Proceedings of the 37th International MATADOR Conference

Srichand Hinduja · Lin Li

Editors

Proceedings of the 37th International MATADOR Conference

 Springer

Editors
Srichand Hinduja
School of Mechanical, Aerospace
  and Civil Engineering
The University of Manchester
Manchester
UK

Lin Li
School of Mechanical, Aerospace
  and Civil Engineering
The University of Manchester
Manchester
UK

ISBN 978-1-4471-4479-3   ISBN 978-1-4471-4480-9   (eBook)
DOI 10.1007/978-1-4471-4480-9
Springer London Heidelberg New York Dordrecht

Library of Congress Control Number: 2012945099

Foreword

It gives us pleasure to introduce the Proceedings for the 37th internal MATADOR conference. The MATADOR Conference was established in 1959, which makes it one of the longest running conferences in the field of manufacturing. It is a truly international conference, which attracts high quality peer reviewed papers from countries all over the world. The Proceedings of this conference include 97 papers.

The MATADOR Conferences in the 1960s and 70s dealt with machine tools and conventional manufacturing processes such as metal cutting, forming and grinding. In the recent conferences, the contents have broadened to include newer areas of manufacturing such as micro- and nano-manufacture, additive manufacturing using lasers and modelling of various manufacturing processes including laser materials processing, with applications in the aerospace, bio-medical, food, energy, automotive, electronics, optics and process industries.

We would like to express our gratitude to all the members of the Organising Committee and to those members of the International Scientific Committee who kindly refereed the papers. Our special thanks to Drs Muhammad Fahad and Robert Heinemann for their assistance in preparing the Conference Proceedings. Finally, we are indebted to Ms Janet Adnams and Andrew Tickle from ConferCare for organising this event.

We hope you find the papers interesting and stimulating.

Professors Srichand Hinduja and Lin Li
Manchester, UK
July 2012

International Scientific Committee

Professor M Brandt, RMIT University, Australia
Professor K Cheng, Brunel University, UK
Professor T H C Childs, University of Leeds, UK
Dr K Davey, The University of Manchester, UK
Professor B Denkena, Leibniz University Hanover, Germany
Professor J R Duflou, Katholieke University Leuven, Belgium
Professor K C Fan, National Taiwan University, Taiwan
Professor F Z Fang, Tianjin University, China
Dr R G Hannam, The University of Manchester, UK
Professor B Hon, University of Liverpool, UK.
Professor X Q Jiang, University of Huddersfield, UK
Dr I Kelbassa, Fraunhofer ILT, Germany
Professor F Klocke, RWTH Aachen, Germany
Professor J P Kruth, Katholieke University Leuven, Belgium
Professor M Kunieda, Tokyo University of Agriculture and Technology, Japan
Professor J Lawrence, University of Lincoln, UK
Professor A Labib, University of Portsmouth, UK
Dr A Leacock, University of Ulster, UK
Professor J A McGeough, University of Edinburgh, UK
Professor S Newman, University of Bath, UK
Dr T S Sudarshan, Materials Modification Inc., USA
Professor F J A M Van Houten, University of Twente, Netherlands
Professor F Vollertsen, BIAS GmbH, Germany
Professor K Watkins, University of Liverpool, UK
Dr Z Wang, University of Bangor, UK
Professor M L Zhong, Tsinghua University, China

Proceedings of
The Thirty-Seventh International

MATADOR

Conference

Organising Committee
Professor S. Hinduja *(Co-Chairman)*
Professor L. Li *(Co-Chairman)*
Dr J. Atkinson
Dr J. Francis
Dr R. Heinemann
Dr P. Mativenga
Dr J. Methven
Dr A. Pinkerton
Dr M. Sheikh

The University of Manchester
School of Mechanical, Aerospace and Civil Engineering

Contents

3 ECM and EDM

4 Machine Tools

5 Machining

6 Manufacturing Systems Management and Automation

7 Metrology

8 Rapid Prototyping

9 Welding

10 Laser Technology – Additive Manufacturing

11 Laser Technology – Modelling

12 Laser Technology – Surface Engineering

1

Casting, Moulding and Forming

Process capability study of hybrid investment casting

Rupinder Singh[1]

[1] Dept. of Production Engineering, Guru Nanak Dev Engineering College, Ludhiana-141006, India

Abstract. The purpose of the present study is to investigate process capability of hybrid investment casting (HIC) for industrial applications. Starting from the identification of component, prototypes were prepared as replicas of plastic patterns (by hybridization of fused deposition modelling and conventional investment casting). Some important mechanical properties were also compared to verify the suitability of the components. Final components produced are acceptable as per ISO standard UNI EN 20286-I (1995). The results of study suggest that HIC process lies in ±4.5sigma (σ) limit as regard to dimensional accuracy of component is concerned. This process ensures rapid production of pre-series technological prototypes and proof of concept at less production cost and time.

Keywords: Hybrid investment casting, process capability, dimensional tolerance.

1. Introduction

The process of casting metal into a mould produced by an expendable pattern with a refractory slurry coating (sets at room temperature), after which the wax pattern is removed through the use of heat prior to filling the mould with liquid metal is known as investment casting (IC)/lost wax process [1,2]. It is one of the economical, mass-production casting processes [2,3].

With the industry growing at a very fast pace reduced production time is a major factor in industries to remain competitive in the market place [4]. Therefore traditional product development methodology has given way to rapid fabrication techniques like rapid prototyping [5]. HIC (Hybridization of fused deposition modelling (FDM) and IC) is one of the rapid manufacturing technique which enables near net shaped metal parts containing complex geometries and features from a variety of metals. FDM forms three-dimensional objects from CAD generated solid or surface models. In this process, FDM material like ABS, elastomer, polycarbonate etc. feeds into the temperature-controlled FDM extrusion head, where it is heated to a semi-liquid state [5]. The head extrudes and deposits the material in thin layers onto a fixtureless base. The head directs the material into place with precision, as each layer is extruded, it bonds to the

previous layer and solidifies [6]. The designed object emerges as a solid three-dimensional part/pattern without the need for tooling [7]. Fig. 1 shows schematic of FDM.

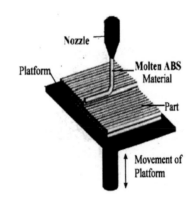

Fig.1. FDM schematic [7]

Individual patterns are attached onto a common wax sprue system to form a cluster in the form of tree. The cluster is repeatedly dip coated in investment slurry containing graded suspensions of refractory particles and followed by stucco application to build shell thickness and strength [8]. When dried, the ABS pattern is melted out via autoclaving to reveal the internal cavities of the ceramic shell. The first layer is normally a fine coating so that a good surface finish on the casting will be obtained. Subsequent layers are made up of ceramic slurry and refractory granules. The shell will normally be made up of between five and eight layers depending on the cooling rate required and the subsequent metallurgical properties [9-10].

The literature review reveals that lot of work has been reported on IC, its applications by different researchers [10,12]. But very few authors have reported on hybridization of IC and FDM process in ordere to get good quality casting. The main objective of this research paper is to study the process capability of HIC. In order to accomplish this objective, 'plain carbon steel casting' has been chosen as a benchmark (Ref. Fig. 2). The chemical composition of plain carbon steel was C=0.43%, S=0.017, P=0.023, Si= 0.35 and Mn=0.75. The component selected is a outer cover of memory stick.

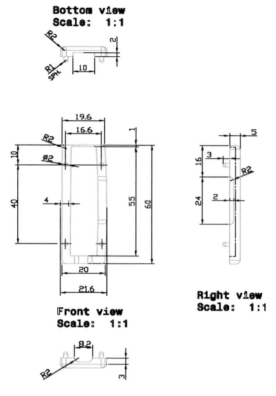

Fig.2. Benchmark dimensions

CAD model (Fig. 2) of the bench mark was made on UNIGRAPHICS software. The 3D CAD model was converted into the STL format which was fed into the computer attached to the FDM machine. The machine was cleaned and the benchmark was set in various orientations. After setting of orientations the component was sliced, layer by layer construction was done and thereafter the components were cleaned. Fig. 3 shows ABS pattern prepared by FDM.

Fig.3. ABS pattern prepared by FDM

2. Experimentation

As a usual practice in industry, for IC three different types of layers; primary (1°), secondary (2°) and tertiary (3°) are applied on shells (as fine, semi coarse and coarse grains of slurry). It should be noted that the number of layers (5-8) are dependent on size/volume/weight of casting and can't be generalised. Based upon pilot experimentation it has been observed that for present case study five layers are providing sufficiently good results (acceptable as per industrial requirements) in terms of surface finish (Ra5.51-6.02μm) and hardness (195-205HV) etc. On the basis of the results obtained as above it was decided to prepare the shell with minimum five layers.

Finally castings produced from plain carbon steel were measured for dimensional accuracy with the help of CMM (Ref. Fig. 2). The observations are shown in Table 1 for four critical dimensions (D1 = 60mm, D2 = 3mm, D3 = 10mm and D4= 21.6 mm). The dimensional data has been collected and analyzed; for the critical dimensions D1, D2, D3 and D4 (Ref. Table 1) by preparing 16 samples of HIC. Table 2 shows the classification of different IT grades according to ISO UNI EN 20286-I (1995) for D1, D2, D3 and D4.

Now for a generic nominal dimension D_{JN}, the number of the tolerance units is evaluated as [13,14]:

$$n= 1000(D_{JN} - D_{JM})/ i,$$

Where D_{JM} is a measured dimension

Tolerance factor $i = 0.45 (D)^{1/3} + 0.001D,$

Where, D is the geometric mean of the limiting values. In present case for nominal dimension (12.46mm):

$D = (10x 18)^{1/2} =13.416$ mm
$i = 0.45 (D)^{1/3} + 0.001D$
$= 0.45 (13.416)^{1/3} + 0.001(13.416)$
$= 1.082964$

For casting obtained from 12mm shell thickness:

$n = 1000(D_{JN} - D_{JM})/ i$
$= 1000(12.46-12.301)/1.082964$
$=147$

The component selected in present study is actually prototype of memory stick cover and dimensions (D1, D2, D3 and D4) was chosen for demonstration over the rest of the dimensions for the measurement and comparison purpose because it is actually mating/functional dimension in assembly. It should be noted that the results are based upon study performed on a simple geometry (Ref. Fig.2), but the same results are applicable to any complex geometry of similar volume as because solidification time depends upon ratio of volume to surface area.

Table 1. Measured dimensions for process capability analysis

Sample No.	D1=10 mm	D2=3 mm	D3=60 mm	D4=21.6 mm
1	9.9216	3.0203	59.8100	21.7129
2	9.9506	2.9710	59.8575	21.7534
3	10.0791	3.0508	59.8901	21.6734
4	10.0335	3.0510	59.8248	21.5921
5	9.9713	2.9559	59.9080	21.6939
6	9.9689	3.0754	60.0500	21.6578
7	10.0276	3.0102	59.9222	21.6718
8	10.0738	3.0247	59.8851	21.5934
9	10.0324	3.0300	60.0223	21.6820
10	9.9470	3.0503	59.8313	21.5977
11	10.0279	2.9776	59.9782	21.6119
12	10.0113	3.0152	59.8908	21.6988
13	10.0228	3.0318	59.9178	21.6366
14	9.9813	2.9812	59.8534	21.6191
15	10.0332	3.0777	59.9575	21.6225
16	9.9786	3.0258	59.8974	21.6811

Table 2. IT grades for measured dimensions

Sample No.	IT Grade for D1	IT Grade for D2	IT Grade for D3	IT Grade for D4
1	IT11	IT10	IT10	IT10
2	IT10	IT11	IT10	IT11
3	IT11	IT12	IT9	IT8
4	IT9	IT12	IT10	IT9
5	IT9	IT12	IT9	IT9
6	IT9	IT13	IT7	IT7
7	IT9	IT9	IT8	IT8
8	IT11	IT10	IT9	IT9
9	IT9	IT11	IT5	IT8
10	IT10	IT12	IT10	IT9
11	IT9	IT10	IT5	IT8
12	IT7	IT9	IT9	IT9
13	IT9	IT60	IT9	IT7
14	IT8	IT10	IT10	IT7
15	IT9	IT13	IT7	IT7
16	IT8	IT10	IT9	IT8

3. Results and Discussion

Based upon observations in Table 1, capability analysis calculations have been made to assess whether a system is statistically able to meet a set of specifications or requirements. The comparison is made by forming the ratio of the spread between the process specifications (the specification "width") to the spread of the process values,

as measured by 6 process standard deviation units (the process "width"). The capable process is one where almost all the measurements fall inside the specification limits. The Capability Indices (Cp) index is used to summarize a system's ability to meet two-sided specification limits (upper and lower). However it ignores the process average and focuses on the spread. If the system is not centered within the specifications, Cp alone may be misleading.

$$Cp = \frac{USL - LSL}{6\sigma} \qquad Cpk = min\left[\frac{USL - \mu}{3\sigma}, \frac{\mu - LSL}{3\sigma}\right]$$

Where: USL = upper specification limit, LSL= lower specification limit, σ = standard deviation, μ = mean of data.

The higher the Cp value the smaller the spread of the system's output. Cp is a measure of spread only. A process with a narrow spread (a high Cp) may not meet customer needs if it is not centered within the specifications. Cp should be used in conjunction with Cpk to account for both spread and centering. Cp and Cpk will be equal when the process is centered on its target value. If they are not equal, the smaller the difference between these indices, the more centered the process is. Cpk is a capability index that tells how well as system can meet specification limits. Since it takes the location of the process average into account, the process does not need to be centered on the target value for this index to be useful. If Cpk is 1.0 the system is producing 99.73% of its output within specifications. The larger the Cpk, the less variation you will find between the process output and specifications. If Cpk is between 0 and 1.0 not all process output specifications meets. For process capability analysis Table 3 shows summary of statistical analysis for nominal dimension D1, D2, D3 and D4.

Table 3. Statistical analysis for nominal dimensions

Statistical analysis	D1	D2	D3	D4
Cp	1.5658	1.3825	1.5833	1.6631
Cpk	1.5375	1.2239	1.1918	1.5557
Mean of data	10.0038	3.021	59.906	21.65
Lower spec. limit (LSL)	9.79	2.81	59.62	21.39
Upper spec. limit (USL)	10.21	3.19	60.38	21.89
Min. value	9.9216	2.9559	59.81	21.5921
Max.value	10.0791	3.0777	60.05	21.7534
Standard deviation	0.04540	0.0359	0.068	0.0477
Range	0.1575	0.1218	0.24	0.1613

Figs. 4-6 shows R chart, X chart and process capability histogram for nominal dimension D1. As observed from

Figs. 4-6, for Cpk value of 1.5, the area under normal curve is 0.999993198 and non conforming ppm is 6.8016. Similarly Cp and Cpk values for other dimensions (D2, D3 and D4) were calculated. The value of Cpk for all critical dimensions is 1.33. The results of study suggest that RSM process lies in ±4.5sigma (σ) limit as regard to dimensional accuracy of plastic component is concerned.

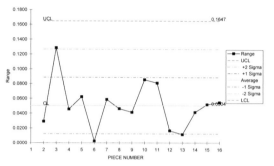

Fig. 4. R chart for nominal dimension D1

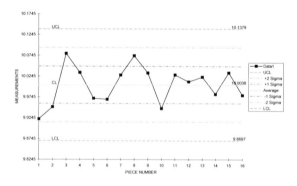

Fig. 5. X chart for nominal dimension D1

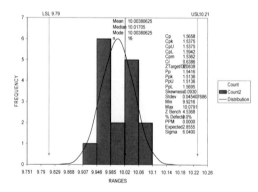

Fig. 6. Process capability histogram for nominal dimension D1

4. Conclusions

On the basis of experimental observations it can be concluded that HIC is highly capable process. It is observed that the 'Cpk value' for all the four critical

dimensions in the present study is 1.33. As Cpk values of 1.33 or greater are considered to be industry benchmarks, so this process will produce conforming products as long as it remains in statistical control. The IT grades of the components produced are consistent with the permissible range of tolerance grades as per ISO standard UNI EN 20286-I (1995). The adopted procedure is better for proof of concept and for the new product, for which the cost of production for dies and other tooling is more.

Acknowledgments: The author is thankful to DST, New Delhi for financial support.

References

[1]   Wang S., Miranda A. G., and Shih C., (2010) A Study of Investment Casting with Plastic Patterns. Materials and Manufacturing Processes 25(12): 1482-88.
[2]   Chattopadhyay H., (2011) Estimation of solidification time in investment casting process. The International Journal of Advanced Manufacturing Technology 55(1-4): 35-38.
[3]   Rafique M.M.A. and Iqbal J., (2009) Modeling and simulation of heat transfer phenomena during investment casting. International Journal of Heat and Mass Transfer 52(7-8): 2132-2139.
[4]   Singh R., (2010) Three dimensional printing for casting applications: A state of art review and future perspectives. Advanced Materials Research 83-86: 342-349.
[5]   Chhabra M. and Singh R., (2011) Rapid casting solutions: a review. Rapid Prototyping Journal 17(5): 328-350.
[6]   Garg H. and Singh R., (2011) A Framework for Development of Pattern for Dies Using Fused Deposition Modeling. International Journal of Advanced Mechatronics and Robotics 53-60.
[7]   Garg H. and Singh R., (2012) Experimental investigations for development of pattern for dies using FDM. Materials Science Forum, Special topic volume: Rapid casting solutions 701: 77-83.
[8]   Mishra S. and Ranjana R., (2010) Reverse Solidification Path Methodology for Dewaxing Ceramic Shells in Investment Casting Process. Materials and Manufacturing Processes 25(12): 1385-88.
[9]   Sidhu B. S., Kumar P. and Mishra B.K., (2008) Effect of slurry composition on plate weight in ceramic shell investment casting. Journal of Materials Engineering and Performance 17: 489–498.
[10] Kumar S., Kumar P. and Shan H.S., (2006) Parametric Optimization of Surface Roughness Castings Produced by Evaporative Pattern Casting process. Mater. Lett. 60: 3048–3053.
[11] Norouzi Y., Rahmati S., Hojjat Y., (2009) A novel lattice structure for SL investment casting patterns. Rapid Prototyping Journal 15(4): 255 – 263
[12] Konrad C.H., Brunner M., Kyrgyzbaev K., Völkl R., Glatzel U., (2011) Determination of heat transfer coefficient and ceramic mold material parameters for alloy IN738LC investment casting. Journal of Materials Processing Technology 211(2): 181-18
[13] UNI EN 20286-1 (1995) ISO system of limits and fits, Bases of tolerances, deviations and fits.
[14] Devor R. E., Chang T., and Sutherland J. W., (2005)Statistical quality design and control contemporary concepts and methods, Pearson Prentice Hall (Second edition) New Jersey.

Enhancing the performance of laboratory centrifuges with carbon fiber rotors

U. Klaeger[1] and V. Galazky[2]

[1] Fraunhofer Institute for Factory Operation and Automation IFF, Magdeburg, Germany

[2] carbonic GmbH, Magdeburg, Germany

Abstract. Centrifuges have multiple uses in medicine and in laboratories, most notably to separate substances such as blood plasma and serums. A centrifuge's rotors hold sample containers. Centrifugal forces equaling 25,000 times the force of gravity act on rotors and samples at speeds of 20,000 rpm. Unlike present conventional solutions (rotors made of special aluminum alloys), the load-bearing structures of the lightweight rotors presented here are made of carbon fiber-reinforced polymers (CFRP). The complex rotor and mold geometries are selective laser sintered. Given the fibers' extremely high load-bearing capacity, lightweight rotors manufactured with this novel technology set new standards for weight, stability and service life compared to current concepts. Moreover, this can reduce the warm-up time typical for aluminum rotors by at least fifty percent.

Keywords: Composite and Polymer Manufacturing, Additive Manufacturing, Forming Processes, Carbon Fiber Material, Centrifuges, Design.

1. Introduction

Centrifugation is the process most frequently applied to separate materials in liquids. Most notably, they are used to separate materials, e.g. blood plasma and serums, and produce genetically engineered substances.

A centrifuge consists of housing, a drive unit with controller, a rotor, a safety enclosure and, frequently, a cooling system. The rotor holds sample containers. There are rotors for different sizes of samples (from the microliter range to one liter) and numbers of sample containers (depending on the task). They are subjected to extreme mechanical loads. Technically this makes them core centrifuge components.

During centrifuging, a solution's solid constituents precipitate under the effect of a stronger gravitational field produced by the centrifugal forces generated by rapid rotation. In this stronger gravitational field, constituents with greater mass displace lighter particles, which are thrust closer to the axis of rotation. The gravitational force increases exponential to the distance from the axis of rotation (radius). Superior centrifuges

operate at high speeds of frequently more than 20,000 rpm. This produces gravitational fields, which exceed normal gravitation several thousands of times over. Conventional rotors are made of special aluminum alloys. They are relatively easy and cheap to manufacture but have drawbacks in terms of their stability and attainable rotational speeds. In addition, undesirable imbalances frequently appear.

Therefore, lightweight rotors are alternatively made of fiber composite materials. A variety of methods exist but, at present, only rotors with resin transfer molded bodies and wound highly stressed annular shells are commercially available.

Admittedly, lightweight rotors made entirely or partially of carbon fiber-reinforced polymers or mixed are more expensive than conventional aluminum rotors since they are predominantly manufactured by hand. However, in addition to having a substantially lower density, they weigh far less and are approximately six times more stable than aluminum rotors.

Lower weight cuts centrifuges' energy consumption and ramp-up times and also reduces an overall centrifuge system's mechanical loads. In addition, lighter CFRP rotors simplify handling since aluminum rotors often weigh up to 50 lbs.

2. Motivation and aims

The authors jointly developed a method of manufacturing lightweight rotors from pre-molded woven carbon fibers. Carbon fibers have high tensile strength, provided the fibers are aligned with the direction of load. They are thusly processed relatively easily when profiles are long and shapes are flat or cylindrical.

However, tapering and freeform surfaces like those of centrifuge rotors are more complicated since the fibers are unable to adhere to these surfaces and slip easily. Hence, rotors are only wound at present.

The drawback of this is that the fibers cannot be aligned with the direction of load. Further, the manufacture of wound rotors requires extremely expensive multi-axis winders and wound surfaces are never really smooth. This affects a rotor's running smoothness adversely.

The new methods of positioning fibers aligned with the direction of load on conical surfaces employs carbon spiral tapes that exact match the winding of a rotor's tapered surface geometrically. In a first step, a base body that holds sample containers and a hub with a conical exterior shape are manufactured. Then, the spiral tape is placed around the base body.

Since the geometry of the spiral tape and base body corresponds, the spiral tape stays in place and only the two ends of the tape have to be secured with some spray adhesive. The base body layered with spiral tape is placed in a second mold and impregnated with resin by RTM. This design reduces the moment of inertia by more than half.

The complicated design principal considered by these authors requires suitable forming tools that reproduce the complex geometries (e.g. undercuts). Therefore, generative (laminate) methods of geometry generation are used to make molds. Given their practically unlimited freedom of design, these technologies can, for instance, already produce close-contour cooling channels during mold making. Furthermore, selective laser sintered, geometrically complex inserts reduce the weight of rotor bodies.

When this novel concept is successfully implemented, a lighter weight CFRP rotor will be at least 10% more stable than a lightweight wound rotor. At the same time, it can be expected to have a permissible speed that is at least 10% higher than that of lightweight rotors in the same class. First, comprehensive physical models and new calculation algorithms for the FEM analysis were developed, which ensure that lightweight rotors can be manufactured reproducibly for future implementation in practice.

3. Conceptual design of the manufacturing technology

First, the CAD models of the two sizes of CFRP rotor analyzed (14 x 50 ml and 6 x 500 ml) were generated. They served as the basis for subsequently calculating the shell design and potential failure criteria (maximum stress, maximum strain, etc.) based on the finite element method (FEM), Fig. 1.

The ANSYS analysis package was used for the calculations since it contains tools that are especially effective for the calculating the structures of fiber composite materials.

This tool was also used to identify and calculate the principal stresses produced during centrifugation and the resultant critical zones in the rotor geometry (Fig. 2).

Taking the properties required of the rotor as the point of departure, the energy of three different rotor geometries was analyzed in order to draw conclusions about the rotational energy of each. The goal targeted for the overall system was 94,000 Nm. Only the variant "Core 2" met this goal (see Table 1).

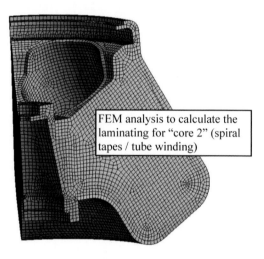

FEM analysis to calculate the laminating for "core 2" (spiral tapes / tube winding)

Fig.1. Rotor design calculation and simulation

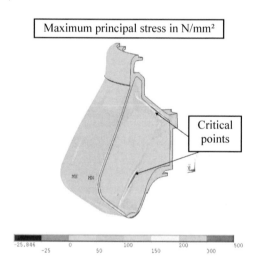

Maximum principal stress in N/mm²

Critical points

Fig 2. Critical rotor geometry zones.

Since the geometry had not been optimized, designing the rotor geometry as a monolithic CFRP block even worsened the initial values.

The simulation determined that CFRP rotors weigh up to 44% less because they have a lower density (approximately 1.5 g/cm³ at a content of approximately 60% fiber by volume) than aluminum (ca. 2.7 g/cm³) rotors.

This weight advantage shortens acceleration and deceleration time while retaining a centrifuge's performance. Consequently, cycle times are shorter.

Table 1. Energy analysis of a 6 x 500 ml CFRP rotor compared with an aluminum rotor

Rotational energy [Nm]			
Type	Rotor	Target	Difference
Aluminum	154,246	94,000	60,246
Monolithic CFRP	158,913	94,000	64,913
Core 1-foam	135,208	94,000	41,208
Core 2-SLS	86,426	94,000	-7,574

In addition, higher speeds can be run, thus increasing the relative centrifugal acceleration. Moreover, this weight reduction makes such rotors easier to handle. Their resistance to corrosion and enhanced fatigue strength are additional advantages.

4. Forming tool design

Taking the calculation results as the starting point, so-called displacers (as inserts) for the rotor casting mold were designed and laser sintered (Fig. 3a).

Fig 3. Selective laser sintered master models displacersas the basis for silicone molds (a). Rotor variant Core 2 with 14 inserted SLS cores (b).

In addition to cutting weight, such laser sintered inserts significantly reduced the rotational energy. Arranged in position, the inserts are intended to assure Core 2 has the rotational energy desired.

The prototyping technologies of selective laser sintering and vacuum casting were combined in order to deliver the large quantities required. The overall design was simultaneously optimized for fiber composites. The prototyping technologies of selective laser sintering and vacuum casting were combined in order to deliver the large quantities required. The overall design was simultaneously optimized for fiber composites.

5. Laboratory prototype and mold making

The results of these tests entered into the development of the manufacturing technology to properly design the future forming tool for casting.

Taking the theoretical calculations as the starting point, initial rotor prototypes were subsequently produced to verify the variance analysis. This entailed producing laminating molds to assure the reproducibility of the manufacturing in certain quantities. One of the first rotor prototypes is pictured in Fig. 3b.

The complete lightweight rotor consists of nine different components, including the aerosol ring, hub, rotor and six filling elements. The filling elements are hidden in the rotor housing between the sample container holders and are made of a lightweight plastic, thus reducing the lightweight rotor's weight.

The hub is force fit with the lightweight rotors while it is being manufactured/layered. The use of two different hubs is planned at present. The aerosol ring is separately made of plastic and subsequently bonded to the rotor once it has been manufactured. The greatest challenge during development was reconciling the design of forming tools with the layering technology to be developed (Fig. 4).

Fig 4. Elements of the casting mold for the first rotor prototypes.

The forming tool must have the rotor's geometric complexity and its design must facilitate the defined fiber layering, which is crucial to facilitating full impregnation and a uniformly high content of fibers by volume.

These are essential for the manufacture of extremely stable rotors. Therefore, the development partners employed simulation methods to optimize the layering technology and to design the forming tools.

The tests executed made it possible to implement design modifications in a matter of hours, thus enabling the development partners to rapidly find and test solutions to the most complicated layering steps. The first proposed solution for the carbon fiber rotor is depicted in Fig. 5.

Fig 5. First solution for the carbon fiber rotor.

6. Conclusion

The extensive work to design and develop the layering technology delivered findings that enabled carbonic GmbH to completely engineer its manufacturing processes without having to modify the mold's design.

Rotors manufactured with the new methods combine the smooth surface of aluminum rotors with the advantages of wound rotors, e.g. lower weight and better fracture characteristics. The new lightweight rotors weigh up to fifty percent less than aluminum rotors and can withstand up to twenty percent higher loads.

Compared with wound rotors, the new methods can produce smaller quantities more cost effectively. Further, rotors manufactured with the new methods are more stable and have a smoother surface, which enables rotors to operate smoothly.

Acknowledgements: Support for this project came from the Investitionsbank of Saxony-Anhalt, for which the authors express their sincere thanks.

References

[1]   Beckwith, S., Hyland, C., SAMPE Journal, Vol 34, No. 6, (1998) Resin Transfer Moulding: A decade of technology advances. 7-19

[2]   VDI-Richtlinie 3404, Beuth Verlag Berlin (2009) Additive fabrication: Rapid technologies (Rapid Prototyping)

[3]   DIN EN 3783, Beuth Verlag Berlin (1992) Aerospace series: fibre composite materials; normalization of fibre dominated mechanical properties

[4]   DIN 58970-1, Beuth Verlag Berlin (1996) Laboratory centrifuges-Part 1: Definitions, testing, marking

[5]   DIN 58970-2, Beuth Verlag Berlin (1998) Laboratory centrifuges; centrifuge tubes for relative centrifugal acceleration up to 4000.

Composite grid cylinder vs. composite tube: Comparison based on compression test, torsion test and bending test

Lai C.L., Liu C. and Wang J. B.
Shaanxi Engneering Research Center for Digital Manufacturing Technology, Northwestern Polytechnical University, Xi'an 710072, China

Abstract: The use of composite grid cylinder, which is known for its very high strength/stiffness-to-weight ratio, gives large weight savings over metallic. However, the mechanical property comparison between composite grid cylinder and traditional composite tube is lake of research. To better quantify the advantages or disadvantages between the composite grid cylinder and the composite tube, this paper describes the fabrication and the test of composite grid cylinder and composite tube. Composite grid cylinders are fabricated manually guided filament winding. And composite tube is fabricated by autoclave moulding. In order to evaluate structure mechanical property, comparison including compression test, torsion test and bending test, are carried out. Finally, comparison results show that strength to weight ratio of composite grid cylinder is lower than composite tube, but stiffness to weight ratio is 2.7 times higher than composite tube in bending test.

Keywords: Composite grid cylinder, Composite fabrication, Mechanical property comparison

1. Introduction

The Composite grid cylinder is characterized by a lattice of rigid, interconnected rids.It consists of a dense system of unidirectional composite helical, circumferential or axial ribs and mainly derive its strength and stiffness form ribs. Because composite grid cylinders are attractive structures for application in aerospace from their high strength/stiffness to weight ratio, high impact resistance and low fabrication cost, they have been successfully applied to interstages, payload adapters[1-4] and fairings of launch vehicles [5,6], and expanded to fuselage section [4]. But all the applications only exploit high load carrying capacity of the structure in axial direction. The potential of beam application which carry torsional load and bending load is lack of research.

Composite tubes, which have more second axial moment of area and can provide higher bending stiffness and torsion stiffness than other solid structure, have been used in spacecraft truss and solar unmanned air vehicle beams [7, 8]. However, in contrast to the composite laminate tube with finite thickness whose mass cannot be reduced beyond some minimum value determined by the finite thickness of the ply, the mass of gride cylinders is governed by the rib spacing and cross-sectional area and can be readily reduced to a desirable value.

In this research, composite grid cylinders and composite tubes for comparison testing are fabricated using the same materials. Comparison testing schemes, including axial compression testing, torsion testing and bending testing, are modified in consideration of test specimen features and test situations. Finally, their compression property, torsion property and bending property are compared and analyzed.

2. Specimen manufacturing

Composite tubes are made by prepreg sheet, cured in autoclave. Layup is [45/-45/45/-45/0_4]s, and the thickness of each ply is about 0.125 mm. Carbon fiber is CCF300-3K carbon fiber, and resin matrix is high temperature epoxy resin. The measured data of the main parameters are shown in Table 1.

Table 1. Composite tube specimen parameter

Specimen	Length (mm)	Outer diameter (mm)	Inner diameter (mm)
Compression	98	40.5	36.3
Bending	373	40.8	36.6
Torsion	355	40.6	36.4

Composite grid cylinders are made by continuous wet winding during which impregnated carbon tows are placed into helical and circumferential grooves formed on the surface of the mandrel with the aid of:

- Machining of the foam core covering the mandrel.
- Forming grooves in the silicon rubber elastic coating that is pulled out of the structure after curing.
- Forming grooves in thin metal or wooden panels mounted on the mandrel surface [2].

The second method, using the silicon rubber elastic coating, provides two additional advantages. The silicon rubber elastic coating is typically a high thermal expansion. It provides lateral compaction to the ribs during curing. In addition, after the structure is cured, it is easy to remove the mandrel and the rubber coating can be pulled out facilely. In this research, composite grid cylinders are made by continuous wet winding, using the silicon rubber elastic coating, and cured in high temperature stove. The materials for composite grid cylinders are the same as composite tube. CCF300-3K carbon fiber is adopted, and high temperature epoxy resin is used for the resin matrix. The typical manufacturing process involves the following operations

- Forming of the silicon rubber elastic coating
- Wet winding of ribs.
- Curing.
- Machining of the rings.
- Removal of the mandrel and the elastic coating.

In general, composite grid cylinders are haracterized with eight design variables (shown in Fig. 1), i.e.,

- D-structure outer diameter.
- L-structure length.
- a_h-spacings of the helical ribs.
- a_c-spacings of the circumferential ribs.
- b_h-widths of helical rib.
- b_c-widths of circumferential rib.
- φ-angle of helical ribs.
- H-height of ribs cross-sections.

The measured data of the main parameters are shown in Table 2. In Table 2, widths of helical ribs cross-sections is equal to widths of hoop ribs cross-sections, namely b_h = b_c.

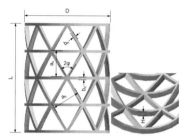

Fig. 1. Grid cylinder variables

3. Test and comparison

Test scheme should take structure features and loading conditions into consideration. Firstly, structure grippedends should be designed to avoid grippedends being destroyed in advance in loading. Secondly, load acting and load bearing should be improved to prevent local points or places of structures destroying in advance.

3.1. Compression test

The ends of the composite tube should be polished smoothly, clean, dry, to provide parallel and level face for compression test. Also, to ensure composite grid cylinder carrying compression load symmetrically, two ends of the composite grid cylinder are cured with epoxy resin. Compression test scheme are shown in Fig. 2.

The damage of composite tubes appears on the end of strcture. It is crushed (shown in Fig. 3).The damage of composite grid cylinder is complex in compression test. The nodes of composite grid cylinder, points at which ribs cross, appear delamination at first. After delamination, the load carrying capacity of ribs declines and the ribs begin buckling. Finally, with increase of compression load, delamination becomes more and more severe and some ribs break off (shown in Fig. 3).

In general, there are two main reasons that cause nodes of the composite grid cylinder delamination in advance.

- The nodes of the composite grid cylinder are points of rib intersections. The intersections cause fiber accumulation, bending and weaken strength and stiffness at nodes. So the nodes are

Table 2. Composite grid cylinder specimen parameter

Specimen	Quality (g)	L (mm)	Φ (°)	D (mm)	H (mm)	a_h (mm)	a_c (mm)	b_h=b_c (mm)
Compression	8.8	95	30	76.7	1.71	24.6	23.5	1.5
Bending	35.2	352	30	76.2	1.90	24.4	24.0	1.7
Torsion	31.5	350	30	77.3	1.61	25.2	24.5	1.7

the weakest place in the cylinder and they are damaged at first in compression test.

- During curing, composite grid cylinders are placed statically without rotating in high temperature stove. The resin flows into the lowest place in composite grid cylinder and cause some nodes resin loss.

The stiffness to weight ratio of the composite grid cylinder is about 45% of the composite tube. And the strength to weight ratio of the composite grid cylinder is about 19.4% of the composite tube. There are three main reasons weakening the strength/stiffness to weight ratio of the composite grid cylinder.

- Fiber volume fraction of composite grid-stiffened cylinder is low.
- Carbon fiber is damaged in wet winding.
- Fiber accumulation and resin loss weaken strength and stiffness of nodes.

Fig. 2. Compression test scheme

Fig. 3. Compression specimen damage

3.2. Torsion test

Composite tube grippedends are the weakest place in torsion testing and can be damaged easily by testing machine. To make matter worse, composite tube grippedends are so slipped that it is very hard to clamp them in load acting. In test preparation phase, to ensure the grippedends are enough burliness to avoid damaging in advance, composite tube grippedends are cured with resin inside and outside of composite tubes.

The ends of Composite grid cylinder can not carry clamping load from radial direction. In addition, the structure outer diameter is larger than inner diameter of testing machine grip. In test preparation phase, two ends of the composite grid cylinder are cured with resin to form grippedends. Torsion testing scheme are shown in Fig. 4.

The damage of composite tube is delamination of laminate in torsion testing (shown in Fig. 5). the damage of the composite grid cylinder also appear at nodes at first. It is delamination of nodes. With increase of torsion load, some ribs are broken off. Causes of nodes delamination are the same as in compression testing.

Torsion strength to weight ratio of the composite grid cylinder is about 31% of composite tube. Reasons which weaken torsion strength to weight ratio of the composite grid cylinder are the same as above mentioned in compression testing.

Fig. 4. Torsion test scheme

Fig. 5. Torsion specimen damage

3.3. Bending test

Three point bending test was not adopted to analyze bending resistance by material testing machine. Because the test specimens are cylinders and tubes, loading area is so small that it is easy to cause local failure in three point bending test. It is worse that support area, between test specimens and test equipments, is too small to finish three point bending test. In three point bending test, a

Table 3. Compressive property comparison

Specimen	Quality (g)	Stiffness (KN/mm)	Stiffness to weight ratio (KN/mm)/kg)	Ultimate load (KN)	Strength to weight ratio (KN/kg)
Tube	39.4	86.92	2206.09	73.58	1867.51
Grid cylinder	8.8	8.74	993.18	3.19	362.50

simple test device is adopted to measure structure stiffness. In the test scheme, two ends of specimens are cured with resin to form support ends. And weight is up in the middle of specimens, as loading. Deflection is measured by dialgage. The bending test scheme are shown in Fig. 6.

The stiffness to weight ratio of composite grid cylinder is about 2.7 times of the composite tube (shown in Table 5). When grid cylinder quality is the same as tube, grid cylinder can provide bigger outer diameter than tube. The composite grid cylinder with bigger outer diameter, have more second axial moment of area to improve bending stiffness.

Table 4. Torsional property comparison

Specimen	Quality (g)	Ultimate torque (kg·m)	Strength to weight ratio (kg·m/kg)
Tube	145.6	110.0	755.49
Grid cylinder	31.5	7.55	239.68

Fig. 6. Bending test scheme

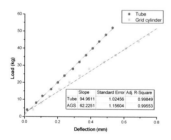

Fig. 7. Load vs. deflection

Table 5. Bending property comparison

Specimen	Quality (g)	Stiffness (kg/mm)	Stiffness to weight ratio ((kg/mm)/kg)
Tube	149.2	94.96	636.46
Grid cylinder	35.2	62.22	1767.61

4. Conclusions

The present work has provided effective reference for selection of structures, especially ultra-lightweight structure. Some conclusions, based on fabricating process and test scheme mentioned above, are in the following.

- Compression strength/stiffness to weight of grid cylinder is lower than tube; torsion strength to weight of grid cylinder is lower than tube; bending stiffness to weight of grid cylinder is higher than tube.
- Because of fiber accumulation, bending and resin loss,the joint, points of ribs intersections, is the weakest place in composite grid cylinder.
- Volume fraction, fiber damage in wet winding and fiber accumulation and resin loss at nodes, weaken strength and stiffness of the composite grid cylinder.

Refenrences

[1] V. V. Vasiliev, V. A. Barynin, A. F. Rasin, Anisogrid lattice structures-survey of development and application[J], Composite Structures 2001 54: 361-370

[2] Yu. O. Bakhvalov, V. P. Molochev, S. A. Petrokovskii, Proton-M Composite interstage structures: design, manufacturing and performance European conference for aerospace sciences (EUCASS)

[3] V. V. Vasiliev, A. F. Razin, Anisogrid composite lattice structures for spacecraft and aircraft applications[J], Composite Structures 2006 76: 182-189

[4] V. V. Vasiliev, V. A. Barynin, A. F. Razin. Anisogrid composite lattice structures-Development and aerospace applications [J], Composite Structures, 2012 94: 1117-1127

[5] Wegner, P.M., Ganley, J.M., Huybrechts, S.M., and Meink, T.E., "Advanced Grid Stiffened Composite Payload Shroud for the OSP Launch Vehicle", 2000 IEEE Aerospace Conference Proceedings.

[6] Peter M. Wegner, John E. Higgins, Barry P. VanWest, Application of Advanced grid-stiffened structures technology to the minotaur payload fairing, AIAA 2002-1336

[7] G. Frulla, E. Cestino, Design, manufacturing and testing of a HALE-UAV structural demonstrator[J], Composite Structures, 2008, 83: 143-153

[8] Giulio Romeo, Giacomo Frulla, Enrico Cestino, and Guido Corsino, HELIPLAT: Design, Aerodynamic, Structural Analysis of Long-Endurance Solar-Powered Stratospheric Platform[J], Journal of Aircraft 2004, 41(6).

Thermal laser micro-adjustment using ultra-short pulses

J. Griffiths, S. P. Edwardson, G. Dearden and K. G. Watkins
Laser Engineering Group, School of Engineering, University of Liverpool, Liverpool, L69 3GH, United Kingdom

Abstract. MEMS manufacturing requires accurate positioning and high reproducibility. Lasers can be utilised in accurate post-fabrication adjustment, allowing for manufacturing processes with relatively large tolerances. Laser micro forming (LµF) is a process for the precision adjustment, shaping or correction of distortion in micro-scale metallic components through the application of laser irradiation without the need for permanent dies or tools. The non-contact nature of the process is also useful in accessing specific micro-components within a device which may be highly sensitive to mechanical force. As such it has potential for widespread application in both the manufacturing and microelectronics industry. In this work, the micro-adjustment of 1000x300µm stainless steel actuator arms using a mode locked fibre laser with a maximum pulse energy and duration of 3 µJ and 20 ps respectively was conducted. The effect of pulse overlap, laser power, irradiation strategy and number of irradiations on the net bend angle is presented.In addition, the mechanism of thermal deformation is investigated through numerical modeling and the potential for its application in the micro-adjustment of MEMS scale components is discussed.

Keywords: Micro-adjustment, Laser Forming, Temperature Gradient Mechanism (TGM), MEMS, FEM

1. Introduction

Micro-electronic systems often comprise functional components which require highly accurate micro-scale adjustment after fabrication [1]. Such functional components are typically difficult to access and highly sensitive to mechanical force. The application of lasers offers the potential for controlled and repeatable micro-adjustment of these components in a contact free process.

At the macro-scale the LF process involves generating thermal stresses within a substrate using a defocused beam. Depending on the desired effect, the process parameters can be altered to either induce elastic-plastic buckling or plastic compressive strains. The most commonly employed mechanism is the Temperature Gradient Mechanism (TGM), which bends the sheet metal out of plane towards the beam. A steep thermal gradient is generated locally along the irradiation path, inducing more thermal expansion on the upper surface of the substrate. Upon cooling, providing the temperature was raised enough to cause sufficient thermal strain, plastic contraction occurs in this upper surface, creating a bend angle of 1-2° per pass [2, 3].

When scaling down the LF process, limits to conventional thermal forming techniques become evident, such as excessive, non-localised heating of the substrate and long thermal relaxation times [4]. Research has been conducted on non thermal LµF techniques, such as utilising shockwaves generated through the breakdown of air to induce compressive stresses in the workpiece upper surface [5, 6]. Providing the fluence (Φ) is below or close to the ablation threshold (Φ_{th}) of the material, ultra-short pulses have the potential to be used to form materials in a thermal process. When the pulse duration is shorter than the lattice interaction time, as is often the case with sub nanosecond pulses, there is little conductive heat transfer into the bulk material. This confines the heating effect to the surface layer of the material, thereby selectively inducing plastic compressive stresses and avoiding thermal damage of the substrate, as investigated in this paper.

2. Experimental

2.1. Design and manufacture of micro-scale actuators

The investigation initially consisted of the design of a suitable micro-scale component and selection of an appropriate laser source for their manufacture. 1000x300µm MEMS scale actuator-style arms (Fig 1) were designed and micromachined out of 50 and 75µm thick AISI 302 stainless steel sheet (Table 1).

The microactuators were fabricated using a High-Q IC-355-800nm, 50 kHz laser operating at 1064nm. The laser parameters used were 5 kHz repetition rate, 300mW average power, 50mm/s traverse speed, 10ps pulse length and a 35µm spot diameter. 600 irradiations were required to penetrate the material thickness.

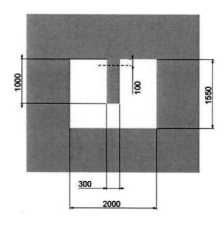

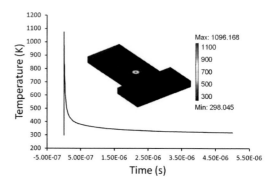

$$-n.q = -n(-k\nabla T) = q_0 + h(T_{inf} - T) + \varepsilon\sigma(T_{amb}^4 - T^4) \quad (2)$$

Fig. 1. Schematic of stainless steel AISI 302 actuator style arms, with dimensions in μm

Fig. 2. Comsol MultiPhysics FE model output for the top surface directly on the laser scan line [1000x300x50μm stainless steel AISI 302, 35μm beam diameter, 500kHz repetition rate, 10mm/s traverse speed]

2.2. Laser forming actuators

An experimental study of LμF was conducted using a 3W Fianium Yb-doped fibre TEM$_{00}$ laser with a pulse length of 20ps, operating at 1059-1069nm and 500kHz. using a Veeco NT1100 white light interferometer to measure the resulting bend angle. The study comprised of adjusting the height of the actuator arms at various fluences and numbers of irradiations. The irradiation path was 100μm from the base of the actuator arm.

2.3 Numerical modelling

To fully understand the thermal aspects of the LF process an FE model was developed. A 2005 element thermal simulation of the laser forming of 300x1000x50μm AISI 302 stainless steel was developed using COMSOL MultiPhysics version 3.5a (Fig 2).

The governing equation for conduction in a solid is the Fourier heat equation (Equation 1).

$$\rho C_p \frac{\partial T}{\partial t} = \nabla.(k\nabla T) \quad (1)$$

Where is the density (kg/m^3), Cp is the specific heat capacity (J/kgK), T is the temperature (K), t is the time (s) and k is the thermal conductivity (W/mK). The term ∇ is the differential or gradient operator (sometimes referred to as the Nabla operator) for three dimensional Cartesian co-ordinate systems. With the exception of the irradiated area the heat flux (q) at a given boundary is governed by Equation 2.

Where h is the heat transfer co-efficient (25 W/m^{2}K) and T$_{amb}$ is the ambient temperature (298.15 K). The incident laser beam was approximated by a Gaussian distributed heat source (Equation 3) with a temperature dependant absorption coefficient [7].

$$I = I_0 e^{-\left(\frac{2r^2}{\omega_0^2}\right)} = \frac{2E_p}{L_p \pi r^2} e^{-\left(\frac{2r^2}{\omega_0^2}\right)} \quad (3)$$

Where E$_p$ is the laser pulse energy (J), L$_p$ is the pulse duration, r is radial distance (m) and ω_0 is the beam radius. The heat source was pulsed through the use of an interpolation function, consisting of a time-dependant Gaussian distributed pulse shape, repeated at regular intervals determined by the repetition rate of the laser.

Material properties were sourced from the ASM Metals Handbook and COMSOL's built in materials library. Thermal expansion coefficient (α_{th}), specific heat capacity (C$_p$), thermal conductivity (k) and density (ρ) were all considered temperature dependant.

The actual geometrical dimensions of the actuator were truncated to leave only the region of interest around the irradiation path. The mesh density was highest along the irradiation path, 100μm from the base of the actuator arm, with the maximum element size being 2.5e^{-6}m. A higher mesh density was assigned to the region of most interest at the centre of the irradiation path where most measurements were taken, where the maximum element size was restricted to 1e^{-5}m (Fig 3).

Table 1. AISI 302 material properties at 298.15K. * Denotes properties considered temperature dependant in numerical simulations

α_{th}^* [1/K]	C_p^* [J/kg.K]	E [Pa]	h [W/m^{2}K]	k* [W/m.K]	ρ^* [kg/m^3]	σ_{ys} [Pa]	v [1]
17.2e^{-6}	500	1.93e^{11}	25	16.2	8000	2.05e^8	0.27

3. Results

3.1. Ablation threshold determination

Prior to thermal LµF being conducted the ablation threshold fluence (Φ_{th}) of the stainless steel substrate was determined experimentally [9] using a 3W Fianium Yb-doped fibre TEM_{00} laser with a pulse length of 20ps, operating at 1059-1069nm and 200kHz. The average power was varied in 50 mW intervals from 100 to 650 mW, with 12 holes drilled at each power. Due to the dependence of Φ_{th} on number of pulses per spot [10] the experiment was repeated for 400 and 600 pulses per drilled spot for validation purposes. The diameter (D) of the ablated craters was measured using a Veeco NT1100 white light interferometer.

The beam radius (ω_0) was determined from a plot of D^2 against E_p and found to be 14.7µm. Using this value the fluence could be determined, and the x-intercept of a logarithmic trend line from a plot of D^2 against Φ_0, plotted on a Log_{10} scale, was taken as Φ_{th}. This was found to be $0.1J/cm^2$ and $0.06J/cm^2$ for 400 and 600 pulses, respectively. Equation 4 describes the relationship between peak fluence (Φ_0) and E_p, whilst D is related to Φ_0 according to Equation 5.

$$\Phi_0 = \frac{2E_p}{\pi\omega_0^2} \qquad (4)$$

$$D^2 = 2\omega_0 \ln\left[\frac{\Phi_0}{\Phi_{th}}\right] \qquad (5)$$

3.2. LµF using Picosecond pulse durations

Multi-pass LµF was conducted using a 3W Fianium Yb-doped fibre TEM_{00} laser with a pulse length of 20ps and spot diameter of approximately 30µm, operating at 1059-1069nm. The irradiation path was 100µm from the base of the arms. Laser micro-adjustment was conducted using both 200 kHz and 500 kHz whilst keeping the pulse energy constant at 3µJ.

Figure 3 reveals little deformation at 200 kHz but significant deformation at 500 kHz, suggesting the latter is the more suitable repetition rate for the laser micro-adjustment process. This can be attributed to the relatively larger cumulative build up in temperature at higher repetition rates due to reduced dwell time between pulses (Fig 4).

Figure 4 reveals a larger cumulative increase in temperature over the same period of time at constant pulse energy for 500 kHz than for 200 kHz. This can be attributed to the shorter interval between pulses.

Operating at 500 kHz, and with increasing multiple irradiations, a cumulative increase in bend angle was observed, as shown in Fig 5.

In addition to a cumulative increase in bend angle, an ablated groove was also observed along the length of the scan path, the aspect ratio of which became too large to resolve the depth by using white light interferometry with multiple irradiations (Fig 6).

This ablated groove can be attributed to the use of multiple irradiations in conjunction with fluences above those of the experimentally determined values of Φ_{th}. An experiment was therefore conducted in which the traverse speed was increased and the ablated groove depth and bend angle was monitored (Fig 7).

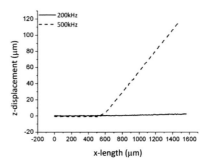

Fig. 3. Profile view of z-deformation after 10 irradiations at 200 and 500 kHz repetition rate and 3µJ pulse energy [1000x300x50µm stainless steel AISI 302, 35µm beam diameter, 10mm/s traverse speed]

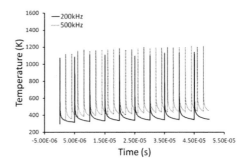

Fig. 4. FE simulation of temperature directly beneath the beam for a stationary laser heat source at 200kHz and 500kHz [1000x300x50µm stainless steel AISI 302, 3µJ pulse energy, 35µm beam diameter]

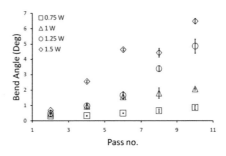

Fig. 5. Cumulative bend angle variation with successive irradiations [1000 x 300 x 50µm stainless steel AISI 302, 35µm beam diameter, 500kHz repetition rate, 10mm/s traverse speed]

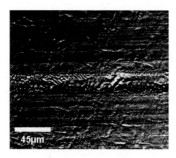

Fig. 6. Scanning electron microscope image of ablated groove after 1 scan [1000 x 300 x 50μm stainless steel AISI 302, 1500mW, 35μm beam diameter, 500kHz repetition rate, 10mm/s traverse speed]

From Fig 7 it is evident that the ablated groove has a detrimental effect on bend angle up to a point, in this instance when deeper than ~2μm. This detrimental effect can be attributed to the relatively large ablated groove depth when compared to the thickness of the thin section steel sheet. The optimum traverse speed at which a suitable combination of bend angle z-depth was achieved was found to be 35mm/s.

Whilst it was possible to limit the ablation depth for single scans, multiple irradiations caused the ablation depth to increase in an exponential fashion. This phenomenon could be attributed to a conditioning of the irradiated surface after an initial irradiation [9], increasing the absorption for subsequent scans. As such an alternative method to multiple irradiations was required to obtain a variation in bend angle. One such method investigated involved a combination of varying power and a hatched scan strategy. The hatch consisted of four single irradiation paths, scanned sequentially in alternate directions, each 30μm apart allowing for little or no overlap (Fig 8).

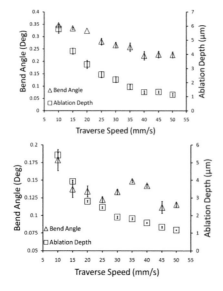

Fig. 7. Bend angle and ablation depth variation with increasing speed, 50 and 75μm thickness left and right respectively. [1000x300μm stainless steel AISI 302, 35μm beam diameter, 500kHz repetition rate, 1.5W average power]

Through variation of laser power with a single line scan strategy, controlled and repeatable micro-adjustment was achieved. The application of a hatched scan strategy increased the range over which micro-adjustment could be achieved whilst keeping the ablated groove to within~2μm depth (Fig 9).

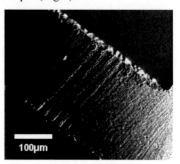

Fig. 8. Scanning electron microscope image of ablated groove after hatch irradiation strategy [1000 x 300 x 50μm stainless steel AISI 302, 1500mW, 35μm beam diameter, 500kHz repetition rate, 35mm/s traverse speed]

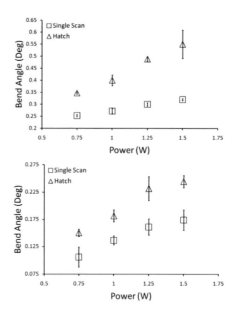

Fig. 9. Bend angle variation with increasing power, 50 and 75μm thickness top and bottom respectively. [1000 x 300μm stainless steel AISI 302, 35μm beam diameter, 500kHz repetition rate, 35mm/s traverse speed]

4. Conclusions

This investigation has demonstrated thin sheet LμF for the micro-adjustment of actuator style components conducted using picosecond duration pulses, with no absorptive or tamping layers required. This novel technique combines short pulse durations with high repetition rates and offers a method of generating localised heat build-up on the top surface of micro-scale components, allowing for controlled and repeatable micro-adjustment. Through a combination of irradiation strategies a large range of deformation is achievable. Empirical and numerical studies were conducted, with an emphasis on the use picosecond pulse durations in LµF. The pulse duration must be sufficiently short as to limit the heat diffusion depth to within a suitable range (i.e. half the sheet thickness) yet not so short that the intensity causes significant material removal. Additionally the repetition rate or degree of pulse overlap must be high enough to ensure a build up in temperature on the surface of the component suitable for thermal forming. The process has significant potential for the post-fabrication micro-adjustment of functional components in micro-electronic devices. Challenges include minimising ablation along the irradiated scan path and obtaining a larger range of deformation through optimised process parameters.

References

[1]   G. Dearden, S. P. Edwardson, Some recent developments in two- and three-dimensional laser forming for 'macro' and 'micro' applications, J. Opt. A: Pure Appl. Opt. 5 (2003) 8–15

[2]   M. Geiger and F. Vollertsen, The Mechanisms of Laser Forming, CIRP ANNALS 42 (1993), pp. 301–304

[3]   J. Magee, K. G. Watkins, W. M. Steen, Advances in Laser Forming, J. Laser Appl. 10, 235 (1998)

[4]   M. Dirscherl, G. Esser, M. Schmidt, Ultrashort Pulse Laser Bending, JLMN- Journal of Laser Micro/Nanoengineering, Vol. 1, No. 1, 2006

[5]   J. Zhou, Y. Zhang, X. Zhang, C. Yang, H. Liu, J. Yang, The Mechanism and Experimental Study on Laser Peen Forming of Sheet Metal, Key Engineering Materials Vol. 315-316 (2006) pp607 611

[6]   Kenneth R. Edwards Stuart P. Edwardson Chris Carey Geoff Dearden Ken G. Watkins, Laser Micro Peen Forming Without a Tamping Layer, Int J Adv Manuf Technol (2009)

[7]   T. J. Wieting, J. L. DeRosa, Effects of surface condition on the infrared absorptivity of 304 stainless steel, J. Appl. Phys. 50(2), February 1979

[8]   P. Mannion, J Magee, E. Coyne, G. O'Conner, Ablation thresholds in ultrafast micro-machining if common metals in air, Proc. Of SPIE, Vol. 4876 (2003)

[9]   P. Mannion, J. Magee, E. Coyne, G. O'Connor, T. Glynn, The effect of damage accumulation behaviour on ablation thresholds and damage morphology in ultrafast laser micro-machining of common metals in air, Applied Surface Science 233 (2004) 275–287.

Precision mould manufacturing for polymer optics

M. Speich[1], R. Börret[1], A.K.M. DeSilva[2] and D.K. Harrison[2]
[1]   Aalen University, Aalen, Germany
[2]   Glasgow Caledonian University, Glasgow, UK

Abstract. Ophthalmic lenses, sensors and sunglasses, for example, are commonly made of polymer materials. Therefore polymer optics is a growing market mainly for medium quality optics but also for precision optics. Polymer optics are usually produced in an injection moulding process. Due to high quality requirements on the final product the manufacturing process of the moulds has to be very accurate and precise. Thus the master tool has to be fabricated to a high quality with regard to shape accuracy and roughness. The focus of this paper is on the process chain for the fabrication of moulds for polymer optics. The typical existing process chains contain steps like to nickel plating, diamond machining and manual polishing. Some of these steps are expensive and require very experienced Staff. In the new process chain an industrial robot with special tools is used for lapping and polishing. Robot polishing replaces the former nickel plating, diamond machining and manual polishing steps. The lapping / polishing process was newly developed to achieve appropriate results. Different tool materials, polishing agents and process parameters have been tested to obtain excellent results directly on hardened steel moulds. Aalen University and UVEX ARBEITSSCHUTZ GMBH have started this project together with Saarland University Faculty of Medicine. The goal of this work is to achieve a ready to use steel mould with a roughness better than 5nm rms and an overall shape accuracy better than 4µm; with just one process step after grinding.

Keywords: robot polishing, mould manufacturing, mould fabrication, precision polishing

1. Introduction

This work focuses on the manufacturing of steel moulds for polymer optics. Polymer optics is a growing market. Applications using polymer optics can be found in almost every area of daily life moreover they are of fundamental use for military issues and medical applications f.e. [1].

Polymer injected optics are usually produced in high numbers. From low cost products with several million polymer parts per mould up to precision optics with quantities of some hundreds the demands on the moulds are very similar [2].

The topic of this work arose out of a research project with UVEX ARBEITSSCHUTZ GMBH. UVEX, a producer of safety goggles, is also using plastic injection moulding for their optics. Moulds are of different sizes

and the radius of curvature also varies. UVEX also uses this process to produce plastic visors for helmets. So there is need of spherical as well as of cylindrical moulds. Another problem is that wearers or workers who are supposed to wear safety goggles often criticize the image quality of the goggles. Fig. 1. shows a barcode seen through goggles in accordance to DIN standards [3].

Fig. 1. left: image definition without safety glasses; right: image definition with safety glasses conform to DIN EN 166 (simulation)

1.1.   Present process chain

Present process chains for mould manufacturing usually consist of 5 steps. These 5 steps can be seen in the following Fig. 2.

The first step is to shape the work piece, therefore usually milling or grinding machines are used. The goal of this first step is to get the best possible surface with a high rate of material removal. [4-6] All following processes have to deal with the results of the first material removal and sometimes structures from grinding remain on the surface until the last polishing. So the surface roughness should be already really good after this first process step. [7]

After grinding, the steel mould is usually nickel plated. Nickel plating can lead to lifetime issues because the surface can get small cracks after some hundred injection runs. So it would be interesting to avoid this step. But this step is necessary to enable the third process step, diamond machining. Normally diamond machining cannot be used for steel parts but there are several attempts to avoid nickel plating and use diamond machining directly on steel surfaces. [8] Usually diamond machining is used to bring the nickel plated surface to a good shape and to produce good roughness values. Not forgetting diamond machining is a rather expensive process, the purchase of diamond machining equipment and machines are almost impossible for small and medium sized enterprises.

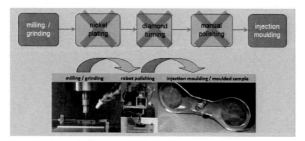

Fig. 2. Present vs. new process chain; 3 instead of 5 steps

After diamond machining the mould is normally manually polished. Manually polishing is a very labour intensive and experience requiring process step. [9] Manual polishers have to be really experienced and skilled to polish steel moulds in specifications for polymer optics. The moulds are not supposed to show any visible defects to achieve good results on polymer parts.

1.2. New process chain

The goal of this work is to find a new process chain that is also feasible for small and medium enterprises. Therefore expensive and difficult to handle process steps are to be superseded with an easy to handle cheap and stable process step. This can be seen in Fig. 2, robot lapping / polishing is used instead of nickel plating, diamond machining and manual polishing. One big advantage of the usage of an industrial robot is independence on machine restrictions. As long as the mould is in the working range of the robot it can be worked with the new process.

One robot step will not be sufficient to achieve the desired results starting from a ground sample. To obtain real results the sample material was chosen according to current mould steels. The steel used for first experiments have been hardened to approx. 59 HRC before grinding. [10]

2. Process development

Process development was started on plane steel samples to determine the best working tool material and polishing agent. The robot moved the steel sample in a constrained movement on the tool. The hardness of the tools was varied for different working steps; first lapping was done with hard tools and then the tool grew softer with every step. For polishing soft tools of polyurethane were chosen.

There have been some experiments to determine an appropriate polishing agent. Different kinds of diamond powder in all sorts of grain sizes have been investigated; starting with cheap monocrystalline diamond powder up to expensive polycrystalline powders. In first tests the cheap, edgeless diamond powder led to pores and pitting on the surface. Pitting is an undesirable defect; carbides are pulled-out of the steel matrix by diamond grain as can be seen in Fig. [11].

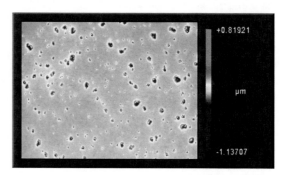

Fig. 3. Pitting after 2 polishing steps
(field of view 4.94 x 3.70 mm, PV 1956 nm, rms 80 nm)

Synthetic diamond turned out to be an appropriate polisihing agent that leads to good material removal rates on hardened steel samples.

With this polishing agent different tool materials have been investigated. Tests showed that hard tools lead to high material removal rates with a big diamond grain size; softer tools lead to better roughness values but the material removal rate was substantial smaller.

The first experiments for process development were realised on plane samples with large, plane tools. To work on spherical or cylindrical moulds or even on freeform shaped moulds it is necessary to develop a process with small tools with different radii of curvature. Therefore the knowledge gained in the experiments with large tools and plane samples was transferred to small tools and different mould shapes.

Fig. 4 shows a polishing head that was developed in Aalen University.

Fig. 4 left: Industrial robot with polishing head, right: Tool with soft cloth [12]

The polishing head rotates the tool and generates the pressure on the surface and is fixed to the industrial robot. The movement on the surface is performed by the industrial robot. The robot movement is generated with an offline programming tool that allows different robot tool paths.

2.1. Tool path

After an appropriate combination of tool and polishing agent was found the influence of the working tool path was investigated. Usually the experiments have been done with linear tool paths and two working directions. But the results showed that the tool footprint is not symmetrical, this is why the tool path was specially investigated. Fig. 5 shows the results after a polishing step with linear tool path and two working directions. It is clearly visible that the tool path is directly reproduced on the surface; this leads to a rough structure on the surface with a PV of 97 nm which is not required.

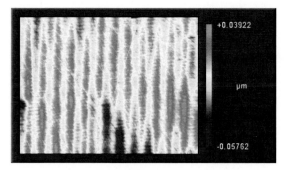

Fig. 5. Tool path visible after polishing in both meander directions (field of view 4.94 x 3.70 mm, PV 97 nm, rms 14 nm)

To remove these structures the tool path was changed and the surface was just worked in one direction. The robot movement was the same as before but the tool was not in contact with the surface on its way back. Fig. 6 shows that there is a substantial difference between polishing in both linear directions or just in one linear direction. Several other tool paths have already been investigated. [13]. Later samples have always been polished either in just one linear direction or with spiral tool path. Especially spherical moulds are often polished like this, due to rotational symmetry. The CNC code for the robot control has been generated in proprietary software from the Centre of Optical Technologies of Aalen University.

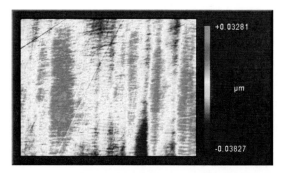

Fig. 6. Surface more homogenous with just one working direction (field of view 4.94 x 3.70 mm, PV 71 nm, rms 10 nm)

3. Conclusions

The results so far show, that the new process chain can be used to polish moulds to a ready to use state for standard moulds. Depending on tool material and mould material several steps are necessary to achieve good surfaces free of visible defects. The number of required lapping and polihsing steps is also heavily dependent on the state of the mould after grinding. The polishing agent has to be adapted to the tool material. But roughness values of rms 10nm can be obtained very steadily with different combinations of polishing agent, tool material and other process parameters. So the process is very stable.

4. Outlook

Up to now it is not certain if the polished moulds are good enough for the production of polymer optics. This will be evaluated in injection moulding tests.
Furthermore the contact area between the tool and the mould surface will be simulated with finite element method. The results of these simulations are inteded to help in characterising new materials with different surface layer properties to select an appropriate tool footprint. Furthermore the finite element analysis should help to understand the problems with the toolpath visible after polishing. The material removal is to be simulated with this new approach. Results will be used for trajectory planning of totally new tool paths.

Acknowledgments: The authors would like to thank the Bavarian Research Foundation for funding IVOS research project.

References

[1]   Beich, W.S., (2002) Plastic Optics - Specifying Injection-Molded Plastic Optics. Photonics Spectra, 2002. 36(3): 127-132.

[2]   Hering, E. and Martin, R., Fertigung optischer Komponenten und Systeme (Manufacturing of optical components and systems), in Photonik - Grundlagen, Technologie und

Anwendung (Photonics - Basics, technology and application), 2005, Springer: Berlin. 105-128.

[3]   Speich, M. and Börret, R., (2011) Mould fabrication for polymer optics. Journal of the European Optical Society: Rapid Publications, 2011. 6.

[4]   Brinksmeier, E., Mutlugünes, Y., Klocke, F., Aurich, J.C., Shore, P., and Ohmori, H., (2010) Ultra-precision grinding. CIRP Annals - Manufacturing Technology, 2010. 59(2): 652-671.

[5]   Comley, P., Walton, I., Jin, T., and Stephenson, D.J., (2006) A High Material Removal Rate Grinding Process for the Production of Automotive Crankshafts. CIRP Annals - Manufacturing Technology, 2006. 55(1): 347-350.

[6]   Mandina, M.P., Design, Fabrication, and Testing; Sources and Detectors; Radiometry and Photometry, M. Bass, Editor 2009.

[7]   Rebeggiani, S. and Rosén, B.-G., High gloss polishing of tool steels – step by step, in Proceedings of The 4th International Swedish Production Symposium 2011. p. 257-262.

[8]   Dong, J., Glabe, R., Mehner, A., Brinksmeier, E., and Mayr, P., Method for micromachining metallic materials, 2006.

[9]   Altan, T., Lilly, B., and Yen, Y., (2001) Manufacturing of Dies and Molds. CIRP Annals - Manufacturing Technology, 2001. 50(2): 404-422.

[10] Rockwell, S.P., Hardness Testing Machine, U.S.P. Office, Editor 1924.

[11] Klocke, F., Rosén, B.-G., Behrens, B., Rebeggiani, S., and Zunke, R., Towards robust polishing strategies for moulds and dies, in Proceedings of the 3rd Swedish Production Symposium, Göteborg, Sweden, 2-3 December 2009 (SPS09) 2009, Swedish Production Academy.

[12] Börret, R., Klingenmaier, J., Berger, U., and Frick, A., (2008) Minimized process chain for polymer optics. Proceedings of SPIE 2008. 7061(Novel Optical Systems Design and Optimization XI): 706118-706118-8.

[13] Tam, H.-y. and Cheng, H., (2010) An investigation of the effects of the tool path on the removal of material in polishing. Journal of Materials Processing Technology, 2010. 210(5): 807-818

Gradient vector method for computing feed-paths and hot-spots during casting solidification

M. Sutaria and B. Ravi
Mechanical Engineering Department, Indian Institute of Technology Bombay, Powai, Mumbai, India

Abstract. In this work, we present a computationally efficient approach to visualise 3D feed-paths and identify hot-spots inside a solidifying metal casting. The proposed Gradient Vector Method (GVM) involves computing the interfacial heat flux vector using an analytically derived solution to the heat transport equation, assuming virtual adjunct of material equivalent to metal-mould interface resistance. The casting geometry is divided into a finite number of radial segments and the resultant flux vector indicates the direction of the highest temperature gradient, which is normal to liquid-solid interface. The feed-path is computed by continuously tracking the successive approximate solutions. The GVM is found to be an order of magnitude faster than Level-set and other numerical methods for casting simulation. Its capability in identifying multiple hot-spots has been validated by casting experiments. Owing to its accuracy and speed, GVM is better suited to optimizing the feeding systems of castings, compared to conventional simulation methods.

Keywords: Casting, feed-path, hot-spot, shrinkage, simulation, solidification.

1. Introduction

During casting solidification and resultant volumetric contraction, hotter metal from adjacent locations flows to the solidifying region along 'feed-paths'. The feed-paths propagate along the maximum temperature gradients, and converge at the last solidifying points, the 'hot-spots'. These are the most probable locations for defects such as shrinkage cavity, porosity, and centre-line shrinkage as shown in Fig. 1(a), (b), and (c). These defects can be minimized by designing an appropriate feeding system to ensure directional solidification from thin to thick sections in the casting, finally leading to feeders.

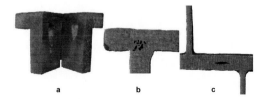

Fig. 1. Shrinkage. a. concentrated; b. distributed; c. centre-line

Over the years, a considerable amount of experimental, analytical, and numerical work has been carried out to understand casting solidification and feeding mechanism. Early experimental work to estimate the feeding distance during solidification was carried out by Myskowski et al. [1]. The feeding guidelines are empirical in nature and do not account for boundary conditions. This limitation can be overcome by the use of analytical solutions of casting solidification. Most of the early analytical work was carried out based on variants of Neumann [2] solution. Garcia et al. [3] proposed and discussed virtual adjunct method to account for resistance at metal-mould interface. Analytical solutions are however, limited to unidirectional and axisymmetrical solidification problems.

More generalised solutions use numerical methods involving finite approximation to the solution of the governing equations. The enthalpy method [4], volume of fluid method [5] and level-set-method [6] are some of the most widely used physics-driven methods to compute temperature and solid fraction during solidification. Sutaria et al. [7] proposed level-set-method for feed-path and hot-spot computation using combined Eulerian-Lagrangian framework. All these methods are based on manipulation of huge matrices and iterative calculations, making them computationally intensive [8]. Ravi and Srinivasan [9] proposed a computationally efficient geometry-driven approach called Modulus Vector Method (MVM), using geometric modulus (ratio of volume to heat transfer area) to compute feed-paths.

From the literature, four major conclusions can be drawn. First, feeding distance rules in general do not account for complex geometries, as well as variation in material properties and boundary conditions. Second, analytical solutions are applicable to only simple unidirectional and axisymmetric problems and are of limited use in practice. Third, numerical techniques can overcome above limitations and are versatile, but require accurate values of material properties and boundary conditions and are computation intensive. Fourth, the MVM takes less computation time but is essentially a geometry-based technique and appears to lack in the accuracy of feed-path profile, especially near the metal-mould boundary. The aim of the present work is to evolve an improved method to compute hot-spots and feed-paths, suited to automatic evaluation and optimization of feeding systems. The accuracy of the proposed method is benchmarked with MVM as well as level-set-method.

2.   Gradient Vector Method (GVM)

A casting solidifies progressively from the mould boundary until it converges to last solidifying points in the mould-cavity, known as hot-spots. Feeding at any given point (compensation for volumetric contraction) occurs in the direction of the highest temperature gradient along the liquid-solid interface.

2.1. Interfacial heat flux

The principle of the proposed GVM is illustrated with a control volume (Fig. 2). Consider a point P_i on liquid-solid interface at time τ. The control volume is divided into n radial segments from P_i. Taking infinitesimally small segment angle, the interfacial heat flux due to evolution of latent heat in radial segment s is given by:

$$q_s = \rho_c L (\partial r / \partial \tau) \qquad (1)$$

where, ρ_c is density of cast metal, L is latent heat evolved, and $\partial r / \partial \tau$ represents liquid-solid interface velocity (solidification velocity). The expression for interfacial velocity is derived based on following assumptions:

- The heat flow is unidirectional in each segment, along the radial direction.
- The heat transfer coefficient at metal-mould interface remains constant during solidification.
- Temperature diffusivity of the metal and mould material is independent of the temperature.

Solidification time is computed for each segment by adding a virtual layer of solid metal and mould as suggested by Garcia et al. [3]. The basic notion involves consideration of interface resistance as equivalent to pre-existing adjunct of material. The thermal contact between metal and mould is then assumed perfect and represented by an infinite heat transfer coefficient. The expression for solidification time is:

$$\tau_{sol} = \frac{1}{4\varphi^2 \alpha} \left(\frac{r}{2} \right)^2 + \frac{k_c}{hSt\alpha} \left(\frac{r}{2} \right) \qquad (2)$$

where, r is average radius of segment, φ is solidification constant, k_c is thermal conductivity of the metal, α is thermal diffusivity of cast metal, h is heat transfer

coefficient at metal-mould interface and St is Stefan number.

$$St = C_c \left(T_f - T_m \right) / L$$

where, T_f is freezing temperature of the metal, T_m is mould temperature, and C_c is specific heat of cast metal.

Solidification constant φ is computed by deriving thermal balance at liquid-solid interface [Eq. (3)].

$$\sqrt{\pi}\varphi e^{\varphi^2} \left(Dt + erf\left(\varphi \right) \right) = St \qquad (3)$$

where, $Dt = \sqrt{k_c \rho_c C_c / k_m \rho_m C_m}$. Here, ρ_m, k_m, and C_m are density, thermal conductivity, and specific heat of mould material, respectively.

Interface (solidification) velocity is computed by differentiating Eq. (2) with respect to r:

$$v_{sol} = \frac{\partial r}{\partial \tau} = \frac{8\varphi^2 \alpha StBi}{r\left(StBi + 4\varphi^2 \right)} \qquad (4)$$

where, Biot number, $Bi = hr/k_c$. Interfacial heat flux for segment s is computed by substituting Eq.(4) into Eq.(1)

$$q_s = \rho_c L \frac{8\varphi^2 \alpha StBi}{r\left(StBi + 4\varphi^2 \right)} \qquad (5)$$

The expression given by Eq. (5) shows that interfacial heat flux is a function of casting geometry, material properties and boundary conditions. At a hot-spot, the vector sum of these segmental interfacial heat fluxes is zero; here the liquid-solid interface vanishes, forming a singularity. If the vector sum is non-zero, the resultant vector indicates the direction of the highest temperature gradient, which is normal to liquid-solid interface profile. Interface movement occurs in the opposite direction as shown in Fig. 2. The resultant interfacial heat flux is directed outwards (solid phase) where as interfacial velocity vector is directed inwards (liquid phase).

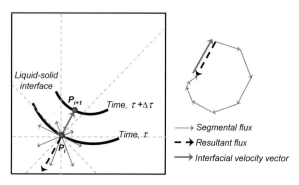

Fig. 2. Computation of interfacial heat flux and velocity vector

2.2. Computation of hot-spot and feed-path

Feed-path is computed by continuously tracking the direction of the resultant interfacial heat flux. The methodology is illustrated with a control volume as shown in Fig. 3. Let us assume that the hot-spot lies at position P_i on liquid-solid interface as an initial solution. Consider a mass-less particle lying at P_i and compute the direction of resultant interfacial heat flux by dividing the casting geometry into infinitesimal number of segment as explained in previous section. Direction of interfacial velocity vector and resultant interfacial heat flux will be opposite to each other as shown in Fig. 3. The mass-less particle is marched along the interfacial velocity vector direction with a small step Δs as shown by thin arrows in Fig. 3. The new position of mass-less particle is:

$$P_{i+1} = P_i + \vec{v}\Delta s \qquad (6)$$

where, $\vec{v}$ is the unit vector along interfacial velocity.

The procedure is repeated at P_{i+1} and continued until it reaches a position where the solution converges (the resultant interfacial heat flux is negligible). This last position P_{i+m}, correspond to the point of local maxima of temperature, and represent the local hot-spot. During solidification, when temperature T_i of the molten metal at position P_i reaches the solidus temperature, the feed metal to compensate volumetric shrinkage is supplied from P_{i+1}, since it has the highest temperature gradient toward P_i. Thus feeding occurs in opposite direction to the mass-less particle movement, that is, from P_{i+m} to P_i as shown by thick solid arrow in Fig. 3.

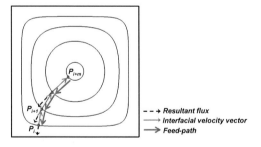

Fig. 3. Hot-spot and feed-path computation

3. Benchmarking and validation

Analysis of two sand castings has been carried out to demonstrate and evaluate the performance of GVM. The first one is a 4.5 Kg 'C' shaped cast steel (0.2% C) part with thickness 0.1 m. The second one is a 2.5 Kg multiple junction ductile iron (500/7) part with thickness 0.04 m. They have different geometric features like re-entrant corners and junctions. The feed-path and hot-spot results are compared with MVM [9] and benchmarked with result obtained by LSM [7]. Finally, validation is carried out by performing experiments.

3.2. Benchmarking with MVM and LSM

Propogation of feed-paths in the right and left arm of the 'C' shaped casting, computed using GVM, is shown in Fig. 4 (a). The GVM feed-paths starting from all part boundaries are shown in Fig. 4 (b). The feed-paths from the same starting points, computed using MVM, are shown in Fig. 5 (a). The GVM feed-paths are (correctly) normal to the part boundary, whereas MVM feed-paths are slightly inclined. The GVM (correctly) exhibits a series of hot-spots, whereas MVM exhibits a single hot-spot in the right bottom junction. The feed-paths computed using LSM are shown in Fig. 5 (b). The direction of the feed-paths as well as their convergence points (hot-spots) obtained by LSM are in close agreement with those obtained from GVM.

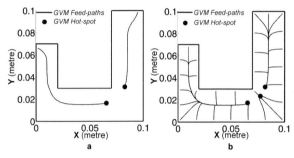

Fig. 4. 'C' shaped casting. a. GVM feed-paths in arms; b. GVM feed-paths from part boundary

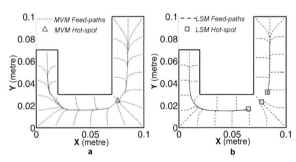

Fig. 5. 'C' shaped casting. a. MVM feed-path; b. LSM feed-path

In the second example, both GVM and MVM feed-path converge to the nearest hot-spots (in the junctions) as shown in Fig. 6. The MVM hot-spots are slightly pulled toward the arm sections. The GVM and MVM hot-spot location differ by about 1% of overall domain dimension. The overlapping plots of feed-paths by GVM and LSM shown in Fig. 7 reveal a close match between feed-path profiles and location of hot-spots.

The computations were performed on a 64-bit Win-XP computer equipped with 2.6 GHz processor and 8 GB RAM. The LSM results took about 105 minutes (excluding pre and post-processing) for a mesh size of 1 mm. Both MVM and GVM results took less than 11 minutes with a step size of 1 mm.

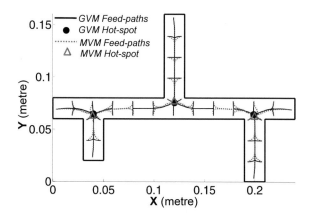

Fig. 6. Multiple junction casting: GVM and MVM results

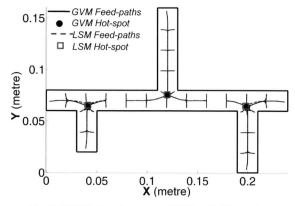

Fig. 7. Multiple junction casting: GVM and LSM results

4. Conclusions

Feed-paths provide a better visualization of feeding directions inside a casting compared to the traditional concept of feeding distance, which is essentially based on experimentally-derived empirical expressions for specific metal-mould combinations and limited to simple shapes. The proposed Gradient Vector Method has been shown to accurately compute feed-paths in complex geometries, as well as identify multiple hot-spots. The GVM could correctly identify multiple hot-spots, and its feed-paths are more reliable than those produced by Modulus Vector Method. This is attributed to the fact that the direction of the maximum temperature gradient in GVM is computed from interfacial heat flux, as a function of geometric parameters, material parameters and boundary conditions. It can thus potentially handle material properties and various thermal-boundary conditions encountered in foundries, being based on analytical solution for heat transport equation. Further, the GVM is shown to be about ten times faster than the numerical techniques (level-set-method). This becomes important when different configurations of casting feeders (location, shape, size, etc.) need to be analysed to determine the most optimum one that gives the desired quality at the highest possible yield.

3.3. Validation by casting

The two castings were produced in sand moulds with the metals mentioned earlier to validate the simulation results. The convergence location of GVM feed-paths in 'C' shaped casting correctly matches with the extended shrinkage cavity observed in the cut-section [(Fig. 8 (a)]. In the multiple-junction casting, three distinct convergence points of feed-paths were identified during simulation. These correctly match the location of shrinkage porosity observed in cut-section [Fig. 8(b)], proving the capability of the method to predict multiple hot-spots.

References

[1]   Myskowski ET, Bishop HF, Pellini WS, (1952) Application of chills to increasing the feeding range of risers. AFS Transactions 60:389-400

[2]   Carslaw HS, Jaeger JC, (1959) Conduction of heat in solids. Oxford University, 2nd edition

[3]   Garcia A, Clyne TW, Prates M, (1979) Mathematical model for the unidirectional solidification of metals: II. Massive molds. Metallurgical Transactions B 10:85-92

[4]   Voller VR, Cross M, (1981) Accurate solutions of moving boundary problems using the enthalpy method. International Journal of Heat and Mass Transfer 24:545–556

[5]   Welch SWJ, Wilson J, (2000) A volume of fluid based method for fluid flows with phase change. Journal of Computational Physics 160:662-682

[6]   Chen S, Merriman B, Osher S, Smereka P, (1997) A simple level set method for solving Stefan problems. Journal of Computational Physics 135:8-29

[7]   Sutaria M, Gada VH, Sharma A, Ravi B, (2012) Computation of feed-paths for casting solidification using Level-Set-Method. Journal of Material Processing Technology, doi: 10.1016/j.jmatprotec.2012.01.019

[8]   Viswanathan S, Duncan AJ, Sabau AS, Han Q, Porter WD, Riemer BW, (1998) Modeling of solidification and porosity in aluminum alloy castings. AFS Transactions 106:411-417

[9]   Ravi B, Srinivasan MN, (1996) Casting solidification analysis by modulus vector method. International Cast Metals Journal 9(1):1-7

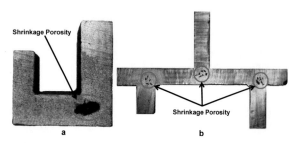

Fig. 8. Cut-section. a. 'C' shaped casting; b. Multiple junction casting

Effect of casting orientation, thickness and composition on solidification structure and properties of ductile iron castings

V. D. Shinde [1], B. Ravi [2] and K. Narasimhan [3]

[1]Department of Metallurgical Engineering & Materials Science, Indian Institute of Technology Bombay, Mumbai-400076 (India), shinde@iitb.ac.in

[2]Professor, Department of Mechanical engineering, Indian Institute of Technology Bombay, Mumbai-400076 (India)

[3]Professor, Department of Metallurgical Engineering & Materials Science, Indian Institute of Technology Bombay, Mumbai-400076 (India)

Abstract. In this work, the effect of casting orientation (horizontal, vertical), thickness (4-16 mm) and composition (Cu, Mn) were investigated on the mechanical properties (tensile strength, yield strength, elongation, hardness) of hypereutectic ductile iron castings. Thermocouples were placed in the middle of four sections of a step casting. Overall horizontal castings were found to cool faster than vertical castings. Thermal analysis (cooling curves) shows a wide difference among the four sections. Thinner sections exhibited significant undercooling and thereby carbide formation, leading to poor ductility. The combined effect of Cu and Mn gives higher strength with a drop in ductility especially in thin sections, which can be used to balance carbide formation.

Keywords: Thin wall casting, ductile iron, solidification, inoculation, microstructure

1. Introduction

The transportation industry faces three major challenges: control of emissions, improvement of fuel efficiency and reduction of costs. One solution to all these challenges is to reduce vehicle weight; a reduction of 100 kg is commonly equated to a fuel efficiency improvement of 0.4 km per liter [1]. Since castings constitute a significant proportion of vehicle weight, manufacturers are exploring weight reduction by redesigning the castings with thinner walls [2, 3]. It is however, a challenge to produce thin wall ductile iron casting with the desired properties.

Solidification of hypoeutectic ductile iron proceeding below liquidus temperature causes nucleation of austenite; graphite spheroids nucleate on pre-existing austenite and grow in the interdendritic regions. In hypereutectic melts, solidification starts with graphite nodules [4, 5], with subsequent decrease in carbon % in the liquid. Upon further cooling, austenite grows dendritically, allowing new graphite spheroids in interdendritic regions [6]. Graphite nodules nucleate on

small inclusions [7] but further growth solely depends on foreign particles or solutes which are added as inoculant [4]. Austenite formed during solidification undergoes solid state transformation while cooling below eutectic temperature, overlapping the solidified structure and making solidification mechanisms more complex [7, 8, 9, 10]. Since it is preferable to have more graphite nodules, especially in thinner sections, hypereutectic ductile irons are preferred for such castings.

The chemical composition, melt treatment and cooling rate are important process parameters which determine the final properties of ductile iron. The graphite nodule count, nodularity (deviation from spherical shape) and the amount of phases are important control parameters to achieve better properties. Melt treatment includes modification and inoculation, in which initially the melt is treated with magnesium alloy (for changing graphite shape from flake to spheroidal) followed by further inoculation (for increasing the nodule count or to suppress carbide formation) to facilitate heterogeneous nucleation [6]. Sufficient graphite nucleation is required in order to avoid formation of carbides due to higher solidification rates in thin wall ductile iron castings [11, 12].

Heat transfer behavior during solidification of actual castings can be studied by inserting suitable thermocouples in the casting cavity. The measured cooling curve reflects the effect of solidification variables such as chemical composition, inoculation and its efficiency [13]. The melt quality is controlled by the chemical composition, pouring temperature, efficiency of inoculation and shrinkage tendency [14].

The commonly measured mechanical properties for ductile iron are tensile strength, yield strength, percent elongation and Brinell hardness. Because of the fairly consistent influence of spheroidal graphite and structure of its base matrix, the tensile properties and the Brinell hardness of ductile iron are well related. In the matrix, the softer ferrite gives higher ductility but lower strength than pearlite. Also the graphite morphology plays an important role, as the graphite shape deviates from the ideal spherical shape, the ductility and strength reduce [15]. The time after spheroidal treatment has significant effect on elongation, but less effect on the tensile strength and hardness of castings.

Chemical composition plays a vital role in solidification processing of ductile iron. Even small changes of the elements show significant increase or decrease in mechanical properties of ductile iron [16]. Silicon is a strong solid solution strengthener; it reduces undercooling and avoids carbide formation by nucleating graphite. It segregates negatively. Copper and Manganese are pearlite stabilizers. Chromium is a pearlite former as well as a carbide former which leads to segregation. Molybdenum is pearlite stabilizer and promotes hardenability [17, 18]. Arsenic and Tin are subversive elements and hence are kept within controlled limits [4].

During nodularizing, numerous inclusions are formed with a sulphide core and an outer shell containing complex magnesium silicates. After inoculation with a ferrosilicon alloy containing Ca, Ba or Sr, the surface of magnesium silicate micro-particles is modified and other complex silicate layers will be produced [8]. Such silicates have the same hexagonal crystal lattice structure as graphite. Due to very good lattice match they act as effective nucleation sites for graphite nodules to grow from the melt during solidification. The particles with good epitaxial fit between nucleant and graphite embryo are the most potent catalysts to grow graphite [9].

As some of the measured Magnesium is in the form of MgS, the final level of Sulfur affects the Magnesium needed to result in nodular graphite. Maximum nodularity can be achieved by keeping magnesium residual just enough, but decreasing magnesium below 0.02% will deteriorate the nodule shape (spheroicity). Nodule count can be maximized by sound base iron melting along with good inoculation practice [18]. Nodule count and nodularity, both are affected by cooling rate. Thin sections (due to fast cooling) result in better nodule shape than slower cooling sections for the same magnesium residuals.

Previous work shows that for improving the strength of a casting, Copper is an important constituent. Manganese is used in this study for strengthening the casting by promoting pearlitic matrix, but increasing manganese may alter the structure by promoting carbides in different section thicknesses. This effect can be studied by varying the amount of manganese and copper, which balance strength and ductility. Further, the effect of different section thickness and casting orientation for a given composition of ductile iron also needs to be studied. These have been taken up in the present investigation.

2. Experimental work

In the current study, experiments were designed and conducted to study solidification behavior in varying thickness ductile iron castings as shown in Table 1. A step casting was designed with four sections having thickness 4, 8, 12 and 16 mm respectively as shown in Fig. 1. Each step is 50 mm long, making the total length of casting 200 mm. The width of the casting is 100 mm, so as to avoid end freezing effects in all sections. Multiple gates were provided (one in each section of the casting) for rapid and uniform filling. A wooden pattern was fabricated, and used to prepare the molds in green sand. K-type thermocouples were inserted in the middle of each step to record the thermal history of casting solidification. Two castings each with vertical and horizontal orientation are poured and four castings with varying Cu and Mn composition are also cast (Fig. 2). A 16-channel data logger (Ambetronics), capable of storing 2995 readings for each port, was used.

Table 1. Experiments conducted and composition

Melt code	Casting orientation	Composition
V1	Vertical	Trace Cu, 0.2% Mn
V2	Vertical	Trace Cu, 0.2% Mn
H1	Horizontal	Trace Cu, 0.2% Mn
H2	Horizontal	Trace Cu, 0.2% Mn
A	Horizontal	Trace Cu, 0.2% Mn
B	Horizontal	0.2% Cu, 0.3% Mn
C	Horizontal	0.4% Cu, 0.4% Mn
D	Horizontal	0.5% Cu, 0.5% Mn

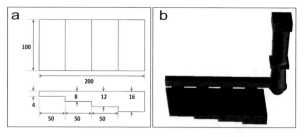

Fig. 1. (a) Details of the step casting, (b) 3D model of step casting

Fig. 2. Horizontal Castings with composition (A, B, C, D)

The melt charge consisted of 50 kg pig iron, 150 kg cold rolled steel scrap and balance foundry returns with suitable chemical composition. The charge mix was melted in 300 kg capacity coreless medium frequency induction furnace. The molten metal was tapped into a preheated ladle containing Ferro-silicon- magnesium (FeSiMg) alloy granules of size 10-15 mm at the bottom covered with steel scrap (sandwich process). The tapping temperature was 1450 °C.

The inoculant was added in the melt stream while pouring for proper mixing. Inoculant particles were of 2 to 3 mm in size so as to dissolve easily and dust free to avoid oxidation losses. The spectroscopic analysis of the melt samples was carried out just before pouring into the mold, using a spectrometer (BRUKER, model Q-4 Tasman).

The treated iron was poured into mold cavity at a temperature of 1380 °C for all castings. Four experiments (two vertical and two horizontal castings) were conducted for studying the effect of casting orientation with the same chemical composition. The measured tensile and Brinell hardness readings on each steps of casting H1 were reported in Table 2. The thermocouples were inserted in the castings H2 and V2.

Table 2. Tensile and hardness test results in step casting H1

Thickness (mm)	T. S. (MPa)	0.2% Y.S. (MPa)	Elongation %	BHN
4	426.33	338.83	12.30	207
8	418.43	326.23	13.75	192
12	412.51	317.47	14.87	179
16	394.38	310.72	14.95	177

Another four experiments studied the effect of chemical composition for a given orientation (horizontal). In all the four melts (labeled A, B, C, and D), C=3.6% and Si=2.5-2.78% gave a carbon equivalent of 4.33-4.43, which is in hyper-eutectic range. Cu varied from 0.035 to 0.512% and Mn varied from 0.216 to 0.518% as shown in Table 3. Other elements present in the melt were Pb<0.01, Al=0.006, Cr=0.015, Mo<0.002, Ni<0.002 and Ti=0.02.

Table 3. Spectroscopic analysis of the castings A, B, C, D

Melt No.	C	Si	Cu	Mn	Mg
A	3.62	2.51	0.035	0.216	0.035
B	3.63	2.69	0.214	0.310	0.034
C	3.68	2.68	0.401	0.392	0.032
D	3.61	2.78	0.512	0.518	0.039

3. Results and discussion

The solidification temperature history recorded using thermocouples and stored in the data logger was used to plot the cooling curves for each section thickness (Fig. 3). The amount of undercooling was noted in each graph. In horizontally oriented sections, the under-cooling varied from 18 °C in thinnest to 5 °C in thickest sections. In vertically oriented sections, the corresponding variation was from 16 to 8 °C.

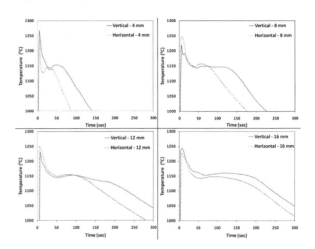

Fig. 3. Cooling curve for vertical and horizontal castings at the center with different wall thickness

Tensile test specimens were prepared from each casting as per ASTM standard E8M-04. The brinell hardness is measured on the samples taken from the middle portion of each casting. The tensile and hardness results are shown in Table 4. The samples for micro-structure studies were taken from the middle portion of the casting and polished. These were etched with 2% Nital (2% concentric Nitric acid and 98 ml Methanol solution). Optical micrographs were taken using a camera attached to a Leintz microscope (Fig.4). The polished samples were studied using an Image Analyzer (Pro-metal-11) for nodule count, nodularity and percentage of ferrite and pearlite content (Table 5).

Table 4. Tensile testand hardness results of the castings

Melt No.	T.S. (MPa)	0.2%Y.S.(MPa)	Elongati on %	Hardness (BHN)
A	412.53	307.34	14.87	179
B	456.58	327.73	10.23	187
C	532.46	382.32	7.35	224
D	614.61	417.23	2.85	247

Table 5. Microstructure image analysis of the castings

Melt	Pearlite %	Ferrite %	Nodularity %	Nodule count
A	11.80	88.19	91.07	395
B	24.56	75.40	91.23	330
C	60.12	39.89	88.52	207
D	80.28	19.72	94.68	199

The thermal analysis of the four sections within the horizontal castings indicates faster cooling rate compared to corresponding sections in vertical casting. Irrespective of the orientation, approximately 5% carbides are found in thinner sections. In vertical casting, the metal is not stagnant during mold filling delays the nucleation leading to lesser amount of carbide formation.

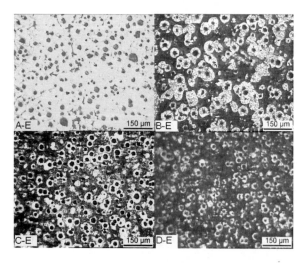

Fig. 4. Microphotographs of ductile iron castings with 2% nital A to D in 12 mm sections etched, labled A-E to D-E respectively

This is more clearly reflected in the cooling curve of 4 mm thin vertically oriented casting. Copper and Manganese promote pearlite and need to be controlled to achieve the desired mechanical properties of ductile iron castings. The effect of copper is to increase the amount of pearlite; manganese stabilizes the pearlite but also promotes carbides. The simultaneous increase of both Cu and Mn enhances both tensile and yield strengths without a significant decrease in ductility (as compared to that observed by an increase in Cu alone). The microstructure study indicates no traces of carbides in 16, 12 and 8 mm sections and only 4 to 5% carbides in 4 mm thick sections.

The nodule size distribution affects shrinkage tendency, since it reflects graphite formation and expansion throughout the entire solidification sequence. Small nodules and uniform distribution indicates an early graphite nucleation which may lead to micro-porosity and wide range of graphite nodule sizes indicating continuous nucleation of graphite while solidification. The amount of under-cooling, which is measure of the energy barrier against heterogeneous nucleation increases with lattice disregistry in the structure of respective regions in the casting.

4. Conclusions

The properties of ductile iron as indicated by their grades are largely determined by their microstructure, which in turn is affected by section thickness of the casting and chemical composition of the melt. Thin wall (4 mm) sections are more prone to deep undercooling and carbide formation, especially in horizontal orientation. The microstructure and thereby mechanical properties (especially tensile strength and ductility) can be improved in thin wall ductile iron castings by simultaneously increasing the amount of copper and manganese. The combined addition of Cu and Mn varying from 0.1 to 0.5% increased the amount of pearlite from 15 to 80% in the ductile iron castings, which in turn increased the strength from 413 to 615 MPa. The corresponding fall in ductility (percentage elongation) was 15% to only 3%.

Acknowledgement: Authors gratefully acknowledge the assistance of Ganesh Foundry, Magna Industries, and S.S. Industries, Ichalkaranji for supporting melting trials, thermal analysis and microstructure analysis.

References

[1]   Dogan O, Schrems KK, Hawk JA (2003) AFS Transactions, 111: 949–959.
[2]   Bockus S, Venckunas A and Zaldarys G (2008) Materials Science, 14(2): 1392-1320.
[3]   Javaid A., Davis K.G. and Sahoo M. (2000), Modern Casting, 6:33-41.
[4]   Flemings MC, (1974) Solidification Processing. McGraw-Hill Book Company, New York.
[5]   Skaland T (2005) Nucleation mechanisms in cast iron, Proceedings of the AFS Cast Iron, September 29-30, Schaumburg, Illinois,13-30.
[6]   Rivera GL, Boeri RE and Sikora JA (2003) AFS Transactions, 111, 979-989.
[7]   Skaland T, Grong Ø and Grong T (1993) Metallurgical and Materials Transactions A, 24:2321-2345.
[8]   Elliot R, (1988) Cast Iron Technology. Butterworth-Heinemann, Oxford.
[9]   Rio Tinto Iron and Titanium, Ductile Iron Databook for Design Engineers, Montreal, 1999.
[10] Campbell J (2004) Castings, Elsevier, Amsterdam.
[11] Stefanescu DM, Ruxanda R and Dix LP (2003) Int. J. Cast Metals Research, 16 (3): 319-324.
[12] Fras E, Gorny M (2007) Archives of Foundry Engineering, 7(4):57-62.
[13] Pedersen KM and Tiedje NS (2008) Materials Characterization, 59:1111-1121.
[14] Sparkman D, (1994) Modern Casting, 84(11):35.
[15] Bockus S, Dobrovolskis A (2004) Materials Science (Medziagotyra), 10(1):1392-1320.
[16] Soinski MS and Derda A (2008) Archives of Foundry Engineering, 8(3):149-152.
[17] Gonzaga RA and Carrasquilla JF (2005) Journal of Materials Processing Technology, 162: 293-297.
[18] Gonzaga RA, Landa PM, Perez A and Villanueva P (2009) Journal of Achievements in Materials and Manufacturing Engineering, 23(2):150-158.

Laser beam shaping for manufacturing processes

Duncan Hand

Institute of Photonics and Quantum Sciences, Heriot-Watt University, Edinburgh EH14 4AS

Abstract. The standard laser beam shape (typically Gaussian or similar) is not always optimal for a particular process. In this presentation the use of adaptive optics (both deformable mirrors and spatial light modulators) to dynamically alter the beam shape is investigated for laser precison machining processes.

Keywords: laser precision machining, laser beam shaping, laser-based manufacturing processes

1. Summary

A particular laser can in general provide a single beam shape, whilst laser-based manufacturing processes often require many different illumination strategies and/or patterns. These can be generated by rapidly moving the focused laser beam relative to the workpiece by means of a high speed optical scanner, trepanning optic, or motorised stages. However, there are applications where this kind of beam manipulation is not sufficient, where it is instead preferable to use spatial beam shaping optics. Such shaping can be achieved with fixed optics, including: masks, diffractive optical elements (DOEs) [1], and phase plates [2]; however the ability to dynamically alter the beam shape offers much increased process flexibility, as well as enabling implementation of speckle reduction techniques (speckle is typically a problem with DOEs). There are various kinds of adaptive optics which can be used in this way, splitting into 2 key types: (i) deformable mirrors and (ii) spatial light modulators. Both can be applied to laser manufacturing processes [3,4], with liquid crystal spatial light modulators having the distinct advantage of being able to operate as either controllable phase plates, DOEs, or amplitude masks. A wide variety of complex beam shapes can therefore be generated at the workpiece (examples are shown in Fig. 1), and these can be dynamically manipulated.

In this presentation the application of dynamic beam shaping techniques to laser precision machining processes will be described, for both nanosecond and picosecond lasers, using green and IR light. The extension of such processing to UV wavelengths will also be discussed.

Applications described include structuring of metals, glasses and ablation of thin films.

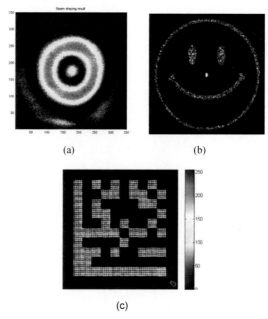

(a)   (b)

(c)

Fig. 1. Example beam shapes produced using liquid crystal spatial light modulators, (a) as a phase plate to generate donut beam at laser focus; (b) and (c) as dynamically variable diffractive optics.

2. Biography

Duncan has been a member of academic staff at Heriot-Watt since 1997. His work on laser-based manufacturing includes laser precision machining; the use of adaptive optics in laser manufacturing processes; and laser joining of microsystems. In this work he collaborates with a range of companies including GE Aviation, Renishaw, BAE Systems and Selex. He has further research activity on the delivery of high peak power laser light through novel optical fibres (with applications in manufacturing and medicine), including a collaboration with the University of Bath on photonic bandgap fibres. He also has an interest in optical sensing, with current activity in optically-addressed fibre optic micro-cantilever sensors. Duncan is also currently Director of Research for the School of Engineering and Physical Sciences.

References

[1]   E. Neiss, M. Flury, J. Fontaine, Diffractive optical elements for laser marking applications - article no. 70032L, in Optical Sensors 2008, F. Bergmans, et al., Editors, Spie-Int Soc Optical Engineering: Bellingham. p. L32 (2008)

[2]   A. Bich et al, Multifunctional micro-optical elements for laser beam homogenizing and beam shaping' article no. 68790Q, in Photon Processing in Microelectronics and Photonics Vii, A.S. Holmes, et al., Editors, SPIE-Int Soc Optical Engineering: Bellingham. pp. Q8790-Q8790A (2008)

[3]   R.J. Beck, J.P. Parry, W.N. MacPherson, A. Waddie, D.T. Reid, N. Weston, J.D. Shephard, D.P. Hand. Adaptive optics for optimization of laser processing. in Proceedings of LAMP2009 – the 5th International Congress on Laser Advanced Materials Processing. 2009. Kobe, Japan.

[4]   Z. Kuang, D. Liu, W. Perrie, S. Edwardson, M. Sharp, E. Fearon, G. Dearden, K. Watkins, Fast parallel diffractive multi-beam femtosecond laser surface micro-structuring. Applied Surface Science, 255(13-14): pp. 6582-6588 (2009)

Computer Aided Engineering

A haptic approach to computer aided process planning

C.A.Fletcher[1], J.M.Ritchie[1] and T.Lim[1]
[1] School of Engineering & Physical Sciences, Heriot Watt University, UK

Abstract. Current solutions in Computer Aided Process Planning (CAPP) can be time consuming, complex and costly to employ and still require the input of an experienced planner. Implementations can require a high degree of configuration, particularly when pre-existing knowledge within the company needs to be incorporated. This means due to the time and expense many companies, particularly smaller ones, do not employ CAPP systems. Within the process planning domain of machining, decisions need to be made regarding routing and processes, with each choice making a significant impact on the final cost and quality of the product. Existing CAPP systems need to be carefully configured as they rely on artificial intelligence or existing data to find the best routes and processes. These systems tend to be very specialised and their configuration can be tper presents a prototype haptic virtual machining application where machining operations are simulated whilst visual and tactile information is fed back to the operator to enhance their experience. As they generate their maching sequence the operator input is logged. By logging their activities it is shown that specialist knowledge can be accumulated unobtrusively and formalised such that the system can immediately generate usable process plans without the need for lengthy configuration and formalisation. Experimental findings show how using a virtual reality (VR) environment can clearly represent a machining task and that relevant knowledge and data can be quickly captured during a simulation. Simulating machining tasks in this way offers a unique non-intrusive opportunity to collect important information relevant to a machining process; this information can then be used further downstream during manufacturing.

Keywords: CAPP, haptics, virtual environment, process planning

1. Introduction

Computer Aided Process Planning is the link between Computer Aided Design and manufacturing, A process plan comprises the selection and sequence of operations and associated processes to transform a chosen raw material into a finished product [1]. A plan includes routing sheets, operation lists and tool lists which detail machining operations, machine scheduling and tooling requirements.

Traditionally process plans were carried out by a senior engineer and by combining their expertise with the part requirements a process plan would be generated that could then be passed to the shop floor for manufacture. The problem with plans generated this way is that they suffer from excessive clerical content, lack consistency and are dependent on the knowledge and experience of the planner.

Early evolutions of PC software aimed at addressing these problems were variant CAPP systems. A variant CAPP system is knowledge based system, which draws from previous designs to generate new or partial plans, through the use of a classification code [2]. These plans can then be taken as a base line by the process planner for further refinement into the final plan.

The next evolutions of CAPP systems were termed generative. These are based on artificial intelligence algorithms to automatically create unique plans for each new design. The intention was to create a fully automated process plan generated directly from the design files. This would allow a fast and seamless transition direct to manufacturing. However, these plans invariably needed to be finished off via some manual input, or would be limited to a very specific part family of components. A further review of papers written from 2009 onwards reveals that much of the research is still focused on trying to create automated systems [3-6], are based on feature recognition [7-9] are knowledge based systems [10,11] are based on a genetic algorithms.

In spite of these extensive research efforts, the uptake of automated CAPP systems into industry has been slow [12] and they tend not to be used in small to medium enterprises (SMEs) at all [13]. With considerable potential savings.

A short survey of 19 SMEs (Fig. 1 Survey of reasons for non implementation of CAPP systems) indicated some of the reasons for CAPP systems not being used: (i) there is no expected time saving; (ii) they require considerable training; and (iii) the final results are not always optimal.

S. Hinduja and L. Li (eds.), *Proceedings of the 37th International MATADOR Conference*,
DOI: 10.1007/978-1-4471-4480-9_2, © Springer-Verlag London 2013

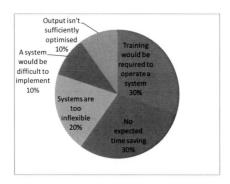

Fig. 1 Survey of reasons for non implementation of CAPP systems

In order to address some of these issues this paper proposes a different approach using a novel virtual reality (VR) paradigm to create a simulated machining environment via haptic routing. In this environment operators will be able to generate the cutting sequence and simulate the machining of a part in real time. Whilst the part is being machined relevant information is automatically logged and used to generate a unique process plan.

Virtual reality can be defined as the creation in software of a real world situation [14]. The benefits a virtual environment can bring to the human experience include: (i) the ability to rapidly prototype different solutions; (ii) a better understanding of problems through improved visualization; and (iii) more collaborative approaches due to the ability to share complex information more easily.

Central to this application is the inclusion of a haptic device. This is an electro-mechanical device which introduces a sense of feel into the virtual environment allowing the operator to interact with models in a more natural way [15].The phantom omni haptic device by Sensable was used in this instance since the movements required for simulating routing cutting operations are similar to several types of part programming and planning approaches used in a typical workshop. By creating an abstract simulation of a machining environment, it is thought that a semi-automated process planning tool can be developed; which will give the process planner the support that is needed, to quickly and efficiently evaluate cutting sequences, and automatically generate plans. The latter feature addresses the issues found in traditional process planning by removing excessive clerical content and the lack of plan consistency. Further, because the system uses a virtual environment similar to a machining environment it should be intuitive and allow the development of highly optimized easily editable solutions. Indeed, the virtual system aims to integrate the intelligence of the process planner and the haptic interface by allowing the human to be immersed in, and experience the process rather than being involved in post editing.

This paper presents a pilot study of a virtual haptic machining application, addressing whether or not a virtual haptic approach can be used to generate plans more efficiently than in the traditional manner. After an initial explanation of the system implementation, the experimental method is explained followed by a short discussion of results, conclusions and future work.

2. Implementation

The virtual environment (Fig. 2. Virtual Environment including Haptic Device) consists of commercially available hardware and an in-house developed software application.

The hardware includes a Sensable Phantom Omni haptic device and a Dell T1500 Workstation, which includes an INTEL CORE I5-750(2.66GH) CPU, 4GB 1333MHZ DDR3 RAM, NVIDIA Quadro FX580 GPU.

The software consists of a multi-threaded application, comprising of a graphical interface, models with physical attributes and a haptic module to provide sensory feedback. The haptic application requires 1 kHz rendering frequency to reproduce forces convincingly. Software libraries used include OpenSceneGraph, ODE, osgModelling, OpenHaptics [16, 17, 18, 19].

Fig. 2. Virtual Environment including Haptic Device

The graphical interface shows a virtual machine table, a selection of tools including drills, end mills, clamps and an inspection probe. The operator can pick up and manipulate the billet, clamp it and then carry out milling and drilling operations based on a routing methodology similar to existing CAD/CAM packages. Meanwhile, in the background, the object being manipulated – such as a cutting tool - is logged and monitored, its position, velocity and virtual cutting force captured. This data is recorded in a text file (Fig. 3 Sample of Logged data.) which is then post-processed to automatically generate routing, operation and tool list sheets (Fig. 4 Sample of data from generated route plan.) the reasoning for which is automatically inferred from the interactive haptic cutting sequence activities.

Time	Tool	L_Position(x)	L_Position(y)	L_Position(z)
1	Start			
5267	Billet	-111.5	6.212	3.954
46886	Billet	-111.5	6.212	3.953
88630	Billet	-111.5	6.211	3.955
132791	Billet	-111.5	6.21	3.958
176890	Billet	-111.5	6.207	3.963
219134	Billet	-111.5	6.203	3.971

Fig. 3. Sample of logged data.

Operation List						
Operation Number	Description	Machine Tool	Tooling	Stop PosX	Stop PosY	Stop PosZ
10	Move Billet			-0.2747	0.4444	0.3644
20	Move Clamp A			-0.04591	0.2923	0.2826
20	Drill	Drilling Station	Drill 5mm	-0.6301	-0.07457	0.6367
30	Drill	Drilling Station	Drill 10mm	-0.6199	-0.09375	0.6457

Fig. 4. Sample of data from generated route plan.

3. Experimental method

Six test subjects comprising of a mixture of novice and intermediate process planners were asked to carry out two process planning tasks in a random order, one task was to be carried out in a traditional manner (Task 1) and the other using the haptic virtual environment (Task 2).

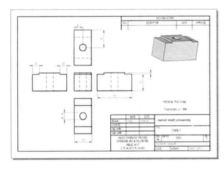

Fig. 5. Clamp

For Task 1 the test subject was presented with a mechanical drawing of a clamp (Fig. 5 Clamp), paper templates for routing sheets, operations lists and tool lists and photographs with descriptions of the tools available to them in the workshop. The tools included a pillar drill, a milling machine, a work bench and a clamp. The tooling family comprised two drill bits of 5mm and 10mm diameter and a 16mm end mill. The test subject was then asked to generate a process plan using the resources available and the task timed.

For task 2 the test subject was requested to simulate the machining of the same object in the virtual environment and told all the information would be automatically logged in order to generate the process plan automatically. The test subject was given around 10

minutes to become accustomed to the environment before beginning the task. The time taken to complete the second task was also timed.

4. Experimental results

The Task Completion Time (TCT) for Task1(blue) and Task 2 (red) for each test subject are shown in Fig. 6 Time to complete process plan.. Trend lines have been added to each set of results for clarity.

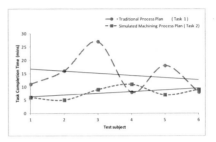

Fig. 6. Time to complete process plan.

The TCT has been plotted against the Plan Data Quality (PDQ) in Fig. 7. Process plan completion time and quality. PDQ is expressed as a % compared to a correct plan. The experience of the test subject is also highlighted in different colours. Operator experience was defined as follows: (i) a novice has no mechanical engineering background; (ii) the intermediate level has a mechanical background and some process planning training; and (iii) the expert level carries out process planning on a daily basis. The TCT and PDQ of the process plans generated by Task 2 are plotted in purple for comparison to Task 1.

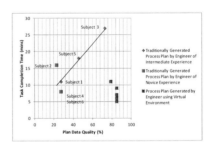

Fig. 7. Process plan completion time and quality.

Figure 8 shows examples of plans automatically generated by the virtual environment and those generated in a traditional manner.

5. Discussion of results

It can be seen from Figure 6 Time to complete process plan that 66.6% of the users developed a process plan in the virtual environment more quickly than in a traditional process planning environment; however, in cases 4 and 6 the time taken to generate the process plan was about equal. Further scrutiny shows a large difference between times taken to generate the plan in task 1. This can be shown to be directly related to the quality of the process plan and the experience of the process planner. It seems the more knowledge the process planner has, the better quality the plan but more time is required to document it. It would be interesting to extend the study to include more process planners, particularly experts, to determine if this trend continues.

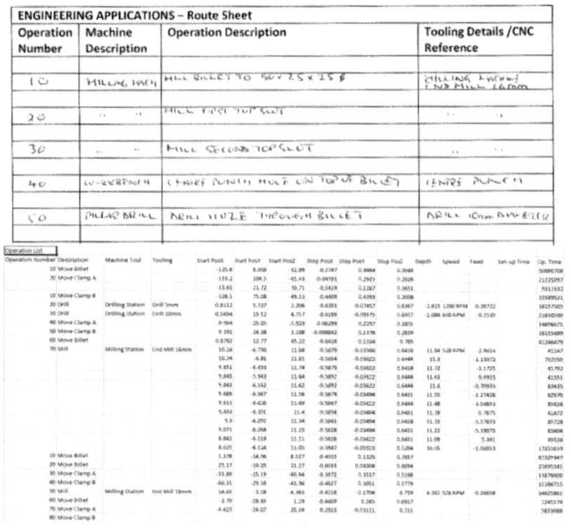

Fig. 8. Examples of traditionally and automatically generated process plans

6. Conclusion and future work

It can be seen that even in early stages of development this prototype virtual haptic environment can be used to generate process plans more efficiently than traditional methods and in a more intuitive way. Even though the data size is limited results encourage further investigation. Future work will include improving the environment to cope with more complicated processes where, it is supposed, that the results could be expected to be more pronounced.

References

[1] Scallan, P., 2003. Process planning: the design/manufacture interface, Butterworth-Heinemann.

[2] Leo Alting & Hongchao Zhang, 1989. Computer Aided Process Planning: the state-of-the-art survey. International Journal of Production Research, 27(4), p.553.

[3] Kriangkrai Waiyagan (∗) & E.L.J. Bohez, 2009. Intelligent feature based process planning for five-axis mill-turn parts. Computers in Industry, 60, pp.296–316.

[4] Butdee, S., Noomtong, C. & Tichkiewitch, S., 2009. A process planning system with feature based neural network search strategy for aluminum extrusion die manufacturing.

[5] Petrzelka, J.E. & Frank, M.C., 2010. Advanced process planning for subtractive rapid prototyping. Rapid Prototyping Journal, 16(3), pp.216–224.

[6] Wang, Q.-H. & Gong, H.-Q., 2009. Computer-aided process planning for fabrication of three-dimensional microstructures for bioMEMS applications. International Journal of Production Research, 47(21), pp.6051–6067.

[7] Yi-Lung Tsai et al., 2010. Knowledge-based Enginering for Process Planning and Die Design for Automotive Panels. Computer-Aided Design & Applications, 7(1), p.75.

[8] Xu, H.-M., Yuan, M.-H. & Li, D.-B., 2008. A novel process planning schema based on process knowledge customization. The International Journal of Advanced Manufacturing Technology, 44(1-2), pp.161–172.

[9] Popma, Houten & Universiteit Twente, 2010. Computer aided process planning for high-speed milling of thin-walled parts : strategy-based support. University of Twente. Available at: http://purl.utwente.nl/publications/71548.

[10] Xie, S.Q., Gan, J. & Wang, G.G., 2009. Optimal process planning for compound laser cutting and punch using Genetic Algorithms. International Journal of Mechatronics and Manufacturing Systems, 2(1), pp.20–38.

[11] Kafashi, S. & Shakeri, M., 2011. Application of genetic algorithm in integrated setup planning and operation sequencing. AIP Conference Proceedings, 1315(1), p.1413.

[12] Ahmad, N., Haque, A. & Hasin, A., 2001. Current trend in computer aided process planning. In Proceedings of the 7th Annual Paper Meet and 2nd International Conference. pp. 81–92.

[13] Denkena, B. et al., 2007. Knowledge Management in Process Planning. CIRP Annals - Manufacturing Technology, 56(1), pp.175–180.

[14] Sherman, W.R. & Craig, A.B., 2003. Understanding Virtual Reality: Interface, Application, and Design 1st ed., Morgan Kaufmann.

[15] Lin, M.C. & Otaduy, M., 2008. Haptic Rendering: Foundations, Algorithms and Applications, A K Peters.

[16] Osfield, R., 2006. osg. Available at: http://www.openscenegraph.org/projects/osg [Accessed April 11, 2011].

[17] Smith, R., 2002. Open Dynamics Engine - home. Available at: http://www.ode.org/ [Accessed April 11, 2011].

[18] Rui, W., 2008. osgmodeling - A modeling library for OpenSceneGraph, creating kinds of parametric curves and surfaces, e.g. the NURBS. - Google Project Hosting. Available at: http://code.google.com/p/osgmodeling/ [Accessed August 21, 2011].

[19] Sensable, 2004. Sensable. Available at: http://www.sensable.com [Accessed April 11, 2011].

Integrated tactile-optical coordinate system for the reverse engineering of complex geometry

F. Li, A. P. Longstaff, S. Fletcher and A. Myers
Centre for Precision Technologies, School of Computing and Engineering, University of Huddersfield
Queensgate, Huddersfield, HD1 3DH, UK.

Abstract. To meet the requirement of both high speed and high accuracy 3D measurement for reverse engineering of artefacts, an integrated contact–optical coordinate measuring system is proposed in this paper. It combines the accuracy of contact measurement using a co-ordinate measuring machine (CMM) and the efficiency of full field of structured light optical scanning methods using a projector and two CCD cameras. A planar target printed with square patterns is adopted to calibrate the projector and cameras while three calibration balls are used to unify two coordinate systems. The measurement process starts from cameras around the volume to capture its entire surface, then the CMM's probe is used to re-measure areas of the object that have not been adequately scanned. Finally the combined data from both systems is unified into the same coordinate system. In this paper the hybrid measurement of a guitar body proves the feasibility of this method.

Keywords: Integrated system, Structured light measuring, CMM, 3D measurement, Data fusion

1. Introduction

In recent years, extensive attention has been given to different methodologies of reverse engineering (RE) aimed at developing more effective measurement methods which provide both high speed and high accuracy. The existing CMM methods are widely used for industrial dimensional metrology, but the digitisation process is very time-consuming for the acquisition of the first set of points on complex, freeform surfaces. An alternative approach is represented by non-contact digitisation of surfaces based on optical techniques, such as time-of-flight lasers [1], laser scanning [2, 3], stereovision [4] and structured light [5]. These optical instruments can efficiently capture dense point clouds in terms of speed and reduces the human labour required. However it is usually difficult for the optical sensors to digitize the non-surface objects, such as slots or holes, due to occlusions and obscuration of these artefacts.

The reduction of the lead time in RE, and the increased requirements in term of flexibility as well as accuracy have resulted in a great deal of research effort aimed at developing and implementing combined systems for the RE based on cooperative integration of in homogeneous sensors such as mechanical probes and optical systems [6-9]. Particular features of a workpiece can be measured with the most suitable sensor, and these measurements with low uncertainty can be used to correct data from other sensors which exhibit systematic errors but have a wider field of view or application range.

However, a limitation of the prosed systems is that the integration of the optical system with the CMM generally takes place but is limited at the physical level, flexibility level and usability level [7, 8]. This paper describes a flexible and effective approach for the integration of a 3D structured light system with dual CCD cameras and a CMM to perform the reverse engineering of freeform surfaces. The system does not need the physical integration of the two sensors onto the CMM, but includes their combination at the measurement information level. The aim is to measure a wider range of objects of complex geometry rapidly with higher accuracy than any individual measurement system alone. The structured light sensors are applied to scan the profile of a part from different views, while the trigger probe is used to measure key features, the edge and blind area of the structured light sensor on a part. The limitations of each system are compensated by the other and measurement results from all of the optical sensors and the CMM probe head are combined into one set. The main characteristics of the methodology are given in the following sections.

2. System configuration

2.1. Elements of the integrated system

The integrated system [see Fig. 1 (a)] is designed and manufactured with the following components:

- Two CCD cameras: IDS UI-1485LE-M-GL, the CCD array resolution is 2560(H) x 1920(V), the dimension of a pixel on CCD array is 2.2um × 2.2um [see Fig. 1 (b)].
- Lens: Fujinon HF12,5SA-1/1,4 5 Megapixel C-Mount Lens, the focal length is 12.5 mm.
- Projector: Panasonic PT-LB60NTEA projector with 1,024 x 768 pixels.
- Co-ordinate measuring machine (CMM): Renishaw cyclone.
- FlexScan3D PRO 3D scanning software.
- Calibration board [12W × 9H × 15 mm Squares, see Fig.1 (c)] for optical scanner calibration and three calibration balls (nominal diameter 52 mm).
- PC Workstation.

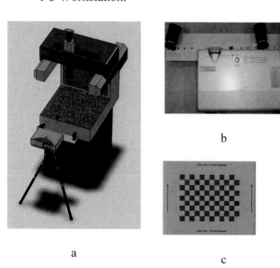

a

b

c

Fig. 1. Elements of the integrated system a. the integrated system; b. cameras and projector ; c. calibration board

2.2. The principle of optical system

The optical system is based on the structured light measurement technique [10-14] with digital sinusoidal fringe and Gray code projection and consists of a digital projector and two CCD cameras that provide redundancy to reduce the effect of obstructions and improve accuracy.

The measurement process is performed in three steps: Generation of stripe patterns, phase measurement (based on the analysis of deformed stripe patterns) and calibration of the phase values in each

pixel of the camera to the real-world (x, y, z) Cartesian coordinates.

3. Measurement process of the integrated system

3.1. Calibration of the optical system

There are many studies of modelling and calibration of CCD cameras. Camera calibration in the context of 3D optical measurements is the process of determining the transformation from 2D CCD image to 3D world coordinate system. The parameters to be calibrated include intrinsic parameters and extrinsic parameters. The intrinsic parameters involve: (1) effective focal length -the distance between the projection centre and the image plane, (2) principle point-the intersection of the optical axis and the image plane, (3) lens distortion and (4) aspect ratio-the length-width ratio of each pixel. The extrinsic parameters involve the transformations between the world coordinate system, the camera coordinate system and the image coordinate system. Extrinsic parameters are needed to transform object coordinates to a camera-centred coordinate system. In many cases the overall performance of a machine vision system strongly depends on the accuracy of the camera calibration procedure [15-19]. To solve all of the intrinsic and extrinsic parameters simultaneously, at least six non-coplanar points in the world coordinate system and their correspondences on the image are required. Zhang [20] proposed a flexible technique for camera calibration by viewing a plane from different unknown orientations. Accurate calibration points can be easily obtained using this method.

3.2. Data fusion of integrated system

Multisensor data fusion in dimensional metrology can be defined as the process of combining data from several information sources (sensors) into a common representational format in order that the metrological evaluation can benefit from all available sensor information and data [9]. The optical scanner and the CMM work in their own separate coordinate systems. If the integrated system is to produce useable results, these two coordinate systems have to be unified.

Coordinate transformation of 3D graphics includes geometric transformations of translation, proportion, rotation and shear. In this first step, the multi-sensor data fusion in this paper assumes no systematic measurement error and only involves translation and rotation transformation. Optical scanner and CMM data fusion can therefore be seen as a rigid body transformation. Since three points can express a complete coordinate frame, data transformation in two systems will be achieved simply with three different reference points and a three-point alignment coordinate transformation method can be used to deal with data fusion.

Since the error of each measuring reference point can be seen as equal weight value, the data fusion errors can be seen as average distributed errors [21]. It is usually very difficult to obtain the same reference point from two different sensors (CCD cameras and CMM in this case) because of different measurement principles and methods of two systems as well as different point cloud density. If we take a reference feature point as the calibration reference point every time, the possibility of occurrence of system error, human errors and accidental errors will increase greatly. Because three points can establish a coordinate, we can calculate the centroid of a standard calibration ball and then use the sphere centre coordinate as the datum reference point coordinate to achieve data fusion and reduce fusion errors. The data fusion of 3D measurement data from different systems will be achieved through the alignment of three datum sphere centre points. In fact, the data fusion problem is, therefore, converted to a coordinate transformation problem. The transformation is determined by comparing the calculated coordinates of the centres of the calibration balls obtained in measurement conducted by the optical system [8]:

$$W_{CMM} = R \cdot W_s + T \qquad (1)$$

where W_{CMM} is the points' coordinates in CMM alignment; R is the rotation matrix; W_S is the point's coordinates in optical alignment; and T is the translation vector.

Specific methods are as follows:

- Calibrate the optical system (section 3.1).
- Use CMM to measure the surfaces of three standard balls.
- Use structured light scanner to measure the surfaces of the same standard balls.
- Use optical scanner and CMM measure the workpieces separately.
- Calculate the sphere centre coordinates measured in two systems and use the sphere centre as reference points to achieve data fusion.

This measuring ball operation has to be performed prior to any measurement, after the calibration of the optical system. It is carried out only once before a series of measurements. Change of configuration of any of systems results in the need of repeating the unification process. The result of this process is a

transformation matrix, which modifies (rotates and translates) the point's coordinates from the optical scanners' relative coordinate system to the absolute system of the CMM.

4. Experimental results

After building the integrated system, an electrical guitar body [see Fig. 2(a)] has been used to demonstrate the feasibility of the proposed method. The surface of guitar body includes a freeform surface and a pick-up slot which is difficult to measure for an optical scanner because of occlusions and obscuration of artefacts. Therefore we can use optical scanner to capture the freeform surface 3D data of guitar body then digitize the pick-up slot surface by using the CMM.

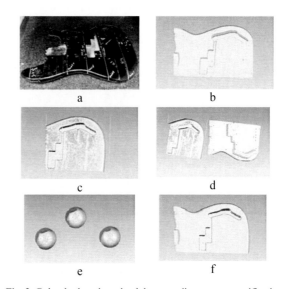

Fig. 2. Guitar body points cloud data coordinate system unification a. optical scanner measurement; b. data measured by optical scanner; c. data measured by CMM; d. data measured by two systems; e. calculation of ball sphere centre; f. data fusion of two systems.

Specific steps are as follows:
- Calibrate the optical scanner by using a calibration board, then use both CMM and optical scanner to measure three calibration balls mounted on the CMM granite bed.
- Use structured light scanner to capture dense point clouds data of freeform surface efficiently [see Fig. 2(b)]. However the cameras cannot capture the surface information of pick-up slot because of ambient occlusion and obscuration of guitar surface, and then we use CMM's contact head to re-measure the slot area that has not been registered [see Fig. 2(c)]. As the scanning

speed of structured light scanner is very fast (An average scan takes between 1 to 6 seconds), while a CMM only needs to digitize the slot surface, compared to performing a full surface measurement by using a CMM alone, the data acquisition speed of the integrated system improves greatly. In this example, the CMM-contact digitisation of one guitar surface need more than 8 hours, while the hybrid system can finish the measurement within one hour. The point cloud data obtained in both systems [Fig. 2 (d)].

- Calculate the sphere centre coordinates calibration balls measured in two systems [see Fig. 2 (e)].
- Use the sphere centre coordinates to unify the two coordinate systems, the unified guitar point clouds data as shown in Fig. 2 (f).

5. Conclusions and discussions

To meet the requirement of measuring complex geometry of workpieces with high accuracy and speed, a full field of integrated scanning system, which mainly consists of a CMM, two CCD cameras and a DLP based standard projector, has been developed in this paper. The unification of contact and non-contact systems are fulfilled by using three calibration balls mounted on a CMM granite bed. The hybrid measurement of guitar body showed this approach is simple, convenient, efficient and reliable. However, in this paper we do not verify the accuracy of the integrated system. No visual separation between the data sets from different measuring systems indicates a fit accuracy within 0.5 mm, proving the feasibility of this approach. Theoretically, the integrated system accuracy and resolution depend on both separate systems, but should be biased towards the contact method, and can be improved by improving the specifications of hardware, including choosing a higher accuracy CMM, choosing more precise calibration balls and a higher resolution projector which can provide sharper stripe. Reducing the measuring range can also improve the system accuracy. Further research will include more extensive experiments to test the accuracy of the integrated system and more sophisticated design of the algorithm to reduce the manual intervention.

Acknowledgements: The authors gratefully acknowledge the UK's Engineering and Physical Sciences Research Council (EPSRC) funding, of the Centre for Advanced Metrology under its innovative manufacturing program.

References

[1]   A. Ullrich, N. Studnicka, J. Riegl et al., Long-range high-performance time-of-flight-based 3d imaging sensors, in: 3D Data Processing Visualization and Transmission, Padova, Italy 2002, pp. 852–855.

[2]   K.-C. Fan, A non-contact automatic measurement for free-form surface profiles, Computer Integrated Manufacturing System 10 (1997) 277–285.

[3]   G. Wang, B. Zheng, X. Li, Z. Houkes, Modeling and calibration of the laser beam scanning triangulation measurement system, Rob. & Aut.Sys. 40 (2002) 267–277.

[4]   D. Gorpas, K. Politopoulos, D. Yova, A binocular machine vision system for three-dimensional surface measurement of small objects, Computerized Medical Imaging and Graphics 31 (8) (2007) 625–637.

[5]   J. Salvi, J. Page's, J. Batlle, Pattern codification strategies in structured light systems, Pattern Rec. 37 (4) (2004) 827–849.

[6]   Carbone, V., et al., Combination of a Vision System and a Coordinate Measuring Machine for the Reverse Engineering of Freeform Surfaces. The International Journal of Advanced Manufacturing Technology, 2001. 17(4): p. 263-271.

[7]   Chan, V.H., C. Bradley, and G.W. Vickers, A multi-sensor approach to automat. co-ordinate measuring machine-based reverse engineering. Comp. in Ind. 2001. 44(2): 105-15.

[8]   Sladek, J., et al., The hybrid contact-optical coordinate measuring system. Measurement, 2011. 44(3): p. 503-510.

[9]   Weckenmann, A., et al., Multisensor data fusion in dimensional metrology. CIRP Annals - Manufacturing Technology, 2009. 58(2): p. 701-721.

[10] V. Srinivasan, H. C. Liu, and M. Halioua, Automated phase-measuring profilometry of 3-D diffuse objects. Applied Optics, Vol. 23, 1984, pp. 3105-3108.

[11] Halioua, M. and H.-C. Liu, Optical three-dimensional sensing by phase measuring profilometry. Optics and Lasers in Engineering, 1989. 11(3): p. 185-215.

[12] Wei Pan, Yi Zhao, Xueyu Ruan, Aplication of phase shifting method in projection grating measurement, Journal of Applied Optics. 2003. 24(4): pp.46-49.

[13] Peisen S. Huang and Song Zhang, Fast three-step phase-shifting algorithm, App. Optics, 2006. 45, pp. 5086-5091.

[14] Song Zhang and Shing-Tung Yau, High-speed three-dimensional shape measurement system using a modified two-plus-one phase-shifting algorithm, Optical Engineering. 2007, 46(11), 113603

[15] E.M. Mikhail, J.S. Bethel, J.C. McGlone, Introduction to Modern Photogrammetry, John Wiley and Sons, Inc., 2001.

[16] O. Faugeras, Three-dimensional Computer Vision: A Geometric Viewpoint, The MIT Press, Fourth Printing, 2001.

[17] Micheals RJ, Boult TE. On the Robustness of Absolute Orientation. In: Proceeding of the International Association for Science and Technology Development (IASTED) Conference on Robotics and Automation. 2000.

[18] Sheng-Wen, S., H. Yi-Ping, and L. Wei-Song. Accuracy analysis on the estimation of camera parameters for active vision systems. in Pattern Recognition, 1996., Proceedings of the 13th International Conference on. 1996.

[19] Sagawa, R. and Y. Yagi. Accurate calibration of intrinsic camera parameters by observing parallel light pairs. in Robotics and Automation, 2008. ICRA 2008. IEEE International Conference on. 2008.

[20] Z. Zhang, A flexible new technique for camera calibration, IEEE Transactions on Pattern Analysis and Machine Intelligence 22 (2000)1330–1334.

[21] Tao, J. and K. Jiyong, A 3-D point sets registration method in reverse engineering. Computers & Industrial Engineering, 2007. 53(2): p. 270-276.

Adapting STEP-NC programs for interoperability between different CNC technologies

M. Safaieh, A. Nassehi and S. T. Newman
Department of Mechanical Engineering, University of Bath

Abstract. Since the 1990s, complex multi-axis and multi-process machines with complex controllers have been commonplace for the efficient batch production of precision parts that incorporate various technologies such as milling, turning, grinding and laser cutting. With this level of complexity, interoperability between different resources and technologies in CNC metal cutting machines has always been difficult. For dealing with this difficulty in interoperability, researchers started to identify an efficient way of realising interoperability between CNC machine tools with different technologies for machining the same component on various CNC machines. STEP-NC is a rich information standard developed to replace the traditional programming languages used for CNC machines. The aim of this paper is to propose a methodology based on the use of STEP-NC to develop a framework for cross technology interoperability. A prototype implementation of the framework is tested by converting a STEP-NC programme written for a 4-Axis milling machine to one suitable for a turn-mill machine.

Keywords: Interoperability, CNC machining, STEP-NC

1. Introduction

In the 21th century, CNC machines with advanced technological capabilities have become the state-of-the-art tools for manufacturing complex parts with tight tolerances. Due to the availability of CNC machines with different technologies, it has become possible to produce complex products with high speed and high precision [1, 2]. To fully enable interoperability, it is essential to have a system, which is not dependent on a machining process. This is because by having such a system, generating NC codes for different CNC machines will be based on manufacturing information not just the machining process [3]. Due to the development of CNC machines with multiple axes and capability of carrying out multiple processes, interoperability between these machines has become a challenging issue for researchers with the aim of finding the way for communication between these machines without the necessity for human resources.

As illustrated in Fig.1, machining a part with two CNC machines with different technologies without the need

for an expert to generate the code is still a gap in interoperability.

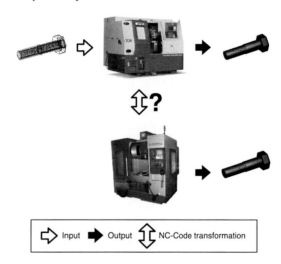

Fig. 1. Machining a part with two CNC machines with different technology

To solve this problem, the authors have designed a system to read G&M codes as a source code from one machine and analyse the necessary information for generating NC code for another machine [3]. The aim for this paper is to enhance the system by identifying the elements in a STEP-NC code and importing them to the intelligent translator to generate a STEP-NC code for another CNC machine with a different technology (destination machine).

2. Background

Interoperability in manufacturing using CNC machines has been a major area of research over the last 10 years [4]. In 2007, Nassehi presented the barriers and issues of incompatibility between the various CAD/CAM/CNC systems and proposed a new framework to overcome these barriers in achieving interoperability in the CAD/CAM/CNC chain [5]. Newman provided a strategic view of how interoperability can be implemented across the CAx chain with the range of standards used to regulate the flow of information [1]. Yusef described and illustrated a STEP compliant CAD/CAPP/CAM system for the manufacturing of rotational parts on CNC turning centres [2]. In 2010, Zhang developed a new software tool to demonstrate the feasibility of interoperable CNC manufacturing based on STEP-NC [6].

STEP-NC (ISO 14649) defines methods for incorporation of a variety of production information in a new type part program which is different from G&M code based on axis movement [7].

The authors proposed the framework for a new translator to allow G&M codes to be translated from one technology to another [3]. In the framework as illustrated in Fig. 2, the adapter read a series of G-codes form the source machine and analyses the imported information for generating new code for a destination machine [8].

The adapter itself communicates with two databases: First the "Manufacturing Dictionary" which consists of machine reference information, cutting tool reference information, operation meta-data, feature meta-data and CNC STEP-NC meta-data. Second the "Manufacturing Process Database" consists of cutting tool information, manufacturing feature information and machining operation information.

The adapter it-self consists of three major elements: A Reader, which imports the code. An Analyser, which analyses the imported information for generating a new code, And A Writer, which generates the STEP-NC code based on the information provided by the analyser.

3. Methods

In this paper the framework in Fig. 2 is extended to use STEP-NC as input from the source to generate new code for the destination machine.

STEP-NC code consists of general information such as filename, author, date and organization, and detailed information such as working steps, work plans, machining operations, manufacturing features and their geometry.

The role of the reader is to identify the tools, features and operation in the source machine STEP-NC code and store them in the manufacturing process data-base.

As illustrated in Fig. 3, the reader will identify the machining working steps and form the information in the machine working step determining manufacturing feature and machining operation.

Tools and technology information will be determined from machining operation and feature geometry form manufacturing features, after identifying the information for the adapter the reader will store all the information to the manufacturing process data-base.

The analyser in the adapter will process the data in the manufacturing process data-base based on information in the manufacturing dictionary and then the writer will generate the STEP-NC code for the destination machine.

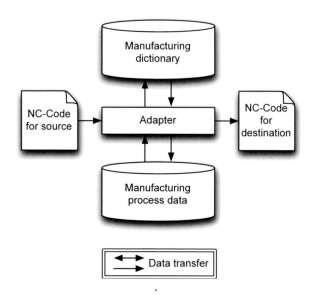

Fig. 2. Framework for information flow in the CNC machining chain

The method used in the analyser is converting machining operation and machining features information from source to destination based on the availability of tools and operations in the destination machine.

To realise such a system, the analyser should categorise the machining operations and features to different sublevels and then start to find the operations and features from the manufacturing process database that are suitable for the destination CNC machine based on the source machining information.

As illustrated in Fig. 4, the writer starts to read the destination machine information from the Manufacturing Dictionary for generating the Header and Data section of STEP-NC file. The information for the header is similar to the information of the source machine. For data section the writer reads the first operation from manufacturing process data-base, (which is carried from analyser decision) and then reads the feature (which is chosen for the current operation).

After reading the operation and the feature, the writer will generate the code based on information in manufacturing dictionary and write the STEP-NC code to a file. The writer finishes the process at the last operation.

For testing the system, a part with milling and turning features has been chosen. The part consists of five different features: Planar Face, Profile Feature, Step, Closed Pocket and Slot. Which can be machined with two operations: Planer_milling and Bottom_and_side_milling? This information will be identified from the input code. An excerpt of the input programme is shown in Table 1.

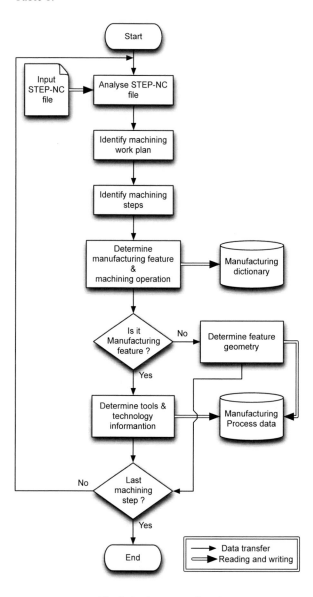

Fig. 3. Reader process flowchart

This programme was fed into the adapter in order to obtain a turn-mill programme to machine the same component. The header of the new programme will be similar to the source and for the data; features and operations are the most codes, which are changing during the conversion. Table 2 illustrates the new STEP-NC code for the destination machine with the differences highlighted. Machining the part on both machines produced identical results.

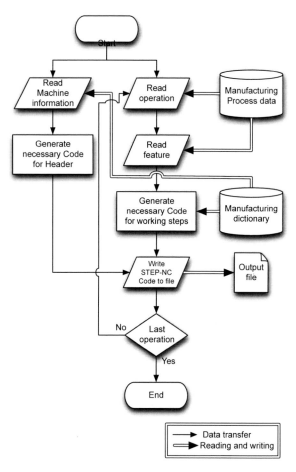

Fig. 4. Writer process flowchart

Figure 5 illustrates the part with five features. For testing the experimental part a 4 axis CNC milling machine and a C-axis turn-mill machine were chosen.

Fig. 5. Experimental part

Table 1. An excerpt of the STEP-NC milling programme written to machine the component

```
DATA;
#1= PROJECT('MATADOR CASESTUDY',#2,(#4),$,$,$);
#2= WORKPLAN('MAIN
WORKPLAN',(#10,#11,#12,#13,#14,#15,#16,#17,#18,#19,#20,#21,#22,#23,
#24,#25),$,#8,$);
#4= WORKPIECE('SIMPLE WORKPIECE',#6,0.010,$,$,$,(#66,#67,#68,#69));
#6= MATERIAL('ST-50','STEEL',(#7));
#7= PROPERTY_PARAMETER('E=200000N/M2');
#8= SETUP('SETUP1',#71,#62,(#9));
#9= WORKPIECE_SETUP(#4,#74,$,$,());
                                                    WORKPIECE DEFINITION
```

```
#10= MACHINING_WORKINGSTEP('ROUGH PLANAR FACE',#62,#30,#40,$);
#11= MACHINING_WORKINGSTEP('FINISH PLANAR FACE',#62,#30,#41,$);
#12= MACHINING_WORKINGSTEP('ROUGH PROFILE',#62,#31,#42,$);
#13= MACHINING_WORKINGSTEP('FINISH PROFILE',#62,#31,#43,$);
#14= MACHINING_WORKINGSTEP('ROUGH STEP',#62,#32,#42,$);
#15= MACHINING_WORKINGSTEP('FINISH STEP',#62,#32,#43,$);
#22= MACHINING_WORKINGSTEP('RUOGH POCKET',#62,#33,#42,$);
#23= MACHINING_WORKINGSTEP('FINISH POCKET',#62,#33,#43,$);
#24= MACHINING_WORKINGSTEP('ROUGH SLOT',#62,#34,#42,$);
#25= MACHINING_WORKINGSTEP('FINISH SLOT',#62,#34,#43,$);
                                                            WORKING STEPS
```

```
#30= PLANAR_FACE('PLANAR FACE1',#4,(#40,#41),#70,#63,#24,#25,$,());
#31= PROFILE_FEATURE('PROFILE',#4,(#42,#43),#71,#63,#24,#25,$,());
#32= STEP('STEP',#4,(#42,#43),#72,#63,#24,#25,$,());
#33= CLOSED_POCKET('POCKET',#4,(#42,#43),#73,#63,#24,#25,$,());
#34= SLOT('SLOT',#4,(#42,#43),#74,#63,#24,#25,$,());
                                                                 FEATURES
```

```
#40= PLANE_ROUGH_MILLING($,$,'ROUGH PLANAR
FACE',10.000,$,#39,#40,#41,$, #60,#61,#42,2.500,$);
#41= PLANE_FINISH_MILLING($,$,'FINISH PLANAR
FACE',10.000,$,#39,#40,#41,$, #60,#61,#42,2.500,$);
#42= BOTTOM_AND_SIDE_ROUGH_MILLING($,$,'ROUGH
POCKET1',15.000,$,#39,#50,#41 ,$,$,$,#51,2.500,5.000,1.000,0.500);
#43= BOTTOM_AND_SIDE_FINISH_MILLING($,$,'FINISH
POCKET1',15.000,$,#39,#52, #41,$,$,$,#53,2.000,10.000,$,$);
                                                               OPERATIONS
```

Table 2. An excerpt of the STEP-NC turn-mill programme written to machine the same component

```
DATA;
#10= MACHINING_WORKINGSTEP('ROUGH FACE',#62,#30,#40,$);
#11= MACHINING_WORKINGSTEP('FINISH FACE',#62,#30,#41,$);
#12= MACHINING_WORKINGSTEP('ROUGH PROFILE',#62,#31,#42,$);
.
.
                                                            WORKING STEPS
```

```
#30= REVOLVED_FLAT('FACE',#4,(#40,#41),#70,#63,#24,#25,$,());
#31= GENERAL_REVOLUTION('PROFILE',#4,(#42,#43),#70,#63,#24,#25,$,());
.
.
                                                                 FEATURES
```

```
#40= FACING_ROUGH($,$,'ROUGH FACE',10.000,$,#39,#50,#51,$,
#60,#61,#42,2.500,$);
#41= FACING_FINISH($,$,'FINISH FACE',10.000,$,#39,#50,#51,$,
#60,#61,#42,2.500,$);
#42= CONTOURING_ROUGH($,$,'ROUGH PROFILE',10.000,$,#39,#52,#53,$,
#60,#61,#42,2.500,$);
.
.
                                                               OPERATIONS
```

4. Conclusion and future works

This paper reviewed the interoperability in manufacturing and standards in interoperability. The major contribution of the paper is the use of ISO14649 (STEP-NC) in a cross technology environment to realise an interoperable system. This system allows a part to be manufactured without dependency on a specific process or resources with two CNC machine centre with very different technologies; provided that they both have the capability to make the part. The future work will be focused on extending the framework to work with more CNC machines with a wider variety of technologies at the same time and also more complex part with additional features.

References

[1] S. T. Newman, A. Nassehi, X. Xu, R. Rosso, L. Wang, Y. Yusof, L. Ali, R. Liu, L. Zheng, and S. Kumar, (2008) Strategic advantages of interoperability for global manufacturing using CNC technology, Robotics and Computer Integrated Manufacturing, vol. 24, pp. 699-708.

[2] Y. Yusof, S. Newman, A. Nassehi, and K. Case, (2009) Interoperable CNC system for turning operations, pp. 941-947.

[3] M. Safaieh, A. Nassehi, S. T. Newman, (2011) Realization of interoperability between a C-axis CNC turn-mill centre and 4-axis CNC machining centre, Flexible Automation and Intelligent Manufacturing, FAIM2011, Taichung, Taiwan.

[4] D. Elias, Y. Yusri, and M. Mohamad, (2012) A Framework for a Development of an Intelligent CNC Controller Based on Step-NC, Advanced Materials Research, vol. 383, pp. 984-989.

[5] A. Nassehi, (2007) The realisation of CAD/CAM/CNC interoperability in prismatic part manufacturing, Thesis (PhD), University of Bath.

[6] X. Zhang, R. Liu, A. Nassehi, and S. Newman, (2010) A STEP-compliant process planning system for CNC turning operations, Robotics and Computer-Integrated Manufacturing.

[7] Y. Yusof, N. D. Kassim, and N. Z. Z. Tan, (2011) The development of a new STEP-NC code generator (GEN-MILL), International Journal of Computer Integrated Manufacturing, vol. 24, pp. 126-134.

[8] M. Safaieh, A. Nassehi, S. T. Newman, (2011) Cross Technology Interoperability for CNC Metal Cutting Machines, presented at the 21th International Conference on Production Research, Stuttgart Germany.

A heuristic approach for nesting of 2D shapes

Gianpaolo Savio[1,2], Roberto Meneghello[2] and Gianmaria Concheri[2]

[1] ENDIF (ENgineering Department In Ferrara) - University of Ferrara - Via Saragat, 1 - 44100 - Ferrara - Italy

[2] DICEA-Lin (Laboratory of Design Tools and Methods in Industrial Engineering) - University of Padova - Via Venezia, 1 - 35131 - Padova - Italy

Abstract. In several manufacturing processes, the cutting of 2D parts from sheets is an important task. The arrangement of the parts in the sheets, supported by computers, is called nesting and is addressed to   minimize the wasted material. In literature some approaches are proposed, based on genetic or heuristic algorithms which emphasize different characteristics, e. g. the time complexity or the wasted material. In shipbuilding the parts to be arranged have significantly different sizes, which are often difficult to pack in a fast way using the standard methods in literature. In this work an approach is proposed, able to arrange parts with very different dimensions, which is based on the identification of a suitable starting rotation that ensures a solution in a reasonable time. The main steps are: a) importation of the model files of the parts to be packed, b) identification of a preliminary orientation and sorting of the parts (starting position), c) optimization of the position of the parts, ensuring a minimum distance between them. For the starting rotation, three different orientations are considered: i) the original orientation, ii) the x axis coincident with the minimum inertia axis, iii) the x axis aligned with the maximum edge. The orientation is selected in order to obtain the minimum area of the bounding box. The implementaiton of the method has been investigated and the results show the advantages of the approach: reduction of waste material and time for performing the nesting.

Keywords: CAD/CAM, CAPP, Nesting, shipbuilding, packing, cutting-stock.

1. Introduction

The task of finding an efficient layout for cutting 2D parts out of a given 2D sheet with minimum waste material is called "nesting". The number of applications, e.g. shipbuilding, textile and furniture industry, is very large.

Many numerical approaches are available in literature, which involve different strategies and solution techniques as genetic or swarm intelligence algorithm [1-3]. Several authors evaluate the no fit polygon techniques (NPF) [2, 4-5] which is based on the aggregation of 2 polygons and can be used to determinate the overlaps. This approach needs to convert the boundary of the surfaces into polygons. Another approach is based on the subdivision of the sheet in pixels [2, 6] adopting different coding schemes to denote the empty space. Other methods are based on the sliding of the shells in the sheet as the bottom left (BL) [1, 7-8] or bottom-left-fill (BLF) [1, 9] approaches.

In actual industrial applications, several techniques (e.g. NPF) are not suitable and simpler methods, as the BLF approach, are preferred.

In the product development of a ship it is necessary to split the geometric model of the hull into shells. The shells must be unrolled in 2D surfaces having dimensions less than the sheet dimensions (e. g. 6000x2000 mm). Then, the shells are nested in metal sheets for the cutting process. Finally the shells should be bended and welded.

In this paper a part of a project for the shipbuilding industry, developed in IronPython - Rhinoceros® V.5 software (from McNeel), is described. More in detail, a plug-in for the naval carpentry management has been developed, which include:

- definition of the properties of the shells,
- creation of shells database
- tracking the welding lines and the reference frames,
- unrolling shells,
- nesting shells, described in this work.

The proposed nesting method is based on the identification of a preliminary optimal orientation of each shell, which is defined by the minimum area of the bounding box. This approach can be easily integrated with the traditional nesting algorithms: bottom-left [1, 7], minimum total potential or lowest-gravity-center and raster [2, 6-7, 9].

The results show a reduction of the waste material, with a minimum rise in computational time. Also using the Rhinoceros® built-in functions, the no-overlap condition is simple to check, the reduction of the surface to a polygon and the implementation of the NPF algorithm is not required, the welding lines and the reference frames can be moved together with the shell, and the respect of the distance between the shells can be obtained by offsetting the boundary.

2. Proposed nesting approach

As previously mentioned, the geometrical model of the hull is split into shells which are developed in 2D surfaces; each of them is saved in a separate model file.

Then, the nesting phase, summarized in Fig. 1, can take place. The first step is the importation and the preliminary rotation of several model files (i.e. shells) in the nesting model file. Afterward, a list of the shells is created and sorted according to the height of the bounding-box from largest to smallest. In order to allow the cutting process, the shells must be spaced (typical distance is 20 mm); this result can be obtained nesting the offset curve of the boundary of each shell, moving together the shell and its offset and then erasing the offset. Following a nesting procedure, selected among four options, the shells are placed in the sheet starting from the top right position and moving left and down until no-overlap occurs. Details are described in the following sections.

2.2. Preliminary orientation

Three orientations of the shell are investigated:
- the original orientation, i.e. the same orientation of the shell reference frame,
- the orientation based on the alignment of the maximum edge lenght with the x-axis,
- the orientation of the principal axes of inertia about the centroid with the sheet reference frame.

If necessary the shell is rotated of 90° in order to align the maximum edge of the bounding box together to the x-axis. The orientation which assures the minimum bounding box area of the shell is selected for the nesting procedure.

Figure 2 shows the application of the procedure to a shell having an hole and irregular boundaries. The original bounding box area is 3.79 m^2, the bounding box area with maximum edge is 3.38 m^2 and the bounding box area with the principal axes of inertia aligned together to the reference frame is 2.61 m^2: the last one is assumed as the optimal orientation for the nesting.

2.3. Nesting rule

Four standard nesting approaches are tested in order to evaluate the improvement due to the preliminary orientation and to the sorting.

2.3.1. BL Algorithm
One of the most widley adopted methods for nesting (especially in industry) is the bottom-left BL alghorithm [1, 7-8]. Starting from the top-right corner of the sheet each item is slided to the left and then down until the item is close to another nested part or to the sheet contour. Adopting the pre-orientation and the sorting of the items the major disadvantage of the method (creation of empty areas, when larger items block the following ones) is reduced.

Similar to literature the sliding method can be described by the following pseudo code, first applied to left and then to down directions:

```
for each shell:
    s=(base bounding-box)/2, t=0.1 mm
    offset shell
    move shell and offset-shell to top-right of sheet
    if no-overlap occurred:
        while s<t:
            slide to left (or down) shell and    offset-shell by the step s
            if overlap:
                slide to right (or up) shell and offset-shell by the step s
                s=s/2
```

The overlap condition is checked by the Rhinoceros V.5 built-in function PlanarClosedCurveContainment between the boundary of the sheet, of the previously nested shells and of the actual offset-shell.

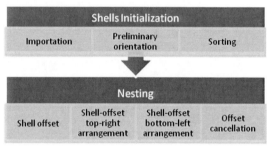

Fig. 1. Flow chart of the nesting proedure.

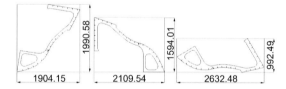

Fig. 2. Different preliminary rotation: original (left), maximum edge lenght (midle), principal axes of inertia (right).

2.3.2. BL & HAPE integrated algorithm
Minimum total potential energy (HAPE) or lowest-gravity-center principles are based on the minimization of the distance beetwen an edge of the sheet and the centroid of the shell.

This approach can be implemented testing different rotations and applying the BL algorithm. Considering the preliminary orientation of the parts, a rotation at 90° steps may be sufficient.

The free part covering the full height of the sheet (free length) is not considered in the evaluation of the wastage. Consequently, in the last sheets, this approach can be useful if there are shells of elongated shape: vertically aligning these shells, the free length increases and the wastage decreases.

2.3.3. BL & Raster integrated algorithm

Raster methods are based on the subdivision of the sheet in points or pixel [2, 6-7, 9]. The simplest coding scheme uses the value 1 to code the area of the sheet covered by a shell, and 0 to code the free area.

In this work a variation of this method is applied to improve the nesting of shells with holes. Initially the shells are nested as in the BL method. When the area of the bounding box is greater than 80% of the shell area, the bounding box of the nested shell is covered by a grid of points with a constant step along the x and y-axis. If the points fall offside the shell, they are put in a list. The shells having smaller area are tested. Each shell is set on each point in the list until no-overlap happens. When this condition occurs the BL procedure is applied and the points inside the shell are deleted from the list.

The left-bottom vertexes of the shells with smaller area are put in the first point of the list. For each shell if no-overlap occurred, the BL procedure is applied and the points inside the shell are removed from the list.

2.3.4. BL, HAPE & Raster combined algorithm

This approach is obtained combininig the methods descibede in the paragraphs 1.2.3.2-1.2.3.3.

3. Results and discussion

The proposed approaches have been tested on the nesting of 152 shells with and without the preliminary rotation.

The shells initialization task has been performed in 6 s, 5 of which were used to import the file.

The results summarized in Table 1 have been obtained assuming:
- sheets dimension 6000x2000 mm,
- shell-shell minimum distance 20 mm,
- rotation step 90° for the HAPE approach,
- points step 100 mm for the raster approach.

For the nesting of all the shells, three sheets are needed. The BL approach is the faster, while the combined is slower. It can be seen that the wastage material is smaller adopting the preliminary orientation, while the time necessary to perform the nesting has no clear behaviour

in relation to the preliminary orientation. The "BL & Raster" method performs all shells in a short time. The high value of wastage material in the last sheet is due to the high number of shell with small area and to the minimum distance between the shells. The best result is obtained with the combined method; in this case more than 4 meters of the last sheet remains unused (free length).

Figure 3 shows the results of the nesting on the first sheet. As it is possible to see, sometime the combined used of the raster and HAPE approach don't give any improvement (in sheet 1 the best results is given by the BL & raster approach, while in sheet 2 and 3 the best results are given by the BL and HAPE method). Effectively in the first nesting step the HAPE method can create empty areas when larger items block the succesive ones, while in the final step the HAPE method can be useful to get better free length in the last sheet. Thus it can be useful to start with the raster approach and to follow with the HAPE. Figure 4 shows the result of a nesting with BL & Raster in the first sheet and BL & HAPE for the second sheet. In this way the total time for the nesting procedure is about 2 minutes and the total waste material decrease to 23 %, with a free length of 4459 mm.

4. Conclusion

In this paper a modification of the standards nesting procedures is proposed introducing a preliminary orientation of the parts based on the minimization of the bounding box area and on the sorting by the height of the bounding box. The results show that the approach reduces the time necessary for the nesting and decreases the waste material.

Table 1. Nesting of 152 shells: calculation time, waste material** and free lenght in the last sheet performed with (Y) or without (N) preliminary orientation.

Algorithm	Time [s]				Wastage [%]								Free lenght last sheet [mm]	
	1 sheet		All shells		Sheet 1		Sheet 2		Sheet 3		Total			
	N	Y	N	Y	N	Y	N	Y	N	Y	N	Y	N	Y
BL	6.4	2.8	26.2	45.7	52.7	38.3	35.3	19.5	47.1	47.1	47.5*	33.1	4147*	2416
BL & HAPE	14.5	19.4	141.4	155.7	49.6	33.7	17.4	16.5	43.6	40.7	36.1	27.7	1672	3570
BL & Raster	71.0	23.5	137.3	44.3	36.2	25.4	30.0	18.8	48.0	54.7	37.3	27.5	1382	3616
Combined	85.6	54.1	159.2	170.4	31.5	28.7	20.8	16.5	54.6	41.3	32.4	25.2	2565	4057

* Four sheets are needed. ** 100 (Sheet_area - Area_of_ nested_shells) / Sheet_area

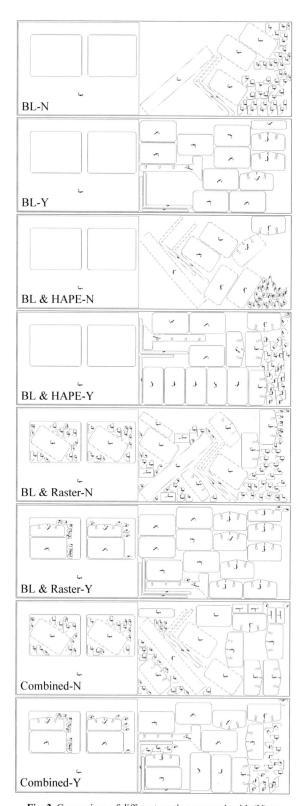

Fig. 3. Comparison of different nesting approach with (Y) or without (N) preliminary orientation.From top to down: BL-N, BL-Y, BL & HAPE-N, BL & HAPE-Y, BL & Raster-N, BL & Raster-Y, Combined-N, Combined-Y.

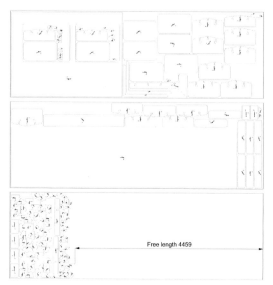

Free length 4459

Fig. 4. Mixed Approach: the first sheet is nested with the BL & Raster, while the second and the third sheets are nested with the BL & HAPE method.

Acknowledgments: The authors are grateful to 3jyachting srl for the financial support.

References

[1] Hopper E, Turton BCH, (2001) An empirical investigation of meta-heuristic and heuristic algorithms for a 2D packing problem. European Journal of Oper. Research 128:34-57.

[2] Bennell JA, Oliveira JF, (2008) The geometry of nesting problems: A tutorial. European Journal of Operational Research 184:397-415.

[3] Xu Y, Yang GK, Bai J, Pan C, (2011) A Review of the Application of Swarm Intelligence Algorithms to 2D Cutting and Packing Problem. Lecture Notes in Computer Science - Proceedings of ICSI 2011, Chongqing, China, June 12-15, 2011, Part I - 6728:64-70.

[4] Albano A, Sapuppo G, (1980). Optimal allocation of two dimensional irregular shapes using heuristic search methods. IEEE Transactions on Systems, Man and Cybernetics, 10(5):242-248.

[5] Liu HY, He YJ, (2006) Algorithm for 2D irregular-shaped nesting problem based on the NFP algorithm and lowest-gravity-center principle. Journal of Zhejiang University SCIENCE A 7(4):570-576.

[6] Ramesh Babu A, Ramesh Babu N, (2001) A generic approach for nesting of 2-D parts in 2-D sheets using genetic and heuristic algorithms. Computer-Aided Design 33:879-891.

[7] LIU X, YE JW, (2011) Heuristic algorithm based on the principle of minimum total potential energy (HAPE): a new algorithm for nesting problems. Journal of Zhejiang University-SCIENCE A 12(11):860-872.

[8] Lee WC, Ma H, Cheng BW, (2008) A heuristic for nesting problems of irregular shapes. Computer-Aided Design 40:625-633.

[9] Burke E, Hellier R, Kendall G, Withwell G, (2006) A New Bottom-Left-Fill Heuristic Algorithm for the Two-Dimensional Irregular Packing Problem. Operations Research 54(3):587-601.

STEP-NC compliant manufacturing cost estimation system for CNC milled part component

Abayomi B. O. Debode. Aydin Nassehi, Linda B. Newnes and Stephen T. Newman
Department of Mechanical Engineering, University of Bath, Bath, BA2 7AY

Abstract. Manufacturers depend on cost estimation system, at the early stage of product development, to supports engineers decision on design alternatives toward reducing final product cost. Lack of detail information about a product and its manufacture to generate fast and accurate cost estimate is a critical challenge in integrated manufacturing environment today. Most cost estimation systems relies on expert opinion and nominal information about the product to generate cost estimate of questionable accuraccy. High level integration of cost estimation system with manufacturer' other computer aided systems for seamless exchange of actual design and process information about a product can potentially improve the accuraccy of cost estimate. However, each computer aided system, including current integrated cost estimation systems, uses vendor-specific programming for unidirectional communication with other manufacturere systems. This hinders seamless flow of actual information about the product that is require for fast and accurate cost estimation during product development. Implementation of a vendor-neutral communication language will facilitate bidirectional sharing of quality information for interoperability and accurate cost estimate. In this paper the evolving ISO 14649 standard (STEP-NC) is implement as a vendor-neutral mediun for high level integration of estimation system with other computer aided systems for fast and accurate cost estimation.

Keywords: Cost estimation, Integrated manufacturing, STEP-NC

1. Introduction

In recent years cost estimation systems have been developed and used by manufacturers to support cost reduction decision during product development.

Existing cost estimation systems used in providing cost related information to design engineer for design alternatives assessment in a traditional product development were reported to have suffered from little or no integration capability with manufacturer's computer aided systems (CAx). They rely on manual data input and depend on expert opinion as well as, historical database. The estimation process also involves time consuming and costly data gathering and analysis [1].

In the past decades, computer aided systems (CAx), such as CAD, CAM, CAPP and CNC are widely used in manufacturing industries and international efforts has led to the development of Standard for the Exchange of Product data model (STEP) that enable total integration of CAx. The use of STEP for the exchange of design data is gaining momentum in manufacturing industry. In the same way, ISO 14649 also referred to as, STEP compliant Numerical Control (STEP-NC), is evolving as a viable solution to bi-directional exchange of information in integrated manufacturing environment [2].

Product design, process and resource models are represented in a STEP-NC data file. This provides an opportunity for the development of high level integrated cost estimation system that utilises these actual product information to generate improved cost estimate. This paper presents a methodology that utilises part design and process infromation available in STEP-NC file, to generate product cost estimate for a prismatic part. The STEP-NC compliant methodology facilitate high level integration with other CAx for improve accuracy of cost estimate and reduces expensive and time consuming data gathering and analysis activities for cost estimation.

2. Manufacturing cost estimation

There are a number of cost estimation techniques that have been reported in literature, Parametric, Analytical and Analogical methods are widely use in manufacturing industry [3]. These methods were applicable to early stages of product development were there is lack of detailed information about a product. Integration of cost estimation system for gathering of available data to support product costing can improve accuracy of estimate.

Various methodologies for integrated cost estimation have been reported in the literature [4]. Reviewed integrated cost estimation methodologies applied a set of rules to generate process plan from a feature based CAD model, calculate the expected material volume removed and use the itterrated value to generate product cost estimate. Being rule based, reported integrated cost estimation lack capability to adabt to manufacturing process innovation. Also, they require extensive manual data input and are therefore susceptible to human induced errors. Furthermore, they do not take into account, secondary fininshing process that may be required to achive a specified surface finish, to improve cost estimate accuracy.

Accuracy of cost estimates improves with the quality of data used to generate it. In an integrated manufacturing environment cost relevant data are available as output at the multiple computer aided systems such as CAD, CAPP, CAM, and CNC that were used to perform product development activities. Utilization of these available data to generate cost estimate require high level of costing system integration with other integrated systems but the replacement of the current uni-directional communication language with a unified bi-directional communication medium for information exchange between manuafcturing systems is a critical challenge that require solution. STEP-NC provides an opportunity for the development of a compliant cost estimation system that can utilise CAx outputed data to generate product cost estimate.

3. Product development and STEP-NC

A STEP-NC compliant adaptive manufacturing platform has been developed by the AMPS team at the University of Bath's Mechanical Engineering Department [5]. The high level integration of the component part of the platform is illustrated in Fig. 1. Based on this platform a STEP-NC compliant estimation system for manufacture (SCES-M) is proposed.

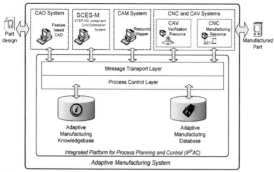

Fig. 1. STEP-NC compliant integrated platform   (adapted from [5])

The proposed SCES-M benefits as a component part of the STEP-NC compliant integrated platform. The platform associates the data exchanged with each component with the semantics of the information handled by that component. As a result, the CAD, SCES-M (costing), CAM, and CNC, utilise for product development communicate through specific interfaces in

the platform instead of having to translate and transfer information individually. These interfaces are STEP-NC compliant for bi-directional information exchange. The CAD interface for example interprets the information received from a CAD package and associates the highest level of semantics that are available for that specific system with the received data, such as geometric features, tolerance and surface finish. These are then passed to the CAM interface where domain specific knowledge is utilized to transform the information to a suitable format for a specific CAM package. The CAM package manipulates the information and sends additional data in the form of manufacturing information to the CAM interface. The interface parses the information and creates the necessary links to maintain the semantic validity of the product model. Complete manufacturing information is then passed on to the CNC interface where the knowledge driven system utilizes the available data to create a suitable set of instructions for part machining. As semantics are retained in the platform, every single piece of information is captured and linked to the rest of the available data in a synchronized manner. This maintains information integrity at all times and ensures that the quality information is made available at all stages of product development.

The result is that SCES-M that is plugged-in to the platform can access and utilise actual design and process information to quickly generate part cost estimate with improved accuracy. This STEP-NC compliant methodology ensures that cost based support for decision on the alternative, during product development, is not confined to the early stage but extends right down to the manufacturing stage.

4. The proposed system

The functionality of SCES-M systems can be described using a flow diagram shown in Fig. 2 .

The process of estimating the cost of a machined part begins with a STEP-NC data file input. STEP-NC file contains design, processes and resources models representation that provide semantically structured cost relevant parameters for machining time and cost estimation. Based on the high level integrated structure discussed above, the SCES-M process for estimating the cost of a machined part is illustrated in Fig. 2 below. The process can be summaries as follows:

1st – The system read STEP-NC files and extract the machining feature and process parameters from the data file.

2nd – The manufacturer database is interrogated for available resources and a machine tool specific process plan is generated based on expected finished features. The process plan parameters are then extracted for use at the next stage of the estimation process.

3rd – An algorithm utilises the extracted machining parameters to generate machining time estimate for each operations performed to realized a finished part. The following expression is used to calculate the machining time for each operation:

$$t_i = \sum_{j=1}^{n} t_j \qquad (1)$$

Where ti is the machining time for each feature (min) and tj is the machining time for each operation (mim).

4th – The sum of estimated time for the machined features are then caculated to obtain the overall machining time estimate. The following expression is used:

$$t_p = \sum_{i=1}^{m} t_i \qquad (2)$$

Where: t_p is machining time estimate for part (mim) and t_i is the machining time for each feature (min).

5th – The system interrogates the manufacturer accounting database for relevant shop and labor rates which are use for machining cost calculation. The machining cost is calculated using expression below:

$$C_m = t_p \times (C_h + L_r) \qquad (3)$$

where, C_m is the part machining cost (£); t_p is the estimated machining time (min); C_h is the hourly cost of a machine and; L_r

5. Case study in Java

The proposed system has been implemented to estimate the machining time for prismatic part [6]. The part on which time estimation was based is shown in Fig. 3.

There are three machining features (facing, a hole and a pocket) requiring a total of five machining operations, namely, facing, drill hole, ream hole, rough pocket and finish pocket. The STEP-NC information category for the milling test component showing the working steps that represent the process plan for the part is shown in Fig. 4. This provides the design and machining data that is required to generate time and thus cost estimate.

Accounting for these machining operations a time estimate of 0.045 hour per part was reported for the example milled part [6]. If a manufacturer uses a CNC machine tool usage cost of £45 per hour that require personnel with labour rate of £14 per hour, the estimated hourly machining cost (C_m) for the example milled part is:

$$C_m = (0.045) * (45+14) = £2.67$$

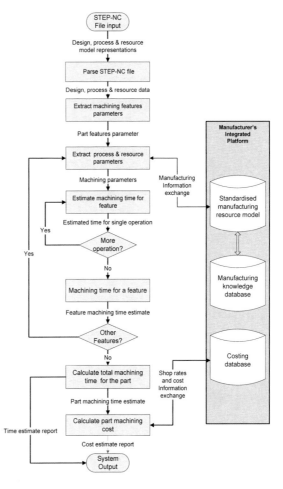

Fig. 2. SCES-M process flow diagram

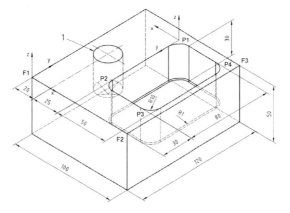

Fig. 3. Milling test component [2]

Based on reported machining time estimate for the example milled part and generic manufacturer's shop rate the above cost per hour has been calculated for the part using STEP-NC file as the information source for the cost estimate. High percentage of medium to high volume manufacturers based their cost estimate on combination of machine hour and production time [7] and the implementation of STEP-NC for high level integrated cost estimation provide the basis for fast cost estimation.

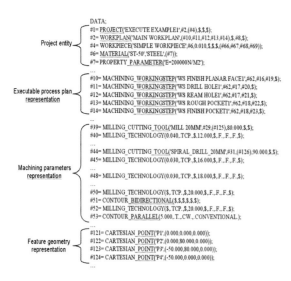

Fig. 4. STEP-NC design and process information category for the example test part [2]

6. Conclusion

With the changes in manufacturing technologies a systemic approach to integrate concurrent design of product and their related processes, for shorter product development cycle, better quality, and reduction of manufacturing cost, have become industry standard. Cost estimation systems are used by manufacturers for timely decision support on reducing manufacturing cost. It is then useful and beneficial approach to reduce the time taken to generate accurate estimate. This paper has presented a STEP-NC compliant estimation system with high level integration capability for machined part cost estimation. The presented estimation system obtains detailed cost parameters from a STEP-NC file and implements algorithm to generate machining cost estimate. This paper demonstrates the feasibility of the ISO standard [2] to facilitate high level integrated cost estimation in a manufacturing environment. . This paper demonstrates the benefit to be had from such an approach. Furthermore, the presented methodology provides manufacturer with actual product and processes information to generate fast and accurate cost estimate. Further work will focus on demonstrating the potential application of such system to support decision on alternatives at all stages of product development and extending this STEP-NC compliant methodology to support decision on optimum manufacturing resources

selection by generating and comparing production cost estimate (for a given part) for a number of machine tools (CNC). The result will be used to support decision on optimum manufacturing resource alternative.

References

[1]   Souchoroukov, P., R. Roy, and K. Mishra, Data and Information in Cost Estimating, 2002, Conference of The Society of Cost Estimating & Analysis. p. 44-45.

[2]. ISO 14649-11, Industrial automation systems and integration - physical device control - data model for computerized numerical controllers, part 11: Process data for milling. ISO.

[3]   Niazi, A., et al., Product cost estimation: Technique classification and methodology review. Journal of Manufacturing Science and Engineering-Transactions of the Asme, 2006. 128(2): p. 563-575.

[4]   Shehab, E.M. and H.S. Abdalla, Manufacturing cost modelling for concurrent product development. Robotics and Computer-Integrated Manufacturing, 2001. 17(4): p. 341-353.

[5]   Nassehi, A., S.T. Newman, and R.D. Allen, STEP-NC compliant process planning as an enabler for adaptive global manufacturing. Robotics and Computer-Integrated Manufacturing, 2006. 22(5-6): p. 456-467.

[6]   O Debode, A., et al., Integrated machining time estimation using standardised micro-process planning information, 2011, Conference: Flexible Automation and Intelligent Manufacturing.

[7]   Brierley, J.A., C.J. Cowton, and C. and Drury, How product costs are calculated and used in decision making: a pilot study", 2001, Managerial Auditing Journal. p. 202-206.

Automated design and STEP-NC machining of impellers

Arivazhagan Anbalagan[1], Mehta NK[2] and Jain PK[2]

[1] Indian Institute of Information Technology Design & Manufacturing, Kancheepuram, Chennai – 600 048, INDIA.

[2] Indian Institute of Technology Roorkee, Roorkee – 247 667, INDIA.

Abstract. This paper presents the four stage approach followed for automated design and STEP-NC based machining of impellers. In the first stage, the design calculations are performed to construct the 'Meridional representation' of the radial impeller. Then 3D curves are projected from the 'Meridional representation' and 3D model is generated using UG-NX software. In the second stage, the process planning activities including tooling & setup plan are completed. Here, ball end mill cutters with suitable diameter and length are selected and appropriate process parameters as suited to 5 axis milling are considered. In the third stage, the tool path data based on contour area milling is generated and verified in the UG NX software. Finally, in the fourth stage, the model with the complete data is imported to STEP-NC software and the AP-238 format is generated. In this article the design procedure adopted for construction of 'Meridional Section' of a radial turbine is discussed with the general methdology to automate the process planning and tool path generation. A test case of radial impeller is presented with the results obtained by adopting STEP-NC format.

Keywords: Impellers / Blades, Modelling & Automation, CAPP, STEP-NC Integration

1. Introduction

Automated design & STEP-NC machining of impellers is considered to be a crucial task as it involves integration of complex design procedures and 5 axis manufacturing process plan data. Impellers which are free form in nature are adopted to pump the flow of gas or fluid in centrifugal & axial compressors/turbines/pumps belonging to oil and gas (O&G), aviation and power generation domains. Generally, these are first casted and then finish machined using a 5 axis milling machine and sometimes completely milled in a 5 axis milling machine. In either case, a manufacturing drawing sheet must be generated from a parametrically strong and geometrically precise 3D CAD models. These 3D CAD models are designed by sweeping the basic curves namely (i) B-Spline and (ii) NURBS which follows recursive blending mathematical representations. The construction procedure of these curves and surfaces are well known [1] and implemented in many CAD/CAM packages. From an automated manufacturing point of view, these 3D CAD models

should contain error free feature data, as even a minor change leads to improper process plan and tool paths. Further, process plan independent CL data generated from these models consumes more time for post processing in a CNC machine. In the present scenario, CL data alone is not sufficient to go ahead with the machining process. Addition details such as tooling, setup and fixture is required to proceed with a robust machining. As regards, researchers adopt STEP/STEP-NC technology owing to the advantage of integrating product life cycle and manufacturing process planning data. Also, it reproduces error free 3D CAD models and reduces the transfer time to a major extent. Even though there are many advances in this domain, automated design and STEP-NC machining of impellers needs attention owing to the complexity encountered while automatic feature recognition, design calculations and generation of process plan with tool paths. HT Young et al. [2] generated tool paths for rough machining centrifugal impeller using a five axis milling machine. They introduced two concepts namely (i) residual tool path and (ii) cutting tool path for removing the material which are closer and away from the blade tip. Pyo Lim [3] presented an approach to optimize the rough cutting factors of impeller with a 5 axis machining using 'response surface methodology'. In his work, the roughing operation is divided into five portions to machine the fillets between blade surfaces and hub surfaces. Julien Chaves-Jacob et al. [4] presented an optimal strategy for finish machining the impeller blades by adopting a 5 axis milling machine. Here, point milling and flank milling strategies are developed to reduce the machining time. Li-Chang Chuang & Hong-Tsu Young [5] presented an integrated rough methodology to manufacture centrifugal impeller. While rough machining constant scallop height is maintained to improve the quality of machining process. They analyzed a theoretical model and developed process plan for machining the part in a 5 axis milling machine. Toh [6] developed a strategy for cutter path calculation in high-speed milling process. He focused on rough machining of moulds and tested the tool

paths using a vertical high-speed-machining centre. An algorithm for parametric tool path correction in a 5 axis machining has been proposed by Gabor et al. [7]. In their approach, machine dependent and independent data is developed to store the prescribed tool path. A machining strategy for milling a set of surface which is obtained by the technique of cross sectional design is performed by Sotiris & Andreas [8]. The surfaces are formed by sliding the Bezier Curve (Profile curve) along another Bezier Curve (Trajectory) and tool-paths are generated by offsetting the boundaries of the profile curve matching with the trajectory curve. He used data point models and produced LOD models and obtained adaptive rough-cut and finish cut tool-paths. Brecher et al. [9] tested STEP-NC program and inspected the feed back in a closed loop CAPP/CAM/CNC process. In their work, they modelled the component in a CAD package and generated the process planning details and validated in a STEP-NC based milling machine. A frame work to interpret the data in AP-238 is done by Liu et al. [10]. In their work, a PC based STEP-NC prototype for STEP compliant CNC is developed to interface and to extract the details required for processing the AP-238 format. After analyzing the literatures, the following points are noticed:
•Machining is conducted without addressing the design calcualtion of impellers
• 5- axis milling ignores the integration of process planning and tool path data in a single format
•There is still a complexity on roughing out the excess material in between the blades.
•While machining, there is a necessity for most efficient tool path, where the tool spends only a minimum amount of time in air.
•The tool length needs to be kept to the minimum to avoid vibration and to prolong tool life
• Focus must be given for integration of tooling, setup and fixturing aspects
•STEP-NC integration focuses on simple rotational, prismatic and sheet metal parts and not for impellers

Based on the above points, it is decided to proceed with an automated design and STEP-NC machining of impellers. As the first step, the design procedures adopted in impellers are analysed. It is noticed there are more than 20 design parameters involved in impeller design process. The next section presents the design calculation and its automation carried out in this research.

2. Design calculation of impellers

The design of an impeller is considered to be most complicated and crucial as there are more than 20 design parameters. These parameters are related to various flow parameters of compressor/pumps and is to be checked in accordance with the desired output. Fig.1(a) shows an impeller with few basic parameters namely (i) a leading edge-as pointed at its top (ii) trailing edge-as pointed at its end; (iii) hub diameter

(iv) hub height (v) shroud (vi) hub & shroud surface and (vii) blade thickness.

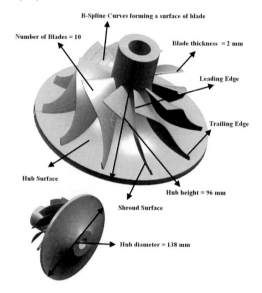

Fig.1(a). Radial Impeller cross section

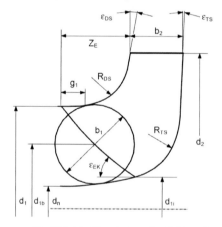

Fig.1(b). Meridional view of radial impeller

In order to draw a 3D impeller it is indeed necessary first to draw the section through the impeller called 'Meridional Representation'. Basically, the leading and trailing edges of a blade are projected into the drawing plane through 'circular projection' and the initial blade profile is drawn [12]. A 'Meridional representation' with various basic parameters required for construction if shown in Fig. 1(b). It consists of parameters namely (i) b_1-impeller inlet width (leading edge) = ½ (d_1- d_n); (ii) b_2-impeller outlet width (trailing edge); (iii) d1-Impeller inlet diameter ; (iv) d_2-Impeller outer diameter ; (v) d_{Ii}-Blade inlet diameter at the inner streamline; (vi) d_{Ib}-stream line diameter; (vii) R_{DS}-Radius of curvature - front shroud = (0.6 to 0.8) b_1; (viii) R_{TS}-Radius of curvature - rear shroud or hub; (ix) z_E-Axial Extension; (x) ε_{DS}- angle of front shroud; (xi) ε_{TS}-angle of rear shroud or hub; (xii) ε_{EK}-axial inlet angle; (xiii) d_n - Hub diameter; (xiv) g_1-

Short section length = (0.2 to 0.3) b_1; (xv) e-Blade thickness; (xvi) d_w- Shaft diameter; (xvii) z_{La}- Impeller blade number; (xviii) β_{1B}.Impeller blade inlet angle; (xix) β_{2B} - Impeller blade Outlet angle and ; (xx) A_{1q}.Throat area.

Further to the above design formulas the following points are also considered: (i) In order to achieve a flatter pressure, the radius RDS should not be tangent to the point defined by z_E, but a short section g_1 = (0.2 to 0.3)×b_1 should be introduced with only a minor increase in radius (ii) For short axial extension of the impeller, smaller values are selected for z_E and R_{DS} than calculated from Eq. (1) (iii) Specific speed is used to find the angle ε_{DS} (iv) ε_{DS} is increased to 15 to 20° with higher specific speeds (v) Positive or negative angle for ε_{TS} can be chosen and (vi) The outer streamline is drawn with d_2, b_2, d_1, z_E g_1, ε_{DS} and R_{DS} defined by a free curve or assembled from straight lines and circular arcs or by Bezier functions. To proceed with the calculation of the basic parameters namely, d_1, d_2, d_{1opt}, Z_E etc. the Equations from Eq.1 to Eq.6 are adopted.

$$d_1 = 2.9 \sqrt[3]{\frac{Q_{La}}{f_q \, n \, k_n \tan\beta_1}\left(1+\frac{\tan\beta_1}{\tan\alpha_1}\right)} \qquad \text{Eq. (1)}$$

$$d_2 = \frac{60}{\pi n}\sqrt{\frac{2g\,H_{opt}}{\psi_{opt}}} = \frac{84.6}{n}\sqrt{\frac{H_{opt}}{\psi_{opt}}} \qquad \text{Eq. (2)}$$

$$d_w = \left(\frac{16 P_{max}}{\pi \omega \tau_{al}}\right)^{\frac{1}{3}} = 3.65\left(\frac{P_{max}}{n\,\tau_{al}}\right)^{\frac{1}{3}} \quad \begin{array}{l} P_{max}\text{ in W}\\ n\text{ in rpm}\\ \tau_{al}\text{ in N/m}^2 \end{array} \qquad \text{Eq. (3)}$$

$$z_E = (d_{2a}-d_1)\left(\frac{n_q}{n_{q,Ref}}\right)^{1.07} \quad \begin{array}{l} R_{Ds}=(0.6\text{ to }0.8)\,b_1\\ b_1=\frac{1}{2}\,(d_1-d_n)\\ n_{q,Ref}=74 \end{array} \qquad \text{Eq. (4)}$$

$$d_{1,opt} = \sqrt{d_n^2 + 10.6\left(\frac{Q_{La}}{f_q\,n}\right)^{\frac{2}{3}}\left(\frac{\lambda_c+\lambda_w}{\lambda_w}\right)^{\frac{1}{3}}} \qquad \text{Eq. (5)}$$

$$\beta_{1B} = \beta'_1 + i_1 = \arctan\frac{c_{1m}\,\tau_1}{u_1 - c_{1u}} + i_1 \qquad \text{Eq. (6)}$$

To find the various parameters initially, the values of the first 7 parameters are assumed. The remaining are calcuated accordingly with their specific fomulas. Further, due to page restriction, partial calculation is shown with the basic parameters assumed for few dimensional parameters. The author can be emailed for the complete calculation part of the impeller. (i) d_n = 1.36 m; (ii) α_1= 60°; (iii) α_2= 35°; (iv) β_1= 30°; (v) β_2= 37°; (vi) β_1'= 45°; (vii) β_2'= 52°; i_1'= 15°; i_2'= δ'= 10°; β_{1B}= i_1'+ β_1'= 60°; β_{2B}= i_2'+ β_2'=60°; $\delta = \beta_{2B} - \beta_2$ = 25°; H_{opt} = 10 m; n = 3000 rpm; λ_c =1.2 to 1.35; λ_w= 0.42; C_{1m} = Q_{La} / f_q A_1 ; A_1 = (π/4 $(d_1-d_n)^2$); C_{1u} = C_{1m} / $\tan \alpha_1$; Q_{La} = Q_{opt} + Q_{sp} + Q_E ; Q_{opt} = 8.9 m^3/s; Q_{sp} = 1.9 m^3/s; Q_E = 0; K_n =0.2;The calculated values are given below: Q_{La} = Q_{opt} + Q_{sp} + Q_E = 10.8 m^3/s; d_1 – based on Eq.1 = 0.241m; $d_{1,opt}$ – based on Eq.1 = 1.30m; d_1 – based on Eq.2 = 0.0893m; z_E – based on Eq.4 = 0.684; R_{DS} – based

on Eq.4 = 0.84m ; After making all the basic calculations the "Meridional Section" is drawn using UG NX software. The 3D representation is also drawn in the UG-NX software from the 'Meridional section' by adopting a similar set of calculation.

3. General methodology adopted in automation process

Step1: Design the radial impeller and model the part in UG NX CAD package

Explanation to Step1: In this step, the part is modelled and parameterized in the UG NX CAD package. Geometric dimensioning and tolerances (GD&T), information of datum's are added to the model. Then drawing sheets associated with the parts are manually generated and checked.

Step2: Using UG/UFUNC functions extract the geometrical and topological data of the model.

Sub step2.1: Ask the tag (number) of part (specific to UG)

Sub step2.2: Using the tag, cycle all the objects in the part and count the number of features/ objects.

Sub step2.3: Get the ID's of all features/objects

Sub step2.4: Extract the data and store it in a text file.

Explanation to Step2/Sub steps 2.1-2.4: Generally, a UG part model will have a single tag in the form of a number. This is extracted and the tags of various sub features / objects are found by cycling the part model through a UG/UFUNC function "UF_OBJ_cycle_objs_in_part". Using these tags the geometry and topological data of the sub features / objects are extracted which is used to find the closeness index with Bezier /B-Spline curves. Some of the other used functions are: (i) UF_CURVE_ask_spline_data (ii) UF_CURVE_edit_spline_feature(iii) F_b_curve_bezier_subtype.

Step3: Match the data with the basic B-Splines / Bezier curves / surfaces and calculate the closeness index

Explanation to Step3: In this step, the extracted data is matched and a closeness index (CI) "0(0-not matching)-10 (10-exact match)" is generated. It is done by calculating the control points, degree of meridional curve, and various parameters (as shown in Fig.1(b)) required for Bezier and uniform/ cubic/open/non-uniform B-Spline curves.

Step4: Calculate the blending functions and identify the machinable area of the impeller / blade features.

Explanation to Step4: After finding the closeness index blending functions are calculated using convolution theorem. Using the blending function data, the rough and finish cut machinable volumes are calculated.

Step5: Specify the process plan details and Adopt the Z-level contour area milling to generate tool paths

Explanation to Step5: Here, ball end mill cutters with appropriate radius and length are used for machining.

Table 1. Process Plan details of the radial impeller

Roughing	Finishing
Ball End mill	Ball End mill
Diameter = 8 mm	Diameter = 5 mm
Length=75 mm	Length=75 mm
Flute length= 50 mm	Flute length= 50 mm
Feed rate	23 mm/min
Spindle speed	2500 rpm

Appropriate process parameters for 5 axis contour area milling as shown in Table1 is adopted for machining.

The work piece is rotated to make cutting surfaces of tool tangent to ideal part features. Two methods namely (i) fixed and (ii) variable contour machining methods are used to finish areas formed by free form surfaces. Intricate contours are machined by controlling tool axis & projection vector. A schematic representation of the impeller machining process is shown in Fig. 2. The tool path is simulated for both roughing & finishing operations and CL data is obtained after post processing.

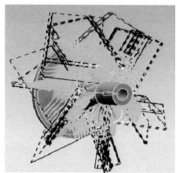

Fig.2 Tool paths simulated with GD&T data

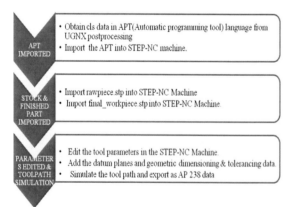

Fig.3. Steps followed to obtain a AP-238 data.

Step6: Integrate and verify with STEP-NC format
Explanation to Step6: Finally, the impeller is machined using standard method of tool path generation available in STEP-NC Machine as shown in Fig.3. The tool path is finally simulated & output file is obtained as AP238 format.

4. Conclusions and future work

The whole process is automated through a software namedFree_Form_Blades_Impeller_Automation F²BIM). It consists of four modules namely (i) Design Module (DM) (ii) Process Planning Module (PPM) (iii) Tool Path Generation Module (TPGM) and (iv) STEP-NC generation Module (STM). All these modules are linked with the main GUI of the software. A user can select/ modify various blades / impellers as suited for industrial needs and can generate the complete set of data required for machining. Presently, cross sectional details of 3 radial impellers are automated. Work is in progress to upgrade the whole software with more than 50 different types of profiles collected from various engineering domains.

Acknowledgements: The author gratefully acknowledge Department of Science and Technology (DST), Government of India for funding this research under SERC-Fast Track Scheme for Young Scientists (FAST).

References

[1]   Donald Hearn, Pauline Baker M, (1996), Computer Graphics, Prentice Hall.
[2]   Young HT, Chuang LC, Gerschwiler K, Kamps S, (2004) A five-axis rough machining approach for a centrifugal impeller. International Journal of Advanced Manufacturing Technology 23:233- 239.
[3]   Pyo Lim, (2009) Optimization of the rough cutting factors of impeller with five-axis machine using response surface methodology 45:821-829.
[4]   Julien Chaves-Jacob, Gérard Poulachon, Emmanuel Duc, (2011) Optimal strategy for finishing impeller blades using 5-axis machining. International Journal of Advanced Manufacturing Technology DOI 10.1007/s00170-011-3424-1.
[5]   Li-Chang Chuang, Hong-Tsu Young (2007) Integrated rough machining methodology for centrifugal impeller manufacturing. International Journal of Advanced Manufacturing Technology 34:1062–1071.
[6]   C.K. Toh (2006), Cutter path strategies in high speed rough milling of hardened steel. Materials& Design 27: 107-114.
[7]   Gabor Erdosa, Matthias Muler, Paul Xirouchakis (2005), Parametric tool correction algorithm for 5-axis machining. Advances in Engineering Software 36 :654-663.
[8]   Sotiris L Omiroua, Andreas C. Nearchoub, (2007) A CNC machine tool interpolator for surfaces of cross-sectional design. Robotics and Computer Integrated Manufacturing 23 : 257-264.
[9]   Brecher C, Vitr M, Wolf J, (2006) Closed loop CAPP/CAM/CNC process chain based on STEP and STEP-NC inspection tasks. International Journal of Computer Integrated Manufacturing 19: 570-580.
[10] Liu R, Zhang C, New man ST, (2006) A Frame work and data processing for interfacing CNC with AP-238 International Journal of Computer Integrated Manufacturing 19: 516-522.
[11] Unigraphics NX 7.5 Help Documentation.
[12] Johann Friedrich Gulich , (2010), Centrifugal pumps, Second Edition, Springer.

LACAM3D, CAM solution for tool path generation for build up of complex aerospace components by laser powder deposition

J. Flemmer[1], N. Pirch[2], J. Witzel[1], A. Gasser[2], K. Wissenbach[2], I. Kelbassa[1,2]
[1] Chair for Laser Technology, RWTH Aachen University, Germany
[2] Fraunhofer Institute for Laser Technology ILT, Steinbachstr. 15, Aachen, Germany

Abstract. This The success of Laser Metal Deposition (LMD) for repair or fabrication of near-net shaped metals components directly from CAD solid models without use of time and cost-prohibitive conventional techniques, such as five-axis milling, linear friction welding and electro chemical machining depends on the availability of a close CAD/CAM chain. Distortion and defects of a worn part implies that the nominal CAD model from the design stage is no longer suitable for the representation of the part geometry. This means that first of all the actual and the target geometry have to be constructed from which the deficit volume is determined. Only with this volume being available the tool path programmed. The original data format of laser scanned data is a polygonal modeling approach for surface representation. The further step of creating a NURBS representation of the geometry requires a major commitment in time, both in instruction and in amount of work. For this reason the ILT designed a process chain for the LMD whose CAM module, LACAM3D, is based on a polygonal modeling approach for geometry representation. NURBS are only used for the reduction of data noise in the scanned data. The target geometry of worn parts is derived by a best fit with the CAD model from the design stage. LACAM3D supports the generation of all sorts of path pattern within a layer which for example may be derived from a distortion reduction analysis. The functionalities of LACAM are exemplified by a repair application and the near-net fabrication of a BLISK (Blade-Integrated diSK) blade direct from the CAD model.

Keywords: Laser powder deposition, reverse engineering, CAM

1. Introduction

LMD is under development in the last few years and just becoming an advanced manufacturing technology for repair or fabrication of near-net shaped metals components. LMD is an additive–layered technique that uses a laser beam to melt injected powder or wire along the tool path and thereby to form the layer by adjoining clad tracks. Due to various defects such as distortion, wear or manufacturing tolerance the most components don't correspond to the geometry data specified in der nominal CAD model. For this reason the actual and the target geometry have to be constructed in order to establish a sound CAD basis for tool path generation.

Almost all CAD programs have the ability to transfer geometry data based on NURBS modeling to polygon representation in a desired accuracy. Reverse engineering derives in the very first step a polygonal geometry representation from the point cloud of scanned data. So polygon modeling i.e. the so called stl format is supported by CAD and reverse engineering. Concerning the geometry construction from scanned data there is a need for research for shape preserving data noise reduction and automatic edge detection.

Because polygon modeling is the very first step in reverse engineering the ILT decided to use a geometric modeling kernel for the CAM module, LACAM3D, which uses exclusively the stl data format. Thereby the complex procedures for NURBS representation of scanned data can be avoided. The polygon modeling approach is both quicker and involves less complexity with a view to rapidly restore of worn parts. In this paper the process chain for LMD and thereby the functionalities of LACAM3D is presented for a repair application and the near-net fabrication of a BLISK (Blade-Integrated diSK) blade direct from the CAD model.

2. LMD repair

Most parts no longer correspond to the geometry data specified in the CAD data record after usage or due to their manufacturing tolerance. On that account and because there is a need to handle manual prepared welding areas the feasibility to construct a CAD model of the welding area of the part in the clamped state is required. For this reason a laser line scanner was integrated in the LMD system to use it as measuring machine (Fig.1). From the resulting point cloud a polygon mesh is derived and made available in the so called stl format as input for LACAM3D. In general the mesh is unstructured and not cross-linked. In order to realize short computing time for the different CAD functionalities on the mesh a suitable topology was introduced. This topology allows a fast access on the neighborhood of any point on the geometry.

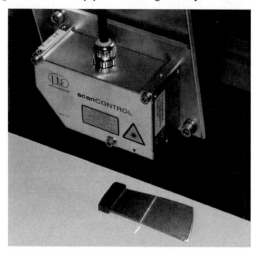

Fig. 1. Laser line scanner

The goal of the LMD repair process is a dimensional accuracy of about 0.5 mm concerning the target geometry. That reduces the post finishing processing substantially. In order to achieve that objective the tool paths of the different layers have to be trimmed at the target geometry i.e. at the so called Master geometry. The Master geometry is a CAD surface model of the non-worn part which includes the welding area. The Master geometry is updated with outsize by a best fit to the digitized part model. The part geometry and the Master limit geometrically the deficit volume with outsize of about 0.5 mm.

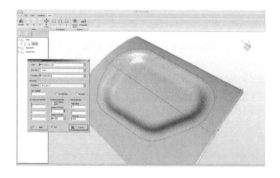

Fig. 2. Boundary line of welding area (blue), first line (red) and the group of equidistant lines on the free form surface derived from the first line.

The digitized part and the Master geometry are provided for LACAM3D in the stl format. For the generation of tool path pattern for the first layer the welding area have to be geometrically limited by a boundary line (Fig.2). The fill pattern with equidistant lines is derived from a start line. This line may be a part of the boundary line or has to be defined by LACAM3D. In the simplest way this

line is defined by two points. This line segment is then projected onto the part geometry whereby the projection direction can be freely selected. The equidistant lines are then generated on the part geometry relative to this projected line in both directions. The equidistant lines are trimmed with the boundary line. The number of points per line can be reduced by i.e. thinning out in a desirable precision.

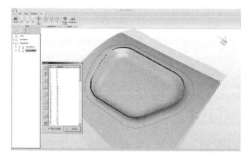

Fig 3. Welding area (trough machined into shaft), Master geometry (25% Opacity) and tool path of the 8. Layer.

For this LACAM3D provides a menu from which the different option for tool path generation can be selected (Fig.2). Even at before equidistant line generation the user can determine a uni- or bidirectional path pattern. Via a window menu LaCam3D offers functionalities to change the numbering and direction of each track and to configure the laser energy and scanning velocity for the tool paths (Fig.3).

Fig 4. Welding area (trough machined into shaft), Master geometry (25% Opacity) and tool path of the 8. Layer.

The generation of the tool paths of the next layers is based on an offset calculation. In order to fill the trough by LMD with a high dimensional accuracy the displaced tool paths from the first layer are trimmed with the Master geometry.
LACAM3D provides a simulation tool for the inspection for potential collisions between laser head and part geometry (Fig.4). For the generation of cnc code LACAM3D uses a post processor which is customized for the different machining systems.
  Fig. 5 shows the trough and the process during LMD repair on basis of the ILT process chain

Fig. 5. Part, welding area und process photo.

3. LMD direct manufacturing

For the LMD manufacturing of turbine components (Fig.5) the CAM program is provided with special module.

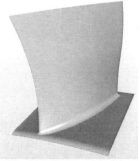

Fig. 6 (a). CAD model of the blade

Fig. 6 (b). Part of the outer periphery of a rotor disk.

Due to the considerably varying blade thickness the LMD process requires in dependence of part height different welding strategies. In the root area the blade exhibits a mean thickness of about 8 mm. The outer periphery of a rotor disk, on which the blade is fixed, represents a free form surface (Fog. 6b). For the manufacturing of the blade by LMD layer by layer the blade geometry has to be sliced with the in z-direction displaced surface part of the outer periphery of the rotor disk to determine the welding boundary (Fig.7). To do this LACAM3D offers a functionality to calculate the intersection of any two free form surfaces.

In the following a midline is derived for the boundary line (Fig.7 left) and a group of equidistant lines is generated which are trimmed with the boundary line (Fig. 7 right). The numbering of the lines and the starting point of the contour path can be modified. The root area requires a meander-like filling pattern (Fig.7 right)
For the upper layers the boundary line becomes continuously smaller [Fig. 8 (a),(b)]. The variation of the blade thickness becomes lower than 4 mm [Fig. 9 (a),(b)]. In preliminaries studies /1/ process diagrams (Fig.19) have been developed which describe the laser power, scanning velocity and beam diameter along the tool path in order to realize tracks with varying thickness but

constant track height. For the generation of cnc code this process diagrams have to be provided to LACAM3D as a discrete measure curve. Depending on the local blade thickness LACAM3D adjusts the process parameters along the tool path according to the processing diagrams.

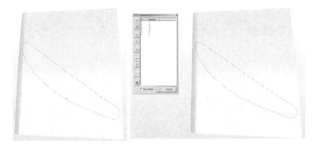

Fig. 7. Contour line as cutting line between blade surface and the outer periphery of a rotor disk and midline in the blade root area (left). Equidistant lines to midline trimmed with layer contour (right).

Fig. 8. Part of the outer periphery of a rotor disk and it's vertically offset (left). Contour of two different layers. (right).

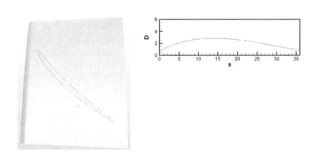

Fig. 9. Contour line as cutting line between blade surface and the outer periphery of a rotor disk and midline out of the blade root area (left). Blade thickness along arc length of the midline (right).

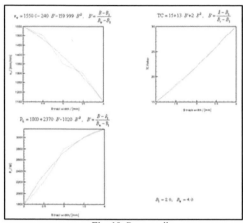

Fig. 10. Process diagrams

Reference

[1]   Witzel, J., Schrage, J., Gasser, A., Kelbassa, I.: Additive manufacturing of a blade-integrated disk by laser metal deposition. ICALEO. 30. Int. Congr. on Applications of Lasers and Electro-Optics, October 23-27, 2011. Paper 502 (7 S.), 2011

Figure 11 shows the resulting blade which has to be finished by milling. The LMD process was adapted in such way that the blade exhibits an oversize of 0.5 mm.

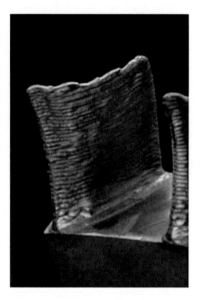

Fig. 11: Blade bevor post finishing.

4. Summary and conclusion

LACAM3D is an offline cam program for LMD equipped with functionalities which enables the user to transfer also complex welding patterns in cnc code just in time. The CAM program has a modular structure so that new welding strategies can be easily added by customer inquiry. At time a prototype version of the program exists and is used for the cnc code generation for different applications for LMD and is under continuously development.

3

ECM and EDM

A comparative study of the ANN and RSM models for predicting process parameters during WEDC of SiC$_p$/6061 Al MMC

Pragya Shandilya[1], P.K. Jain[1] and N.K. Jain[2]

[1] Department of Mechanical and Industrial Engineering, Indian Institute of Technology, Roorkee, 247667, India

[2] Department of Mechanical Engineering, Indian Institute of Technology, Indore, 452017, India

Abstract. Metal matrix composites (MMCs) have found many successful industrial applications in recent past as high-technology materials due to their properties. Wire electric discharge cutting (WEDC) process is considered to be one of the most suitable processes for machining MMCs. Lot of research work has been done on WEDC, but very few investigations have been done on WEDC of MMCs. This paper reports work on the analysis of material removal rate (MRR) and cutting width (kerf) during WEDC of 6061 Al MMC reinforced with silicon carbide particles (i.e. SiC$_p$/6061 Al). Four WEDC parameters namely servo voltage (SV), pulse-on time (T_{ON}), pulse-off time (T_{OFF}) and wire feed rate (WF) were chosen as machining process parameters. Artificial neural network (ANN) models and response surface methodology (RSM) models were developed to predict the MRR and kerf using Box-Behnken design (BBD) to generate the input/output database. It was observed that prediction of responses from both models closely agree with the experimental values. The ANN models and RSM models for WEDC of MMC were compared with each other on the basis of prediction accuracy which shows that ANN models are more accurate than RSM models for MRR and kerf because the values of percentage absolute errors are higher for RSM models than ANN models.

Keywords: Artifical neural network, Response surface methodology, Wire electric discharge cutting, Metal matrix composites, Material removal rate, kerf.

1. Introduction

Concerning industrial applications, MMCs now have a proven record of accomplishment as successful high-technology materials due to the properties such as high strength-to-weight ratio, high toughness, lower value of coefficient of thermal expansion, good wear resistance, and capability of operating at elevated temperatures [1]. MMCs are fabricated using several processes such as casting, forging and extrusion. However, cutting and finishing operations of MMCs are not well understood. The use of traditional machining processes to machine hard composite materials causes serious tool wear due to abrasive nature of reinforcing particles thus shortening tool life [2]. Although, nontraditional machining techniques such as water jet machining (WJM) and laser beam machining (LBM) can be used but the machining equipment is expensive, height of the workpiece is a constraint, and surface finish obtained is not good [3]. On the other hand, some techniques such as electric discharge machining (EDM) and wire electric discharge machining (WEDM) or wire electric discharge cutting (WEDC) are quite successful for machining of MMCs. EDM has limited applications as it can be used only for drilling purpose. WEDM which is a derived process of EDM seems to be a better choice as it conforms to easy control and can machine intricate and complex shapes. The setting for the various process parameters required in WEDC process play crucial role in achieving optimal performance. According to Patil and Brahmankar [4], during WEDC an accurate and efficient machining operation without compromising machining performance is achievable. Effective and economical WEDC of MMCs will open new areas of applications for MMCs. The most important performance measures in WEDC are MRR and kerf. In WEDC, the material removal is by melting and/or evaporation of electrically conductive phase of SiC$_p$/6061 Al MMC. kerf determines the dimensional accuracy of the finishing part. Extensive experimental work is therefore needed to analyze and optimize the process parameters to understand their effect on product quality. The current investigation aims at investigating the suitability of MRR and kerf predictive models based on RSM and ANN models during WEDC of SiC$_p$/6061 Al MMC.

2. Experimentation

The experiments were conducted on the ECOCUT WEDM machine from Electronica India Pvt Ltd. 6061 aluminum based MMC, made by stir casting technique having 10% SiC particles (by weight) as reinforcement were used as the workpieces. The workpieces were of rectangular shape having a thickness of 6 mm. The

S. Hinduja and L. Li (eds.), *Proceedings of the 37th International MATADOR Conference,*
DOI: 10.1007/978-1-4471-4480-9_3, © Springer-Verlag London 2013

deionized water was used as dielectric. The dielectric temperature was kept at 20°C. A diffused brass wire of 0.25 mm diameter was used as the cutting tool. The four input process parameters namely servo voltage (SV), pulse-on time (T_{ON}), pulse-off time (T_{OFF}) and wire feed rate (WF) were chosen as variables to study their effect on the quality of cut in SiC_p/6061 Al MMC during WEDC. The ranges of these parameters were selected based on literature survey, machining capability and preliminary experiments conducted by using one-variable-at-a-time approach [5]. Table 1 gives the levels of various parameters and their designation.

To calculate the MRR, the following equation [6] is considered:

$$MRR = \frac{Mf - Mi}{\rho t} \qquad (1)$$

where, Mi, are masses (in gm) of the work material before and after machining respectively, ρ is the density of workpiece material and t is the time of machining in minutes. An electronic weighing machine with an accuracy of 0.001 mg is used to weight the material. The kerf was measured using the stereo microscope, and is expressed as sum of wire diameter and twice of wire-workpiece gap.

Table 1. Levels of process parameters

Process parameters	Levels		
	-1	**0**	**+1**
Voltage (V)	70	80	90
Pulse-on time (μs)	1	2	3
Pulse-off time (μs)	6	8	10
Wire feed (m/min)	5	7	9

The WEDC process was studied according to the Box-Behnken design (BBD). In this investigation, total 29 experiments were conducted. Levels and values of four process parameters for 29 experimental runs were given in Table 1.2. The 'Design Expert 6.0' software was used to establish mathematical models for optimization of the parameter settings to achieve the required MRR and kerf during WEDC of SiC_p/6061 Al MMC.

3. RSM based predictive models

Mathematical models based on RSM for correlating responses such as MRR and kerf with various settings of process parameters during WEDC of SiC_p/6061 Al MMC have been established, and are represented in the following regression equations

$$MRR = +2239.042 - 3.650SV - 12.115T_{OFF} - 9.427WF + $$
$$0.0134SV^2 + 0.248T_{OFF}^2 + 0.228WF + 0.0980VT_{OFF} + {}^{(2)}$$
$$0.077VWF$$

$$kerf = -6.016 + 0.097SV + 0.095T_{OFF} + 0.272T_{OFF} + $$
$$0.325WF - 0.004SV^2 - 0.022T_{ON}^2 - 0.006T_{OFF}^2 - $$
$$0.008WF^2 + 0.009SVT_{ON} - 0.001SVT_{OFF} - 0.008SVWF - $$
$$0.007T_{ON}T_{OFF} - 0.004T_{ON}WF - 0.007T_{OFF}WF \qquad (3)$$

Table 2 shows the RSM predicted values for MRR and kerf for the 29 experimental runs.

Table 2. BBD with four parameters and experimental MRR and kerf

Exp. No.	SV	T_{ON}	T_{OFF}	WF	MRR (mm³/min)	Kerf (mm)
1	0	1	0	-1	4.200	0.381
2	-1	0	0	-1	9.965	0.266
3	0	0	1	1	4.931	0.328
4	-1	0	1	0	5.188	0.287
5	0	-1	-1	0	4.586	0.359
6	1	0	0	-1	3.679	0.415
7	1	0	-1	0	3.026	0.438
8	0	0	0	0	3.243	0.424
9	0	0	0	0	3.943	0.387
10	1	0	0	1	5.072	0.308
11	0	0	1	-1	3.897	0.407
12	0	0	0	0	3.939	0.394
13	0	-1	0	1	4.740	0.342
14	0	0	0	0	3.170	0.426
15	0	1	0	1	5.075	0.302
16	0	0	0	0	3.293	0.422
17	1	0	1	-1	4.702	0.352
18	0	1	-1	-1	4.974	0.322
19	0	0	-1	1	4.340	0.368
20	0	-1	0	-1	4.251	0.372
21	0	0	-1	-1	4.851	0.334
22	0	-1	1	0	3.901	0.401
23	1	1	0	0	2.590	0.446
24	-1	0	-1	0	11.354	0.261
25	-1	1	0	0	5.991	0.274
26	-1	0	0	1	5.126	0.294
27	-1	-1	0	0	5.325	0.282
28	1	-1	0	0	3.132	0.432
29	0	1	1	0	4.997	0.316

4. Development of ANN models for prediction of responses

An ANN is an information-processing system that has certain performance characteristics in common with biological neural networks. Generally, an ANN is made up of some neurons connected together via links. Among various neural network models, the feedforward neural network based on back-propagation is the best general-purpose model [7]. The network has four inputs of servo voltage (SV), pulse-on time (T_{ON}), pulse-off time (T_{OFF}), wire feed rate (WF) and two outputs of MRR and kerf. The training of the ANN for 29 input-output patterns has been carried out using the Neural Network Toolbox available in MATLAB software package. The network consists of one input layer, one hidden layer and one output layer. In the proposed model, there were four input variables and two outputs. Hence the number of input neurons was taken as four and the number of output neurons was two. The selection of number of neurons in the hidden layer is usually model dependent. The numbers of hidden layers neurons are decided by trial and error method on the basis of the improvement in the error with increasing number of hidden nodes [8]. Hence, there were fifteen neurons in hidden layer. To train each network, an equal learning rate and momentum constant (α) of 0.05 and 0.9 respectively were used, the activation function of hidden and output neurons was selected as a hyperbolic tangent, and the error goal (mean square error, MSE) value was set at 0.0001, which means the training epochs are continued until the MSE fell below this value. To calculate connection weights, a set of desired network output values is needed. Desired output values referred as training dataset, obtained with the help of design of experiments (DOE). MRR and kerf values corresponding to training data were obtained from experimental runs generated by BBD based on RSM. The dataset generated by BBD shown in Table 2. Table 3 shows ANN predicted values for MRR and kerf for the 29 training set.

5. Comparison of the RSM and ANN models

An attempt was made to compare the RSM and ANN predicted models on the basis of their prediction accuracy. The RSM and ANN models were tested with 29 data sets of BBD of experiments. For each input combination, the predicted values of responses were compared with the respective experimental values and the absolute percentage error is computed as follows:

$$\text{percentage absolute error} = \frac{Yj,expt - Yj,pred}{Yj,expt} * 100 \qquad (4)$$

Where $Yj,expt$ is the experimental value and $Yj,pred$ is the predictive value of the response for the jth trail by the RSM and ANN models. The absolute percentage error were found for ANN and RSM models and maximum values of error for MRR and kerf were tabulated in Table 4. From this table it can be concluded that ANN predictions are more accurate than RSM predictions because the values of maximum percentage absolute error are less for ANN models than RSM models for each response parameter. Figure 1 illustrate the comparison of error profile for responses, for the 29 data set of the training patterns. From this figure it has been observed that ANN predictions are better than the RSM predictions.

In order to test the interpolation of the prediction from the developed models, experimetnal MRR and kerf values were compared with the predicted values of the ANN and RSM models. This comparison is shown in Fig. 2. It is observed from this figure that the prediction of

responses from both models closely agree with that of experimental values.

An attempt was also made to compare the RSM and ANN models on the basis of correlation coefficient. Table 4 lists the values of correlation coefficients for ANN and RSM models of each response parameter. The correlation coefficient between the experimental values and predicted values is a measure of how well the variation in the predicted values is explained by the experimental values. The value of correlation coefficient equal to 1 indicates the perfect correlation between the experimental values and the predicted values. In most cases, the value of correlation coefficient is more closer to unity for ANN models for response parameters. Which clearly indicates that prediction accuracy is higher for ANN model compared to RSM model.

Table 3. List of ANN predictions and RSM predictions for MRR and kerf

Exp no.	MRR (mm^3/min)		kerf (mm)	
	ANN prediction	RSM prediction	ANN prediction	RSM prediction
1	4.2685	4.358	0.3753	0.371
2	9.9197	8.9887	0.2697	0.2549
3	5.0303	4.763	0.3285	0.2989
4	5.185	4.9591	0.2959	0.3049
5	4.6744	5.0357	0.3628	0.353
6	3.6631	2.4133	0.4283	0.4481
7	2.957	2.6882	0.4379	0.4196
8	3.5045	3.4473	0.4077	0.4077
9	3.5045	3.4473	0.4077	0.4077
10	4.9932	5.5341	0.3162	0.3322
11	3.8746	4.7583	0.4134	0.4038
12	3.5045	3.4473	0.4077	0.4077
13	4.6485	4.3631	0.3605	0.3516
14	3.5045	3.4473	0.4077	0.4077
15	4.9805	4.3631	0.2922	0.3124
16	3.5045	3.4473	0.4077	0.4077
17	4.6683	4.7738	0.3419	0.4317
18	5.0936	5.9467	0.3254	0.3271
19	4.385	5.9515	0.3758	0.3586
20	4.2335	4.3583	0.3715	0.3755
21	4.8143	5.9467	0.3266	0.3362
22	4.0098	3.8472	0.3853	0.388
23	2.682	3.0603	0.445	0.3989
24	11.2925	10.0686	0.265	0.2522
25	5.9127	6.5197	0.2727	0.2614
26	5.217	5.8775	0.2907	0.2883
27	5.3344	6.5197	0.276	0.3022
28	3.1771	3.0603	0.4242	0.4018
29	4.9813	3.8472	0.32	0.335

Table 4. Percentage absolute error and correlation cofficients for between RSM and ANN model

Percentage absolute error	MRR		kerf	
	ANN	RSM	ANN	RSM
	11	35	6	15
Correlation coefficient	MRR		kerf	
	ANN	RSM	ANN	RSM
	0.989	0.869	0.939	0.984

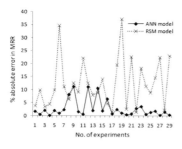

a

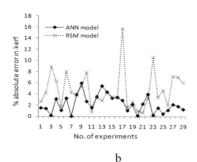

b

Fig. 1. Comparison of error profile for RSM and ANN model a) for MRR b) for kerf

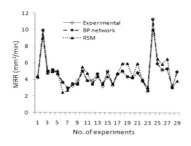

a

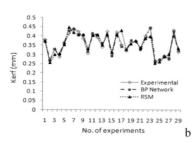

b

Fig. 2. Comparison of experimental measurements with predicted results from RSM and ANN model a) for MRR b) for kerf

6. Conclusions

Present study reports prediction of MRR and kerf through RSM and ANN techniques. RSM predicted values and ANN predicted values of MRR and kerf were compared with the experimental values to decide the nearness of prediction with the experimental values. Good agreement was obtained between the predicted values of both models and experimental measurements. Comparison of RSM models with ANN models on the basis of their prediction accuracy shows that ANN models are more accurate than RSM models for each response parameter because the values of percentage absolute error is higher for RSM models than ANN models. Comparison of RSM models with ANN models on the basis of correlation cofficients shows that in most cases the values of correlation coefficients are more closer to unity for ANN models. Which clearly indicates that ANN models are more accurate than RSM models because the value of correlation coefficients are almost equal to 1. This indicates a perfect correlation between the experimental values and the predicted values.

References

[1]   Rosso M (2006) Ceramic and metal matrix composites: routes and properties. Journal of Materials Processing Technology 175: 364-375.

[2]   Yan BH and Wang CC (1993) Machinability of SiC particle reinforced aluminum alloy composite material. Journal of Japan Institute Light Metals 43: 187-192.

[3]   Lau WS and Lee WB (1991) Comparison between EDM wire cut and laser cutting of carbon fiber composite materials. Materials and Manufacturing Processes 6: 331-342.

[4]   Patil NG and Brahmankar PK (2010) Some studies into wire electro-discharge machining of alumina particulate reinforced aluminum matrix composites. International Journal of Advanced Manufacturing Technology 48: 537-555.

[5]   Shandilya P, Jain NK, Jain PK (2011) Experimental studies on WEDC of SiCp/6061 Al metal matrix composite, key Engineering Materials 450: 173-176.

[6]   Neto JCS, Silva EM and Silva MB (2006) Intervening variables in electrochemical machining. Journal of Materials Processing Technology 179: 92-96.

[7]   Hassoun MH (1995) Fundamentals of artificial neural networks. MIT Press.

[8]   Benardos PG and Vosniakos GC (2003) Predicting surface roughness in machining: a review, International Journal of Machine Tools and Manufacture 43: 833-844.

Improved surface properties of EDM components after irradiation by pulsed electrons

J. Murray[1] and A.T Clare[1]
[1]Precision Manufacturing Centre, Department of M3, University of Nottingham, Nottingham, UK, NG7 2RD

Abstract. Electrical discharge machining (EDM) is a useful process for producing high aspect ratio features for components such as mold tools. The recast layer produced by rapid quenching however is often undesirable in engineering components due to its brittleness, the occurrence of cracking as well as high surface roughness, which contribute to reducing the part's overall fatigue and corrosion resistance. In the case of high aspect ratio geometries, recast layer removal via chemical etching then mechanical grinding is not always possible. In this study we investigate the use of the large-area electron beam irradiation technique to improve the properties of the EDM'd surface of a stainless steel so that a more desirable recast layer may be produced. As well as improvement in surface roughness, repair of EDM induced surface cracks was observed the and mechanism of this repair is discussed.

Keywords: EDM, recast layer, crack repair, polishing, electron beam melting

1. Introduction

EDM is a widely used process for the production of mould tools and engineering components involving high-aspect ratios and conformal geometries. During the process much of the material melted during discharge on-time is not removed but resolidified as a recast layer, with properties often undesirable in engineering applications. These include high roughness, brittleness as well as surface cracks. Reducing the extent of this layer is a continued goal of EDM research, given faster machining parameters usually yield a larger recast layer.

It has been shown that the presence of surface cracks induced by the EDM process can reduce the fatigue life of EDM'd components [1]. This is due to reducing the 2-stage fatigue failure process of crack formation then propagation into one stage of crack propagation. As well as the presence of cracks, surface features such as asperities and surface cavities which are stress concentrators are also known to have a strong influence on fatigue life [2, 3]. Uno et al. [4] showed that pulsed electron beam treatment of an EDM'd surface can reduce its roughness from 6μm Rz to under 1μm as well as improve its corrosion resistance. Okada et al. have more

recently improved the corrosion resistance and blood repellency of surgical steel by this method [5]. The process has also been used to improve the wear resistance of aluminium based alloys [6]. TEM analysis of grain size by Zuo et al. [7] has shown grain size reduction to approximately 50nm can occur at the surface after irradiation, explaining the improved mechanical properties of such treated surfaces.

The potential for the elimination of surface cracks by this process as well as changes in phase and orientation of the EB treated surfaces of EDM'd components has not been assessed. This study therefore investigates the use of large-area electron beam irradaition for the potential of improving the fatigue life of EDM'd components via the suppression of surface cracks as well as the reduction of surface roughness of EDM'd surfaces.

2. Experimental

AISI 310 stainless steel was used as a workpiece material in this study. It is an austenitic, general purpose steel used widely in corrosive and high temperature applications. Surface roughness was measured by white-light interferometry (WLI) with a Fogale nanotech "Photomap 3D", and the Sa roughness parameter was used. XRD analysis was performed with a Bruker AXS D8 Advance" diffractometer using CuK (α) monochromatic radiation. Angles used were between 40° and 100°, with a step size of 0.04°. Electron microscopy was performed with a Hitachi S-2600 SEM.

2.1. EDM

A Sodick AP1L micro die-sink EDM machine was used to machine shallow slots in 310 stainless steel and induce a typical recast layer. Positive electrode polarity was used to best represent represent a typical EDM setup. Due to carbon adhesion onto tool electrodes in EDM using hydrocarbon dielectrics, minimal electrode wear is associated with positive tool polarity [8]. A copper electrode was sunk 500μm into the workpiece as shown in Fig.

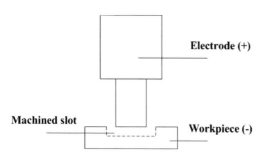

Fig. 1 Schematic of EDM operation

The chosen EDM parameters were based upon literature for EDM of features with dimension of the order of hundreds of microns [9], as well as previous experimental tests. The parameters used to produce the surface to be subject to electron irradiation are shown in Table 1.

Table 1 EDM Parameters

Electrode polarity	On-time (μs)	Off-time (μs)	Main current (A)	Gap voltage
+	30	3	4.5	90

2.2. Electron beam irradiation

Electron beam irradiation was performed using a Sodick PF32A "EBM". The process is carried in vacuum chamber into which Argon is supplied at a pressure of 0.05Pa. To produce the electron beam of 60mm diameter, a solenoid coil firstly produces a magnetic field, and at its maximum intensity a pulsed voltage is applied at the anode. Penning ionisation causes electrons to be generated which then move towards the anode. Argon atoms are then ionised by repeated collisions with electrons, generating plasma near the anode. When the intensity of this plasma is at a maximum, a pulsed voltage is applied to the cathode and electrons from the plasma are accelerated towards the workpiece. The bombardment of the electrons with the workpiece causes its surface to heat then rapidly quench.

Table 2 Electron beam parameters

Cathode voltage (kV)	No. of Shots	Anode voltage (kV)	Argon Pressure
25,35	1,5,10,20	5	4.5

The chosen EB parameters are shown in Table 2. Both the cathode voltage and the number of shots are vital parameters in this process, since the voltage determines the acceleration of electrons and therefore the energy

density, while further shots repeat the process. Two cathode voltages were chosen based on previous experimental tests to represent medium and high power irradiation, with 1, 5, 10 and 20 shots for both settings.

3. Results and discussion

3.1. Surface roughness

The surface finish after irradiation was noticeably improved from the initial EDM surface, dominated by asperities and cracks. A before and after image with irradiation by 20 shots at 35kV is shown in Fig. 2.

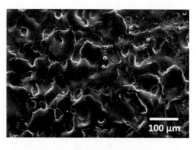

a

b

Fig. 2 EDM surface a) before and b) after irradiation

After measurement by WLI, the largest reduction in surface roughness occured after the highest number of shots of 20 at highest power of 35kV cathode voltage. A pre irradiation average roughness of 3.1μm Ra could be reduced to 0.9μm.

The trends in roughness according to increasing numbers of shots and the 2 cathode voltages are presented in Fig. 3. With both voltages there is a trend downwards for Sa roughness with increasing numbers of shots, although as expected the higher voltage parameter produces a more rapid improvement.

Crater formation commonly associated with the processing of materials by pulsed electron irradiation, and is well discussed in the literature was observed in our experiments. Under irradiation of 5 shots at 25kV as well as 20 shots at 35kV, craters were present. This phenomenon occurs due to melting which occurs below the surface, and the subsequent expansion and eruption from the surface. In steels this has been shown to occur at the location of carbides, which serve as nucleation sites for sub-surface melting [10]. The frequency of crater formation as a result of this process is necessary to understand since they are known to accelerate pitting corrosion [11].

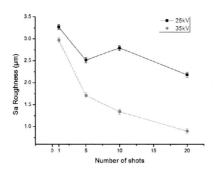

Fig. 3 Reduction in surface roughness with 25kV and 35kV cathode voltage irradiation

As well as the improvement in surface finish by the process, microscopy revealed a significant improvement in the proliferation of surface cracks. The next section quantifies and discusses this phenomenon.

3.2 Crack density

The change in surface crack density was quantified by SEM imaging of a statistically significant total area of 3.15mm^2, imaging then tracing the length of each individual crack in "ImageJ" image processing software. The total length of cracks for each sample was then divided by the total area to give the crack density. The results are shown in Fig. 4.

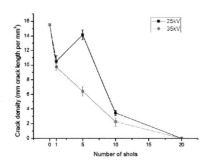

Fig. 4 Change in surface crack density after 25kV and 35kV cathode voltage irradiation

With 1 shot at both 25kV and 35kV cathode voltage, crack density was reduced. After 5 shots however, 25kV cathode voltage irradiation increased crack profileration to near control levels, which is likely due to the exposure of cracks previously repaired at the near surface by irradiation induced evaporation. Increasing numbers of shots at the higher cathode voltage of 35kV only reduced crack density, suggesting any evaporation taking place at the surface is not enough to compensate for the repairing

effect of the irradation process. Further work is to be undertaken to assess the mass loss induced by the process. Surface images (15kV, 5 shots) were obtained tp elucidate the mechanism of repair. These can be seen in Fig. 5.

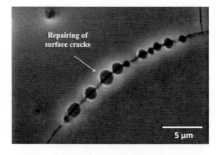

a

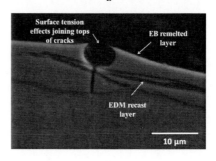

b

Fig. 5 Mechanism of repair of EDM surface cracks after irradiation

The rounded morphology of cracks at the surface subjected to low voltage irradiation suggests surface tension effects when both sides of the cracks are molten and are responsible for the merged features that are observed. This is supported by cross-sectional imaging, revealing the rounding of cracks into the recast layer with a narrow section joining at the very top surface. After 20 shots at 35kV voltage, there is no longer evidence of this effect, and no cracks are present in the newly re-melted layer as observed in the cross-section. The re-melted layer after 35kV and 20 shots, compared to the EDM'd material can be seen in Fig. 6.

It was observed that with these particular EDM settings, some of the initial recast layer remained unaffected at 35kV voltage and highest number of shots. This was beneath the uniform, newly re-melted layer produced by irradiation. In the unaffected EDM recast layer some cracking was observed, although penetration neither of the surface nor into the bulk occurred. It is thought that with improvements to the current irradiation process, specifically higher acceleration voltages, larger EDM layers can be re-melted, thus guaranteeing full depth of crack repair.

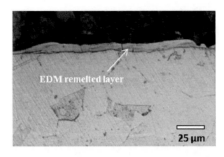

a

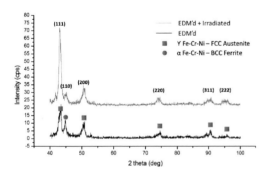

b

Fig. 6 Cross-section of a) EDM recast layer and b) uniform remelted layer after irradiation

3.3. XRD analysis

To assess any phase changes or changes in crystalline orientation at the surface, XRD analysis was performed on an EDM'd sample and a sample irradiated with 20 shots at 35kV. The XRD plot can be seen in Fig. 7.

Fig. 7 XRD plot of EDM'd and irradiated surface. Significant crystalline texture is introduced after irradiation.

Significant crystalline texture is introduced by the irradiation process, with the (111) planes of the FCC phase orienting parallel to the surface. The ratio of intensities of the (111) to the (200) peak increases by a factor of 2.2. There is also a reduction in the strength of the BCC ferrite peak after irradiation. Crystalline texture at the surface has potential implications for the introduction of anisotropic layers for corrosion resistance, as has been shown by Shahryari et al. [12] whereby

pitting corrosion resistance was improved in 316LVM stainless steel under (111) and (100) planar orientation.

4. Conclusions

Pulsed electron irradiation with a diameter of 60mm can be used to improve the surfaces of EDM'd stainless steel in approximately 15 minutes. Sa roughness of surfaces above 3μm can be reduced to below 1μm. Repair and elimination of surface cracks can occur in the newly remelted surface. Crystalline texture can be introduced at the surface.

References

[1] Tai, T.Y. and S.J. Lu, Improving the fatigue life of electro-discharge-machined SDK11 tool steel via the suppression of surface cracks. International Journal of Fatigue, 2009. 31(3): p. 433-438.

[2] Andrews, S. and H. Sehitoglu, A computer model for fatigue crack growth from rough surfaces. International Journal of Fatigue, 2000. 22(7): p. 619-630.

[3] Rokhlin, S.I. and J.Y. Kim, In situ ultrasonic monitoring of surface fatigue crack initiation and growth from surface cavity. Int. Journal of Fatigue, 2003. 25(1): p. 41-49.

[4] Uno, Y., et al., High-efficiency finishing process for metal mold by large-area electron beam irradiation. Precision Engineering, 2005. 29(4): p. 449-455.

[5] Okada, A., et al., Surface finishing of stainless steels for orthopedic surgical tools by large-area electron beam irradiation. CIRP Annals - Manufacturing Technology, 2008. 57(1): p. 223-226.

[6] Walker, J., et al., Dry Sliding Friction and Wear Behaviour of an Electron Beam Melted Hypereutectic Al–Si Alloy. Tribology Letters, 2011: p. 1-10.

[7] Zou, J.X., et al., Microstructures and phase formations in the surface layer of an AISI D2 steel treated with pulsed electron beam. Journal of Alloys and Compounds, 2007. 434-435(0): p. 707-709.

[8] Kunieda, M. and T. Kobayashi, Clarifying mechanism of determining tool electrode wear ratio in EDM using spectroscopic measurement of vapor density. Journal of Mat. Processing Technology, 2004. 149(1-3): p. 284-288.

[9] Liu, H.-S., et al., A study on the characterization of high nickel alloy micro-holes using micro-EDM and their applications. Journal of Materials Processing Technology, 2005. 169(3): p. 418-426.

[10] Zou, J.X., et al., Cross-sectional analysis of the graded microstructure in an AISI D2-steel treated with low energy high-current pulsed electron beam. Applied Surface Science, 2009. 255(9): p. 4758-4764.

[11] Zhang, K., et al., Improved pitting corrosion resistance of AISI 316L stainless steel treated by high current pulsed electron beam. Surface and Coatings Technology, 2006. 201(3-4): p. 1393-1400.

[12] Shahryari, A., J.A. Szpunar, and S. Omanovic, The influence of crystallographic orientation distribution on 316LVM stainless steel pitting behavior. Corrosion Science, 2009. 51(3): p. 677-682.

Modeling and experimental study of electrical discharge diamond cut-off grinding (EDDCG) of cemented carbide

S. K. S. Yadav[1] and V. Yadava[2]

[1] Assistant Professor, MED, H.B.T.I Kanpur, India, sanjeevyadav10@rediffmail.com
[2] Professor, MED, MNNIT Allahabad, India, vinody@mnnit.ac.in

Abstract. This paper reports the development of neural network model and experimental study of electrical discharge diamond cut-off grinding (EDDCG) during machining of cemented carbide for average surface roughness (Ra). EDDCG is combination of diamond grinding and electrical discharge grinding. This process has been developed for machining of electrically conductive difficult to machine very hard materials such as cemented carbide, Ti-alloy, super alloys, metal matrix composites etc. ANN model was development for EDDCG process, to correlate the input process parameters such as current, pulse-on time, duty factor and wheel RPM with the performance measures namely, surface roughness. The experiments are carried out on a self developed electrical discharge diamond grinding setup in cut-off mode. The range of machining parameters was decided on the basis of pilot experiments. A total of 81 experiments were performed based on full factorial design of experiments. After experimentation the data set were divided into a training set and testing set for ANN modeling. Seventy present data of total available data set were used for training the network and remaining set were used for testing the network. The developed architecture can predict average surface roughness (Ra) with 0.0140 APE for training and 0.0042 APE for testing. Further, variation of surface roughness is plotted against different process parameters such as wheel RPM and pulse current.

Keywords: EDDCG, Ra, Cemented carbide, ANN Hybrid process.

1. Introduction

The Hybrid machining processes is defined as combination of two or more processes for shaping and finishing machine parts, it combining various physical and chemical processes acting on workpiece material into one machining process. These hybrid machining processes are developed to enhance advantages and to minimize potential disadvantages associated with an individual technique. Electrical discharge diamond grinding (EDDG) is a one of the hybrid machining process. It is a combination of diamond grinding and electrical discharge grinding (EDG). In this process metal bonded diamond grinding wheel is used. There are three basic configuration by which the combination of grinding

and EDG can be classified. (1) Electro-Discharge Diamond Surface Grinding (EDDSG) (2) Electro-Discharge Diamond Face Grinding (EDDFG) (3) Electro-Discharge Diamond Cut-off Grinding (EDDCG). EDDSG is used to machine flat surfaces by using periphery of the metal bonded diamond grinding wheel. Since the work is normally held in a horizontal orientation, peripheral grinding is performed by rotating the grinding wheel about a horizontal axis. The relative motion of the workpiece is achieved by reciprocating the workpiece. EDDFG is used flat face of the metal bonded diamond grinding wheel. In this mode, the metal bonded diamond grinding wheel rotates about vertical spindle axis and fed in a direction perpendicular to the machine table. While machining, the rotating wheel is fed downwards under the control of servo system.

Fig.1. Electrical Discharge Diamond Grinding (EDDG) setup in cut-off grinding

EDDCG performed using periphery of the thin metal bonded diamond grinding wheel. While machining, the rotating wheel is fed downward using servo control, for material removal in cut-off configuration. Fig. 1 shows the self developed electrical discharge diamond grinding (EDDG) setup in cut-off mode. In this process metal bonded diamond grinding wheel was used. Sparking takes place between metallic bonding material and work piece. Heat generated during sparking and due to heat softens the work material and hence machining by diamond abrasive particles becomes easier. Sparking in the inter electrode gap during EDDCG, results in continuous dressing of the grinding wheel and hence the wheel doesn't clog also. As a result cutting properties of the grinding wheel are stabilized. The projection of diamond grain on the bonding material of wheel is called protrusion height and gap between the work piece and bonding material is called gap width. For proper machining of work piece material, the protrusion height should be more then inter electrode gap [1].

2. Experimentation

Experiments were conducted on an EDM, attached with self developed grinding attachment of EDDCG. The set-up consists of a metal bonded diamond grinding wheel, D.C motor, shaft, V-belt and bearing, mounted on the ram of the machine to rotate the metal bonded grinding wheel about an axis. The rotating wheel is fed downwards under servo control, for material removal in the cut-off configuration. The thin metal bonded grinding wheel of 5.7mm and the work surface are physically separated by a gap, the magnitude of which depends on the local breakdown strength of the dielectric for a particular gap voltage setting [2]. Experiments were performed on cemented carbide workpiece. Four input parameters, such as current, pulse-on time, duty factor and wheel RPM and one output parameter, Ra chosen for modiling. According to the size of attachment, size of dielectric tank and requirement for machining of cemented carbide material, specification of diamond wheel shown in Table 1

Table 1. EDDCG Wheel specification

Wheel diameter	100 mm
Thickness of wheel	5.7 mm
Grit size	200/230
Concentration	75
Bonding	Bronze
Work material	Cement Carbide
Workpiece thickness	5.7 mm

It was decided to use full factorial design. The number of experiments to be performed using full factorial design can be given by formula [3].

$$N = F^k$$

where, N= is number of experiments, F= number of levels, k= number of factors

In the present work total of 81 (3^4) experiments have been performed and the value of Ra were taken. After experimentation the data set were divided into a training set and testing set for ANN modeling. The variation of Ra is plotted against different process parameters such as wheel RPM and pulse current.

3. ANN based modelling of EDDCG

The present work was aimed at establishment of correlation between input process parameters such as current, pulse-on time, duty factors and wheel speed with output parameter Ra. There are several algorithms in a neural network and the one that has been used in the present study is the back-propagation training algorithm. The back-propagation neural networks are usually referred to as feed forwarded, multilayered network with number of hidden layers. The error back–propagation process consists of two passes through the different layer of the network: a forward pass and a backward pass. In the forward pass, an activity pattern (input vector) is applied to the sensory nodes of the network, and its effect propagates through the network layer by layer. Finally, a set of output is produced as the actual response of the network. During the backward pass, all synaptic weights are adjusted in accordance with the error correction rule. Specifically the actual response of the network is subtracted from the desired (target) response to produce an error signal. The synaptic weights are adjusted so as to make the actual response of the network move closer to the desired network [4].

The steps of the ANN calculation during training using back propagation algorithm are as follows.

Step 1: The network synaptic weights are initialized to small random values.
Step 2: From the set of training input/output pairs, an input pattern is presented and the network response is calculated.
Step 3: The desired network response is compared with the actual output of the network, and all the local errors to be computed.
Step 4: The weights preceding each output node are updated according to the following update formula:

$$\Delta wij\,(t) = \eta\,\delta_i\,o_i + \alpha\,\Delta wij(t\text{-}1)$$

Where, η the learning rate, δ the local error gradient, α the momentum coefficient, o_i the output of the ith unit wij represents the weight connecting the ith neuron of the input vector and the jth neuron of the output vector. The local error gradient calculation depends on whether the unit into which the weights feed is in the output layer or the hidden layers. Local gradients in output layers are the product of the derivatives of the network's error function and the units' activation functions. Local gradients in hidden layers are the weighted sum of the unit's outgoing weights and the local gradients of the units to which these weights connect.

Step 5: The cycle (step 2 to step 4) is repeated until the calculated outputs have converged sufficiently close to the desired outputs or an iteration limit has been reached.

The various signals are individually amplified, or weighted, and then summed together within the processing element. The resulting sum is applied to a specific transfer function, and the function value becomes the output of the processing element. Transfer function used in the back-propagation network is known as 'sigmoid function', which is shown below [5].

$$f(x) = \frac{1}{(1+e^{-x})}$$

where, X is the sum of the node input

The absolute prediction error (APE) is calculated using following equation [6].

$$APE(\%) = \left| \frac{Experimental\ result - ANN\ Predicted\ result}{Experimental\ result} \right| \times 100$$

4. Training and testing of Neural network

Experiments were performed on cemented carbides workpiece. For the training and testing of the neural network, it is decided to use full factorial design, considering four input process parameters (current, pulse on-time, duty factor and wheel RPM) at three different levels, total of 81 (3^4) experiments have been performed and the value of Ra taken. After experimentation the data set were divided into a training set and testing set for ANN modeling. 70% data of total available data set were used for training the network and remaining used for testing the network. Before training and testing the total input and output data were normalized for increase accuracy and speed of the network. The datasets were normalized using the following equation.

$$X_n = \frac{X - X_{min}}{X_{max} - X_{min}}$$

Where, X_n is the normalized value of variable X, X_{min} and X_{max} minimum and maximum value of X in total data sets.

First, neural network architecture has been decided, as there are four inputs and one outputs in the present problem, the number of neurons in the input and output layer has to set to four and one, respectively. As the number of hidden layer in the network increase, the complexity increases. Further single hidden layer gave comparatively better result with an optimum training time [5]. Hence, in the present case only one hidden layer has been considered.

In present problem the number of neurons in the hidden layer is changed and the total average prediction error for training and validation was calculated for each case.

Figure 2 shows the variation of APE with number of neurons in hidden layer in which the number of hidden neurons was varied from 3 to 19. The number of neurons in hidden layer is 18, for which the average prediction error is minimum.

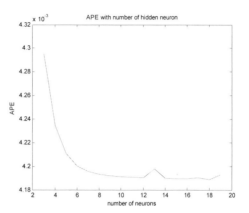

Fig. 2. Plot for determining the number of neurons in hidden layer

5. Results and discussions

The data were normalized to lie between 0 and 1. The model developed is a back-propagation network having 18 neurons in the hidden layer. Figure 3 shows the ANN architecture proposed for the present problem. It considers four inputs current, pulse on-time, duty factor, wheel RPM and one outputs Ra. So, 4-18-1 (4 input neurons, 18 hidden neurons and 1 output neurons) is the most suitable network selected for the present task with the help of self developed ANN based MATLAB code.

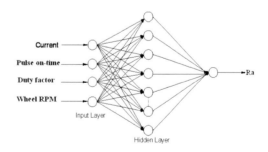

Fig. 3. The 4-18-1 ANN model for the Ra

The developed model was used to predict the output. During the training of network the learning rate taken as 0.01 to successfully train the network with maximum number of epochs 2000. The developed architectures 4-18-1 can predict Ra with 0.0140 APE for training and 0.0042 APE testing respectively shown in Fig. 4.

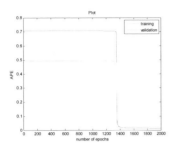

Fig. 4. The variation of APE with number of epochs

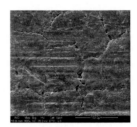

Fig. 7. Micrograph of machined surface observed under SEM

Figure 5 shows the variation of Ra with wheel RPM for different value of gap current. Here, Pulse on-time of 50 µs and duty factor of 0.5 was taken. It is observed that Ra decreases with increase in wheel RPM for all current values. This is due to increase in wheel RPM effective flushing of inter-electrode gap occurs and therefore, the adherence of resolidified eroded particles on the work surface is reduced and the resulting surface presents a better finish. It is also observed that for a particular wheel RPM, with increase in gap current, the R_a value is increases.

6. Conclusions

From the above results and discussion it can be concluded that: The developed architectures 4-18-1 can predict Ra with 0.0140 APE for training and 0.0042 APE testing respectively. Thus, it is apparent that the ANN model can be reliably used for prediction of output responses Ra in close conformity to the actual experimental data. An improvement in surface quality was found by 70% when wheel RPM is increase by 700 to 1300 RPM. For RPM range 700-1300 RPM if current is increased two times (from 3 to 7A).The Ra increase almost by a factor of 1.5 here the quality of surface detoriates for higher current. The experiments shows that better Ra achieve with lower range of pulse current with higher value of wheel RPM.

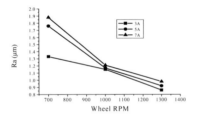

Fig. 5. Effect of wheel RPM on Ra for different currents during EDDCG

Figure 6 shows the variation of Ra with gap current for different values of wheel RPM. Here it is observed that the Ra increases with an increase in gap current for all value of wheel RPM. This is due to the simultaneous occurrence of flushing and efficient sparking material removal rate will increase as the energy input increases, resulting in bigger craters on the work surface and poor surface finish. At higher current increases in maximum protrusion height of grains also leads to deterioration of surface finish [7]. The SEM micrograph of machined surface at 1000 RPM and 5A are shown in Fig.7.

References

[1]   Choudhary SK, Jain VK, Gupta M, (1999) Electrical discharge diamond grinding of high -speed steel : Mach. Sci. Tech. 3: 91-105

[2]   Koshy P, Jain VK, Lal GK, (1997) Grinding of cemented carbide with electrical spark assistance. J. Mater. Process Technol. 72: 61-68

[3]   Jain VK, Mote RG, (2005) On the temperature and specific energy during electro- discharge diamond grinding (EDDG): International Journal of Advanced Manufacturing Technology. 26: 56-67

[4]   Jain RK, Jain VK, (2000) Optimum selection of machining condition in abrasive flow machining using neural network: J. Mater. Process Technol..108: 62-67

[5]   Karunakar DB, Datta GL, (2008) Prevention of defects in castings using back propagation neural networks:.Int J Adv Manuf Technol. 39:1111-1124

[6]   Dhara SK, Kuar AS, Mitra S, (2008) An artificial neural network approach on parametric optimization of laser micro-machining of die-steel: J.Adv. Manuf. Technol. 39:39-46

[7]   Kumar S, Choudhury SK, (2007) Prediction of wear and surface roughness in electro-discharge diamond grinding: J. Mater. Process. Technol..191:206-20.

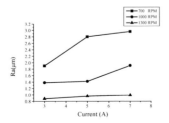

Fig.6. Effect of current on Ra for different RPM during EDDCG

Experimental investigation on material removal rate in wire electrical discharge turning process for Al/SiC$_p$ metal matrix composite

M. Rajkumar[1], M. Kanthababu[2,*] and S. Gowri[3]
[1]PG Student, [2]Assistant Professor, [3]Professor
Department of Manufacturing Engineering, College of Engineering Guindy, Anna University, Chennai – 600 025,India.
*Corresponding author email: kb@annauniv.edu

Abstract. An attempt is made for the first time to machine metal matrix composite (MMC) consisting of aluminium alloy 360 (A360) reinforced with silicon carbide (SiC) particulate with 15 % volume fraction using wire electrical discharge turning (WEDT) process. WEDT is carried out by incorporating a rotary spindle attachment in the existing wire electrical discharge machine (WEDM). The experiments are carried out by Response Surface Methodology (RSM) using Box-Behnken method. The input machining parameters varied at three levels are pulse-on time (T_{ON}), gap voltage (V) and spindle speed (SS). The significant WEDT machining parameters and their levels are obtained for higher material removal rate (MRR) by using ANOVA and response surfaces. From this study, it is observed that high T_{ON}, medium V and low SS are resulted in higher MRR. Regression equation is established for the MRR for predication. The results obtained from this study will be useful to engineers in selecting the appropriate input WEDT parameters and their levels for cylindrical turning of MMC involving Al/SiC$_p$.

Keywords: Wire electrical discharge turning; Metal matrix composite; Response surface methodology; Material removal rate; ANOVA.

1. Introduction

Manufacturing industries today realise the need for generating different cylindrical forms in hard and difficult to machine materials especially composites. Hence, current researchers have made attempts to generate precise cylindrical forms using WEDT process in the conventional as well as hard to machine materials [1-9]. WEDT is type of hybrid machining processes generally carried out in WEDM by incorporating a rotary spindle attachment. WEDT process is found to have advantages over conventional turning process such as good repeatability, less deflection of workpiece due to non-contact process, elimination of stresses during machining, etc. In WEDT, the electrically charged wire is controlled by the x-y table of the WEDM and the rotational movement of the workpiece is controlled by the attached rotary motor which results in the removal of work

material and generates desired cylindrical forms. Fine surface finish, less roundness error and enhanced material removal are possible by using significant WEDT process parameters.

Literature review indicates that researchers have studied the performance of WEDT process considering T_{ON}, pulse-off time (T_{OFF}), V and SS on MRR, surface roughness (R_a) and roundness error for different materials. It is observed that the selection of appropriate WEDT parameters for each material is a difficult task and it is material specific. It is also observed that there has been no attempt made to identify significant input WEDT process parameters for MMCs. MMCs have gained importance in various fields like aerospace, automobiles, defense, etc due to their superior properties [10]. The most commonly used MMC in different applications is found to be the combination of aluminium alloy reinforced with SiC (Al/SiC). The Al/SiC MMC in various cylindrical forms are expected to replace some of the existing materials in different applications. Therefore, in this work, an attempt has been made to indentify siginificant WEDT process parameters for generating cylindrical form in Al/SiC MMC.

2. Experimental details

Electronica make WEDM is used in this work, in which the WEDT setup is installed (Fig. 1). The WEDT set-up consists of various components such as self centering chuck, spindle shaft, deep groove single row ball bearings, spindle housing, DC motor (12V, 6W), timing belt, flanged gears and lock nut. The straight turning configuration is used in this work. The gap between the wire and workpiece is constantly maintained in the ranges between 0.075 mm to 0.1 mm by a CNC positioning system. The brass wire of 0.25 mm diameter is used as the electrode material. The depth of cut and the length of the cut are maintained as 0.2 mm and 6 mm respectively. T_{OFF} and wire tension of 6 μs and 1.2 kgf respectively were kept statistically constant. Deionized water is used as dielectric fluid. The workpiece material used in this work is MMC consisting of A360 reinforced with 15% volume fraction of SiC particles (Al/SiC_p), which is prepared by stir casting process. The particle size of the reinforcement material is around 30 μm. Experiments are carried out by RSM [11] using Box-Behnken method in order to establish relationship between the WEDT process parameters and MRR (response). The important WEDT parameters such as T_{ON}, V and SS are varied at three levels (Table 1). The allocation of WEDT parameters in the RSM table and experimental results are shown in Table 2. Typical machined WEDT component is shown in Fig. 2. ANOVA and response surfaces are used to determine the effect of input WEDT parameters, which are obtained from Design-Expert software. The response MRR is calculated using the following formula:

$$MRR = \frac{\text{Total volume removed from the workpiece}}{\text{Time taken}} \, mm^3 / min. \quad (1)$$

Fig. 1. Photograph of the WEDT setup installed in the WEDM

Fig. 2. Photograph of the typical WEDT workpiece

Table 1. Input process parameters and their levels

Parameter	Low	Centre	High
Pulse on time [T_{ON}] (μs)	6	8	10
Gap voltage [V] (Volts)	45	50	55
Spindle speed [SS] (rpm)	125	150	175

3. Results and discussion

Table 3 shows ANOVA results obtained for the MRR. It indicates that among the input process parameters studied in this work, the individual effect of T_{ON} is found to be significant, while V and SS are found to be insignificant. However, the quadratic effect of the V is also found to be significant (Table 3). The relationship between the input parameters and the response MRR is expressed in the form of regression equation and it is given below:

$$MRR = -175.29 + 7.24*T_{ON} + 6.45*V - 0.144*SS - 0.043*T_{ON}*V - 0.014*T_{ON}*SS - 0.006*V*SS - 0.14*T_{ON}^2 - 0.051*V^2 + 0.002*SS^2 \quad (2)$$

Table 2 WEDT parameters and experimental results

Ex. No	T_{ON} (μs)	V (Volts)	SS (rpm)	MRR (mm^3/min)
1	6	45	150	6.99
2	10	45	150	10.17
3	6	55	150	8.38
4	10	55	150	9.84
5	6	50	125	8.97
6	10	50	125	14.26
7	6	50	175	9.71
8	10	50	175	12.15
9	8	45	125	8.28
10	8	55	125	11.56
11	8	45	175	10.93
12	8	55	175	11.36
13	8	50	150	10.81
14	8	50	150	10.76
15	8	50	150	10.45

Table 3 ANOVA table of MRR

S	SS	Dof	M	F	P
*Model	39.80	9	4.42	5.57	0.036
*T_{ON}	19.13	1	19.13	24.08	0.004
V	2.84	1	2.84	3.58	0.117
SS	0.15	1	0.15	0.18	0.686
T_{ON} x V	0.74	1	0.74	0.93	0.378
T_{ON} x SS	2.03	1	2.03	2.56	0.170
V x SS	2.03	1	2.03	2.56	0.170
T_{ON}^2	1.09	1	1.09	1.38	0.293
*V^2	6.09	1	6.09	7.67	0.039
SS^2	4.83	1	4.83	6.08	0.056
Residual	3.97	5	0.79	-	-
Lack of Fit	3.90	3	1.30	34.14	0.028
Pure Error	0.08	2	0.04	-	-
Cor Total	43.77	14	-	-	-

*Significant, S- Source, SS-Sum of square, Dof- Degree of freedom, M- Mean Square, F- F value, P- P value.

Fig. 3 indicates response surface of MRR by varying the SS and V, while T_{ON} is held constant at low level [Fig. 3

(a)], medium level [Fig. 3 (b)] and high level [Fig. 3 (c)]. Fig. 3a indicates that with low T_{ON}, higher MRR is achievable with low SS and V in between medium and high level (50 V to 55 V). The maximum MRR achievable with this combination is found to be 10.3 mm^3/min. Fig. 3 (b) indicates that with medium T_{ON}, higher MRR is achievable with V in between medium and high level (50 V to 55 V) and low SS. The maximum MRR is found to be 11.9 mm^3/min. Fig. 3c indicates that with high T_{ON}, higher MRR is achievable with V in between medium and high level (50 V to 55 V) and low SS. The maximum MRR is found to be 13.4 mm^3/ min. By comparing the influence of different levels (low, medium and high) of T_{ON} from Fig. 3, it is observed that high T_{ON} results in higher MRR. High T_{ON} generates high discharge energy and therefore creates wide and deep craters in the workpiece surface, which leads to higher MRR [2,4,12,13].

(a)], medium level [Fig. 4 (b)] and high level [Fig. 4 (c)]. Figure 4 (a) indicates that with low V, higher MRR is achievable with T_{ON} at its high level and low SS. The maximum MRR is found to be around 11.4 mm^3/ min. Fig. 4 (b) indicates that with medium V, higher MRR is achievable with low SS and high T_{ON}. The maximum MRR is found to be around 13.4 mm^3/ min. Fig 4 (c) indicates that with high V, higher MRR is achievable with low SS and high T_{ON}. The maximum MRR is found to be around 13 mm^3/ min. By comparing the influence of different levels (i.e. low, medium and high) of V from Fig. 4, it is observed that medium V results in higher MRR. The high V also results in nearly similar value. However, the use of high V is not advisable because it will increase the manufacturing cost (high wear rate of the wire), lead to arcing and also increase the rate of deposition of resolidification/recast layer on the workpiece [12,13]. Hence, medium V is preferable for higher productivity and continuous machining without breakage of wire electrode.

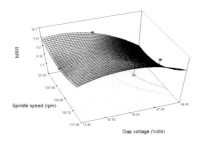

a) Spindle speed Vs Gap voltage (at low T_{ON})

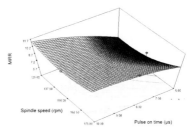

a) Spindle speed Vs Pulse on time (at low V)

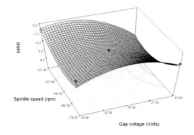

b) Spindle speed Vs Gap voltage (at medium T_{ON})

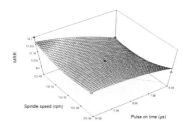

b) Spindle speed Vs Pulse on time (at medium V)

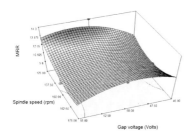

c) Spindle speed Vs Gap voltage (at high T_{ON})

Fig. 3 Response surface of MRR at various T_{ON} levels

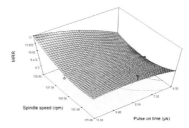

c) Spindle speed Vs Pulse on time (at high V)

Fig. 4 Response surface of MRR at various V levels

Fig. 4 indicates response surface of MRR by varying the SS and T_{ON}, while V is maintained at low level [Fig. 4

Fig. 5 indicates response surface of MRR by varying the T_{ON} and V, while SS is held constant at low level [Fig. 5

(a)], medium level [Fig. 5 (b)] and high level [Fig. 5 (c)]. Fig. 5 (a) indicates that with low SS, higher MRR is achievable by maintaining the V between medium level to high level and the T_{ON} at its high level. The maximum MRR is found to be around 13.1 mm^3/min. Fig. 5 (b) indicates that with medium SS, higher MRR is achievable with high T_{ON} and V between medium and high level. The MRR achievable with this combination is found to be around 11.7 mm^3/min. Fig. 5 (c) indicates that at with high SS, higher MRR is achievable between medium and high V and T_{ON} at its high level. The maximum MRR is found to be 12.2 mm^3/min. By comparing the influence of different SS levels from Fig. 5, it is observed that low SS results in higher MRR. Low SS may increase the temperature concentration on the workpiece surface during machining compared to that of medium and high SS and hence results in the increased MRR. Low SS may also effectively improve the circulation of the dielectric fluid in the spark gap and enhances the MRR [2,4].

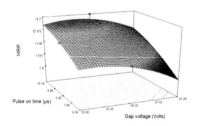

a) Pulse on time Vs gap voltage (at low SS)

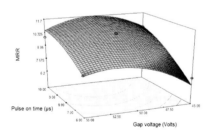

b)   Pulse on time Vs gap voltage (at medium SS)

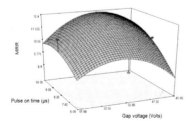

c) Pulse on time Vs gap voltage (at high SS)

Fig. 5 Response surface of MRR at various SS levels

4. Conclusion

The influence of WEDT process parameters such as T_{ON}, V and SS on MRR are analysed for machining of Al/SiC$_p$ MMC. The experiments are carried out as per RSM using Box-Behnken method. The significant parameters and their levels are identified for achieving higher MRR with the help of ANOVA and response surfaces. It is found that high T_{ON}, medium V and low SS leads to higher MRR and effective machining. Hence, these combinatons are recommended for WEDT of Al/SiC$_p$ MMC in order to achieve higher MRR. Regression equation is established for MRR for easier predication.

References

[1]   Haddad MJ, Alihoseini F, Hadi M, Hadad M, Tehrani AF, Mohammadi A, (2010) An experimental investigation of cylindrical wire electrical discharge turning process. Int J Adv Manuf Tech 46:1119-1132
[2]   Qu J, Shih AJ, Scattergood RO, (2002) Development of the cylindrical wire electrical discharge machining process Part 1: Concept, design, and material removal rate. J Manuf Sci and Engg 124:702-707
[3]   Qu J, Shih AJ, Scattergood RO, (2002) Development of the cylindrical wire electrical discharge machining process Part 2: Surface integrity and roundness. J Manuf Sci and Engg 124: 708-714
[4]   Mohammadi A, Tehrani AF, Emanian E, Karimi D, (2008) Statistical analysis of wire electrical discharge turning on material removal rate. J Mat Proc Tech 205:283-289
[5]   Mohammadi A, Tehrani AF, Emanian E, Karimi D, (2008) A new approach to surface roughness and roundness improvement in wire electrical discharge turning based on statistical analyses. Int J Adv Manuf Tech 39:64-73
[6]   Haddad MJ, Tehrani AF, (2008) Material removal rate (MRR) study in the cylindrical wire electrical discharge turning (CWEDT) process. J Mat Proc Tech 199:369-378
[7]   Haddad MJ, Tehrani AF, (2008) Investigation of cylindrical wire electrical discharge turning (CWEDT) of AISI D3 tool steel based statistical analysis, J Mat Proc Tech 198:77-85
[8]   Matoorian P, Sulaiman S, Ahmad MMHM, (2008) An experimental study of optimization of electrical discharge turning (EDT) process. J Mat Proc Tech 204:350-356
[9]   Janardhan V, Samuel GL (2010) Pulse train data analysis to investigate the effect of machining parameters on the performance of wire electro discharge turning (WEDT) process. Int J Mach Tools Manuf 50:775-788
[10] Chawla N, Chawla KK (2006) Metal matrix composites, Springer, Newyork, USA
[11] Montgomery DC (2001) Design and analysis of experiments. John Wiley & Sons, Singapore
[12] Satishkumar D, Kanthababu M, Vajjiravelu V, Anburaj R, Thirumalai Sundarrajan N, Arul H (2011) Investigation on wire electrical discharge machining characteristics of Al6063/SiCp composites. Int J Adv Manuf Tech 56: 975-986
[13] Garg RK, Singh KK, Sachdeva A, Sharma VS, Ojha K, Singh S (2010) Review of research work in sinking EDM and WEDM on metal matrix composite materials. Int J Adv Manuf Tech 50:611-624.

Study on electrolyte jet machining of cemented carbide

K. Mizugai[1], N. Shibuya[2] and M. Kunieda[1]

[1] Department of Precision Engineering, The University of Tokyo, Tokyo, Japan
[2] Department of Mechanical Systems Engineering, Tokyo University of Agriculture and Technology, Tokyo, Japan

Abstract. In this study, electrolyte jet machining (EJM) was attempted on cemented carbide. $NaNO_3$ aqueous solution was used as the electrolyte instead of NaOH aqueous solution, which is normally used for the electrochemical machining of cemented carbide, because of its hazardous characteristics. The machining was carried out with alternating current (AC) because it enables more localized dissolution under the jet than direct current (DC), however an insulator nozzle was additionally employed during AC machining, and was found to be effective for preventing nozzle wear often seen in AC machining as well as for obtaining more localized dissolution area even in DC machining.

Keywords: Electrolyte jet machining, ECM, micro machining, cemented carbide, insulator nozzle

1. Introduction

In electrolyte jet machining (EJM) [1, 2], the workpiece is machined only in the area which is hit by the electrolyte jet by applying an electrical current through the jet. By translating the jet on the workpiece, intricate patterns can be fabricated without the use of special masks [3], since the distribution of current density is localized under the jet [4]. Given that electrolyte jet machining is an electrochemical process, no burrs, cracks, or heat affected zones are generated on the machined surface. Kunieda et al. [3] has therefore been using this method for texturing micro indents on the surface of rolling bearings to extend fatigue life. Moreover, because most electrically conductive materials can be machined regardless of their hardness, this method can be applied to the machining of cemented carbide. However, use of pure $NaNO_3$ or NaCl aqueous solution results in the formation of a tungsten oxide layer on the workpiece surface which hinders the further dissolution of tungsten carbide (WC). Although WC grains can be dissolved by an electrolyte containing NaOH [5], this hazardous electrolyte cannot be used in EJM, because electrolyte mist is generated when the electrolyte is hitting the workpiece surface. On the other hand, Masuzawa et al. [6] succeeded in the electrochemical surface finishing of cemented carbide using harmless $NaNO_3$ solution as the electrolyte with

bipolar pulse current, based on the fact that NaOH is generated on the cathode surface during the process. For this reason, the authors attempted the machining of cemented carbide by EJM using $NaNO_3$ aqueous solution in this study. To obtain higher machining accuracy, choice of jet nozzle materials and machining current conditions were also investigated.

2. Electrolyte jet machining (EJM)

2.1. Principle of EJM

Electrolyte jet machining is carried out by jetting electrolytic aqueous solution from a nozzle towards the workpiece while applying voltage between the nozzle and workpiece. When the electrolyte jet hits a plate at a sufficiently high velocity, it flows radially outwards in a fast thin layer which suddenly increases in thickness. Equipotential surfaces in the jet at this time are shown in Fig. 1 (a) and distribution of the current density is concentrated in the area under the jet as shown in Fig.

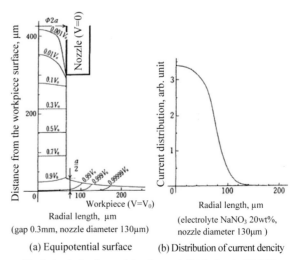

(a) Equipotential surface (b) Distribution of current dencity

Fig. 1. Analysis of potential and current distribution in EJM [4]

1(b) [4]. Thus, electrolytic dissolution is limited to the jet impinging area. The following is the chemical reaction on the anode surface in the electrochemical machining of cemented carbide using a mixed solution of NaOH and NaCl as the electrolyte [5]. Co is oxidized into Co^{2+}.

$$Co \rightarrow Co^{2+}2e^- \quad (1)$$

WC is oxidized into WO_3, thereby forming a thin oxide film on the anode surface. Then NaOH reacts with WO_3 as:

$$WO_3+2NaOH \rightarrow Na_2WO_4+H_2O \quad (2)$$

Since the oxide layer of WO_3 is removed, machining can progress. If NaOH is not mixed into the $NaNO_3$ solution, WO_3 film cannot be removed, thus machining is prevented. On the other hand, the following reaction occurs on the cathode surface in parallel to the reaction of hydrogen gas generation.

$$2H_2O+2Na+2e^- \rightarrow 2NaOH+H_2\uparrow \quad (3)$$

In EJM using direct current (DC), since NaOH formed on the inner surface of the cathode nozzle is transferred to the surface of the anode workpiece by the convection of the jet, NaOH need not to be added to the electrolyte prior to machining. Using alternating current (AC), as Masuzawa et al. [6] found in the electrochemical surface finishing of cemented carbide with $NaNO_3$ aqueous solution, the WO_3 film formed on the workpiece surface when the workpiece serves as the anode can be removed by NaOH which is generated when the workpiece serves as cathode. Hence in EJM, cemented carbide can be machined using either DC or AC current.

2.2. Experimental equipment

Figure 2 shows the experimental setup. The workpiece was fixed in a work tank whose position was controlled horizontally using an XY table. The nozzle was installed on the Z table to adjust the gap width between the nozzle

and workpiece. The electrolyte was supplied from the pressure tank pressurized by an air compressor.

3. Pit machining of cemented carbide

3.1. DC machining

NaOH generated on the inner surface of the cathode nozzle is transferred to the workpiece surface in EJM. Based on this idea, pits were machined on a cemented carbide plate using a standing jet of $NaNO_3$ aqueous solution, through which DC current was supplied. The machining conditions are shown in Table 1. A metallic cylindrical nozzle φ400μm in inner diameter was used. The cross-sectional view of a machined pit is shown in Fig. 3(a). The diameter of the machined area was over four times larger than the nozzle inner diameter. This is because the NaOH concentration distributed widely. From the equipotential surfaces in the jet shown in Fig. 1(a) calculated by Yoneda et al. [4], it is found that NaOH is mostly generated at the inner edge of the nozzle outlet where the potential gradient is steep. Hence, the NaOH concentration is highest on the side surface of the cylindrical jet but lowest at the center. As shown in Fig. 4, since NaOH is transferred to the workpiece surface

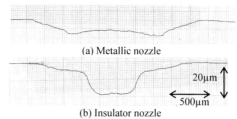

(a) Metallic nozzle

(b) Insulator nozzle

Fig. 3. Cross-sectional view of pits machined using

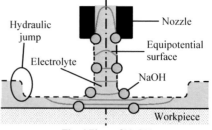

Fig. 4 Flow of NaOH

Table 1. Machining conditions in pit machining

Nozzle inner diameter [μm]	φ 400
Gap width between nozzle and workpiece [mm]	4.0
Tank pressure [MPa]	0.5
Machining current [mA]	30
Machining time [s]	30
Work material	Cemented carbide V30 (WC particle size : 1.0-2.5μm)

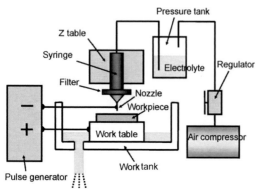

Fig. 2. Experimental equipment

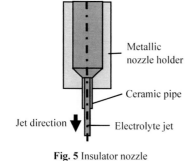

Fig. 5 Insulator nozzle

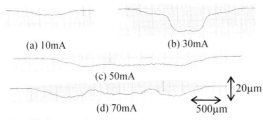

(a) 10mA   (b) 30mA

(c) 50mA

(d) 70mA

Fig. 6 Influence of current on pit shape with DC current

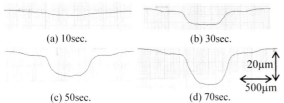

(a) 10sec.   (b) 30sec.

(c) 50sec.   (d) 70sec.

Fig. 7. Influence of machining time on pit shape with DC current

without mixing into the center of the jet, machining mostly occurs in the periphery of the jet colliding area. This finding led to the idea of using the insulator nozzle shown in Fig. 5. Since NaOH is generated in the hollow space of the metallic nozzle holder, NaOH concentration tends to become uniform due to the turbulent flow before the electrolyte flows into the ceramic pipe. Figure 3 (b) shows the cross-section of a pit machined using the insulator nozzle under the same conditions as those shown in Table 1. It was found that the insulator nozzle is effective for realizing selective machining under the jet. Figures 6 and 7 show the effects of machining current and time on the pit shape, respectively. The other conditions used were the same as those in Table 1. Fig. 6 shows that excessive currents larger than 30mA cause enlargement of the machined pit area with decreased pit depth. Fig. 7 shows that the pit becomes deeper proportionally to the machining time under the conditions used in the experiment.

3.2. AC machining

Even if NaOH is not mixed in the $NaNO_3$ aqueous solution, NaOH can be generated on the workpiece surface while the polarity of the workpiece is negative using AC current. Masuzawa et al. [6] employed this

reaction and succeeded in the electrochemical surface finishing of cemented carbide using AC current. With AC current, the NaOH generated when the workpiece polarity is negative dissolves the WO_3 film formed over the workpiece when the workpiece polarity is positive. Hence, cemented carbide can be machined by EJM using AC current. To prove this idea, we machined pits on cemented carbide using the same cylindrical metallic nozzle as that described in Section 1.3.1 with an AC rectangular pulse current of ±30mA at a frequency of 5Hz. Fig. 8 (a) shows the cross-section of the pit machined. The comparison between Fig. 8 (a) and Fig. 3 indicates that the diameter of the pit machined with AC current is smaller than that with DC current. With the metallic nozzle however, the inner edge of the nozzle outlet wears out when the polarity of the nozzle is positive, resulting in deformation of the jet. Fig. 9 shows the cross-sections of a metallic nozzle at the outlet before and after the machining time of 15 min. The inner edge of the nozzle outlet wore preferentially because the current density was highest at the edge as indicated by Fig. 1(a). To solve this problem, the insulator nozzle in Fig. 5 was applied to the AC machining. As shown in Fig. 8 (b), the diameter of the pit machined using the insulator nozzle was twice as large as the nozzle inner diameter, a little larger than the case of the metallic nozzle shown in Fig. 8 (a). This is probably because the jet shape was not parallel in the case of the ceramic nozzle because it was not built specifically for this study. We investigated the effects of peak current, machining time, and current frequency on the shape of the pits machined using the insulator nozzle, as shown in Figs. 10, 11, and 12, respectively. Machining conditions other than these were

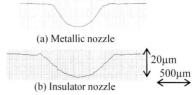

(a) Metallic nozzle

(b) Insulator nozzle

Fig. 8 Cross-section of pit machined with AC (5Hz)

(a) Before machining   (b) After machining

Fig. 9. Wear of metallic nozzle at outlet in AC machining

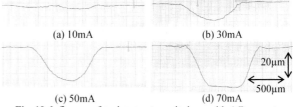

(a) 10mA   (b) 30mA

(c) 50mA   (d) 70mA

Fig. 10. Influence of peak current on pit shape with AC current

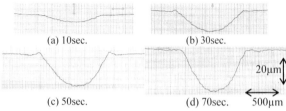

(a) 10sec.   (b) 30sec.

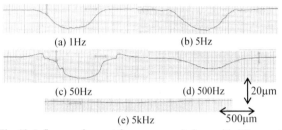

(c) 50sec.   (d) 70sec.   500μm

20μm

Fig. 11 Influence of machining time on pit shape with AC current

(a) 1Hz   (b) 5Hz

(c) 50Hz   (d) 500Hz   20μm

(e) 5kHz   500μm

Fig. 12. Influence of current frequency on pit shape with AC current

the same as those used in Fig. 8. Although higher machining current resulted in deeper pit depth, excessive current led to saturation in the pit depth. The machined depth increased proportionally to the machining time. Figure 12 shows that the optimal frequency of the AC current was 5 Hz.

4. Observation of machined surface

Figure 13 shows the SEM images of the micro structures on the bottom surface of the pits machined by EJM under the machining conditions in Table 1. It can be seen from Figure 13 (c) that WC grains were removed more preferentially than Co binder with AC current of 5Hz, where machining rate was comparatively high. On the other hand, Figs. 13 (b) and 13 (d) show that the Co binder was dissolved selectively leaving WC grains on

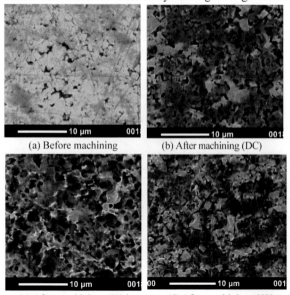

(a) Before machining   (b) After machining (DC)

(c) After machining (5Hz)   (d) After machining (50Hz)
Fig. 13 Micro structure of surface machined by EJM

the surface in machining using DC current and AC current with 50 Hz, where machining rate was low.

5. Conclusions

In the EJM of cemented carbide with DC current, NaOH generated in the cathode nozzle is transferred by the jet to the workpiece surface, where WO_3 film is removed, thereby allowing material removal to progress. The newly developed insulator nozzle enables more localized dissolution under the jet impinging area than metallic nozzle. Although AC current realizes more localized machining than DC current, the inner edge of the metallic nozzle wears out. Thus, the insulator nozzle is useful for preventing nozzle wear when machining with AC current.

Acknowledgements: This work was supported by the Grant-in-Aid for Challenging Exploratory Research from the Japan Society for the Promotion of Science (23656096).

References

[1]   Ippolito R, Tornincasa S, Capello G (1981) Electron-Jet Drilling. Annals of the CIRP 30(1):87–89.

[2]   Kozak J (1989) Some Aspects of Electro Jet Drilling. 4th International Conference on Developments in Production Engineering Design & Control, 363.369.

[3]   Kunieda M, Yoshida M, Yoshida H, Akamatsu Y (1993), Influence of Micro Indents Formed by Electro-chemical Jet Machining on Rolling Bearing Fatigue Life, ASME, PED-Vol64, 693-699.

[4]   Yoneda K, Kunieda M (1996) Numerical Analysis of Cross Section Shape of Micro-Indents Formed by the Electrochemical Jet Machining. Journal of JSEME 29(63):1-8. (in Japanese)

[5]   Maeda S, Saito N, Haishi Y (1967), Principle and Characteristics of Electro-Chemical Machining. Mitsubishi Denki Giho, 41(10): 1267-1279 (in Japanese)

[6]   Masuzawa T, Kimura M (1991), Electrochemical Surface Finishing of Tungsten Carbide Alloy, Annals of the CIRP, 40(1):199-202.

The material removal rate increases during machining of the steel St 37 (DIN 17100) in the electrical discharge machining process

S. Santos, A. Gomes and J. D. Marafona
Departamento de Engenharia Mecânica, Faculdade de Engenharia da Universidade do Porto
Rua Dr. Roberto Frias, 4200-465 Porto, Portugal

Abstract. In this article is shown that the material removal rate increases with machining time, until a peak in the material removal rate be reached for a depth of cut, followed by its decrease with the increase of the depth of cut. This statement is opposite to that says that material removal rate decreases with machining time followed by a stabilization. The increase of the material removal rate depends more on the pulse duration, than of the current intensity. There is a fast increase in the material removal rate until the peak be reached when small pulses are used, in opposition to a slow increase when long pulses are used. The material removal rate is low for small machining times independently of the pulse duration and current intensity. This behavior of the material removal rate was found for the current intensities and duration of pulses used. Thus, the research shows not only that the material removal rate increases during machining time, in opposition to the idea that it decreases, but also that the material removal rate decreases for a depth of machining due to the degradation of the machining conditions. Therefore, the increase of the material removal rate is due to the machining conditions (debris particles in the gap) and/or occurrence a metallurgical modification.

Keywords: Electrical discharge machining (EDM); machining depth; debris particles; gap flushing; Material removal rate (MRR).

1. Introduction

Electrical discharge machining (EDM) is a non-traditional manufacturing process where the material is removed by a succession of electrical discharges, which occur between the electrode and workpiece that are submersed in a dielectric fluid, such as, kerosene or deionised water. The electrical discharge machining process is widely used in the machining of hard metals and its alloys in the aerospace, automobile and moulds industries.

The flushing of debris particles and the cooling of dielectric fluid in the gap are critical characteristics to prevent the same localization and concentration of discharges [1]. The debris particles play a role in the ocurrence of the discharge in the electrical discharge machining (EDM) [2]. So, the stability of electrical discharges for small depths of cut is easily achieved by replacement of the contaminated dielectric fluid with a fresh one.

The material removal rate increases due to the interactions of workpiece hardness and EDM input parameters [3]. The workpiece hardness is obtained with heat treatments, so, being the electrical discharge machining characterized by a electric-thermal phenomenon, leads to the ocurrence of metalurgical modification of the base material of the workpiece [4].

On the one hand, it is known that the material removal rate decreases when there is a decrease in the surface roughness of the workpiece. On the other hand, the material removal rate increases simultaneously with the reduction of surface roughness of workpiece, according to Abbas et al. [5]. Only, the metallurgical modifications of the base material of the workpiece explains the latter statement. Therefore, the authors decided to investigate the effect of cut depth on material removal rate and electrode wear rate in a low carbon steel – St 37 (DIN 17100) using the flushing of debris with dielectric fluid at a pressure of 1 bar.

2. Experimental methodology

This methodology was designed and performed in a die-sinking EDM machine, AGIE COMPACT 3, equipped with adaptive control facilities. The adaptive control optimization (ACO) system enables the process to be optimized automatically and it was switched off so that the results can be generalized to all machines. The electrode and workpiece materials are electrolytic copper and steel DIN 17100-St 37, respectively. The workpiece was a parallelepiped with dimensions of 300x60x25 mm3. The electrodes used were copper rods 16 mm in diameter and a length of 100 mm. The EDM performance is related to the efficiency, which is determined in the EDM process by the material removal rate and tool wear rate. Quality is determined by the accuracy and surface roughness.

Three values of current intensity together with three pulse duration values were used in the experiments and the remaining input parameters were taken from the handbook of the AGIE manufacturer. So, the experiments were realized with the current intensities of 19.3A, 25.4A and 37.1A, together with duration pulses of 18μs, 75μs and 420μs.

This experimental methodology enables determine the cut depth as a significant contributor to the material removal rate for different cut depths with flushing of debris with dielectric fluid at a pressure of 1bar.

3. Experimental results

The experimental results obtained in the research, using the die-sinking EDM machine AGIE COMPACT 3, were analyzed to check the removing of material during the cut depth. On one hand, it is well accepted that material removal decreases with machining time due to the large number of debris in the gap, causing instability in the machining. On the other hand, some authors point out that the cleaning conditions of the dielectric in the beginning of the machining are not favorable to a good machining.

The authors of the research intend to demonstrate that the material removal rate in the beginning of the machining is smaller than during machining.

3.1. Effect of machining depth on material removal rate

Large cut depths are a factor of instability, because the debris particles that are in the gap are difficult to flush with fresh dielectric under pressure. However, the flushing of debris particles can be done without difficulties for small depths of cut and so, it would be expected to increase in the material removal rate for small depths of cut. Marafona and Araújo [3] pointed out that the workpiece hardness and its interactions with input parameters affect the EDM performance. So, the authors studied these effects on steel with low content of carbon St 37 (DIN 17100). Therefore, the material removal rate was evaluated for the cut depths of 1mm, 6mm and 12mm.

The effect of cut depth on material removal rate for the current intensity of 19.3A is analyzed in Fig.1. From the Figure is demonstrated that the material removal rate is smaller for the cut depth of 1mm, independently of the pulse duration used. The maximum removal rate seems to be achieved between the depth of 6mm and 12mm, follow by a degradation of the machining conditions. Figure 2 and 3 show that the material removal rate found in similar conditions of machining for the current intensities of 25.4A and 37.1A. The results are very similar to the current intensity of 19.3A presented in Fig. 1.

The current intensities of 19.3A, 25.4A and 37.1A show a maximum rate of removal material of 45.0mm3/min,

54.1mm3/min and 83.2mm3/min respectively. These maximum values were obtained using the pulse duration of 420μs and the cut depth of 12mm. So, one can conclude that the material removal rate increases with the increase of the cut depth and after there is a degradation of the machining conditions that leads to a decrease in the material removal rate.

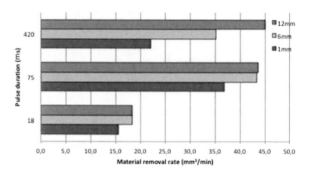

Fig. 1 Effect of machining depth on MRR for the current intensity of 19.3A.

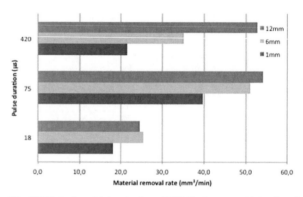

Fig. 2 Effect of machining depth on MRR for the current intensity of 25.4A.

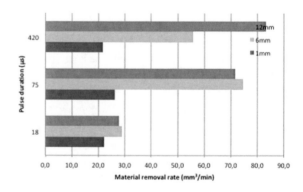

Fig. 3 Effect of machining depth on MRR for the current intensity of 37.1A.

In Fig. 1 is shown that the material removal rate is higher for the pulse duration of 75μs and the cut depths of 1mm and 6mm. This behaviour is followed by other current intensities as shown in Fig. 2 and 3, being this behaviour according to the trend of the material removal rate with the pulse duration.

Therefore, the research demonstrates that the material removal rate increases with the increase of cut depth,which is opposite to the idea that the material removal rate decreases and tends to a constant value during the machining.

4. Conclusion

The research results show that the material removal rate is affected by machining cut depth. This is demonstrated by the experimental results using the steel, St 37 (DIN 17100) that has a very small amount of carbon, and so, passible to have great changes in its metallurgical structure. As, main conclusion the authors can to say that there is an increase in the material removal rate with the increase of machining cut depth, followed by a degradation of the machining conditions, which lead to a decrease in the material removal rate, when is used the steel St 37 (DIN 17100).

References

[1]   Kunieda M., Lauwers B., Rajurkar K.P.,Schumacker B.M., (2005) Advancing EDM through Fundamental Insight into the Process, Annals of the CIRP, 54 (2) 599-622.

[2]   Schumacker B.M., (1990) About the Role of Debris in the Gap During Electrical Discharge Machining, Annals of the CIRP, 39 (1) 197-199.

[3]   Marafona J.D., Araújo A., (2009) Influence of workpiece hardness on EDM performance, International Journal of Machine Tools & Manufacture, 49 (9) 744-748.

[4]   Soni J.S., Chakraverti G., (1996) Experimental investigation on migration of material during EDM of Die steel (T215Cr12), Journal of Material Processing Technology, 56 439-451.

[5]   Abbas N.M., Solomon D.G., Bahari Md. F., (2007) A review on current research trends in electrical discharge machining, International Journal of Machine Tools & Manufacture, 47 1214-1228.

4

Machine Tool

Idealising mesh modelling for haptic enabled services and operands

E. Govea-Valladares[1]*, H. I. Medellín-Castillo[1], C. Fletcher[2], T. Lim[2], J. Ritchie[2], Xiu-Tian Yan[3], Victor Arnez[4] and Ernesto Hernandez[4]

[1] Facultad de Ingeniería, Universidad Autónoma de San Luis Potosí, S.L.P. México
[2] Innovative Manufacturing Research Centre, Heriot-Watt University, Edinburgh, UK
[3] Design, Manufacture & Engineering Management, University of Strathclyde, Glasgow, UK
[4] Facultad de Ingeniería, Universidad Nacional Autónoma de México, D.F. México
* Corresponding author: eder.govea@gmail.com

Abstract. Communicating the knowledge and science of product engineering, analysis and manufacturing planning is an area of continued research driven by the digital economy. Virtual Reality (VR) is a generally accepted interactive digital platform which industry and academia have used to model engineering workspaces. Interactive services that generate a sense of immersion, particularly the sense of touch to communicate shape modelling and manipulation, is increasingly being used in applications that range from Design For Manufacturing and Assembly (DFMA) and Process Planning (PP) to medical applications such as surgical planning and training. In simulation, the natural way for solid modelling is the use of primitive geometries, and combinations of them where complex shapes are required, to create, modify or manipulate models. However, this natural way makes use of Booleans operands that require large computational times which make them inappropriate for real time VR applications. This work presents an insight on new methods for haptic shape modelling focused on Boolean operands on a polygon mesh. This is not meant as a contrast to point/mesh-editing methods, instead it is focused on idealising polygonal mesh modelling and manipulation for use with haptics. The resulting models retain a high level of geometric detail for visualisation, modelling, manipulation and haptic rendering.

Keywords: Mesh modelling, Boolean operations, Haptic rendering, Process Planning.

1. Introduction

Through the years, computer simulation has been a tool used to model real life situations by using a computer program. Traditionally, system modelling uses a mathematical model, which attempts to find analytical solutions to problems, trying to predict the behaviour of a system [1]. There are many different types of computer simulations; the common feature that they all share is the attempt to generate a sample of representative scenarios for a model and its behaviour [2].

Object modelling is a tool used in mechanical engineering to design parts [3]. Simulations play an important role in the product design process, reducing the need for expensive prototyping and reducing the product development cost [4]. Most of commercial modelling software focuses on the visual editing, but when haptics is added the user experience through tactile engagement is superior. Precision and accuracy are desired characteristics in these systems, but it is directly dependent on the computational capabilities and the data size and complexity. The more data, the more specific becomes the model [5]. The time performance of the modelling and simulation process is a current subject under study. Current methods such as performing Boolean Operations on Polygon Meshes and implicit functions [6] are usually very slow.

This paper presents a new algorithm developed for mesh modelling and manipulation with haptics. Union, intersection and difference Boolean operands are 'localised' to the contact area between the two objects to speed up processing to move closer to a real-time performance. The implementation is done in VTK [7] with collision libraries of VTKBioEng [8] programmed in C++.

2. Related Work

The editing and manipulation of 3D models has seen constant research activity particularly in the area of CAD. Several techniques relevant for physics-based simulations are listed in [7]. In this work the modelling techniques for fully-automatic or semi-automatic simplification of CAD models are also characterized. In [8] a set of interactive free-form editing operators for direct manipulation of level-set models to support the creation and removal of surface detailed by operators for volumetric implicit surfaces is presented.

3D modelling with triangular mesh has become increasingly popular in engineering where fast

S. Hinduja and L. Li (eds.), *Proceedings of the 37th International MATADOR Conference*,
DOI: 10.1007/978-1-4471-4480-9_4, © Springer-Verlag London 2013

modifications of the mesh models means critical problems can be solved quickly without going back to the CAD model. An algorithm that removes the intersecting faces in an n-ring neighbourhood is presented in [9]. The algorithm produces triangles whose sizes smoothly evolve according to the possibly heterogeneous sizes of the surrounding triangles. Regarding the formation of new models from primitive objects, techniques for rendering implicit surfaces using point based primitives were presented in [10]. A method for real time modelling was presented in [11], where a modelling approach using signed distance functions for objects and complex surface manipulations with immediate visual feedback was described. A new method for implicit modelling was presented in [12]. It was proposed a method to describe sharp features (edges and vertices) applying a new surface modelling representation.

In [13] Boolean operations to construct heterogeneous material objects were introduced. Boolean operations use Boolean algebra to model more complex objects. The operands of addition, subtraction and common are used in this modelling technique [14]. Boolean modelling can be readily implemented in CAD/CAE/CAM software. An algorithm to calculate intersection, union and difference was proposed in [15]. The algorithm is valid for general planar polygons based on algebraic operations to calculate the intersection between general polygons. Similarly, an algorithm for Boolean operations on polyhedral solid representations using approximate arithmetic was described in [16].

Haptic editing of 3D models was presented in [17], where the integration of Virtual Reality (VR) and Computer-Aided Design (CAD) was investigated. The proposed system made possible the intuitive and direct 3D edition of CAD objects through B-Rep modelling in CATIA and haptic aided by grid and extrude commands. Some commercial haptics 3D modellers are based on point clouds that define surfaces, essentially for applications of sculpture or artistic modelling. For example, the Splodge software of Sensable [18] or the Cre8 of Novint [20] allow 3D modelling of objects in real time.

From the literature review, it can be concluded that simulation with mesh models, haptic fast edition and manipulation of the 3D models using open source libraries and multi haptic devices, is not an area that has been explored by researchers. The purpose of this work is to model 3D mesh objects via a haptic interface and using enhanced Boolean operations. The proposed method is based on dividing the main piece (Object 1) into regions and manipulates the haptic cursor or tool (Object 2) to modify the main piece and create a new model. The aim of the method is the optimization of the processing speed when a Boolean operation is applied.

3. Methodology

Figure 1 presents the proposed methodology for haptic modelling using localised Boolean. This methodology has been implemented using C++, Visualization Toolkit libraries (VTK 5.6.1), and collision detection with VKTBioEng v5.0.1. Open Source Haptics H3D v2.1.1 is also used for haptic rendering, which in this case is carried out by the Phantom Omni from Sensable and the Falcon from Novint haptic devices. The system has been implemented in a PC with a 1.73 GHz processor, 2.0GB of RAM and Windows XP.

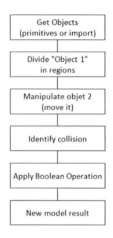

Fig. 1. Proposed Boolean localised methodology

1. - Get objects: The modelling process start with the definition of the objects in the scene (objects 1 and object 2). These objects can be created with VTK libraries (primitives) or imported from CAD systems as STL files.

2. - Divide "Object 1" in regions: In this step Object 1 is divided into smaller parts and named with an ID for identification. The algorithm divides the mesh using a filter that separates the cells of a dataset into spatially aggregated pieces using an Oriented Bounding Box (OBB) method. The division can be done by specifying the number of elements or the number of points in one area.

3. - Manipulate "Object 2": Object 2 can be either the haptic cursor or an object being controlled with the haptic cursor. Object 1 and 2 must be in contact to allow Boolean operand being used.

4. – Identify collision: The algorithm takes only the regions of object 1 that are in contact with object 2 and creates a "polydata" variable. The data like points and elements of these regions can be extracted. The elements that are not in contact with the cursor will not be selected and will not be included in the Boolean operation procedure.

5. - Apply Boolean operation: The selected "polydata" is sent to the Boolean operation function to perform Union, Intersection or Difference.

6. - New model result: Once the Boolean operand has been done, the new section will be added to object 1.

Figure 2 shows an example of Object 1 after being divided into regions (colors) and the object 2 (white sphere) being the haptic cursor, before the Boolean operation.

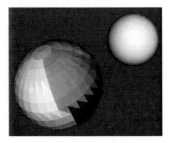

Fig.2 Objects for the haptic modelling

4. Implementation

Four case studies were selected to test and evaluate the proposed localised approach to perform Boolean operands (intersection, union and difference) in virtual haptic modelling. For comparison purposes, Boolean operations were applied to these case studies using both, the proposed and the conventional, Boolean methods.

Case 1: Two primitive objects with regular curved surfaces (spheres) were used. Each comprises a mesh size of 780 elements. The results of each Boolean operand are shown in Fig. 3.

Case 2: In this case, a primitive cube comprising 12 triangular elements, and a sphere cursor with 780 elements were selected. The purpose of this case was to observe the behaviour of the Boolean operations when applied to an object with low resolution. Fig. 4 shows the results of the Boolean operands.

Case 3: A more complex model i.e., a gear with 1767 triangular elements was used. The objective is to test the response of the Boolean operands in areas with elements of different sizes. Fig. 5 shows the results.

Case 4: The last case study is shown in Fig. 6. A jaw model with 9230 elements and spherical cursor were selected. The objective is to assess the performance of Boolean operands in high-resolution models of irregular geometry.

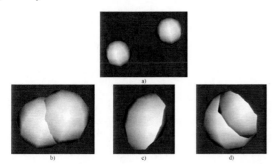

Fig. 3 Objects for the case 1: (a) source objects, (b) union, (c) intersection and (d) difference.

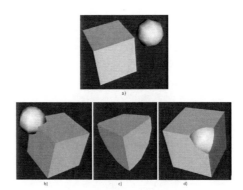

Fig. 4 The regular cube and cursor: (a) source objects, (b) union, (c) intersection and (d) difference.

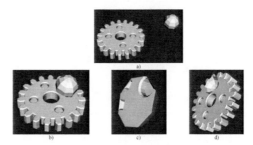

Fig. 5 Gear and tool for the case study 3: (a) source objects, (b) union, (c) intersection and (d) difference.

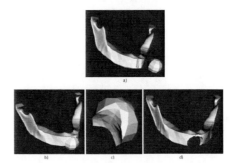

Fig. 6 High resolution jaw and cursor: (a) source objects, (b) union, (c) intersection and (d) difference.

5. Results

In order to compare the time performance of the two Boolean methods, localised and conventional, processing time was measured using the CPU processor clock. Table 1 present the time performance results of the Boolean operands using the conventional approach, whilst Table 2 shows the time performance results using the proposed localised based approach. Each Boolean operand was applied ten times to each case study; the results reported in Table 1 and 2 correspond to the average values of each test.

From the results presented in Table 1 and 2 it can be observed that the proposed localised Boolean approach is approximately 25% faster than the conventional Boolean approach. This is due to the fact that the proposed approach uses less data (localised data) than the conventional approach where the whole mesh is used in the Boolean operand. Since the proposed algorithm separates into parts the complete mesh of each object and uses a collision detection filter to localise and select the area of interest, the time performance is improved. By enabling discreet haptic region selection, only the elements and associated vertices of the selected regions will be affected, thus avoiding the need of procesing the whole mesh.

It can be observed that as the number of elements of the mesh models increases (more complex models), the processing time increases. The quality of the models was not affected when using the localised proposed method.

Table 1. Conventional Boolean operation, milliseconds

Haptic cursor with	UNION	INTERSECTION	DIFFERENCE
Sphere	5.715	5.766	4.422
Cube	4.922	4.798	4.098
Gear	6.948	6.630	5.778
Jaw	7.896	7.577	7.487

Table 2. Localised Boolean operation, milliseconds

Haptic cursor with	UNION	INTERSECTION	DIFFERENCE
Sphere	4.458	4.497	3.449
Cube	3.741	3.646	3.114
Gear	5.211	4.973	4.334
Jaw	5.685	5.455	5.391

6. Conclusions and future work

A new method to improve the time performance of Boolean operands has been proposed. The proposed method is based on a collision detection approach to localise contact areas. It has been proved that by 'localising' a Boolean operand on mesh models, it is possible to reduce the execution time compared with Boolean conventional operations that do not discriminate elements. It has to be mentioned that in haptic virtual reality systems, haptic rendering may be slower than Boolean operands and therefore the simulation may not be perform in real-time. Future work considers the optimization of haptic virtual reality applications using the proposed Boolean operations approach based on localised collisions.

Acknowledgements: The authors wish to thank to Consejo Nacional de Ciencia y Tecnologia (CONACyT), Universidad Autónoma de San Luis Potosí (UASLP) and Heriot-Watt University (HWU) for the invaluable support to carry out this research work.

References

[1] Oliver Rübel, Sean Ahern, E. Wes Bethel, Mark D. Biggin, Hank Childs, Estelle Cormier-Michel, Angela DePace, Michael B. Eisen, Charless C. Fowlkes, Cameron G. R. Geddes, Hans Hagen, Bernd Hamanna, Min-Yu Huang, Soile V. E. Ker¨anen, David W. Knowles, Cris L. Luengo Hendriks, Jitendra Malik, Jeremy Meredith, Peter Messmere, Prabhat, Daniela Ushizima, Gunther H. Weber, Kesheng Wu. "Coupling visualization and data analysis for knowledge discovery from multi-dimensional scientific data", Procedia Computer 1 (2010) 1757-1764.
[2] ChaoliWang and Han-Wei Shen, "Information Theory in Scientific Visualization", Entropy 13 (2011), 254-273
[3] Junji Nomura, Kazuya Sawada. "Virtual Reality Technology and its Industrial Applications", Annual Reviews In Control 25 (2001) 99-109
[4] Jesús David Cardona, Miguen Ángel Hidalgo, Héctor Castán, Fabio Rojas, Diego Borro, Héctor Jaramillo, "Realidad Virtual y Procesos de Manufactura" ISBN 978-958-8122-51-9. (2007) Colombia, primera edición.
[5] Atul Thakur, Ashis Gopal Banerjee, Satyandra K. Gupta. "A survey of CAD model simplification techniques for physics-based simulation applications. Computer-Aided Design 41 (2009) 65-80
[6] H. Masuda, "Topological operators and Boolean operations for complex-based nonmanifold geometric models", Computer-Aided Design 2 (1993) 119 - 129.
[7] VTK - The Visualization Toolkit, http://www.vtk.org/
[8] Vtkbiocng, http://www.bioengineering-research.com/
[9] Atul Thakur, Ashis Gopal Banerjee, Satyandra K. Gupta. "A survey of CAD model simplification techniques for physics-based simulation applications", Computer-Aided Design 41 (2009) 65-80
[10] Manolya Eyiyurekli, David Breen, "Interactive free-form level-set surface-editing operators", Computers & Graphics 34 (2010) 621-638
[11] Ruding Loua, Jean-Philippe Pernot, Alexei Mikchevitch, Philippe Vérona, "Merging enriched Finite Element triangle meshes for fast prototyping of alternate solutions in the context of industrial maintenance", Computer-Aided Design 42 (2010) 670-681
[12] Emilio Vital Brazil, Ives Macedo, Mario Costa Sousa, Luiz Velho, Luiz Henrique de Figueiredo, "Shape andtonedepictionforimplicitsurfaces", Computers & Graphics 35 (2011) 43-53
[13] Tim Reiner, Gregor Muckl, Carsten Dachsbacher, "Interactive modeling of implicit surfaces using a direct visualization approach with signed distance functions", Computers & Graphics 3 5(2011) 596-603
[14] Xinghua Song, Bert Juttler, "Modeling and 3D object reconstruction by implicitly defined surfaces with sharp features", Computers & Graphics 33 (2009) 321-330
[15] W. Sun, X. Hu, "Reasoning Boolean operation based modeling for heterogeneous objets", Computer-Aided Design 34 (2002) 481-488
[16] M. Rivero, F.R. Feito, "Boolean operations on general planar polygons", Computers & Graphics 24 (2000) 881-896
[17] J.M. Smith, N.A. Dodgson, "A topologically robust algorithm for Boolean operations on polyhedral shapes using approximate arithmetic", Computer-Aided Design 39 (2007) 149-163
[18] P. Bourdot, T. Convard, F. Picon, M. Ammi, D. Touraine, J.-M. Vézien, "VR-CAD integration: Multimodal immersive interaction and advanced haptic paradigms for implicit edition of CAD models", Computer-Aided Design 42 (2010) 445-461
[19] Sensable, http://www.sensable.com/
[20] Novint, http://www.novint.com/

Analysis of dynamic properties of a multi-stage gear system using the flexible multi-body system modelling technique

M. Sulitka[1], Z. Neusser[2] and J. Veselý[1]

[1] Research Center for Manufacturing Technology, Czech Technical University in Prague, Faculty of Mechanical Engineering, Horská 3, Praha 2, 128 00, Czech Republic

[2] Department of Mechanics, Biomechanics and Mechatronics, Czech Technical University in Prague, Faculty of Mechanical Engineering, Technická 4, Praha 6, 166 07, Czech Republic

Abstract. The paper presents a model of a tooth gear system which allows a complex analysis of dynamic properties of the feed drive. The model, comprising a description of shaft, bearings, wheel and gearbox compliances, is assembled in state space as a coupled system of finite element models and includes a description of the stiffness of tooth contact. A measurement for verification of the proposed model is done using a single gearbox unit, as well as the entire system of a machine tool feed drive.

Keywords: Tooth gear system, Feed drive model, Machine tool

1. Introduction

The development of big machine tools with large working ranges has been intensively expanding, using rack-pinion feed drive systems increasingly as a substitute of common ball screw drives. A suitable choice of feed drive mechanism is closely linked to the interaction between the feed drive and the structure of the machine tool. An effectively optimised feed drive design can be achieved by using coupled models combining the description of feed drive and machine tool frame.

A model of a multistage tooth gear system derived from a discrete description is shown e.g. in [1]. In the present paper, a FE-based model is introduced which was developed for application in machine tool feed drive simulations. In addition to a detailed description of tooth contact and the compliance of the mechanical structure of the gear system, special attention has been paid to the contact between shaft, key and wheel.

2. Gear system model

For simulations of dynamic properties of the gear system both in frequency and time domain, a model was created which includes a description of all the important elements of the mechanical gear system including tooth contact, shaft and wheel mechanics, bearings and gear box.

2.1. Tooth contact model

The tooth contact model assumes an ideal contact of a wheel couple. Calculation of the contact uses a modified Hertz theory [2], using the more precise parameters of tooth tilting derived in [3] and tooth bending in [4] (see Fig. 1). The resulting force is dependant on the load of the coupled gears and the rotation, with a different number of teeth in contact. Stiffness used to substitute tooth contact has been calculated as an average of instantaneous flexibilities. Dynamic properties of the tooth contact model and comparison with measurements introduces [5].

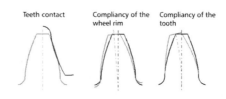

Teeth contact   Compliancy of the wheel rim   Compliancy of the tooth

Fig. 1. Components of the compliance of tooth contact

2.2. Contact stiffness

The calculation of the stiffness of the gear contact is based on the two tooth penetration formula, in which the total penetration is dependant on the bending of both teeth caused by contact force F_N, the tilting of both teeth and the transformations of the sides of the teeth in contact, according to formula 1.

$$\delta = F_N g_{tilting1} + F_N g_{bending1} + F_N g_{contact}(F_N) + \\ + F_N g_{tilting2} + F_N g_{bending2} \tag{1}$$

Compliances g_x are dependant on the line of contact, tooth profile and material properties of the gears. Variable $g_{contact}$ is described in a non-linear dependence featuring contact force F_N. Therefore, the formula (1) must be solved iteratively. The stiffness of the contact is determined by dividing the contact force with total penetration.

2.3. Shaft and wheel model

An important component of the overall compliance of the gear system is represented by the system of shafts and wheels. A model was designed, enabling simulations of the operation of the gear system over time and the creation of the machine tool feed drive coupled models. An FE model, including the contact problem between shaft, key and wheel, is used for the calculation of static analysis, delivering the stiffness k_{CS} [N/m] between a wheel W on which a tangent force F is applied and a pinion P, fixed on the circumference by the tooth contact line (Fig. 2). The shaft is fixed in bearing locations.

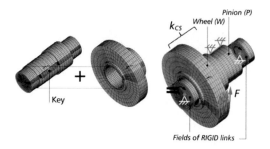

Fig. 2. The FE model of the shaft and wheel for a static analysis with the shaft, key and wheel contact problem.

This type of model, however, cannot be used for efficient simulations in time domain. For this reason, another FE model was created, where the shaft is united with the wheels.

Through modal decomposition, this model is transformed into state space with interface nodes i, j, k, l selected to define the force inputs and outputs of positions and rotations at points, representing the bearings and wheels (Fig. 3). This model can then be used to determine the torsion stiffness k_{SS} [Nm/rad] between the wheel W and pinion P.

2.4. Coupled shaft and gearbox model

The coupling of the shaft models with the wheels and the gearbox is carried out in state space (Fig. 4). The coupling force between the shaft wheels 1 and 2 is determined using the total stiffness of one wheel pair

$$k_{12} = \frac{1}{\frac{1}{k_t} + \frac{1}{k_{cs}} - \frac{r_w^2}{k_{ss}}} \text{ [N/m]}, \tag{2}$$

where k_t [N/m] is the contact stiffness derived from (1) and k_{CS} [N/m] is the stiffness between the wheel and the pinion of the FE model considering the contact between shaft, key and wheel (Fig. 2). Subtracted is the element with the torsion stiffness k_{SS} [Nm/rad], transformed to its axial representation using the wheel radius r_w. On interface nodes j, k (Fig. 4), there is a torque of

$$M_{1,2} = k_{12}(r_1\varphi_1 + r_2\varphi_2)r_{1,2}^2 \tag{3}$$

where φ is the rotation of the teeth and r is the appropriate gear diameter.

To determine the coupling force between the shaft and the gearbox, the stiffness values of the bearings are used.

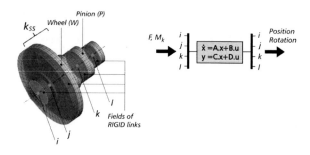

Fig. 3. A full FE model of the shaft with wheels for export in state space (a schematic is shown on the right)

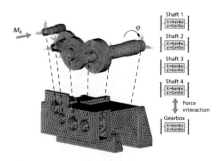

Fig. 4. Coupled model of shaft with wheels and the gearbox

3. Measuring the stiffness of the gear system

The stiffness of the pinion's mesh with the rack and the overall stiffness of the gear system is determined via measuring on a multifunctional test bench. The considered gear system has three stages, straight teeth and a pinion on the output shaft meshing with a diagonal-toothed rack. The kinematic path of the gear system corresponds to the model in Fig. 4.

The arrangement of the test bench and tested gear system for the measurement can be seen in Fig. 5. In addition to the rotation (1) on the input shaft, rotation (2) on the output pinion and the torque on the input, deviations of the gearbox relative to the rack were also measured in directions X, Y and Z.

The gear system has been loaded by a controlled alternating torque of the servomotor, with three selected amplitude levels of the load torque. The information from rotation sensors φ_1, φ_2 and torque M_k on the output shaft are used to determine the overall stiffness of the gear system relative to the input as,

$$k_T = \frac{\varphi_1 - p_R \cdot (\varphi_2 - \Delta\varphi_2)}{M_k},$$ (4)

where p_R is the overall transmission. For the additional rotation of the pinion $\Delta\varphi_2$ caused by the gearbox being moved relative to the rack because of the compliance of the consoles, the following can be derived:

$$\Delta\varphi_2 = \frac{\Delta y \cdot \tan(\beta) \cdot \cos(\alpha)}{r_b},$$ (5)

where α is the angle of the sides of the rack's teeth, β the angle of the teeth and r_b is the radius of the pitch circle of the pinion.

The measurements are used to derive not only the stiffness of the entire transmission system, but also of the pinion's mesh with the rack.

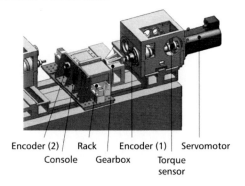

Fig. 5. Picture of gear system measuring arrangement

4. Verfirication of the gear system stiffness

Hysteresis loops of the dependence of load torque on rotation are determined from the measurements, with an example of shown in Fig. 6.

In the range of used torque amplitudes, stiffness values have been determined which correspond to central lines of the hysteresis loops, with a range of 300 to 500 Nm/rad (). These points at obviously non-linear characteristic of the gear system. The simulation model evaluates linearised stiffness values, which are compared

with the measured ones in the chart in Fig. 7. Well match of the model with measurements can be seen.

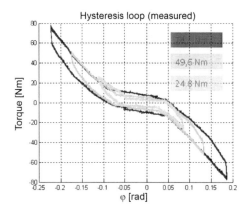

Fig. 6. Measured hysteresis loops of the gear system

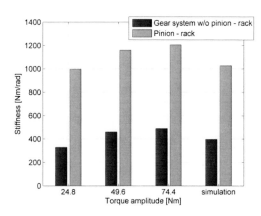

Fig. 7. Stiffness of the gear system relative to input. Comparison of the measurements and simulation

5. Verification of the feed drive model of the machine tool motion axis

The tested gear system is used on a large portal vertical milling machine as a feed unit of the X motion axis. For the analysis of the feed drive dynamic properties, a coupled model of the drive of X axis is created, including an FE model of the machine tool structure and a simplified two-mass substitution of the gear system mechanics. The gear system simplified model uses the value of the overall torsion stiffness determined using the detailed model described above.

5.1. Feed drive coupled model

A coupled model of the feed drive and machine tool structure is created using the procedure described e.g. in [6], [7]. The machine tool structure is modelled by FE and using the modal decomposition method, the FE model is transformed into state space. A schematic picture of the model is shown in Fig. 8.

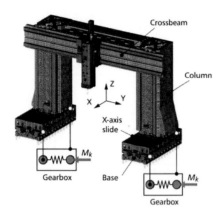

Fig. 8. FE model of the machine and the connection with simplified models of the X-axis gear system

6. Feed drive dynamic properties

Dynamic properties of the feed drive mechanical structure are expressed with the frequency transfer φ_M/x_L between the rotation of motor φ_M (encoder) and a linear movement of the column x_L (linear ruler). This transfer can be evaluated with a combination of functions φ_M/M_k and x_L/M_k

$$G_{ML} = \frac{\varphi_M}{x_L} = \frac{\varphi_M}{M_k} \cdot \frac{M_k}{x_L} = \frac{\varphi_M}{M_k} \cdot \left(\frac{x_L}{M_k}\right)^{-1} \quad (6)$$

A comparison between the measured and simulated characteristics is shown in a chart in Fig. 9. It can be seen that the simulation corresponds to the measurements very well, both in frequency and amplitude. It is particularly important to note the very good correspondence of the value of the first anti-resonance frequency at the first point the characteristic's amplitude drops.

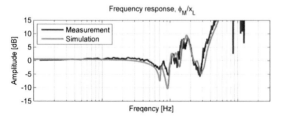

Fig. 9. Frequency transfer of the machine's feed drive. Comparison of measurements and simulation

Regulation theory [8] has proven that setting of the position control loop gain is limited primarily by the first anti-resonance frequency. In addition to this value, however, there are also higher natural frequencies and

their oscillation amplitudes entering the feed drive control dynamics, and the model shows very good correspondence with the measurements in this area as well. The resulting model can thus be beneficially used for relevant simulations of feed drive control dynamics.

7. Summary

A detailed model of a gear system was created, including a description of the stiffness of the teeth mesh, the transmission path and the gearbox. The model is designed as a flexible multi-body system, enabling time and frequency domain simulations. The proposed shaft modelling approach also considers the contact of the shaft – key – wheel system. The model is verified with measurements on a real gear system using a specialised testing bench, determining a very good correspondence between the simulated and measured values. The gear system model is also verified on a real machine tool motion axis. The model shows a very good correspondence with the real dynamic properties of the feed drive and therefore it can well be applied in machine tool feed drive coupled models for simulating and optimizing the feed drive control dynamics.

Acknowledgements: This research has been supported by the 1M0507 grant of the Ministry of Education of the Czech Republic.

References

[1]   Tanaka E, Tanaka N, Ohno k, (2001) Vibration analysis of a multi-stage gear system including drive mechanism elements. JSME Int Journal, Ser C. Mech Systems, Mach Elem Manuf, Vol. 44, No. 2

[2]   Petersen D, (1989) Auswirkung der Lastverteilung auf die Zahnfusstragfähigkeit von hoch überdeckenden Strinradpaarungen. Dissertation, Hamburg, Germany

[3]   Sainsot P, Velex P. (2004) Contribution of Gear Body to Tooth Deflections - A New Bi-dimensional Analytical Formula. Journal of Mechanical Design, Vol. 126: 748-752

[4]   Weber C, Banaschek K, (1953) Formänderung und Profilrücknahme bei Gerad-und Schrägverzahnten Rädern. Heft 11, F. Vieweg und Sohn, Braunschweig, Germany.

[5]   Neusser Z, Sopouch M, Schaffner T, Priebsch H-H, (2010) Multi-body Dynamics Based Gear Mesh Models for Prediction of Gear Dynamics and Transmission Error. SAE konference, Detroit, USA.

[6]   Sulitka M, Strakoš P, (2007) Complex model of a real machine tool feed drive axis with ball screw. Conference VIDA 2007; Poznan, Poland,

[7]   Vesely J, Sulitka, M, (2008) Machine Tool Virtual Model. International Congress MATAR 2008, Part 1: Drives & Control, Design, Models & Simulation; Prague; Czech Republic;

[8]   Souček, P, (2004) Servomechanism in machine tools (in Czech). CTU of Prague, Praha.

An efficient offline method for determining the thermally sensitive points of a machine tool structure

N. S. Mian, S. Fletcher, A. P. Longstaff and A. Myers
University of Huddersfield, Queensgate, HD1 3DH, UK

Abstract. Whether from internal sources or arising from environmental sources, thermal error in most machine tools is inexorable. Out of several thermal error control methods, electronic compensation can be an easy-to-implement and cost effective solution. However, analytically locating the optimal thermally sensitive points within the machine structure for compensation have been a challenging task. This is especially true when complex structural deformations arising from the heat generated internally as well as long term environmental temperature fluctuations can only be controlled with a limited number of temperature inputs. This paper presents some case study results confirming the sensitivity to sensor location and a new efficient offline method for determining localized thermally sensitive points within the machine structure using finite element method (FEA) and Matlab software. Compared to the empirical and complex analytical methods, this software based method allows efficient and rapid optimization for detecting the most effective location(s) including practicality of installation. These sensitive points will contribute to the development and enhancement of new and existing thermal error compensation models respectively by updating them with the location information. The method is shown to provide significant benefits in the correlation of a simple thermal control model and comments are made on the efficiency with which this method could be practically applied.

Keywords: Finite element analysis, FEA, Matlab, Thermal error, Thermal error compensation, Thermally sensitive locations.

1. Introduction

Thermal errors have been identified as a major contributor to the overall volumetric error of a machine tool, in many cases up to 70% [1]. Several techniques based on analytical, empirical and numerical methods have been established to control the effect of thermal errors. These techniques are widely used and applied with a basic ideology to establish a thermal model based on relationships between the measured temperature of the machine from various locations, used as temperature inputs and the displacement at the tool [2]. The temperature inputs however in some cases may be difficult to identify if propagation of the temperature gradients is complex due to the combined effect of internal and external heat sources and perhaps due to the

complexity of the machine structure. These ambiguities therefore add complexities to identify sensitive locations within the structure and stand out to be a challenging task with a limited number of temperature inputs. It has been observed that the performance of the conventional empirical and statistical approaches such as Artificial Neural Network (ANN) and Linear Regression [3, 4] heavily rely on the data from sensitive location within the machine structure for effective and robust thermal compensation such as varying environmental conditions. Kang et al. [5] used a hybrid model consisting of regression and NN techniques to estimate thermal deformation in a machine tool. The total of 28 temperature sensors were placed on (18) and around (10) the machine to acquire internal heating and environmental data. The training time for the model was 3 hours. Yang et al. [6] tested INDEX-G200 turning centre to model thermal errors. Temperature variables were selected using engineering judgement as temperature sensors were placed on or near the possible heat sources and Multiple Linear Regression technique was used to model thermal errors. Training time for the thermal model however was not mentioned. Krulewich [7] used the Gaussian integration method using polynomial fit to identify the optimum thermal points on the machine spindle. The spindle was put through heating and cooling cycles providing 3.5 hours of training data to locate three optimum measurement points where the results correlated to 96%. The author compared this method with a statistical technique and found that the Gaussian integration method requires significantly less training data.

It has been observed that a significant amount of data is generally required to identify sensor locations and train models which inevitably require machine downtime; therefore, such methodologies can be impractical for general application. It is also the fact that machine structures are sensitive to environmental changes which means that the training data acquired in the first instance may not respond well to the new conditions and therefore a new set of training data may be required [7]. This paper presents an offline technique based on FEA. The technique provides the ability to identify optimised sensitive locations within the machine structure offline for any set of data either from internal heating or external environmental conditions. Being software based, using the Graphical User Interface (GUI) of the FEA software, this technique integrates the visual aspect to aid reviewing the location of the sensitive areas and the practicality for sensor installations. The application of this technique requires minimal machine downtime as any set of the measured thermal conditions can be assessed offline to obtain the thermal behaviour of the machine. This means that new sensitive areas inside the machine structure may be located according to the new thermal conditions. Satisfactory correlations between the measured and the FEA simulated results are a pre-requisite to the application of this technique. In this paper, this technique is applied on the results from simulation case study previously conducted.

2. Case Study

This study was conducted on a 3 axis Vertical Machining Centre (VMC) located on the shop floor with uncontrolled environmental temperature. The FEA model of machine was created in Abaqus/Standard 6.7-1 software [10] using manufacturer provided engineering drawings. Fig. 1 shows the generated CAD model of the machine. The model of the machine was simplified by cutting into half because of the symmetry in the X axis direction and complex structures such as fillets and chamfers were simplified and represented using simple corners to avoid complexity of meshing and nodes.

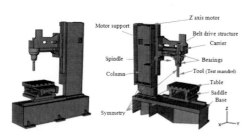

Fig. 1. Generated CAD model of the machine assembly

Mian et al. [8, 9] conducted tests to exploit the thermal behaviour of the VMC when subjected to the spindle heating and varying environmental conditions. Mian et al. [8, 9] proposed a technique in which only one short term data set obtained during one hour internal heating is required to obtain thermal parameters and simulate the heat transfer within structures. This short term data set is used to create the FEA thermal model to simulate the machine for a variety of real world testing regimes. The results showed good correlation between the experimental results and the FEA simulated results typically between 70% and 80%. Mian et al. [9] also conducted environmental tests where the machine was tested for

three continuous days in two seasons (winter and summer). The aim was to achieve good correlation in results from one season test and validate the methodology with good correlation results in different environmental conditions i.e. in a different season. Both tests successfully validated the FEA environmental thermal model with good correlations typically above 60%. This technique in effect can remarkably reduce the machine downtime by creating the CAD model of the machine in the FEA software and simulate it to create an environmental thermal model that is able to simulate the effect of any set of varying environmental conditions.

This method therefore provides a platform to use FEA modelling as an offline tool to determine not only machine behaviour, but also help with the development of compensation models by determining the location of sensitive nodes/areas. The case study by Mian et al. [8, 9] was therefore used for differentiating between areas sensitive to internal heating and environmental temperature fluctuations.

The remainder of the paper details a method and the developed software for the offline assessment of the FEA data and help determine the temperature-displacement sensitive nodes based on search parameters and their physical locations within the FE model. The information can be used to retrofit sensors for compensation; however there can be practical limitations to their attachment.

3. Nodal data extraction

Abaqus simulation software provides the facility to extract surface and sub-surface nodal data within the FEA model. Since the model has to be meshed for FEA analysis, nodes from the mesh can be used to represent individual points on the structure. Therefore, using this facility, the nodal data was extracted to find nodes of interest. The predicted error is obtained as the difference in displacement between a node on the table and a node on the tool. In this case the dependant parameters are slope and hysteresis.

The slope is simply the magnitude of displacement for any given change in temperature ($°C/\mu m$). Hysteresis is caused by the time lag involved with typical surface temperature measurement which is related to the distance between the temperature sensor and the true effective temperature which is causing the distortion. A node location with high slope sensitivity will require lower resolution in the measurement of temperature and induce less noise when applied in models, as described later. The lowest hysteresis will represent that area that relates well to thermal displacement and responds in a linear fashion whether the machine is being heated or cooled. The nodal data is extracted from Abaqus and the files are converted and imported into Matlab software. Matlab functions were written to calculate the slope and hysteresis for each node and return the best ones with respect to an axis.

3.1. Matlab program routines

The function imports the nodal data from the FEA software and extracts the error between the tool and workpiece in each direction, and the temperature of all the nodes. Then it calculates the slope (°C/µm) using a linear least square fit and hysteresis (µm), using deviation from the straight line, for all nodes. These are compared against a predefined set of ranges to filter out the best nodes. The range may be set based on the resolution of the temperature sensors and required accuracy for compensation. There can be thousands of nodes depending on the mesh density of the machine model. If no nodes are found then the range must be widened. The nodes are filtered for slope and hysteresis separately to maintain flexibility so that different nodes can be used for different jobs, not always both. The final node numbers satisfying both filters are then used to locate their positions in the CAD model of the relevant structure. Fig. 1 shows the function calls where comparison takes place using a specified range, in this case the range for the slope sensitivity is from 0.17 °C/µm (min) to 0.20 °C/µm (max) and 5.44 µm (min) to 8 µm (max) for the hysteresis. The first and second lines filter out node numbers for the slope sensitivity and hysteresis respectively using the range. The third line is then used to match node numbers in both arrays and obtain the matched nodes numbers. Fig. 2 shows the Matlab array editor displaying 8 nodes filtered out from the total of 4113 from the carrier (Fig. 5) structure mesh. The first column shows node number, the second column shows slope sensitivities and the third column shows the hysteresis values. These 8 nodes have shown to have the highest slope sensitivities (Fig. 3) and the lowest hysteresis values and will effectively be used to place permanent temperature sensors for use in error compensation systems. It can also be observed that nodes 738 and 739 possess the highest slope sensitivity among the other filtered nodes and a slightly higher hysteresis values relative to the other filtered nodes, however an agreement can be obtained to prioritize the selection of nodes that were located at the surface for practical installation of temperature sensors. This priority may not be the case if slope sensitivities and hysteresis values are significant at node positions inside the structure.

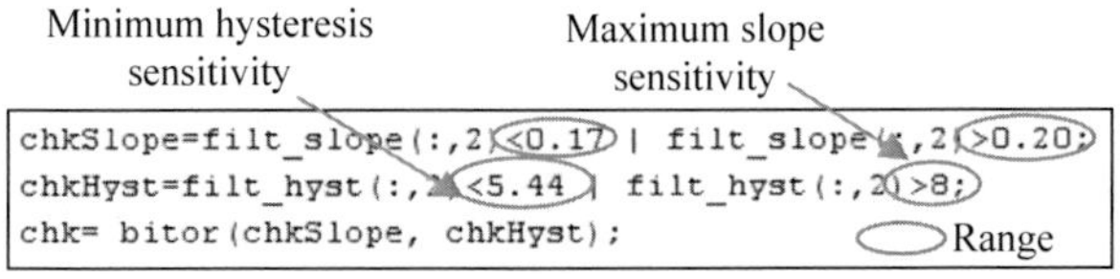

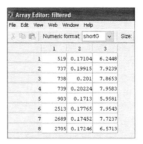

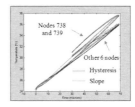

Fig. 2. Filtered nodes

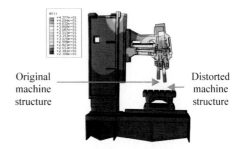

Fig. 3. Slope and hysteresis plot

4. Internal heating test – Carrier sensitivity against the Y axis and Z axis displacement

Since the carrier holds the spindle in place, it is the most affected structure as the heat from the spindle flows directly into it. Therefore, this structure was analysed to locate the temperature-displacement sensitive nodes for internal heating. Fig. 4 shows the visual representation of the simulated deformation of the machine.

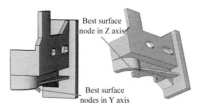

Fig. 4. Simulated visual representation of deformation of the machine due to internal heating

Fig. **5** shows the best surface nodes found using the Matlab search routine. Other visible nodes are inside the structure.

```
chkSlope=filt_slope(:,2)<0.17 | filt_slope(:,2)>0.20;
chkHyst=filt_hyst(:,2)<5.44 | filt_hyst(:,2)>8;
chk= bitor(chkSlope, chkHyst);
```

Minimum hysteresis sensitivity — Maximum slope sensitivity — Range

Fig. 1. Part of Matlab program code for assigning range

Best surface node in Z axis — Best surface nodes in Y axis

Fig. 5. Nodes sensitive to spindle heating on the carrier

5. Validations

Using the similar approach shown in section 3.1, the best identified surface node (Fig. 6) was checked which give the sensitivity of 0.20°C / µm and hysteresis of 7µm. This linear fit gives a simple model for the Y axis of $5\Delta t_{int}$ - 106.5. This was applied to measured temperature data from a sensor fitted to the machine surface close to the identified node position, with correlation to measured displacement of 84% as shown in Fig.7.

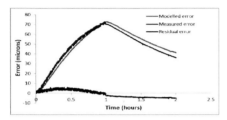

Fig. 6. Validation of the FEA model against measured error due to internal heating

5.1. Environmental sensitive nodes inside the full machine structure

Using the similar procedure the nodes sensitive to the varying environmental conditions, including different seasons, were found in the machine structure. During this preliminary work, each structure was analysed individually for efficiency to locate sensitive nodes with the higher slope and lowest hysteresis approach. Further to consider the full machine structure as one component to locate the set of sensitive nodes. Fig. 7 shows the full machine FEA model with highlighted environmental sensitive nodes individually located on components.

6. Conclusions

It has been observed that the simulation of thermal behaviour of complex machine structures using FEA can provide a solid platform for offline assessment of the machine error and model identification. FEA results from previously conducted case studies were used to locate nodes in the structural elements of a 3 axis VMC that were sensitive to temperature change and movement of the machine structure in Y and Z axes. Matlab functions were used to manipulate the extracted data from the FEA software, calculate the hysteresis and slope for any given node and filter out the best node locations by using a range of highest slope sensitivity and lowest hysteresis value. The location of the filtered nodes were analysed using the Abaqus GUI. The priority is given to surface nodes rather than the internal nodes for practical temperature sensor installation on the machine. The validation result showed the predicted sensitive nodal location correlated to better than 84%. By determining the best linear relationships, simple models are available

and compatible with the common thermal compensation methods available in most modern NC controllers.

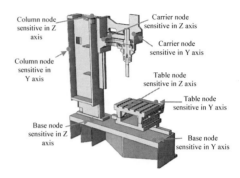

Fig. 7. Environmental sensitive nodes within the full machine

Acknowledgements: The authors gratefully acknowledge the UK's Engineering and Physical Sciences Research Council (EPSRC) funding of the Centre for Advanced Metrology under its innovative manufacturing program.

References

[1]   Bryan, J., International Status of Thermal Error Research (1990). CIRP Annals - Manufacturing Technology, 1990. 39(2): p. 645-656.

[2]   Chen, J.S. and G. Chiou, Quick testing and modeling of thermally-induced errors of CNC machine tools. International Journal of Machine Tools and Manufacture, 1995. 35(7): p. 1063-1074.

[3]   Tseng, P.C., A real-time thermal inaccuracy compensation method on a machining centre. International Journal of Advanced Manufacturing Technology, 1997. 13(3): p. 182-190.

[4]   Yang, S., J. Yuan, and J. Ni, The improvement of thermal error modeling and compensation on machine tools by CMAC neural network. International Journal of Machine Tools and Manufacture, 1996. 36(4): p. 527-537.

[5]   J Yang, J.G., et al., Testing, variable selecting and modeling of thermal errors on an INDEX-G200 turning center. International Journal of Advanced Manufacturing Technology, 2005. 26(7-8): p. 814-818.

[6]   Kang, Y., et al., Estimation of thermal deformation in machine tools using the hybrid autoregressive moving-average - Neural network model. Proceedings of the Institution of Mechanical Engineers, Part B: Journal of Engineering Manufacture, 2006. 220(8): p. 1317-1323

[7]   Debra A, Krulewich., Temperature integration model and measurement point selection for thermally induced machine tool errors. Mechatronics, 1998. 8(4): p. 395-412.

[8]   Mian, N.S., et al., Efficient thermal error prediction in a machine tool using finite element analysis. Measurement Science and Technology, 2011. 22(8): p. 085107.

[9]   Mian, N, Fletcher, S, Longstaff, A.P., Myers, A and Pislaru, C, Efficient offline thermal error modelling strategy for accurate thermal behaviour assessment of the machine tool. In: Proceedings of Computing and Engineering Annual Researchers' Conference 2009: CEARC'09. University of Huddersfield, Huddersfield, pp. 26-32. ISBN 9781862180857.

[10] ABAQUS/CAE: Hibbitt, Karlsson and Sorensen, Inc., Pawtucket, RI 02860-4847, USA.

Development of an Abbé Error compensator for NC machine tools

K. C. Fan, T. H. Wang, C. H. Wang and H. M. Chen
Department of Mechanical Engineering, National Taiwan University, Taiwan, ROC

Abstract. Abbé error is the inherent systematic error in all numerically controlled (NC) machine tools due to the fact that the scale measuring axis is not in line with the cutting axis. Any angular error of the moving stage will result in the position offset from the commanded cutting point. In this report, a new concept of multi-sensor feedback system of the NC controller is presented. A miniature three-axis angular sensor is embedded in each axis to real-time detect angular errors of the moving stage. An error compensator is developed to calculate induced volumetric errors and fed back to the NC controller. This feedback error compensation system automatically corrects the Abbé error of the machine tool. Experiments show that the volumetric accuracy can be improved significantly by employing the proposed Abbe error compensator.

Keywords: machine tools, Abbé error, multi-sensor feedback, volumetric error compensation.

1. Introduction

Abbé error is the inherent systematic error in all numerically controlled (NC) machine tools. the Abbé principle is regarded as the first principle in the design of precision positioning stages, machine tools, and measuring instruments [1]. It defines that the measuring apparatus is to be arranged in such a way that the distance to be measured is a straight-line extension of the graduation used as a scale. Bryan further made a generalized interpretation with that if the Abbé principle is not possible in the system design, either the slideway that transfer the displacement must be free of angular motion or the angular motion data must be obtained to compensate the Abbé error by software [2, 3].

Nowadays, most commercial machine tools and CMMs still cannot comply with Abbé principle because the scale axis is always parallel to the moving axis. A very popular way to improve the accuracy is to store the positioning or volumetric errors through prior measurement or calibration process and then compensate for the error budget with software, which is called the feed-forward compensation [4, 5]. It, however, can only compensate for the mean systematic errors. The angular errors are subject to the time-varied temperature changes. It is known that if the Abbé principle is not possible in the system design, one effective method is to obtain the

real time angular data and compensate for the volumetric error in real-time [6].

Techniques of non-contact angle measurement find applications in many fields. Autocollimators are commonly used optical tools for straightness calibration [7]. Some multi-degree-of-freedom (MDOF) measurement systems have been developed for measuring angular and straightness errors of precision machines but did not feed back to the controller for real-time compensation [8-10]. Laser interferometer, with its superiority in accuracy and resolution, also has been applied for angle measurement [11, 12]. By counting interference fringes the tiny displacement of objective point can be detected and converted into angle value. The resolution can be improved by techniques of phase subdivision to very fine [13]. The author's group has developed a miniature interferometer system for holographic gratings with good performance in measuring uncertainty and signal quality [14].

This paper presents a new approach for real-time Abbé error compensation on the machine tools by hardware. A low-cost three-angle sensor is developed that can embed in each axis of the machine tool. With an appropriate interface connection with the NC controller, this system can successfully compensate for the Abbé error during machine running condition. Experimental results show that the positioning errors within the working volume can be significantly reduced.

2. Abbé error in machine tool

Current NC controller in the machine tool feeds back the scale reading position, which is offset from the real commanded position, as shown in Fig. 1. The straightness error of the slideway will cause angular motion (θ) of the moving table yielding inevitable positioning error (δ) at the cutting point, which is offset from the scale reading position by L.

$$\delta = L \tan(\theta) \tag{1}$$

From the 3D point of view, the moving table has three angular errors, namely pitch, yaw, and roll. Any of these angles will induce positioning errors at the cutting points in three dimensions, as shown in Fig. 2. The corresponding errors can be expressed by the following equation.

$$\begin{bmatrix} \delta_X \\ \delta_Y \\ \delta_Z \end{bmatrix} = \begin{bmatrix} -\theta_Z \cdot L_Y + \theta_Y \cdot L_Z \\ \theta_Z \cdot L_X - \theta_X \cdot L_Z \\ -\theta_Y \cdot L_X + \theta_X \cdot L_Y \end{bmatrix} \qquad (2)$$

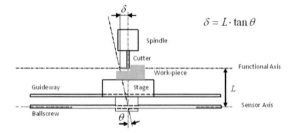

Fig. 1. Abbé error in 1D stage

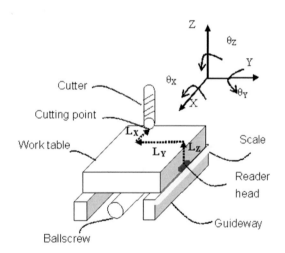

Fig. 2. Abbé error in the 3D space of a linear stage

It is known that to eliminate δ by letting L or θ zero is impossible. The only way is to compensate for the positioning errors by sensing both L and θ and correct the cutting position through the controller. The design of low cost three-angle sensor is necessary.

3. Principle of miniature angular interferometer

The optical structure of the proposed system is shown in Fig. 3. The principle is based on the classic model of Michelson interferometer. The approximately linear polarized beam from the laser diode is split by the polarization beam splitter PBS1. The P-polarized beam

passes through and the S-polarized beam is reflected to the left. With careful rotation of the PBS1 these two beams will have equal intensity. Then, the reflective mirrors M1, M2 and M3 guide these two beams to the object mirror in parallel and equal path distance. When the object mirror has an angle displacement, the change of the optical path difference will cause interference of two returned beams after joining together, which can be converted into corresponding angle value. After passing through the quarter waveplate Q1 twice, the left-arm beam will be converted into P-polarized beam and pass through PBS1. The right-arm beam has the similar feature. This design is to avoid the beam returning back to the laser diode. After passing through Q3 the left-arm beam and right-arm beam will be converted into right-circularly and left-circularly polarized beams, respectively. The NPBS divides both beams into two split beams of equal intensity. These four beams will be separated by 0–90–180–270 degrees by PBS2 and PBS3 (set fast axis to 45 degrees) and interfere with each other. Four photo detectors (PD) will convert the beam intensity to corresponding current. A proper sinusoidal signal processing circuit can reach 0.1 arc-sec resolution. Fig. 4 is the compact size of this developed yaw angle sensor.

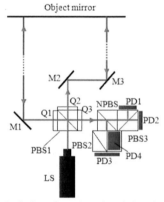

Fig. 3. Optical configuration of angle interferometer

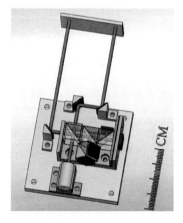

Fig. 4. Design drawing of yaw sensor

4. The Three-angle Sensor

A miniature three-angle sensor has been developed for this purpose, as shown the schematic diagram in Fig. 5. One laser diode splits the beam into two angle interferometer modules set in orthogonal directions, one for the yaw and another for the pitch measurements. The second laser diode also splits the beams to two parallel paths and each one is reflected by a corner cube reflector (CCR) and collected by a quadrant detector. The relative up and down straightness motions of two CCRs reflect the roll angle motion of the stage. Fig. 6 shows the physical size (about 160 mm x 130 mm) of the developed three-angle sensor on one axis of the machine tool. After calibration, the pitch and yaw sensors can reach ±0.3 sec accuracy for the range of ±100 sec, and for roll angle it is ±1 sec accuracy for the range of ±150 sec. These performances are good enough for machine tool use.

controller. By this way, the cutting point can be automatically adjusted in real-time with the amount of Abbé errors in space.

A test trial has been carried out on a small NC machine tool. The experimental setup is shown in Fig. 7 for the X-axis motion. A laser interferometer of HP5529 was amounted at different Z heights of the spindle head as a calibration reference. Same procedure can also be conducted for the Y-motion. Figures 8 and 9 show the comparison of positioning errors with and without the Abbé error compensation in X- and Y-axis respectively. The kinematic error of the table can be regarded as a rigid body motion. It is clearly seen that the positioning errors can be significantly reduced when the Abbé error compensation scheme is activated at any position.

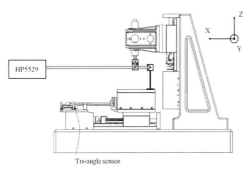

Fig. 7. Experimental setup for positioning test

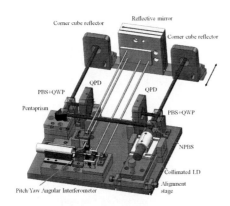

Fig. 5. The integrated structure of a three-angle sensor

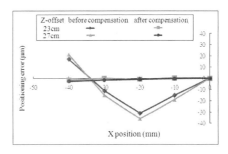

Fig. 8. Experimental results of X-positioning error calibration.

Fig. 6. Photo of a three-angle sensor on the machine tool

5. Experiments

Fig. 7 shows the schematic diagram of mounting the three-angle sensor module on each axis of the machine tool and the integration with the PC-based NC controller. A microprocessor that processes the angle signals and calculates Eq. (2) is called the Abbé error compensator, which can dynamically acquire the current three coordinate positions from the NC controller and, after processing, send the compensated command into the

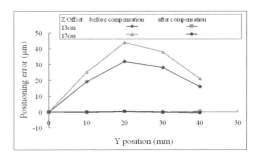

Fig. 9. Results of Y-positioning error calibration

6. Conclusions

In this paper, a developed low cost miniature three-angle sensor module is presented. It is able to embed in the machine tool structure and compensate for the positioning errors within the working zone. The developed Abbé error compensator can be equipped to any machine tool for real-time Abbé error compensation. Experimental results show the effectiveness of this system. Since this is a real-time feedback error compensation system, it can realize the goal of time-variant volumetric error compensation on any NC machine tool.

References

[1]   Abbé E, (1890) Meßapparate für physiker. Zeitschrift für Instrumentenkunde 10: 446–448

[2]   Bryan JB, (1979) The Abbé principle revisit: An updated interpretation. Precision Engineering 1: 129–132

[3]   Wright DA and Bryan JB, (1979) Letters. Precision Engineering 2: p2.

[4]   Ni J and Wu SM, (1993) An on-line measurement technique for machine volumetric error compensation. ASME J of Engineering for Industry 115: 85–92.

[5]   Okafor AC, Ertekin YM, (2000) Derivation of machine tool error models and error compensation procedure for three axes vertical machining center using rigid body kinematics. International J of machine Tools & Manufacture 40: 1199–1213.

[6]   Huang PH and Ni J, (1995) On-line error compensation of coordinate measuring machines. International J of Machine Tools 35: 725–738.

[7]   Yoder PR, Schlesinger JE, and Chickvary JL, (1975) Active annular beam laser autocollimator system. Applied Optics 14: 1890–1895.

[8]   Fan KC, Chen MJ and Huang WM, (1998) A six-degree-of-freedom measurement system for the motion accuracy of linear stages. Int. J. of Machine Tools & Manufacture 38: 155–164.

[9]   Liu CH, Hsu CC, Jywe WY and Hsu TH, (2005) Development of a laser-based high-precision six-degrees-of freedom motion errors measuring system for linear stage. Review of Scientific Instruments 76: 055110-1–055110-6.

[10]   Gao W, Arai Y, Shibuya A, Kiyono S and Park CH, (2006) Measurement of multi-degree-of-freedom error motions of a precision linear air-bearing stage. Precision Engineering 30: 96–103.

[11]   Jablonski ER, (1986) Interferometric measurement of angles. Measurement 4 (4): 148-153.

[12]   Ikram M and Hussain G, (1999) Michelson interferometer for precision angle measurement. Applied Optics 38: 113–120.

[13]   Brich KP, (1990) Optical fringe subdivision with nanometric accuracy. Precision Engineering 12: 195-198.

[14]   Cheng F and Fan KC (2011) A linear diffraction grating interferometer with high alignment tolerance and high accuracy. Applied Optics 50: 4550-4556.

A novel haptic model and environment for maxillofacial surgical operation planning and manipulation

X-T Yan[1], E Hernandez[2], V Arnez[2], E Govea[3], T Lim[4], Y Li[1], J Corney[1] and V Villela[2]

[1] Design, Manufacture & Engineering Management, University of Strathclyde, Glasgow, UK
[2] Facultad de Ingeniería, Universidad Nacional Autónoma de México, D.F. México
[3] Centro de Investigación y Estudios de Posgrado, Facultad de Ingeniería, Universidad Autónoma de San Luis Potosí, S.L.P. México
[4] Innovative Manufacturing Research Centre, Heriot-Watt University, Edinburgh, UK
arnezvictor@comunidad.unam.mx[2]

Abstract. This paper presents a practical method and a new haptic model to support manipulations of bones and their segments during the planning of a surgical operation in a virtual environment using a haptic interface. To perform an effective dental surgery it is important to have all the operation related information of the patient available beforehand in order to plan the operation and avoid any complications. A haptic interface with a virtual and accurate patient model to support the planning of bone cuts is therefore critical, useful and necessary for the surgeons. The system proposed uses DICOM images taken from a digital tomography scanner and creates a mesh model of the filtered skull, from which the jaw bone can be isolated for further use. A novel solution for cutting the bones has been developed and it uses the haptic tool to determine and define the bone-cutting plane in the bone, and this new approach creates three new meshes of the original model. Using this approach the computational power is optimized and a real time feedback can be achieved during all bone manipulations. During the movement of the mesh cutting, a novel friction profile is predefined in the haptical system to simulate the force feedback feel of different densities in the bone.

Keywords: Haptic surgical planning, mesh model, bone cutting, DICOM images, process planning, friction model.

1. Introduction

During the last decade there have been several research efforts [1-4] to achieve an interface with which the user could do practices in medicine, without the need to use a human body or animal, all inmersed in a virtual environment. In order to enhance this work a haptic property is added, having said that, many researchers have been driven to develop a system that is realistic in both feeling and visual [5-6].

The increasing research interests into tactile displays and haptic feedback systems to augment virtual reality in the last five years have led to trial haptic devices that aid the training of hand-based skills in applications such as medical training. There are commercially available devices such as the PHANToM by Sensable Technologies [7] and the Cyberglove by Immersion Inc. [8]. Research has already been conducted into simulating jaws, body parts and so on using such devices. However, these devices are designed to give haptic feedback for relatively lower resolution and large surface areas to give the impression of interacting with large volumes, and cannot provide the correct force feedback. The devices were therefore found to be unsuitable for haptic feedback. In addition, the extremely high cost of these devices makes them inaccessible to mass medical training.

This paper describes findings of research work undertaken in collaboration among authors from several institutions on the application of haptic technology and development of a haptic model for maxillofacial operation planning and training. This haptic environment has been used in conjunction with medical images obtained from computerized tomography scanners, looking to train medical doctors and in some cases to plan surgical operations.

The solutions in physics are implemented in an environment that generates a friction and stiffness on the body, both forces have been studied and reported several times, but the solutions in visual environments are mainly going in two ways, one using polygonal meshes and the other using voxels. Both types of solutions have their own difficulties when interacting with the physics solutions. When using meshes, the computational power tends to be exhaustive when detecting collisions and doing a cutting function, but the resolution can be good enough for medical purposes. In more recent work, the use of voxels is becoming more present since they need less computational power to handle cutting functions and detecting collisions, but in the counter part they are not very accurate for some particular applications.

The need to develop a method that achieves a precise cutting function with a computational power enough to maintain a haptic rendering is paramount at this stage of the study. It is because of this that the objective of the present paper is to design and develop a haptic interface that uses few computational resources, allowing having a real-time feedback and a good resolution when cutting the model.

2. Image capture of a subject's head

2.1. DICOM images processing

CT scans are commonly used to scan a patient and capture the tissues, bone void etc in many slices of images. A CT scan model is provided for this research to develop a haptic model for geometry representation of a patient head. At this first stage the images captured by the CT are counted, and these images are in an unrecognizable format for the computer without the specific software. They consist in a series of files inside a folder where each file contains the information that describes the material density of a transversal section of the patient's body: in other words, is the radiography taken of a specific plane of the patient, in this manner, by joining all the images together a 3D image of the body can be reconstructed.

To achieve this image compilation and the representation as a volumetric body, the VTK (Visualization ToolKit) libraries were used in this research. These libraries have functions that allow an easy management of the images and the graphic resources of the computer, and they are built on top of the open software OpenGL, which is the standard in graphics management for various operating systems such as Microsoft Windows.

To compile a series of images stored in the folder, it is essential to develop an appropriate algorithm to construct the three-dimensional model of the patient. The algorithm of the program to read the DICOM images (images taken from the CT) is as follows:

- Create a VTK variable to read DICOM files. Load the address where the images are allocated in the computer memory.
- Make sure that the model is in three dimensions.
- Apply a reduction factor.
- Create a variable to allocate the volume information.
- Add the properties of colour and volume opacity.
- Apply filters to the variable that contains the information of the DICOM images. These filters are: contour filter, triangle filter to generate the mesh, decimation, cleaning the mesh of non-connected points.
- Generate a STL file.

- Draw the volume on screen to visualize the result.
- End program.

Codifying the above algorithm it is possible for the program to create a window to visualize the DICOM images and a file with in STL format. The generated image represents a skull for this example, and the program adds some properties to make it look like bone shown in Fig. 1.

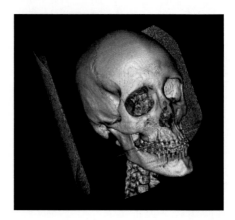

Fig. 1. DICOM images representation.

2.2. Partition the jaw part from the head model

As the research is to focus on the support of surgical maxillofacial operations, it is more efficient to extract the jaw model only in order to minimise the computational burden for haptic modelling. For this phase no filter generated by computational algorithms is needed, the reason is that an image processing based on the human anatomy is required, and to do so the human knowledge is needed. Because of this, the software MeshLab is used, which in essence reads the STL file and shows it on the screen. In addition, it gives the freedom to edit the mesh file manually: that is, it supports the manipulation of the geometry by selecting the points and faces with the mouse. This provides freedom for more accurate and smooth operations of the model if necessary.

Since the jaw is required to be isolated from the skull (Fig. 2), a computational tool could be used to save time and effort in some circumstances. This tool is a filter implemented in VTK that allows to delete from the mesh all of the points and faces that have no connection with the biggest continuous volume in the file. This tool can be executed as many times as needed during this phase.
The algorithm to apply this filter is as follows:

- Create a VTK variable to read the STL file. To load the address of the file in the variable.
- Apply the connexion filter to the previous variable.
- Create a STL file with the resulting information after the filter.
- End program.

Fig. 2. Jaw obtained.

For the purpose of demonstration of concept and for future tests, a portion (Fig. 3) of the previous mesh is selected and used, and it is the right side of the bone.

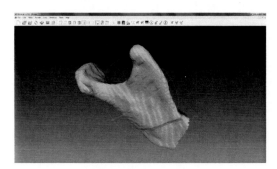

Fig. 3. Portion of the jaw.

3. Haptic model

3.1. Preparing the cut

This third phase is the main objective of the work in this project, because it involves adding the cutting function to a haptic interface. All the load of the work goes directly over C++ using VTK libraries.
The algorithm for the cut is as follows:

- Create a VTK variable to read STL files. To load the address of the STL file to use.
- Call a cutting function giving the Cartesian coordinates of the position where the cut is required.
- Inside the cutting function, use a VTK function to extract geometries from de VTK variable that contains the mesh information.
- Create a widget using the data received in the cutting function as arguments, and using a thickness proportional to the surgical tool.
- Filter the result to eliminate all the points and faces that have no continuity with the larger volume.
- Obtain the resulting mesh after retrieving the specified geometry.

- Apply the filter to remove just the points that have no connexion within the mesh.
- Display the result on screen.
- End program.

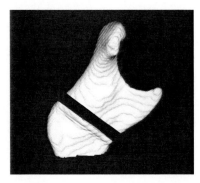

Fig. 4. Cut jawbone by the middle with a box with thickness of 5 graphic units, the faces of the box are coplanar with the planes xy, xz and yz.

Fig. 5. Cut jawbone with thickness of 4 graphic units, the faces of the box are oriented 30 degrees from the planes xy, xz and yz.

3.2. Haptic interface

To create the interface a tool was added to the omnicursor. The reasson is to obtain a point from which the cutting function will be applied and also calculate an orientation and depth of the cut. Having all this information set, the function can create a friction profile to interact with the piece of the bone that has been cut and give the feeling of being removing material from the original mesh.

Once the omnicursor form the haptic device is in place and the user clicks the primary button of the haptic tool, two main things happened after, the position and oritentation of the tool is obtained and the farthest point of the mesh in the orientation of the tool is calculated. The next thing to operate is that the mesh is cut graphically and the movement of the tool is then restricted only to push the cut piece inside the mesh, as if the user would be actually cutting it. In order to give a more realistic feeling, a novel friction profile is created to the movement of the tool.

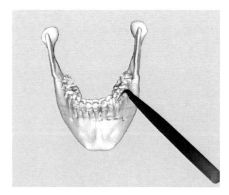

Fig. 6. Haptic interface.

3.3. Friction and stiffness profile

In this section the physical characteristics are given to the cutting body, such as the dynamic friction and the stiffness, a surgeon can adjust both in order to make the feel of operation more realistic.

The work with friction (Fig. 7) needs to define a maximum and a minimum of this variable, the first represents the bone and the second the bone marrow. The importance of these two friction values is that they are not the friction felt when moving the omnicursor in the surface of the mesh, but when the omnicursor is moving inside the mesh, so the feeling can be interpreted as cutting the bone.

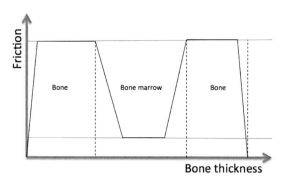

Fig. 7. Friction profile.

What the user will see is that the mesh was cut at the moment the user pushes the button and then the user is holding the cut piece and pushes it inside the bone. This makes the user feel like cutting. Once the movement is complete and the omnicursor is on the other side of the bone, the cut piece disappears and the user is ready to cut the bone in other part.

4. Conclusions

A new method to achieve real-time force feedback with a cutting function of polygonal meshes in a virtual jaw

model environment was developed. This solution has enabled the real feeling of bone friction and stiffness, even during the cut function. It can maintain stability in the haptic device since the movement is restricted and no collision detection is required. The main significance and implication of this research are obtaining a precise cut in the polygonal mesh, which is more precise than a voxel model and in real time and less requirement on computational resources, which has been the main problem when using polygonal meshes. Future work consists of generating different shapes of cut, with different tools, to create different suitable resistance to emulate the real time cutting experience. It is also planned to try the system with surgeons and evaluate the degrees of real feel and to create a better cutting algorithm in order to obtain a sharper mesh after applying the function.

Acknowledgements: We would like to thank CONACyT in Mexico for the support given in the form of scholarship during the work of this project. Another thanks for Universidad Nacional Autónoma de México in Mexico City, Mexico, the Universidad Autónoma de San Luis Potosí in S.L.P., Mexico and the University of Strathclyde in Glasgow, UK.

References

[1] Xia J.J., Phillips C.V., Gateno J., Teichgraeber J.F., Christensen A.M., Gliddon M.J., Lemoine J.J., Liebschner M.A.K., Cost-Effectiveness Analysis for Computer-Aided Surgical Simulation in Complex Cranio-Maxillofacial Surgery, Journal of Oral and Maxillofacial Surgery, 64(12):1780-1784, 2006

[2] G. Moy, U. Singh, E. Tan, and R. S. Fearing, "Human psychophysics for teleaction system design," Haptics-e, The Electronic Journal of Haptics Research, vol. 1, 2000.

[3] K. McGovern, and R. Johnston, "The Role of Computer-Based Simulation for Training Surgeons," Studies in Health Technology and Informatics: Medicine Meets Virtual Reality: Health Care in the Information Age, Vol. 29, pp. 342-345, 1996.

[4] A. Crossan, S. A. Brewster, S. Reid, and D. Mellor, "Multi-Session VR Medical Training – The HOPS Simulator," British Computer Society Human Computer Interaction, London, UK, pp. 213-226, 2002.

[5] R. L. Williams, J. N. Howell, M. Srivastava, and R. R. Conaster Jr, " The Virtual Haptic Back Project," IMAGE 2003 Conference, Scottsdale, Arizona, USA, 2003.

[6] G. Burdea, G. Patounakis, and V. Popescu, "Virtual Reality-Based Training for the Diagnosis of Prostate Cancer," IEEE Transactions on Biomedical Engineering, pp. 1253-1260, 1999.

[7] http://www.sensable.com

[8] http://www.immersion.com

"LCA to go" – Environmental assessment of machine tools according to requirements of Small and Medium-sized Enterprises (SMEs) – development of the methodological concept

R. Pamminger, S. Gottschall and W. Wimmer
Institute for Engineering Design, Vienna University of Technology, Austria

Abstract. The goal of the "LCA to go" project is spreading the use of LCA across European SMEs. For the sector of machine tools a webtool will be developed to help SMEs with conducting environmental assessment. First the SMEs requirements regarding environmental assessment were gathered in a survey. Second LCA case studies and third standards and legislation were studied. Out of these three sources a simplified environmental assessment methodology is developed. This resulted in a two-step approach including the life cycle phase raw materials and use phase of the machine tool. In the first step of the methodology the Cumulative Energy Demand have to be calculated to analyse the environmental hot spots. Depending on the hot spots a detailed environmental assessment using the CED or an Energy Efficiency Index is proposed in the second step. The methodology should be kept simple but lead to useful data for environmental communication. Next steps in the project are the detailed specification of the methodolgy, the data collection and the tool development.

Keywords: Environmental assessment methodology, machine tools, SMEs

1. Introduction

Environment is one of the leading concerns of our industrialized life. The increasing interest in environmental impacts of products over the whole life cycle is reflected by the numerous standards and activities. Large sized companies have enough budget and workforce to cope with these environmental necessities, but what about SMEs? The objective of the project "LCA to go" is to develop open source webtools for SMEs to perform a sector specific life cycle based environmental assessment. It provides tailor-made solutions to integrate simplified life cycle approaches into daily business processes. Industry machines and more specifically machine tool are in focus of the "LCA to go" project next to other sectors.

2. Approach

This paper shows the development of the methodology concept for the environmental assessment of machine tools. To name an appropriate assessment method (LCA, Carbon Footprint, Energy Efficiency Index etc.) a SMEs needs assessment in form of a survey, a research for current case studies of environmental assessments of machine tools and an analysis of current and future legislation and standards have been conducted (Fig. 1).

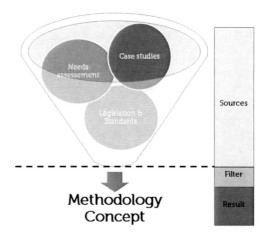

Fig. 1. Development of methodology concept [1]

2.1. Needs assessment

A survey, where 22 SMEs specialized in machine tool manufacturing responded, helped to define the needs of the European SMEs. It can be recognized that environmental issues are already anchored in SMEs but often just in form of cleaner production and theoretical knowledge around environmental assessment methods.

Only 2 companies have practiced LCA once and just 36% of respondents know that machine tool are use-intensive products (because of energy consumption). 23% think that disposal is the most problematic life cycle (Fig. 2).

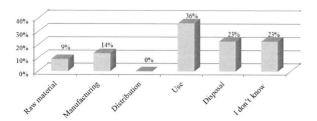

Fig. 2. Survey answers to the question "What is the most problematic life cycle phase of a machine tool?" [1]

Nevertheless the SMEs pointed out that the software tool should focus on energy aspects and it should support in fulfilling legal requirements. As environmental communication instrument a voluntary environmental label focusing on energy efficiency is of most interest. Additionally the tool should help improving product quality, product environmental performance should support in reducing manufacturing costs and helps to be prepared for future requests according to customers' demands. It should be possible to assess innovative products, without complete life cycle data sets.

2.2. Case studies

Case studies about environmental assessment of machine tools give the scientific perspective via providing environmental profiles where the most environmental aspects can be derived.

For conducting an LCA of machine tools different methods (CML, Ecoindicator 99, Cumulative Energy Demand) have been used in the case studies. Machine tools have usually a high weight (> 5 tons) and average lifetime of about 10 years running on a 2-3 shift basis. The result of the environmental assessment pointed out that the energy consumption during use phase causes 55-90% of the total environmental impact. Fig. 3 shows the environmental profile for an injection moulding machine. Additionally the energy consumption in the use phase is broken down into its main consumer, where the tempering unit is of main importance consuming nearly 50% of the total energy consumption.

Further the case studies showed that only for least intensive used machines e.g. one shift operation of a press brake, the raw material use has a relevant environmental impact with 40% of the total. This indicates that the environmental impact of a machine tool is quite sensitive according to the use scenario.

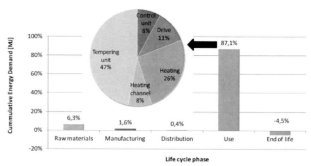

Fig. 3. Environmental Profile of an injection moulding machine [2]

2.3. Legislation and Standards

To ensure a high application rate of the developed web tool, the methodology should be in line with the actual and future legislation and standards.

Machine tools are already in focus of different environmental driven legislative initiatives. The European Commission has started a Product Group Study related to the Ecodesign Directive [3] with the aim to identify and recommend ways on how to improve the environmental performance of machine tools. This pressure from the EC resulted in two further initiatives driven by the industry. The ISO 14955 [4] and the Self-Regulation Initiative [5] concentrate on the environmental assessment of machine tool. The SRI and the ISO focus on the evaluation of energy efficiency of machine tools, where according to the functions of a component their energy consumption is allocated and the most relevant machine components can be identified. Further the German Society on Numeric Control (NCG) [6] proposed a method to measure the energy usage to calculate an Energy Efficiency Index (EEI).

All mentioned initiatives and standards are focusing on the energy consumption during the use phase of a machine tool. Additionally, also Product Category Rules for preparation of an Environmental Product Declaration are under development, where the whole life cycle is considered [7].

2.4. Methodological concept

According to some leading questions the key aspects of the methodological concept is identified (see Table 1). These questions are answered on the basis of the three defined sources needs assessment, the case studies and the legislation.

For machine tools the most relevant environmental aspect is the energy consumption during use as it is of major importance according to all three sources. The energy consumption represents up to 90% of the total environmental impact of a machine tool, already first attempts in legislation are seen and also the companies get more and more aware of this issue.

Table 1. Main questions leading to methodological key aspects (excerpt) [1]

Main questions	Sources	Key aspects for methodology
On which **environmental aspect(s)** the assessment should focus?	Case studies, Needs assessment,	Energy consumption, materials consumption
What kind of **environmental assessment** should be provided?	Case studies, Needs assessment	Cumulative Energy Demand (MJ), Energy Efficiency Index (kWh/production unit)
What kind of **environmental communication** instrument should be used?	Needs assessment, Legislation	Energy savings, EEI, CED

Additionally the raw material use is the second relevant environmental aspect to consider as they are relevant for the companies (needs assessment) and also due to some environmental assessment case studies. The relative environmental impact of the raw materials rises, especially if the machines running just a few hours a week and less energy are used during the whole life of the machine. Materials become also important, when large quantities or rare elements are addressed within machine tools or if the customer's request a material declaration. Moreover increasing costs of materials or future legislation could bring these aspects more in focus.

Therefore, the environmental assessment method as well as the developed software tool will be limited on the relevant life cycle phase raw materials and use phase. Parameters such as auxiliary materials or energy consumption during manufacturing of the machine tool are excluded as this causes minor environmental impact over the full life cycle of the machine tool. For environmental communication a voluntary environmental label, focusing on energy efficiency, is of most interest. SMEs want to inform their business clients about energy savings of machines compared to reference products.

To fulfil these key requirements a two-step assessment is proposed. In the first step a hot spot assessment with applying the Cumulative Energy Demand (CED) will be conducted. Just the impacts of indicators for energy and for materials are in focus of the assessment. In the second step a detailed assessment will be conducted, either an EEI or a more specific CED will be calculated.

2.4.1. First step: Hot spot assessment with CED

The goal in the first step is to find the environmental hot spot of the machine tool. With only limited data input the dominant environmental life cycle phase can be highlighted. According to these results a detailed assessment can be conducted in the second step.

The CED is an appropriate approach to assess impacts due to energy and material consumption leading to aggregated results in MJ. The results are easy understandable for SMEs. In the material section a rough estimation of the CED will be calculated. Just knowing

the total weight of the machine tool the CED is calculated using a general material data set, including an average material mix. Then the main focus lies in the definition of the use scenario. Therefore the operating hours of the machine tool over the full life time have to be assessed. It has to be declared if the machine is used in 1-shift, 2-shift or 3-shift operation and what is the targeted lifetime. Within one shift different machine modes (operating, stand-by, idle) have to be considered. For all this machine modes the energy consumption has to be measured according to a defined measuring standard. As result the environmental performance just focusing on raw materials and the use phase, calculated in MJ is given. If the CED shows a significant environmental impact of both raw materials (>10% of the total CED) and the use phase than the CED should be used for further environmental considerations – step two. If the CED of the materials represents less than 10% of the total energy the detailed environmental assessment should focus just on the use phase. In this case an Energy Efficiency Index (EEI) will be calculated.

2.4.2. Second step: Detailed assessment with CED or EEI

To get accurate results in the second step the environmental assessment will be conducted in more detail. If the accuracy rises to a certain level e.g. 95% of the environmental impact, the results can be used for environmental communication as well. Additionally the results should help and give advice on how to environmentally improve the machine tool.

In case of significant environmental impact of both the materials and the use phase the CED will be calculated in more detail. Therefore the specific materials have to be declared. For each material a dataset is available. The more materials are declared the higher the accuracy of the results. In the use phase the energy consumption is measured according to the energy measurement standard giving the energy consumption for all main components. This will help at the analysing stage when it comes to product improvement.

In comparison to other environmental assessment methods the advantage of the CED methodology is manageable data and time effort, which was a main criterion for SMEs using an environmental assessment method. Moreover it delivers easy to understand results, even for users which have low experience in the field of environmental assessment.

If the use phase is dominating an EEI should be calculated instead of the CED. An EEI has the purpose to assess the energy efficiency of products and to show the efficiency performance in comparison to other products. The EEI is also very much favoured as business to business communication from the SMEs, as it provides clear and short information about energy consumption of a machine tool during the use phase. In comparison to other communication instruments like the product carbon footprint (PCF) the EEI methodology is easy to calculate and to understand. Moreover the value of the EEI is the same for a specific product in every country. Considering a PCF (calculated with CO_2-equivalents) the value for one and the same product is different due to the different energy mixes.

In developing an EEI, the challenge is to get comparable results. This has to be secured by defining a suitable energy measurement standard. For example, NCG has proposed a standard where the machine has to run through a 15 min test cycle without producing a work piece. This leads to a method applicable for a broad range of machine tools, but on the other side the productivity and the energy consumption during production are not included. Another approach to define specific test pieces like it is foreseen in the ISO/CD 14955-1 Part 3. This lead also to comparable results but a test piece for each product type is needed.

In Fig. 4 a model on how to display the energy efficiency of a product is shown. The energy efficiency is defined as the relation of the energy consumption to the production unit per hour. Within this model the energy efficiency of a machine tool can be compared with other machine tools and additionally the energy class (A, B, C, etc.) can be defined. For example machines with the Best Available Technology (BAT) represent the energy class B or machines with Best Not Yet Available Technology (BNAT) are defined as class A. This energy classes can then also be used within an Energy Efficiency Label for environmental communication.

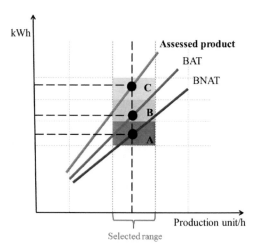

Fig. 4. Model of Energy Efficiency Index

3. Outlook

In the next step a simplified operating method is generated including compiling environmental profiles and developing Product Category Rules (PCR). To define the EEI the challenges will be the definition of the measurement standards and the data collection for reference products (BAT, BNAT). More aspects to clarify are the definition of the use scenarios and the energy measurement standard. Defining the details of the tool and the methodology will be conducted in collaboration with the later users, the SMEs.

References

[1] Pamminger R, (2012) Deliverable 1.2: Methodology concepts for LCA support of SMEs, EU-FP7 Project, LCA2go Project, GA Nr. 265096
[2] Pamminger R, Wimmer W, Winkler R (2009) Entwicklung von Kriterien zur Kommunikation der Energieeffizienz von Kunststoff verarbeitenden Maschinen. Berichte aus Energie und Umweltforschung 5/2010
[3] Schischke K, (2010) Energy-Using Product Group Analysis, Lot 5: Machine tools and related machinery, Executive Summary – Version 2
[4] ISO/CD 14955-1 (2011) Machine tools - Environmental evaluation of machine tools Part 1: Design methodology for energy-efficient machine tools
[5] CECIMO (2010) Concept Description for CECIMO's Regulatory Initiative (SRI) for Sector Specific Implementation of the Directive 2005/32/EC (EuP Directive), PE International GmbH
[6] Kaufeld M, (2011) Energieeffizienz – eine Kennziffer a' la NCG für den Anwender, NC transfer, Nr. 49
[7] PCR 44214 (2011) Product Category Rules CPC 44214: Machine-tools for drilling, boring or milling metal, open consultation version, 29.09.2011, EPD International System.

Analytical modeling of the machine tool spindle dynamics under operational conditions

O. Özşahin[1], E. Budak[2], H. N. Özgüven[1]
[1] Middle East Technical University, Department of Mechanical Engineering, Ankara, Turkey
[2] Sabanci University, Faculty of Engineering and Natural Sciences, İstanbul, Turkey

Abstract. Chatter is an important problem in machining operations, and can be avoided by utilizing stability diagrams which are generated using frequency response functions (FRF) at the tool tip. In general, tool point FRF is obtained experimentally or analytically for the idle state of the machine. However, during high speed cutting operations, gyroscopic effects and changes of contact stiffness and damping at the interfaces as well as the changes in the bearing properties may lead to variations in the tool point FRF. Thus, stability diagrams obtained using the idle state FRFs may not provide accurate predictions in such cases. Spindle, holder and tool can be modeled analytically; however variations under operational conditions must be included in order to have accurate predictions. In authors previous works Timoshenko beam model was employed and subassembly FRFs were coupled by using receptance coupling method. In this paper, extension of the model to the prediction of operational FRFs is presented. In order to include the rotational effects on the system dynamics, gyroscopic terms are added to the Timoshenko beam model. Variations of the bearing parameters are included by structural modification techniques. Thus, for various spindle speeds, and holder and tool combinations, the tool point FRFs can be predicted and used in stability diagrams.

Keywords: Machine Tool Dynamics, Chatter, Gyroscopic Effects

1. Introduction

In high speed machining operations, stability diagrams can be used to avoid chatter [1-3] and accurate tool point FRFs are needed for determination of the stability diagrams which is usually obtained for the idle state of the machine tool [4-6]. Inconsistent results, on the other hand, are frequently observed between the actual and the predicted stability especially at high spindle speeds which can be attributed to the changes of the dynamic properties of the structures during cutting. With the development of the noncontact measurement devices such as Laser Doppler Vibrometer (LDV), variation of the machining center dynamics during cutting operations has been investigated experimentally and significant deviations have been observed [7-8].

At high speeds gyroscopic moments, centrifugal forces and thermal expansions cause variations in machine dynamics. In addition to the structural variations due to the rotational effects, bearing properties are also affected by the gyroscopic moments and centrifugal forces [9-11]. In order to analyze these effects Finite Element Modeling (FEM) has been used [12-13].

In this paper, a complete model for a machining center under operational conditions is presented by extending the previously developed analytical spindle-holder-tool assembly dynamics model by the authors [5], including the variations under operational conditions. The spindle, holder and tool subassemblies of the machining center are modeled analytically by using the Timoshenko beam model including gyroscopic effects. The subassembly FRFs are coupled using receptance coupling method with the contact parameters at the spindle – holder and holder – tool interfaces. In addition to the structural dynamics, bearing properties are also added to the system with structural modification techniques. Since bearing properties mainly affect the spindle modes, and these properties vary during cutting, speed dependent bearing properties are adapted to the model. Finally speed dependent tool point FRF, and thus stability diagrams are obtained, and variations of the chatter stability under operational conditions is investigated.

2. Model development

2.1 Component modeling

The Timeshenko beam model was used for modeling of the spindle-holder-tool dynamics by Erturk et al. [6]. In order to include the gyroscopic effects rotary inertia should be included in the beam model. Therefore, Euler-Bernoulli beam model cannot be used for modeling a rotating structure. Furthermore, for low slenderness ratios, shear deformation becomes important at high frequencies. The Rayleigh beam model which includes rotary inertia effects but neglects shear deformation does not provide accurate results. Therefore, for accurate modeling of the system, the Timoshenko beam model is used.

Equation of motion for the rotating Timoshenko beam can be written as follows:

$$EI_x \frac{\partial^4 u_y}{\partial z^4} + \rho A \frac{\partial^2 u_y}{\partial t^2} - \rho I_y \left(1 + \frac{E}{kG}\right) \frac{\partial^4 u_y}{\partial z^2 \partial t^2}$$
$$+ \frac{\rho^2 I}{kG} \frac{\partial^4 u_y}{\partial t^4} + 2\rho I_y \Omega \left(\frac{\partial^2}{\partial z^2} \left(\frac{\partial u_x}{\partial t} \right) - \frac{\rho}{kG} \frac{\partial^3 u_x}{\partial t^3} \right) = 0 \quad (1)$$

where ρ is the density, A is the cross sectional area, I is the area moment of inertia of the beam cross section about neutral axis, G is the shear modulus, k is the shear coefficient and Ω is the spin speed of the beam.

As seen from equation 1, due to the gyroscopic effects, motions in two orthogonal planes are coupled. Therefore, classical solution methods cannot be applied for the rotational Timoshenko beam equations. However, since the element is axially symmetric, it is known that the mode shapes of the beam in two orthogonal planes will be related to each other by the following relations:

$$U_x(z) = iU_y(z) \quad U_x(z) = -iU_y(z) \quad (2)$$

Modes given by equations 2 correspond to the forward and backward modes in rotor dynamics, respectively. For the harmonic forcing case and axially symmetric geometry, linear displacements in two orthogonal planes can be decoupled and equation for the backward and forward motions in each orthogonal plane can be obtained. Thus, the solution procedure given by Aristizabal [14] can be applied to the Timoshenko beam model with the additional gyroscopic terms, and the mode shapes can be determined for the free–free boundary conditions. Finally, since the rotating Timoshenko beam model is a non-self adjoint system, by using the right and adjoint left eigenvectors, biorthonormality can be applied [15-17] and the receptance functions of the beam element can be obtained as follows:

$$H_{ij}(\omega) = \sum_0^\infty \left(\frac{U_{y_r}(x_i)\bar{U}^a_{y_r}(x_j)F(t)}{i\omega - \lambda_r} + \frac{\bar{U}_{y_r}(x_i)U^a_{y_r}(x_j)F(t)}{i\omega - \bar{\lambda}_r} \right) \quad (3)$$

In order to check the accuracy of the proposed method, the analytical predictions are compared with the FEM results. End point FRF of a 1 m long beam with 60 mm diameter is determined using ANSYS and compared with the analytical solution in Fig. 1 which shows very good agreement.

2.2 Receptance coupling

Spindle-holder-tool assembly dynamics can be modeled using the receptance coupling method where the subassembly components are modeled using the Timoshenko beam model presented in the previous section along with the contact parameters at the spindle-holder and holder - tool interfaces [5].

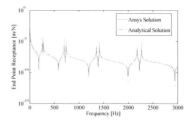

Fig.1. End Point Receptance of a 1m long beam obtained by Ansys and proposed analytical model

Front and rear bearings are also added to the model using structural modification techniques [18]. The coupling procedure is shown in Fig. 2.

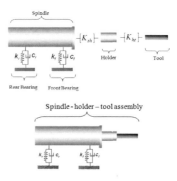

Fig. 2. Spindle Holder Tool assembly

In Fig. 2, k_f and c_f are the linear displacement – to – force stiffness and damping values of the front bearing, respectively; k_r and c_r are linear displacement – to – force stiffness and damping values of the rear bearing, respectively. $[K_{sh}]$ and $[K_{ht}]$ are the contact parameters at the spindle – holder and holder – tool interface, respectively.

2.3 Contact parameters

In dynamic modeling of the spindle–holder–tool assembly the contact parameters play a crucial role [6]. Since there is no theoretical model for obtaining these parameters, Orkun et al. [19] proposed an experimental identification procedure. The identified parameters were used to construct an artificial neural network [20], so that for different spindle, holder and tool combinations, contact parameters can be predicted.

2.4 Bearing parameters

Bearing properties mainly affect the spindle modes of a machine center [7]. During cutting, centrifugal and gyroscopic forces acting on bearings may lead to decrease in their stiffness [9]. In order to determine these variations of bearing properties, Li and Shin [11] proposed a thermo mechanical model. In a recent study, Orkun et al. [21] used spectral measurement techniques to identify the bearing properties during cutting. Their results showed that bearing properties change significantly during cutting. Thus, variation in the bearing dynamics should also be considered for accurate predictions of FRFs under operational conditions.

3. Case Studies and results

In order to investigate the variation of the tool point FRF under operational conditions several cases are presented in this section.

3.1 Effect of gyroscopic forces

A spindle-holder-tool assembly given by the authors [6] is modeled with the proposed Timoshenko beam model for 30 000 rpm spindle speed. The subassembly FRFs are coupled with receptance coupling method using the translational and rotational stiffness at the holder-tool interface as 2.5×10^7 N/m and 1.5×10^6 N/rad, respectively. The stiffness for the front and rear bearings are taken as 7.5×10^5 N/m and 2.5×10^6 N/m, respectively. The calculated tool point FRF presented in Fig. 3 shows that the spindle modes located at 64 Hz and 190 Hz are not affected by the gyroscopic terms whereas there is a small variation in the tool modes.

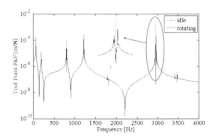

Fig. 3. Tool point FRFs for idle state and at 30000 rpm.

In order to investigate the effect of the contact parameters on the tool point FRF variation, the translational and rotational stiffness at the holder–tool interface is increased to 7.5×10^7 N/m and 7.5×10^6 N/rad, respectively. The predicted tool point FRFs at 30000 rpm given in Fig. 4 which indicates that more stiff connection at the holder – tool interface causes separation of the backward and forward modes at the third and fourth tool modes and the gyroscopic effects become more crucial. The contact parameters also affect the tool modes. Thus, accurate identification of the contact parameters becomes an important key point in the prediction of the tool point FRF in operation.

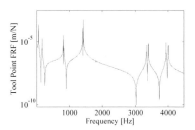

Fig. 4. Tool point FRFs at 30 000 rpm for stiffer connection

3.2 Effect of variations in the bearing properties

The variation of the bearing parameters during operation were identified using the milling force and vibration signals at different speeds [21]. In order to investigate the effects of bearing parameter changes during operation, the tool point FRF is calculated using the bearing properties the idle (5×10^7 N/m and 8×10^7 N/m for the front and rear bearings, respectively), and operating conditions. Assuming similar speed dependent behavior given in references [11, 21], the bearing properties are updated for the rotating case as 4×10^7 N/m and 6×10^7 N/m for the front and rear bearings, respectively 5000 rpm. The FRFs for both cases are shown in Fig. 5. As seen from Fig. 5, with the updated bearing properties even at a moderate speed of 5000 rpm, the variation of the bearing properties causes significant changes in the spindle modes.

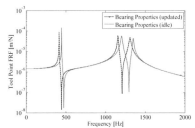

Fig. 5. Tool point FRF variation due to the variation of the bearing properties.

4. Conclusion

In this paper, a complete model for the dynamics of spindle-holder-tool assembly on machining centers under operational conditions is presented. This is done by extending the model developed by the authors for the idle state of the machine to the operational conditions. First a new solution procedure for the rotating Timoshenko beam is proposed and used for the modeling of spindle, holder and tool subassemblies. Obtained subassembly FRFs are coupled using the receptance coupling method with the contact parameters, and the bearing properties are added using the structural modification techniques. Finally, the tool point FRF is obtained for the operational conditions.

Using the analytical modeling approach presented the effects of the gyroscopic moments, contact parameters and bearing parameter variations on the tool point FRF are investigated. Results show that, variations in the structural dynamics due to the gyroscopic moment have negligible effects on the tool point FRF for the case studied. However when the contact parameters of the interface dynamics change, gyroscopic effects become more important. In addition, the effect of the speed dependent bearing parameters on the tool point FRF is investigated and it is observed that the variation in the bearing properties causes significant changes in the spindle modes, and thus in the stability diagrams. Therefore, accurate modeling of the speed dependent bearing properties plays a crucial role in the chatter stability prediction.

References

[1]   Tlusty, J, Polacek M, The stability of machine tools against self-excited vibrations in machining, (1963) Proceedings of the ASME International Research in Production Engineering, Pittsburgh, USA, 465-474

[2]   Altintas Y, Budak, E., (1995) Analytical prediction of stability lobes in milling, Annals of the CIRP, 44:357-362

[3]   Budak E., Altintas Y., (1998) Analytical prediction of chatter stability in milling – part I: general formulation; part II: application to common milling systems, Transactions of ASME, Journal of Dynamic Systems, Measurement, and Control, 120:22-36

[4]   Schmitz T., Donaldson R., (2000) Predicting high-speed machining dynamics by substructure analysis, Annals of the CIRP 49: 303-308

[5]   Ertürk   A., Özgüven H.N., Budak E., (2006) Analytical modeling of spindle-tool dynamics on machine tools using Timoshenko beam model and receptance coupling for the prediction of tool point FRF, International Journal of Machine Tools and Manufacture 46: 1901-1912

[6]   Ertürk   A., Özgüven H.N., Budak E., (2007) Selection of design and operational parameters in spindle-holder-tool assemblies for maximum chatter stability by using a new analytical model, International Journal of Machine Tools and Manufacture 47:1401-1409

[7]   Tatar, K., Gren, P., (2007) Measurement of milling tool vibrations during cutting using laser vibrometry, International Journal of Machine Tools and Manufacture, 48:380-387

[8]   Zaghbani, I., Songmene, V., (2009) Estimation of machine-tool dynamic parameters during machining operation through operational modal analysis, International Journal of Machine Tools and Manufacture, 49:947-957

[9]   Rivin, E., (1999) Stiffness and Damping in Mechanical Design, N.Y., USA, Marcel Dekker Inc

[10]   Stone B. J., (1982) The state of the art in the measurement of the stiffness and damping of rolling element bearings, CIRP Annals Manufacturing Technology, 31:529-538

[11]   Li, H. Shin Y. C., (2004) Analysis of bearing configuration effects on high speed spindles using an integrated dynamic thermo-mechanical spindle model, International Journal of Machine Tools & Manufacture, 44:347-364

[12]   Xiong, G. L., Yi, J. M., Zeng, C., Guo, H. K., Li, L. X., (2003) Study of gyroscopic effect of the spindle on the stability characteristics of the milling system, Journal of materials Processing Technology, 138:379-384

[13]   Movahhedy, M. R., Mosaddegh, P., (2006) Prediction of chatter in high speed milling including gyroscopic effects, International Journal of Machine Tools & Manufacture, 46: 996-1001

[14]   Aristizabal-Ochoa J.D., (2004) Timoshenko beam-column with generalized end conditions and nonclassical modes of vibration of shear beams, Journal of Engineering Mechanics 130:1151–1159

[15]   Wang W., Kirkhope J., (1994) New eigensolutions and modal analysis for gyroscopic/rotor systems Part 1, Journal of Sound and Vibration, 175(2):159-170

[16]   Lee C. W., Katz R., Ulsoy A. G., R. A. Scott, (1988) Modal analysis of a distributed parameter rotating shaft, Journal of Sound and Vibration, 122(1):119-130

[17]   Lee C. W., Jei Y. G., (1988) Modal analysis of continuous rotor-bearing systems, Journal of Sound and Vibration, 126(2):345-361

[18]   Ozguven H.N., (1990) Structural modifications using frequency response functions Mechanical Systems and Signal Processing 4 (1):53-63

[19]   Özşahin O., Ertürk A., Özgüven H. N., Budak E., (2009) A closed-form approach for identification of dynamical contact parameters in spindle-holder-tool assemblies, , International Journal of Machine Tools and Manufacture, 49:25-35

[20]   Özşahin O., Budak E, Özgüven H. N., (2008) Estimation of Dynamic Contact Parameters for Machine Tool Spindle-Holder-Tool Assemblies Using Artificial Neural Networks, Proceedings of the 3rd International Conference on Manufacturing Engineering (ICMEN), pp. 131-144

[21]   Özşahin O, Budak E, Özgüven H. N., (2011) Investigating dynamics of machine tool spindles under operational conditions, 13th CIRP Conference on modeling of machining operations

5

Machining

Optimization of cutting parameters for drilling Nickel-based alloys using statistical experimental design techniques

L. J. Zhang, T. Wagner and D. Biermann
Institute of Machining Technology (ISF), Baroper Str. 301, Technische Universität Dortmund, 44227 Dortmund, Germany

Abstract. Nickel-based alloys are frequently applied in the aerospace and power generation industries due to their excellent material properties, such as high temperature strength and high corrosion resistance. These advantageous material properties, however, result in challenges for cutting operations. Contrary to turning, where good results for the machining of nickel-based alloys have been obtained, drilling processes are less investigated until now. In general, coated cemented carbide drills have been proven to show good performances in drilling operations based on their higher strength compared to high speed steel (HSS) tools. Hence, they can be utilized with more efficient process parameters, whereby tool life will likely be reduced as a consequence of the higher loads. In order to find reasonable trade-offs between efficiency and tool life, a multi-objective optimization based on both criteria is presented in this article. The optimization of the cutting parameters is performed for drilling the popular nickel-based alloy Inconel 718. It is assisted by empirical models based on statistical experimental design techniques. By these means, the trade-off surface between process efficiency and tool wear can be approximated within a small experimental effort. In addition, the dominant mechanisms behind the tool wear for different process parameters are discussed.

Keywords: Nickel-based alloys, Inconel 718, Drilling, Optimization

1. Introduction

High temperature materials, such as nickel-based alloys, are often used to produce components in the aerospace and gas turbine industries [1]. In particular, they are appropriate for the hot sections of turbine parts due to their good tensile, fatigue, creep, and rupture strength at high temperatures up to 700 °C [2]. During machining processes, however, the low thermal conductivity and high tendency to work hardening of nickel-based alloys could make the cutting process problematic. High levels of tool wear and a poor surface finish are typical issues for the machining of these materials [3-5]. In comparison to turning processes, drilling operations are even more challenging. The cutting zone of those processes is more closed than the one of turning operations. As a consequence, high cutting temperatures result in high levels of tool wear and an inefficient chip evacuation [6].

TiAlN (or TiAlN multi-layered) coated cemented carbide drills have commonly been used to deal with nickel-based alloys because of a good cutting performance by the cemented carbide and a superior oxidation protection under high temperatures by the TiAlN (based) coating due to a protective aluminium oxide layer at the surface [7]. In general, the amorphous aluminium oxide forms on the surface of the TiAlN coating realize a higher operating temperature during the machining in comparison to most other coatings, such as TiN or TiCN. However, the range of the process parameters was still limited in the case of drilling high temperature materials with TiAlN coated tools, as the high cutting temperatures result in higher levels of tool wear compared to the one of other workpiece materials like steels or iron-based alloys [8-9]. Increasing the process parameters can enhance the cutting performance, but the tool life will be reduced correspondingly. Lower process parameters can reduce the tool wear, but increase the operating time. In this paper, the trade-off surface between these two criteria, tool life and performance, is empirically approximated. A statistical experimental design is employed to obtain the process parameters resulting in optimal trade-offs.

2. Experimental setup

Drilling experiments were carried out to build up a statistical empirical model for the optimization of the process parameters. These experiments were conducted on a 4-axis machining centre GROB BZ 600. Whereas the efficiency of the process parameters could be directly calculated by means of the material removal rate Q_w, circular arrays of holes were drilled for each experimental setup (cf. Table 1) in order to evalaluate the tool life. After each drilled hole, the tool wear at specific positions of the drills, namely the primary cutting edge, the minor cutting edge, the chisel edge and the flute, were inspected optically and recorded using a light optical microscope. An experiment was stopped as soon as an average flank

S. Hinduja and L. Li (eds.), *Proceedings of the 37th International MATADOR Conference*,
DOI: 10.1007/978-1-4471-4480-9_5, © Springer-Verlag London 2013

wear of VB = 0.25 mm or a maximum flank wear of VB_{max} = 0.5 mm was reached. The corresponding tool life volume V_w was recorded as wear criterion.

2.1. Workpiece

Inconel 718, one of the most important nickel-based alloys, was supplied as face-milled cylindrical plates with a diameter of d = 130 mm and a thickness of δ = 35 mm. The plates were solution heat-treated and have a hardness of 382 HBW 10/3000 (42 HRC). The chemical composition of the material is shown in Table 1.

Table 1. Chemical composition of the workpiece material

Ni+Co	Fe	Cr	Cb	Mo	Ti+Al
54.01	18.43	17.83	5.19	2.93	1.49

2.2. Tool

Cemented carbide drills (K10/20) having TiAlN coatings were employed. The drill diameter and the maximum cutting length were d = 8.5 mm and l_c = 47 mm, respectively. The drill point angle was designed as σ = 120°. The cutting edge is composed of a concave edge near the periphery and a straight edge near the chisel edge. For all drills considered in the experiments, the cutting edges were prepared and measured by a structured light microscope. Basic statistics of the cutting edge parameters are detailed in Table 2. The values Sα and Sγ indicate the asymmetry of the curvature. Their ratio k = 1 means a symmetrically rounded edge. The distance Δr indicates the general magnitude of the rounding. A sharper edge has a smaller Δr value.

Table 2. Cutting edge geometry of the new drilling tools

radius			k-factor			Δr		
mean	min	max	mean	min	max	mean	min	max
9	8	10,1	0,798	0,491	1,111	15,5	9,8	21,4
9	7,9	10	0,717	0,405	1,056	19,4	15,1	23,6

Sα			Sγ			chipping	
mean	min	max	mean	min	max	mean	max
20,9	11	31,6	24,3	13,5	35,5	1,5	5,1
24,9	18,5	32,3	31,1	13,6	49	1,2	4

2.3. Design of experiments

The cutting speed and the feed rate were varied from v_c = 20 m/min to v_c = 60 m/min and from f = 0.04 mm/rev to f = 0.16 mm/rev, respectively, in order to analyze a wide spectrum of tool wear levels. In order to maximize the number of factor steps, the first six experiments were designed using Latin hypercube sampling [10]. The two additional experiments were sequentially planned based on the available evaluations.

Using the criterion of the SMS-EGO [11], the new experiments aimed at improving the approximation of the trade-off surface between efficiency and tool wear. The direction of the feed rate was horizontal for a center drilling. The lubricant used was a water-based 6% emulsion supplied by an internal cooling setting with a pressure of p = 20 bar. The experimental design and the observed response values are provided in Table 3.

Table 3. Experimental designs and the corresponding observations of the considered response values

v_c [m/min]	f [mm/rev]	n [rev/min]	V_w [mm³]	Q_w [mm³/min]
20	0.15	748	29791.13	6366.79
45	0.1	1685	7944.30	9561.54
60	0.11	2246	3972.15	14019.42
35	0.16	1310	9930.38	11893.76
50	0.14	1872	1986.08	14871.73
30	0.12	1123	25818.98	7646.96
25	0.08	936	77456.95	4249.07
35	0.04	1392	39721.51	3159.56

3. Results and discussion

3.1. Tool wear

The drilling operations of Inconel 718 are more complex than the ones of steels with the same hardness. Hence, the drilling tools of all experiments suffered from high stresses, high cutting temperatures in the cutting zone, and high forces during the process. These effects result in a strong wear of the drilling tool. The empirical measurements of the tool life volume V_w were used as basis for computing an empirical model which can be used for the optimization of the tool wear. In order to not assume any functional form of the response, modern empirical models of the design and analysis of computer experiments were used [12]. In these models, no trend function and a Gaussian correlation kernel were applied. A leave-one-out cross-validation resulted in a determination coefficient of R2 = 0.85. The model could thus be used to predict the tool life volume V_w. The model of the tool life volume is shown in Fig. 1. It rapidly decreases with increasing cutting speed v_c and feed rate f. The mechanisms behind this decrease are explained in the following.

3.1.1. Flank wear
The flank wears after experiments with different cutting speeds v_c and almost the same feed rate f are shown in Fig. 2. It is visible that the higher cutting speed of v_c = 50 m/min (right) induced more heat in the cutting zone, which caused a high thermal load on the drilling tool. In conjuction with mechanical stresses in the flow zone of the tool-workpiece-interface, the flank wear of

the drilling tool for the experiment with the higher cutting speed is thus significantly higher leading to the decrease in the tool life volume. In contrast, the variation of feed rates did not have a significant influence on the flank wear.

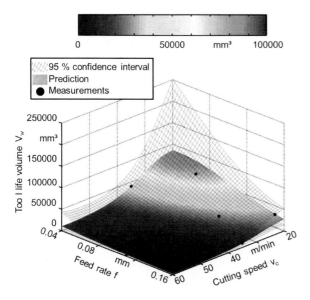

Fig. 1. Observed values of the material removal volume V_w (black dots) and model prediction of V_w (colored surface) over the considered ranges of the feed rate f and the cutting speed v_c

Chipping on the cutting edges of the drills was found in all the experiments except in the test with the lowest cutting speed of $v_c = 20$ m/min and the feed rate of $f = 0.15$ mm/rev.

Fig. 2. Flank wear comparison of the process parameters $v_c = 20$ m/min, $f = 0.15$ mm/rev (left) and $v_c = 50$ m/min, $f = 0.14$ mm/rev (right) for a cutting length of $L = 35$ mm

3.1.2. Wear of the chisel edge and the rake face
The effect of the feed rate f on the wear of the chisel edge and the rake face are analyzed in Fig. 3 and 4. For lower values of the feed rate f (left), the wear was marginal. However, it could be found that the highest feed rate $f = 0.16$ mm/rev led to a rather large plastic deformation of the chisel edge compared to the test with the lower feed rate $f = 0.04$ mm/rev. This is due to the higher feed rate, with which more material is pushed and extruded under the chisel point during one revolution. In conjuction with that, the rake face in the tests with higher feed rates was also more stressed as shown in Fig. 4 and the tool life volume V_w is decreased.

Fig. 3. Chisel edge wear comparison of the process parameters $v_c = 35$ m/min, $f = 0.04$ mm/rev (left) and $v_c = 35$ m/min, $f = 0.16$ mm/rev (right) for a cutting length of $L = 35$ mm

3.2. Multi-objective optimization

The material removal rate Q_w is calculated using the simple formula $Q_w = f * n * \pi * (d/2)^2$. It is thus linearly increasing with the feed rate f and the cutting speed v_c. As a consequence, the effects of cutting speed v_c and feed rate f on tool life volume V_w and material removal rate Q_w are opposing. It seems not to be possible to optimize both criteria at the same time. Nonetheless, a model-based multi-objective optimization of the empirical models of both quality indicators was performed. To accomplish this, a regular grid of points with steps of 1 m/min in the cutting speed v_c and steps of 0.01 mm of the feed rate f was evaluated on the empirical models. Interestingly, not all parameter settings resulted in a trade-off in which no criterion can be improved without deteriorating the other. There were possibilities to simultaneously improve both of them by varying the cutting speed v_c and the feed rate f. This is shown in Fig. 5 (right), in which the gray points depict all evaluations of the grid, whereas the red points highlight the ones which cannot be improved in both objectives. These points represent the optimal trade-off surface, from which the final parameter setting can be chosen, e. g. based on the current price of a machine hour and a drilling tool.

The corresponding parameter vectors are also shown (left). The desired trade-offs, again highlighted using red color, are obtained for values of the cutting speed from $v_c = 20$ m/min to $v_c = 45$ m/min and for $f = 0.04$ mm/rev to $f = 0.16$ mm/rev, whereby these parameters have to be increased accordingly.

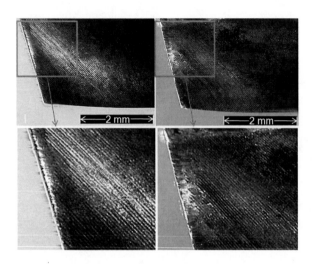

Fig. 4. Rake face wear comparison of the process parameters $v_c = 35$ m/min, $f = 0.04$ mm/rev (left) and $v_c = 35$ m/min, $f = 0.16$ mm/rev (right) for a cutting length of $L = 35$ mm

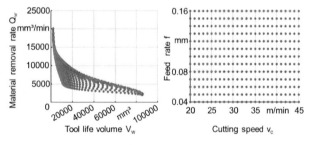

Fig. 5. Visualization of the possible and optimal trade-offs (left) and the corresponding process parameters (right)

4. Conclusion

In this paper, drilling operations of Inconel 718 using TiAlN coated cemented carbide drills were carried out. The experiments conducted with varying process parameters have shown significant effects of these parameters on the tool life volume and the material removal rate. Higher feed rates reduced the operating time, but led to more plastic deformation on the chisel edge. The same holds for higher cutting speeds for which the tool cutting edges encounter higher thermal loads and mechanical stresses. By using modern techniques of statistical experimental design and principles from multi-objective optimization, it was found that cutting speeds

from $v_c = 20$ m/min to $v_c = 45$ m/min and appropriately chosen settings of the feed rate f result in optimal trade-offs between both criteria.

Acknowledgments: The work presented in this paper was supported by the Graduate School of Energy Efficient Production and Logistics, North-Rhine Westphalia, Germany and is based on investigations of the project D5 "Synthesis and multi-objective model-based optimisation of process chains for manufacturing parts with functionally graded properties" as part of the collaborative research center SFB/TR TRR 30, which is kindly supported by the Deutsche Forschungsgemeinschaft (DFG).

References

[1]   Adam P, (1998) Fertigungsverfahren von Turboflugwerken. Birkhäuser, Basel

[2]   Loria EA, (1988) The status and prospects of alloy 718. Journal of Metal 40:36–41

[3]   Choudhury IA, El Baradie MA, (1998) Machinability of nickel-based super alloys: A general review. Journal of Materials Processing Technology 77:278–284

[4]   Thakur DG, Ramamoorthy B, Vijayaraghavan L, (2009) Machinability investigation of Inconel 718 in high-speed turning. International Journal of Advanced Manufacturing Technology 45:421–429

[5]   Arunachalam R, Mannan MA, (2000) Machinability of nickel-based high temperature alloys. Machining Science and Technology 4:127–168

[6]   Ezugwu EO, Lai CJ, (1995) Failure modes and wear mechanisms of M35 high-speed steel drills when machining Inconel 901. Journal of Materials Processing Technology 49:295–312

[7]   Sharman ARC, Amarasinghe A, Ridgway K, (2008) Tool life and surface integrity aspects when drilling and hole making in Inconel 718. Journal of Materials Processing Technology 200:424–432

[8]   Klocke F, Gerschwiler K, Frisch R, Lung D, (2006) PVD-coated tools and native ester – an advanced system for enviromentally friendly machining. Surface and Coatings Technology 201:4389–4394

[9]   Chen YC, Liao YS, (2003) Study on wear mechanisms in drilling of Inconel 718 superalloy. Journal of Materials Processing Technology 140:269–273

[10] McKay MD, Conover WJ, Beckman, RJ, (1979) A comparison of three methods for selecting values of input variables in the analysis of output from a computer code. Technometrics 21:239–245

[11] Wagner T, Emmerich MTM, Deutz A, Ponweiser W, (2010) On expected-improvement criteria for model-based multi-objective optimization. In: Schaefer R, Cotta C, Kolodziej J, Rudolph G (eds.), Proc. 11th Int'l Conf. Parallel Problem Solving from Nature (PPSN XI). Springer, Berlin

[12] Biermann D, Weinert K, Wagner T, (2008) Model-Based Optimization Revisited: Towards Real-World Processes. In: Michalewicz Z, Reynolds RG (eds.), Proc. 2008 IEEE Congress on Evolutionary Computation (CEC 2008). IEEE press, Los Alamitos, CA.

Machining performance of a graphitic SiC-Aluminium matrix composite

V. Songmene, R. D. Njoya and B. T. Nkengue
Department of Mechanical Engineering; Ecole de Technologie Supérieure (ÉTS); Université du Québec,
1100 Notre-Dame Street West, Montreal, QC, H3C 1K3, Canada

Abstract. Graphitic SiC-reinforced aluminium matrix composites were developed for high wear resistance application as replacement for cast iron, where lower part weight, high thermal conductivity and diffusivity are desired. Original graphitic metal matrix composites consisted of an aluminium matrix reinforced with both hard particles (SiC) for wear resistance and soft particles (nickel-coated graphite particles) for improved friction and machinability. Coating the graphite with nickel improves the wettability of the coated particle and thus facilitates its incorporation into the aluminium alloy, but adversely, also leads to the formation of nickel-based intermetallic precipitates (Al_3Ni), which can have adverse effects on the machinability of the composite. Also, the machining of this type of composite, as that of most metals, generates fine metallic particle that can be detrimental to the machine-tools parts' reliability and to occupational health and safety. This paper investigates the machining strategies to cost effectively machine a graphitic SiC-aluminium matrix composite made of an A356 aluminium alloy reinforced with 10vol%SiC and 4vol%Gr. The machinability is evaluated through tool wear, tool life and fine metallic particle emission. Empirical models governing the tool life and the fine metallic particle emission are developed to help determining the machining conditions leading to economical, ecological and occupational safe machining practices.

Keywords: Composite, dry milling, tool life, fine metallic particle.

1. Introduction

The strength and the physical properties of the composite materials are improved by the presence of the reinforcing particles. However, the rapid tool wear and the associated poor surface finish during machining MMCs are the leading drawbacks of the existence of the reinforcement particles as well [1]. In order to improve the machinability and the friction properties of particulate composites, aluminium MMC reinforced with both soft lubricating graphite particles and hard silicon carbide particles was developed [2,3,4,5]. Rohatgi, Bell and Stephenson [2] demonstrated the beneficial effects of a dispersion of graphite within the composite, which acts as a solid lubricant in aluminium silicon alloys. The introduction of both soft graphite particles and hard silicon carbide particles into the aluminium matrix improves its wear resistance while the graphite lowers the coefficient of friction of the resultant composite.

The first composite of this family, GrA-Ni 10S-4G, consisted of the A356 aluminium matrix reinforced with 10vol% SiC and 4vol% nickel-coated graphite [7]. By coating the graphite with nickel, the wettability of the coated particle is improved and this facilitates its incorporation into the aluminium alloy [3], but adversely, also leads to the formation of nickel-based intermetallic precipitates (Al_3Ni). Both the Al_3Ni precipitates and the SiC particles can have adverse effects on the machinability of the composite. Hard and abrasive particles such as SiC reinforcing particles and Al_3Ni intermetallics abrade the tool flank face, causing excessive wear. While studying the performance of the GrA-Ni 10S4G when machined with diamond tools, Songmene and Balazinski [8] found that diamond-coated carbide tools outperform polycrystalline diamond tools in terms of productivity while polycrystalline diamond tools produce a better surface finish on the part.

During the machining of this type of composite, fine metallic particle are emitted. These fine particles can be detrimental to the machine-tools parts' reliability and to occupational health and safety. Some researchers [9, 10-16] have investigated the emission of metallic particle during machining but very limited work was done for metal matrix composites. It was found that during machining of SiC-reinforced MMC, a tool with smooth coating emits more dust than rough coating with sharp structure [19]. The authors [19] attributed it to the wavier shear plane during machining with s of a rough surface coated tool as they also identified the friction at the shear plane to be the main source of dust emission.

The purpose of this work is to examine the performance of the GrA-Ni 10S-4G composite during dry milling process using carbide cutting tools. The machining process performance indicators used include the tool wear, the tool life and the emission of fine metallic particle.

2. Experimental procedure

2.1. Workpiece material

The workpiece used for this work is a GrA-Ni 10S.4G composite. It consists of an aluminium matrix reinforced by SiC (10-15 μm) and nickel coated graphite particles (100-150 μm). Its microstructure (Fig. 1) consists of a matrix of aluminium-silicon (grey) with a 10 vol% dispersion of SiC particles (small black phase) and 5 vol% coarser graphite particles (large black phase). Also found in the matrix are 6 vol% Ni-based intermetallic precipitates (Al_3Ni), which are formed on solidification of the alloy as a result of the dissolution of the nickel coating in molten aluminium. Graphite improves the machinability by lubricating the cutting and facilitating chip breaking [3, 5]. It also improves the composite tribological properties [4, 6]. GrA-Ni 10S 4G composite is designed for high wear resistance applications (cylinder liners and brake rotors) as replacement for grey cast iron [3].

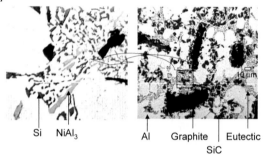

Si NiAl₃ | Al Graphite | Eutectic
SiC

Fig. 1. Description of GrA-Ni 10S 4G composite [7]

2.2. Machine-tool and cutting tools

The machining operation was performed on a CNC milling machining tool (10,000 rpm, 17 kW), using an Iscar ADK-D1.50 helimill cutter (Clearance angle : 11°; Lead angle : 90°; Rake angle : 8°; Number of teeth: 3; Tool diameter: 30.48 mm). The inserts used were Iscar ADKR 1505 PDR-HM (clearance angle: 15°; wiper clearance angle: 15°; entering angle: 90°; corner radius: 0.8 mm), TiCN-coating.

In order to increase the metallic particle emission measurement efficiency a plexiglas box is added on the table so that the machining process is carried out in a closed and limited environment (Fig. 2). Polluted air within the closed box is dragged into the dust measurement unit through a 10 mm diameter polyester tube (about 1 foot) and kept straight to minimize dust lost in the tube. The dust monitoring unit used (Fig. 2) is the TSI8520 Dust track laser aerosol monitor capable of measuring mass concentration ranging from 0.1 to 100 mg/m³ and particle size ranging from 0.1 to 10 μm. The air dragged passes through a 2.5 μm impactor (filter) before entering into the dust measurement unit (laser photometer). The system flow rate was set to 1.7 L/min.

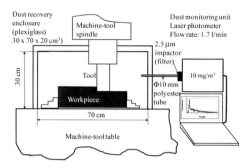

Fig. 2. Schematic representation of the experimental set-up used

3. Results and discussions

The main wear pattern observed was regular flank wear (Fig. 3). The wear of the flank face is similar to that observed in other studies [17, 18]. Flank wear is due to the abrasive action of the reinforced particles within the composite on the flank face of the tool. The harder SiC particles (2700-3500 HV), Figure 4, grind the face of the tool on similar way of a grinding wheel.

The flank wear (VB) was measured and the data plotted to show the evolution of the flank wear as a function of the volume of chip removed (Fig. 5). As expected, the tool flank wear rate increased with the increase in cutting speeds. Curves similar to those presented in Fig. 6 were constructed for other machining parameters (feed, depth of cut and tool immersion) and based on a 0.3 mm flank wear criterion, the tool life curves as a function of these machining parameters were constructed (Fig. 6)

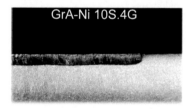

Fig. 3. Image of a worn insert after 10 minutes of cut. Cutting speed: 61 m/min, Feed rate: 0.254 mm [5]

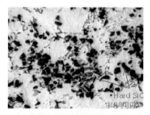

Fig. 4. SEM image of GrA-Ni 10S-4G composite (200 x) showing hard and abrasive SiC-reinforcing particles

The initial wear rate (Fig. 5) increases when the cutting speed is increased. Similar results were found for the effects of feed rate, and that of the width of cut on tool wear. This is the result of the increased normal force on the flank face of the tool.

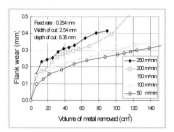

Fig. 5. Flank wear as function of volume of metal removed

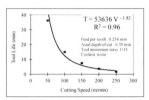

a) Effect of cutting speed

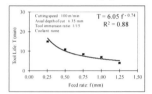

b) Effect of feed rate

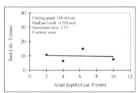

c) Effect of depth of cut

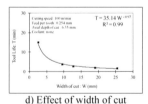

d) Effect of width of cut

Fig. 6. Tool life as function of cutting parameters (Tool life critera: 0.3 mm flank wear)

From Fig. 6, it appears that the speed, the feed rates and the width of cut have similar impact on tool life while the axial depth of cut seemed not to influence the tool life or the wear rate. It did increase the total worn area on the flank face on the tool without increasing the flank wear length. This can be explained by the fact that in side milling the increase of depth of cut plays a key role on radial load on the tool. This load impacts more in deflecting the tool than increasing the normal load, responsible for abrasion wear on the flank face of the tool. The use of higher feed, depth of cut and width of cut will increase the tool life and the tool performance since the metal removal rates will be increased and the tool will

have limited contacts with abrasive particles for a given volume of material to remove.

Figure 7 shows the progression of the total metallic particle emission (PM2.5) as a function of the cutting parameters. In general, increasing the cutting speed, the depth of cut or the tool immersion led to increased amount of dust produced. A statistical analysis of the dust generation led to the modeling of the dust generation (Dg), Eq. 1 as a function of the cutting speed (V), the feed rate (f), the width of cut (W) and the depth of cut (p). This model (Eq. 1) explained 89 % of the variability found in dust generation (Dg).

$$D_g = \frac{p^{0.645} \cdot W^{0.072}}{V^{0.107} \cdot f^{0.79}} \qquad R^2 = 89\% \quad (1)$$

The product of the depth of cut and the width of cut present in the numerator of the Eq.1 describe the chip section and the surface of the chip generated at each advance of the tool. The terms on the numerator (depth of cut (p) and width of cut (W) of the Eq.1 are related to chip geometry while those in the denominator (the cutting speed (V) and the Feed rate or the feed per tooth (f)) are related to the speed at which chip is being generated Let us define:

$C_p = d\,W$ the chip parameter. It represents the volume of chip removed per unit length of workpiece.
$S_p = f\,V$ the tool speed parameter,

From Eq. 1, it comes that the total dust emission is proportional to the chip parameter and inversely proportional to the speed parameter. A statistical analysis of the dust emission as a function of C_p and S_p has led to the two regression models that follow (Eq. 2 and 3):

$$e^{D_g} = A \times \frac{C_p^{1.31}}{S_p^{0.42}} \qquad R^2 = 83\% \quad (2)$$

$$D_g = k\,\frac{C_p^{0.44}}{S_p^{0.182}} = k\,\frac{(p \cdot W)^{0.44}}{(V \cdot f)^{0.182}} \qquad R^2 = 74\% \quad (3)$$

Where Dg (µg) is the total dust, A and k are constants depending on material ($A = 0.307$ and $k = 0.60$ for the GrA-Ni 10S 4G composite tested), C_p (mm^2) the chip parameter, S_p (mm^2/s) the speeding parameter

According to Eq. 1, 2 and 3, low depth of cut and width of cut combined with higher cutting speed and feed rates will result in very limited dust emission. This strategy is very practical as the low chip volume is combined to higher speed, thus maintaining the productivity at the same time that the dust emission is reduced. The use of high feed rate will also limit the wear of the cutting tools [5]. Higher feed rates will limit the amount of time the tool is in contact with the abrasive workpiece material.

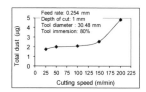

a) Effect of the cutting speed

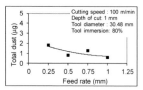

b) Effect of feed rate

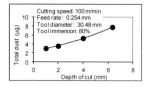

c) Effect of depth of cut

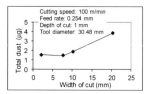

d) Effect of tool immersion (width of cut)

Fig. 7. Metallic particle emission (total dust) as function of cutting parameters

4. Conclusions

This work has shown that during the milling of graphitic SiC-reinforced aluminum matrix composite with carbide tools, the wear mechanism is pure abrasion. The wear and the tool life are determined by the speed, feed and with of cut used. The axial depth of cut does not influence the flank wear land nor the tool life. The metallic particle emission during this milling process was found to be proportional to the chip parameter (product of depth of cut and width of cut) and inversely proportional to the speed parameter (product of the feed rate and the cutting speed). Empirical models describing the relationship between the dusts produced and the cutting parameters were proposed. A strategy for reducing the dust production while maintaining the machining process productivity was recommended: use of small chip section associated with higher cutting speed and feeds. The use of higher feed rate is also favorable for the tool life as a large volume of metal will be removed before the tool wears out completely.

References

[1]   Balazinski M., Songmene V. and H. Khishawi, (2011), Analyzing the machinability of metal matrix composites, in

H. Hocheng: Machining Technology Composite Mat.: Principles &Practice, Woodhead Publishing Ltd, 394-411.

[2]   Rohatgi, PK, Bell J.A., Stephenson T.F., Aluminium-Base Metal Matrix Composite, European Patent Ep0567284A2, Inco, April 20 & October 27, 1993.

[3]   Bell J.A., Stephenson T.F., Warner A.E.M, Songmene V., 1997, Phy. Prop.of Graphitic Silicon Carbide Aluminium Metal Matrix Composites, SAE technical paper 970788.

[4]   Rohatgi P.K., Narendranath C.S., 1993, Tribological Properties of Al-Si-Gr-SiC Hybrid Composite, proc. of ASM Materials Congress, Pittsburgh, Pennsylvania, 17-21, 21-25.

[5]   Songmene V., Balazinski M.,1999, Machinability of Graphitic MMCs as a Function of Reinforcing Particles, Annals of CIRP, 48/1, 77-80

[6]   Ames W., Alpas A.T., 1993, Sliding Wear of an Al-Si Alloy Reinforced with Silicon Carbide Particles and Graphite Flakes, proc. of ASM Mat. Congress, Pittsburgh, Pennsylvania, Oct. 17-21: 27-35.

[7]   Azzi, L.; Ajersch, F., Stephenson, T.F., 2000, Rheological Characteristics of Semi-Solid GrA-Ni® Composite Alloy, 6th international conference on Semi-Solid Processing of Alloys and composites, Sept. 27-29, Turin, Italy. Pre-print

[8]   Songmene V., Balazinski M., 2001, Machining of graphitic SiC aluminium MMC with Diamond Tools, Proc. 1st Int. conf. On Progress in Innovative Manufacturing Engineering, Genoa, Italy, June 20-22, 2001, 73-76.

[9]   Sutherland JW, Kulur VN, N.C. King, (2000), An Experimental Investigation of Air Quality in Wet and Dry Turning. CIRP Ann, Manufacturing Technology 49(1):61-64

[10] Arumugam, P. U., Malshe, A. P., and Batzer, S. A., Bhat, D. G., (2002), Study of airborne dust emission and process performance during dry machining of aluminium-silicon alloy with PCD and CVD diamond coated tools NAMRC. Society of Manufacturing Engineers MR02-153, 1-8

[11] Khettabi R., Songmene V., Zaghbani I. and Masounave J (2010), Modeling of fine and ultrafine particle emission during orthogonal cutting, Mat. Eng. & Perf. 19, 776–789.

[12] Khettabi R., Songmene V., and Masounave J, (2010), Effects of cutting speeds, materials and tool rake angles on metallic particle emission during orthogonal cutting, Materials Engineering & Performance, 19, 767–775.

[13] Khettabi, R. Songmene, V. (2009), Particle emission during orthogonal and oblique cutting, Int. J. Advances Machining and Forming Operations, v. 1, N.1, Jan-June 2009, 1-11.

[14] Zaghbani, I., Songmene, V. and Khettabi, R., (2009), Fine and Ultra fine particle characterisation and Modeling In High Speed Milling of 6061-T6 Aluminium Alloy; Materials Eng. & Performance, Vol.18, I. 1, 38-49.

[15] Songmene V., Balout B. and Masounave J. (2008), Clean Machining: Experimental Investigation on Dust Formation, Int. J. of Environmentally Conscious Design and Manufacturing, vol. 14, N. 1, 1-33.

[16] Balout B., Songmene V. et Masounave J., (2007) An Experimental Study of Dust Generation during Dry Drilling of Pre-cooled and Pre-heated Workpiece Materials, Manuf. Processes, SME, vol. 9; No 1, 23-34.

[17] Tomac, N., Tonnessen, K.. and Rasch, F.O. 1992, Machinability of Particulate Aluminium Matrix Composites, Annals of CIRP, 41/1, 55-58.

[18] Weinert, K.; Köning, W.; 1993, A consideration of tool wear Mechanism when Machining Metal Matrix Composites (MMC), Annals of CIRP, 42/1, 95-98.

[19] Kremer, A.; El Mansori, (2009), Influence of nanostructured CVD diamond coatings on dust emission and machinability of SiC particle-reinforced metal matrix composite; Surface and Coatings Technology, 204, n 6-7, Dec. 2009, 1051-1055.

Surface integrity of Al6061-T6 drilled in wet, semi-wet and dry conditions

Y. Zedan, V. Songmene, R. Khettabi, J. Kouam and J. Masounave
Department of Mechanical Engineering, École de Technologie Supérieure (ÉTS), 1100 rue Notre-Dame West, Montréal, Québec, Canada.

Abstract. The quality of the surface dictates the functional performance and service-life of produced parts. The main objective of the present paper is to investigate the effect of different lubrication types and modes on the integrity of the surface produced during high speed drilling of 6061-T6 aluminum alloy. Surface integrity index investigated include surface roughness, dimentional accuracy, and microhardness of the subsurface of the drilled hole. Exit burr height is also investigated as a function of the lubrication modes. Results of this study revealed that, in general lubrication types and their interaction with cutting parameters have significant effects on the surface quality and integrity of the drilled holes. It is also found that dry and semi-wet also called mist lubrication machining can produce parts with surface quality comparable to that obtained in wet drilling conditions. Some loss of part mechanical properties have been observed from the microhardness analysis of the drilled holes, for all tested lubrication conditions.

Keywords: aluminum alloys, drilling, cutting fluid, Mist, dry drilling, surface integrity, burr height

1. Introduction

Aluminum alloys have been the most widely used structural materials in the aerospace and automotive industries for several decades. Currently, one of the most commonly used aluminum alloys is the 6000-series (Al-Mg-Si) [1]. This attributes to superior mechanical properties such as a high strengh/weight ratio, good corrosion resistance, weldability, and deformability. The functional behavior and dimensional stability of a finished component is greatly influenced by the surface intergrity induced during machining [2]. To improve the part surface finish and limit the tool wear, the machining process has been performed for long times using metal working fluids. The use of cutting fluid in machining however degrades the environment and increases the machining cost. Therefore, a special interest is given to dry machining since last few years [3]. Dry drilling is one of the most difficult machining processes because of difficulties associated with chip removal and the resulting high cutting temperatures [4]. To limit this elevated temperature, it is recommended to use minnimum quantity lubrication machining (MQL). This new technique consist of applying only a few millimeter of fluid in form of mist to the tool's cutting edge [5]. The cutting fluid was found to have a significant effect on thermal deformation and dimensional error [6].

Very limited studies so far, dealt with the influence of lubrication modes and its interaction with machining parametes on surface integrity of non-ferrous alloys such as drilled aluminum alloys. While studying the effect of lubrication (dry, mist and flooded lubrication) on drilling of AA1050 aluminum, Davim *et al.* [7] found no difference in hole surface texture between mist and flooded drilling.

The objective of this reserach study is to investigate the effects of lubrication and machining parameters on surface texture, dimentional accuracy, burr formation and microhardness of the subsurface of drilled holes performed on 6061-T6 aluminium alloy.

2. Experimental work

A set of experiments was performed to investigate the effect of different lubrication method in drilling of aluminum alloy 6061-T6 (95BHN). Drilling experiments were carried out on CNC high speed machining center (28000 rpm, 50 N.m) using 9.525 mm high speed steel twist drills. The machining conditions used are shown in Tables 1. and 2.

The surface roughness of holes was evaluated using Mitutoyo SJ 400 profilometer. The surface roughness was taken at four locations (90° apart) and repated twice at each point on the face of the machined surface. The sample from last holes drilled for each condition was sectioned in parallel to feed direction. These samples were ultrasonically cleaned in ethanol bath in order to investigate the surface texture of drilled hole using SEM for each case. The hole diameter dimension was measured in order to appreciate the component accuracy using Coordinate Measuring Machine CMM.

Table1. Lubrication modes

Type of Machining	Quantity and description of lubricant
Dry	0 m/h
Mist	Delivery pressure is 6 bar gauge; flow rate is 50 ml/h. The lubricant was a vegetable oil
Wet	Water miscible mineral oil at concentration of 5% ; flow rate of 5000 ml/h

Table 2. Machining parameters

Parameters	Condition
Material	6061-T6 (200 mm x40 mm x3 mm)
Tool	HSS twist drill-9.525 mm diameter , 118° point angle
Speed	30, 60, 120, 180, 240 , 300 m/min
Feed	0.15, 0.25, 0.35 mm/rev
Depth of cut	3 mm

The burr height was obtained using Mitutoyo Height Gauges with a sensitivity of 0.0005 in (13 μm). To measure the burr height, the gauge indictor was first placed on the datum surface at the hole exit, and then move on the top of the burr. The distance between the two measurements was thus the burr height. The measurements were done four times and then the average value was used for analysis. The burr form was also captured using optical microscopy.

Microhardness measurements of the workpiece and chip were undertaken with a Vickers indenter (Digital Microhardness Tester, FM-1) using a load of 50 g and a loading time of 10 s. Ten indentation loads were carried out on the middle of the sample to obtain the baseline value of the bulk microhardness. A series of five readings were taken at equispaced distances in the direction normal to the hole surface, and to depth aproximely 350 μm below the surface, as shown in Figs. 1.a, b and c.

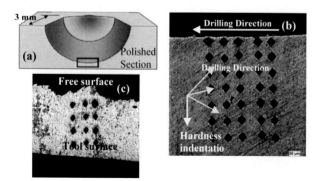

Fig. 1. (a) Schematic diagram of the quarter section of drilled hole showing the plan which is the mid-section of the hole parallel to the workpiece; (b) an optical micrograph of this section with the Vickers indentation impressions; and (c) optical micrograph showing microhardness variations beneath the chip produced.

3. Results and discussions

Fig. 2 (a) represents the vartions in the microhardness as function of depth below the ground surface. The results obtained from microhardness measurement of Al6061-T6 in dry, mist and wet application did not indicate significant sub-surface modification. Microhadness variation is gradually increasing with increase depth below the machined surface. This figure exhibts also an increasing in the micro hardness values ranging from 15-20 $HV_{0.05}$ confining to depth of cut around 50 μm from the machined surface before the reaching its stable bulk microharness values. This could be due to the fact that, depending upon the temperature of cutting process, annealing of the work-piece may occur during machining, causing softening close to the finished surface.

Fig. 2 (b) compares the average microhardness value of the produced chips with diffrent lubrication mode at the same cutting conditions. For both the dry and mist cutting, the microhardness measurments show higher values compatred to wet application. This result is mostly likely due to the higher cutting temperature induced into produced the chips during dry and mist applications. This in turn leads to higher flow stress resulting in an increase in surface deformation that ultimately led to an increase in the work hardening of the produced chips dominated by plastic deformation.

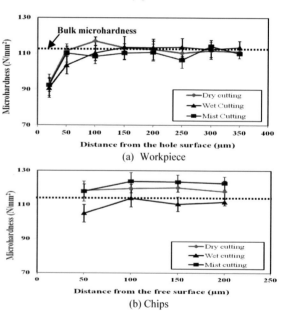

(a) Workpiece

(b) Chips

Fig. 2. a. Microhardness variations beneath the surface produced for dry, mist and wet conditions; (b) Microhardness variations beneath the chip produced for various lubrication. Machining condition: cutting speed = 120 m/min, feed rate 0.25 mm/rev). The dashed line indicates the average bulk hardness.

Cutting fluid application was found to have an effect on the average diameter of holes drilled with different cutting speeds and feed rates. Fig. 3 (a) shows the effect of the feed rate on the mean diameter with different cutting speeds during wet drilling. It was observed that the mean diameter decreases with increased feed rates. This observation is more pronounced at high cutting speeds (300 m/min), however the effect of feed rate on the mean diameter is insignificant with the low cutting speed i.e. 60 m/min. Fig.3 (b) presents a comparison of mean diameter values obtained against cutting speed with various cutting fluid application methods. It was found that the wet condition exhibits a high increase in mean diameter compared to mist and dry applications. One explanation of the larger mean diameter seen with the cutting fluid [Fig.3(b)] may be attributed to the thermal expansion and heat-removal properties of the cutting fluid. This would suggest that, without cutting fluids, enough heat is retained in the workpiece to allow expansion to occur. After machining, the workpiece cooled down, and the diameters of the holes become smaller in comparison to the holes drilled with cutting fluids [8]. The effect of drill temperature can be ignored because of the depth of cut used in this study is very small about 3 mm and the time the drill was in contact of the material very limited.

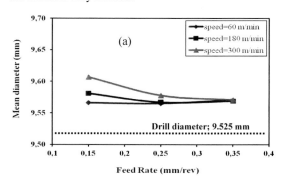

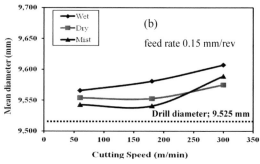

Fig.3. a. Showing the effect of feed rate on the mean diameter; (b) comparison of average values of machined hole diameters obtained with wet, dry and mist lubrication versus cutting speed

Fig. 4 shows SEM photographs of typical sectioned drilled hole surface and corresponding surface for three drilling conditions with the same cutting speed (120 m/min) and feed rate (0.25 mm/rev). As shown clearly, the dry and mist drilled holes exhibit a smoother surface while the wet conditions produce heavily deformed zones on the side-wall with significant feed marks resulting in increased roughness. The surface roughness measurements show that the Ra values of holes drilled in

the wet condition are almost one-and-half times higher than those drilled with mist and dry conditions. The adverse surface effects found with wet drilling applications may be explained by the following phenomena: post-cut (drill removal) occurrences, being which chips dragged against the side wall of the hole as the drill is being retracted, this phenomene was aleready observed by [4]. The severity of this effect was high with the external pressurized cutting fluid application. The denoting that external high pressure fluid application might have trapped the chip within the tdrill flute.

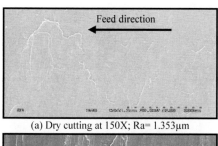

(a) Dry cutting at 150X; Ra= 1.353µm

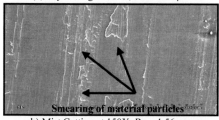

b) Mist Cutting at 150X; Ra = 1.56 µm

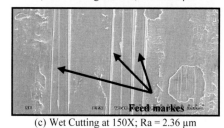

(c) Wet Cutting at 150X; Ra = 2.36 µm

Fig. 4. SEM of surface texture was observed on the holes drilled with various methods of cutting fluid application with the same cutting speed and feed rate.

Burrs generated during machining may cause part reliability problems and deterioated the performance. Fig. 5 show the two burr types created on the exit surface of the drilled holes under dry, mist and wet lubrication modes. Two types of burr form namely the transient and the uniform burrs were observed. The transient type was created for dry condition with low cutting speed and feed rate [Fig. 5(a)]. The uniforms burr with a uniform height and thickness was found to form at low cutting speed and feed rate for wet and mist lubrication modes [Figs. 5 (b) and (c)]. The most common burr type observed in this study was the uniform burr.

Fig. 6 exhibits that the burr height decreases significantly with increase in feed rate. In addition lubrication conditions have shown a great influence on drilling burr size. This is due to the intense effect of cutting fluid on temperature reduction in cutting zone area which greatly facilitate the cutting process and chip evacuation in drilling process; therefore less burr height was expected by using wet cutting condition. Fig. 5 proves also that at lower cutting speed, lubricated machining application (mist and wet) produces smaller-size burrs compared to dry machining; however the burr height is not influenced under various cutting fluid applications at high cutting speed as shown in Fig. 6 (b).

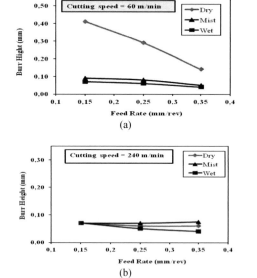

(a) Dry cutting   (b) Mist cutting

(c) Wet cutting

Fig. 5. Optical microscopy image showing typical burr formation for 6061-T6 for various cutting fluid applications

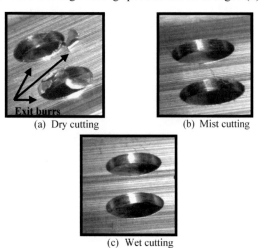

(a)

(b)

Fig. 6. A comparison of average values of burr height obtained with wet, dry and mist lubrication versus feed rate at differen cutting speed

4. Conclusions

The effects of lubrication modes on machining performance were investigated. It was found that with a proper selection of cutting parameters, mist and dry

drilling can led to advantageous performances in terms of hole surface finish and burr height, compared to wet drilling. More specifically, it can be concluded that:The results obtained from microhardness measurements indicate a loss in hardness values from 110 to 90 $HV_{0.05}$ at 20 μm beneath the surface produced for dry, mist, and wet conditions. The cutting fluid applications has an effect on the average diameter of the drilled hole, The wet condition produces an increase in mean diameter as compared to mist and dry applications. Dry and mist drilled holes exhibit smoother surfaces while wet conditions produce heavily deformed zones on the side wall, with feed marks resulting in increased roughness. The burr height decreased significantly with an increase in the cutting speed and feed rate, independently of the various methods of cutting fluid applications. However, at lower cutting speeds, lubricated machining applications (mist and wet) produced smaller size burrs as compared to dry machining.

References

[1]   Lee SH, Saito Y, Sakai T, Utsunomiya H, (2002) Microstructures and mechanical properties of 6061 aluminum alloys processed by accumulative roll-bonding. Materials science and engineering A 325: 228-235

[2]   Toh CK, Kanno S, (2004) Surface integrity effects on turned 6061 and 6061-T6 aluminum alloys. J. Materials Science 39:3497-3500

[3]   Dasch, JM, Ang, CC, Wong, CA, Cheng, YT, Weiner, AM, Lev, LC, Konca, E (2006) A comparison of five categories of carbon-based tool coatings for dry drilling of aluminum. Surface & coatings technology 200: 2970-2977

[4]   Kalidas S, DeVor RE, Kapoor SG (2001) Experimental investigation of the effect of drill coatings on hole quality under dry and wet drilling conditions. Surface & coatings technology. 148:117-128.

[5]   Weinert K, Inasaki I, Sutherland JW, Wakabayashi T. (2004) Dry machining and minimum quantity lubrication. CIRP Annals-Manufacturing Technology. 53: 511-537

[6]   Lopez de Lacalle LN, Lamikiz A, Sanchez, JA, Cabanes I (2001) Cutting conditions and tool optimization in high speed milling of aluminum alloys. Proceedings of the Institution of Mechanical Engineers, Part B: Journal of engineering manufacture. 215: 1257-1269

[7]   Davim, JP, Sreejith, PS, Gomes, R, Peixoto C (2006) Experimental studies on drilling of aluminum (AA1050) under dry, minimum quantity of lubricant, and flood-lubricated conditions. Proceedings of the institution of mechanical engineers, Part B: J.Eng. Manuf. 220: 1605-1611

[8]   Haan DM, Batzer SA, Olson WW, Sutherland JW (1997) An experimental study of cutting fluid effects in drilling. Materials processing technology. 71: 305-313.

Dry machining of aluminum alloys and air quality

A. Djebara and V. Songmene
Products, Processes, and Systems Engineering Laboratory (P2SEL), Department of mechanical Engineering
Ecole de technologie Supérieure, ÉTS, 1100 Notre-Dame West, Montréal (Qc), Canada.

Abstract. Most manufacturing and especially metal working activities generate aerosols (dry or wet) that can be harmful or degrade the environment due to the use of new processes and advanced materials such as materials containing nanoparticles. Dry machining is an environmentally conscious process, but under certain conditions, it can produce significant quantities of metallic particles. A new problem appears concerning the risk related to the exposure to metallic particles dispersed in air. To limit the metallic particles generation, it is essential to know under what conditions they are formed as well as the mechanisms underlying their formation. The main objective of this study was to evaluate the impact of machining conditions on metallic particles emission during dry machining. This work was carried out in order to minimize dust emission and thus preserve the environment and improve air quality in machine shops. Microscopy observations of particles produced during dry machining show that there are a great heterogeneity in the particles shape and a large dispersion for the size (a few nanometers to a micrometer). It is also found that during high speed machining, the emission of metallic particles decreases with the increase of cutting speed. This result is very encouraging from a practical standpoint. It is thus possible to machine parts at very high speeds, which ensures high productivity, good quality parts and limited metallic particles emission.

Keywords: Dry machining, aluminium alloys, ultrafine particles, fine particles.

1. Introduction

Indoor air quality is an emergent issue, directly related to worker exposure to polluted indoor air [1]. Machining occupies a privileged place in the shaping process because of its necessity and its large application domain, but it also represents a potential danger to health and the environment due to the aerosol generated. These aerosols may be liquid (from cutting fluids) or solid (metal particles emitted during cutting). The subject has gained a lot of interest in scientific and governmental circles in the last decade [2, 3] because of high risk associated with exposure metallic particle.

It is accepted that exposure to fine or ultrafine metallic particles can be responsible for diseases ranging from simple irritation of the lung airways to cancers [3]. Some countries have set regulations and standards for dust emission in workplace. In Germany for example, the new limit value is 0.3 mg/m^3 for respirable dust [4]. The National Institute for Occupational Safety and Health (NIOSH) recommended exposure limits of 1.5 mg/m^3 for fine particles and 0.1 mg/m^3 for ultrafine particles [5]. Air quality control in the industrial environment is usually carried out by sampling particulate matter smaller than 2.5µm (PM$_{2.5}$) or by gas receptors, in situations where the main polluters are gases [6]. If the liquid aerosols can be reduced by eliminating the use of cutting fluid, it remains solid particles emitted during metal cutting. For example, aluminum, titanium and composite materials used in aerospace can decompose into dangerous and explosive dust [7].

Djebara et al. (2010) show that the metallic particles produced when milling at high speed have different shapes [8, 9]. Kouam et al. (2011) found that friction produces more ultrafine particles than fine particles [10]. Various research studies have been conducted to identify the major factors influencing exposure to metalworking aerosols [1, 6, 8, 11-13]. A committee of risk prevention and control of the working environment of the World Health Organization held in Switzerland in 1999 wished there be research relating to the dust production process parameters, which would help assess the reliability and cost of changing systems to improve control of dust [14].

The experimental setup used in this work, the sampling method and the analyses proposed address some of those challenge. The main objective of this study is to show the effects of machining processes parameters, workpiece materials and tooling on metallic particle emission during a slot milling process. Such knowledge is needed by design and process engineers in selecting appropriate ventilation systems and controlling strategies to minimize exposure to indoor metallic particles. Only dry machining is considered because of its beneficial effects on environment and machining costs.

2. Experimental procedure

The experiments were performed on CNC milling center (Power: 50kW, Speed: 28000 rpm Torque: 50 Nm). Fine and ultrafine particles size were measured using Scanning Mobility Particle Sizer (SMPS). The experimental setup used is presented in Fig.1.

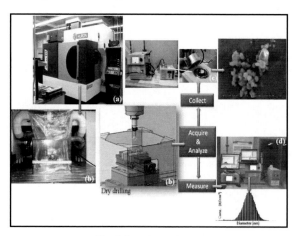

Fig. 1 Experimental setup

The workpiece materials used and their mechanical properties are presented in Table 1. The machining tests were based on a multi-level full factorial design (DOE), Table 2. The cutting tool used was a 19 mm diameter three flutes end mill cutter. The inserts has the following characteristics:

IC328: 0.5 mm nose radius, TiCN coating
IC928: 0.83 mm nose radius, TiAlN coating
IC4050: 0.5mm nose radius, TiCN/Al$_2$O$_3$/TiN coating

Table 1 Mechanical properties of the three aluminium alloys

Material	Brinell - Hardness	Yield Strength
Al 6061-T6	95 HB	275 MPa
Al 2024-T351	120 HB	325 MPa
Al 7075-T6	150 HB	505 MPa

Table 2 Factors and levels used in DOE

Factors	min	midle	max
Cutting speed (m/min)	300	750	1200
Feed rate (mm/rev)	0.03	0.165	0.3
Depth of cut (mm)	1	-	2
Workpiece materials	6061-T6	2024-T351	7075-T6
Tools (inserts)	IC328	IC908	IC4050
Cutting fluid	None		

3. Results and Discussions

3.1. Typical machining particle generation: direction, morphology and size distribution

Prior to performing the machining tests, a preliminary test was performed using an ice block to determine the direction taken by the generated aerosol. When machining a long and deep slot (Fig.2) most of the dust produced is ejected from the slot machined in the opposite direction of the feed rate. Therefore the appropriate location of the dust or metallic particle sampling is at the back side, opposed to tool path direction.

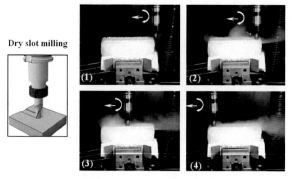

Fig. 2. Direction taken by the particles emitted during milling process: (750 m/min, 0.165 mm/rev, 19 mm tool diameter)

An analysis of the chip obtained during the machining process gives an indication on one of the possible mechanism of metallic particle generation (Fig. 3). The formation of particles during the machining is caused by different phenomena such as: macroscopic and microscopic friction, plastic deformation and chip formation mode. The friction of the chip micro-segments between themselves produces particles of micrometric and nanometric in size. Similarly, the friction at the tool rake face with the chip also produces particles. The outer surface irregularity of the chip observed (Fig. 3) shows that the separation degree between the segments coincides with the particles generation for each material. These results reflect the high segmentation number and the spacing between them generated much metallic particle. It is likely that much particle emission comes from this area.

Figure 4 shows typical particle emission results as a function of the particle diameters obtained using SMPS. All the distribution curves obtained have the same profiles (Fig.4). Table 3 summarizes the distribution of metallic particles as a function of three selected intervals. After collecting the ultrafine particles with the NAS (Nanometer Aerosol Sampler), observations with transmission electron microscope and scanning electron microscope showed ultrafine particles to be heterogeneous (Fig.5) and agglomerate (Fig.6).

This morphology depends on the nature of the material and the mechanism that produced it. Similarly, the agglomeration of ultrafine particles does not lead to spherical particles (Fig.6).

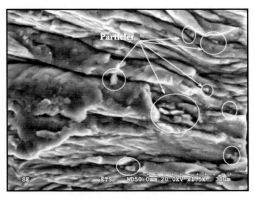

Fig. 3. SEM image of chip

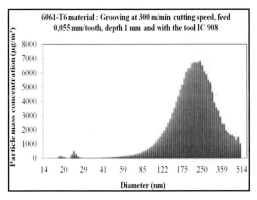

Fig. 4. Mass concentration from SMPS of 6061-T6 at cutting speed 300 m/min, feed 0,165 mm/rev

Table 3 Particle number fraction distribution

Diameters range	Percentage of total particle
Φ between 0.1 and 0.5 μm	15% of total particles generated
Φ between 0.02 and 0.1 μm	20 % of total particles generated
Φ < 0.02 μm	65 % of total particles generated

While the equivalent sphere concept used to represent particles is widely used, in some cases, it is however too simplistic. For particles with irregular shapes, it is necessary to refine the size and description in order to characterize the particle, rather than using a single parameter. This characterization should be done by attaching additional parameters that attempt to quantify

the extent to which the studied particle moves away from the sphere model [11].

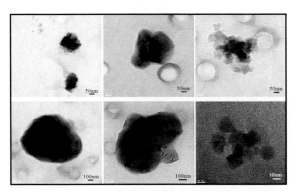

Fig. 5. TEM images of particles emitted during machining of aluminum alloy 6061-T6

Fig. 6. SEM image of particles emitted during machining of aluminum alloy 6061-T6

3.2. Statistical analysis of effects of process parameters, tooling and workpiece materials on particle emission

A series of tests was conducted to analyze the contributions of the milling process parameters (speed, feed and depth cut), workpiece materials and cutting tool (geometry and coating) on particle emission during a. dry slot milling process. A multi-level full factorial design was used in this study (Table 2). The total number of experiments performed was 162 tests.

The diagram of direct effects (Fig.7) of the total particles mass generated highlight the important factors, namely, the material and the tool. The 7075-T6 material generates less particles than the material 2024-T351 in the same cutting conditions. Reducing the tool nose radius decreases the dust emission. Factors such as depth, cutting speed and feed rate had small influences.

The Pareto diagram (Fig.8) compares the relative importance and statistical significance of main factors and interaction effects between factors. The reference line in the Pareto chart indicates the statistically significance level at a confidence level of 95%. The Pareto diagram shows the predominance of the material factor for the response mass concentration. The material and tool alone explain more than 90% of the variation found in the response. The contributions of feed rate and cutting speed are very small and not significant. Therefore, the tool and the workpiece material factors appear as those controlling the reduction in particle emission for the studied process and conditions.

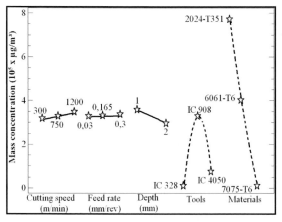

Fig. 7. Direct factors effects for the particles generation

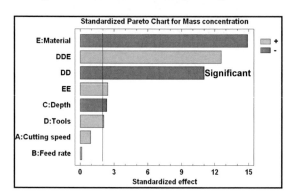

Fig. 8. Pareto diagram of mass concentration

Figure 9 presents the comparison results of the total mass concentration obtained during dry slot milling tests for 6061-T6 and 7075-T6. For 6061-T6, the particles emission are higher compared to 7075-T6 material at low cutting speed (< 400 m/min), but at high speed the particles emission are higher for 7075-T6 material. This observation can be explained by the 7075-T6 hardness is higher than the 6061-T6 one. On the other hand, their mechanical properties as well as their hardness (150 HRB for 7075-T6 and 95 HRB for 6061-T6) could have played an important role.

The response surfaces (Fig.10) embody the change in the dust generation as a function of cutting speed and feed rate. Figure 10 (c) shows that the airborne swarf of the aluminum alloy 7075-T6 is uniform across the cutting speed and feed ranges selected. Conversely, for aluminum alloy 2024-T351, there is a wide variation. Eventually, the response surface identified a region of the experimental field in which the dust emission is maximal (to be avoided). This maximum is given by the combination of a critical cutting speed and feed rate. For the 6061-T6 material, the experimental field in which the ultrafine particles emission is maximal to correspond to a feed rate less than 0.06mm/rev at low cutting speed or to

a feed rate less than 0.03mm/rev when using a high cutting speed.

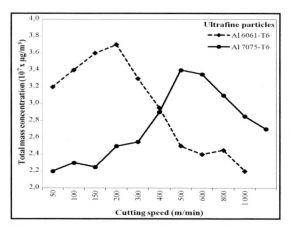

Fig. 9. Total mass concentration at different speeds obtained from SMPS at 0,165 mm/rev of feed and indexable inserts IC908

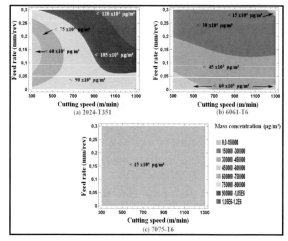

Fig. 10. Contours of estimated response surface for the mass concentration

4. Conclusion

In this work it was shown that dust emission in machining is affected by the combination of material and tool geometry. Statistical analysis has shown for a full slot milling of wrought aluminum alloys, the particle emission is controlled by the cutting tool geometry and coating, and the workpiece material rather than the cutting parameters. Using special tool geometry during dry cutting, the machining can be advantageous and sustainable. The results also confirm the existence of ranges cutting speed and feed rate to what the dust emissions is minimal.

This examination indicates that is possible to machine parts at very high speeds, which ensures high productivity, good quality parts, and without producing harmful metallic particles. Implementing this should represent a very great step in industry because manufacturers could become more productive and more competitive without compromising workers' health.

References

[1] Sutherland JW, Kulur VN, N.C. King, (2000) An Experimental Investigation of Air Quality in Wet and Dry Turning. CIRP Annals Manufact. Technology 49(1):61-64

[2] EPA US, (1995) Characterizing Air Emissions from Indoor Sources. U.S. Environmental Protection Agency, Washington, DC. EPA 600/F-95/005

[3] Ostiguy C, Roberge B, Ménard L, Endo C, (2009) A good practice guide for safe work with nanoparticles: The Quebec approach. IOP Publishing 012037

[4] Aronson RB, (1995) Why dry cutting? Manufacturing Engineering 2(1):33–36

[5] Sreejith PS and Ngoi BKA, (2000) Dry machining, machining of the future. Journal of Materials Processing Technology 101:287–291

[6] Arumugam, P. U., Malshe, A. P., and Batzer, S. A., Bhat, D. G., (2002) Study of airborne dust emission and process performance during dry machining of aluminum-silicon alloy with PCD and CVD diamond coated tools NAMRC. Society of Manufacturing Engineers MR02-153, 1-8

[7] Agrawal JP, Hodgson RD, and Corporation E, (2007) Organic chemistry of explosives: Wiley Online Library

[8] Djebara A, Songmene V, al. e, (2010) Experimental investigation on ultrafine particles emission during dry machining using statistical tools. Proceedings of the International Conference on Nanotechnology: Fundamentals and Applications 490(1):1-10

[9] Djebara A, Khettabi R, Kouam J and Songmene V, (2011) Comparison of the Capability of Peak Function in Describing Real Condensation Particle Counter Profiles. Advanced Materials Research 227:96-100

[10] Kouam J, Songmene V, Djebara A, Khettabi R (2011) Effect of Friction Testing of Metals on Particle Emission. Journal of Materials Engineering and Performance:1-8

[11] Djebara, A., (2011), Métrologie des particules ultrafines d'usinage : optimisation de la caractérisation et de la mesure, Ph.D thesis, École de technologie supérieure, ÉTS, Montréal, Canada, 200 pages (In french).

[12] Songmene, V., Balout, B., Masounave, J. (2008), Clean machining : Experimental investigation on dust formation, International Journal of Environmentally Conscious Design & Manufacturing 14(1): 1-34

[13] Khettabi, R., Songmene, V., Masounave, J. (2010). Effects of Speeds, Materials, and Tool Rake Angles on Metallic Particle Emission During Orthogonal Cutting, Journal of Materials Engineering and Performance 19 (6): 767-775

[14] WHO (1999). Global Air Quality Guidelines. Geneva. WHO.

Study on hybrid magnetic force assistant abrasive flow machining process

Ramandeep Singh and R.S. Walia
PEC University of Technology, Chandigarh, India

Abstract: Abrasive flow machining is a non-conventional machining process and was developed in late 1960's as a method to deburr, polish and radius difficult to reach surfaces such as intricate geometries by flowing a semi-liquid paste over them. Abrasion occurs wherever the medium passes through the highly restrictive passage. The key components of AFM process are the machine, tooling and abrasive medium. The AFM is capable of economically producing high surface finish. One serious limitation of this process is its low productivity in terms of rate of improvement in surface roughness. Till now limited efforts have been done towards enhancing the productivity of this process with regard to better quality of work piece surface. In recent years, hybrid-machining processes have been developed to improve the efficiency of such processes. This paper discusses magnetic force as a technique for productivity enhancement in terms of percentage improvement in surface roughness (Ra) and material removal (MR). The magnetic force is generate around the full length of the cylindrical work piece by applying DC current to the solenoid, which provides the magnetic force to the abrasive particles normal to the axis of work piece. The result shows that magnetic force assisted AFM gives better results in terms of material Removal and percentage improvement in surface roughness compared to conventional AFM..

Keywords: Abrasive Flow Machining (AFM), Magnetic Force, MFAAFM. MAFM

1. Introduction

Abrasive flow machining (AFM) is one of the latest non-conventional finishing processes, which possesses excellent capabilities for finish-machining of inaccessible regions of a component. It has been successfully employed for deburring, radiusing, and removing recast layers of precision components by extruding an abrasive laden polymer medium with very special rheological properties. High levels of surface finish and sufficiently close tolerances have been achieved for a wide range of components [6]. The polymer abrasive medium which is used in this process possesses easy flowability, better self deformability and fine abrading capability. A special fixture is generally required to create restrictive passage or to direct the medium to the desired locations in the workpiece.

The basic principle behind AFM process is to use a large number of random cutting edges with indefinite orientation and geometry for effective removal of material. The extremely thin chips produced in abrasive flow machining allow better surface finish upto 50nm, close tolerances in the range ± 0.5μm, and generation of more intricate surface [2]. In this process tooling plays very important role in finishing of material.

In order to cater to the requirement of high-accuracy and high-efficiency finishing of materials, AFM is gaining importance day by day. The AFM process has a limitation too, with regard to achieving required surface finish. With the aim to overcome the difficulty of longer cycle time, the present paper reports the findings of a hybrid process, which permits AFM to be carried out with additional centrifugal force applied onto the cutting media.

The elements required for AFM process are the machine, workpiece fixture (tooling) and media. The machine used in AFM process hydraulically clamps the work-holding fixtures between two vertically opposed media cylinder. These cylinders extrude the media back and forth through the workpiece(s).Two cylinder strokes, one from the lower cylinder and one from the upper cylinder, make up one process cycle. Both semiautomatic machines and high-production fully automated system are widely used. The extrusion pressure is controlled between 7-200bars, as well as the displacement per stroke and the number of reciprocating cycles are controlled. AFM process is an efficient method of the inner surface finishing process. Generally speaking, the control parameters of the AFM process are extrusion pressure, media flow volume, number of working cycle etc. The result of the surface quality will be harmfully affected if improperly control the parameters in the process. It is necessary that an engineer must work on accumulating experience of the test results and does his/her best to understand the control parameters of the process such as engaging in the parameter experiment of the AFM process in order to identify the dominant factors, supporting the on-sight operation as consultant, promoting the efficiency of manufacture process and reducing the variables.

A number of studies [3, 5, 7, 8, 9, 10, 12, 13, 14] show that the material removal rapidly increases during the initial cycles and there after it stabilized at higher number of cycles. This is due to the fact that higher peaks are removed during the initial process cycles when abrasive particles abrade these peaks; later the peaks become somewhat flatter and the rate of material removal and that of ΔRa reduce. Increased extrusion pressure, with all other parameters remaining constant, has significantly affected the work surface roughness [3,4,11]. Jain and Jain [9] reported that at higher pressure the improvement in material removal just tends to stabilize probably due to localized rolling of abrasion particles. The media flow rate has been reported to be a less-influential parameter in respect to material removal [2]. It has also been observed that greater the reduction ratio the more is the material removal from the work piece for a specified number of cycles. It has been noted that the fine grain size of the abrasive particles results in greater improvement of surface finish and the material removal decreases. The reason for this seems obvious as the fine grains are expected to make finer but large number of cuts on the high spots on the work surface, thus generating smoother surface. There exists the possibility of using a large range of concentration of abrasive particles in carrier media (2–12 times the weight of carrier media) [14]. However, it has also been suggested that abrasive grain to base material ratio (by weight) should vary from 4:1 to 1:4 with 1:1 as the most appropriate ratio [1]. The media viscosity and geometrical shape of the work piece also affect the flow pattern.

2. Magnetic Force Assisted Abrasive Flow Machining (MFAAFM)

The fixture employed for the MFAAFM process is shown in Fig. 1. In the current investigation the work piece was placed in between the media cylinders to create an artificial dead zone and increase the pressure required for extruding the media. The fixture was made in three parts. The work piece and the attachment were placed in central part. During the operation, the media containing the abrasive particles was made to flow from one cylinder to the other cylinder through the central hole in the work piece. The work piece was surrounded by an attachment specially designed to give necessary magnetic force through the whole length of the work piece resulting in pulling of the abrasive particles on the internal surface (normal to the axis) of the work piece. Thus the media was subjected to the extrusion pressure as well as to the additional magnetic force.

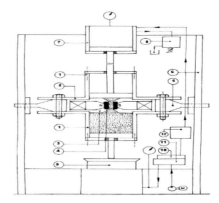

Fig. 1. Schematic illustration of the MFAAFM setup: 1 Cylinder Containing Medim; 2 Flange; 3 Nylon Fixture; 4 Workpiece; 5 Eye Bolt; 6 Hydraulic Press; 7 Auxiliary Cylinder; 8 Modular Relief Valve; 9 Piston of Hydraulic Press; 10 Directional Control Valve; 11 & 12 Manifold Blocks; 13 Electromagnet.

2.1. The Electromagnet

The electromagnet was designed and fabricated for its location around the cylindrical work-piece. It consists of two poles that are surrounded by coils arranged in such a manner as to provide the maximum magnetic field near the entire internal surface of the work-piece.

2.2. Media

Media used for present investigation consists of silicon based polymer, hydrocarbon gel and abrasive and iron particles. A Polymer-to-Gel (PGR) 1:1 has been taken. Abrasives-to-media ratio is also one. The abrasive was of Aluminium Oxide and Iron Powder, both of grit size 200, in a ratio of 3:2 have been used.

2.3. Work-piece

In the present investigation, brass as work-piece material was used. The cavity to be machined in the test specimen was prepared by drilling operation followed by boring to the required size. The test work piece is shown in Fig. 2. The internal cylindrical surface was finished by AFM process. Each work-piece was machined for a predetermined number of cycles. The work-piece was taken out from the setup and cleaned with acetone before the subsequent measurement.

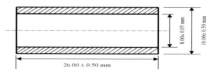

Fig. 2. Test Piece

The selected parameters and their range for the detailed experiments are shown in Table 1.

Table 1. Selected Process Parameters and their Range.

S. N.	Process Parameter	Range	Unit
1	Magnetic Flux	0.2-0.7	Tesla
2	Extrusion Pressure	5	N/mm2
3	Number of cycles	4	No.
4	Abrasive particle size	60-65	Micron
5	Media Flow Volume	290	cm3
6	Abrasive to media concentration	1:1	% by weight
7	Polymer-to-Gel Ratio	1:1	% by weight
8	Aluminium Oxide - to- Iron Powder	3:2	% by weight
9	Temperature of media	32 ± 2	°C
10	Reduction Ratio	0.90	---
11	Initial Surface Roughness	0.6-1.1	μm

3. Scheme of experiments

The magnetic flux density was selected as independent variable keeping the other parameters constant as described in table-1 and eight experiments were conducted. MR and percentage improvement in surface roughness value (ΔRa) were taken as the response parameter. The experimental specimens were chosen from a large set of specimens in such a way that selected specimens had inherent variation in their initial surface roughness values in a narrow range. It was not possible to remove this variability completely; therefore percentage improvement in surface roughness (ΔRa) has been taken as the response parameter.

4. Results and discussion

Eight experiments were conducted with magnetic flux as only variable. Figs. 3 and 4 show the simultaneous effect of magnetic flux density on MR and ΔRa respectively. From these figures it can be observed that material removal and percentage improvement in surface roughness both increase with the increase in applied magnetic flux density. The effect of magnetic field is typically more prominent beyond a magnetic flux density of 0.2T. The simultaneous increase in MR and ΔRa

indicates a unique behaviour of AFM when compared with other machining processes. In AFM, the material removal takes place first from hills or peaks of the surface profile. More material removal produces a smoother surface. This holds good until all of the high hills are removed and quite a smooth surface is produced. It is also clear from the trend of the surface obtained in Figs.3 and 4 which indicates that whereas the ΔRa to improve continues with increase in magnetic field, material removal appears to start stabilising at higher densities of magnetic field. When a strong magnetic field is applied around the workpiece, the flowing abrasive particles experience a sideways pull that causes a deflection in their path of movement to get them to impinge on to the work surface with a small angle, thereby resulting in microchipping of the surface. The particles that otherwise would have passed without striking the surface now change their path and take an active part in the abrasion process, thus causing an enhancement in material removal. It is to be mentioned here that although the mechanical pull generated by the magnetic field is small, it is sufficient to deflect the abrasive particles, which are already moving at considerable speed. Therefore it appears that, by virtue of the application of the magnetic field, more abrasive particles strike the surface. Simultaneously, some of them impinge on the surface at small angles, resulting in an increased amount of cutting wear and thereby giving rise to an overall enhancement of material removal rate.

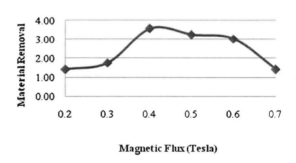

Fig. 3. Effect of magnetic flux density on MR

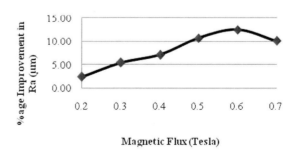

Fig. 4. Effect of magnetic flux density on (ΔRa)

It is further seen from Figs.3 and 4 that both MR and ΔRa decreases at higher values of magnetic flux density. This is a consequence of the fact that, initially, the total peaks available upon the surface of the workpiece are more. The greater the number of peaks, the more will be the material removal. However, as the surface is subjected to repeated cycles, the number of peaks and their heights continue decreasing, and hence the material removal declines after certain value of magnetic flux density. This may be explained as when a very strong magnetic field is applied around the workpiece, the abrasive particles get *ploughed* instead of *impinge* on to the work surface, thereby resulting in macro-chipping of the surface and enhancing the surface roughness.

5. Conclusions

A magnetic field has been applied around the workpiece being processed by MFAAFM and an enhanced rate of material removal and increase in percentage improvement in surface roughness were achieved. It can be concluded from study that; Magnetic field significantly affects both MR and ΔRa. The slope of the curve indicates that MR increases with magnetic field more than ΔRa. At 0.4 Tesla magnetic field densities MR is maximum and then there is marginal variation upto 0.6 Tesla. Thereafter the MR reduces sharply. ΔRa increases almost linearly upto 0.6 Tesla and thereafter after it reduces sharply.

References

[1] Siwert, D.E. (1974), "Tooling for the extrude hone process", SME International Engineering Conference, pp.302–315.

[2] L.J. Rhoades (1985), "Abrasive flow machining and its use". Proceedings of Non Traditional Machining Conference, Cincinnati, OH, December, pp. 111–120

[3] Przyklenk (1986) "Abrasive flow machining: a process for surface finishing and deburring of workpiece with a complicated shape by means of an abrasive laden medium", Advances in Non-traditional Machining, PED, ASME,22:101–110.

[4] Williams, R.E., Rajurkar, K.P. and Rhoades, L.J. (1989) "Performance Characteristics of Abrasive Flow Machining", SME Technical Paper No. FC 89-806

[5] Williams, R.E. and Rajurkar, K.P. (1992) "Stochastic modeling and analysis of abrasive flow machining", Trans. ASME, J. Engg. for Ind., 114:74–81.

[6] Loveless T.R., Williams R.E., Rajurkar K.P.(1994), "A study of the effects of abrasive-flow finishing on various machined surfaces", Journal of Materials Processing Technology 47:133-151.

[7] Shan, H.S. and Dubey, A.K. (1997). "Micro Machining by Flow of Abrasives". Proceedings 17th AIMTDR Conference, Warrangal, India, 269–275.

[8] Jain R.K. and Jain V.K. (1999), "Abrasive fine finishing processes- a review", Int. J. of Manufacturing Science and Production, 2(1):55-68

[9] Jain, R.K., Jain, V.K. and Dixit, P.M. (1999) "Modelling of material removal and surface roughness in abrasive flow machining process", Int. J. Machine Tool and Manufacture, 39:1903–1923.

[10] Rajurkar, K.P. and Kozak, J. (1999), "Hybrid Machining Process Evaluation and Development", Online, www.unl.com

[11] Singh S. ,Shan H.S., and Kumar P. (2001),"Effects of magnetic field on abrasive flow machining processes", Proc. 12th Int. DAAAM Symposium Jena, Germany, October.

[12] Walia R.S., Shan H.S., Kumar P. (2006), "Parametric optimization of centrifugal force assisted abrasive flow machining (CFAAFM) By the Taguchi Method". Journal of Materials and Manufacturing Processes, USA. Vol. 21, Issue 4,375-382.

[13] Walia R.S., Shan H.S., Kumar P. (2006b) "Multi-response optimization of centrifugal force assisted abrasive flow machining (CFAAFM) process through Taguchi method and Utility concept", Journal of Materials and Manufacturing Processes, 21:907-914.

[14] Sankar R.M., Monda, S., Ramkumar J., Jain V.K. (2009), "Experimental investigations and modelling of drill bit-guided abrasive flow finishing (DBG-AFF) process", International Journal of Advanced Manufacturing Technology, 42:678–688.

The effect of in-plane exit angle on burr minimization in face milling of medium carbon steels in dry environment

P.P. Saha, A. Das and S. Das
Department of Mechanical Engineering, Kalyani Govt. Engineering College, Kalyani- 741325, West Bengal, India

Abstract. Machining burr is a plastically deformed material adhered to the edge of the machined surface. It usually creates problem in production lines in terms of product quality and injury to the operator. So, burr prevention and its minimization are essential. An additional, non-productive deburring process is generally required for removal of burr. A number of experimental observation was performed in the past on burr formation and its minimization. The present experimental investigation has been performed on medium carbon steel (45C8) specimens to observe the extent of burr formation with different in-plane exit angles with and without an exit edge bevel. Encouraging results have been found when an exit edge bevel angle of 15° and in-plane exit angle is 60° are provided on the workpiece leading to negligible burr.

Keywords: Burr, exit edge bevel, in-plane exit angle

1. Introduction

Burrs are often observed in milling operation at the exit edge of the workpiece. Presence of burr in the finished component creates problem in assembly, work handling, etc. So, a finished component should be free from any burr. For this, an additional deburring operation may be employed to remove burr. It adds an extra cost that may be as high as 30% of the production cost. To understand the mechanism of burr formation and its minimization, several research works were carried out [1- 3].

Nakayama and Arai [1] classified machining burrs and reported that burrs might cause grove wear and accelerate burr growth. Gillespie [4, 5] observed that cost of deburring and edge finishing for precision components might constitute as much as 30% of the part cost. Chern and Dornfeld [6] did experiments on burr formation mechanism in orthogonal cutting and found out that negative shear plane tends to form when steady state chip formation stops as the tool moves to the end of the cut. Lee and Dornfeld [7] investigated micro-burr formation in machining. They studied the type and size of burr created on stainless steel 304 and also developed a control chart to minimize burr size. On the other hand, Luo et al. [8] did experimental investigation on the

mechanism of burr formation in slot milling of aluminium alloy workpiece in feed direction. They found out an increasing trend of exit burr height with the increase in exit angle, and at 90° exit angle, large burr was observed. Biermann and Heilmann [9] did experiments on improvement of workpiece quality in face milling of aluminium alloys. They noted that the modification on machining process along with the use of cutting fluid is a promising approach to avoid deburring. On the other hand, Pekelharing [10], Hashimura et al. [11] and Das et al. [12] tried to explore burr formation mechanism using finite element method, and validated experimental findings. Classification of milling burrs according to the shape, location and burr formation mechanism based on fractography was also proposed [10]. Heisel et al. [13] observed that the influence of minimum quantity lubrication had minor effect on burr formation in face milling.

Saha et al. [14], Das et al. [15, 16], Saha and Das [17-19] and Das et al. [12] did experiments on burr formation. They reported that burr formation depends on various parameters, like exit edge bevel angle, in-plane exit angle, cutting velocity and feed. They investigated the machining burr formation in milling and shaping on different steels and aluminium alloy, and observed that burr became minimum at 15° exit edge bevel angle of the workpiece. This may be due to the need of less backup support material along the bevel.

The aim of the present experimental work is to obtain minimum or negligible burr formation on medium carbon steel (45C8) workpiece under different machining conditions considered. The influence of in-plane exit angle on burr formation of 45C8 steel workpiece without edge bevel and at 15° exit edge bevel in face milling using coated carbide tool insert is experimentally investigated.

2. Experimental investigation

Experiments are done in a vertical axis CNC milling machine on medium carbon steel, with a single titanium titride (TiN) coated carbide tool insert mounted on a 54 mm diameter cutter to observe the nature of burr formation. Two set of experiments (set 1 and set 2) are carried out in dry environment. Constant cutting velocity vc = 339 m/min, depth of cut of ap = 3 mm and feed of fz of 0.1 mm/tooth are considered during experiments performed in dry condition. In-plane exit angle is varied from 30° to 150° at a step of 15° without any exit edge bevel in experiment set 1 to find out the condition for minimum burr formation.

Table 1. Experimental set up in CNC milling

Machine Tool	Vertical axis CNC milling machine Make: Bharat Fritz Werner (BFW) Ltd., Bangalore, India Model: Akshara VF30CNC, Sl. No 5081/269
Cutting Tool	Face milling cutter with a single TiN coated carbide insert Cutter specification: 490-054Q22-08M241259, Diameter: 54mm Make: Sandvik Asia Ltd., India Insert specification: 490R-08T308M-PM, Make: Sandvik Asia Ltd., India Condition and type of tool: Sharp new negative rake tool Normal clearance angle = 15°, Principal cutting edge angle = 75°
Job Material	Medium carbon steel (45C8) Size : 63 mm x 55 mm x 21 mm, Hardness: 175 HB Composition (wt%): C (0.44%), Si (0.29%), Mn (0.85), S (0.033%), P (0.028%), Fe (remainder)

In-plane exit angle is the angle between the cutting velocity vector at the exit point of the cutter leaving the workpiece and the cutter feed direction which is along the exit edge of the workpiece as considered in this work. To provide an exit edge bevel angle, the end portion of the workpiece where the cutter leaves it is beveled to have an inclination downward. Definition of in-plane exit angle and exit edge bevel angle was shown with schematic presentation in references [18, 19] of the first and corresponding author of this paper.

Next, in-plane exit angle is varied from 60° to 150° at a step of 15° with 15° exit edge bevel angle in experiment set 2. An exit edge bevel angle of 15° is chosen in this work following the observation reported earlier [12, 14, 17- 19]. The bevel is made of a height of 3 mm. After machining, the exit edge of workpiece is observed under a Mitutoyo, Japan make tool makers microscope, and the

respective height of burr is measured. Details of the experimental set up and machining conditions are shown in Table 1 and Table 2 respectively.

3. Discussion on experimental results

After conducting milling experiments in experimental set 1, burr height of the specimen is measured using a tool makers microscope. Measured burr heights are presented in Fig. 1. It is noted from Fig. 1 that burr height is increased with increase in in-plane exit angle up to 105°, and has a decreasing tendency after that. Quit low burr height is observed at 30° in-plane exit angle. Low tool engagement may be the reason for small amount of burr height (68 μm) at 30° in-plane exit angle. Similar observation was also made by Avila and Dornfeld [20]. At 105° and 120° in-plane exit angle, burr height at exit edge is observed to be more than 300 μm.

Table 2. Machining conditions

Variables	Experimental set 1	Experimental set 2
Exit edge bevel angle (°)	0	15
In-plane exit angle (°)	30, 45, 60, 75, 90, 105, 120, 135, 150	60, 75, 90, 105, 120
Cutting velocity vc = 339 m/min, Depth of cut of ap = 3 mm, Feed of fz 0.1 mm/tooth, Environment: Dry		

Microscopic views of typical burrs for in-plane exit angle of 30° and 90° with no exit edge bevel angle of the workpiece are shown in Fig. 2 (a) and (b). It is clear from the photograph that at 30° in-plane exit angle, notably less burr is there, when at 90° in-plane exit angle, distinct saw tooth type burr is present.

In experiment set 2, face milling of medium carbon steel (45C8) has been performed with an exit edge bevel angle of 15°; in-plane exit angle is varied from 60° to 120° at a step of 15° for this experiment. As beyond an in-plane exit angle of 45°, average burr height of more than 100 μm occurs, in the experiment set 2, burr height is tried to reduce substantially with 15° exit edge bevel corresponding to in-plane exit angle of 60°-120°. Other machining conditions are given in Table 2.

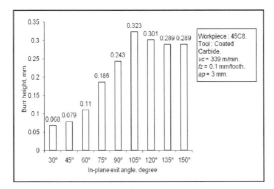

Fig. 1. Variation of average burr height with different in-plane exit angles of medium carbon steel specimens without beveled exit edge for experiment set 1

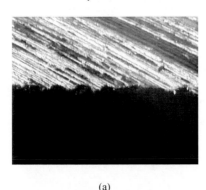

(a)

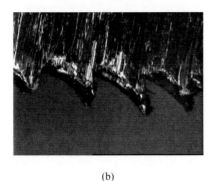

(b)

Fig. 2. Microscopic view of burrs of a face milled specimen for in-plane exit angle of (a) 30° and (b) 90° for medium carbon steel specimens with no edge bevel angle for experiment set 1 (View in the microscope is taken under x20 magnification)

Variation of average burr height with different in-plane exit angles under 15° exit edge bevel angle is presented in Fig. 3. It shows that burr height increases with an increase in in-plane exit angle up to 105° and then decreased at 120°. On the whole, burr height observed at these conditions is lesser than that observed without an exit edge bevel.

Gradual reduction in cutting force along the beveled edge results in less need of back up support, and thus reducing the height of burr. Quite low burr height of 16 μm is observed at 60° in-plane exit angle. Low tool engagement and less requirement of back up support

material along the beveled exit edge of the workpiece may be the reason behind formation of negligible burr.

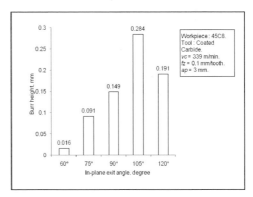

Fig. 3. Variation of burr height with different in-plane exit angles under 15° exit edge bevel of medium carbon steel specimens for experiment set 2

Figure 4 (a) and (b) show microscopic view of typical burrs for in-plane exit angle of 60° and 90° with 15° exit edge bevel of the workpiece. Presence of almost no burr is observed from Fig. 4 (a) at 60° in-plane exit angle, while saw tooth type burr with lesser frequency is there as seen in Fig. 4 (b) at 90° in-plane exit angle with the provision of an exit edge bevel angle of 15°.

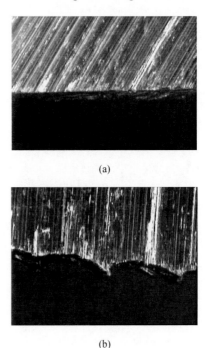

(a)

(b)

Fig. 4. Microscopic view of burrs of a face milled specimen for in-plane exit angle of (a) 60° and (b) 90° for medium carbon steel specimens with 15° edge bevel angle for experiment set 2 (View in the microscope is taken under X20 magnification)

4. Conclusions

From the experimental observation on milling burr formation at the exit edge of medium carbon steel (45C8) specimens under dry environment, following conclusions may be drawn; Low burr height of 68μm is observed using coated carbide tool at 30° in-plane exit angle with no exit edge bevel of the workpiece at the selected machining condition. When an exit edge bevel angle of 15° is provided, burr height observed at these conditions is lesser than that observed without an exit edge bevel. Negligible burr of only 16 μm height is also found corresponding to an in-plane exit angle of 60°. To obtain negligible burr, hence, either quite small in-plane exit angle is to be provided, or an exit edge bevel angle of 15° is to provide with a suitable in-plane exit angle.

References

[1]   Nakayama K, Arai M, (1987) Burr formation in metal cutting, Annals of the CRIP, 36(1): 33-36.

[2]   Gillespie LK, Bolter PT, (1976) The formation and properties of machining burr, Transactions of the ASME, Journal of Engineering for Industry, 98(1): 66-74.

[3]   Ko SL, Dornfeld DA, (1991) A Study on burr formation mechanism, Transactions of the ASME, Journal of Engineering Materials and Technology, 113(1): 75-87.

[4]   Gillespie LK, (1985) The academic challenge of burr technology, Allied bendix aerospace, KC division, SME paper no. BDX 613-3151.

[5]   Gillespie LK, (1979) Deburring precision miniature parts, Precision Engineering, 1(4): 189-198.

[6]   Chern GL, Dornfeld DA, (1996) Burr/Break out model development and experimental verification, Transactions of the ASME, Journal of Engineering Materials and Technology, 118(2): 201-206.

[7]   Lee K, Dornfeld DA, (2005) Micro-burr formation and minimization through process control, Precision Engineering, 9: 246-252.

[8]   Luo M, Lin G, Chen M, (2008) Mechanism of burr formation in slot milling Al-alloy, International Journal of Materials and Product Technology, 31(1): 63-71.

[9]   Bierman D, Heilmann M, (2009) Burr minimization strategies in machining   operations, Proceedings of the CIRP International Conference on Burrs- Analysis, Control and Removal, Part-1, University of Kaiserslautern, Germany: 13-20.

[10]   Pekelharing AJ, (1978) The exit angle failure in interrupted cutting, Annals of the CIRP, 27(1): 5-10.

[11]   Hashimura H, Chang YP, Dornfeld DA, (1999) Analysis of burr formation mechanism in orthogonal cutting, Transactions of the ASME, Journal of Manufacturing Science and Engineering, 121(1): 1-7.

[12]   Das A, Saha PP, Das S, (2011) Burr minimization in shaping En25 steels: using experimental and stress analysis technique, International Journal of Manufacturing and Industrial Engineering, 2(2): 61-65.

[13]   Heisel H, Schaal M, Wolf G, (2009) Burr formation in milling with minimum quantity lubrication, Production Engineering Research and Development, 3: 23-30.

[14]   Saha PP, Das D, Das S, (2007) Effect of edge beveling on burr formation in face milling, Proceedings of the 35th International MATADOR Conference, Taipai, Taiwan: 199-202.

[15]   Das A, Mondal P, Samanta S, Das S, Mahata S, (2011) Burr minimization in milling: through proper selection of in-plane exit angle, Journal of the Association of Engineers, India, 81: 38- 47.

[16]   Das A, Saha PP, Das S, (2011) Minimization of burr formation in milling of nickel chrome alloy steels: through appropriate selection of in-plane exit angle, Indian Science Cruiser, 25(5): 43-49.

[17]   Saha PP, Das S, (2010) A study on the effect of process parameters and exit edge beveling on foot and burr formation during machining of medium carbon steels. Proceedings of the National Conference on Recent Advances in Manufacturing Technology and Management, Jadavpur University, Kolkata: 30-35.

[18]   Saha PP, Das S, (2011) A simple approach for minimization of burr formation using edge beveling of alloy steel workpieces, International Journal of Mechatronics and Intelligent Manufacturing, 2(1/2): 73-84.

[19]   Saha PP, Das S, (2011) Burr minimization in face milling: an edge beveling approach, Proceedings of the Institution of Mechanical Engineers, Part B: Journal of Engineering Manufacture, 225(9): 1528-1534.

[20]   Avila MC, Dornfeld DA, (2004) On the face milling burr formation mechanism and minimization strategies at high tool engagement, Proceedings of the International Conference on Deburring and Edge Finishing, University of California, Berkeley: 191-200.

A surface generation model for micro cutting processes with geometrically defined cutting edges

Jost Vehmeyer[1], Iwona Piotrowska-Kurczewski[1] and Sven Twardy[2]

[1] University of Bremen, Center for Industrial Mathematics, 28359 Bremen, Germany
[2] University of Bremen, Laboratory for precision machining, 28359 Bremen, Germany

Abstract. Conventional machining techniques like milling and turning are well established techniques in manufacturing. In the last years considerable progress in miniaturisation of components has been made. Micro-machining operations have made the production of small components flexible, but scaling down these processes is extremely challenging. In the production of dies and molds special attention is paid to the forming zone where surfaces interact under heavy stress. Optimization of the tribological behaviour has been investigated over the last few years due to high rejection rates. The simulation of micro-structures and topographical characterisation are essential factors for the study of the functional performance of surfaces. Our contribution is a numerical surface generation model, which is able to simulate the micro-topography for machining operations with geometrically defined cutting edges. In the presented model the tool-workpiece kinematics are described by static and dynamic motions, including tool run-out and tool deflection caused by cutting forces. In the material removal algorithm, the minimum chip thickness and tool wear are considered. Experiments for ball-end milling are carried out for different process parameters to verify the simulation results.

Keywords: surface generation, micro-topography, ball-end milling, tool deflection

1. Introduction

Mechanical micro-machining is commonly used in the production of micro-components. Besides the geometric specification, there is a growing demand for textured or structured surfaces to provide a desired functionality [1]. Therefore the understanding of surface generation mechanisms and prediction of micro-topography is an essential step to make the production chain as effective and functional as possible.

In micro-milling, helical ball-end tools have been developed, due to their flexible abilities in production of sculptured surfaces. Round-nosed tools have a more complicated tool-workpiece engagement than cylindrical tools, which is recognizable by non-constant machining variables like cutting-depth a_p or uncut chip thickness h and comma-shaped chips. Therefore, even under ideal conditions, the kinematic surface topography for ball-

ends is quite complex. In [2] the term pick-scallop and feed-scallop are introduced to describe the surface topography under ideal conditions. Both scallops are the result of the curvilinear shape of ball-ends and will not occur for cylindrical or insert tools. In [3], topography of face milled surfaces is simulated and verified by experiments.

This paper is organized in two parts, first we introduce the surface generation model, followed by some experimental verifications.

2. Surface generation model

Mechanically machined surfaces are the result of a highly complex material removal process and the first question that arises is on which scale the surface is to be simulated.

In macro-scale machining, surface simulation is widely used for virtual prototyping or tool path optimization. When surface topography becomes the main object of study, the degree of idealization may be reduced. This especially concerns tool sweeping and material removal. Swept volumes are used in terms of computer graphics to statically represent the space that is traversed by an object in motion. To study the global form of a workpiece, form deviation, or waviness, one option would be to occupy the swept volume of the tool by moving a simplified model of the tool, e.g. an implicit representation of the volume of revolution, along the tool path. Afterwards, this volume can be subtracted by a Boolean operation from the workpiece to get the machined surface.

To study micro-structures, the above method must be improved in two points. An obvious improvement is to uncouple translational from rotational motion by sweeping the cutting edge rather than the volume of rotation. Especially for rotating round-nosed tools this advancement is crucial. The Boolean material removal is an accepted idealization at micro-scale; it is equivalent to an ideal sharp cutting edge. At macro-scale the edge

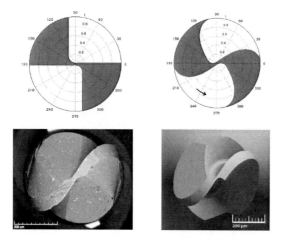

Fig. 1. top: crosssectional area and projected area, bottom: ball-end SEM micrograph (left), computer model (right)

radius of a cutting tool is insignificant small compared to the tool radius. It is a fact that the edge radius cannot be downscaled like other tool dimensions; this phenomenon is known to be a size effect in micro machining [4]. Therefore a modified material removal is used in the presented model, which is presented in the following subsections.

3. Tool geometry

The cutting edge geometry directly affects surface topography. Helical ball-end millers have mostly two curvilinear cutting edges, which are a result of the spheric tool tip and helix angle. Geometrically, they are built by the intersection line of flank face and rake face. One option for calculation is to project the cross-section area onto a sphere, taking the helix angle α into account, exemplarily shown in Fig. 1. For the simulation, a parametrisation of the first cutting edge is needed, it is formulated with tool length l and radius r. For small cutting depths (e.g. 15µm) the helical curvature of the active part of the cutting edge is hardly visible, so an idealisation for the y-component in (1), setting it to a small offset, is acceptable. We investigate this offset with the help of microscopic examination; Fig. 1 shows a SEM micrograp of a ball-end of radius 250µm.

$$(x, y, z, 1)^{T} = \varepsilon(\varphi) =$$
$$(r \cos \varphi, y(\varphi), r \sin \varphi - 1 + r)^{T}, \qquad \varphi \in [0, {}^{\pi}\!/_{2}]$$

Furthermore, it is important to note that the edge geometry changes during operation. The effect of wear is clearly visible in micrographs and due to the discrete nature of the model, simple to include by flattening the edge according to the width of wear mark.

4. Kinematic model

For the tool sweeping algorithm, we need a unique configuration of the tool in the coordinate system, which is given by 3 rotations and 3 translations. This is provided by the kinematic model, which takes process-defined kinematics, as well as positioning errors into account. Even in fields of precision machining, positioning errors can never be avoided entirely. To the contrary, the effect of a misaligned tool on the surface generation increases, as downscaling continues. Positioning errors are classified into static ones, e.g. tool run-out, which results if tool and spindle location are not precisely concentric, and dynamic ones, like vibrations or cutting forces.

The tool trajectories are composed by the rotational and translational motion of the tool. As proposed in [5], a concatenation of homogeneous matrices is used to formulate the tool trajectories, leading to the parameter to state operator

$$\Phi : P \to \Re^{4 \times 4},$$

where P denotes the parameter space. The static part of Φ consists of the rotational and translational tool motion, furthermore a run-out is included by using a parallel axis offset $\rho = (\rho_x, \rho_y, 0)^{T}$.

In order to include dynamical changes into the generation of tool trajectories, we implemented a dynamic force model, which interacts with the process parameter cutting depth a_p and feed velocity v_f. Cutting force can be formulated as a function of a_p and v_f and gives the deflection of tool tip $\delta \sim F$. On the other hand a_p and v_f are functions of δ. This interaction leads to a system of ordinary differential equations, described detailed in [5].

5. Material removal

At macro scale, usually an ideally sharp cutting edge is assumed. Realizing that the cutting edge is built by a cylindric shell, the material removal mechanism becomes more complicated. It has been known for a long time that a threshold for the undeformed chip thickness exists, below which chip formation is disturbed or impossible [6]. For round nosed tools this is not only critical when

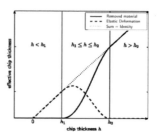

Fig. 2. Truncation of chip thickness

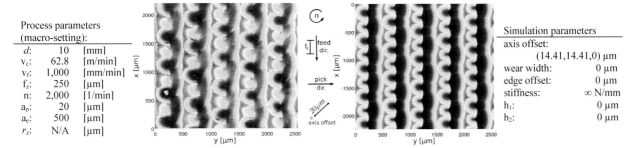

Process parameters (macro-setting):		
d:	10	[mm]
v_c:	62.8	[m/min]
v_f:	1,000	[mm/min]
f_z:	250	[μm]
n:	2,000	[1/min]
a_p:	20	[μm]
a_e:	500	[μm]
r_ε:	N/A	[μm]

Simulation parameters	
axis offset:	(14.41,14.41,0) μm
wear width:	0 μm
edge offset:	0 μm
stiffness:	∞ N/mm
h_1:	0 μm
h_2:	0 μm

Fig. 3. Process parameters for the macro setting, resulting experimental and simulated surface.

small cutting depths are used. Surface topography is affected by the minimum chip thickness for arbitrary cutting depths, caused by the complex tool-workpiece engagement for ball-ends. The chip thickness h takes values in $[0,h_{max}]$, depending on the location of the cutting edge and will always go below the minimum chip thickness on a small part of the cutting edge. Therefore, the minimum chip thickness is an important factor for the topography of ball-end machined surfaces.

It has been established to divide the chip formation in micro-scale machining into three parts, compare [7]. Below the minimum chip thickness, material is elastically deformed, no material is removed. When undeformed chip thickness is approximately equal to the minimum chip thickness, material is both deformed and removed. At significantly higher chip thickness compared to the minimum chip thickness deformation still exists, but is negligible compared to the material removed.

To model the effect of minimum chip thickness for surface generation, we apply a truncated chip thickness to the material removal procedure. To avoid abrupt scallops, we use a continuous truncation; modeled with help of the function $f(h)$, see Fig. 2. The truncation function is demarcated by two points $h_1 < h_{min} < h_2$ into the three schematic chip formation sections:

For $h < h_1$ pure elastic material deformation followed by complete recovery is assumed. In this case no material will be removed. In middle section $h_1 \leq h \leq h_2$ both elastic recovery and material removal take place. This interval represents the critical phase $h \approx h_{min}$, where chip formation is unstable. Rake angle will be negative and chips are sheared off the material. For $h > h_2$, ideal chip formation takes place and deformation is reduced to an insignificant value.

6. Numerical results and experimental verification

The surface model describes in the last section has been implemented in MATLAB. The algorithm is similar to the Z-buffer method, commonly used in computer graphics. The essential modification is the concatenation of the truncation function.

Experimental verification was carried out on a precision milling machine tool (Ultrasonic 20 linear, Sauer GmbH). For first milling experiments common 2-fluted ball-end shaped uncoated HSS cutting tools with a diameter d of 10 mm were used in combination with aluminum alloy (AlMg3), which is known for good machineability. This setup should exclude as many influences as possible and show the true kinematic profile of the tool engagement. Uncut chip thickness is very high (f_z=250μm) compared to the cutting edge radius of an uncoated cutter. Furthermore, tool deflection is also negligible, because of the ratio between shank diameter and cutting forces. Cutting forces were recorded using a 3-axis piezo-dynamometer (MiniDyn 9256C2, Kistler Holding AG). After milling experiments were carried out, all machined surfaces were measured by an optical profilometer (Plµ 2300, Sensofar Technology GmbH).

After the macro test, applying a 10 mm cutter on aluminum, the experimental validation for micro milling of hardened steel was carried out on the same machine tool. For micro milling experiments 2-fluted ball-end shaped tungsten carbide cutting tools with a diameter d of 500 μm and TiAlN coating (VHMSK, Van Hoorn Hartmetaal BV) were used. A spray formed HSS Alloy (C1Cr8Mo2V0.4Al1, 67 HRC) was used as workpiece material. This Alloy is of great interest for micro cold forming tools, because of its outstanding mechanical

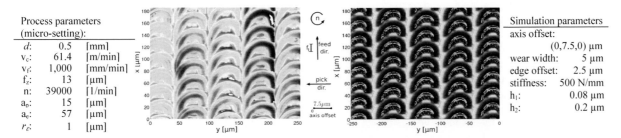

Process parameters (micro-setting):		
d:	0.5	[mm]
v_c:	61.4	[m/min]
v_f:	1,000	[mm/min]
f_z:	13	[μm]
n:	39000	[1/min]
a_p:	15	[μm]
a_e:	57	[μm]
r_ε:	1	[μm]

Simulation parameters	
axis offset:	(0,7.5,0) μm
wear width:	5 μm
edge offset:	2.5 μm
stiffness:	500 N/mm
h_1:	0.08 μm
h_2:	0.2 μm

Fig. 4. Process parameters for the micro setting, resulting experimental and simulated surface.

properties and was also showing sufficient machinability in former tests [8].

Cutting conditions and the resulting surfaces of experiment and simulation for the macro and micro setting are presented in Fig. 3 and Fig. 4 respectively.

7. Discussion

To verify the pure kinematic surface generation, experiments at macro scale for up- and down-milling mode and an alternating combination of both are carried out. In all cases simulation and experiment can be identified with high visual conformity and the Abbott curves show a good agreement of high distribution. These results are achieved by choosing almost ideal simulation parameters, i.e. ideally sharp cutting edge in the sense of no truncation of chip thickness and ideal edge geometry. The run-out effect is too dominant to be neglected and is clearly visible in force signal. Absolute value and direction of the axis offset is determined by manual minimization of the error between measured and simulated surface.

In a next step the milling process is scaled down from $d=10mm$ to $d=0.5mm$. The surface topography of the micro process can hardly be identified as a downscaled version of the macro-sized surface. The structure shows a complete different characteristic. In particular, the feed scallops are more or less symmetric along tool path and have a distance of the twice of feed per tooth, instead of feed per tooth. This means that cutting action is not even, anymore. In the simulation this effect is reproduced with the run-out in combination with the effect of minimum chip thickness. The demarcation points h_1 and h_2 are chosen the same as above by visual inspection. This technique may be extended to a new method for determination of the minimum chip thickness value. Mathematically this is a parameter identification problem. In general, the determination of minimum chip thickness is still challenging, influencing factors are complex and indirect methods are required [9].

8. Conclusion

Overall, this study shows that the presented surface generation model can control the basic mechanisms of surface generation and provides an accurate prediction of surface topography for macro-scaled processes. Taking the minimum chip thickness into account, the simulation is qualified to identify the significant surface structures in micro-sized ball-end milling, too. To meet the exact high distribution, further optimizations of the kinematic or material removal model are necessary.

Furthermore, the presented surface simulation has concrete potential for deeper mathematical investigations. Due to growing requirements in precision and functional performance, micro-cutting processes have become subject of mathematical optimization [10]. Surface generation is the last and most important step in modeling a cutting process and combines the results of all involved sub-models like force and vibration model.

Acknowledgments: The work presented in this article was supported by the German Foudation (DFG) within the SFB 747, subproject C2 and the grant SPP 1480 German Research Foundation (DFG).

References

[1]   Bruzzone, A.A.G. ; Costa, H.L. ; Lonardo, P.M. ; Lucca, D.A. Advances in engineered surfaces for functional performance, CIRP Annals - Manufacturing Technology. 57(2008) 2 - p. 750-769 , 2008

[2]   J.S.A Chen, Y.K. Huang, and M.S. Chen. A study of the surface scallop generating mechanism in the ball-end milling process. International Journal of Machine Tools and Manufacture , 45(9):10771084, 2005.

[3]   M.A Arizmendi, J. Fernandez, L.N.L.d. Lacalle, A. Lamikiz, A. Gil, J.A. Sanchez, F.J. Campa, and F. Veiga. Model development for the prediction of surface topography generated by ball-end mills taking into account the tool parallel axis offset. experimental validation. CIRP Annals - Manufacturing Technology, 57(1):101104, 2008.

[4]   F. Vollertsen. Categories of size effects. Product Eng, 4(2):377#383, 2008

[5]   I. Piotrowska-Kurczewski, J. Vehmeyer. Simulation model for micromilling operations and surface generation. Advanced Materials Research, 223:849858, 2011.

[6]   P. Stockinger. Die Bedeutung der Mindestspandicke in der spanabnehmenden Fertigung. Feinwerktechnick, 73 Heft 6:253256, 1969.

[7]   A. Aramcharoen and P.T. Mativenga. Size effect and tool geometry in micromilling of tool steel. Precision Engineering, 33(4):402#407, 2009.

[8]   Brinksmeier, E.; Riemer, O.; Twardy, S.: Surface Analysis of Micro Ball End Milled Cold Work Tool Steels. Proceedings of the 2nd Nanoman, International Conference on Nanomanufacturing, Tianjin (2010) #173.

[9]   Mian A.J, Driver N., and Mativenga P.T: Estimation of minimum chip thickness in micro milling using acoustic emission, Proceedings of the Institution of Mechanical engineers, Part B: Journal of Engineering Manufacture, vol. 225 no. 9 1535-1551, doi: 10.1177/0954405411404801.

[10] Brandt, P. Maaß, I. Piotrowska-Kurczewski, S. Schiffler, O. Riemer, E. Brinksmeier: Mathematical methods for optimizing high precision cutting operations. To appear in the International Journal of Nanomanufacturing.

An investigation into abrasive water jet machining of TRIP sheet steels using Taguchi technique and regression models

J. Kechagias[1], I. Ntziantzias[2], N. Fountas[3], N.M. Vaxevanidis[3]

[1] Department of Mechanical Engineering, Technological Educational Institute of Larissa, Larissa GR 41110

[2] Department of Mechanical and Industrial Engineering, University of Thessaly, Volos 38334, Greece

[3] Department of Mechanical Engineering Educators, School of Pedagogical & Technological Education (ASPETE), N. Heraklion Attikis GR 14121, Greece

Abstract. An experimental investigation of abrasive water-jet machining (AWJM) of transformation induced plasticity (TRIP) multi-phase sheet steels using design of experiments was carried out. The quality characteristics selected for examination were the Kerf mean width and average surface roughness. The process parameters were the nozzle diameter, the stand-off distance and the traverse speed. For the design of experiments the Taguchi methodology was applied. Optimal process parameter values were identified and regression models were applied to the experimental results and were tested by using evaluation experiments. All the predictions are reasonable and compares well with the experimental values. The experimental design indicated that the nozzle diameter is the most important parameter that affects the mean Kerf width and surface roughness followed by the stand-off distance. The proposed methodology could be easily applied to different materials and initial conditions giving reliable predictions, resulting in process optimization and providing a possible way to avoid time- and money-consuming experiments.

Keywords: AWJM, Taguchi design, Kerf width, Surface roughness

1. Introduction

The AWJM belongs to the category of non-conventional material removal methods and it is used in industry to machine different materials ranging from soft, ductile to hard and brittle materials. This process does not produce dust, thermal defects or fire hazards. It is a preferable process for shaping composite materials and imparts almost no surface delamination, see [1].

The primary interests in TRIP sheet steel processing by AWJM are the Kerf shape (Kerf width and Kerf taper) and surface quality (surface roughness of cut), as well as burrs which may be formed at the jet exit (Figure 1). Kerf shape and quality in slotting sheet materials by AWJM and the resulting surface roughness have been studied in recent research works [2-4].

The innovation of the present work relies on the use of a hybrid modeling approach based on the Taguchi method and the regression analysis for the modeling of

cutting surface quality characteristics in AWJM. The experiments were performed on TRIP 800 HR-FH steel sheets which were processed using AWJM with three different diameters of the nozzle (nozzle diameter), three different distance values between the nozzle and the sheet steel (stand-off distance) and three different traverse speeds (also known as cutting speed or travel speed).

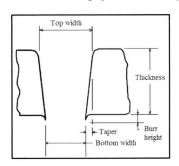

Fig. 1. Schematic representation of a typical cut in AWJM.

An L_9 (3^4) orthogonal array was used in order to reduce the experimental effort, following the Taguchi design of experiments [5]. Taguchi design uses an orthogonal matrix experiment for the parameter design of a process and exploits the orthogonality of the matrix in order to predict the performance of a quality characteristic according to parameter levels.

The selected quality characteristics (also known as performance measures) were the mean Kerf width (Kerf) and the arithmetic mean surface roughness (Ra). Analysis of means (ANOM) and analysis of variances (ANOVA) were applied on the experimental results and the best combination of parameter levels is obtained and the effect of each process parameter on the Kerf and Ra are revealed. Finally, interactions between the process parameters are examined and regression models are adopted and evaluated for Kerf and Ra parameters predictions.

2. Experimental design and procedures

Details concerning the experimental procedure and the materials are given elsewhere; see [6]; therefore only the main features are summarized below.

The TRIP steel tested was designated as TRIP 800 HR-FH. Specimens of materials were of square form (10x10 cm^2). Machining was performed on a SIELMAN HELLENIC HYDROJET industrial AWJM system. In each specimen a slot of 3 cm in length was cut. Each slot corresponds to different machining conditions. After processing, each specimen was separated in order to allow surface roughness measurements to be performed on the machined surface.

The pressure at which a water jet operates is about 400 MPa, which is sufficient to produce a jet velocity of 900 m/s. Such a high-velocity jet is able to cut materials such as ceramics, composites, rocks, metals etc [1].

Based on Taguchi design, the standard orthogonal array L$_9$ (3^4) has been selected in order to perform the matrix experiment [5]. Three levels for each factor were selected (Table 1). Following the L$_9$ (3^4) orthogonal array nine experiments were performed (Table 2). Columns 1, 2, and 3, are assigned to nozzle diameter (A, mm), stand-off distance (B, mm), and traverse speed (C, mm/min), respectively.

The performance measures were the arithmetic mean surface roughness (Ra, μm) and the mean Kerf width (Kerf, mm).

Ra measurements were performed with a Surtronic 3+ profilometer. As it is illustrated in Figure 1 the Kerf is of tapered form and to evaluate this characteristic, the semi-sum of the upper area width and the lower area width were measured by a stereoscope [7].

3. Experimental results and analysis

The Taguchi design is in fact a simple technique for optimizing the process parameters. The main parameters are located at different rows in a designed orthogonal array [5]. Thus, an appropriate number of statistically important experiments are conducted. Generally, the signal to noise (S/N) ratio (η, dB) represents the response of the data observed in the Taguchi design of experiments. Both, the arithmetic mean roughness, Ra and the mean Kerf width are characterized as 'the smaller the better' quality characteristics since lower values are desirable. The objective function for such quality characteristics is defined as follows [5]:

$$\eta = -10 \log_{10}\left[\frac{1}{n} \sum_{i=1}^{n} y_i^2 \right] \qquad (1)$$

In Eq. (1) y_i is the observed data at the ith trial and k is the number of trials [5]. By analyzing the observed data (S/N ratios), the effective parameters having an influence on the quality response can be seen and the optimal levels of the process parameters can be obtained.

Table 1. Parameter Design

Process Parameters		Units	Levels		
			1	2	3
A	nozzle diameter	mm	0.95	1.2	1.5
B	stand-off dist.	mm	20	64	96
C	traverse speed	mm/min	200	300	600
D	vacant	-	-	-	-

Table 2. Matrix experiment

	Proc. Parameters				Obj. Functions	
No	A	B	C	D	η$_{Kerf}$ (dB)	η$_{Ra}$ (dB)
1	0.95	20	200		0.19	-13.1
2	0.95	64	300		-1.25	-15.8
3	0.95	96	400		-0.68	-17.0
4	1.2	20	300		-1.40	-16.0
5	1.2	64	400		-1.22	-16.4
6	1.2	96	200		-3.21	-16.8
7	1.5	20	400		-2.829	-15.3
8	1.5	64	200		-3.528	-16.5
9	1.5	96	300		-3.862	-16.7
mean					-1.979	-15.9

According to the L$_9$ orthogonal array, nine experiments were performed with each experiment producing a 3 cm long slot in which the mean Kerf width and the average surface roughness (Ra) were measured. Then, each objective function η$_{Kerf}$, and η$_{Ra}$ was calculated according to the negative logarithmic formula (Table 2):

$$\eta_{Kerf} = -10 \log_{10}\left[Kerf^2 \right]$$
$$\eta_{Ra} = -10 \log_{10}\left[Ra^2 \right] \qquad (2)$$

For each of the three process parameters (A, B, and C) an average value, m$_i$ for every level i, was calculated for each of the two performances measures (Table 3).

Based on the average values, ANOM diagrams (Fig. 2) were drawn indicating the impact of each factor level on the performance measures η$_{Kerf}$ and η$_{Ra}$ of the parts fabricated. Thus, based on the ANOM, one can derive the optimum combination of process variables, with respect to performance. The optimum level for a factor is the level that gives the maximum value of η.

Table 3. Mean parameter values

	η_{Kerf}			η_{Ra}		
	1	**2**	**3**	**1**	**2**	**3**
m_{Ai}	-0.58	-1.94	-3.40	-15.2	-16.3	-16.1
m_{Bj}	-1.34	-2.00	-2.58	-14.7	-16.2	-16.8
m_{Ck}	-2.18	-2.17	-1.57	-15.4	-16.1	-16.2

Table 4. ANOVA analysis

Par.	Deg. of Fr.	Kerf			Ra		
		Sum of Sq.	Mean Sq.	%	Sum of Sq.	Mean Sq.	%
A	2	11.9	5.98	74.7	1.99	0.99	17.1
B	2	2.3	1.15	14.4	6.64	3.32	56.9
C	2	0.72	0.36	4.5	1.07	0.53	9.2
Tot	8	16.0			11.6		
Er	2	1.01	0.50		1.95	0.97	

According to the ANOM diagrams mean Kerf width is affected mainly by nozzle diameter. On the other hand, the stand-off distance and nozzle diameter mostly affect the surface roughness Ra.

Moreover, it was evident that when the traverse speed increased, the mean Kerf width decreased and the surface roughness increased.

Taguchi design performs an analysis of variances (ANOVA) of the experimental results in order to evaluate the relative importance of the process parameters and error variances. The ANOVA analysis results can be seen in Table 4 for the mean Kerf width and surface roughness.

According to the ANOVA analysis the nozzle diameter affects the mean Kerf width and the surface roughness by about 75 % and 17 % respectively. The stand-off distance affects the surface roughness at most by approximately 57 %.

Interaction charts between nozzle diameter and standoff distance can be seen in Fig. 3 concerning Kerf and Ra, respectivelly.

4. Predictive modeling and evaluation

Assuming that: (i) the process parameters are continuous and controllable within the range of machining experiments, (ii) there are "synergistic" interactions between the process parameters as indicated by the interaction charts, (iii) the dominant parameter for the

mean Kerf width variation is the nozzle diameter, (iv) the dominant parameters for the surface roughness variation are the stand-off distance and the nozzle diameter; the response regression model for each of the quality indicators can be expressed as follows:

$$y = b_1 + b_2 A + b_3 B + b_4 C + b_5 AB \pm e \qquad (3)$$

In Eq. (3) y is the corresponding response of each objective function, bi, coefficients which should be determined and e is the error. Interaction terms are used due to 'synergistic' interactions between process parameters as it is evident from the plots in Fig. 3.

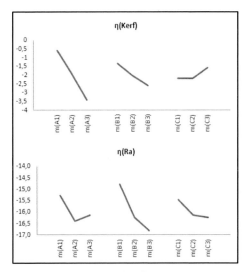

Fig. 2. ANOM diagrams

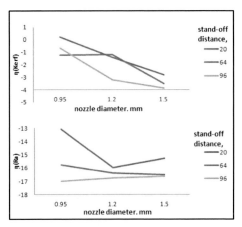

Fig. 3. Interaction charts

In general, Eq. (3) can be written in a matrix form as:

$$Y = bX + E \qquad (4)$$

In Eq. (4), Y is the matrix of the measured values for the y response and X the matrix of process parameters and their products. The matrices b and E consist of coefficients and errors, respectively. The solution of Eq. (4) can be obtained by a matrix approach [8].

$$b = (X^T X)^{-1} X^T Y \qquad (5)$$

In Eq. (5) X^T is the transpose of matrix X and $(X^T X)^{-1}$ is the inverse of the matrix $(X^T X)$.

From the data listed in Table 2 and Equation (5), b_i coefficients for each objective function were calculated and then tabulated in Table 5.

Table 5. b_i coefficients

	b_1	b_2	b_3 $(x10^{-2})$	b_4 $(x10^{-3})$	b_5 $(x10^{-2})$
η_{Kerf}	6.3	-7.2	-5.9	4.9	3.5
η_{Ra}	-8.5	-4.4	-8.8	-1.2	5

Three evaluation experiments were carried out in order to compare the actual valueswith the predicted ones given from Eq. (5). The results are presented in Table 6.

The proposed regression models can be used for the optimization of the cutting parameters during AWJM of TRIP 800 HR-FH steel sheets. This can be done by testing the behavior of the response variable (Kerf and Ra) under within a range of values of the nozzle diameter, the stand-off distance and the traverse speed (Fig. 4).

Table 6. Evaluation experiments

Exp. No	Kerf (mm)	Ra (µm)
A|B|C	Pred|Act|Dev%	Pred|Act|Dev%
0.95|20|300	0.95|0.87|9.4	4.96|5.3|-6.3
1.2|64|200	1.32|1.28|3.1	6.24|6.8|-8.2
1.5|96|400	1.43|1.51|-5.1	7.04|6.9|2.0

5. Conclusions

The mean Kerf width and the mean surface roughness (Ra) have been selected as the quality indicators for AWJM process multi-parameter optimization using Taguchi design.

The experimental design indicated that the nozzle diameter is one of the most important parameters that affect the mean Kerf width and surface roughness by 75% and 17% respectively. The stand-off distance was the second most important parameter within the experimental range tested and affects the mean Kerf width and surface roughness by 14%, and 57% respectively. Traverse speed also affect the mean Kerf width and surface roughness but by lower degree than the nozzle diameter and stand-off distance. The mean Kerf width decreases when nozzle diameter decreases or stand-off distance decreases. The

surface roughness Ra decreases when distance decreases and/or nozzle diameter decreases stand-off.

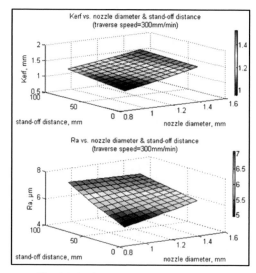

Fig. 4. Behaviour of the response variables

In addition, regression models were applied to the experimental results and three verification experiments were carried out in order to compare the measured (actual) and the predicted values. All the predictions are reasonable and close to the actual values.

Results obtained indicate that both the proposed approaches could be effectively used to predict the Kerf geometry and the surface roughness in AWJM, thus supporting the decision making during process planning.

References

[1] Momber AW, Kovacevic R (1997) Principles of abrasive waterjet machining. Springer.

[2] Gudimetla P, Wang J, Wong W (2002) Kerf formation analysis in the abrasive waterjet cutting of industrial ceramics. J Mater Process Tech 128:123-129.

[3] Hascalik A, Ulas C, Gurun H (2007) Effect of traverse speed on abrasive waterjet machining of Ti–6Al–4V alloy. Mater Design 28(6):1953-1957.

[4] Valicek J et al (2007) Experimental analysis of irregularities of metallic surfaces generated by abrasive waterjet. Int J Mach Tool Manuf 47(11):1786-1790.

[5] Phadke MS (1989) Quality Engineering using Robust Design. Prentice-Hall, Englewood Cliffs, NJ.

[6] Kechagias J, Petropoulos G, Vaxevanidis N (2011), "Application of Taguchi design for quality characterization of abrasive water jet machining of TRIP sheet steels", Int J Adv Manuf Tech, DOI 10.1007/s00170-011-3815-3.

[7] Petropoulos G, Tsolakis N, Vaxevanidis NM, Antoniadis A (2009) Topographic description of abrasive waterjet machined surfaces, Proceedings 2nd European Conference on Tribology (ECOTRIB 2009), Pisa, Italy, p.309-314.

[8] Box GEP, Hunter WG and Hunter JS (1978) Statistics for Experimenters. Wiley.

SelecTool: Software tool for the search and comparison of cutting tools depending on standard, geometric and cutting properties and user's criteria

Juan Arjona Montes[1], Joaquim Minguella i Canela[1], Joan Vivancos i Calvet[2]
[1]Fundació Privada Centre CIM
[2]Dept. d'Enginyeria Mecànica, Universitat Politècnica de Catalunya

Abstract. Sometimes it is difficult to choose one or several cutting tools for specific or general uses between the myriad of tools that exist today in the market. It is also necessary to spend lots of time analyzing costs and other important properties between the models that are available. To overcome this, there exist some software applications that might help, but normally with clearly commercial purposes that can distort the best selection. For this reason, trying to adopt a multilateral approach, it is not easy neither quick to gather partial or total information about a single tool searching it directly from the manufacturers website or massive *pdf* files. The study of the different manufacturer's catalogues reveals lots of common points by presenting the technical information required. The pair material-operation represents the basis of the data base structure with which the software application presented is developed. Despite the heterogeneity of the catalogues formats, all the properties presented are classified in only three groups: general, geometric and mechanistic properties; the later one depending on the pair material-operation selected. The data base can contain tool references of different manufacturers for materials and operations and it is totally upgradable by the user. *SelecTool* is developed to guide the user step by step to specify the search properties. A different range of values of selected properties will modify the results that appear on screen. An important aspect supported is that the user defines the exact hierarchy with which the application searches and shows the results. Finally, the application presented is able to compare graphically the selected tools and generate other type of files for further analysis.

Keywords: Software, data base, cutting tools, MS Excel, search, select, compare, graphical comparison, reports.

1. Introduction

From the point of view of the user, it represents a big deal to select the most appropriate cutting tools from several cutting tool manufacturers and thousands of tool models. Only the data search and comparison of a small group of them usually mean lots of dedication in energy and time. Besides, an inappropriate selection can imply a substantial delay and implicit costs.

Also, many manufacturers are interested in approaching their products to their regular and potential customers by using other strategies. Many of them have already not developed it yet and they continue offering their customer service as it was ten years ago.

Inspired by the *Kendu*'s software [1], developed by *Iketek* [2], *SelecTool* helps the user from the early stages of the cutting tool selection to the last ones. It establishes a formal communication channel between the manufacturers and the customers.

2. Design

The software's design and requirements are based on the *Kendu*'s software design and capabilities [1][2][3][4]. The previous software analysis has shown specific aspects to be implemented, improved and ignored.

2.1. Design specifications

The main aspects concerning the design specifications, the requirements and terms of usage of the program are explained as follows:

- The application is designed only for information and evaluation issues. It is highly recommended to ask the manufacturer before purchasing any tool.
- The data contained in the program is strictly based on the values presented by the manufacturer in the catalogues.
- *Selectool* offers information only about cutting tools. The information about other accessories is not included. However, the data base is designed to be upgradable and so to include further information related about new models, materials or operations.
- Roughing or finishing cutting conditions are not considered directly on the program. Default cutting properties are only for roughing operations.
- Due to the awkwardness relative to the interaction tool-workpiece, *SelecTool* does not enforce any particular criteria for workpiece geometry.

- The program is scalable; making it possible to integrate other functions and capabilities than the initial one. The program code permits the upgrade of the application interface and functions.
- In the database contained in the first release, there are only two materials supported: austhenitic stainless steel *AISI316* and *Ti6Al4V*-alloy. The operations available are side milling, slotting and *3D* surface milling.
- The application needs to be installed before being run on the computer. Because of the computer requirements, this program can run on a standard PC. *SelecTool* v1.0. is only for *Windows*® *OS x86* and *x64* available.
- In order to generate graphical comparisons and print reports, it is necessary to install *MS Excel*® *2007* or later.

2.2. Parameters

There are some common tool parameters: material, operation and type (square-end mills, ball-end mills and other). The standard, geometrical and mechanistic parameters are shown in the table below (Table 1.):

Table 1. Available parameters on the first release (milling).

Standard	Geom. Solid	Geom. Indexed	Cutting props.
Model	*Model*	*Model*	*Model*
Manufac-turer	Diameter 1	Form	*Material*
Type 1	Radius	Radius	*Operation 1*
Type 2	Diameter 2	Teeth	*Operation 2*
Refrige-ration	Teeth	Width	Feed per tooth
Coating 1	Subjection	Height	Feed per revolution
Coating 2	Helix height	Deep	Cutting speed
Price	Total height	Rake angle	Feed
	Neck height	Clearance angle	Rpm
	Radius ε	Edge length	Chip flow
	Tolerance	Subjection	
	Helix angle	Subjection diameter	
		Chip breaker	

This application orders automatically the results by the same hierarchy specified by the user. The hierarchy itself is considered as an additional parameter in the program to present the results in the way the user asks for.

It is also possible that the user knows exactly which tool model has to use, nevertheless, he ignores, for example, on which material could this tool work properly. This software is able to gather all the information requested by the reference of the searched model and present it on the screen.

2.3. Software development platform

Microsoft Visual Studio® [4][5] has been used for the development of the software tool. The *T-SQL* (*Transact - Structured Query Language*) for the data base queries has been used [6]. The *Object Oriented Programming* (*OOP*) philosophy has been employed [8].

3. Design

The developed software searches and gathers all the information from a data base by executing *SQL* queries. The data base was implemented on *Microsoft Access*® *2007* and its format is *mdb*.

The data base structure is based on a series of tables whose fields correspond to the above presented parameters: *General table* with common information, *Geometrical table* (2 tables) with geometrical information and *Material-Operation table* with cutting information. Figure 1 below presents the current structure.

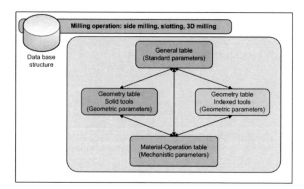

Fig. 1. Data base structure (milling).

The same data base structure is used to store information about drilling, turning, threading and other operations not available in this version yet [9]. Selected parameters on Table 1. mark *identity fields* on each data base table. New brand tool models information is processed from manufacturers' handbooks by a *MS Excel*® template to calculate exact cutting information. Later, this data is imported and added to the *MS Access*® database file. Database updates can be performed in a relative short time depending on the amount of new available tool models.

3.1. Application wizard

The task of searching tools is not a trivial one: it hides a big complexity in itself. The program has to deal with the previous experience of the user by comparing and purchasing tools [7].

In order to guide the user through the searching task, the application has a search wizard in which the user selects sequentially the parameters and its values with which to construct the appropiate *SQL* query.

3.2. User interface

The aspect of the user interface's controls is the common look of a standard application based on *Windows*®. The application's interface has been built to perform the visibility of the results giving the user a clear feedback [7]. Its central workspace is divided in two parts: numerical & graphical results (*left side*) and selected parameters and values information & options for graphic generation (*right side*). The different interface types *Graphic User Interface*, *Menu User Interface* and *Command User Interface* are combined for developing the software interface.

The main menu is localized above the screen. In the main menu, the complete amount of functions and options implemented in the program are ordered by common main menu elements: *File*, *Edit*, *View*, *Tools*, *Graphs*, *Windows* and *Help*. The status bar is fixed on the bottom. It gives real time feedback to the user related about the current status of the application.

All the employed interface elements are considered programming standard elements. In relation to the software's accessibility, text format and overall style keep constant and ease a comfortable reading of the results. For this version, there's no multilingual support. Hence, the interface is not customizable.

3.3. Graphics

The ability to generate graphic comparisons takes the difference within other software of the same kind. The software establishes communication with *MS Excel*® to build custom graphs [5]. Then it imports them via clipboard. Some presentation options, graphic designs and other *MS Excel*® graphical properties are available on this application.

3.4. Printing reports and file format

Sometimes dealing with big amounts of information on the screen can lead the user to confusion. Therefore this application includes a printing function based on *MS Excel*® [5].

It's possible to save the results in different file formats. Supported formats are given by *MS Excel*® itself [5], because the saving procedure is driven by this program: *xls*, *xlsx*, *txt* and etcetera. Saved files content selected parameters and specified values, all founded results, the last graphical comparison generated and the selected data to generate this last one.

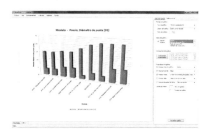

Fig. 2. Graph generation of *SelecTool* based on *MS Excel*®.

3.5. User's help

One of the most common problems on this kind of applications seems to be the lack of detailed information, examples and general user assistance. A clear and complete help form will answer the user the major part of the questions he/she has regardless of the degree of knowledge on the subject.

Furthermore, it is recommended that help is available through all the searching process so the user has full access to the help contents [9]. The overall content includes conceptual information for non-technical users.

Finally, some related links are exposed which leads the way for finding more useful information about the viewed topics.

3.6. Software update

The main feature to maintain and update is the data base inserting new models, modifying some values of existent models or deleting obsolete ones. By now, the application has the possibility of editing the data base manually. Nevertheless, it's possible to have a new release of it by simply downloading *SelecTool* from the *internet*.

Finally, the characteristics of a newer version could install updates automatically, correct and debug program exceptions, develop an own graphics engine, improve existing functions and create new ones.

4. Assessment and validation

In order to ensure the quality envisaged, *SelecTool* development process has been submitted to intensive testing, demonstrating the achievement of the major critical aspects.

Regarding the information database, the user has to choose from which database they want to get the cutting tool information. Database selection and-or modification is permitted at any time.

The user has different ways to proceed with the searching task: either by the selection wizard or by the main menu. In addition, at the beginning of the process, the user decides with which method wants to proceed: search by parameters (*default*) or by known model reference. It is possible to choose every combination and a high range of values for the selected parameters. This aspect will obviously take to a large or a small results list. Finally, the user can specify the hierarchy or accept merely the default one (order of appearance of the parameters).

Example of tool search:
- Material: *AISI 316*
- Operation: *Milling, Side milling*
- Tool type: *flat-end mill, solid mill*
- Parameters: *Price* (0-85€), *Diameter 1* (0-16mm), *Radius* (0-8mm), *Teeth* (1-4)
- Hierarchy: *Price, Teeth, Diameter 1, Radius.*

The cutting tool selection steps are sequentially presented and the software permits to change the order of the parameter selection at any time. Nevertheless, it is recommended to fill all the required data before.

The search function takes proportional time to the length of the data base: a custom search in a *3000* cutting tool models data base finishes in less than *8 seconds*. Results (a total of *137 results*) are shown on screen in the same hierarchy order. The software's interface shows the results in a grid: rows are for tool models and columns for specified parameters. Number of results (rows) shown on screen is controlled by the user (*default:* all the results).

The graphical capabilities of *SelecTool* exploit the major part of *MS Excel*®'s graphical engine offering to the user different options by making custom comparisons: graph type, design (*default:* custom design) and style (*default:* style 2) are available options to be modified. Graph format can be modified by updating the current graph. It takes about *25 seconds* for every graphical generation or graphical update.

Import, export and zoom (25-800%) functions are already embedded in the application. This software could be used as a common picture viewer by importing any picture in standard image format: *bmp, jpg, png, tiff* and so on.

Report printing function is supported, but the printed results don't have an appropriate format to show them all.

The program gives real time feedback, accepts hotkeys and shortcuts for triggering its options and functions and help button keeps always visible and enabled.

User's help includes program functions' explanations and examples, related links and additional technical data related with the viewed topic and materials engineering. Help information is printable, but not multilingual.

5. Conclusion

Data base contents are appropriate for giving an overall view of the program's search performance; which has been constructed with the aim of being a trans-manufacturer complete portfolio. The application offers a big amount of parameters with which to complete the search task, different search methods and the chance to specify personal criteria. Elapsed search time is low and

found results are coherent. Roughing and finishing cutting conditions must to be included in later versions.

The tool search shown before demonstrates an objective and custom tool searching method, always available and totally independent from any particular opinion. Unlike *Kendu*'s software [1], it avoids clearly some commercial issues by comparing tools from different manufacturers.

The use of *MS Excel*®'s graphics engine for graphical comparisons implies a quick development solution, but make the application rely in the dependency towards non-free software. Data transfer from the application to *MS Excel*® takes too much time and must to be reduced.

For this reason, in the first release, the printing reports function has not been completely developed, as it is envisaged to utilize a further implementation based on free software solutions. In any case, many different file formats are supported for further data manipulation with other applications.

Selectool interface demonstrates a clear appearance in relation to the presented numerical and graphical data. The selection tree indicates the user selected and non-selected parameters and values. The menu structure and right panels show any supported option or function in a logical order. By its simplicity, it promotes an intuitive and progressive learning.

The user's help form and included information is correct and very useful for experts and beginner users with technical and non-technical knowledge about material or manufacturing engineering.

Online support and update seem to be strictly necessary for newer releases of the program. This application could support scientific disclosure.

References

[1]   http://www.kendu.es/ Kendu's company web site.
[2]   http://www.iketek.com/ Iketek's company web site.
[3]   E. Petroutsos; La Biblia de Visual Basic 2008; Ediciones Anaya Multimedia 2009.
[4]   http://msdn.microsoft.com/en-us/vstudio/aa718325 MSDN - Microsoft support on Visual Studio®.
[5]   B. "MrExcel" Jelen, T. Syrstad; Excel Macros y VBA, Edición revisada y actualizada 2010; Ediciones Anaya Multimedia, 2010.
[6]   http://msdn.microsoft.com/en-us/library/ms189826%28v=sql.90%29.aspx MSDN - Microsoft support on T-SQL.
[7]   http://www.w3.org/WAI/ - Web on standards on Accessibility
[8]   I. Horton; Ivor Horton's Beginning C++: The Complete ISO/ANSI Compliant (Wrox Beginning Series); Third Edition, Appress, 2004.
[9]   B. Boehm, C. Abts, A. Winsor Brown, S. Chulani, B. K. Clark, E. Horowitz, R. Madachy, D. J. Reifer, B. Steece; Software Cost Estimation with CoCoMo II; EngleWood Cliffs, Prentice Hall, 2000.

Computational fluid dynamics analysis for predicting the droplet size in MQL during grinding of super-alloy

Balan A.S.S, Vijayaraghavan L, Krishnamurthy. R
Department of Mechanical Engineering, IIT Madras Chennai, India 600036

Abstract. Increasing concerns for environmental and health hazard in processing lead a change from wet cutting to dry/near-dry cutting is highly recommended. In recent years, Minimum Quantity Lubrication (MQL) machining is regarded as a promising method for reducing machining cost by way of minimizing power,floor space and oil comsumption, while improving cutting performance; thus, it has been investigated vigorously. Very few investigations have been carried out on the influence of MQL parameters, such as oil flow rate and air pressure on droplet size and consequent perfomance.The present work aims to develop a simulation model to replicate the mist formation in MQL grinding using Fluent based Computational Fluid Dynamics (CFD) flow solver. The MQL parameters considered for the study are air pressure and the mass flow rate. Simulation of the atomization under turbulent conditions was done in Discrete Phase Model (DPM) owing to the fact that oil mass flow rates are very low and oil acts as a discrete medium in air. Droplet diameters were obtained under different inputs to find the optimum value of air pressure and mass flow rate of oil to achieve the desired results (lower cutting force and surface roughness) in MQL grinding of superalloy (Inconel 751). It is seen that medium size (around 16.3 µm) of droplet plays a sigificant role in improved performance by the way of reduction of cutting force and surface roughness.

Keywords: MQL, CFD, Droplet size, Surface roughness, cutting forces.

1. Introduction

Grinding operation, without using sufficient grinding, normally leads to thermal damages and dimensional inaccuracy on the workpiece surface. Hence methods of dry grinding have not yet been fully successful in industrial applications. Thus to avoid possible quench cracking (wet grinding) and thermal softening (dry grinding), an attractive alternative is the minimum quantity lubrication (MQL) grinding and it is referred as near dry grinding [1]. During MQL grinding, cutting fluid droplets with the gas jet dash against walls of the workpiece. When the droplets arrive at high-temperature zones, they absorb much heat and become oil vapour rapidly [2]. Tawakoli et al. [3] investigated the significance of the workpiece material hardness and grinding parameters in the MQL grinding process. Based on the results of their investigations, considerable improvement can be achieved by MQL grinding of hardened steel relative to dry grinding process. The boundary layer of air rotating with the grinding wheel can results in fluid starvation in the contact region. The boundary layer acts a barrier to fluid penetration and prevents fluid reaching the contact region and the fluid is deflected elsewhere. Such a situation is inefficient and wasteful. So the proper air supply pressure is needed for MQL grinding [4]. The main objective of this work is to simulate the atomization of oil mist using Computational Fluid Dynamics (CFD) model using Fluent 6.3 flow solver. The parameters varied being the air pressure and the mass flow rate. Simulation of the atomization under turbulent conditions was done in discrete phase model (DPM). Sauter Mean Diameter (SMD) is most widely used in these types of models where aerosol sprays are simulated. SMD is defined as the diameter of a spherical droplet which has same ratio of surface area/volume as that of the whole spray. A 2D numerical discrete phase simulation has been developed to study the droplet diameters under different inputs conditions to find the optimum value of air pressure and mass flow rate of oil to achieve the best results in MQL grinding of INCONEL 751 superalloy.

2. Experimentation

Straight surface grinding experiments were carried out on Inconel 751 super alloy with resin bonded diamond wheel. The schematic of the experimental setup is shown in Fig. 1. The details of machining parameters and the grinding wheel specification used in this present study are summarized in Table 1. In this MQL system, the compressed air (pressure) and lubricant flow rate can be adjusted separately and mixed in the special nozzle- (air atomizer) to make micro droplets of cutting oil jetting to the cutting zone by the compressed air.

Table 1 Experimental conditions

Parameters	Conditions
Grinding Machine	Tool and cutter grinder (schuette make)
Grinding wheel	D126 C75 (resin bonded Diamond wheel of 150 mm diameter; 13 mm width; grit size 126)
Wheel velocity (V_s)	2826 m/min
Work feed rate (V_w)	0.9 m/min
Depths of cut (a)	30 µm
MQL oil flow rate (Q)	60, 80 ,100 ml/hr
Air pressure(P)	2, 4 , 6 bar
MQL oil	Cimtech D14 MQL oil with viscosity =5 cst, and ρ= 1080 kg/m3

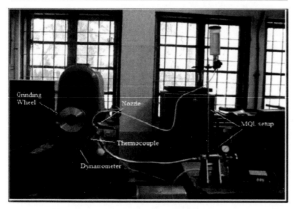

Fig. 1. Experimental Setup.

The workpiece was mounted on a three component piezoelectric dynamometer (Make: Kistler, Model: 9265A) in order to monitor both the tangential (F_x) and normal (F_z) components of grinding force. Temperature at the tool-work interface was measured using a k-type (Chromel / Alumel) thermocouple. . Surface finish of the machined surface was measured using a Talysurf CCI - Lite Non-contact 3D Profiler.

3. Experimentation

A 2D CFD simulations were performed with ANSYS FLUENT version 6.3. Air atomizer geometry was modeled and meshed using GAMBIT 2.3. Size functions were used to further reduce mesh size. The air phase and particle tracking were performed in steady state. The compressed air and oil (cutting fluid) were set as the primary and secondary phase, respectively. Throughout

all simulations the Realizable $k - \varepsilon$ model and Discrete Phase model (DPM) were used for oil mist formation. FLUENT simulate the discrete second phase in a Lagrangian frame of reference. This second phase consists of spherical particles which may be taken to represent droplets of oil dispersed in the continuous phase. FLUENT computes the trajectories of the droplets by solving the time-averaged Navier-Stokes equations for air, while the oil droplets is solved by tracking a large number of droplets through the calculated flow field. A fundamental assumption made in this model is that the dispersed second phase occupies a low volume fraction, that perfectly fits the model of MQL in which the mass flow rates are of the order of micro liters. The governing equations for mass and momentum for the present model can be shown in tensor form as

$$\frac{\partial p}{\partial t} + \Delta(\rho \vec{v}) = 0 \qquad (1)$$

$$\frac{\partial}{\partial t} + (\rho \vec{v}) + \Delta(\rho \vec{v} \vec{v}) = -\Delta p + \Delta(\overline{\overline{\tau}}) + \rho \vec{g} + \vec{F} \qquad (2)$$

4. Geometry and boundary conditions

According to the structure of nozzle and spray characteristic, the model of simulation was built in 2D for computational simplification. The boundary conditions for different regions were decided based on the constant inputs that were given for that particular region. Figure 2 shows different boundary conditions and geometry of the model.

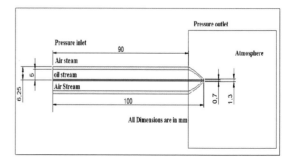

Fig. 2. Geometry and boundary conditions

The air stream input boundary was selected as 'pressure inlet' owing to the fact that air is supplied at constant pressure from that boundary. Similarly oil stream input and boundaries of wall were selected as mass flow inlet and pressure outlet respectively. Different domains such as air/oil streams, mixing area, nozzle outlet were separated using 'interior' boundary condition.

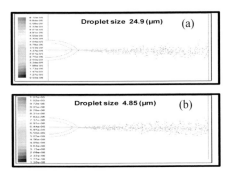

Fig. 3. Contours of (a) Maximum and (b) Minimum Droplet size

5. Results and discussion

5.1. MQL droplet size prediction

The three levels for the two input conditions (air pressure and mass flow rate) were used to create simulation results for all the possible combinations (full factorial experiment). The standoff distance is considered to be constant for all the case. Summary after post processing for DPM was obtained for all the simulations. Sauter Mean Diameter (SMD) was taken and tabulated in Table 2.

Table 2. SMD values for Simulated and experimental conditions

Pressure (bar)	Flow rate (ml/hr)	Simulated SMD (µm)	Experimental SMD (µm)
2	60	24.9	23.7
2	80	14.2	12.9
2	100	8.1	7
4	60	20	18.3
4	80	9.3	7.5
4	100	6.6	4
6	60	18.3	16.3
6	80	8.23	6.8
6	100	4.85	2

In two-phase (gas-liquid) jet of MQL, droplet characteristics are mostly affected by the primaryphase (compressed air). Increase in the air supply pressure and mass flow rate leads to a decrease in droplet size. Larger droplet size with high pressure can easily penetrate the boundary layer to lubricate the grinding zone effectively and this can be achieved when the oil mist velocity is substantially greater than the tangential velocity of the wheel. The maximum droplet size of (24.9µm) was found in the case of 2 bar air pressure and 60 ml/hr mass flow rate whereas, the minimum droplet size (4.85µm-Fig. 3) was found in the case of 6 bar air pressure and 100 ml/hr.

5.2. Experimental validation

The Malvern Analyzer (manufactured by Malvern Instruments, England) was used to measure the exact droplet size. The Setup of the Malvern Analyzer is shown in Fig. 4. It is rapid, nondestructive and requires no external calibration. The extremely fast, 10 KHz, data acquisition rate of the new system produces real-time particle size distributions with a 100 microsecond resolution.

Fig. 4. Malvern Analyzer (Malvern Instruments, England)

The Sauter Mean Diameter for different MQL conditions was measured and tabulated in table 1.2. The experiments were repeated to check the repeatability. The comparisons between numerical and experimental data of droplets were carried out and it is shown in Fig. 5. The numerically simulated values are matching with experimental values at lower and medium mass flow rate. At higher mass flow rate and air pressure of (4 & 6 bar), the predicted Sauter Mean Diameter is higher.

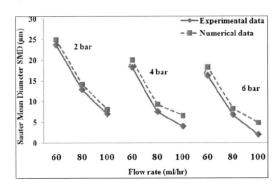

Fig. 5. Comparison of numerical vs. Experimental data

5.3. Effect of droplet diameter on grinding performance

Grinding experiments have been performed in different MQL oil flow rates and air pressures with maximum cutting velocity (V_s= 2826 m/min), feed rate (V_w=0.9m/min) and depth of cut (a=30 µm). Fig. 6 illustrates the significance of Droplet size on grinding force components, surface roughness and temperature. As the droplet size increases, there was a steep rise in the normal force and then tends to decrease gradually up to 16.3 µm droplet size and then increases drastically. The same trend was observed with surface roughness. The diameter of an oil droplet has the most significant influence on the relative flight distance and is proportional to the square of the diameter, it is very difficult for a very small oil droplet to move freely in the air flow [5]. So it can easily changes its track away due to the cutting velocity and then leading to insufficient lubrication condition with higher normal force and poor finish. Larger droplet with lower pressure also leads to insufficient lubrication, owing to the fact that cannot be carried to the grinding zone because of its higher mass. Therefore medium size droplet with higher pressure can easily penetrate the boundary layer and provide effective lubrication, thereby reducing the normal force, temperature and surface roughness. This phenomenon occurs due to the reduction in the sliding friction of grinding and the adhesion of the chip on the rake face of the grit (loading). As the lubrication effect (film forming tendency) increases, the energy formation of the work material drops down (due to Rehbinder effect) facilitating elastic–plastic deformation under the cutting edge of the abrasive grain, resulting in reduction of cutting forces [6]. The function of the high cooling air is to flush away the chips from the cutting zone (flushing effect) and maintain a clean cutting edge, which in turn reduces the surface roughness. There is no variation in the tangential force between (7.5 -16.3 µm) droplet size because of effective lubrication. The temperature in the cutting zone decrease with decreasing droplet size. This is due to both the cooling and the lubrication effects of the fluid provided by MQL, as lubrication reduces the cutting forces and cutting energy, while convection heat transfer and/or boiling carries away some of the heat from the cutting zone. From the results it is clearly seen that the droplets ranging from (7.5 -16.3 µm) gives effective lubrication, while grinding of Inconel 751 superalloy.

6. Conclusions

The potential of the numerical modeling combined with the experimental results is used as a powerful tool to obtain a better understanding of the phenomena observed during the MQL grinding of Inconel 751 alloy and to analyze the influence of droplet size on grinding performance. The significant findings derived from this work are summarized below.

- Droplet size plays a key role in MQL grinding. Medium droplet sizes with higher pressure can easily penetrate the boundary layer to lubricate the grinding zone effectively.
- The numerically simulated values are matching with experimental values at lower and medium mass flow rate at various air pressure.
- Effective transportation of the droplets to the grinding zone, enhance boundary lubrication environment that develops around the wheel-work interface, which enhances to better grinding performance.
- The droplets ranging from (7.5 -16.3 µm) give effective lubrication that reduces normal force, and surface roughness values.

Acknowledgements: Authors want to express their sincere gratitude to Tamal Mukherjee, Application Specialist, Malvern - Aimil Application Lab for providing the Malvern Analyzer used in this work.

References

[1]   Malkin S (2008) Grinding Technology, Theory and Applications of Machining with Abrasives, second ed., Industrial Press Inc.

[2]   Silva R, Bianchi EC, Fusse RY, Catai RE, Franca TV, Aguiar PR, (2007) Analysis of surface integrity for minimum quantity lubricant–MQL in grinding , International Journal of Machine Tools & Manufacturing,47;412–418.

[3]   Tawakoli T, Hadad MJ, Sadeghi MH, Daneshi A, Stockert S, Rasifard A,(2009) An experimental investigation of the effects of workpiece and grinding parameters on minimum quantity lubrication–MQL grinding, International Journal of Machine Tools & Manufacturing,49;924–932.

[4]   Wu1 H, Morgan MN, and Lin B, (2009), Investigation of the Grinding Wheel Air Boundary Layer Flow, Advanced Materials Research,76-78;113-118.

[5]   Brinksmeier E, Heinzel C, Wittmann M, 1999, Friction cooling and lubrication in grinding, CIRP Annuals. Manufacturing Technology, 48; 581–598.

[6]   Obikawa T, Asano Y and Kamata Y, (2009), Computer fluid dynamics analysis for efficient spraying of oil mist in finish-turning of Inconel 718, International Journal of Machine Tools & Manufacture, 49; 971–978.

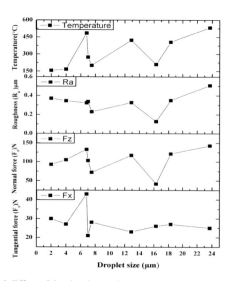

Fig. 6. Effect of droplet size on Cutting force, Surface roughness and Temperature

Cryogenic CNC machining of individualised packaging

V.G. Dhokia[1], A.Nassehi[1], S.A.Wolf[1,2] and S.T.Newman[1]

[1] Department of Mechanical Engineering, University of Bath, UK, BA2 7AY
[2] WZL, RWTH Aachen, Manfred-Weck Haus, 52056 Aachen, Germany

Abstract. With the continuing growth in the high value product manufacturing sector there is a need to provide new ways to store and transport high value precision parts without damage and contamination. Traditional packaging is produced from foams and elastomers using forming techniques. This provides difficulty for producing individualised packaging solutions as individual moulds and forming tools would be required. In addition present solutions do not provide adequate packaging as they can cause damage to the part in some cases. This paper explores the design and manufacture of precision machined individualised neoprene foam packaging using cryogenic CNC machining technology. Cryogenic CNC machining is discussed and an example case study is presented showing the viability and efficacy of the proposed design and manufacturing method.

Keywords: Cryogenic, CNC, Packaging

1. Introduction

Packaging is a major issue for a variety of different industrial and consumer based products. The fundamental aim is to prevent damage and part contamination during storage and transportation. Current methods primarily consist of forming technologies such as thermoforming to produce the required geometrically accurate packaging. For lower value products, simple, cut to fit foam inserts, polystyrene, and bubble wrap packaging is used and this can in a number of cases prove to be inadequate. Producing specific packaging using current methods is difficult as individual moulds are required, which subsequently leads to increased cost and manufacturing process chain. Figure 1 shows an example of a soft foam package.

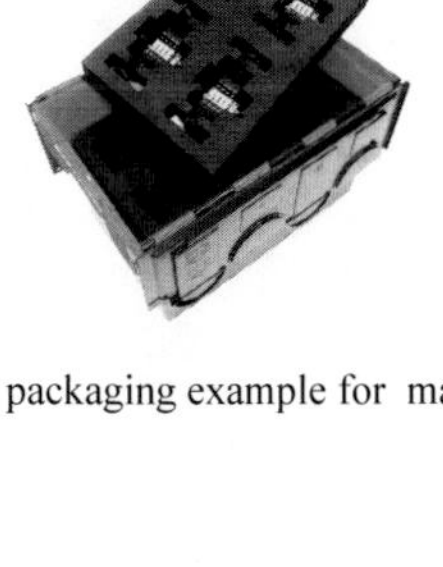

Fig. 1. Soft material packaging example for mass produced parts

Correct packaging is vital, particularly if parts are being sent to customers as replacement or as a consignment of 'bits' for onsite assembly. It is also useful when parts need to be transported for further processing. In addition, as these types of parts are often expensive, packaging takes on added significance, not only for product security against damage, but also from a business perspective. A well-packaged part is likely to procure added enhancement to the product, thus increasing its customer appeal and value. For example, perfume packaging can represent 22% of the manufacturer's gross costs and can contribute up to 40% of the value of the product [1]. It is not merely the case that a product or part should be stored or transported in adequate standardised packaging, particular if the part being packaged is of high value.

A process being pioneered at the University of Bath is that of direct cryogenic CNC machining of soft polymers [2, 3]. Using this process it becomes possible to produce product specific packaging that can be directly machined from different materials depending on the end application and on the amount of part protection required. This paper illustrates the concept through a case study for design and manufacture of individualized packaging for specialist bicycle parts.

2. Cryogenic CNC machining of elastomers

The concept of soft elastomeric material machining is based on the need to remove the moulding process, which can be expensive and does not allow for constant design change and inhibits customised and individualised design [4, 5]. Direct CNC machining of polymers provides the ability to change designs instantaneously emphasizing the realistic opportunity to produce individualised product packaging. However the major challenge of machining polymers and soft elastomeric materials is the inability to conventionally impart a sufficient chip bending moment, which can result in significant deformation, tearing and burning of the material. Figure 2 presents example images of different machining conditions for soft materials. The first image is captured from dry machining and it is clear to see the tearing and serration marks imparted on the material. The second image illustrates a chip sample from cryogenic machining of neoprene foam and it is clear to see the clean-cut edges and full chip formation, indicating a significantly improved machined part and surface.

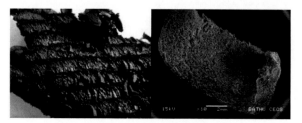

Fig. 2. Surface from dry machined rubber and a fully formed chip from cryogenically machined neoprene

The cryogenic CNC machining facility [2] has been designed to machine a range of soft elastomer materials, which include, neoprene foam and ethylene vinyl acetate (EVA). The major aim of the cryogenic CNC machining process is to freeze a soft material to below its glass transition temperature (T_g) value and then to directly machine it using standard conventional CNC machine tools and tooling. The T_g is the temperature at which a material shows similarities to that of a glass type structure. Only after this temperature has been achieved and is maintained can the material be successfully machined. Each material has its own T_g value, which is determined using dynamic mechanical thermal analysis methods (DMTA) [6]. Machining above this temperature can lead to deformed features and material tearing as illustrated in Fig. 2 leading to inferior parts, products and potentially reduced tool life.

The cryogenic CNC machining facility consists of a cryogenic fixture designed to securely clamp a test part sample, an 18mm diameter vacuum jacketed piping that feeds directly from a 180-litre high pressure Dewar into a custom designed fixture and a multi nozzle spray jet unit. The temperature of the fixture is monitored using a series of low temperature thermal probes. The material block is securely located inside the fixture and is cooled directly using liquid nitrogen (LN_2), with the use of the spray jet unit. The spray jet unit activates using a timer mechanism, so as to regulate the amount of LN_2 used during a given machining cycle. Fig. 3 provides a schematic of the cryogenic machining setup.

The developed cryogenic CNC machining facility is designed to be retrofitted to any type of vertical machining centre. The control unit for controlling the rate of flow of LN_2 at any given time in the machining process is totally independent of the machine tool and was developed as part of another study [4]. This is a timer-based system that uses the time taken to achieve the material specific T_g and the time taken to maintain this T_g as the input factors.

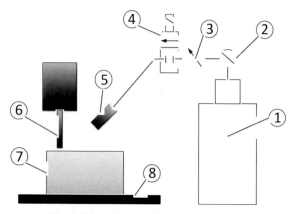

Fig. 3. Schematic setup for cryogenic machining, 1-Liquid nitrogen Dewar, 2- pressure gauge, 3- gate valve, 4- solenoid on/off valve, 5- specially designed nozzle, 6- cutting tool, 7- workpiece, 8- machine table

3. Individualised packaging design

There are different types of packaging solutions that can be used depending on the part / product required to be packaged, transported and or stored. Mass produced products typically tend to use thermoformed packaging manufactured using specifically designed mould tools and soft low-density, impact attenuating materials. The following Fig. 4 illustrates an example of the current method for mass produced packaging.

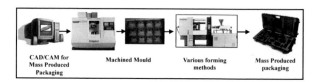

Fig. 4. Mass produced packaging

The design of individualised packaging requires a totally new approach. In order to produce the correct packaging geometry a CAD representation of the part is required. This can be either obtained from the CAD part file or from using scanning reverse engineering techniques to produce an accurate 3D digital representation of the part features and geometry. Using the CAD part, fully individual male type negative moulds of the part can be replicated. In addition, these computer-generated designs can then be altered, adjusted and manipulated to increase packaging support for the part and part numbers corresponding to the part can be engraved directly into the material.

Using the generated packaging design moulds CAM techniques can be used to generate the machine specific CNC code. The Cryogenic CNC machining process can then be used to freeze the material and machine the required packaging from a variety of different soft materials, which depend on the end application and on the amount of material support required for the part.

4. Methodology

The methodology for designing and manufacturing of a fully individual product packaging solution consists of using scanning technology to capture the part geometrical information and then using the developed cryogenic CNC machining technology to produce the required packaging. The methodology used for the individualised packaging is shown in Fig. 5 through an IDEF0 representation. The three main areas for producing a fully individualized soft material package consist of scanning, designing and CNC machining. The scanning output is a point cloud of the part data, which can be meshed together. The output of the design phase is the mould packaging design and the output from the third phase is the individualised packaging solution.

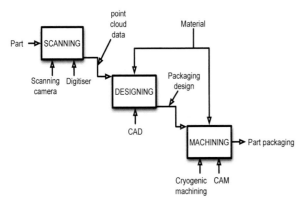

Fig. 5. IDEF0 representation of the methodology

4.1. Case study and results

In order to demonstrate the efficacy of the developed cryogenic CNC machining process and the individualised packaging design a case study example was developed and tested for a high value product.

Moulten Bicycles based in the UK are a company that specialize in hand made custom bicycles retailing in the price range of £9000 to £14000 [7]. A handle bar holder was chosen as the case study example, as this part requires safe damage free transportation around the production site for different manufacturing and finishing operations. In addition, this part is often sent to the customer directly as part of a consignment of bits to be assembled by the customer or as a replacement part. Fig. 6 illustrates an example of the bicycle and handle bar holder that is used as the case study example.

The process for designing of the product specific packaging first begins with capturing of the part data digitally. The handle bar holder is scanned using a 3D laser-stripe non-contact scanner to capture the complete geometry of the part in the form of a digitised point cloud. The part requires two scans, which are then meshed together. The density of the point cloud is controlled so as not to capture large amounts of redundant points, which would increase the processing and design

time. The orientation of the part throughout the scanning process is critical in order to capture all the necessary data. Anomalous data, captured from the scanning, such as background points can be filtered out and removed in the design process. The point cloud data is converted into curves and surfaces and this provides a functional CAD model with which to design the part packaging around. Fig. 7 shows the scanned part and the final CAD model of the part.

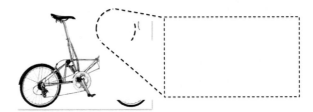

Fig. 6. The Moulten bicycle and the case study part

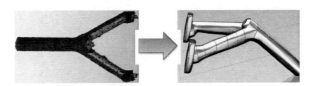

Fig. 7. Digitised part and full CAD part

The part packaging solution is then designed and in this case a sandwich type configuration is used in order to encapsulate the part in the central sections. This provides full part protection. For this particular case study the part is located centrally so as the degree of impact resistance is consistent around the part. Also, as the part has two additional assembly items, these are positioned within the complete handle bar holder packaging solution. Figure 8 illustrates an example of the designed packaging.

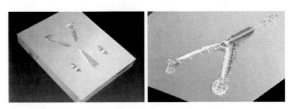

Fig. 8. The CAD packaging design

Using the cryogenic CNC machining facility and the correct material dependent machining parameters each segment of the package is machined. Tooling is selected based on the geometry, required feature size and generated process plans. The machining strategies are also devised based on the material properties and the complexity of the part.

Figure 9 depicts the finished neoprene foam packaging solution for the Moulten Bicycle handle bar.

Fig. 9. The final machine packaging

5. Conclusions

This paper has illustrated and demonstrated a totally new process for rapidly machining elastomeric material on conventional CNC machine tools using liquid nitrogen freezing technology. The concept of individualised packaging is discussed and an example case study is presented and machined using the cryogenic CNC machining process. Using this type of design and manufacturing process it makes it increasingly cost effective and efficient to produce high quality precision machined packaging for bespoke and high value products and parts.

References

[1]   Poor packaging design adding to industry footprint. http://envirowise.wrap.org.uk. [Web page] 2011 31/01/2012].

[2]   Dhokia V.G., Newman S.T., Crabtree P. and Ansell M.P., (2011), Adiabatic shear band formation as a result of cryogenic CNC machining of elastomers. Proceedings of the Institution of Mechanical Engineers Part B-Journal of Engineering Manufacture. 225(B9): 1482-1492.

[3]   Dhokia V. G., Newman S. T., Crabtree P. and Ansell M. P., (2010), A methodology for the determination of foamed polymer contraction rates as a result of cryogenic CNC machining. Robotics and Computer-Integrated Manufacturing. 26(6): 665-670.

[4]   Dhokia V.G., Newman S.T., Crabtree P. and M.P. Ansell, (2011), A process control system for cryogenic CNC elastomer machining. Robotics and Computer-Integrated Manufacturing. 27(4): 779-784.

[5]   Dhokia V.G., Crabtree P., Newman S.T., M.P. Ansell and R.D. Allen. (2009), Rapid craft CNC machining of personalsied orthotic insoles for medical and sporting applications. in Asia-Pacific Conference on Sports Technology. Hawaii, USA.

[6]   Herzog B., Gardner DJ., Lopez-Anido R. and Goodell B., (2005), Glass-transition temperature based on dynamic mechanical thermal analysis techniques as an indicator of the adhesive performance of vinyl ester resin. Journal of Applied Polymer Science. 97: 2221-2229.

[7]   Company Moulten Bicycle. [cited 2012 20/01/2012]; Available from: http://www.moultenbicycles.co.uk.

Modeling and simulation of shot peening process

X J Zhang [1], J B Wang [1], A Levers [2] and K K B Hon [3]

[1]   School of Mechanical Engineering, Northwestern Polytechnical University, Xi'an, Shaanxi 710072, China
[2]   School of Applied Sciences, Cranfield University, Cranfield, Bedfordshire MK43 0AL, UK
[3]   School of Engineering, University of Liverpool, Liverpool L69 3GH, UK,

Abstract. Shot peening is a widely used mechanical surface processing method for fatigue strength enhancement. It is a complex process as multiple nonlinear factors are involved. In this paper, a novel method is presented to model and simulate shot peening process in 3D system. Workpieces and tools were modelled in 3D system to simulate the dynamic characteristics of the shot peening process. A vectorial routing method was introduced to define the movement of nozzles in 3D space. Shot peening impacts over the surface of the specimen was simulated with a segmented field function. The deformation of specimen was simulated instantaneously with a combined analytical/numerical method by dividing the shot peened area of the specimen into segments according to the segmented field function of coverage. Verification of simulation results by shot peening experiments showed that this approach could facilitate process design and predict operational behaviours.

Keywords: modeling, simulation, shot peening

1. Introduction

Shot peening is a widely used mechanical surface processing method in forming of large scale thin component for its fatigue strength enhancement and low cost. However, it is difficult to model and simulate the shot peening process because the final state of the workpiece is the summative result of a large amount of elasto-plastic impacts on the surface.

Numerical methods have been used effectively for modeling and simulation of shot peening process. A direct way is to simulate the impacts against the workpiece surface one by one incrementally.

Meguid et al. [1] modeled and simulated the single and twin spherical indentations using the finite element (FE) method. Majzoobi et al. [2] obtained the compressive residual stress profiles introduced in the shot peening process by modeling and simulating multiple shot impacts on a target plate at different velocities.

However, it is impossible to model and simulate the shot impacts over the whole surface of a large scale thin workpiece because very fine mesh is required which makes the numerical model too large in size to be solved efficiently.

Indirect methods, i.e. equivalent loads of the shot impact action on the target show efficiency at reducing the problem size of the numerical models of the shot peening process.

Wang [3] applied pre-stresses and pre-strains to layers of the static finite element model to reduce the problem size. Gardiner and Platts [4] proposed to simulate the action of the residual stresses on large surface areas by imposing some level of prestress on the part. Levers and Prior [5] simulated the effect of peening on large flexible panels by applying varying thermal load to section points through the thickness. Homer and VanLuchene [6] introduced equivalent stretching and bending loads to element nodes to simulate the shot peening effect on the shot peened structures with the FE method.

In this paper, a novel combined analytical/numerical method of modeling and simulation of single-sided shot peening process in 3D space is investigated. An analytical loading model is built based on the concept of linear density of indentations. An equivalent loading method is proposed by scheduling the shot peening process with vectorial routing method and dividing the routes into segmented strips. FE models are built to simulate the deformation of large scale thin component in 3D space based on narrow strip shot peening experiments.

2. Models of shot peening routes and coverage

2.1. Vectorial shot peening routing method

In a typical system for shot peening large scale thin components, the workpiece mounted on the fixture of the shot peening machine moves relatively to the nozzles in 3D space. The projection of the relative motion path of the nozzles on the plane of the workpiece forms a series of routes which indicates the shot peening routes of the process. Therefore, the shot peening routes can be defined with a series of data sets A^i as follows:

$$A^i = \left(\vec{Q}^i, \vec{Q}^{i+1}, \vec{Vec}^i, v^i, sta^i \right) \tag{1}$$

where A^i defines the ith data set of the shot peening routes; $\vec{Q}^i$ and $\vec{Q}^{i+1}$, define the starting point and terminal point of the relative motion; $\vec{Vec}^i$ is a unit vector indicating the direction of the relative motion; v^i is a scalar that defines the relative speed of the motion; and sta^i is a status variable indicating whether there is a shot flow or not: if $sta^i = 0$, there is no shot flow; else if $sta^i = 1$, there is a shot flow.

Consequently, the shot peening routes of the process for a workpiece in a simulation system are defined by a series of vectorial data sets A^i.

2.2. Segmented field function model of coverage

Assuming the nozzles are perpendicular to the surface of the workpiece, the relationship between the width of the shot peened strip and the parameters of the nozzle can be formulated as,

$$r_c = kr_n + kd_n \tan\alpha \tag{2}$$

where r_c is the width of the shot peening strip, k is the number of nozzles, r_n the radius of the nozzle, d_n the distance of the outlet of nozzle to the surface of the shot peened surface, and α the scattering angle of shot flow.

Assuming the distribution of the shot peened indentations is uniform, when the shot flow is given, i.e. the number of shots from the nozzle per minute, the number of indentations covered by unit length in the shot peened area, namely, linear density of indentations is given in equation. 3:

$$c_l = \sqrt{1/\left[r_c \left(v/f_s + \xi \right) \right]} \tag{3}$$

where c_l is the linear density of indentations, ξ a compensational factor for saturated shot peening condition.

The linear density of shots on the ith segment of the shot peening routes can be given by a segmented field function as

$$c_l(x,y) = c_l^i\left(x_{Q^i}, y_{Q^i} \right) = 10^4 \sqrt{6 f_s^i / \left(\pi\rho v_n^i r_c^i d^3 \right)} \tag{4}$$

when,

$$(x,y) = \left(x_{Q^i}, y_{Q^i} \right) + \lambda^i \vec{Vec}^i + \eta^i \vec{Vec}^i \cdot \begin{bmatrix} 0 & 1 \\ -1 & 0 \end{bmatrix} \tag{5-1}$$

$$\lambda^i \in \left[0 \ \left\| \vec{Q}^{i+1} - \vec{Q}^i \right\| \right], \ \eta^i \in \left[-r_c/2, r_c/2 \right] \tag{5-2}$$

where f_s^i is the flow of shots on the ith segment of the shot peening route, v_n^i the speed of nozzles, r_c^i the width of the shot peened strip.

3. Combined analytical/numerical method

3.1. Analytical representation of shot peening loads

Assuming a cylindrical coordinate system $o\rho z$ on the cross section of an indentation, the bending moment $\bar{M}$ introduced by the compressive residual stresses to a line with a depth of δ can be expressed approximately as

$$\bar{M} = \int_{-W/2}^{W/2} \int_0^T \sigma_\rho \left(\delta - z \right) \mathrm{d}l \mathrm{d}z \tag{6}$$

Consider an array of indentations distributed along a line AB in the shot peened area, the bending moment to introduced by the residual stresses of the indentations covered by the line AB can be formulated as follows

$$M \cong n\bar{M} \tag{7}$$

where n is the equivalent number of the indentations cut across by the line AB.

3.2. Finite element simulation of shot peening

Analytical equivalent bending loads were first determined with narrow strip single-sided shot peening experiments and then applied in the FE models.

2024T351 aluminum alloy specimens, with sizes of 0.40m(length)*0.20m(width)*0.008m (thickness), were used in the experiments with a narrow shot peened strip covering along the symmetrical line of each specimen as depicted in Fig. 1. The specimens were shot peened on a pneumatic shot peening machine with five different feed rates: 0.5m/min, 1.0m/min, 2.0 m/min, 3.0 m/min, 5.0 m/min, and 8.0 m/min while the other shot peening parameters were kept at: shot peening distance 0.30m, shot peening air pressure 0.40MPa, shot flow 6kg/min, and shot type ASH660.

The equivalent bending moment for the action of the indentations on the shot peened specimen is calculated through the following formula

$$M^* \cong D^* l/r \tag{8}$$

where D^* is the equivalent rotational inertia corresponding to the loading moment applied to the specimen in FE model as shown in Fig. 2.

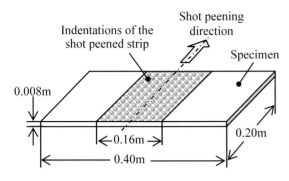

Fig. 1. Schematic diagram of narrow strip shot peening of the specimen

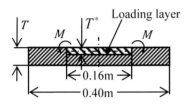

Fig. 2. Cross section of loading layer and the specimen in FE model. The loading layer corresponds to the compressive layer, which deforms the specimen, of the shot peened strip of the specimen.

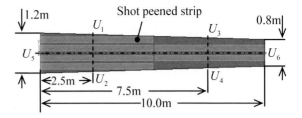

Fig. 3. Shot peening pattern of the FE model of the workpiece

Two different FE models of the shot peening process are built with Abaqus/Standard procedure: a verification model and a process model. Linear quadrilateral shell element S4 with five integration points in thickness is adopted in both models.

The verification FE model, the cross section of which is shown in Fig. 2, is built to verify the validity of the modeling method. The thickness of the loading layer T^* is assigned 0.002m for the model, and the equivalent bending moments applied to the edges of the loading layer in the lateral direction are calculated with Eq. (8) as 717.70 Nm, 584.49 Nm, 490.01 Nm, 452.30 Nm, 408.91 Nm, and 248.95 Nm corresponding to the respective feed rates.

The process FE model is built to simulate the deformation of the shot peened workpiece in 3D space. The shot peening pattern of the FE model corresponding to the defined shot peening routes is shown in Fig. 3. Equivalent bending moments corresponding to various feed rates of the shot peening routes are applied to the FE model via loading layer along the shot peened strips. Three cross section lines U_1U_2, U_3U_4, and U_5U_6 as illustrated in Fig. 3, are deployed to examine the deformation of the workpiece in the simulation.

The material of the specimens and the workpiece is aluminum alloy 2024T351 with Young's modulus 6.8×10^{10}Pa, Poisson's ratio 0.3, and mass density 2.8×10^3Kg/m^3.

4. Results and discussion

4.1. Verification experiments

The bending radii of the shot peened specimens in experiments and simulation are plotted against the feed rate as shown in Fig. 4. The maximum tolerance of the radii obtained from simulation compared with those obtained from experiments is no more than -11.4%.

4.2. Simulation of large scale thin workpiece

The deformed shape of the workpiece in 3D space under the action of the shot peening strips is given schematically in Fig. 5 with a contoured distribution of Von Mises stresses of the shot peened surface. The distribution of the stresses indicates that while higher lever stresses are introduced in the area of the shot peened strips, lower lever stresses or deformation are resulted in the non-shot-peened area because of the radial action of the residual stresses introduced by shot peening.

The deflection radius of the cross section curve U_1U_2 and U_3U_4 are plotted against the normalized point position measured from U_1 and U_3 respectively as shown in Fig. 6.

The distribution of the deflection radius along the cross section curves U_1U_2 and U_3U_4 shows that the shot peening strips deform the workpiece mainly at the shot peened area although related deformation is introduced in the surrounding non-shot-peened area.

Deflection in the longitudinal direction along the mid-cross section curve U_5U_6 is observed as well with a value of 338.44mm and an arc length of 71.04mm. The distribution of deflection radius is plotted against the normalized point position along the curve U_5U_6 measured from U_5 as given in Fig. 7.

Figure 7 shows that the deflection radius along the longitudinal mid-cross section curve U_5U_6 are generally more than 10 times larger than those along the lateral curves that covered by the shot peened strips. The reason is that the non-shot-peened area, which reduces the linear density of shot peened indentations n/l as defined in Eq. (11), resists the bending deformation of the workpiece in the longitudinal direction.

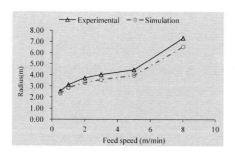

Fig. 4. Bending radius of the specimens versus feed speed.

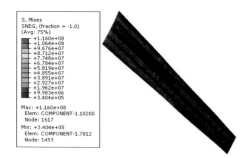

Fig. 5. Deformed workpiece in 3D space

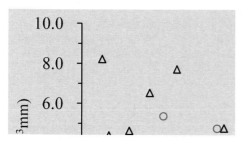

Fig. 6. Deflection radius versus normalized point positon along cross section curve U_1U_2 and U_3U_4.

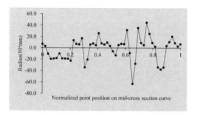

Fig. 7. Deflection radius versus normalized point positon along mid-cross section curve U_5U_6

4.3. Time efficiency

The problem size of the FE model of the workpiece shown in Fig. 3 is 14709 4-node shell elements with 15856 nodes, which took a total CPU time of 753.30 seconds and a memory of 114 Mbytes to solve the

problem on a personal computer with an Intel Core Duo CPU of 2.4GHz. Therefore, it is a time-efficient way of simulating the shot peening process with the analytical/numerical method.

5. Conclusions

A novel analytical/numerical method of modeling and simulation of single-sided shot peening process is proposed by combining analytical formulation of bending moment based on the concept of linear denstiy of shot-impact indentations and applying bending loads via loading layer strips.   The main conclusions are summarised as; Vectorial shot peening routing method defines the feed rates of each segment of shot peening route, which provides   auxiliary information for the design of shot peening strips. Segmented shot peened strips covered with uniform distribution of impacted indentations forms an equivalent uniform bending moment applied to the edges of the strips under small deformation   condition.   The   combined analytical/numerical method is demonstrated as time efficient in modeling and simulation of the deformation of sigle-sided shot peened large scale thin workpiece.

References

[1]   Meguid, S., Shuagal, G., Stranart, J. and Daly, J. 1999. Three-dimensional dynamic finite element analysis of shot-peening induced residual stresses. Finite Elements in Analysis and Design, 31:179-191.

[2]   Majzoobi, G., Azizi, R. and Alavi Nia, A. 2005. A three-dimensional simulation of shot peening process using multiple shot impacts. Jour. Materials Processing Technology, 164-165:1226-1234.

[3]   Wang, T. 2003. Numerical simulation and optimisation for shot peen forming processes. PhD Thesis, University of Cambridge.

[4]   Gardnier, D. and Platts, M. 1999 Towards peen forming process optimisation. ICSP-7. Warsaw, Poland.

[5]   Levers, A. and Prior, A. 1998. Finite element analysis of shot peening. Jour. Materials Processing Technology, 80-81:304-308.

[6]   Homer, S.E. and Van Luchene, R.D. 1991. Aircraft Wing Skin Contouring by Shot Peening. Jour. Materials shaping Technology, 9/2:90-101.

Evaluation of tool performance and hole quality when drilling C355 aluminium alloy using diamond coated and PCD drills

R. Rattanakit[1], S.L. Soo[1], D.K. Aspinwall[1], B. Haffner[2], Z. Zhang[3], D. Arnold[4], P. Harden[5]

[1]   Machining Research Group, School of Mechanical Engineering, University of Birmingham, Edgbaston, Birmingham, B15 2TT, UK

[2]   Doncasters Sterling, Colliery Lane, Exhall, Coventry, CV7 9NW, UK

[3]   Doncasters Group Ltd., Millenium Court, First Avenue, Burton-upon-Trent, Centrum 100, DE14 2WH, UK

[4]   MAPAL Ltd., Old Leicester Road, Swift Park, Rugby, CV21 1DZ, UK

[5]   Element Six Ltd., Shannon Airport, Shannon, Co. Clare, Ireland

Abstract. The paper details experimental work to investigate the influence of tool material/coating (uncoated carbide, chemical vapour deposited (CVD) diamond coated carbide and polycrystalline diamond (PCD)) and cutting speed (130 and 260m/min) on drill wear, cutting forces/torque and hole quality/integrity, when drilling 4.8mm diameter blind holes in cast, heat treated C355 aluminium alloy (~ 4.7% Si content). Tool flank wear after ~1680 holes was up to ~6 times higher with uncoated carbide (WC) drills compared to PCD, while thrust forces correspondingly showed a two-fold increase with the former at a cutting speed of 260m/min. Hole surface roughness produced using PCD drills did not exceed ~0.3µm Ra for all conditions tested, while it varied between 0.6 – 1.8µm Ra along the length of the hole when employing uncoated and coated WC tools, due to scoring and smearing of the workpiece. Hole roundness was below 7µm in all trials assessed. Hole cross sectional micrographs showed no significant damage or deformation of the material microstructure, with the exception of some minor pitting on surfaces produced using the uncoated WC drills.

Keywords: Drilling, CVD diamond, PCD, aluminium alloy

1. Introduction

Aluminium alloys are widely used in industry not least because of their exceptional strength-to-weight ratio, the tensile strength and density of a typical heat-treatable cast material (LM16 equivalent) being 240MPa and 2.71g/cm^3 respectively [1, 2]. Silicon is added to Al alloys to further improve casting ability and strength while minimising shrinkage during solidification and thermal expansion [2]. The presence of Si however can significantly reduce tool life during machining due to its abrasive properties [3], whilst the softer Al phase has a tendency to adhere to the surface of tools and form a built-up edge (BUE) especially at low cutting speeds, leading to poor component quality [4].

High-speed steel (HSS) and uncoated tungsten carbide (WC) tools have traditionally been utilised with cutting fluid for the machining of Al alloys [5], while low-friction coatings such as diamond-like carbon (DLC) and molybdenum disulphide (MoS2) based products, have been successfully used to minimise BUE and workpiece material adhesion when drilling Al-Si alloys under dry conditions or using minimum quantity lubrication (MQL) [6, 7]. Since their introduction in the early 1970's, polycrystalline diamond (PCD) products and more recently thin/thick film chemical vapour deposited diamond (CVD diamond) tools, have seen increased application in the automotive and aerospace industries. The significant advantages of PCD compared to standard WC cutters when machining non-ferrous materials have been well documented [8], and include significantly higher tool life, superior workpiece surface finish, reduced tool changing and increased productivity. Although less commonly employed, thick film CVD diamond (500µm thick) tooling has been shown to match or exceed the performance of PCD and WC when turning abrasive materials such as 20% SiC reinforced Al metal matrix composites and hypereutectic Al-Si alloys [9, 10]. Edge preparation costs aside, thick film CVD diamond generally results in longer tool life compared to the thin-film variant ($\leq$ 30µm thick), despite the former being prone to chipping from the deposition process induced residual stresses which can weaken the cutting edge [11].

The drilling of blind holes is a key operation in high volume production of aluminium compressor impellers for automotive turbocharger assemblies, prior to finish reaming. The following work was undertaken to investigate the influence of tool material/coating and cutting speed on drill wear, cutting forces/torque and hole quality/integrity when drilling 4.8mm diameter blind holes in cast C355 aluminium alloy.

2. Experimental work

The C355 aluminium alloy blocks containing ~4.7% Si were produced by die casting and heat treated to give a bulk hardness of 140HV. Individual test pieces measuring 340x180x30 mm were used for mainstream tool wear/life trials while smaller samples of 20x30x180 mm were prepared for force measurement, hole roundness, cylindricity, surface roughness, microhardness and microstructure assessment. Uncoated fine grain WC (K30F with 10% Co binder) tools were benchmarked against multi-layer CVD diamond coated products (~8μm thick applied on an indentical WC substrate) and two different brazed PCD grades, which involved a unimodal (CTB010; 10μm average grain size) as well as a mixed/multimodal (CTM302; 2-30μm grain size) formulation. The uncoated and CVD diamond coated twist drills were 3 flute with helix and point angles of 15° and 150° respectively whilst the PCD tools had 2 straight flutes (0° helix) and a corresponding 130° point angle.

Blind hole (4.8mm diameter) drilling tests to a depth of 28mm were performed on a Matsuura FX-5 high speed machining centre involving 2 variable factors of tool material (4 levels) and cutting speed (2 levels), see Table 1. Feed rate was kept constant at 0.1mm/rev while tool overhang was typically ~36mm with corresponding runout of <0.01mm. All tests were performed in a flood environment (3bar, 25l/min) using a water based emulsion containing mineral oil at a concentration of 7-10%. Tool life criterion was either 1680 holes (840 holes per block) or a maximum flank wear (VB_{Bmax}) of 0.3mm.

Response measures were recorded following the first hole and at intervals of 280 holes thereafter. Tool wear was measured with an optical microscope while thrust forces and torque were recorded using a four-component drilling dynamometer fitted with a bespoke workpiece fixture and connected to charge amplifiers. Mean hole diameter was evaluated at the top, middle and bottom locations with a digital indicator attached to a Diatest split ball probe (0.001mm resolution). Hole roundness and cylindricity measurements were performed on a Talyrond 300 unit, each with a total of 15 profiles.

Table 1. Test matrix and variable factors

Test No.	Tool material/coating	Cutting speed, m/min (rpm)
1	Uncoated WC	130 (8,500)
2	Uncoated WC	260 (17,000)
3	CVD diamond coating	130 (8,500)
4	CVD diamond coating	260 (17,000)
5	PCD CTB010	130 (8,500)
6	PCD CTB010	260 (17,000)
7	PCD CTM302	130 (8,500)
8	PCD CTM302	260 (17,000)

Drilled hole surface roughness (Ra) was assessed with a Form Talysurf 120L having a 2μm radius diamond stylus using a 0.8mm cut-off and 4.0mm evaluation length. Measurements were taken at ~2mm and 25mm from the hole entry and averaged. Cross-sectioned specimens (using wire EDM down the hole centreline) were cold mounted in an epoxy resin mixture followed by grinding and polishing prior to the measurement of microhardness depth profiles, which was carried out on a Mitutoyo HM 124 tester. These were performed on the last hole of each test with a Knoop diamond indenter (25g load) at a depth of ~12mm from hole entry. The polished samples were subsequently etched in Keller's reagent to enable optical investigation of microstructure and surface/subsurface damage. No trial replications were performed due to limited tool availability and cost constraints.

3. Results and discussion

Figure 1 shows optical micrographs of drill edges in the new and worn condition for each trial. Significant adhesion of workpiece material and BUE was evident on the flank and chisel edges of uncoated drills irrespective of cutting speed. In contrast, this was not observed when employing CVD diamond coated tools, although peeling of the coating layer occurred after hole 282 when operating at the higher cutting speed, which led to premature termination of the test. Similarly, no signs of material adhesion or BUE was visible on any of the PCD blanks, but there was accumulated material on the carbide body/chisel edge of the tools.

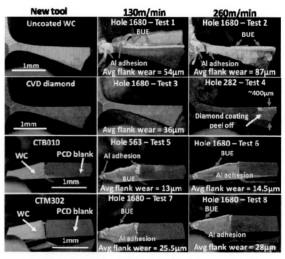

Fig. 1. Micrographs of drills in new and worn condition

The drill in Test 5 suffered catastrophic breakage due to a malfunction in the tool changer during the experiment and hence the test had to halted after 563 holes. The flank wear progression of each tool over the test duration is detailed in Fig. 2. In general, the PCD drills showed the lowest wear rate, with the CTB010 grade (Test 6) having a VB_{Bmax} of only 14.5μm after 1680 holes. Conversely, wear on the uncoated WC drills at low (Test 1) and high (Test 2) cutting speeds was 55μm and 87μm respectively at test cessation (~4 and ~6 times higher respectively).

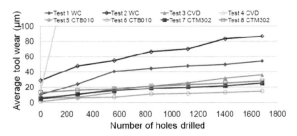

Fig. 2. Drill flank wear curves

Similar results were recorded for thrust force data, where a significant reduction was observed when employing PCD drills. Despite a straight fluted geometry, the PCD products (Test 5 to Test 8) generated ~70 to ~125% lower average thrust forces compared with uncoated and CVD diamond coated twist drills (Test 1 to Test 4) under equivalent machining conditions, see Fig. 3. Additionally, forces in the former remained stable throughout the trials. Although not shown here, torque values generally showed an increasing trend with the number of holes drilled, due to tool wear progression. A number of tests also exhibited high initial torque levels (Tests 2, 3, 4 & 8), which was possibly due to Al material adhering on drill corners at trial commencement.

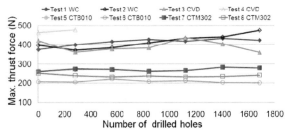

Fig. 3. Maximum thrust forces

Hole diameter in general decreased as cutting progressed due to tool wear, see Fig. 4. All holes produced with the CVD diamond coated drills were oversized, which was a result of the coating thickness. Although initially oversized by 4-10μm, the diameter of holes machined with uncoated WC tools steadily decreased by ~15μm at the end of the trial. Despite being undersized, holes drilled with PCD tools showed minimal diametral variation over the entire experiment.

Hole roundness and cylindricity results are presented in Fig. 5. The average out of roundness did not exceed 7μm and was comparable for each of the different tool materials evaluated. In terms of cylindricity, the mean deviation was marginally higher when using PCD drills, although this did not exceed 22μm. This was most likely

due to the difficulty in chip/swarf evacuation as a result of the straight fluted tool geometry. In addition, cylindricity tended to improve when operating at lower cutting speed.

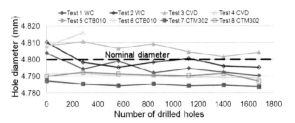

Fig. 4. Hole diameter

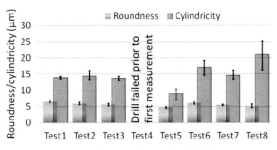

Fig. 5. Hole roundness and cylindricity

Surface roughness near the hole entry was generally lower compared to the bottom location, see Fig. 6 and 7 respectively, most probably due to the increased difficulty of swarf evacuation and possible tool vibration as drilling depth increased [12]. In general, carbide drills produced holes with the poorest surface finish particularly towards the exit position, which reflects the greater tool wear rate and high levels of BUE as evident from previous tool micrographs. The PCD drills generated the lowest surface roughness irrespective of cutting speed, which was below 0.3μm Ra even after 1680 holes and consistent over the entire hole length.

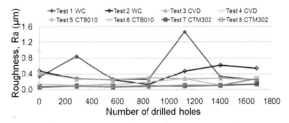

Fig. 6. Hole surface roughness near entry location

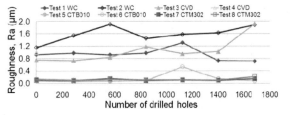

Fig. 7. Hole surface roughness near exit location

Other factors which compromised Ra was the formation of 'scuff' marks similar to that illustrated in Fig. 8(a), which were caused by ploughing/re-deposited material, and the occasional presence of workpiece defects (inclusions, voids, etc.) as shown in Fig. 8(b). This was probably the reason for the unusual peaks/spikes in the surface roughness profiles observed in Tests 1 and 6.

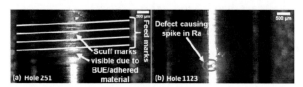

Fig. 8. Hole cross section from; (a) Test 1-entry, (b) Test 6-exit

Figure 9 shows workpiece microhardness results from trials after 1680 holes (Tests 1-3, 6-8). A hardened region up to a maximum of 169HV (~21% above bulk hardness) and extending to a depth of ~100µm from the machined surface was observed in the majority of specimens analysed. No major variation in hardness was detected with changes in drill speed. Investigation of corresponding cross-sectional micrographs detailing drilled hole surface/subsurfaces, see Fig. 10, showed no obvious evidence of microcracks, surface damage or microstructural alterations, irrespective of tool condition. Minor pitting however was seen in the last and first holes from Tests 1 and 2 (uncoated WC tools) respectively.

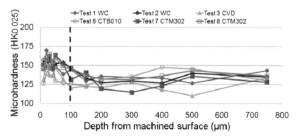

Fig. 9. Microhardness depth profiles at hole 1680

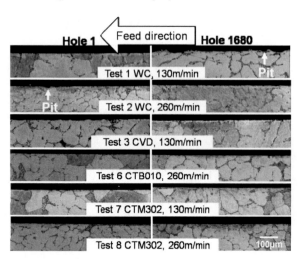

Fig. 10. Cross-sectional micrographs of hole microstructure

4. Conclusions

Substantial amounts of workpiece adhesion and BUE were present on the uncoated WC drills, which led to comparatively high thrust forces (up to ~500N). In contrast, evidence of adhered material on the CVD diamond and PCD tool surfaces was minimal. While the CVD diamond coating generally increased drill life compared to uncoated tools, the relatively poor bonding between the coating and substrate led to peeling/delamination at the higher cutting speed level. A hardened region measuring up to ~29HV above the workpiece bulk hardness to a depth of ~100µm was typically recorded in tests which achieved 1680 holes. Sample hole cross-sectional assessment of microstructure revealed no microcracks or microstructural alterations. While small surface pits were observed in trials using the uncoated WC tools, such flaws would easily be removed by any subsequent finishing/reaming operation.

References

[1]   Bauccio M, (1993) ASM Metals Reference Book. ASM International

[2]   Polmear I, (2006) Light alloys: From traditional alloys to nanocrystals. Butterworth Heinemann

[3]   Roy P, Sarangi SK, Ghosh A, Chattopadhyay AK, (2009) Machinability study of pure aluminium and Al-12% Si alloys against uncoated and coated carbide inserts. Int J Refract Met Hard Mater27(3):535-544

[4]   Yousefi R, Ichida Y, (2000) A study on ultra-high-speed cutting of aluminium alloy: Formation of welded metal on the secondary cutting edge of the tool and its effects on the quality of finished surface. Precis Eng 24(4): 371-376

[5]   Smith GT, (2008) Cutting tool technology. Springer

[6]   Bhowmick S, Alpas AT, (2008) Minimum quantity lubrication drilling of aluminium-silicon alloys in water using diamond-like carbon coated drills. Int J Mach Tools Manuf 48(12-13):1429-1443

[7]   Wain N, Thomas NR, Hickman S, Wallbank J, Teer DG, (2005) Performance of low-friction coatings in the dry drilling of automotive Al-Si alloys. Surf Coat Technol 200(5-6):1885-1892

[8]   Jennings M, Clark I, (1995) PCD in the automotive industry. Ind Diamond Rev 55(2):51-53

[9]   Sussmann RS, Brandon JR, Collins JL, Whitehead AJ, (2001) A review of the industrial applications of CVD diamond. Ind Diamond Rev 61(4):271-280.

[10]   Uhlmann E, Friemel J, Brucher M, (2001) Machining of hypereutectic aluminium-silicon alloy. Ind Diamond Rev 61(4):260-265

[11]   Kanda K, Takehana S, Yoshida S, Watanabe R, Takano S, Ando H, Shimakura F, (1995) Application of diamond-coated cutting tools. Surf Coat Technol 73(1-2):115-120

[12]   El-Khabeery MM, Saleh SM, Ramadan MR, (1991) Some observations of surface integrity of deep drilling holes. Wear 142(2):331-349.

Real-time monitoring, control and optimization of CO_2 laser cutting of mild steel plates

E. Fallahi Sichani, J. De Keuster, J.-P. Kruth and J. R. Duflou
Dept. of Mechanical Engineering, Katholieke Universitieit Leuven, Celestijnenlaan 300 B, B-3001 Heverlee (Leuven), Belgium

Abstract. This paper presents a real-time monitoring and control system for oxygen-assisted CO_2 laser cutting of thick plates of mild steel. The proposed system consists of two subsystems, namely a process monitoring and a control and optimization module. The process monitoring module evaluates the cut quality based on a set of sensing parameters, derived from analysis of data provided from two optical sensors (photodiodes and a NIRcamera). These sensing parameters are chosen to be well correlated with different quality characteristics of the cut surface. An overview of the most suitable set of sensing parameters and how they correlate with different quality deteriorations are mentioned as well as the reasons for these correlations. The real-time control and optimization module is implemented as an expert system. It compares the sensing parameters with predefined thresholds and diagnoses the quality defect. Furthermore the system modifies the cutting parameters based on the predefined set of rules corresponding to the identified quality defect. The obtained results prove the effectiveness of the system in terms of increased autonomy, productivity, and efficiency of the process, as well as elimination of the need for manual quality control and the possibility to automatically generate quality reports.

Keywords: laser cutting, monitoring, adaptive control, optimization

1. Introduction

In the case of cutting thick plates, the very narrow process window of CO_2 laser cutting necessitates using a real-time monitoring and adaptive control system which is not commercially available so far. As state of the art prior to the reported contribution, the following efforts can be referred to.

Jorgensen [1] used a Si photodiode and a CCD camera to monitor the laser cutting process. The mean value and the variance of the photodiode signal proved to correlate well with dross formation and the roughness of the cut surface, respectively. As for the CCD camera, the mean pixel value of the recorded images proved to correlate well with dross formation, whereas the variance could be correlated to the roughness of the cut surface. Furthermore, based on the images, the cut kerf width could also be measured. Authors [2] have reported on successful performance of a photodiode-based monitoring

and control system being capable of detecting and correcting typical quality deteriorations in cutting thick plates of mild steel. Chatwin et al. [3-5] developed a knowledge-based adaptive control system for laser cutting, based on a photodiode and a CCD-camera that monitors the spark cone ejecting from the bottom of the plate. A photodiode was used to measure the temperature/irradiance, emitted by the cut front. Reportedly a bad cut quality was typically characterized by a larger, more diffuse spark cone. Decker et al. [6] developed a collinear photodiode set-up using a partially transmitting mirror. The presented results are in agreement to those of Jorgensen. Kaplan et al. [7] measured the thermal radiations emitted from the cutting front using a selective ZnSe mirror and Si photodiode. The roughness proved to be proportional to the detected striation period. The occurrence of burning defects could be detected by a quality factor combining the mean and variance of the photodiode signal.

Poprawe et al. [8, 9] developed a universal coaxial process control system, a modular system that can be equipped with different sensors, like CCD-cameras and photodiodes, for monitoring different laser manufacturing processes. The major advantage of the photodiode-based monitoring system is high monitoring frequencies thanks to the very short response time and relatively fast signal processing techniques. The main advantage of camera-based monitoring systems is the capability of providing spatial information which offers more insight into the process.

2. Experimental setup

The setup used for the experiments reported in this paper was a 2D laser cutting machine with flying optics architecture, a 6 kW CO_2 laser source ($M^2 = 4.24$) and a Fanuc controller. A monitoring direction coaxial with the laser beam was chosen. As shown in Fig. 1, the optical setup consists of:

- a dichroic mirror transmitting the laser beam and reflecting the NIR process emission towards the camera setup
- a diaphragm and a focusing lens to focus the process radiation
- a beam-splitter folding mirror to assure optimal process light transmittal to reflect a part of the process radiation towards the camera and to transmit the remainder towards a photodiode sensor (not shown on Fig. 1).

Taking the melt temperature of mild steel and the laws of Planck and Wien into account, it can be calculated that the peak wavelength of the process emission spectrum is around 1.7 μm. Therefore a NIR camera with an InGaAs detector was chosen.

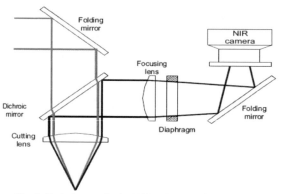

Fig. 1. Optical setup for the NIR camera monitoring system

3. Quality criteria and sensing parameters

In compliance with relevant standards [10-12], the following cut quality characteristics were selected for monitoring purposes:

- dross attachment: quantified by the height of the dross (resolidified material) roughness: quantified by R_z, measured by a non-contact profilometer at a height as specified in [10, 11]
- drag of the striations: quantified as the distance between the start and the end point of the striation, orthogonally projected to the plate surface (measured by optical microscope)
- burning defects: quantified as the average number of burning defects per unit length
- squareness of the cut edge: quantified as per [10], measured by a tactile probe
- kerf width

Figure 2 shows NIR images recorded during the laser flame cutting of ST52-3 15 mm at cutting speeds 100-120% of the standard value. Offline analysis of the cut surface revealed that the drag of the striations and the dross attachment are directly proportional to the cutting

speed. In NIR images, this is clearly visible as a significant enlargement of the central hot zone, particularly along the cutting direction. This can be attributed to the elongation and slightly higher location of the process front which is the result of depleted energy input per unit length.

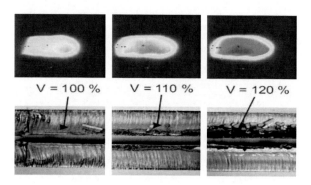

Fig. 2. NIR images vs. cut quality for laser flame cutting 15 mm thick ST52-3 with increasing speed

At low cutting speeds (e.g. 60-100% of the standard speed), the drag of the striations decreases and no dross attachment was observed. However the roughness, number of burning defects and cut kerf width all increase. In the NIR images captured while cutting at moderately low speed, the process zone appears less elongated and with a low intensity (as illustrated in the image corresponding to V=80% in Fig. 3). This is the result of the surplus energy per unit length, leading to a straighter and steeper process front and thus easier melt removal. When the cutting speed is further reduced, the width of the tail of the process front increases and varies significantly (as illustrated in the image corresponding to V=60% in Fig. 3). Such an image is suggestive of localized increases of kerf width (burning defect) because of a too high energy input per unit length.

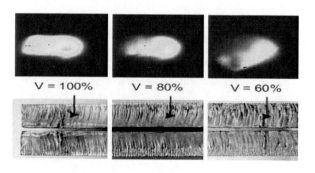

Fig. 3. NIR images vs. cut quality for laser flame cutting 15 mm thick ST52-3 with decreasing speed

The kerf width was determined based on the calculation of the intensity gradient of the NIR images (Fig. 4). This method is computationally expensive and is therefore not implementable in a high-frequency monitoring algorithm.

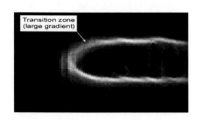

Fig. 4. Kerf width indicated by the transition zone between solid and molten material

No sensing parameter could be identified that correlates to the squareness of the cut edge because of the view point of the camera. Table 1 offers an overview of sensing parameters correlated with different cut quality aspects in terms of Arbituary Digital Units (ADU).

4. Adaptive control and optimization

The adaptive control was implemented as a rule-based expert system. During the real-time monitoring, the values of different sensing parameters are compared with well-chosen threshold values and a predefined quality class is assigned to the current cut quality.

- Quality class 0: acceptable cut quality;
- Quality class 1: too high drag of the striations, indicating a high risk of dross formation;
- Quality class 2: presence of burning defects, therefor increasing the roughness of the cut edge;
- Quality class 3: combination of Quality class 1 and 2 ,typical for preheated workpieces;
- Quality class 4: loss of cut.

The rule-based expert system associates each quality class with a quantitative corrective strategy. based on which cutting parameters are overwritten in real time to salvage that specific quality deterioration. For a detailed explanation on the corrective strategy corresponding to each quality class see [13].

The performance of the adaptive control and optimization algorithms was evaluated on different material/thickness combinations. Fig. 5 shows the real-time alteration of the process parameters during a linear cut in HARDOX-400 25mm starting from standard process settings. While the cut quality remained acceptable, the optimization resulted in approximately 21% increase of cutting speed (i.e. 168 mm/min) due to an increase of the duty cycle and the gas pressure.

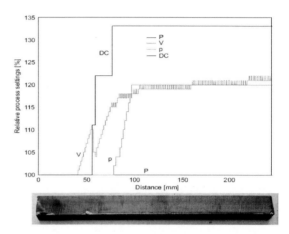

Fig. 5. Control and optimisation of linear cutting of HARDOX-400 25 mm starting from standard process settings

Figure 6 illustrates the real-time alteration of the process parameters during a linear cut in ST52-3 15 mm, starting with a cutting velocity 20% higher than the standard value. It can be noted that at the outset of the cut, the shortage of energy input per unit length was sensed by the control and optimization system and therefore the velocity and the duty cycle were drastically changed in order to correct the cut quality.

A typical problem in laser cutting of complex contours is the heat accumulation at sharp corners. This preheating effect drastically increases the oxidation rate. The performance of the control and optimisation system was verified during cutting of complex contours including all possible geometrical features: linear segments, smoothly curved segments, corner segments including both obtuse and sharp angles (Fig. 8).

In this test the outer contour was cut before the inner one. This unusual cutting sequence was opted to create the worst case scenario with respect to heat occumulation by prohibiting the heat removal from the central zone of the workpiece. Fig. 7 shows the real-time evolution of the

Table 1. Overview of identified sensing parameters

Quality criterion	Sensing parameter(s)
Dross attachment	Length of the hot process zone (ADU > 3000), Area of the hot process zone (ADU > 3000)
Roughness	Variation of the width of the global process zone (ADU > 1000)
Drag of striations	Length of the hot process zone (ADU > 3000), Area of the hot process zone (ADU > 3000)
Burning defects	Variation of the width of the global process zone (ADU > 1000)
Kerf width	Calculation based on the intensity gradient of the NIR image

process parameters during cutting of the above-mentioned part. Fig. 8 compares the resultant cut quality with and without the real-time control and optimization system.

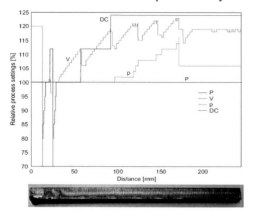

Fig. 6. Control and optimisation of linear cutting of ST52-3 15 mm, starting from V=120 % of $V_{standard}$

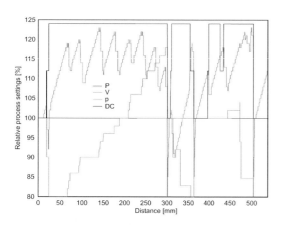

Fig. 7. The performance of the control and optimisation system for a complex workpiece

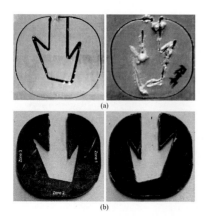

Fig. 8. Top and bottom view of the complex workpiece, (a) without the control and optimization system; (b) with the control and optimization system

5. Conclusion

The developed NIR-camera-based monitoring system is capable of sensing the main quality parameters in CO_2 flame laser cutting. Dross attachment, roughness, drag of striations and the occurrence of burning defects can be monitored in real time; whilst the kerf width can be calculated at a sufficiently high sample frequency. The developed real-time monitoring system was linked with a rule-based adaptive control system for automatic process quality assurance. The functional performance of this control system was illustrated by means of a series of tests on different material/thickness/geometry combinations.

References

[1]   Jorgensen H. , PhD thesis, Technical University of Denmark, Lyngby, Denmark, 1990

[2]   De Keuster J., Duflou J. R., Kruth J.-P., in Proceedings of the LANE 2007, Vol. 2, 979-992, 1999

[3]   Huang M.Y. and Chatwin C.R., Optics and Lasers in Engineering 21, 273-295, 1994

[4]   Huang M.Y. and Chatwin C.R., Lasers in Engineering 3, 125-140, 1994

[5]   Lim S.Y. and Chatwin C.R., Lasers in Engineering 3, 99-112, 1994

[6]   Decker I., Heyn H., Martinen D., and Wohlfahrt H., in Proceedings of the SPIE, volume 3097, 29-37, 1997

[7]   Kaplan A.F.H., Wangler O., and Schuöcker D., Lasers in Engineering 6, 103-126, 1997

[8]   Poprawe R. and Konig W., in Annals of the CIRP 50, 137-140, 2001

[9]   Abels P., Kaierle S., Kratschz C., Poprawe R., and Schulz W., in Proceedings of the ICALEO 1999, Vol. 87, E99-108, 1999

[10]   ISO 9013: Thermal cutting - Classification of thermal cuts – Geometrical product specification and quality tolerances

[11]   VDI 2906: Blatt 8, Quality of cut faces of (sheet) metal parts after cutting, blanking, trimming or piercing - Laser cutting

[12]   DIN 2310: Thermal cutting - Part 30: Classification of thermal cuts, principles of process, quality and dimensional tolerances

[13]   Fallahi Sichani E., De Keuster J., Kruth J.-P., Duflou J. R. Physics Procedia, Volume 5, Part 2, 2010, Pages 483-492.

The stress-strained state of ceramic tools with coating

Grigoriev[1] S., Kuzin[1] V., Burton[2] D., Batako[2] A. D.
[1] Moscow State Technological University «Stankin», Vadkovsky per. 3a, Russia.
[2] Liverpool John Moores University, Byrom Street, Liverpool , L3 3AF, UK.

Abstract. This paper presents some fundamental investigations into the physical laws that govern the wear of coated ceramic tools. A mathematical model of a stress-strained state of ceramic tools with coating has been developed to analyze this process in details. This mathematical model is built on the finite elements method that is based on multi-dimensional theories of heat transfer and elasticity. In this work, the developed model uses the solutions of two-dimensional heat conductivity and elasticity to predict the wear of coated ceramic tools in stress-strained state. Results of the simulated stress-strained state of ceramic tools with coating are presented. A new generation of coating for ceramic tools have been developed using of these results.

Keywords: Ceramic Tools, Coating, Stress-Strained State, Mathematical Model.

1. Introduction

The ceramics is a very promising tool material. Ceramic tools considerably intensify production processes and improve product precision and quality [1]. However, such tools are of limited applicability on account of their poor reliability.

Effective method of improving of ceramic tools is the application of functional coatings. The composite nature of such an arrangement, allows to combine exploiting efficiently the properties of the internal bulk ceramic material and the coating. Functionality of these coatings consists of the following; the coating eradicates technological defects at the surface layer of the tools. The dense structure of a coating increases the resistance of the ceramic tool to the formation of cracks. Layers of coating become a barrier for the growth of cracks generated from the inside bulk of ceramic material.

However, operating experience of ceramic tools shows that coatings from various refractory materials do not always affect the operational characteristics. This is because coated tools become a complex system «ceramics – coating» in which each component has its own physical mechanical and thermal properties. An unfavourable combination of these elements can lead to the formation of localized high stresses at the upper layer of ceramics and to the generation of cracks.

Therefore, the identification of the relationship of the coating properties with a stress-strained state of the surface layer of ceramic tools is an actual scientific task. This article is devoted to the solution of this task.

2. The Mathematical model

The stress-strained state of various structures may be effectively studied by the finite element method, which is well adapted to machine calculations. It produces a complete picture of the stress-strained state of complex structures, with high precision, and the results are presented in useful form, including graphical display. This method is used in the present work.

A mathematical model was developed to study the stress-strained state of coated ceramic tools. At model creation it was assumed that the flat problem is considered; no plastic deformation occurs at the cutting edge of the tool; the structure of ceramics and coating is the faultless so there are neither pores nor cracks.

Initially, a micro-structure model of a coated ceramic plate was developed [2]. The algorithm of realization of this approach is given in Fig. 1 (a). The cutting plate is presented as a repeating elementary fragment of coated ceramic plate consisting of structural elements of ceramics (grain, inter-grain phase and matrix); the coating and the workpiece material. This allows using the heterogeneous properties of the ceramic and the workpiece materials in the model. Then on the basis of the allocated elementary fragment of coated ceramic plate the calculation scheme was generated. It is presented in the form of a design [Fig. 1 (b)]. This design consists of a single grain having ellipse form with semi-major axis (a) and semi-minor axis (b). This grain is closed up in a matrix through inter-grain phase of a thickness δ_f.

The free surface of the grain, inter-grain phase and matrix is covered by N layers of coating with a thickness d_j, where $j = 1, 2..., N$. The elements of this design is characterised by the following properties: density ρ, modulus of elasticity $E(T)$, thermal conductivity $\lambda(T)$, heat capacity $c_p(T)$, linear expansion α and Poisson ratio μ.

The following scheme of loading of coated ceramic plate is developed. Figure 1 (b) shows the schematic complex loading approach implemented in this model. The external contour of layer N is subjected to the action of time-dependent point forces F_i inclined at angles β_i, to the y axis; distributed forces P_1 and P_2; and heat fluxes Q_1 and Q_2. In the calculations, we take account of the convective heat losses with heat-transfer coefficients h at the sections of the contour with no heat flux.

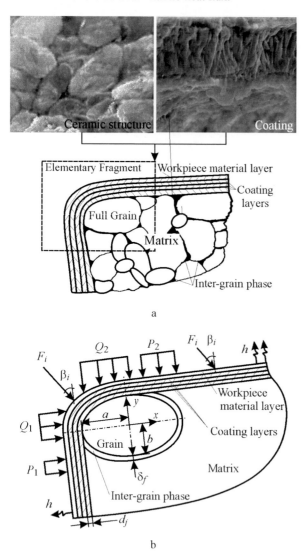

a

b

Fig. 1. Micro-structure model of a coated ceramic plate

The mathematical model of the stress-strained state of coated ceramic plate is based on the solutions of two dimensional problems of thermal conductivity and elasticity. Finite element method is used to implement algorithms to solve stationary and non-stationary problems of thermo-elasticity. The algorithms are realized as an automated system of computing the thermo-strength calculations of ceramic plates [3].

The calculation scheme shown in Fig. 2 was used for identification of the relationship of the coating properties with a stress-strained state of the surface layer of ceramic plate. Silicon nitride (Si_3N_4) plate coated with different refractory material was used in this study. The workpiece material was chromium steel (X32CrMoV12-28). The following loads were used: $F_1 = F_2 = 0.01$ N; $\beta_1 = \beta_2 = 45°$; $P_1 = P_2 = 1 \cdot 10^8$ Pa; $Q_1 = Q_2 = 1.4 \cdot 10^8$ W/m^2 and $h = 1 \cdot 10^5$ W/m$^2 \cdot$K.

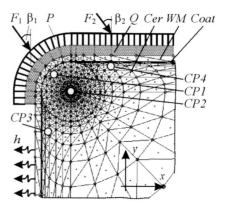

Fig. 2. Calculation scheme for modelling (a = b = 50 μm)

The control points $CP\,1 - CP4$ were used for an analysis of calculations. These control points mean fixed four finite elements of reseaching design. In Fig. 2 the following abbreviations are used: *"Cer"* is the ceramics, *WM* is the workpiece material and *"Coat"* is the coating.

3. Results and discussion

The results of calculation of temperatures taken at the control points are given in Table 1, where the first row shows the temperatures of the ceramic plate without coating. It is observed from this table that coatings have a drastic effect on the thermal state of the ceramic plate. The temperature in the plate with the coatings of various refractory materials varies in the range of 25%. Plates with TiB_2 coating have the lowest temperature whereas BN coating give the highest values. TiB_2, TiN, WC, BeO, Mo_2C and TiC coating provide the reduction in temperature comparing plates without coating. However, the other types of coating including Al_2O_3 increase the temperatures. With plates made of Si_3N_4, the TiN coating gave 8% reduction in temperatures whereas TiC achieved only 1%.

Different refractory materials of coating practically do not influence the character of a temperature field of Si_3N_4 plates. The minimum changes of parities between temperatures in different control points for different ceramic plates testify this.

From the obtained results, the relationship of the thermal conductivity of the coating and the thermal state of the plate was derived. The temperature (T) at the control points $(CP1-CP4)$ can be expressed as follows:

$$T^{1} = 1489.9\ \lambda_{coat}^{-0.0474};\ T^{2} = 986.8\ \lambda_{coat}^{-0.0429},$$

$$T^{3} = 443.78\ \lambda_{coat}^{-0.034};\ T^{4} = 968.4\ \lambda_{coat}^{-0.0428}$$

Table 1. Temperatures at control points on the plate

Coating	Temperatures, T, $^\circ$ C			
	CP1	CP2	CP3	CP4
No coat	1240	836	391	822
TiB$_2$	1103	762	378	762
TiN	1145	789	387	791
WC	1181	810	392	810
BeO	1219	809	364	780
Mo$_2$C	1222	829	393	821
TiC	1235	843	392	831
SiC	1309	875	393	847
TaC	1315	878	400	858
MgO	1340	898	415	882
Y$_2$O$_3$	1340	898	415	882
Al$_2$O$_3$	1347	893	400	867
AlN	1355	899	404	875
ZrO$_2$	1358	908	418	893
BN	1368	908	408	886

Table 2 gives the results obtained in studying the effect of coating the stresses in a silicon nitride plate. It is seen that the coatings have a direct influence on the stresses. The stress intensity at the control points is increased up to 20 times in coated plate.

Table 2. Stresses at control points on the plate

Coating	Stress intensity σ_i, MΠa			
	CP1	CP2	CP3	CP4
No coat	275	45	641	600
WC	237	45	477	510
TiB$_2$	239	28	496	600
TiN	240	17	585	590
Mo$_2$C	240	32	595	540
TiC	264	55	590	520
BeO	270	46	601	632
TaC	322	91	584	485
SiC	405	114	551	520
MgO	600	296	693	201
Y$_2$O$_3$	600	296	693	201
AlN	646	177	590	488
BN	1231	252	423	419
Al$_2$O$_3$	2236	459	660	580
ZrO$_2$	2272	1137	746	1047

The lowest intensity is obtained with *WC* coating and the highest was recorded with the *ZrO$_2$*. It was found that

coating with *WC*, *TiB$_2$*, *TiN*, *Mo$_2$C*, *TiC* and *BeO* on *Si$_3$N$_4$* plate reduce the stresses in the surface layer of ceramic plate whereas the other coatings simply increase the stress level.

It is observed that *TiN* and *TiC* reduced the intensity of stresses at the check point CP1 by 15 and 5% respectively compared to uncoated plate. However the study of an average stress intensity shows that *TiC* has some advantages over the *TiN* with *Si$_3$N$_4$* plate. Coating with *Al$_2$O$_3$* considerably increases stress intensity σ_i at all control points at the surface layer.

The variation of the coating thickness δ_{coat} does not provide a linear effect on the thermal state of the *Si$_3$N$_4$* cutting tips. It was fond that, with the increase of the coating thickness, the temperature at check point *CP1*, *CP2* and *CP4* decreases but at *CP3* the temperature increases. The effect of the coating thickness on the temperature was derived as follows:

$$T^{1} = -11.143\ \delta_{coat} + 1324.1;\ T^{2} = -5.6939\ \delta_{coat} + 883.27;$$

$$T^{3} = 1.602\ \delta_{coat} + 389.26\ ;\ T^{4} = -5.3367\ \delta_{coat} + 868.91.$$

Also we investigated influence of different properties of a coating on a stress-strained state of the surface layer of ceramic plate. The coating used in these experiments was *TiC* with the thickness δ_{coat} = 5 μm. In this study a series of experiments were conducted where one parameter was varied from minimum to maximum for different refractory materials used in this work. For example, the coefficient of thermal conductivity varied between 2.4 and 150 W/m·K when investigating its effect on temperatures and stresses in the plate. Other parameters kept constant (identified *TiC*).

Special attention was paid on influence of coefficient of thermal conductivity of coating on a stress-strained state of ceramic plate. This experiment was carried out with λ_{coat} =2.4 ÷ 50 W/m.K. From the obtained results, it was identified that increasing the thermal conductivity up to 150 W/m.K led to the decrease of temperature at all control points by 1.25~1.63 times. Here, the following dependence of the temperature on the coating thermal conductivity was derived, i.e.

$$T^{1} = -3.53\lambda_{coat} + 1385.6;\ T^{2} = -2.07\lambda_{coat} + 922.2;$$

$$T^{3} = -0.56\lambda_{coat} + 417.4;\ T^{4} = -1.89\lambda_{coat} + 905.2.$$

The influence of coating thermal conductivity on the stresses in the ceramic inserts is not conclusive. However, the increase of coating thermal conductivity from 2.4 to150 W/m.K reduces the stress intensity in the tools at the control point *CP1-CP3* but it considerably increases the intensity of stresses at *CP4* which situated at a midpoint of the action of the loads. Here using plate of *Si$_3$N$_4$* with a *TiC* coating the stress intensity was expressed as function of coating thermal conductivity at respective control points:

$$\sigma_i^1 = -0.11\lambda_{coat} + 268.9; \quad \sigma_i^2 = -0.13\lambda_{coat} + 60.8;$$

$$\sigma_i^3 = -0.93\lambda_{coat} + 614.9; \quad \sigma_i^4 = 1.65\lambda_{coat} + 463.9.$$

The effect of thermal expansion of the coating (α_{coat}) on the stresses in the ceramic plate was investigated by varying the coefficient in the range of $\alpha_{coat} = 1 \div 15 \cdot 10^{-6}$ m/mK. It was found that this effect depends on the location of the control points relative to the heat flux.

Therefore, with the increase in the coating thermal expansion α_{coat} the stress intensity in control point CP1 and CP2 decreases, however the stress intensity increases at CP3 where there is heat convection into the environment and at CP4 which is closer the heat source. The effect of thermal expansion was defined as follows at respective points:

$$\sigma_i^1 = -2.36\alpha_{coat} + 286.5; \quad \sigma_i^2 = -4.07\alpha_{coat} + 89.8;$$

$$\sigma_i^3 = 19.93\alpha_{coat} + 433.9; \quad \sigma_i^4 = 1.29\alpha_{coat} + 510.4.$$

The modulus of elasticity of the coatings E_{coat} had great extent effect on the performance of the ceramic plate. Using values in the range of 150 up to 900 GPa, it was established that the modulus of elasticity considerably affected the stress state of the ceramic plate for any kind of external loads. Under applied forces P and F, the increase of the modulus of elasticity E_{coat} leads to a decrease in the stress intensity at all four control points. This effect on the stress intensity has been established in the following dependency for each control point:

$$\sigma_i^1 = -0.068 E_{coat} + 350.9; \quad \sigma_i^2 = -0.017 E_{coat} + 169.5;$$

$$\sigma_i^3 = -0.026 E_{coat} + 139.8; \quad \sigma_i^4 = -0.095 E_{coat} + 537.6.$$

However under complex mixed loading, with the increase of the modulus of elasticity E_{coat} the intensity of stress decreases at control points CP1, CP2 and CP4, but at the point CP3 the stress intensity increases. The following relationship expresses this effect at respective points:

$$\sigma_i^1 = -0.085 E_{coat} + 304.9; \quad \sigma_i^2 = -0.057 E_{coat} + 85.1;$$

$$\sigma_i^3 = 0.044 E_{coat} + 565.4; \quad \sigma_i^4 = -0.14 E_{coat} + 589.7.$$

Poisson ratio μ_{coat} of the coating had no noticeable effect on the stress in the ceramic plate. However, under force loading only, increasing μ_{coat} from 0.1 to 0.3 an increase in the stress intensity less than 1% was recorded at the point CP1. Under complex mixed leads, increasing Poisson ratio leads to some minor decrease of stress intensity at CP4.

The effect of the specific heat capacity of the coating (c_{pcoat}) on the stress in the ceramic plate was studied. The values of c_{pcoat}, covered the range from 0.3 up to 1.4 kJ/kg.K. Therefore it is concluded that the specific heat capacity of the coating had no effect on the stresses in coated ceramic plate. Similarly, by varying the density of the coating (ρ_{coat}) from 2 to 16 kg/m^3 no effect was identified. Thus there is no dependency of the stresses in coated ceramic plate on the density of the coatings.

4. Conclusion

A systematic approach to the investigation into a stress-strained state of coated ceramic tools has been presented. The developed mathematical model allowed to study the effect of various properties of the refractory materials on the temperature and stress intensity response of a coating ceramic plate under the complex mixed loads. A new generation of coating for ceramic tools have been developed using of these results.

References

[1] Grigor'ev S.N., Kuzin V.V. Prospects for tools with ceramic cutting plates in modern metal working // Glass and Ceramics, 2011, Vol. 68, No. 7-8, pp. 253-257.

[2] Kuzin V.V. Microstructural model of ceramic cutting plate // Russian Engin. Research, 2011, Vol. 31, No. 5, pp. 479-483.

[3] Grigor'ev S.N., Myachenkov V.I., Kyzin V.V. Automated thermal-strength calculations of ceramic cutting plates // Russian Engineering Research, 2011, Vol. 31, No. 11, pp. 1060-1066.

Vibration assisted surface grinding of mild and hardened steel: Performance of a novel vibrating jig design

V. Tsiakoumis and A.D. Batako
Liverpool John Moores University, Byrom Street, L3 3AF.

Abstract. One of the most significant processes in the manufacturing sector is grinding due to its high precision and accuracy. Nowadays, the demand for higher quality products and new technologies has been increased. The importance of grinding lies in the fact that it stands in the final stages of a component's manufacturing chain and therefore, the possibility of errors must be at the lowest levels. In the current work a novel method of vibration-assisted surface grinding of mild and hardened steel using aluminium oxide grinding wheels is examined. Specifically, the design concept along with the static and dynamic characteristics of a simplified vibrating jig is presented. Its purpose was to accommodate and oscillate the workpiece during surface grinding in order to improve the performance of the process in terms of achieving lower grinding forces and thus lower grinding power consumption along with lower material surface roughness values. Two grinding wheels and three workpiece materials with different properties have been ground during conventional and vibration-assisted surface grinding methods and the results are compared. The benefits of this non-conventional, advanced grinding process are clearly shown.

Keywords: vibration, grinding, forces, surface roughness, oscillating jig.

1. Introduction

Grinding accounts for 20-25% of the total expenditure on machining processes in industrialized countries. However, finishing grinding is usually found to be more costly than other processes per unit volume of material removal [1]. Therefore, the accuracy and efficiency of this process is one of the main concerns. Many researchers attempted to develop new techniques in order to improve the performance of the process. One of them was to apply ultrasonic vibrations to the wheel in order to reduce frictional effects due to interrupted contact and consequently reduce the cutting forces during grinding of monocrystal silicon [2] and soft steel [3] workpiece materials. Ultrasonic assisted machining attracted a number of researchers who achieved remarkable results in terms of lower forces and high values of surface quality [4 and 5]. The majority of these works focus on

the application of ultrasonic vibration of the tool and not the workpiece. The approach of the present study is completely different to the ultrasonics. Low frequency vibrations are induced to the workpiece material via a piezoelectric actuator.

A limited number of papers have been published following the principle of low frequency vibrations. In one of these works a piezo-table that consisted of a parallelogram frame and a piezo-electric actuator was used for vibration-assisted grinding of ceramics. The vibration frequency was 200 Hz with 10μm displacement amplitude and applied in a direction coinciding with the wheel spindle axis. The superimposed vibration produced a reduction in both, normal (Fn) and tangential (Ft) force as well as better surface quality [6]. In another study low values of surface roughness were achieved during low frequency (up to 114 Hz) vibration-assisted grinding of silicon samples where vibrations were applied to the workpiece in three directions but not simultaneously [7]. However, in both studies the vibrating mechanisms were complex and the ground materials were not steel.

The aim of the present study is to design, fabricate and test a jig that can accommodate and vibrate the workpiece during shallow surface grinding of mild and hardened steel. The direction of vibration would coincide with the work-speed (Vw). A simplified vibrating jig was introduced to the process which could be used to oscillate the workpiece at its resonance frequency. The reason for selecting this frequency band was that the system could oscillate at high amplitudes with low voltage input. The wheel workpiece system was in compression in order to avoid any hammering process that would lead to rapid wear of the grinding wheel. In order to achieve that, according to vibro-impact theory the following condition must be met [8]:

$$2 \pi \alpha f < Vw \tag{1}$$

Where α is the displacement amplitude of vibration and f the applied vibration frequency.

2. Machine tool frequency response

Firstly, the dynamic characteristics of the machine tool were identified. The purpose was to obtain the spindle response in dynamic loading conditions and to identify its resonance frequencies. These results were obtained with a sweep-sine test where a piezoelectric actuator was used to excite the spindle unit at rising frequencies while a displacement sensor measured the deflection at any time. The target was to avoid the spindle's natural frequency while excite the vibrating jig in order to avoid any resonance phenomena.

The data were recorded using a National Instrument data acquisition system (DAQ) and all the devices were calibrated before use. Fig. 1 depicts the frequency response of the spindle unit of the grinding machine used in this work.

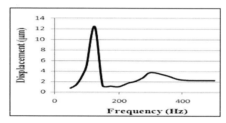

Fig. 1. Spindle unit amplitude-frequency response

The natural frequency of the spindle unit was found to be at 125 Hz with a second peak at around 300 Hz. Therefore, the vibrating jig was designed in such a manner that its natural frequency, and thus the excitation frequency, does not match the natural frequency of the spindle unit.

3. Vibrating Jig

The design concept of the vibrating jig was based on the following specifications:

- Be driven at its resonant frequency (notch of the spindle unit natural frequency).
- Not being very stiff.
- Be subjected to modifications (allow for different stiffness).
- Be simple and adaptable to any machine tool.

After a number of different designs the most appropriate one for this specific application was selected and used for the main volume of experimental work. Figure 2 illustrates the design aspects of a two flat springs vibrating jig.

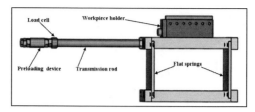

Fig. 2. Two-flat springs vibrating jig.

A piezoelectric actuator was used to vibrate the jig-workpiece and it was connected to the mechanism through a steel connecting rod. At the end of the rod a special arrangement was developed in order to preload the piezoelectric actuator as required by the manufacturer. The actuator was placed outside the working environment in order to avoid any interaction with grinding swarf and debris. In order to identify the frequency response of the vibrating jig the impact test method was employed. During the impact test the jig was impacted with an impulse at one end while at the other end an accelerometer was recording the magnitude of the acceleration and the results were illustrated in Labview software using a FFT. Its natural frequency was found to be at 275 Hz.

4. Experiment

Figure 3 depicts the actual experimental configuration along with all the necessary devices.

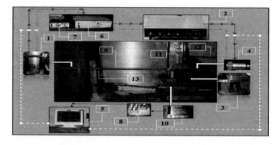

Fig. 3. Experimental Configuration [9].

A series of grinding tests were conducted using the Abwood Series 5025 conventional surface grinding machine. The function generator (7) sets the sine wave, its amplitude and frequency then sends the signal to the power amplifier (6) which drives the piezo-actuator (1). The piezo-actuator drives the oscillating rig (3) through the transmission rod (9) which has a preloading mechanism and a load cell (13). The rig is mounted on the dynamometer (10) connected to the charge amplifier (8) linked to the data acquisition system (5). The workpiece (11) sits on the rig which is driven in translation and vibratory motion under the grinding wheel (12). The closed loop control system (2) and the accelerometer (4) were used to control the amplitude of oscillation as opposed to open loop which is mainly used by most researchers. However the work using closed loop control will be presented in future publication when investigating the effect of frequency on the process [9].

4.1. Experimental Parameters

Two wheels and three workpiece materials were put into test during conventional and vibration-assisted surface grinding. Specific normal and tangential forces, as well as surface roughness, were measured at the end of each trial. The first wheel was an medium grain - medium grade semifriable with coarse to open structure Al_2O_3 (454A 601 L 7G V 3), and the second one was an Al_2O_3 Altos long grain friable wheel with 54% porosity and high aspect ratio. The workpiece materials were Mild steel (BS970 080440, HRB 90.1), hardened En31 steel (BS534 A99, HRC 64.5), and M2 toolsteel (BS BM2, HRC 62). All the grinding experiments were conducted in dry environment in order to understand the effect of superimposed vibration on the process. The overall grinding and vibration parameters are as follows:

- Wheelspeed: 30 m/s
- Workspeed: 50 mm/s
- Vibration frequency: 275 Hz
- Vibration amplitude: 15 μm (Peak)
- Depth of cut: 15 μm

5. Grinding and results

The overall results of all wheels and workpiece materials are presented next.

5.1. Grinding Trials with 454A 601 L 7G V 3 Wheel

In the first set of results the performance of medium grain-medium grade grinding wheel is presented for all three workpiece materials. The graphs in Fig. 4 and 5 show the variation of forces during conventional and vibration-assisted grinding whereas Fig 6 depicts the difference of surface roughness between the two methods.

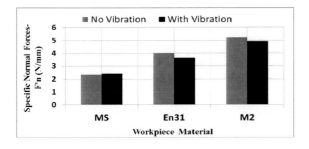

Fig. 4. Specific normal forces for 454A 601 L 7G V 3 wheel.

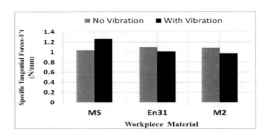

Fig. 5. Specific tangential forces for 454A 601 L 7G V 3 wheel.

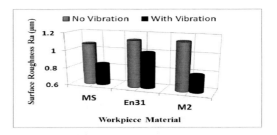

Fig. 6. Surface roughness for 454A 601 L 7G V 3 wheel.

According to the graphs in Fig. 4 and 5 vibration assisted grinding performed very well using a medium grain wheel to grind hard materials. The applied vibration provided a reduction of 10% for En31 normal forces whereas a similar reduction was achieved for M2 tool steel material. However, observation from the mild steel cutting forces shows an increase with the medium grain wheel along with the application of vibration. The reason is that soft materials require harder grade wheels with coarse grain sizes. In this case the wheel clogged easier and resulted in higher forces. Better surface finish, 30.7% for M2 tool steel. In this case the wheel is semi-friable which means that it can produce high surface quality along with high stock removal. In friable wheels the abrasive grains break easier by the impact and generate the self-sharpening process during grinding. The vibration led to a better surface finish in all of workpiece materials due to the induced lapping process.

5.2. Grinding Trials with Altos Wheel

The use of the porous grinding wheel and the applied vibration led to the reduction of cutting grinding forces. An overall reduction in cutting forces is observed. A better performance is seen when machining hardened materials, i.e. 23.6% reduction in tangential forces while grinding M2 hardened steel. However, mild steel did not respond well with the application of vibration as no significant reduction was observed in this test. This may be explained by the ductility and the ability of the material to deform elastically to great extent without actual fracture. The high friability of the wheel causes fast and easy fracture of its grains. This type of wheels performs better with hard steel.

The results of surface roughness in Fig. 8 are not conclusive, however it points to a fact that the surface finish depended more on the wheel surface structure rather than the process. Therefore, surface roughness did not vary and only a 7.6% improvement was observed with the En31 and a smaller decrease in surface roughness for mild steel. High porous wheels with very open structures perform much better in wet grinding applications as they allow better insertion of the coolant and lubricant through their pores. Thus, there is a faster and more efficient delivery of the coolant lubricant to the grinding zone.

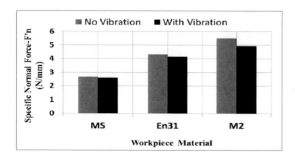

Fig. 7. Specific normal forces Altos wheel.

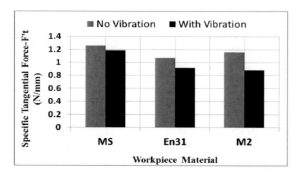

Fig. 8. Specific tangential forces Altos wheel

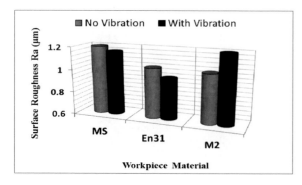

Fig. 9. Surface roughness for Altos wheel

6. Conclusions

The superimposition of vibration into a surface grinding process was achieved using a simplified vibrating jig. A complete study of the frequency response of the machine tool and the vibrating jig was undertaken. Three different materials and two grinding wheels were investigated in terms of process performance in conventional and vibration-assisted grinding. One of the key findings is that the induced vibration improved the performance of the process in terms of workpiece surface quality and the decrease the grinding forces, which was marginal in some cases. Another finding was that the nature and properties of the abrasive wheels play a vital role in specific applications. Grain size, hardness and structure of the wheel should be taken into account. Regardless of the combination wheel and the workpiece, the application of vibration secured some advantages over conventional grinding. Also these findings open a scope to apply this technique to improve fluid delivery in cases such as High Efficiency Deep Grinding.

References

[1]   Malkin, S. (1989) 'Grinding Technology-Theory and Applications of Machining with Abrasives' Ellis Horwood, Chichester.

[2]   Liang, Z., Wu, Y., Wanga, X., Zhao, W. (2010) 'A new two-dimensional ultrasonic assisted grinding (2D-UAG) method and its fundamental performance in monocrystal silicon machining' International Journal of Machine Tools & Manufacture, v50 issue 8 pp728–736.

[3]   Tawakoli, T. & Azarhoushang, B. (2008) 'Influence of ultrasonic vibrations on dry grinding of soft steel' International Journal of Machine Tools & Manufacture, v48 pp1585– 1591.

[4]   Babitsky, V.I., Kalashnikov, A.N., Meadows, A., Wijesundara, A.A.H.P. (2003) 'Ultrasonically assisted turning of aviation materials' Journal of Materials Processing Technology v132 pp157–167.

[5]   Moriwaki, T. & Shamoto, E., (1991) 'Ultraprecision Diamond Turning of Stainless Steel by Applying Ultrasonic Vibration' Kobe University Japan vol.40 No.1

[6]   Zhang, B., Hu, Z., Luo, H., Deng, Z. (2006) 'Vibration-Assisted Grinding – Piezotable Design and Fabrication' Nanotechnology and Precision Engineering Vol.4 No.4 pp283-289.

[7]   Zhong, Z. W., Yang, H. B. (2004) 'Development of a vibration Device for Grinding with Microvibration.' Materials and Manufacturing Processes 19:6, pp: 1121-1132

[8]   Babitsky, V.I., Kalashnikov, A.N., Meadows, A., Wijesundara, A.A.H.P. (2003) 'Ultrasonically assisted turning of aviation materials' Journal of Materials Processing Technology v132 pp157–167

[9]   Batako, A.D.L & Tsiakoumis, V. "A simplified Innovative Method of Vibration-Assisted Grinding of Hardened Steel: Fixture/Machine Tool Response and Process Performance" International Journal of Machine Tools & Manufacture. IJMACTOOL-S-11-00812-1.

Investigation on the effect of cutting parameters on heat partition into multilayer coated tools in HSM

M. Fahad, P.T. Mativenga, M.A. Sheikh

School of Mechanical, Aerospace and Civil Engineering, The University of Manchester, M13 9PL, UK

Abstract. The determination of thermal loads and their distribution during dry or high speed machining can be important in predicting cutting tool performance. In this study, thermal aspects during high speed turning of AISI/SAE 4140 with commercially available coated (TiN, TiCN and Al_2O_3) tools with different coating schemes are investigated. High speed machining was conducted over a wide range of cutting velocities between 314 and 879 m/min. In addition finite element modelling was carried out to evaluate heat load distribution into the cutting tool. The work shows that the tool-chip contact conditions and coating layer deposition schemes are imperative in controlling the heat distribution into the cutting tool.

Keywords: Multilayer coated tools, heat partition, finite element modelling, high speed machining, tool-chip contact, sticking and sliding.

1. Introduction

During machining a large part of the power consumed is converted into heat [1] and the flow of heat into the cutting tool during machining is a critical issue which leads to thermally activated wear mechanism and hence premature tool failure [2]. To overcome the problems which arise in high speed machining (HSM), new tool materials and coating materials/compounds have been and are being developed and deposited on cutting tools to improve tool performance. The use of these thin, hard and sometimes thermally insulating coatings can reduce heat partition into the cutting tool [2, 3].

It is widely accepted that heat generation during machining and its distribution into the cutting tool depends such as cutting forces, tool-chip contact length, chip compression ratio and thermo-physical properties of the cutting tool material [4]. The focus of this article is to study and establish the effect of different cutting parameters/factors on heat partition into the cutting tool. This was conducted by turning AISI/SAE 4140 workpiece with multilayer coated tools. Evaluation of heat partition into the cutting tools was carried out by using experimental data along with finite element modeling (FEM).

2. Evaluation of heat partition into the cutting tool

Evaluation of heat partition into the cutting tool insert was conducted using cutting test data and finite element (FE) transient thermal analysis using commercial code Abaqus/Standard 6.10-1. The modeled inserts were carbide with geometry TCMT 16T308 (restricted contact length groove type). The insert and tool holder were modelled in a CAD software and imported into the FE software. Temperature dependent thermal properties [5] were assigned to each part of the coating tool system (i.e. coatings, shim seat and cutting tool holder). A four-node tetrahedral heat transfer element DC3D4 was used for the whole assembly of the coated tool system. A single element for each layer of coating was modelled because it was found to result in the least element aspect ratio distortion. Initial temperature and the far end surface of the tool holder were set at room temperature (25 $^\circ$C on the day). Convective heat transfer coefficient for the entire model, except for the tool-chip contact area, was specified as $h = 20$ W/m^2 $^\circ$C [6]. All parts of the cutting tool system i.e. coated insert, the shim seat and the tool holder were assumed to be in perfect contact. A constant heat flux was considered in the width of cut direction.

Thermal load q_{st} applied on the tool-chip contact length was calculated using experimental data using equation 1,

$$q_{st} = \tau_{sh} V_{ch} \qquad (1)$$

where, τ_{sh} and V_{ch} are shear stress and chip flow velocity respectively. The shear stress was evaluted from the friction force along the rake face normalised over the contact area. Due to the presence of both plastic (sticking) and elastic (sliding) regions along the tool-chip contact length, a uniform thermal load was applied in the sticking zone and a non-uniform scheme (linearly decreasing to zero) in the sliding zone. This scheme has been used by other reseachers before [3, 7, 8].

Initially 100% of the calculated heat flux (from equation 1) was applied on the tool chip contact area in the FE model for the same duration as the machining contact time. The applied heat flux was successively reduced after each simulation until the minimum acceptable difference was achieved between simulated and experimental temperatures on preselected paths (i.e. Path-1 and Path-2) on the rake face of cutting insert as shown in Fig. 1. The percentage of heat flux at which the FE and experimentally measured temperatures were matched, was considered as the percentage of the heat entering the cutting tool.

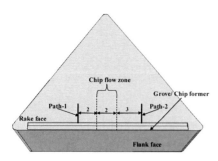

Fig. 1. Temperature measurement locations

3. Cutting tests and results

Nearly orthogonal (with 91° approach angle) dry (without coolant) turning tests were conducted on a semi automated Dean Smith and Grace Lathe machine between the cutting speeds of 314 m/min and 879 m/min. The feedrate and the width of cut were kept constant at 0.16 mm/rev and 2 mm, respectively. A tube of AISI/SAE 4140 (a low carbon high tensile strength alloy steel) with 2 mm thickness and external diameter of 200 mm was used as the workpiece material. Commercially available CVD deposited cutting tool inserts, coated in series and functionally graded, from Iscar (TS-1), SECO (TS-II) and Sandvik Coromant (TG) with ISO specification TCMT 16T308 (restricted contact length groove type) were used. Coating specifications and the thickness values measured using SEM are given in Table 1.

Temperature measurements were obtained by using an infrared thermal imaging camera, FLIR CAM SC3000 equipped with ThermaCAM Researcher package. Emissivity values in the range of 0.4 to 0.65 for Al_2O_3 and 0.2 to 0.3 for TiN [5] were used for the calibration of the camera during the cutting tests.

Table 1. Average thickness values of coatings deposited

Coatings	TS-I (μm)	TS-II (μm)	TG (μm) Rake face	TG (μm) Flank face
TiN	2.5	-	-	2
Al_2O_3	4.5	10	7	7
TiCN	8.2	8	7	7

Cutting forces were measured using a Kistler three-component piezoelectric dynamometer type 9263 along

with data acquisition software. Figure 2 shows the variation of cutting forces with cutting velocity. It can be seen that as the cutting velocity increases, cutting forces decrease for all three cutting tool inserts.

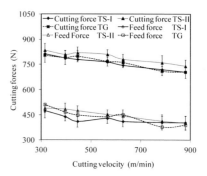

Fig. 2. Variation of cutting forces with cutting velocity for TS-I, TS-II and TG tools

Figure 3 shows the variation of tool-chip contact area on the rake face with cutting velocity which was measured using SEM. TS-I insert has the lowest tool-chip contact area in the entire range of the cutting velocity (i.e. from 314 - 879 m/min). The percentage reduction in the contact area between cutting speed of 314 m/min and 879 m/min for TS-I, TS-II and TG tools is 25%, 31% and 36% respectively.

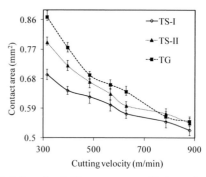

Fig. 3. Variation of tool chip contact area with the cutting velocity

Further investigations were carried out within the contact length to distinguish between sticking and sliding zones. SEM-Energy-Dispersive X-ray Analysis (EDXA) was conducted to evaluate the percentage of Fe (iron) concentration on the tool-chip contact length (Table 2).

Figure 4 shows the variation of heat partition into the cutting tool inserts with cutting velocity. It can be noted that heat fraction flowing into the cutting tool inserts decreases as the cutting speed increases. A reduction in thermal conductivity of Al_2O_3 coating for higher cutting velocities gives a lower heat partition into the cutting tool in case of TS-II and TG. The reduction in heat partition from cutting speed 314 m/min to 879 m/min for TS-I, TS-II and TG inserts is 61%, 68% and 63%, respectively.

Table 2. Percentage of sticking (St) and sliding (Sl)

Cutting velocity (m/min)	TS-I		TS-2		TG	
	St (%)	Sl (%)	St (%)	Sl (%)	St (%)	Sl (%)
314	60	40	55	45	54	46
395	58	42	52	48	51	49
446	55	45	50	50	47	43
565	51	49	48	52	45	55
628	44	56	40	60	38	62
785	40	60	35	65	34	66
879	35	65	29	71	25	75

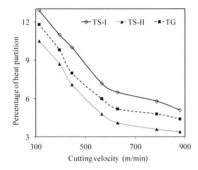

Fig. 4. Variation of heat partition into the cutting tools with cutting velocities

4. Discussion

Figures 5 (a), (b), (c) show the correlation coefficient between heat partition into the cutting tool and cutting parameters; compression ratio, cutting force, feed force and tool-chip contact length. It can be noted that the correlation coefficient (R) of heat partition is above 90% for chip compression ratio and above 88% for cutting forces for all three inserts. The correlation coefficients between heat partition and chip compression ratio and cutting force suggest that heat distribution into the cutting tool is directly related to these two parameters. Thinner chips produce lower chip loads on the tool rake face and in turn reduces the real tool-chip contact area as compared to the apparent area of contact between the tool and the chip. According to Trent, reduction in the real area of contact gives rise to sliding contact between the chip and the tool rake face [1].

Figure 6 shows a decrease in tool-chip contact length as the chip compression ratio decreases. The lighter chips at higher temperatures curl-up and depart early from the tool rake face hence give smaller contact area. From Fig. 5 it is clear that the correlation between the feed force and heat partition is above 88%. This can be attributed to the coefficient of friction between the chip and the tool rake face. Lower value of coefficient of friction gives lower feed forces and hence lowers heat partition into the cutting tool. This coefficient of friction was calculated

from cutting forces, for a 0° rake angle tool, it is simply the ratio between feed forces and cutting forces. It can be noted from Fig. 7 that in case of TS-II and TG inserts, the reduction in the coefficient of friction is 9% and 14% respectively from cutting speed of 314 m/min to 879 m/min

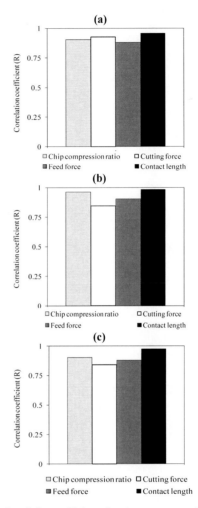

Fig. 5. Correlation coefficient of cutting parameters with heat partition of (a) TS-I (b) TS-II (c) TG cutting tools

The coefficient of friction in case of TS-I at cutting speed of 314 m/min is lower than the other two inserts (i.e. TS-II and TG), this is due to the excellent tibological properties of TiN at lower temperatures. But as the cutting speed increases no significant reduction is observed which is only 1.7% from cutting speed of 314 m/min to 879 m/min. This is due to TiN poor thermal and chemical stability and higher surface roughness at elevated temperatures [9]. Moreover, the poor chemical stability of TiN at higher temperatures allows the chip material to stick on the rake face [5]. This, in turn, allows the uniform shear stress distribution to prevail within the contact area and hence leads to higher heat partition into the cutting tool. A significant reduction in the coefficient of friction for TS-II and TG cutting tools is due to the presence of Al_2O_3 top layer. Al_2O_3 has a self-lubricating ability at higher temperatures [10].

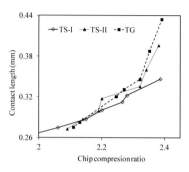

Fig. 6 Variation of tool-chip contact length with chip compression ratio

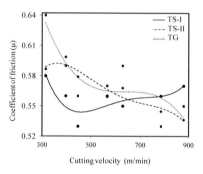

Fig. 7 Variation of coefficient of friction with cutting velocity

It can be noticed from Fig.5 (a), (b) and (c) that the highest correlation coefficient is for the contact length. Therefore, it can be argued that heat partition into the cutting tool is significantly sensitive to contact length. A larger contact area increases the heat distribution into the cutting tool and vice versa [11]. On the contrary, it can be argued from Fig. 3 that despite the contact area in case of TS-I insert being the lowest, the heat partition is highest (Fig. 4) as compared to other cutting inserts (TS-II and TG). The reason for higher heat partition in case of TS-I is the larger length of sticking contact as compared to the other two inserts (Table 2). The lower heat partition in case of TS-II and TG is attributed to smaller sticking zone. Smaller sticking zone is favourable up to some extent but if reduced further, it shifts the higher temperatures to a narrow zone near the cutting edge and leads to premature tool failure [5].

It can be argued that the heat partition into the TS-I tools is the highest due to the presence of TiN top coating. But in case of TS-II and TG tools, all the parameters discussed earlier i.e. tool chip contact, sticking and sliding contact regions are comparable, especially at higher velocities. Heat partition is lower for TS-II insert for all cutting velocities. This could be due to the 10 μm layer of Al_2O_3 and cumulative thickness (18 μm) of multilayers $(TiCN/Al_2O_3)$ compared to 7 μm fo Al_2O_3 and cumulative thickness of 14 μm $(TiCN/Al_2O_3$ on rake face) for the TG insert.

5. Conclusions

In this study the presence of TiN as a top layer in TS-I inserts produced a smaller tool-chip contact length as compared to Al_2O_3 (in inserts TS-II and TG). However, due to the larger sticking zone in case of TS-I the amount of heat transfer was higher. Therefore, it can be concluded that, although, the tool-chip contact length is an important process parameter to heat partition into the cutting tool, a lower contact length does not guarantee lower heat partition. The size of the seizure zone is a critical consideration. This implies that the selection of top coating for a cutting tool insert is crucial for efficient HSM.

References

[1]   Trent EM, Wright P, (2000) Metal cutting. Fourth Edition, Butterworth-Hwineman, Boston, MA. 85.
[2]   Fahad M, Mativenga PT, Sheikh M A, (2011) Finite element methods in manufacturing process. Chap.2: ISTE Wiley, London, 45.
[3]   Akbar F, Mativenga PT, Sheikh MA, (2009) On the heat partition properties of (Ti,Al)N compared with TiN coating in high-speed machining. Proceedings of the Institution of Mechanical Engineers, Part B: Journal of Engineering Manufacture, 223(4): 363-375.
[4]   Akbar F, (2009) An investigation of heat partition in high speed machining using uncoated and coated tools. PhD Thesis, School of Mechanical, Aerospace and Civil Engineering, The University of Manchester.
[5]   Fahad M, Mativenga PT, Sheikh MA, (2011) A comparative study of multilayer and functionally graded coated tools in high-speed machining. International Journal of Advance Manufacturing Technology,DOI 10.1007/s00170-011-3780-x.
[6]   Yen YC, Jain A, Chigurupati P, Wu WT, Altan T, (2004) Computer simulation of orthogonal cutting using a tool with multiple coatings. Machining Science and Technology, 8(2): 305-326.
[7]   Tay AE, Stevenson MG, Davis DG, (1974) Using the Finite Element method to determine temperature distribution in orthogonal machining. Proceedings of Institution of Mechanical Engineers, 188(55): 627-638.
[8]   Wright PK, Mccormick SP, Miller TR, (1980) Effect of rake face design on cutting tool temperature distributions. Transaction of ASME, Journal of Engineering for Industry, (102):123-128.
[9]   Wilson S, Alpas AT, (2000) Tribo-layer formation during sliding wear of TiN coatings, Wear, 245(1-2):223-229.
[10] Xiao H, Yin J, Senda T, (2008) Wear mechanism and self lubrication of engineering ceramics at elevated temperatures. Key Engineering Materials, (368-372):1092-1095.
[11] Abukhshim NA, Mativenga PT, Sheikh MA, (2005), Investigation of heat partition in high speed turning of high strength alloy steel. International Journal of Machine Tools and Manufacture, 45(15):1687-1695.

Effect of key process variables on effectiveness of minimum quantity lubrication in high speed machining

I. H Mulyadi and P.T. Mativenga
School of Mechanical, Aerospace and Civil Engineering, The University of Manchester, Manchester, M13 9PL, UK

Abstract. A great concern about environmental impacts of manufacture as well as the need to reduce cost and cycle times raises interest in dry high speed machining (HSM). Many researchers have reported the feasibility of using minimum quantity lubricants (MQL) in HSM. However, the selection of optimum cutting parameters for MQL application has not received a systematic study. In this paper, the existing knowledge on near dry machining is critically reviewed and this then leads into an experimental evaluation of key process variables. Taguchi experimental design was used to define the dominant process parameters and the optimum setting for minimising tool wear and machined surface roughness. The results show that the selection of MQL quantity is the most dominant factor in improving wear performance and the use of lower cutting speeds and higher chip thickness to cutting edge radius ratio (more efficient cutting) leads to better wear performance when machining tool steel. This can be a strategy for roughing operations. The use of feed per tooth lower than the tool edge radius can help in improving surface finish in finishing operations. The work is an important industrial guide for near dry milling processes and in identifying key areas for process improvement.

Keywords: High speed machining, minimum quantity lubrication, key process variables

1. Introduction

Cutting fluids play a significant role in cutting processes. They have two main objectives. Firstly, cooling down cutting tools, so that tool temperature can be reduced to a certain level, helps minimise thermally activated wear. Secondly, lubricating the contact areas where the friction occurs. These contact areas are the tool-chip interface and tool-workpiece interface. In addition, cutting fluids help by flushing off the chips from the machining zones. However, extensive use of cutting fluids has a negative impact on the environment and human health [1,2]. Machado and Wallbank [3] did a series of cutting tests using a small quantity of cutting fluids. The authors concluded that the machining processes only require a certain amount of cutting fluid quantity depending on the process.
A method that could be employed to mitigate the impact of using cutting fluid is dry machining [4]. For processes such as the intermittent milling process (cyclic cutting hence heating and cooling), dry machining is a promising candidate. Viera et al [5] compared dry machining and wet machining during face milling of H640 steel and reported that dry milling provided the best performance in terms of reducing tool wear and improving surface finish. The effectiveness of cutting fluid when machining at high cutting speed was questioned.

In high speed machining (HSM), working at higher cutting speed leads to increased thermal loads during the cutting process. Consequently, this increases interface temperature and decreases tool life [6]. Sometimes introducing flood coolant leads to thermal shock and hence the interest in machining dry or with minimum cutting fluids. This is particularly interesting to develop dry high speed machining for difficult-to-cut materials [7] where traditionally copious amounts of cutting fluids are used. It has been noted that, more process development is required to achieve widespread use of dry machining in industry [8].

Minimum quantity lubrication (MQL), also known as Near Dry Machining (NDM) can be a favourable machining technology [9, 10, 11]. However, there are contradicting results regarding the application of MQL in machining. For example, Rahman et al [9] reported that MQL was suitable only for low cutting speeds, yet, while, using cutting fluids of different viscosity, Liao et al reported that MQL was beneficial for both high and low cutting speed regimes [11]. In addition, several studies have investigated the use of MQL method [12-18]. Unfortunately, most research on MQL was focused on the benefits and feasibility of this technology [19] and not on performance optimization.

The aim of this study is to identify the significance of key process variables in MQL performance and fill the knowledge gap regarding optimum settings for machining with MQL.

2. Experimental details

The optimisation of cutting variables in MQL application was done for H13 tool steel. In high speed milling the practice is to use lighter cuts and higher table feeds. The use of lower depth of cut and feed per tooth implies that the size effect in relation to the tool edge radius needs to be considered. In this work the depth and feed per tooth are converted into a size effect ratio by relating them to the tool edge radius. Selection of the cutting parameters level was adapted to suit the high speed milling requirements. The volumetric flow rate levels for MQL parameter were selected according to a preliminary test which was done to quantify volumetric output of the supply system on the Mikron HSM 400 machine tool. The MQL was ECOLUB 2032 and this was delivered at 4 bar pressure through 2 nozzles at 45 degrees to the tool feed direction. Taguchi L_9 orthogonal array was selected as an experimental design. The factors and their levels considered are shown in Table 1. Taguchi design has been recognised in engineering and industry research for providing the optimum settings of control parameters [20].

Table 1. Factors and their levels for Taguchi L_9 orthogonal array

Parameters		Level		
Cutting speed (Vc), m/min	A	225	300	400
Feed per tooth-tool edge radius (Rf)	B	0.81	1.13	1.45
Depth of cut-tool edge radius (Rap)	C	1.37	1.82	2.42
Flow-rate (Q), ml/h	D	16.8	22.4	29.9

A down milling, end milling operation was performed to cut a H13 workpiece material on a Mikron HSM 400 milling machine. Test was performed using a tool diameter of 8 mm, mounted with rectangular TiAlN PVD coating tool inserts (IC928) manufactured by ISCAR. The edge radius of the inserts was evaluated to be 62 μm. Every set of experiments were repeated twice.

Surface roughness was measured using a Taylor Hobson Surtronic 3+ Talysurf surface roughness tester. The cut-off length and the evaluation length were set to 0.8 mm and 4mm, respectively. Furthermore, the images of flank wear land and rake face were captured using a Leyca Microscope. Subsequently, the magnitude of flank wear and contact length were measured by using an image processing software, Axio Vision release 4.8. The average flank wear and contact length were calculated from the data taken by 15 lines of measurement on the flank wear land of each worn insert.

3. Results and discussions

3.1. Importance of key process variables on tool life

Table 2 shows the experimental plan, while the results for flank wear and the signal-to-noise ratio (S/N) are shown in Fig. 1.

Table 2. Experimental Design and average flank wear response

Vc (m/min)	Rf	Rap	Q (ml/h)	VB_ average (μm)
225	0.81	1.37	16.80	64
225	1.13	1.82	22.40	63
225	**1.45**	**2.42**	**29.90**	**58**
300	0.81	1.82	29.90	63
300	1.13	2.42	16.80	65
300	1.45	1.37	22.40	67
400	0.81	2.42	22.40	68
400	1.13	1.37	29.90	60
400	1.45	1.82	16.80	61

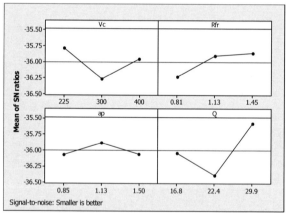

Fig. 1. S/N Ratio analysis for average flank wear

The results in Table 2 and Fig. 1 show that a low cutting speed (225 m/min), high feed per tooth to tool edge radius (Rf=1.45), moderate depth of cut-tool edge radius (Ra_p=1.82) and a high volumetric flow rate of cutting fluid (29.9 ml/h) gave the least flank wear. The low cutting speed is associated with lower cutting temperatures and hence lower average flank wear. This supports previous reports on the performance of MQL over a range of cutting speeds [9, 16, 18]. The higher feed to tool edge radius ratio implies a more positive rake angle and more efficient chip formation process.

In terms of reducing flank wear, the results of ANOVA analysis (Table 3) show that volumetric flow rates have the highest contribution (60%) followed by cutting speed (21%), feed-tool edge radius ratio (15%)

and depth of cut–tool edge radius ratio (of 5%). This dominance of MQL reflects the fact that the lubricant reduces friction hence lowering cutting tool temperature and promoting longer tool life.

Table 3. ANOVA analysis result for average tool wear

Source	Df	SS	MS	F	Cont.(%)	P
Vc	2	37.8	18.88	2017	21%	1E-12
Rfr	2	26.1	13.07	1396	15%	6E-12
Rap	2	8.2	4.11	439	5%	1E-09
Q	2	107.1	53.57	5723	60%	0.000
Error	9	0.1	0.01		0%	1.000
Total	17	179.35	89.64		100.00%	

Moreover, a subsequent analysis was carried out to predict the possible minimum average flank wear that can be achieved by the suggested optimal setting. By utilizing the statistical software MINITAB 15.03, it was found out that the minimum value of average flank wear was 56.60 μm. The result is similar to the statistical prediction calculation that was done manually by using 95% of confidence level. A confirmation test was performed and the actual minimum value of average flank wear at this optimum setting was measured to be 50 μm. Therefore, compared to the initial experiments, there was a 13.8 % decrease in flank wear gained by applying the optimum setting that was predicted by the Taguchi methods.

3.2 Importance of key process variables on surface roughness

Surface roughness is one of the important indicators for quality in a machining process. In this present investigation, the measured values are shown in Table 4 and the S/N ratio is shown in Fig. 2. Both sets of data show that a combination of moderate cutting speed, low feed-tool edge radius ratio, low depth of cut-tool edge radius ratio and low volumetric flow rate contributes to better-machined surface finish.

In finishing operations, in order to produce superior surface finish, the quantity of cutting fluid is less critical as indicated in the ANOVA analysis results given in Table 5. This supports Da Silva and Wallbank [21] who concluded that only small amount of lubricant is required to remove adhering broken off BUE out of the machined surface. The ANOVA analysis result shows that controlling feed-tool edge radius ratio followed by cutting speed and depth of cut-tool edge radius ratio is the hierarchy for improving the surface finish. The need for a low ratio of feed per tooth to the tool edge radius supports the improvement in surface finish that is obtained at a trade off between shearing and ploughing based cutting modes. This is also associated with the polishing effect.

Table 4. Experimental Data of surface roughness response

Vc (m/min)	Feed Ratio	Rap	Q (ml/h)	Ra_average (μm)
225	0.81	1.37	16.80	0.232
225	1.13	1.82	16.80	0.363
225	1.45	2.42	16.80	0.491
300	0.81	1.82	29.90	0.223
300	1.13	2.42	16.80	0.310
300	1.45	1.37	22.40	0.329
400	0.81	2.42	22.40	0.379
400	1.13	1.37	29.90	0.353
400	1.45	1.82	16.8	0.453

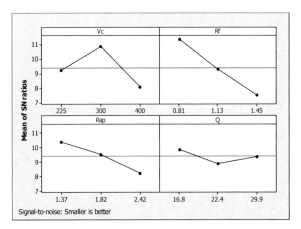

Fig. 2. S/N Ratio analysis of surface roughness response

Table 5. ANOVA analysis result for surface roughness

Source	Df	SS	MS	F	Cont.(%)	P
Vc	2	0.04	0.02	6.3	24%	0.019
Rf	2	0.07	0.03	11.2	42%	0.004
Rap	2	0.02	0.01	4.1	15%	0.055
Q	2	0.00	0.00	0.4	2%	0.666
Error	9	0.03	0.00		17%	
Total	17	0.15	0.07		100.00%	

Similarly, the predicted minimum surface finish was 0.157 μm Ra and the confirmation cutting test result was 0.150 μm Ra. Thus, comparing between the best results from the initial experiment and predicted optimum setting, a 33% improvement in surface finish can be obtained.

4. Conclusions

The motivation for this study was to establish optimum cutting conditions for the effective machining of tool steel under MQL conditions. It can be concluded that;

The rate of flank wear in MQL machining is significantly sensitive to the quantity of MQL. This is more dominant compared to the impact on tool wear of cutting variables of cutting speed, feed per tool and depth of cut. In this study, the highest quantity of MQL gave the lowest wear rate. Where surface finish is a constraint for example, in finishing passes when using MQL, the ratio of feed per tooth to tool-edge radius is the dominant factor to be manipulated in order to significant reduce surface roughness. This is so, when compared to changing cutting speeds or depth of cut-tool edge radius ratio or quantity of MQL. A strategy for MQL machining of tool steel is to undertake roughing passes at higher feed per tooth and moderate cutting speeds using the maximum quantity of MQL. The finishing pass operation can use the lowest feed per tool in order to exploit the size effect for improving surface roughness. It is possible to obtain sub-micro surface roughness of 0.150 µm Ra when machining tool steel. The fact that the optimum MQL quantity reported in the study was at the highest level tested suggests that there could be tool wear reduction benefits for setting MQL flow rate at values more than 29.9 ml/hour.

References

[1]   P.M. Chazal, Pollution of modern metalworking fluids containing biocides by pathogenic bacteria in France: Re-examination of chemical treatments accuracy, European Journal of Epidemiology 11 (1995) 1-7.

[2]   I. A. Greaves, et al., Respiratory health OF automobile workers exposed TO metal-working fluid aerosols: Respiratory symptoms, American Journal of Industrial Medicine 32 (1997) 450–459.

[3]   A.R. Machado, and J. Wallbank, , The Effect of Extremely Low Lubricant Volumes In Machining, Wear 210 (1997) 76-82.

[4]   G. Byrne, and E. Scholta, Environmentally clean machining processes-A strategic approach, CIRP Annals-Manufacturing Technology 42 (1) (1993) 471-474.

[5]   J.M. Viera, et al., Performance of cutting fluids during face milling steels, Journal of Materials Processing Technology 116 (2001) 244-251.

[6]   N. Narutaki, et al., High-speed machining of Inconel 718 with ceramic tools, CIRP Annals-Manufacturing Technology 42 (1) 1993 103-106.

[7]   H. Schulz, and T. Moriwaki, High–speed machining, Annals of the CIRP 41 (1992) 637-643.

[8]   P.S. Sreejit, and B.K.A. Ngoi, Dry machining: Machining of the future, Journal of Materials Processing Technology 101 (2000) 287-291.

[9]   M. Rahman M, et al., Experimental evaluation on the effect of minimal quantities of lubricant in milling. International Journal of Machine Tools and Manufacture 42 (2002) 539–547.

[10]   M.C. Kang, et al., Effect of the minimum quantity lubrication in high-speed end-milling of AISI D2 cold-worked die steel (62 HRc) by coated carbide tools. Surface and Coating Technology 202 (2008) 5621-5624.

[11]   Y.S. Liao, et al., Feasibility study of the minimum quantity lubrication in high speed end milling of NAK 80 hardened steel by coated carbide tool, International Journal of Machine Tools and Manufacture 47 (2007) 1667-1676.

[12]   Cheng-Hsien Wu and Chih-Hsien Chien, Influence of lubrication type and process conditions on milling performance, Proc. IMechE Part B: J. Engineering Manufacture 221 (2007) 835-843.

[13]   P.W. Marksberry, and I.S. Jawahir, A comprehensive tool-wear/tool-life performance model in the evaluation of NDM (near dry machining) for sustainable manufacturing, International Journal Machine Tools and Manufacture 48 (2008) 878–886.

[14]   T. Ueda, et al., Effect of oil mist on tool temperature in cutting, Journal of Manufacturing Science and Engineering, Transaction of the ASME 128 (2006) 130-135.

[15]   J. Sun et al., Effects of coolant supply methods and cutting conditions on tool life in end milling titanium alloy, Machining Science and Technology 10 (2006) 355-370.

[16]   Iqbal A et al., Optimizing cutting parameters in minimum quantity lubrication milling of hardened cold work tool steel, Proceedings of the IMechE Part B: Journal of Engineering Manufacture 223 (2006) 43-54.

[17]   T. Obikawa et al., Micro-liter lubrication machining of Inconel 718, International Journal of Machine Tools and Manufacture 48 (2008) 1605-1612.

[18]   T. Thepshonti et al., Investigation into minimal-cutting-fluid application in high-speed milling of hardened steel using carbide mills, International Journal of Machine Tools and Manufacture 49 (2009) 156-162.

[19]   V.P. Astakhov, Metal cutting theory foundation of near-dry (MQL) machining, Int. J. Machining and Machinability of Materials 7 (1/2) (2010) 1-16.

[20]   Peace G S, Taguchi Methods : A Hands-On Approach to Quality Engineering, Addison-Wesley, Massachusetts, 1993, p. 2.

[21]   Da Silva M.B., and J. Wallbank, Lubrication and application method in machining, Industrial Lubrication and Tribology, 50(4) 1998 149-152.

Development of a free-machining (α+β) Titanium alloy based on Ti 6Al 2Sn 4Zr 6Mo

M. S. Hussain, C. Siemers and J. Rösler
Institut für Werkstoffe, Technische Universität Braunschweig, Braunschweig, Germany

Abstract. Titanium alloys are widely used in the aerospace and power generation industries due to their high specific strength and corrosion resistance. On the other hand, Titanium alloys like the modern (α+β) titanium alloy Ti 6Al 2Sn 4Zr 6Mo (Ti-6246) are difficult to machine due to the formation of long chips which hinders automated manufacturing. In the present study, small amounts of Lanthanum have been added to the standard alloy Ti-6246 to improve its machinability, i. e. to reduce the chips' length. As Lanthanum and Tin form the La_5Sn_3 intermetallic phase, the 2% of Tin had to be replaced by 3% of Zirconium. The matrix of Ti 6Al 7Zr 6Mo 0.9La (Ti-676-0.9) contains pure metallic Lanthanum precipitates which have a relatively low melting point compared to Titanium. During machining of this new free-machining alloy short and strongly segmented chips are observed enabling automation of machining operations. This can be explained by softening of Lanthanum particles during segmented chip formation. The microstructure, phase composition and mechanical properties of the new free-machining alloy were analysed after different thermo-mechanical treatments.

Keywords: Titanium, Ti6Al 4V, Lanthanum, machinability, free-machining alloy

1. Introduction

Titanium alloys have been widely used in aerospace, biomedical and chemeical plant industries because of their good strength to weight ratio and superior corrosion resistance. However, machining of Titanium alloys is difficult due to their poor machinability [1]. This difficulties arise from the physical, mechanical and chemical properties of Titanium. Due to the relatively poor thermal conductivity of Titanium, heat generated by the cutting action cannot diffuse quickly into the chips' material so that heat is concentrated in front of the rake face of tool. Low modulus of elastitcity of Titanium results in springback during the cutting action causing chatter, deflection and rubbing problems. The high strength of Titanium alloys is maintained up to elevated temperature so that a high amount of engery is needed to form a chip [2]. The high chemical reactivity of Titanium limits the number of possible materials for the cutting tools as Titanium has a strong affinity to elements used in cutting tools at operating temparature. This causes galling, welding, and smearing leading to rapid tool destruction and, in addition, decreases the quality of the finished workpiece [2]. Finally, machining operations cannot be automated due to the formation of long chips. These chips have to be removed from the process zone by an operator [3].

In machining, three different types of chips are known to form, namely, continous chips having a constant chips thickness, segmented chips showing a saw-tooth-like geometry and completely separated segments. (α+β)-Titanium alloys like Ti 6Al 2Sn 4Zr 6Mo(Ti-6246) form segmented chips for many different machining process and for a wide range of cutting speeds and cutting depths [4]. Segmented chip formation can be explained as follows: During the beginning of the cut, the tool is penetrating the workpiece and the material is dammed in front of the tool. The plastic deformation is concentrated in a narrow zone, the so-called primary shear zone, Most of the energy used for the plastic deformation is transformed into heat in the primary shear zone which leads to local softeneing of the material and, hence, to the formation of segmented chips, The strain in shear bands can easily exceed 800% at strain rates up to 10^7 s^{-1}. The temperature in the shear band of titanium alloys can easily exceed 1000°C. The chip is afterwards guided along the rake face of the tool, the so-called secondary shear zone. The temperature at the end of the secondary shear zone can also rise to more than 900°C [4].

In the present study, the machinability and the chip formation process of Ti 6Al 2Sn 4Zr 6Mo (Ti-6246) and a modified alloy Ti 6Al 7Zr 6Mo 0.9La (Ti-676-0.9) alloy have been studied in CNC turning experiments. The microstructure of the resulting chips has been investigated by optical microscopy and scanning electron microscopy (SEM) including chemical analysis by energy dispersive X-ray spectroscopy (EDS).

2. Material and experimental procedure

2.1. Material and alloy production

Ti-6246 alloy was produced by the GfE Metalle and Materialien GmbH in Nuremberg, Germany. After 2x vacuum arc remelting (VAR) the alloy was forged from approx. diameter 200 mm to diameter 75 mm in the two-phase field followed by air cool and stress releif anneal.

For alloy modification, 0.9% of Lanthanum (La) have been added to improve the machinability, i. e. to reduce the chips' length [5]. Tin (Sn) was replaced by Zirconium (Zr) as Lanthanum and Tin form intermetallics like La_5Sn_3 during crystallisation. From earlier studies it is known that La_5Sn_3 particles do not lead to improved machinability of Titanium alloys as its dissolution temperature lies above 1500°C [6]. 3% Zirconium have been used instead of 2% Tin to ensure similar solid solution hardening and similar β-transus temperters in both alloys. The final composition of the modified alloy was therefore Ti 6Al 7Zr 6Mo 0.9La (Ti-676-0.9).

To ensure similar starting conditions in the cutting experiments, the two alloys have been produced by plasma arc melting (PB-CHM) in a laboratory furnace with a capacity of about 300 g Titanium [7]. The standard Ti-6246 alloy (remelting of the as-received mateieral) and the modified alloy Ti-676-0.9 were melted and twice remelted to achieve a homogenous alloy. Finally, the alloys have been cast into a water cooled copper crucible of cylindrical shape (fast cooling). The rods had a diameter of 13.2 mm and a length of approx. 90 mm.

2.2. Turning Experiments

The two Titanium alloys have been subjected to turning experiments on a CNC lathe at conventional cutting speeds (v_c) between 30 m/min to 60 m/min. Cutting depths (a_p) of 0.5 mm and 1.0 mm have been used at a constant feed rate (f) of 0.1 mm/rev.

3. Results and Discussion

3.1. Microstructure and Phase Analyses

Both alloys consist of α' martensite only after casting as expected. As the solubility of Lanthanum in Titanium is extremely low at room temperature discrete particles (globular shape) with a high Lanthanum content are present mainly on the grain boundaries in Ti-676-0.9, see Fig. 1. The average particle size is about 2 μm

Fig. 1. Microstructure of Ti-676-0.9: Besides the Titanium matrix Lanthanum-rich particles are visible on the grain boundaries.

Hard X-ray (BW5, HASYLAB, DESY) investigations at modified CP-Titanium, Ti 6Al 4V (Ti-64) and Ti 5Al 5V 5Mo 3Cr (Ti-5553) alloys have shown that the precipitates consisted of metallic Lanthanum [8]. Hence, it is very likely that the particles in Ti-676-0.9 alloy are of metallic nature, too.

The grain size of Ti-6246 has been measured to 147 μm ± 2.1μm whereas the grain size of as cast sample of Ti-676-0.9 was 43 μm ± 1.5μm which shows that the Lanthanum particles act as a grain refiner in the new alloy during crystallization. Due to this observation, heat treatments at different temperatures have been carried out. It has been observed that no grain growth occures in Ti-676-0.9 alloy at temperatures below beta transus whereas grain growth has been observed in the standard Ti-6246 alloy in all heat treatments see Fig. 2. This once more confirms the grain size stabilisation effect of Lanthanum in (α+β)-Titanium alloys.

The β-transus temperature has been investigated by heating the sample to three different temperatures followed by water quenching. It was observed that β-transus of both alloys lie between 900°C and 915°C. As Lanthanum (α-stabiliser) is mostly present in precipitates and not dissolved in the matrix, it is obvious that Lanthanum does not affect the β-transus temperture.

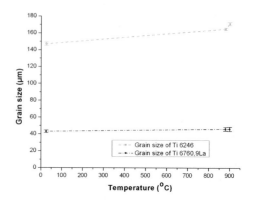

Fig. 2. Grain size with respect to temperature of both alloys.

3.2. Machinability

3.2.1. Chip Morphology
During machining of the standard Ti-6246 alloy long chips devloped in all cutting conditions whereas machining of modified alloy Ti-676-0.9 lead to the formation of short breaking chips, see Fig. 3.

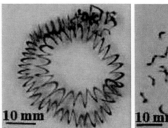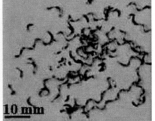

Fig. 3. Chips of Ti-6246 (left) and Ti-676-0.9 (right) alloys. The difference in the length of the chips is clearly visible.

This observation can be explained as follows: During segmented chip formation the temperature in the shear bands reaches or exceeds 1000°C, metallic Lanthanum particles (melting temperature: 918°C) which are present in zone of localized deformation will drastically soften or even melt once the segment starts to form. The adhesion between the segments will be diminished so that the chips fall apart either directly once the shear band forms or due to vibrations during further progress of the tool after the segment is completely developed. The chip separation is shown in Fig. 4.

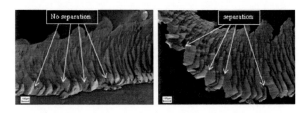

Fig. 4. Chip morphology of Ti-6246 (left) and Ti-676-0.9 (right). The top-down SEM images clearly show the chip separation in case of Ti-676-0.9 alloy.

Several benefiting effect result from the formation of short breaking chips: (1) As the chips are much shorter than in the standard alloy, the contact length between chip and tool decreases so that the temperature of the rake face also decreases. Hence, the tool wear is reduced. (2) As the contact pressure between tool and work piece is diminished, the surface roughness of the finished work piece is reduced. (3) Finally, automated machining processes are now possible as the removal of the chips by cooling fluids is now enabled.

3.2.2. Effect of Cutting Parameters on Chip Geometry
Cross sections of chips of the two Titanium alloys have been analysed by optical microscopy with resepct to their dependencies of the cutting parameters. Two different geometrical features of the chips, namley the degree of segmentation (d_s, $d_s = (h_{max} - h_{min}/h_{max})$) and the segmented shear angle (ϕ_s) have been investigated, see Fig. 5.

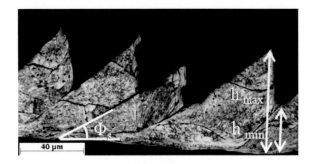

Fig. 5. Definition of geometrical parameters used for the comparison of different chips.

The degree of segmentation increased with increasing cutting depth and cutting speed. In addition, it can be stated that the absolute degree of segmentation was higher ($0.5 \leq d_s \leq 0.8$) in chips of Ti-676-0.9 alloy compared to the standard alloy ($0.2 \leq d_s \leq 0.4$). Material separations (which also increase the degree of segmentation) are visible in the primnary shear zone in case of Ti-676-0.9 alloy, see Fig. 6. The tendency for shear band separations increases with increasing cutting speed. The primary shear zones in Ti-6246 alloy on the other hand do not show any material separations. This once more shows that short chips in case of Ti-676-0.9 alloy are a result of reduced adhesion between the segments.

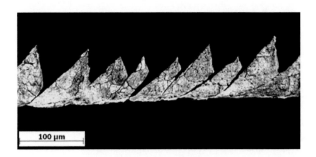

Fig. 6. Microstructure of Ti-676-0.9 alloy chips, $v_c = 40$ m/min, $a_p = 0.5$ mm, f = 0.1 mm/rev. The material separations in the primary shear zones are clearly visible.

In Ti-6246 alloy the segment shear angle has been measured between 40° and 45° independent of the cutting depth and cutting speed. The shear angle for Ti-676-0.9 alloy increases with increasing cutting speed and cutting depth from about 30° to 57°. The results of the geometry measurements are summarised in Fig. 7.

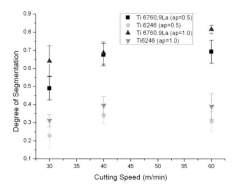

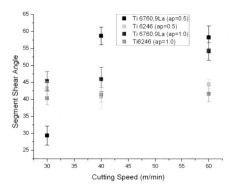

Fig. 7. Degree of segmentation of the chips (top) and segmented shear angle (bottom) for the tow different alloys as a function of cutting speed (v_c) and cutting depth (a_p).

4. Conclusion and Future Work

The machinability of (α+β)-Titanium alloys like Ti-6246 can be improved by the introduction of micrometer-size second-phase particles into the Titanium matrix. The size of the particles should not undercut 1 μm (which is the smallest shear band width observed in the experiments presented here) as otherwise the adhesion between the segments will not be small enough to produce short chips in machining. The melting point of the precipitates should be lower than 1000°C. These two requirements are fulfilled by the two rare earth metal elements Cerium (T_{Melt} = 698°C) and Lanthanum (T_{Melt} = 916°C). In our study, Lanthanum has been chosen to maintain the mechanical properties also to elevated temperature. Benefitting effects of such modified alloys are reduced tool wear, better surface quality and the possibility of automated machining, especially in turning and drilling operations. If in addition, the particles are mainly located on the grain boundaries, an effect of grain refinement is observed as well. Future work will now concentrate on

the investigation of possible deformation routes like extrusion and forging and adequate heat treatment procedures for the modified alloy. Afterwards, physical and static and dynamic mechanical properties of the new modified alloy will be investigated to study the feasibility of industrial applications.

Acknowledgment: The research leading to these results has received funding from the European Union Seventh Framework Programme (FP7/2007-2013) under grant agreement No. PITN-GA-2008-211536, project MaMiNa. Financial support of the European Commission is therefore gratefully acknowledged.

References

[1] Peters M and Leyens C (2002) Titanium and Titanium Alloys, Wiley-VCH, Weinheim, Germany.

[2] Donarchie J (1988.) Titanium – A Technical Guide, ASM International, Metals Park, Ohio, USA.

[3] Lütjering G und Williams J C (2003) Titanium, Springer Verlag, Berlin, Heidelberg, Germany.

[4] Rösler J, Bäker M, Siemers C, (2005) Mechanisms of Chip Formation in: H.-K. Tönshoff and F. Hollmann (eds.), High Speed Machining, VCH-Wiley, Weinheim, Germany

[5] Siemers C, Laukart J, Zahra B, Rösler J, Spotz Z and Saksl Kl, (2011) Materials Science Form Vol. 690 262-265.

[6] Siemers C, Jencus P, Baeker M, Roesler J, Feyerabend F (2007): A new free machining Titanium alloy containing Lanthanum Proceedings of the 11th World Conference on Titanium (Ti-2007), Kyoto, Japan, Vol. I, pp. 709 – 712.

[7] Siemers C, Laukart J, Zahra B, Rösler J, Spotz Z and Saksl K, (2010) Proceedings of the COM2010, Section Light Metals 2010 - Advances in Materials and Processes, Vancouver, Canada, pp. 311 – 322.

[8] Technische Universität Braunschweig, Rösler J, Bäker M and Siemers C, (2004) German Patent DE 103 32 078, DPMA Publications.

[9] Siemers C, Laukart J, Zahra B, Rösler J, Rokicki P, Saksl K, (2011) Advanced Titanium Alloys containing Micrometer-Size Particles, Proceedings of the 12th World Conference on Titanium (Ti-2011), Beijing, China, (in press).

An improved setup for precision polishing of metallic materials using ice bonded abrasive tool

Boopalan M., Venkatarathnam G. and Ramesh Babu.
Department of Mechanical Engineering, Indian Institute of Technology Madras, Chennai, India – 600036

Abstract. This paper presents an improved setup for Ice Bonded Abrasive Polishing (IBAP) of materials with in-built refrigerating unit and temperature controller. Freezing of water into ice serves as a bond that can firmly hold the abrasives and aids to form a flexible tool for polishing. Unlike the earlier setup that made use of liquid nitrogen to prepare a IBAP tool, this work was attempted to develop a refrigerating unit that helps to prepare the tool in-situ and to enhance the life of tool by maintaining its temperature within reasonable limits. Preliminary experiments were attempted on Ti6Al4V alloy, copper and stainless steel in order to study the effectiveness of polishing of metallic materials with the improved setup.

Keywords: Bonded abrasive tool, Refrigeration, Ice.

1. Introduction

The functional performance of a material is directly influenced by the quality of surface generated by finishing process. Production of high precision components, used in various applications like optical systems, silicon wafers, solar cells, micro-electromechanical systems and medical applications, has been under the focus of attention for a long time.

Ultra fine surface generation processes essentially consume enormous time and require extreme care to produce fine and precise surfaces on the components. Hence, several efforts were made to develop efficient processes which can provide ultra-high precision with ease and also provide enough flexibility in handling of simple to complex parts made of high-strength and wear resistant materials.

In view of the complexity and non-deterministic nature of conventional polishing methods, several unconventional polishing processes, like chemical polishing, laser assisted polishing, electrochemical polishing, magneto-rheological polishing and chemical mechanical polishing methods, are developed over the recent times [1]. Among them, chemical mechanical polishing is widely used for planarization of materials to achieve high degree of surface intergrity with tighter tolerances. Certain difficulties in chemical mechanical polishing include precise feeding of slurry into polishing zone, flexibility of the process in polishing of simple to complex profiles, poor material removal rates, deflection of pad surface and periodic reconditioning of the pad.

Some of these drawbacks can be overcome with Ice bonded abrasive polishing (IBAP) process [2]. This process is a cryogenic type chemical mechanical polishing process where ice acts as a bonding medium to hold abrasives together for polishing of simple to complex surfaces. During polishing, certain amount of heat generated due to friction between tool and work surface causes ice to melt continuously thus exposing fresh abrasives for polishing. Studies have shown the production of nano-level surface finish i.e. Ra = 8nm on 304L stainless steel specimen with ice bonded abrasive polishing process [3]. In this setup, liquid nitrogen was poured into the slurry of water and abrasives in order to prepare the tool for polishing of flat specimen. But the tool prepared could be used only for about 15 minutes. Further, the tool prepared has shown certain inconsistency in its composition and structure due to over or under freezing of water in different regions of the slurry in the mould. Maintaining the temperature of the tool has been one of the challenges in this setup and has thus reduced the effective life of tool for polishing of the specimen.

With a view to enhance the life and effectiveness of ice bonded abrasive polishing tool, this paper covers the efforts made to develop a setup containing a refrigerating unit with an appropriate temperature control scheme for polishing of specimen. It presents the details of the refrigerating unit along with kinematic arrangement of IBAP setup. Preliminary studies were attempted to polish copper specimen with the improved setup. Based on the feasibility of setup for polishing of copper specimen, the studies were extended to polish Ti6Al4V and stainless steel specimen in order to demonstrate the effectiveness of this process for polishing different metallic materials.

2. Basis for the development of setup

A setup for in-situ preparation of IBAP tool was felt essential to polish specimen for a longer duration. Several factors influenced the generation of heat during polishing and has thus shortened the life of tool prepared by freezing the mixture of water and abrasives with liquid nitrogen in offline mode. Control over the temperature of the tool plays a crucial role in the preparation of tool and in polishing the specimen. In order to gain insight into various phenomenon that have contributed to thermal degradation of tool, an attempt was made to estimate the energy inputs to frozen tool i.e. heat input due to conduction, convection, radiation and friction between the tool and the work surface, by considering the configuration of existing setup.

Table 1. Energy losses due to different phenomena in IBAP tool

Mode	Theoretical heat addition (J/s)
Friction between tool and workpiece [4]	0.29
Convection (Top, bottom and radial)	40.39
Radial conduction	3.32
Motor heat	53.81

From the estimates listed in Table 1, it is clear that convection over the top, bottom and radial directions of IBAP tool and motor heat input are the major contributors in melting of ice into water and thus shortening the life of IBAP tool. Heat input into the tool due to radiation and friction are seen to be negligibly small. However, the frictional heat liberated between tool and work surface is sufficient to refresh the tool during polishing by way of melting of ice into water. In order to enhance the overall life of tool, several changes were made to bring out a new setup that can minimize the amount of heat input into the system. Modifications made in the configuration of the setup include:

- Refrigerant supply in the annulus between the inner and outer cylinder so as to reduce conduction and convection from the bottom;
- Insulation of outer cylinder to reduce the transfer of heat from the refrigerant to the ambience;
- Belt and pulley drive to offset the motor from IBAP tool so as to avoid the conduction of motor heat into the tool;
- Replacement of swing arm (over hanging beam) type workpiece holder with a straight flat arm (fixed beam) to hold and place the work flat over the tool;
- Removal of tool dressing unit with an appropriate tool preparation technique.

3. Description of the improved IBAP setup

Figure 1 shows the configuration of improved setup made for in-situ preparation of IBAP tool for polishing of flat specimen. It consists of a refrigeration unit, a drive unit and a tool and work holding unit.

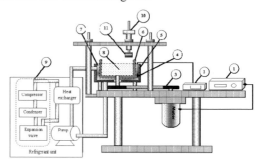

Fig. 1. Schematic of improved setup for flat specimen

[1. Motor speed control, 2. Temperature display unit, 3. Belt and pulley drive, 4. Thermocouple, 5. Stainless steel outer cylinder, 6. Stainless steel inner cylinder, 7. Annulus between the inner and outer cylinder, 8. IBAP tool, 9. Vapor compression refrigeration system, 10. Dead weight loading mechanism, 11. Workpiece holder]

3.1. Refrigeration unit

Figure 2 shows the refrigeration unit that was used to supply the refrigerant at low temperature to IBAP system. The unit basically consists of a primary refrigeration system, a coil-in-coil type heat exchanger and a secondary refrigerant unit. The primary refrigeration system, enclosed by dotted lines in Fig. 2, follows vapor compression refrigeration cycle where R134a is used as the refrigerant. As this refrigerant cannot be supplied directly to the open atmospheric condition prevailing in IBAP system, a secondary refrigerant is used to transfer the cooling effect to the IBAP system at ambient conditions. The heat transfer between the refrigerants occurs in coil-in-coil type heat exchanger. This secondary refrigerant is then pumped to the top of the stationary outer cylinder in IBAP system and is made to flow through the annulus between the outer and inner cylinders.

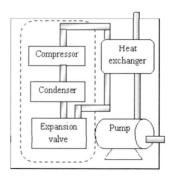

Fig. 2. Schematic diagram of the refrigeration unit employed in IBAP setup

Ethylene glycol, the most commonly used anti-freeze, was employed as the secondary refrigerant, since it is cheaper and readily available in the market. By adding 30% water to 70% ethylene glycol (by weight), the freezing temperature of the solution reduced to -50°C. A commercially available AC monoblock pump (9m head and 500 lph) was used to pump the secondary refrigerant to the IBAP system. This arrangement provided -20°C under no load conditions and -12°C when coupled with the IBAP setup. The extent of insulation needed to maintain this temperature in the setup was determined by trial and error.

3.2. Drive unit, tool and work holding unit

Figure 3 shows the drive unit which comprises of a DC motor and a belt and pulley arrangement, inner cylinder which holds the tool and work holder unit. The belt and pulley system connects the motor to the inner cylinder for rotation of the tool. The tool is firmly held inside the inner steel cylinder with the help of reinforcements that are provided at the bottom of the cylinder. The work holding unit is held flat against the tool by a straight horizontal arm during the polishing process.

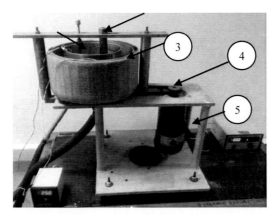

Fig. 3. A photograph of drive unit, tool holder and work holder units in IBAP setup
[1. Work holder, 2. Inner stainless steel cylinder, 3. Outer stainless steel cylinder with insulation, 4. Belt and pulley drive, 5. DC motor]

4. Preparation of IBAP tool

The slurry is prepared by mixing abrasives and water in a fixed concentration by weight, as shown in Table 2. It is then filled layer by layer, into the inner cylinder of the setup, shown in Fig. 3, to form the IBAP tool. To avoid unnecessary use of abrasives in the formation of tool, pure water was poured until the reinforcements submerge in it.

Unlike the previous setup which used liquid nitrogen, the refrigeration unit in the current setup took longer time to freeze the slurry. During this time, certain amount of settlement of abrasives at the bottom of the layer was

observed. To avoid this particular drawback, the volume of slurry poured was limited to 20ml. Through various experiments carried out on the setup, it was observed that the temperature of the tool was 2-5°C less than the temperature of secondary refrigerant. To avoid under or over freezing of tool, the optimum temperature range of secondary refrigerant had to be -5°C to -10°C. The tool prepared was expected to have uniform distribution of abrasives in the ice.

5. Preliminary experimentation

The tool prepared with this improved setup was used to polish electrically condutive copper, 304L stainless steel and Ti6Al4V specimens employing optimum parameters found from the earlier studies [5], and these parameters are listed in Table 2. Each specimen of 22mm diameter and 18 to 20mm thick was first turned in CNC lathe and was then ground on surface grinding machine, before their polishing with IBAP tool.

Table 2. Process parameters of IBAP process

Abrasive	Silicon carbide, 2μm
Concentration by weight	10%
Polishing speed	150rpm
Polishing duration	15 minutes (Cu), 30 minutes (SS, Ti6Al4V)
Polishing load	25kPa

A stylus type roughness measuring (MAHR MARSURF perthometer) instrument was used to measure the finish, in terms of Ra, on the specimen.

6. Results and discussion

Table 3 shows the trends of improvement of finish on copper specimen polished using the earlier and improved IBAP setup. The improvement in surface roughness (Ra) was found to vary in the range of about 50-65% using the improved setup whereas the old setup gave around 52%.

Table 3. Improvement in surface finish (Ra) on copper specimen

Earlier setup [3]	51.95%
Current setup	
Sample 1	63.27%
Sample 2	51.40%
Sample 3	58.53%

Figure 4 shows trend of variation of surface finish for copper specimen polished with the current setup. This trend clearly shows the consistency of improvement in surface finish achieved on copper specimen. Figure 5 shows a similar trend of improvement in finish on

Ti6Al4V and 304L stainless steel specimen polished with IBAP tool.

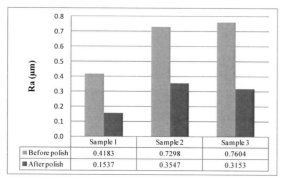

Fig. 4. Improvement in arithmetical mean roughness (Ra) for Copper specimens

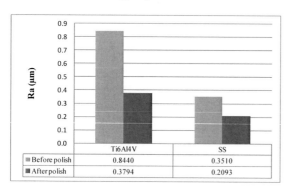

Fig. 5. Improvement in arithmetical mean roughness (Ra) with 8micron SiC abrasives

The percentage improvement in finish was about 55% for Ti6Al4V and 40% for 304L stainless steel. All these trends clearly demonstrate the effectiveness of the improved setup in polishing of different metallic specimen. Apart from this, the current setup made the task of preparing the IBAP tool easier and enhanced the life of the tool. Though the polishing trials conducted with the current setup was limited to 30 minutes, the refrigeration unit provided in the system is capable of maintaining the tool life for much longer period.

7. Conclusion

The developed IBAP setup included a vapor compression refrigeration unit that enabled precise temperature control during tool preparation. This avoided the formation of undesirable cracks and dents in the tool due to over cooling. The secondary refrigerant flow in the annulus between inner and outer cylinder of the IBAP setup helped in maintaining the tool temperature between -5°C to -10°C during polishing. This has considerably improved the life of tool. Polishing studies on copper, Ti6Al4V and 304L stainless steel specimen with the

current setup demonstrated its feasibility to polish different metallic materials.

Acknowledgement: The authors would like to sincerely acknowledge the assistance of Mr.Elangovan and Mr.Srinivas, Refrigeration and Air conditioning laboratory, IIT Madras during the fabrication of refrigeration unit.

References

[1]   Zhong ZW, (2008) Recent Advances in Polishing of Advanced Materials, Materials and Manufacturing Processes, 23(5):449-456

[2]   Mohan R, Ramesh Babu N, (2011) Design, development and characterization of ice bonded abrasive polishing process, International Journal Abrasive Technology, 4(1):57-76

[3]   Mohan R, Ramesh Babu N, (2010) Ultra-fine finishing of metallic surfaces with ice bonded abrasive polishing process, Proceedings of 36th International Conference on Manufacturing Automation and Systems Technology Applications Design Organisation and Management Research (MATADOR), Manchester, UK, 105-108

[4]   De Koning JJ, De Groot G, Van Ingen Schenau GJ, (1992) Ice friction during Speed Skating, Journal of Biomechanics, 25(6):565-571

[5]   Mohan R, Investigations on ice bonded abrasive polishing of metallic materials, PhD Thesis, IIT Madras.

Experimental investigation and FEA simulation of drilling of Inconel 718 alloy

Sureshkumar M.S[1], Lakshmanan D[2] and A. Murugarajan[1]
[1] Department of Mechanical Engineering, Sri Ramakrishna Engineering College, Coimbatore, India
[2] Park College of Technology, Coimbatore, India

Abstract. Drilling of Inconel alloy structures is widely used for aerospace components. Inconel 718, a high strength, thermal resistant Nickel-based alloy, is mainly used in the aircraft industries. Due to the extreme toughness and work hardening characteristics of the alloy, the problem of machining Inconel 718 is one of the ever-increasing magnitudes. This paper discusses the drilling cutting conditions on the machinability of Inconel 718. In this work, an empirical modelling is proposed for predicting the cutting force exerted during the drilling of alloy. The full factorial design is constructed based on the significant drilling process parameters. The cutting force is measured using the dynamometer.. The experiments are conducted on high speed radial drilling machine at a constant depth of cut. Using the experimental results the empirical model is developed for prediction of cutting force and verification results are compared with DEFORM 3D simulation software. The validation experiments are well correlated with empirical model and simulation results for prediction cutting force. However the proposed simulation model further to be considering chatter and vibration of the machine tool structure and different drilling tool geometry.

Keywords: Inconel718, Drilling, Cutting force, DEFORM.

1. Introduction

The knowledge of cutting forces developing in the various machining processes under given cutting factors is of great importance, being a dominating criterion of material machinability, to both: the designer-manufacturer of machine tools, as well as to user. The prediction of cutting force helps in the analysis of optimization problems in machining economics, in adaptive control applications, in the formulation of simulation models used in cutting databases. Also it is useful to study the machinability characteristics of the work materials, to estimate the cutting power consumption during machining and in monitoring the conditions of the cutting tool and machine tool. Cutting force calculation and modeling are one of the major concerns of metal cutting theory [1]. The large number of interrelated parameters that influence the cutting forces (cutting speed, feed, depth of cut, primary and secondary cutting edge angles, rake angle, nose radius, clearance angle, cutting edge inclination angle, cutting tool wear, physical and chemical characteristics of the machined part, cutting tool coating type, chip breaker geometry, etc.) makes the development of a proper model a very difficult task. Although an enormous amount of cutting force related data is available in machining handbooks, most of such data attempt to define the relationship between a few of the possible cutting parameters whilst keeping the other parameters fixed.

In the aerospace industry, drilling accounts for nearly 40% of all the metal-removal operations [2]. About 60% of the rejections are due to the defects in the holes. These defects would create reduction in structural stiffness, leading to variation in the dynamic performance of the whole structure. Many of these problems are due to the use of non-optimal cutting tool designs, rapid tool wear, and machining conditions [3–5]. In conventional machining, the drilling by twist drill is the most applied method though as much as 40% for all materials removal processes [6]. Prediction of critical thrust force was the objective of several studies [7–11] during drilling of composites. With the recent rapid development of industry, the need has increased for precision drilling of special materials like Inconel 718 alloy, which is widely used in Gas turbines, rocket motors, spacecraft, nuclear reactors, pumps.

In this paper, an experimental analysis is carried out for prediction of drilling force in machining of Inconel 718 alloy Titanium Carbide drill bit. The full factorial experimental design is constructed based on the key process parameters includes feed rate and cutting speed. The experiments are carried out and cutting force exerted during machining of alloy is measured using the drill tool dynamometer. The empirical model is developed and verified with experimental results. Furthermore, the proposed empirical model is compared with FEA simulation using Deform 3D software.

Table 1. Input parameters

Parameters	Unit	Notation	Level 1	Level 2	Level 3
Cutting speed	rpm	v	450	640	820
Feed rate	mm/rev	f	0.104	0.211	0.315

2. Experimentation

Factorial designs are used widely in experiments involving several factors on a response. The meaning of factorial design is that each complete test or replications of all the possible combinations of the levels of the factors are investigated [2]. Using full factorial design of experiment, empirical model of cutting force as a function of speed (v), feed (f) have been developed with 95% confidence level. These model equations have been used to predict the cutting force. Table 1 shows the variables and their levels considered for the experimentation.

Inconel 718 alloy of dimensions 110mm x 40mm x 6 mm plate is used for the experiment. Drilling is performed by a radial drilling machine (Model P2/25/GTA).The experiment is carried out by Titanium Carbide drill bit of 6.1mm diameter under MQL condition. The cutting force measurement is acquired by drill tool dynamometer (strain gauge type). Figure 1 shows the schematic view of the experimental setup. Figure 2 shows the photographic view of the setup. The experiments were carried out and the cutting force measurements were observed from the digital output of the drill tool dynamometer.

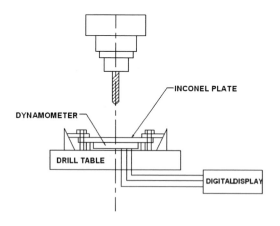

Fig. 1. Schematic view of the experimental setup.

Fig. 2. Photograph of the experimenta setup

3. Emperical cutting force model

The empirical model of cutting force is formed as the relationship between dependant output variable and the two independent input parameters. The functional relationship between output parameter with the input parameters could be postulated using the following equation,

$$F_r = A v^a f^b \qquad (1)$$

The above non-linear equation is converted into linear form by logarithmic transformation and can be written as

$$log\,F = log\,A + a\,log\,v + b\,log\,f \qquad (2)$$

The above equation can be written in the linear form as as

$$y = \lambda_0 + \lambda_1 v + \lambda_2 f \qquad (3)$$

Where, y is the true value of dependent machining output on a logarithmic scale v and f are the logarithmic transformation of the different input parameters λ_0, λ_1 and λ_2 are the corresponding parameters to be estimated. The empirical model of cutting force is formulated using the experimental conditions and measured force with the help of MINITAB software. The measured cutting force values for given experimental conditions are summarized in Table 2. The empirical model was developed based on measured cutting force and it is formulated as,

$$F = 112 + 0.098v + 708\,f \qquad (4)$$

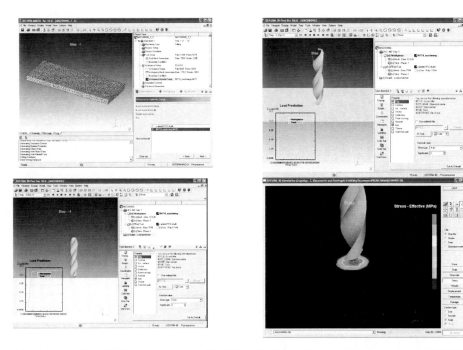

Fig. 3. Screen shot of different steps involved in drilling simulation of Inconel 718 alloy using DEFORM 3D

4. FEA simulation model

DEFORM-3D is a robust simulation tool that uses the finite element method (FEM) to model complex machining processes in three dimensions[12]. One of the most recent processes that has been modeled in DEFORM is drilling. During the drilling simulation, the cutting edges of the drill bit are shearing the workpiece material at different cutting speeds which separate the material from the workpiece by chip formation. Initially, a parting line model was assumed to simplify the simulation process. The stress distribution is simulated and further software can predict the cutting force with help of machining simulation capabilities. A machining specific preprocessor streamlines the setup of routine drilling simulations. Figure 3 shows screen shot of different steps involved in simulation of drilling of Incone 718 alloy using DEFORM 3D. The different runs of trial is conducted for the experimental conditions. The obtained simulated cutting force results are summarized in Table 2.

5. Experimental results

The experiments were carried out using carbide drilling tool for the experimental conditions. The predicted force calculated from empirical formula, measured resultant forces acquired using dynamometer and the forces calculated from simulation model and empirical model are tabulated in Table 2.

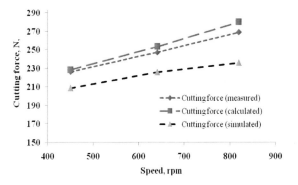

(a) Speed vs Cutting force at feed of 0.104 mm

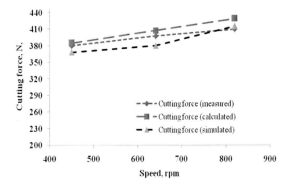

(b) Speed vs Cutting force at feed of 0.315 mm

Fig. 4. Cutting force obtained at different feed rate

Table 2. Measured, calculated and simulated cutting force values at different experimental conditions.

Exp No	Speed (v) rpm	Feed (f) mm/rev	Cutting force, N		
			Measured	Calculated using emprical formula	Simulated from FEA model (DEFORM -3D)
			F_{Ex}	F_{Em}	F_{Si}
1	450	0.104	226.33	228.33	208.59
2	450	0.211	295.45	301.45	256.49
3	450	0.315	380.34	385.34	368.46
4	640	0.104	247.27	253.27	225.95
5	640	0.211	301.34	301.34	265.56
6	640	0.315	397.73	407.73	380.31
7	820	0.104	268.92	279.92	236.05
8	820	0.211	332.48	345.48	304.94
9	820	0.315	410.33	429.33	415.69

It is observed that increasing the feed rate leads to increase in cutting force exerted during drilling of Inconel 718 alloy. Also similar trend is observed when increasing the speed. The results revealed that measured cutting force good agreement with calculated force using empirical model and FEA simulation model as shown in Fig. 4. However, the experiments are to be conducted for full factorial design and considering surface finish and delaminating factor to be useful for further research.

6. Conclusion

The following conclusions are found out from experimental and empirical model results. The cutting force could be effectively predicted by using spindle speed and feed rate as the input parameters for drilling of Inconel 718 alloy using carbide drilling tool. Also apart two input process parameters, feed rate was found to be the most influencing parameter compared with spindle speed. The predicted values from empirical analysis are compared with FEA Simulation model using DEFORM-3D. The results showed are in well agreement with experimental results. It is suggested that a proposed empirical relationship for the measurement uncertainty should always be provided in cutting force measurements, with account being taken of process-related contributions and external disturbances. Furthermore, the model enables the prediction of the uncertainty of cutting force measurements for a defined range of cutting parameters such as considering tool wear, chatter and vibration of the machine tool and data acquisition.

References

[1]   Tamas Szecsi, (1997) Cutting force modelling using artificial networking. Mater. Process Technol 92-93: 344-349.

[2]   Subramanian K, Cook NH. (1997) Sensing of drill wear and prediction of drill life. Transactions of the ASME, Journal of Engineering for Industry. 99:295–301.

[3]   Konig W, Cronjager L, Spur G, Tonshoff HK. (1990) Machining of new materials. Ann CIRP : 39(2):673–80.

[4]   Komanduri R (1997). Machining of fibre reinforced composites. Mach Sci Technol; 1(1):113–52.

[5]   Davim JP, Pedro Reis, Conceicao Antonia. (2004) Experimental study of drilling glass fiber reinforced (GFRP) manufactured by hand lay-up. Comp. Sci Tech.; 64:289–97.

[6]   Brinksmeier, E.(1990). Prediction of tool fracture in drilling. Ann CIRP: 39 (1), 97–100.

[7]   Hocheng H, Dharan CKH. (1990) Delamination during drilling in composite laminates. Trans ASME ;112:236–9.

[8]   Bhatnagar N, Naik NK, Ramakrishnan N. (1993) Experimental investigations of drilling on CFRP composites. Mater Manufact Process ;8(6):683–701.

[9]   Xin W, Wang LJ, Tao JP. (2004) Investigation on thrust force in vibration drilling of fiber-reinforced plastics. J Mater Process Technol : 148:239–44.

[10]   Tsao CC, Hocheng H. (2005) Effect of eccentricity of twist drill and candle drill on delamination in drilling of composite materials. Int J Mach Tools Manufact; 45:125–30.

[11]   Singh I, Bhatnagar N, Viswanath P (2008) Drilling of uni-directional glass fiber reinforced plastics: experimental and finite element study. Mater Des:299(2):546-553

[12]   Gardner JD, Dornfeld, D (2006) Finite element modelling of drilling using DEFORM, Consortium on deburring and Edge finishing.

Performance evaluation of bandsaw using scientific method when cutting tool steels

M. Sarwar[1], J. Haider[2], M. Persson[3] and H. Hellbergh[3]

[1] School of CEIS, Northumbria University, Newcastle upon Tyne, NE1 8ST, United Kingdom
[2] School of Engineering, Manchester Metropolitan University, Manchester M1 5GD, United Kingdom
[3] R&D Saws, SNA Europe, Fiskaregatan 1, Lidkoping, Sweden

Abstract. Bandsawing is a key primary machining operation for cutting off raw material into required dimensions, which is subjected to further secondary machining operation(s) to manufacture a product or a component. Bandsawing is distinctively characterised from other multipoint cutting operations with small depth of cut or feed per tooth (5 μm - 50 μm) compared to the cutting edge radius (5 μm - 15 μm). Bandsawing performance significantly differs depending on workpiece material characteristics. In the current investigation, performance of bandsawing operation has been scientifically evaluated when cutting two different tool steel materials (Orvar Supreme and Sverker). Bandsawing tests were carried out with 6 different cutting speeds ranging from 31 m/min to 90 m/min and four different depths of cut ranging from 1 μm to 4 μm. Cutting forces and thrust forces were continuously measured throughout the cutting tests. Specific cutting energy parameter calculated based on cutting force and material removal data has been used to quantitatively measure the efficiency of the metal cutting process at different feeds and speeds. The bandsaw teeth at the worn condition have been studied under a Scanning Electron Microscope to identify wear modes and mechanisms. The chip characteristics at different feeds and speeds have also been discussed. The sawing community including bandsaw end users and design engineers should find the results presented of interest.

Keywords: Bimetal Bandsaw, Bandsawing, Tool steel, Cutting Force, Specific Cutting Energy, Wear

1. Introduction

Bandsawing is a primary machining operation extensively used by steel stockholders and steel manufacturers to cut off raw materials in a suitable size for secondary machining operations (turning, milling, drilling etc.). Bandsawing process has not been studied by the scientific community to the same depth and extent as in other machining opearations. Several researchers in the past have contributed to the understanding of the bandsawing process [1-8]. Particularly, Thompson, Sarwar and colleagues were very instrumental in distinguishing the bandsawing from other machining operations with the following characteristics: multipoint intermittent cutting action, small feed per tooth (5 μm - 50 μm) compared to the edge radius (5 μm - 15 μm), smaller chip ratios (0.1 as compared 0.3 in turning), variation in number of active cutting edges in contact with the workpiece, restricted chip flow in bandsaw gullets etc.

New materials with improved properties are constantly being developed to meet the demand for different applications. Machining community faces the challenge of efficiently cutting the newly developed materials. Knowledge and understanding of chip formation mechanism, optimum machining parameters, forces developed in the cutting tool and tool wear characteristics are essential to minimise the machining cost, to extend tool life and to improve product quality. Previously the authors scientifically evaluated the bandsawing performance of several steel workpieces (ball bearing steel, stainless steel, Ni-Cr-Mo steel) by studying chip formation mechanism, measuring forces and specific cutting energy (Esp) and establishing wear modes and mechanisms in bandsaw teeth [5, 6]. The aim of this investigation is to assess the performance of bandsaws when cutting tool steel workpieces at different cutting parameters. Esp has been used to quantitatively measure the bandsawing efficiency as it is more sensitive to low depths of cut, which is the case in bandsawing.

2. Experimental procedure

2.1. Workpiece material

Two different tool steels were selected as workpiece materials for the bandsawing investigation. The details of the workpeice materials are given in Table 1. Orvar supreme is particularly useful for casting dies, forging tools and extrusion tools. Sverker is recommended for applications such as blanking and shearing tools, press tools, forming tools and plastic moulds.

Table 1. Workpiece materials details

Material characteristics	Workpiece 1	Workpiece 2
Trade name	Orvar Supreme (BS 4659 BH13)	Sverker (BS 4659 BD6)
Bar dimension	254 mm × 153 mm	254 mm × 153 mm
Vickers Hardness 5 kg (Kgf/mm²)	200	260
Chemical composition (wt%)	C (0.32-0.42); Si (0.85-1.15); Mn (0.40); S (0.035); Cr (4.75-5.25); Mo (1.25-1.75); Ni (0.40); V (0.90-1.10)	C (2.05); Si (0.30); Mn (0.80); W (1.10); Cr (12.70)

2.2. Cutting tool

Variable pitch bimetal bandsaws (M42 High Speed Steel welded to low alloy backing material) in the form of loops were used for the machining tests. The band thickness, width, tooth pitch and loop length were 1.6 mm, 54 mm, 1.4/2 TPI and 8.8 m respectively. The rake angle and clearance angle of each tooth were 10° and 35°.

Fig. 1. Bandsaw (1.4/2 TPI) teeth used in this investigation

Fig. 1 shows a section of the bandsaw used. The bandsaw teeth were set according to the sequence of Right (R) - Left (L) - Neutral (0).

2.3. Machining test

A fully instrumented vertical feed bandsaw machine (NC controlled, Behringer HBP650/850A/CNC) was used to carry out the full bandsaw cutting tests by sawing off small sections from a long workpiece. Fig. 2 shows the experimental set-up and cutting and feed directions during bandsawing. The bandsaw machine was calibrated using a 3-axis Kistler dynamometer and relevant

transducers. Cutting force and thrust force components, cutting speed and feed rate were measured during the bandsawing tests. The test conditions are given in Table 2. The saw blade was flooded with a coolant (Castrol Cooledge) during sawing at a flow rate of 3.8 liters/min. Samples of cutting edges were examined using a scanning Electron Microscope (SEM). Chips were also collected for different test conditions in order to characterise them

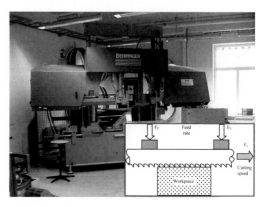

Fig. 2. Experimental set-up used for bandsawing tests with schematic diagram of cutting and feed directions

Table 2. Machining conditions for the selected workpieces

Material	Cutting speed (m/min)	Set depth of cut per tooth or Feed (µm)
Orvar Supreme	31, 45, 60, 75, 90	1, 2, 3, 4
Sverker	31, 40, 50, 60	1, 2, 3, 4

3. Results and discussions

3.1. Forces and specific cutting energy

The cutting and thrust forces generated at different depths of cut and cutting speeds with Orvar supreme and Sverker are presented in Fig. 3 and Fig. 4 respectively. The forces increased with the increase of depth of cut per tooth due to the higher material removal rate. In general, cutting forces (Fv) were higher than the thrust forces (Fp) [8]. However, in some cases Fv was equal to Fp or Fp was even slightly greater than Fv. The combination of cutting parameters and material characteristics could be responsible for this.

The specific cutting energy (Esp) was calculated from cutting force and material removal rate data [7] and presented in Fig. 5 and Fig. 6 for the workpiece materials. All the curves showed exponential nature with the depths of cut. At low depth of cut, the machining efficiency was poor as indicated by higher Esp due to the significant edge radius effect or size effect (edge radius being greater than the depth of cut). The decrease of edge radius effect at higher depth of cut indicated a better machinability characteristic (lower Esp). For Orvar Supreme the general trend of Esp was satisfactory i.e., the cutting efficiency improved with the increase in cutting speed. However, the bandsawing efficiency deteriorated with the increase in cutting speed when cutting Sverker possibly due to a different material characteristic.

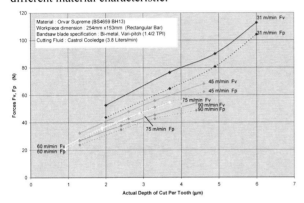

Fig. 3. Influence of cutting speed on forces when cutting Orvar Supreme workpiece

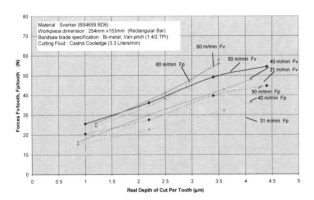

Fig. 4. Influence of cutting speed on forces when cutting Sverker workpiece

It should be noted that the actual depths of cut achieved (Fig. 3 to Fig. 6) by the bandsaw were not equal to the set depths of cut (Table 2). The depth of cut achieved per tooth was influenced by the cutting speed and furthermore by the workpiece material characteristics.

3.2. Cutting edge wear chracteristics

At the new condition, the cutting edges were nominally sharp. However, at high magnification under SEM sign of damage along the cutting edge was visible possibly due to the breaking of very small grinding edge-burr during the production of the teeth. This process created an edge radius of approximately 7 µm.

Bandsaw teeth are worn in such a way that wear flat (flank wear) was produced at the tip of each tooth and the outer corners of the teeth are rounded (corner wear). The SEM pictures of the worn cutting edges are presented in Fig. 7 after cutting 20 sections of Orvar Supreme and Sverker materials. It was established that flank and corner wear were the principal wear modes. A combination of abrasion and adhesion were identified as the underlying mechanisms of the flank and corner wear [5, 6]. However, workpiece adhesion on the bandsaw cutting edge was more severe for cutting Sverker than Orvar Supreme. Rake face wear was generally not as severe as the flank wear. The wear in bandsaw tooth alters the edge geometry and affects the chip formation mechanism [8].

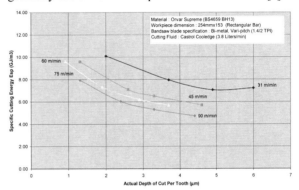

Fig. 5. Influence of cutting speed on specific cutting energy when cutting Orvar Supreme workpiece

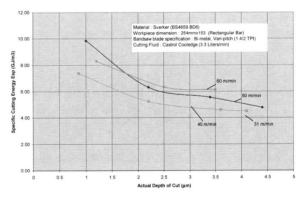

Fig. 6. Influence of cutting speed on specific cutting energy when cutting Sverker workpiece

3.3. Chip chracteristics

Fig. 8 and Fig. 9 present the examples of chips produced during machining of the two workpiece materials at different bandsawing parameters. In general, long chips were found when machining Orvar Supreme material. However, chips produced from Sverker were generally shorter. In case of Orvar Supreme, the chips turned into more curly shape with the increasing speed irrespective of the depth of cut. On the other hand, for Sverker the chips turned into more curly shape with the increasing speed at low depth of cut only. At higher depth of cut, the chips became shorter and straighter.

a

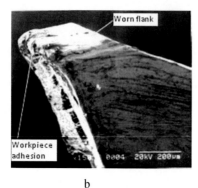

b

Fig. 7. SEM pictures of worn bandsaw teeth after 20 cuts when cutting (a) Orvar Supreme and (b) Sverker workpieces

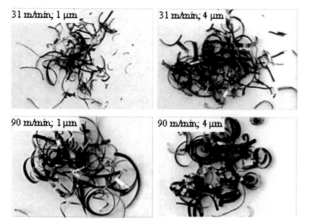

Fig. 8. Chip characteristics when cutting Orvar Supreme material at different combinations of speeds and depths of cut

4. Conclusions

The following conclusions can be drawn from the full bandsawing tests carried out on two different tool steel materials (Orvar Supreme and Sverker). The forces (cutting and thrust) were influenced by the cutting speed

and depth of cut (forces increase). The bandsawing efficiency increased (lower Esp) with the cutting speed when cutting Orvar Supreme. However, with the increase in cutting speed, the bandsawing operation became less efficient for Sverker (higher Esp). Flank and corner wear were caused by a combination abrasion and adhesion. Workpiece adhesion on the worn flank was more prominent for Sverker. The characteristics of the chips produced during bandsawing were influenced by workpiece materials and cutting parameters.

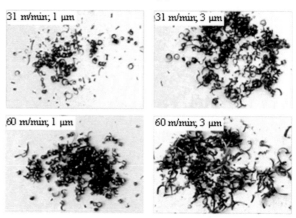

Fig. 9. Chip characteristics when cutting Sverker material at different combinations of speeds and depths of cut

References

[1]   Colwell LV, McKee RE, (1954) Evaluation of bandsaw performance, Transactions of the ASME 77:951–960
[2]   Sarwar, M, Thompson PJ, (1974) Simulation of the cutting action of a single hacksaw blade tooth, The Production Engineer 53:195-198
[3]   Ahmad MM, Hogan B, Goode E, (1989) Effect of machining parameters and workpiece shape on a bandsawing process, International Journal of Machine tools and Manufacture 29:173–183
[4]   Andersson C, Stahl JE, Hellbergh H, (2001) Bandsawing. Part II: detecting positional errors, tool dynamics and wear by cutting force measurement, International Journal of Machine Tools and Manufacture 41:237-253
[5]   Sarwar M, Persson M, Hellbergh H, (2005) Wear and failure modes in the bandsawing operation when cutting ball-bearing steel, Wear 259:1144–1150
[6]   Sarwar M, Persson M, Hellbergh H, (2007) Wear of the cutting edge in the bandsawing operation when cutting austenitic 17–7 stainless steel, Wear 263:1438–1441
[7]   Sarwar M, Persson M, Hellbergh H, Haider J, (2009) Measurement of specific cutting energy for evaluating the efficiency of bandsawing different workpiece materials, International Journal of Machine tools and Manufacture 49:958-965
[8]   Sarwar M, Persson M, Hellbergh H, Haider J, (2010) Forces, wear modes and mechanisms in bandsawing steel workpieces. IMechE Proceedings Part B: Journal of Engineering Manufacture 224:1655-16662

Manufacturing Systems Management and Automation

Development of a robust handling model for foundry automation

Wadhwa, R.S.

NTNU Valgrinda, Department of Production and Quality Engineering, Trondheim, 7051, Norway

Abstract. This paper compares the simulation and experimental results for robust part handling by radially symmetric cylindrical electromagnetic gripper heads, that are used in foundry manufacturing assembly operation. Knowledge of the direct holding force is essential to determine if a given electromagnet is capable of preventing part slipping during pick and place operation. Energy based model and the magnetic circuit model have been described. The latter is developed further and compared with results from a FEA software. It was found that the magnetic circuit model, although simple in form, was limited in its ability to accurately predict the holding force over the entire range of conditions investigated. The shortcomings in the model were attributed to its inability to accurately model the leakage flux and non-uniform distribution of the magnetic flux. A finite element allowed for the ability to couple the mechanical and magnetic models. The finite element model was used to predict the magnetic field based off the solutions to the mechanical (σ) and the magnetic model (B).

Keywords: Handling Electromagent Design, Foundry Automation

1. Introduction

The Robot grippers are used to position and retain parts in an automated assembly operation. Electromagnet grippers offer simple compact construction with no moving parts, uncomplicated energy supply, flexibility in holding complex parts and reduced number of set-ups, and are thus suitable to ferrous metalcasted parts. However, their use is limited to ferrous materials (Iron, Nickel, Cobalt), electromagnet size is directly dependant on required prehension force; residual magnetism in the part when handled when using DC supplies requires the additional of a demagnetizing operation to the manufacturing process. Smart materials, commonly classified according to their energy transduction mode as piezo-electrics, shape memory alloys, and magnetostrictives, have been shown to be useful in low bandwidth application, and micro gripping applications, but they still have limitations in a high volume manufacturing environment [1]. While the choice of material limits application, and demagnetizing is a requirement, the holding force is an important unknown.

An electromagnet consists of at least one pair of north and south magnetic poles that are separated by an airgap. In this way, there is practically no magnetic field present when a current flows through the coil, because air presents a very high reluctance to the magnetic flux. When a part is placed on the surface of the electromagnet in such a way that it connects a north and south pole, the magnetic flux can be established, given that the part is made of a ferromagnetic material. The magnetic flux will produce a force of attraction between the part and the electromagent, as mentioned in the previous section. Two parts made of the same material and having the same geometry and dimensions could experience a different force of attraction on a given electromagnet if the contact conditions between the workpiece and the electromagnet are different for the two of them. Of one of the parts has a rougher surface or has a larger flatness error, the contact interface will have larger airgaps that have to be transversed by the magnetic flux in order to complete the magnetic circuit.

The users of electromagnets in iron foundries know that factors such as material hardness, surface contact conditions, and electromagnet design influence the holding force. Most of the available literature in foundry automation is of a commercial nature [2,3,4]. The author believes that a predictive model for determining the holding force will enable the design of the optimum operating geometry and/or conditions to prevent part slip during robot handling/assembly. Consequently, the need for costly and time intensive experimentation will be minimized.

This paper compares the results from the magnetic circuit model and energy model with available commercial software COMSOL, for a cylindrical radially symmetrical electromagnet head, and substantiates it with experimental analysis.

2. Modeling electromagnetic behaviour

Several energy based models have been created in an attmept to capture the non-linear behaviour or electromagents. Modeling techniques bu Dapino et al [5] include a thermodynamic approach for estimating magnetization to field. Additional modeling techniques have been reported by Sablik and Jiles [6], where internal energy minimization is used to ensure mechanical equilibrium. Analytical methods have also been depevloped, but mainly for predicting magnetostrictive performance [7]. When analyzing complex geometries finite element method generally gives more accurate results.

In the following sections, the energy based model, magnetic circuit model and the FEA model used to simulate experimental data are explained. As a basis for comparison, the analytical method for calculating magnetization factor for cylindrical electromagnets is compared to FEA predictions where the effect of airgaps and varying current through the electromagnet coil on the holding force is investigated. Comsol Multiphysics 4.0 magentostatics (with currents) is the finite element model was used in the research.

To determine the magentization effect via FEA approach the external field is calculated by determining the magnetic field at a point of interest in space at a certain distance from the magnet, in the absence of the sample part. The magnetization factor was calculated for several aspect ratios, where the thickness of the sample always remained 2-inches. The airgap was varied to change the aspect ratio, and the effect of electromagnet coild heating over a period of time was observed.

3. Industrial setup

The six axis ABB ERB 6400 robot used in foundry assembly operation is shown in Fig. 1. A Sony XCG-U100E overhead camera was used for identifying the orientation of the part lying on the conveyor belt, which was internally tracked by the robot. The image captured by the camera was processed by Scorpio Vision System (Tordivel AS) and transferred via closed network Ethernet connection to the Robot. The robot gripper then moved the electromagnets accordingly to pick the part.

Fig. 1. Robot Assembly Cell.

The purpose of the vision system was to recognize the part and extract the orientation. The second purpose of the vision system was to assist in decision making. The part orientation was identified by the markers (Fig. 2)

which were cast in the part. The vision system conveyed the parts orientation to the robot gripper via the Ethernet and the electromagnets were translated to orient towards the grasping the part.

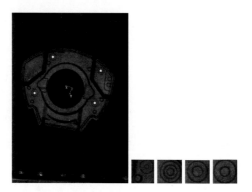

Fig. 2. Markers casted on the part.

4. Magnetic field distributions

This section concentrates on the analysis of different basic electromagnet setups and their effects on the magnetic properties of a system. Magnetic field distributions will be used for each setup to emphasize key differences between different designs.

As a basis for comparison, the analytical method for calculating the magnetization factor of a cylindrical core [34] is compared to FEA predictions,where the effect of aspect ratio on magnetization is investigated. A COMSOL Multiphysics 4.0 magnetostatic (with currents) finite element model was created with a geometry consisting of a rectangular core immersed in a coil, surrounded by an air domain with dimensions of three times the largest dimension of the rod and coil (Fig. 3).

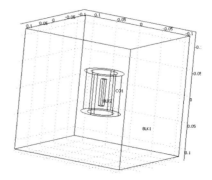

Fig. 3. COMSOL Magentostatics (with currents) geometric rod coil setup.

Equation 1 gives a general expression for determining the effective magnetic field within a sample, with a known value of N_d (magnetization factor) [8]. Since calculation of the effective magnetic field (H_{eff}) requires a knowledge of N_d (geometry dependent), the magnetic field within the sample is normally difficult to calculate (especially for complicated geometries).

$$H_{eff} = H_{ext} - H_d = H_{ext} - N_d M \qquad (1)$$

Ellipsoidal geometries have been shown to have a relatively constant magnetic field distribution, leading to one value of magnetization factor for the entire geometry. An analytical expression for calculating the demagnetization factor as a function of a dimensional ratio (k) was developed by [9] (Equation 2). The dimensional ratio, k, is determined by dividing the length of the semimajor axis by the semiminor axis.

$$N_d = \frac{1}{k^2 - 1}\left[\frac{k}{\sqrt{k^2 - 1}}\log_e\left(k + \sqrt{k^2 - 1} - 1\right)\right] \qquad (2)$$

It was desired to simulate the magnetic field behavior along the radius and length of the rod. A 2D axisymmetric, magnetostatics (with currents) model was utilized. The rod was placed at the r = 0 location and was surrounded by a coil. The rod and coil setup is surrounded by an air domain ($\mu R = 1$) with dimensions that are three times the length of the coil, which is the largest component of the circuit. A current density of 3e6 A/m2 was assigned to the geometry corresponding to the coil. A relative permeability of 50 was assigned to the rod for the experiment.

Figure 4 shows a 2D axisymmetric streamline of the magnetic field when the magnetic sample ($\mu R = 50$) is placed inside the coil. It is evident that due to the demagnetization effect, the magnetic field leaks throughout the length of the rod (i.e., some streamlines fail to travel the full distance of the sample). However, if a magnetic circuit is incorporated into the design of the transducer, then the flux leakage can be drastically reduced. In Fig. 5, a steel flux return path is added to the same setup as in Fig. 4, with $\mu R = 2000$ for the steel. The use of a well defined magnetic circuit will allow for the full use of the material capabilities, as there is negligible field lost due to flux leakage.

Rods having a radius of 0.25-inches were analyzed for aspect ratios of 1, 2, and 4 (0.5, 1, and 2-inch long rods respectively). The magnetic rod is assumed to have a constant permeability of 50. The steel flux return path discussed in the second case has a permeability of 2000.

The current density used here is 3e6 A/m2 for all cases. First, the magnetic field distribution was studied for the no steel flux return case. The radial magnetic field distribution (at the mid-height of the rod) was studied for the no steel flux return case, for cylindrical rods with aspect ratios of 1, 2, and 4, and all with radii of 0.25-inches. The magnetic field data for each aspect ratio was non-dimensionalized according to its maximum magnetic field value. The radial position was also non-

dimensionalized (i.e., max field is one, and outer radius position is one).

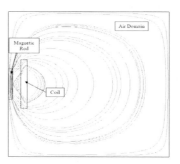

Fig. 4. 2D axisymmetric view of magentic field streamlines showing lines of flux leakage resulting from rod and coil setup.

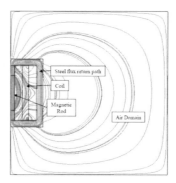

Fig. 5. 2D axisymmetric view of magentic field streamlines showing flux leakage for rod and coil with steel flux return path.

Figure 6 shows the non-dimensional results for the three different aspect ratios. There are two sets of data shown for each aspect ratio. The dashed lines correspond to rods with length and width of half the sample shown by the red lines. Using these dimensions gives the same aspect ratio. It is evident that the non-dimensional magnetic field distribution does not vary for rods of the same aspect ratio. It is important to note that this is only true when comparing magnetic field distributions of the same shape. Also, these distributions are unique to the specific coil design and applied current density. Furthermore, Fig. 6 shows that lower aspect ratio samples experience a larger amount of non-dimensional magnetic field leakage from their centerline to the outer radius. Additionally, it is evident that the magnetic field increases from the center of the rod and reaches a maximum at the end of the rod for all aspect ratio cases. It should also be noted that the relationships shown in Fig. 6 are parabolic. This parabolic magnetic field behavior plays a key role in the element type that is chosen for the mesh.

Next, the magnetic field distribution along the length of the rod was studied for the no steel flux return case, for cylindrical rods with aspect ratios of 1, 2, and 4, and all with radii of 0.25-inches. Again, the magnetic field data for each aspect ratio was non-dimensionalized according to its maximum value. The magnetic field behaviour described in the aboe two cases is summarized in Table 1.

As the aspect ratio is increased, the percentage difference between the maximum and minimum magnetic field through different locations along the radial span decreases. On the contrary, the percentage difference between maximum and minimum magnetic field along the length increases for samples with higher aspect ratio ratios. Differences in magnetic field of 80.9% were seen along the length of a 2-inch, 0.25-inch diameter sample.

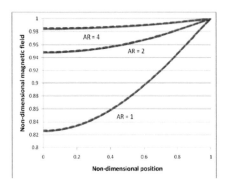

Fig. 6. Nondimensional radial magnetic field verses non dimensional position for a rod and coil setup with varying aspect ratios. The dashed lines show rods with different dimensions that yield the same dimensions as the rods shown in the solid lines. All data is normalized to its respective maximum magnetic field.

Table 1. Percentage difference of magnetic field along radius and length of cylindrical samples with radii of 0.25-inches and lengths of 0.5, 1, and 2-inches (aspect ratio ratios of 1, 2, and 4) for rod and coil setup.

Aspect ratio aspect ratio	% difference in magnetic field along radius	% difference in magnetic field along length
1	17.4	54
2	5.2	70
4	1.6	80.9

The radial magnetic field distribution (at the mid-height of the rod) was studied for the steel flux return case, for the same cases as done for the no steel flux return path studies. It was clear that the presence of the steel flux return path increases the magnetic field within the sample, as well as creates a more uniform distribution of magnetic field. Table 2 summarizes the magnetic field behavior. It can be observed that as the aspect ratio increases, the percentage difference between the maximum and minimum magnetic field through the radial span decreases. In contrast, the percentage difference between maximum and minimum magnetic field along the length increases for samples with higher aspect ratios. Differences in magnetic field of 6.5% were seen along the length of a 2-inch, 0.25-inch diameter sample. However, in comparison to Table 1, the steel flux return path eliminates a large amount of the flux leakage leading to small percentage differences in magnetic field along the radius and length of the sample.

It was clear that the spatial variation of magnetic field varies greatly with the setup. The above parametric study suggested that it is important to include a flux return path.

It was also found that it is desirable to use samples of lower aspect ratio, as the percentage change in magnetic field is much smaller for lower aspect ratio samples. Seeing these, the magnetic circuit was modeled before experimentation.

Table 2. Percentage difference of magnetic field along radius and length of cylindrical samples with radii of 0.25-inches and lengths of 0.5, 1, and 2-inches (aspect ratios of 1, 2, and 4) for rod, coil, and steel flux return path..

aspect ratio	% difference in magnetic field along radius	% difference in magnetic field along length
1	0.9	0.7
2	0.6	3
4	0.2	6.5

5. Magnetic Models

Energy Based Model

In the magnetic field, the energy associated with the system is distributed throughout the space occupied by the field. Assuming no losses, the energy stored in the system per unit volume when increasing the flux density from zero to B is:

$$W_f = \int_0^B H dB \tag{3}$$

From this expression, a relation for the mechanical force can be obtained by the method of energy or coenergy. These two methods are derived from the principle of conservation of energy and are very well documented in the literature such as Sen 1989, Fitzgerald 1985. The expression for the force obtained with the energy method is:

$$F = -\frac{\delta W_m}{\delta x}\Big|_{\lambda=cons\tan t} = \frac{B^2.A}{\mu_0} \tag{4}$$

Where, λ is the flux linkage, which is equal to the magnetic flux (ϕ) in the system times the number of turns in the coil (N) generating the magnetic field. It can be seen from Eq. (4) that the stored energy in the magnetic field, and thus the mechanical force, is a function of the magnetic flux (or flux density) present in the system. Thus, the available force for a specific device with a given MMF is determined by the reluctance of the device.

Magnetic Circuit Model

The magnetic circuit approach is an analytical method, analogous to electric circuit analysis, for modeling electromagnetic devices [10,11]. Cherry et al. in a classic paper [12], demonstrated the duality between electric and magnetic circuits. The driving force in a magnetic circuit is the magnetomotive force (MMF) $\mathfrak{S}$ which produces a magnetic flux against a coil reluctance $\mathfrak{R}$. The reluctance is defined as:

$$\mathfrak{R} = \frac{l}{\mu A}$$

(5)

Where l is the length of the magnetic flux path, A is the cross section area perpendicular to the flux, and μ is the permeability of the material [10].

For a given MMF and $\mathfrak{R}$, the flux ϕ in the circuit can be found from Kirchoff's law for magnetic circuits. The holding force can be computer using the following simple relation:

$$F = \frac{B^2 A}{2\mu_0}$$

(6)

Where B represents the magnetic flux density in the airgap separating the components, A is the cross section area of the airgap and μ_0 is the permeability of air.

The flux depends on the overall reluctance of the system. The reluctance is low when there is perfect contact between the part and electromagnet. However, part form errors, e.g., roughness, and deviation from flatness give rise to air gaps between the part and gripper. Since actual size and distribution of the airgaps in the gripper-part interface are difficult to determine for a gripper directly in contact with the part surface; it is proposed to model a small uniform air gap that can be reproduced in an experiment. When the part rests directly on the magnet gripper surface, a uniform air gap equal to the part out-of-flatness error is could be used. It can be assumed here that the reluctance of this air gap is equivalent to the reluctance of the actual contact.

The reluctances proposed in this model include those of the electromagnet, air gaps, part, and the surrounding air medium. The procedure for modeling the reluctances is described next.

Electromagnet Reluctance $\mathfrak{R}_{Electromagnet}$. To simplify the shape path of the flux lines only half of a cylindrical electromagnet is considered.

Part Reluctance $\mathfrak{R}_{Part}$. The part reluctance is calculated using the following equation [7]:

$$\mathfrak{R}_{Part} = \oint \frac{dl}{\mu(l).A(l)}$$

(7)

To evaluate this line integral, a numerical integration scheme can be used. The mean path is such that it is normal to the radial line representing the cross sectional

area. The variation in part permeability along the flux path is explicitly accounted for in the calculation of the circuit reluctance.

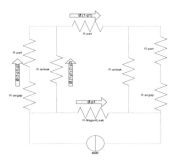

Fig. 7. Equivalent Magnetic Circuit of the Magnet-Part System

Airgap Reluctance $\mathfrak{R}_{Airgap}$. The airgap term applies to the flux lines crossing the magnet-part interface. In reality, the airgap length varies at each point in the interface because of surface roughness and form errors. In this model, an equivalent uniform airgap length is used. The cross-sectional area of the air gap is equal to the magnet-part contact area.

Of the simplifications made above, the use of a mean magnetic flux path is most significant since it implies that the magnetic circuit model cannot predict the distribution of flux in the magnet-part system. However, it can still be used to estimate the total normal holding force and to gain an insight into the effects of magnet and part variables.

Model Solution. Solution of the magnetic circuit model involves determining the flux flowing through each component of the circuit. This is done using Kirchoff's law for magnetic circuits, which states that the sum of MMF in any closed loop must be equal to zero [15]:

$$\sum_i MMF = \mathfrak{I} - \sum_i R_i\phi_i = \mathfrak{I} - \sum_i H_i l_i = 0$$

(8)

where index i represents the ith element of the closed loop. H_i is the magnetic field in the ith element, and l_i is the length of the flux path in the ith element. For the magnet used in the gripper, the equation reduces to:

$$\mathfrak{I} + H_{Work} l_{Work} + (2B_{Airgap} / \mu_0) l_{Airgap} = 0$$

(9)

The factor of 2 in the last term accounts for the crossing of airgap twice, once from the N pole to the part and again from the part to the S pole. For the circuit shown in Fig. 8, the part holding force is produced by the fraction of flux that crosses the magnet-part air gap, which is given by:

$$\phi_P = \phi.(1 - p_1).p_2 \qquad (10)$$

where p_1 is the fraction of ϕ leaking and p_2 is the fraction of $\phi.(1 - p_1)$ entering the airgap and the part. Equation (10) when combined with Equation (4) gives the mechanical force acting on ½ of the model of the part. The results from the above magnetic circuit model were compared with COMSOL FEA model.

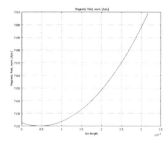

Fig. 8. Field distribution along radius with 6-node element coarse mesh.

COMSOL FEA Model

The magnetic model was created using a 2D axisymmetric magnetostatics (with currents) module, as shown in Fig. 9. The air domain was assigned the dimensions of width and length equal to three times the maximum dimension of the electromagnet. The air and aluminum parts were assigned a $\mu R = 1$, steel casing a $\mu R = 2000$, and the center rod was assigned a variable permeability via $\mu R(B, \sigma)$. The magnetic field is assigned via a current density which has units of current per unit area.

An important boundary condition includes axial symmetry and continuity. Axial symmetry is defined for 2D axisymmetric models at r = 0, and assumes a symmetric condition such that the properties of the system do not vary azimuthally. The continuity boundary condition enforces continuity of the tangential components of the magnetic field via: n x (H2 – H1) = 0. This boundary condition is used at the junction of two different subdomains with different magnetic properties.

When modeling a magnetic circuit, it is important to account for the surrounding atmosphere. Generally, this atmosphere is air, with a relative permeability of one. There have been several ways to model the surrounding air in a magnetic circuit. The two most notable techniques are the use of an air domain and infinite elements. Atulasimha et al. [7] describes how to determine the proper dimensions of an air domain. Another method of modeling the surrounding air is by using infinite elements. Infinite elements are assigned to a small subdomain region which defines the outside of the setup. The outer domain containing infinite elements causes the domain to be stretched to infinity. This allows for the flux lines to flow as far as they normally would without constraints. It was desired to compare the two previously described methods of modeling the surrounding air domain. To do this, the rod and coil was implemented with infinite elements and with air domains of different sizes. The results are shown in Fig. 10, where it was found that an air domain with dimensions of three times the largest geometric dimension of the setup gives same results as infinite elements. It was also desired to determine if larger applied current densities lead to the need for a larger air domain. In Fig. 11, magnetic fields of ~7-7.2 kA/m are seen and the previously specified dimensions of the air domain still give a very good answer as compared to the infinite element result. Fig. 11 shows the same plot as Fig. 10, but at a higher applied magnetic field (~117-120 kA/m). Air domain sizing of three times the largest dimension in each direction still provides an accurate answer.

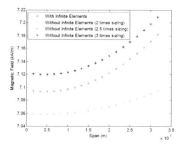

Fig. 9. COMSOL model of radially symmetric electromagnet.

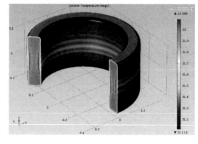

Fig. 10. Radial distribution of magnetic field for rod and coil setup with air domain of different dimensions and infinite elements.

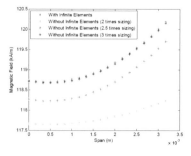

Fig. 11. Radial distribution of magnetic field for rod and coil setup with air domain of different dimensions and infinite elements at high applied magnetic fields.

6. Conclusions and future work

It was shown that a steel flux return path greatly reduces the radial and longitudinal variations of field. The use of infinite elements will generally reduce the total number of elements required in a magnetic model. Seeing this, models using infinite elements will generally prove to be more computationally efficient than using an air domain. However, infinite elements take longer to solve than standard elements, so there is a tradeoff between using the two methods.The part texture attributes (surface roughness and texture) affect the holding forces of an electromagnet gripper. Future effort in this area will present the effect of these attributes on normal and tangential holding forces.

References

[1]   Wadhwa, R.S., Lien,T., Monkman, G.J. Robust Prehension for ferrous metalcasted product families, Proceedings of MITIP 2011

[2]   Anonymous, New shelf robot saves vital space in the foundry environment, The Industrial Robot. Bedford: 2006. Vol. 33, Iss. 2; p. 145

[3]   Anonymous, The Castings Center selects STRIM, Euclid, and Prelude Software, The Industrial Robot. Bedford: 1996. Vol. 23, Iss. 6; p. 6

[4]   Wetzel, S. GM's Iron Finishing Automation, Modern Casting, 2008; 98,1 ABI/INFORM Complete pg.38

[5]   P. G. Evans, M. J. Dapino, and J. B. Restorff, "Bill Armstrong memorial symposium: free energy model for magnetization and magnetostriction in stressed Galfenol alloys," in Proceedings of SPIE, Behavior and Mechanics of Multifunctional and Composite Materials, San Diego, CA, 2007, p. p. 652619.

[6]   M. J. Sablik and D. C. Jiles, "Coupled Magnetoelastic Theory of Magnetic and Magnetostrictive Hysteresis," Ieee Transactions on Magnetics, vol. 29,no. 4, pp. 2113-2123, Jul 1993.

[7]   S. Datta, J. Atulasimha, C. Mudivarthi, and A. B. Flatau, "The modeling of magnetomechanical sensors in laminated structures," Smart Materials &Structures, vol. 17, no. 2, p. 9, Apr 2008.

[8]   J. Atulasimha, "Characterization and Modeling of the Magnetomechanical Behavior of Iron-Gallium Alloys," Phd Dissertation Department of Aerospace Engineering, MD, 2006.

[9]   S. Chikazumi, Physics of Magnetism: John Wiley and Sons, Inc., 1964.

[10]   Hoole, S.R, Computer Aided Analysis and Design of Electromagnetic Devices, 1989

[11]   Law, J.D., Modeling of Field Regulated Reluctance Machines, PhD Thesis, University of Wisconsin-Madison,1991

[12]   Cherry, E.C.: The duality between interlinked electric and magnetic circuits and the formation of transformer equivalent circuits, Proc. Phys. Soc., 1949, 62, p.101.

Concise process improvement - A process variation diagnosis tool

Steven Cox, John A. Garside, and Apostolos Kotsialos
School of Engineering and Computing Sciences, Durham University, Sciences Laboratories, South Road, Durham, DH1 3LE UK
This paper is dedicated to the memory of Prof. Valentin Vitanov.

Abstract. This paper examines the efficiency and objectivity of current Six Sigma practices when at the Measure/Analyse phase of the DMAIC process improvement cycle. A method, known as the Process Variation Diagnosis Tool (PROVADT), is introduced to demonstrate how tools from other quality disciplines can be used within the Six Sigma framework to strengthen the overall approach by means of improved objectivity and efficient selection of samples. From a structured sample of products, PROVADT is able to apply a Gage R&R and Provisional Process Capability study fulfilling pre-requisites of the Measure and early Analyse phases of the DMAIC process improvement cycle. From the same samples a Shainin Multi-Vari study and Isoplot can be obtained in order to further the analysis without additional samples. The latter quality techniques are associated with the "Clue Generation" phase of the Shainin System. The PROVADT method is tested in industry case studies to demonstrate its effectiveness of driving forward a process improvement initiative with a relatively small number of samples, which is particularly important for low volume high value manufacturing. Case studies were conducted at a leading manufacturer of microprocessor based electric motor control systems, a global technology, manufacturing and service company that provide advanced systems in the automotive industry and a furniture manufacturer. Using PROVADT and sample sizes of 20 units it was possible in all cases to validate the measurement system and gain an early objective insight into potential root causes of variation, leading to significant cost savings for both companies.

Keywords: Quality management and control, six sigma, electric motors, automotive components, furniture manufacturing.

1. Introduction

A Six Sigma quality improvement project typically follows the five phase improvement cycle: Define, Measure, Analyse, Improve and Control (DMAIC) [1,2]. Six Sigma texts, [1,2], outline many techniques and tools that can be used at each stage of the quality improvement cycle. When it comes to the Analyse phase it often jumps from extremely subjective approach, using brainstorming and cause-and-effect diagrams to form casual hypothesis, to complex statistical tools to validate casual hypothesis. This paper will introduce techniques to improve the weakness in Six Sigma's "exploration" [3].

It is particularly important when the sampling cost is high or low volume of product is available to test. Thus extremely complex Designs of Experiments (DOE) can be impractical and using less powerful screening techniques, like Fractional Factorials, will reduce the numbers of experiments needed but at the expense of higher order interaction effects. Other approaches to identify important factors such as scatter plots can lead to potentially erroneous results as correlations appear by coincidence or are linked by a related underlying cause [2] and cause-and-effect matrices can be extremely subjective. Importantly the real root cause of a quality problem could be missed if DOE is applied based on casual hypothesis techniques to identify root causes.

This paper outlines a sampling strategy known as the Process Variation Diagnosis Tool (PROVADT). The PROVADT was devised to improve the objectivity of the early analysis of a quality problem when there are a large number of factors in a process to analyse and a relatively low volume of product to sample. It can provide a Gage Repeatability & Reproducibility (GRR) study and a Process Capability study, techniques classically used in the Six Sigma Measure and Analyse phase respectively. These techniques take time to apply, and will shine little light on the root cause of a problem.

The PVDT from the same samples allow a GRR and a Provisional Process Capability (PPC) study to be extrapolated, as well as provide a Shainin Multi-Vari (SMV) study and an Isoplot[SM], techniques associated with the "Clue Generation" phase of Shainin System [4]. Based on them, the "signature of variation" can be found allowing for a more efficient analysis of a quality problem. The Analyse process starts by reducing down the numbers of factors under consideration, eliminating unimportant factors objectively with data driven information. This reduction in factors with SMV significantly reduces the subjectivity of the early analysis. Thus, later application of DOE is more powerful and more meaningfully used. With fewer important factors to analyse, fewer experiments are needed to fully understand the interaction effects.

2. The PROVADT procedure

The PROVADT sampling structure must be defined before implementation to ensure enough data is collected to fulfil the statistical techniques requirements.

A sample size, n, must be selected, with a minimum of 20 units, as this is the minimum number of units needed to calculate a PPC. This sample size depends also on the data collection time period α and the number β of consecutive units sampled from each batch or time period according to (1).

$$n - \alpha\beta \qquad (1)$$

The value for α in a SMV is 3-5 periods, where a sample period could be over a shift, a day, a week if collecting from a flow line or if there is batch production, the periods could correspond to batches. These must be spread out over a time frame in order to capture at least 80% of historical variation. β must be a least 3 consecutive units for a SMV [5]. However n should be greater than 20 to fit the requirements of the PPC, therefore if $\alpha=5$, β must be 4 or more.

The critical-to-quality characteristic on the samples should be measured repeatedly by a minimum of two appraisers. Let r_i is the number of repeats taken by an individual appraiser, then the total number of measurements taken, r_{Total}, is;

$$r_{total} = \sum r_i \qquad (2)$$

The total number of measurements made, φ, is;

$$\varphi = nr_{total} \qquad (3)$$

The value of φ is important for the GRR calculation to be valid and is typically 60 measurements or greater. The first $n/2$ sampled units are measured by appraiser 1 first then by appraiser 2. The second half is measured by appraiser 2 first then by appraiser 1 and the results can be analysed using GRR, Isoplot^{SM}, PPC and SMV.

3. Case Studies of the Practical Implementation of the PROVADT

3.1. Overview

This section explains two case study implementions of PROVADT. Each case has collected information to both validate the measurement system and gain insight into the potential root causes of the quality problems.

3.2. Case One: Edge Banding Trimming

3.2.1. The Quality Problem

Conducted at a leading furniture manufacturer, the most critical quality issue was an Edge Banding process. This process takes Medium Density Fibreboards, which are cut to the correct width; edge banding veneers are glued and

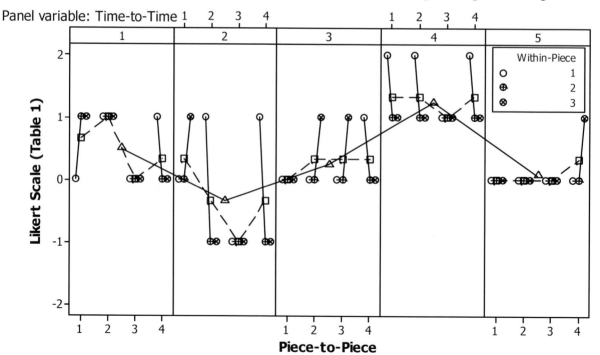

Fig. 1. Shainin Multi-Vari study for Edge 1, showing Red X as within-piece and a possible Pink X Time-to-Time

applied to the freshly cut edges. The boards are then rotated to be cut to the correct length. The final edge banding veneers are then glued and applied to these edges.

All four edges are processed in one run. Output from this process is around 10,000 panels per day and had been subject of a number of quality improvement programmes over many years. At the start of the study the company was seeing around 20% of its output being returned due to edging problems.

3.2.2 PROVADT Implementation
The focus of the process improvement programme was the over- and under- trimming of the edging. In order to use PROVADT, four consecutive panels were acquired from five different time periods, the time periods were selected based on historical data in order to capture 80% of the process variation; each edge was measured three times. This means from a sample of 20 panels (4 panels × 5 time periods), each edge has a total of 60 measurements taken (20 panels × 3 repeated measures). Therefore, the PROVADT parameters: α=5, β=4, n=20, r_{total}=3, r_1=2, r_2=1 and φ=60 are applied. Each of the four edges had the diagnostic tool applied separately as they were trimmed at different points in the machine process. Altough the PROVADT was applied on all edges only the results for edge 1 will be presented, however these results were typical of the four edges.

A quantitative grading score for the problem was introduced. Previously the product was considered as conforming or non-conforming. The modified labelling system, a variant of a five point Likert scale [5], is shown in Table 1. This modification allows the improved expression of the process capability.

Table 1. Modified Labelling System Implemented

2	Reject	(very under trimmed, out of specification)
1	Accept	(under trimmed, within specification)
0	Good	
-1	Accept	(over trimmed, within specification)
-2	Reject	(very over trimmed, out of specification)

3.2.3 Six Sigma Metrics
The results of the GRR experiment, demonstrated there was serious problem with either the measurement system or non-uniformity along the edging. Panel edge 1 had scores of 78%. With a result above 30% classed as inadequate [6].

The PPC of the edge banding process on panel edge 1 is P_p=0.61. The capability study of $P_p \leq 1$ shows extremely low values' indicating the process is failing to produce sufficient products within specification. This is consistent with the high numbers of product returns that were experienced prior to the investigation.

3.2.4 Shainin Multi-Vari Study
A SMV was extrapolated from the PROVADT data. The SMV Study for edge 1, fig. 1, is typical for the four edges. The largest signature of variation or Red X, is a within-piece problem. The within-piece variations are the groups of 3 circles joined by a solid line. This is a typical pattern when there is a measurement system problem or non-uniformity along the edge [5].

The overall SMV investigation clearly showed the Red X was predominantly a within piece problem. This was consistent with the GRR study which highlighted large variation across measures of the same piece. Although the SMV showed there was piece-to-piece and time-to-time variation, it determined that the factors responsible for the within-piece variation had the biggest effect on improving the capability of this process.

3.2.5 Conclusions of the Edge Banding Case Study
From PROVADT the following previously suspected factors were ruled out of the investigation:

- Different size panels were affecting trimming performance; if the edging and trimming machines were affected by the size of the panels there would have to be a significant batch to batch change in variation.
- Settings are being altered between batches; again the effect of changing setting to accommodate different size panels would show up as a batch-to-batch problem.

From the application of PROVADT future investigations should be focused on:
- Full validation of the measuring system using Isoplots to ensure large variation due to a poor measurement system isn't masking another problem.
- If the measurement system is found to be accurate, the investigation should focus on the edging and trimming machines, as the Red X variation results from non-uniformity along the edging.

These follow-up investigations were conducted in-house and the Quality Engineer commented that the project team had *"driven the project further in 2 weeks than it had been in the previous 2 years"*.

3.3. Laser weld mechanical strength

3.3.1 The Quality Problem
Conducted at a global manufacturing company providing advanced systems to the automotive industry, the process scrutinised involved an automated laser welding process used to weld Field-Effect Transistors (FET) to a stamping grid. This component is then used as part of a power steering module. There was concerns over the consistency of the current online destructive shear test method for mechanical strength. A non-destructive peel test was developed in order to establish a statistically valid measurement system. The peel test involves applying a 5N point load on the FET and measuring the deflection.

3.3.2 PROVADT Implementation

In order to validate the experimental non-destructive peel test a PROVADT analysis was planned. The only variation from the previously outline strategy was to collect 24 samples rather than 20. Therefore the PROVADT parameters: $\alpha=6$, $\beta=4$, $n=24$, $r_{total}=3$, $r_1=2$, $r_2=1$ and $\varphi=72$ are applied. This decision was made to evenly sample across the entire shift. Samples were collected to ensure 80% coverage of the historical process variation. In order to asses reproduceability the samples were tested twice on a 500N load cell and once on a 5kN load cell. Although each stamping grid contains 4 FETs, only the results for FET 1 are shown, since.they are representative of all 4 FETs.

3.3.3 Isoplot^SM Interperatation

The Isoplot result in Fig. 2 from testing the FETs demonstrate that the variation present due to the measurement equipment is large compared to the variation present due to the product.

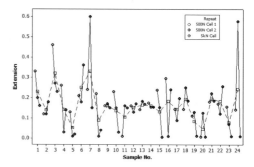

Fig. 3. Shainin Multi-Vari Study for FET 1 showing large with-in piece variation

4. Conclusions

This paper has presented a method for increasing the efficiency and subjectivity of diagnosing a process problem. PROVADT builds on the established methodology and methods of Six Sigma's DMAIC process improvement cycle. Parameters are set out which can capture a GRR study and a PPC Study necessary in the Measure and early Analyse phases of the DMAIC cycle. The parameters also allow for a SMV study and Isoplot^SM to be collected from the same sample. These graphical tools come from the Shainin System, but when applied in the context of PROVADT they can add subjective analysis by narrowing down the root causes of quality problems within a Six Sigma framework. Minimum parameters were demonstrated in all case study material. From 20 units it was possible to fulfil the pre-requisites required to perform a GRR, PPC and SMV Study. It was also reported how the PROVADT method on all occasions drove the improvement projects forward from the samples required to validate the measurement system.

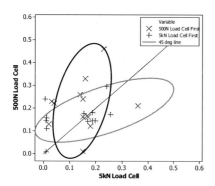

Fig. 2. Isoplot^SM result for FET 1 demonstrating the test effecting the characteristic of the weld

It also demonstrates the results form two distinct groups, shown in red and black, splitting samples tested on the 500N Load cell first from those tested on the 5kN load cell. This is a clear indication that the peel test is effecting the characteristic of the weld and is therefore destructive, counter to the previous belief that it was non-destructive.

3.3.4 Shainin Multi-Vari Study

The SMV study in fig. 3 demonstrates that the with-in piece variation is the Red X. This is consistent with the Isoplot in fig. 2 which demonstrates that there is a large amount of measurement variation.

3.3.5 Conclusionsof the Laser Weld Case Study

GRR was not performed. The graphical Isoplot^SM demonstrated the test was destructive. Thus the common assumption made in GRR calculations that the sample is robust against the measurement process is not valid.

As a result the experimental Peel Test was scrapped and the in-house team estimated a £10,000 saving in labour and resources compared to the classic validation investigations.

References

[1] P. S. Pande, R. P. Neuman, and R. R. Cavenagh, The Six Sigma Way: How GE, Motorola, and Other Top Companies are Honing Their Performance. McGraw-Hill Professional, 2000

[2] M. L. George, D. Rowlands, M. Price, and J. Maxey, The Lean Six Sigma Pocket Toolbook: A Quick Reference Guide to 70 Tools for Improving Quality and Speed. McGraw-Hill Professional, 2005

[3] J. De Mast, "A methodological comparison of three strategies for quality improvement," International Journal of Quality & Reliability Management, vol. 21, no. 2, pp. 198-213, 2004.

[4] P. D. Shainin, "Managing Quality Improvement," in ASQC Quality Congress Transactions, pp. 554-560, 1993.

[5] K. R. Bhote, World Class Quality: Using Design of Experiments to Make It Happen, 1st ed. AMACOM, 1991.

[6] International Automotive Sector Group, QS9000: Measurement Systems Analysis. 1995.

An evaluation of physics engines and their application in haptic virtual assembly environments

G. González-Badillo[1], H. I. Medellín-Castillo[1], C. Fletcher[2], T. Lim[2], J. Ritchie[2], S. Garbaya[3]

[1] Universidad Autónoma de San Luis Potosí, S.L.P., México
[2] Innovative Manufacturing Research Centre, Heriot-Watt University, Edinburgh, UK
[3] Arts et Métiers ParisTech, Le2i, CNRS, Institut Image, 2 Rue Thomas Dumorey, 71100 Chalon-sur-Saône, France

Abstract. Virtual Reality (VR) applications are employed in engineering situation to simulate real and artificial situations where the user can interact with 3D models in real time. Within these applications the virtual environment must emulate real world physics such that the system behaviour and interaction are as natural as possible and to support realistic manufacturing applications. As a consequence of this focus, several simulation engines have been developed for various digital applications, including VR, to compute the physical response and body dynamics of objects. However, the performance of these physics engines within haptic-enabled VR applications varies considerably. In this study two third party physics engines - Bullet and PhysX[tm]- are evaluated to establish their appropriateness for haptic virtual assembly applications. With this objective in mind five assembly tasks were created with increasing assembly and geometry complexity. Each of these was carried out using the two different physics engines which had been implemented in a haptic-enabled virtual assembly platform specifically developed for this purpose. Several physics-performance parameters were also defined to aid the comparison. This approach and the subsequent results successfully demonstrated the key strengths, limitations, and weaknesses of the physics engines in haptic virtual assembly environments.

Keywords: virtual reality (VR); physics engine; Bullet; PhysX; haptics; haptic assembly; virtual assembly.

1. Introduction

Physical based modelling (PBM) uses physics simulation engines to provide dynamic behaviour and collision detection to virtual objects in virtual environments emulating the real world. This results in better appreciation and understanding of part functionality and can also lead to improved training of manual tasks [1,2]. However, there are several challenges when haptics is integrated with physics engines, e.g. synchronization, non-effective collision detection, high computational cost and a negative impact on the performance of the application [3]. This is due to the fact that simulation engines are not adapted to haptic rendering, mainly because the typical frequency of haptics simulations is over 1 kHz and around 100 Hz for physics simulations [4,5].

This work presents an evaluation of two physics engines for haptic environments to assess their performance in haptic assembly tasks. The experiments are aimed to identify the strengths and weaknesses of each simulation engine.

2. Related work

Physics simulation engines have been used in many applications from computer games through to movies. Laurell [6] identified five key points in any physics engine: contact detection, contact resolution, force calculation, integrating motion and the impact of real time constraints (time step) where anything below 25 frames per second (fps) is perceived as slow and stammering. Additionally, the update rate of the whole system, both graphics and physics, must be less than 40 milliseconds per cycle.

Howard and Vance [7] found that while mesh to mesh assembly enabled accurate collision detection, realistic physical response was not demonstrated particularly when objects had continuous contact with each other since excessive surface stickiness and model penetration was observed. The physics update rate was found to be directly related to the number of contacts generated between colliding geometries.

Seth, et al. [8] identified three main challenges that virtual assemblies must overcome to increase the level of realism: collision detection, inter-part constraint detection and physics-based modelling.

Seugling and Rölin [3] compared three physics engines - Newton, ODE and PhysX - against the following run-time executions: friction on a sliding plane, gyroscopic forces, restitution, stability and scalability of constraints, accuracy against real, scalability of contacts (pile of boxes), stability of piling (max number of stacked boxes), complex contact primitive-mesh, convex-mesh and mesh-mesh. According to their results PhysX was the best evaluated simulation engine except in the stability of piling test and the mesh-mesh collision detection due to unwanted behaviour.

Boeing and Bräunl [9] carried out an investigation to compare PhysX (formerly Novodex), Bullet, JigLib, Newton, ODE, Tokamak and True Axis using PAL (Physics Abstraction Layer). Their comparison criteria included: integrator performance, material properties, friction, constraint stability, collision system and the stacking test. They concluded that PhysX had the best integrator method whereas Bullet provided the most robust collision system.

On the other hand Coumans and Victor [10] made a simple comparison analysis of the following physics engines: PhysX, Havok, ODE and Bullet. Collision detection and rigid body features were used as the comparison criteria. According to the authors PhysX was the most complete engine.

Glondu et al. [4] introduced the possibilities of implementing a modular haptic display system that relies on physical simulation and haptic rendering. With this in mind, four physical simulation libraries are evaluated: Havok, PhysX, Bullet and OpenTissue. The performance criterion was based on computation time, stability and accuracy. PhysX showed penetration in some of the tests whilst Havok showed the best average computation time, stability and friction accuracy.

The previous background study has revealed that several research works have been conducted to evaluate different simulation engines. In general, it is concluded that PhysX is the most complete simulation engine. However, these works have not considered the use of haptic rendering in the virtual environment being evaluated. Thus, it can be said that the performance evaluation of simulation engines in haptic enabled virtual environments is still needed. Hence, the objective in this work is to conduct a series of experiments to find the most appropriate simulation engine for a specific haptic application. It is envisaged that the work reported in this paper can contribute to the haptic research community.

3. System overview

A haptic assembly virtual platform, named as HAMMS, has been developed and is shown in Fig. 1. The HAMMS system (Fig. 1) comprises the Visualization Toolkit libraries (VTK 5.8.0) and the Open Haptics Toolkit v3.0. Two physics engines i.e., PhysX™ v. 2.8.4 and Bullet v. 2.79, have been integrated and the user can select between the two during run time. Single and dual haptic is provided using Sensable's Omni haptic device. One of the main characteristics of HAMMS is a control panel where the user can modify in real time simulation parameters; haptic properties like stiffnes, damping and

friction; and physical properties like mass, restitution, tolerance, etc.

Fig. 1. Virtual haptic assembly application

4. Comparative analysis

In order to identify the usefulness and capability of the two physics engines in haptic virtual assembly environments, a set of virtual assembly tasks were defined and carried out using the two physics engines.

4.1. Model representation

Collision detection is a key aspect of assembly analysis and it is directly related to the model representation in the physic simulation engine [11]. Assembly tasks may comprise several objects or components with different shapes. In general, objects can be divided into two groups: convex and concave objects, being the last the most common objects in assembly tasks.

Bullet 2.79 use GIMPACT libraries to calculate collisions for concave objects represented by a triangular mesh, its representation is very similar to the graphic model as shown in Fig. 2. A convex decomposition algorithm such as HACD [12] can also be used to create concave shapes.

Fig. 2. Physic representation of objects using GIMPACT

PhysX v2.8.4 does not support collision detection for triangular meshes; however, an algorithm to create a concave compound object from a triangle mesh, convexFT (CFT), is provided. The algorithm transforms each triangle of the mesh into a convex element, so the final shape has as many convex hulls as triangles in the original mesh.

4.2. Assembly tasks

Five assembly tasks were selected to analyse the performance of each physics engine in HAMMS:

(1) A pile of boxes assembly task was selected to evaluate the manipulation and performance of primitive

shapes, it is also used to analyse the simulation engine performance and stability where multiple and accumulative contacts are considered, Fig. 3 (a).

(2) The packing boxes assembly task, Fig. 3 (b), is useful to identify the physics engine performance using different representation algorithms such as convex decomposition or triangular meshes. The purpose of this task is to observe the collision response and stability when multiple contacts in different directions are present.

(3) The peg and hole assembly task which is commonly used in assembly tests because it represents a generalized case of cylindrical parts' assembly, Fig. 3 (c).

(4) A more complex pump assembly task, Fig. 4a, and (5) a bearing puller [Fig. 4 (b)] are selected as they represent the virtual models of real components with complex shapes.

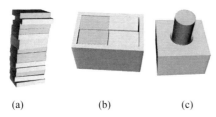

(a)   (b)   (c)

Fig. 3. Assembly tasks: a) Pile of boxes b) Packing box c) Peg & hole

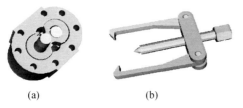

(a)   (b)

Fig. 4. Assembly tasks: a) Pump b) Bearing Puller

Each of these five assembly tasks was performed by an experienced user in both haptics and virtual assembly. Five repetitions were carried out for each task, all the tests were performed using a single haptic device to manipulate virtual objects and the mouse to manipulate the camera.

In a physics simulation engine, the integrator method is refered to the numerical methods that it uses for calculate the new position of the object on each time step during the simulation. In order to assess the integrator performance under different conditions, virtual free fall experiments were carried out, measuring the time on reach the floor when the object was dropped from an elevation of 500 units.

5. Results and discussion

The results of the free fall test are shown in Table 1, where it is shown that when the number of triangles of the model is smaller than 300 the integrator performance of PhysX is not affected, whereas in the case of Bullet an increment of 50% was observed. When the object comprises about 2000 triangles the time performance is

greatly affected (about 100% increase) for both Bullet and PhysX.

Table 1. Free-falling time with respect to shape (number of triangles)

Shapes	Triangles	Bullet (sec)	PhysX (sec)
Box	12	0.993	2.11
Pin	44	1.011	2.105
Big Cog	276	1.445	2.115
Housing	1934	2.918	4.199

Table 2 shows the influence of the model representation on time performance, these results were obtained with the haptic rendering loop on. The results indicate that when primitive shapes are used in PhysX, free fall time is 5.6 seconds compared to Bullet's 0.9 second average. ConvexFT and convex decomposition (HACD) [12] showed similar results and the best performance for PhysX. Bullet showed a time increment related to the increment of the model shape complexity. Model representation using primitives showed the best performance.

Table 2. Free-falling time with respect to model representation

Model	Representation	Bullet (sec)	PhysX (sec)
Box	Box	0.999	5.627
	Trimesh- ConvexFT	0.993	2.11
Pin	Cylinder	0.998	5.608
	Trimesh- ConvexFT	1.011	2.105
Cog	Trimesh- ConvexFT	1.445	2.115
	Convex dec. HACD	1.428	2.09

Table 3 shows the percentage of increase in time in the free-falling test when the haptic rendering loop was running with respect to a situation where only physics and graphics loops were running, Bullet showed a time increment of 50% when the haptic rendering loop was on and the model was complex, compared to PhysX that showed only an increment of 2%; however, the falling time in all test was smaller using Bullet than using PhysX, this suggest that PhysX rendering loop is more adapted to be used together with haptics. The theoretical falling time is 0.316 seconds.

Table 3. Influence of the haptic loop on free-falling time (%)

Model	Representation	Bullet	PhysX
Box	Box	3.65	0.195
Pin	Cylinder	0.91	0.139
Cog	Trimesh	50.42	2.16
	Convex Dec HACD	48.28	1.08

Table 4 presents the task completion time (TCT) in minutes for each assembly tasks, different model representations and each physics engine. It can be observed that for the assembly tasks of pile of cubes, packing box and peg & hole, when primitives or convex decomposed model representation is used, PhysX posted the least TCT, however when using triangular meshes and Bullet, TCT was least in all the tasks, except the packing box due to unnatural collision response. Real and virtual tests were carried out on the pump assembly. A mean

TCT of 37 seconds was obtained in the real assembly task whilst in the virtual platform the TCT value was 58.3 seconds using Bullet (56% more than the real assembly) and 1.21 minutes using PhysX, this difference may be due to several factors such as the manipulation of virtual models trough the haptic device, physical properties of materials (friction, restitution, mass, etc.).

Table 4. Task completion time

Case	Reps	Bullet (min)	PhysX (min)
	Primitives	03:59.8	03:24.6
Pile of cubes	Convex dec. (HACD)	05:11.8	03:32.6
	Trimesh- ConvexFT	02:41.8	03:23.7
Packing Box	Convex dec. (HACD)	04:17.4	02:09.7
	Trimesh- ConvexFT	03:19.2	02:45.5
Peg & hole	Convex dec. (HACD)	00:13.1	00:07.1
	Trimesh- ConvexFT	00:05.4	00:06.5
Pump	Trimesh- ConvexFT	00:58.3	01:21.0
Puller	Trimesh- ConvexFT	01:33.9	n/a

The results obtained for haptic and physics update rates indicate that PhysX offers better update rates when using non complex geometries represented by primitives or by convex decomposition. However Bullet physics showed better update rates when simulating complex parts represented as triangular meshes.

Assembly performance parameters were evaluated using a scale from 0 to 3, where 0 represents the worst performance and 3 the best. Users assign a value to each parameter according to their perception of the assembly task. Performance parameters include: Collision precision (CP) indicates penetration of virtual models when colliding with other virtual objects. Collision response (CR) is the reaction and how natural objects behave. Assembly stability (AS) indicates if the objects are stable once the assembly is completed. Manipulability (M) indicates how easy the models can be manipulated, and the total (T) indicates the overall score of CP+CR+AS+M. The results are shown in Table 5.

Table 5. Performance evaluation

Case	Model Representation	Bullet					PhysX				
		CP	CR	AS	M	T	CP	CR	AS	M	T
Pile of cubes	Primitives	2	3	2	3	10	3	3	3	3	12
	Convex dec. (HACD)	3	1	1	1	6	3	3	3	3	12
	Trimesh- ConvexFT	3	3	3	3	12	3	3	2	2	10
Packing Box	Convex dec. (HACD)	1	2	1	1	5	3	3	3	3	12
	Trimesh- ConvexFT	3	3	1	3	10	3	2	2	3	10
Peg & hole	Convex dec. (HACD)	2	2	3	2	9	3	3	2	3	11
	Trimesh- ConvexFT	3	3	3	3	12	3	2	3	3	11
Pump	Trimesh- ConvexFT	3	3	2	3	11	3	1	2	2	8
Puller	Trimesh- ConvexFT	3	3	3	3	12	3	2	0	2	7
	Overall	23	23	19	22	87	27	22	20	24	93

It is notable that PhysX displayed better performance than Bullet in simple assembly tasks such as the pile of cubes, packing box and peg and hole. However, in complex assembly tasks like the pump and puller assembly, Bullet showed better performance, less assembly time (58.3 seconds) and better evaluation by the user (11 points of 12 possible) than PhysX (assembly time 1:21.0 min and a total evaluation of 8 points). Moreover, in PhysX the puller assembly tasks could not be completed because the puller screw could not be inserted in the puller base, due to a poor model representation.

6. Conclusions and future work

A performance evaluation of two different physics simulation engines for haptic assembly has been presented. The results have suggested that for assembly tasks that involve non complex geometries like boxes and cylinders (primitives), the use of PhysX offers a better performance than Bullet; however when the assembly comprises more complex shape components, Bullet has better performance than PhysX. A more comprehensive study must be carried out including the effect of simulation parameters, the use of a dual haptics configuration, and others physics simulation engines such as Havok or ODE.

Acknowledgments: Authors acknowledge the financial support from CONACYT (National Science and Technology Council of Mexico) and from SIP-UASLP (Research and Postgraduate Secretariat of Universidad Autonoma de San Luis Potosi, Mexico).

References

[1] Zerbato D, Baschirotto D, Baschirotto D, Debora Botturi D, (2011) GPU-based physical cut in interactive haptic simulations. Internetional Journal CARS 6:265-272.

[2] Wang Y, Jayaram S, Jayaram U, (2001) Physically based modeling in virtual assembly. The int. journal of virtual reality vol 5, No. 1, pp. 1-14.

[3] Seugling A, Rölin M, (2006) Evaluation of physics engines and implementation of a physics module in a 3D auhoring tool. Master's thesis in computing science Umea University.

[4] Glondu L, Marchal M, Dumont G, (2010) Evaluation of physical simulation libraries for haptic rendering of contacts between rigid bodies. Proceedings of ASME , WINVR 2010, may 12-14, Ames, Iowa, USA, WINVR201-3726, pp. 41-49.

[5] Ritchie JM, Lim T, Medellin H, Sung RS, (2009) A haptic based virtual assembly system for the generation of assembly process plans. XV congreso internacional SOMIM, 23- 25 September 2009, Cd. Obregon, Sonora. México.

[6] Laurell B, (2008) The inner workings of real-time physics simulation engines. IRCSE '08 IDT workshop on interesting results in computer science and engineering; Mälardalen univerity, Vasteras, Sweden.

[7] Howard BM, Vance JM, (2007) Desktop haptic virtual assembly using physically based mod.. Vir. Rea. 11:207-215.

[8] Seth A, Vance JM, Oliver JH, (2011) Virtual reality for assembly methods prototyping: a review. Virtual rea. 15:5-50.

[9] Boeing A, Bräunl T, (2007) Evaluation of real-time physics simulation systems. Graphite '07 Proc. of the 5th Int. Conf. on Comp. graph and int tech. Aus. and SW Asia, pp: 281-288.

[10] Coumans E, Victor K, (2007) COLLADA Physics. Web3D '07 Proc. of the 12 int. Conf. on 3D web tech., pp. 101-104.

[11] Chen T, (2010) Virtual Assembly of mechanical component and collision detection. Int. Conf. on electronics and information engineering (ICEIE 2010) IEEE vol.1: 443-447.

[12] Khaled M, Faouzi G, (2009) A simple and efficient approach for 3D mesh approximate convex decomposition, 16th IEEE Int. Conf. on image processing (ICIP '09), Cairo, 7-10 November 2009.

Effect of weight perception on human performance in a haptic-enabled virtual assembly platform

G. González-Badillo[1], H. I. Medellín-Castillo[1], H. Yu[2], T. Lim[2], J. Ritchie[2], S. Garbaya[3]

[1] Universidad Autónoma de San Luis Potosí, S.L.P., México
[2] Innovative Manufacturing Research Centre, Heriot-Watt University, Edinburgh, UK
[3] Arts et Métiers ParisTech, Le2i, CNRS, Institut Image, 2 Rue Thomas Dumorey, 71100 Chalon-sur-Saône, France

Abstract. Virtual assembly platforms (VAPs) provide a means to interrogate product form, fit and function thereby shortening the design cycle time and improving product manufacturability while reducing assembly cost. VAPs lend themselves to training and can be used as offline programmable interfaces for planning and automation. Haptic devices are increasingly being chosen as the mode of interaction for VAPs over conventional glove-based and 3D-mice, the key benefit being the kinaesthetic feedback users receive while performing virtual assembly tasks in 2D/3D space leading to a virtual world closer to the real world. However, the challenge in recent years is to understand and evaluate the added-value of haptics. This paper reports on a haptic enabled VAP with a view to questioning the awareness of the environment and associated assembly tasks. The objective is to evaluate and compare human performance during virtual assembly and real-world assembly, and to identify conditions that may affect the performance of virtual assembly tasks. In particular, the effect of weight perception on virtual assembly tasks is investigated.

Keywords: virtual assembly platforms; haptics; assembly task; weight perception.

1. Introduction

According to Howard and Vance [1], a successful virtual assembly environment requires virtual parts to emulate real parts in real world. Two basic methods for simulating physical part behaviour are physically based modelling (PBM) and constraint based modelling. PBM uses Newtonian physics laws to describe the motion of objects and forces acting on bodies and to model realistic behavior by simulating the effect of gravity and collision response between objects in the virtual environment.

Several VR platforms for mechanical assembly have been developed over the years. From these developments, it has been observed that the time required to complete a virtual assembly task is always larger than the time required for the same assembly task in the real world. Several factors that may contribute to this difference include: the interface used (haptic device, glove, 3D mouse, etc), the manipulability of virtual objects, virtual objects shape and weight, camera manipulation, rendering type (stereo-view, 2D screen, head mounted displays), force feedback, etc.

In this paper the influence of weight of virtual objects in virtual assembly task is investigated in order to identify their effect on the completion time of virtual assembly tasks. The investigation is carried out using a virtual PBM assembly platform enabled with haptics.

2. Related work

During the last decade several researchers have proposed different types of virtual assembly environments with some interesting conclusions. Wang et al. [2] analysed methods of dynamic simulation that may affect the assembly task. A scaling factor for gravity was used, which was attained by trial and error. It was concluded that dynamic simulation of parts in virtual environments greatly enhances the realistic feelings of virtual spaces but does not contribute significantly to the assembly task.

On the other hand, Lim et al. [3, 4] investigated the impact of haptic environment on user efficiency while carrying out assembly tasks. It was observed that small changes in shape, the use of full collision detection and the use of stereo-view, can affect assembly times in haptic virtual assembly environments, Similar results were obtained by Garbaya [5], who observed that human operators have superior performance when provided with the sensation of forces generated by the contact between parts during the assembly process.

Huang [6] studied the effects of haptic feedback on user performance when carrying out a dynamic task using a beam and ball experiments. The results showed that user performance is affected by the magnitude of the force feedback and the complexity of the system dynamics.

The problems related to haptic interaction between human operator and virtual environments were investigated by Tzafestas [7]. The results demonstrated that human perception of weight when manipulating objects in virtual environments is similar to the pereception when manipulating real objects. It was also concluded that imperfections and limitations of the mechatronic haptic feedback device may lead to a small decrement on the user performance.

The influence of control/display ratio (C/D) on weigth perception of virtual objects was evaluated by Dominjon [8]. The results showed that when using a C/D ratio smaller than 1, amplification of user movements on virtual environment, the participants perceived the manipulated object ligther than its actual weight, in some cases it was even possible to reverse weight sensation to make a heavy object feel lighter than a light object. This suggested that mass can be added to the list of haptic properties that can be simulated with pseudo-haptic feedback.

Hara [9] examined user weight perception when the heaviness of a virtual object suddenly changes using a haptic device, results indicate that users perceive a change in weight of the virtual object only when the difference between the initial and the actual weight is significative, it was concluded that when the user cannot perceive the change of weight, he/she may unconsciously adjust their muscle command to the weight change.

Uni-manual and bi-manual weight perception when lifting virtual boxes was evaluated by Giachristis [10], who proposed that accurate perception of simulated weight should allow the user to execute the task with high precision.The results indicated that bi-manually lifted virtual objects tended to feel lighter than the same objects unimanually lifted. Users seem to be five times less sensitive to virtual weight discrimination than for real weights.

Vo and Vance [11] examined the context in which haptic feedback affects user performance. In the weight discrimination task the user was asked to idendify the heaviest object of a couple of virtual models showed. The results obtained showed that users compared models in less time using haptic feedback than using only visual perception and that identification of weight is dependent on which hand was used to manipulate the object.

According to the previous research works, a method to evaluate user performance in virtual assembly tasks is by measuring the Task Completion Time (TCT). Several authors have observed that the weight of virtual objects is one of the most important factors that affect the TCT [7, 8, 9]. Thus, this work evaluates the effect of virtual weight on the TCT. An assembly task is used as case of study to compare the performance of virtual assembly vs. the real assembly in terms of the TCT.

3. System overview

A haptic virtual assembly system, named as HAMMS, has been developed and it is shown in Fig. 1. The HAMMS interface has been developed in Visual C++ and comprises the Visualization Toolkit libraries (VTK 5.8.0)

and the Open Haptics Toolkit v3.0. Two physics engines, PhysX™ v2.8.4 and Bullet v2.79, have been integrated in HAMMS and the user is able to select any of them during run time. Single and dual haptic is provided using Sensable's Omni haptic device. One of the main characteristics of HAMMS is a control panel where the user can modify in real time simulation parameters; haptic properties like stiffnes, damping and friction; and physical properties like mass, restitution, tolerance, etc.

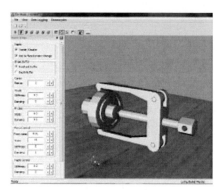

Fig. 1. HAMMS interface

The use of chronocycles [12] is also implemented in the HAMMS platform, allowing the user to graphically observe the path of the assembly once it is completed.

Two interaction phases are identified while the application is running: the first logs when the objects are only touched ("inspect") by the haptic proxy but not manipulated, the second phase is when the objects are being manipulated ("control") with the haptic device. In the "control" phase the physics model is attached to the haptic model by a spring – damper system and the graphic model is updated with the physic model (Fig. 2).

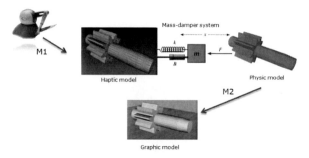

Fig. 2. "Control" phase of virtual objects

The use of the spring-damper model allows the calculation of force feedback that the user will feel when a collision occurs or when virtual objects are manipulated (weight perception). The values of spring constant and damping are determined empirically and adjusted so that smooth and stable movement of the manipulated part is obtained [5].

4. Experimental setup

The assembly task selected in this investigation was a gear oil pump. The pump comprises five parts: the housing, a big cog, a small cog and two figure-eight bearings retainers. The virtual and real pump components are shown Fig. 3. The virtual objects are imported into the HAMMS application as STL files.

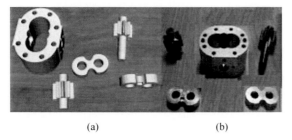

(a)   (b)

Fig. 3. Pump components, a) Virtual and b) real

Eight levels of weight, measured in Newtons (N), are defined for each pump component, L1 to L8, where L1 is the minimum weight and L8 the maximum weight in the virtual scene. The virtual and real weights are presented in Table 1. These virtual weights were obtained by scaling the density of the virtual objects. The maximum force suported by the Sensable Omni Device (3.3 N) was considered when assigning the weight level 8 to the heaviest manipulated object, the big cog. The housing is not considered because during the assembly process, real and virtual, it is the base part and remains static.

The real assembly was performed by 5 persons with each subject performing the task with one and two hands. The virtual tasks were carried out by an experienced user in both haptics and virtual assembly in order to avoid learning. The pump assembly was performed four times for each level of weight and using one and two hands. The virtual assembly tasks were performed using both physics engines, Bullet Physics and PhysX^{tm}, in order to observe the influence of different simulation engines on the assembly process.

Table 1. Levels and weights (N) of pump components

Level	Housing	Big Cog	Small cog	Bearings
L1	0.02	0.02	0.02	0.02
L2	1.3	0.17	0.13	0.1
L3	3.34	0.41	0.34	0.29
L4	>4	0.82	0.66	0.51
L5	>4	1.11	0.9	0.69
L6	>4	1.64	1.31	1.01
L7	>4	2.23	1.81	1.34
L8	>4	3.24	2.71	1.47
Real	16.7	6.7	5.2	1.6

5. Results and discussion

The chronocycles employed in HAMMS offer a graphic representation of the trajectories and user haptic manipulations in the virtual environment. When a virtual object is being manipulated with the haptic device the movements are recorded and once the assembly has been completed chronocycles can be observed to analyse the manipulation of the object. Figs. 4 (a) and (b) show the chronocycles of the pump assembly task using level of weight L1and L8 respectively. The red spheres represent

the path of the virtual objects when they were manipulated by the haptic device, and the distance between each sphere represents the speed of the motion; a low velocity is identified when one sphere is very close to the next one.

From observation, the initial manipulation of the object, when moving from its initial position to the target position, is faster at weight level L1 compared to level L8. This is verified from the chronocycles curves for L8, suggesting that larger inertia influences the assembly operations. With increased inertia in the virtual objects, the manipulation speed decreases, as it occurs in the real world, i.e. heavier objects are more difficult to manipulate.

(a)   (b)

Fig. 4. Chronocycles using A) weight L1, B) weight L8

The task completion time in Fig. 5 shows the results obtained from a single-handed assembly task. In the case of the real assembly task, the TCT mean value for one hand was 37 seconds. The TCT of the virtual assembly task using the PhysX simulation engine is shown as the red dashed line, while the same virtual task but using Bullet physics is shown as the green dotted line. Both red and green lines represent the mean values for each level of weight.

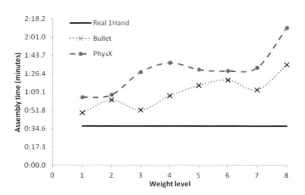

Fig. 5. Task completion time with one hand

From the results (Fig. 5) the minimum assembly time when using one hand corresponds to the weight level L1, where the weight of virtual components is minimal (0.02 N), only enough to keep the system stable. Using Bullet physics the minimum mean time was 49 seconds, with minimum TCT reported at 39 seconds. For PhysX the minimum mean value was 64 seconds and the minimal reported value was 59 seconds. The mean value for the task completion time (TCT) for one hand real assembly was 37 seconds and for two hands 27 seconds.

Figure 6 shows the results obtained with bi-manual assembly. The real process took an average of 27 seconds to complete. In the virtual experiments with two hands the minimum TCT value also corresponded to weight level, L1. Bullet physics posted a minimum mean time of 52 seconds, with a minimum reported assembly time of 44 seconds. In the case of PhysX, the TCT for weight level L1 was 75 seconds and the minimum reported TCT 61 seconds. A dip can be observed in weight level L6 for both simulation engines, Bullet and PhysX, this may be due to compensation of haptic stiffness; at weight level L6, the weight of all the models are above 0.88 N, that is the maximum continuous force for the Phantom Omni device; this value must not be confused with the maximum rendering force.

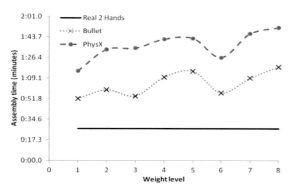

Fig. 6. Task completion time with two hands

The procedure to assemble the pump using two hands in the real world was very similar to the assembly in the virtual environment; two hands were used to align and fit the gears in the housing and bottom bearing. In the real assembly process, the TCT was smaller when using two hands than when using only one hand, however in the virtual environment the TCT using two haptic devices was greater than when using only one device. This difference may be caused by the collision response, such as sticky objects or unnatural reactions, when two objects are being manipulated. It was also observed that the fitting and aligning of the two gears was a difficult task in the virtual environment when using two hands. In general, it was observed that the time to complete a real assembly task is still smaller than the time to complete the same assembly task but in a virtual environment. Weight perception affects the TCT, as the weight of the virtual objects increases, the assembly time increases. Also, it can be said that the physics engine affects the performance of the assembly task.

6. Conclusions and future work

The effect of weight perception on human performance in virtual assembly environments enabled with haptics has been investigated. The results suggested that the performance of virtual assembly tasks, in particular the TCT, is directly affected by the weight of virtual objects; i.e. as the weight of virtual objects increases, the TCT will also increase. The chronocycles analysis

showed that as the weight of the manipulated objects increases, the manipulation velocity decreases. Further investigation considers the analysis of the weight perception effect on the user by measuring muscle and brain activity in order to identify how the user reacts to different conditions on virtual assembly tasks.

Acknowledgments: Authors acknowledge the financial support from CONACYT (National Science and Technology Council of Mexico) and from SIP-UASLP (Research and Postgraduate Secretariat of Universidad Autonoma de San Luis Potosi, Mexico).

References

[1] Howard BM, Vance JM, (2007) Desktop haptic virtual assembly using physically based modeling. Virtual Reality 11:207-215

[2] Wang Y, Jayaram S, Jayaram U, (2001) Physically based modeling in virtual assembly. The international journal of virtual reality, vol 5, No. 1, pp. 1-14

[3] Lim T, Ritchie JM, Dewar RG, Corney JR, (2007) Factors affecting user performance in haptic assembly. Virtual reality 11: 241-252

[4] Lim T, Ritchie JM, Sung R, Kosmadoudi Z, Liu Y, Thin AG, (2010) Haptic virtual reality assembly – Moving towards Real Engineering Applications. Advances in Haptics, pp. 693-721. INTECH. ISBN 987-953-307-093-3

[5] Garbaya S, Zaldivar-Colado U, (2007) The affect of contact force sensations on user performance in virtual assembly tasks. Virtual Reality, Springer, 11: 287-299

[6] Huang F, Gillespie RB, Kuo A, (2002) Haptic feedback and human performance in a dynamic task. Proceedings of the 10th symposium on haptic interfaces for virtual environments and teleoperator systems, (HAPTICS'02), IEE, pp. 24-31

[7] Tzafestas CS, (2003) Whole hand kinesthetic feedback and haptic perception in dextrous virtual manipulation, Transactions on systems, manufacturing and cybernetics, part A: systems and humans, vol 33.no. 1, IEEE, pp 100-113

[8] Dominjon L, Lécuyer A, Burkhardt JM, Richard P, Richir S, (2005) Influence of control/display ratio on the perception of mass of manipulated objects in virtual environments. Proceedings of the IEEE Virtual Reality, pp. 19-25

[9] Hara M, AshitakaN, Tambo N, Huang J, Yabuta T, (2008) Consideration of weight discriminative powers for various weight changes using a haptic device. International conference on intelligent robots and systems, IEEE/RSJ, Acropolis convention center, Nice, France, Sept 22-26, pp. 3971-3976.

[10] Giachristis C, Barrio J, Ferre M, Wing A, Ortego J, (2009) Evaluation of weight perception during unimanual and bimanual manipulation of virtual objects. Third joint eurohaptics conference and symposium on haptic interfaces for virtual environments and teleoperator systems, Salt Lake City, UT, USA, March 18-20, pp. 629-634

[11] Vo DM, Vance JM, Marasinghe M, (2009) Assessment of haptics-based interaction for assembly task in virtual reality. Third joint eurohaptics conference and symposium on haptic interfaces for virtual environments and teleoperator systems, Salt Lake City, UT, USA, March 18-20

[12] Ritchie JM, Lim T, Medellin H, Sung RS, (2009) A haptic based virtual assembly system for the generation of assembly process plans. Memorias del XV congreso internacional anual de la SOMIM, 23 al 25 de septiembre, Cd. Obregon, Sonora. México.

Impact of workers with different task times on the performance of an asynchronous assembly line

Folgado R., Henriques E. and Peças P.
IDMEC, Instituto Superior Técnico, TULisbon, Av.Rovisco Pais, 1049-001 Lisbon, Portugal

Abstract. Assembly lines are often dependable on the human elements when an extended automation is not economically viable, even if technologically possible. Workers have variations in their performances which can mean they have a different average completion time and/or a different dispersion in the time they take to perform a task. Using simulation, these combined effects are tested in a three workstation asynchronous and unbuffered assembly line, in order to understand how the different combinations of performances affect the system output.

Keywords: assembly system, human-centred systems, asynchronous line

1. Introduction

The modelling and simulation of assembly systems have been specially focused on technological and operational aspects, in which operators are represented as resources performing defined tasks with variable availability and/or variable efficiency [1]. Depending on the cases, this approach might be sufficient on highly mechanised processes. However, there are assembly systems in which the tasks are mainly performed by workers, with little or no use at all of automation [2].

Workers are expected to present individual differences among them, represented by variations both on the average time to perform the assigned tasks as well as on the amount of variability on those times. When evaluating an assembly system performance, the task time variability has been modelled as a function of the task content or of environmental effects. Nonetheless previous research studies demonstrate that differences in task times are mainly caused by differences among workers, rather than by minor differences in tasks content or in the time period they are performed [3]. It is more likely to obtain an inaccurate estimate of system productivity if one assumes that all workers come from a pool represented by a single value or a single probability distribution.

Performance analysis was one of the first manufacturing systems applications of discrete event simulation (DES), and has proven to be one of the most flexible and useful analysis tools in this field [4]. In this paper, discrete event simulation is used to model an assembly system and test the influence of the different workers task times.

Data from an industrial setting is used as an example of the different task times, which could be found when performing manual high motor content assembly tasks. The observed performances, categorized in different classes (slower/faster; higher/lower variability), are tested in different scenarios of workers allocation in an unbuffered asynchronous assembly line with 3 workstations and are then compared to a scenario in which all workers perform, according to the "average worker", with the same variability and average task time. An unbuffered system is subject to 'coupling effects', instances of idleness imposed on a workstation when it is 'starved' for available work or 'blocked' from passing completed work to the next station. Nonetheless, many assembly systems are required to hold minimal work-in-process inventories because of space or capital limitations [5]. The assembly line with 3 workstations limits the possible combinations (allocation of the 5 possible performances), while still has a central position to enable symmetrical allocations comparisons. The production of a given number of parts is simulated to assess the impact of the different scenarios in the assembly line performance.

2. Workers task times: Input data

Previous studies performed by the authors [6] indicate that the workers might have significant variations of both average time and variability when performing high motor content assembly tasks. Therefore, the worker performance can be measured in terms of deviations to the workstation expected values (the average time and average variability). A worker might have a task time which is slower or faster than the average and/or more or less dispersed than average. Given that, four types of performance can be considered in terms of deviations to the average values:

- The worker is slower completing the assembly task and the task times are more dispersed than expected;
- The worker is faster completing the assembly task and the task times are more dispersed than expected;
- The worker is faster completing the assembly task and the task times are less dispersed than expected;
- The worker is slower completing the assembly task and the task times are less dispersed than expected.

According the referred study [5], several workers had their times sampled on the industrial setting and their performances were classified according to their deviation to the overall performance - Expected (E). Table 1. presents the 4 established performance classes with the deviations, (x, y), to the overall average and dispersion task time, that characterize the average task time and dispersion within each performance class.

Table 1. Average Performance Deviations to (E)

Class of Performance	Performance deviations (x;y)
Quadrant I (QI)	(+15.9 %; +26.4%)
Quadrant II (QII)	(-4.5%; +13.3%)
Quadrant III (QIII)	(-11.3%; -21.4%)
Quadrant IV (QIV)	(3.8%; -9.1%)

During the data collection it was observed that the workstation with the given type of assembly tasks of high spelling motor content, had an expected average completion time of 15 sec and a standard deviation of 1.95 sec. Based on the (x;y) deviations, the type of workers performance can be assessed and compared to the expected average time and standard deviation. Using a triangular centred task time distribution, the minimum, average and maximum times for each performance is calculated and used as input to the simulation study (Table 2). Depending on the performance class, the time distribution can shift and/or can be wider or narrower than the expected one.

Table 2. Task times input data

Class of Performance	Min. [sec]	Average [sec]	Max. [sec]	Standard Deviation [sec]
E	10.22	15.00	19.78	1.95
QI	11.35	17.39	23.43	2.46
QII	8.91	14.33	19.74	2.21
QIII	9.55	13.30	17.06	1.53
QIV	11.23	15.58	19.92	1.77

3. Simulation model

To test the impact of having significantly different task times, it was considered a serial assembly line, with three workstations (as depicted in Fig.1). Each workstation has one dedicated worker. The part transfer between workstations is done asynchronously. This means that when the worker finishes the assembly tasks on his workstation, transfers it to the next workstation if it's idle (waiting for a part). If the next workstation is not idle (is either working or blocked), then the workstation becomes blocked, the worker has to wait and cannot accept any other part. The first workstation is never waiting and the last station is never blocked. There isn't the possibility to buffer parts between workstations. Also, in order to isolate the performance variations effect on the system, the work content in each workstation is the same, meaning that the line is perfectly balanced. Any unbalance will then be caused by the workers performance variations.

Fig. 1. Simulated assembly line

The simulation model was implemented on MATLAB, using Object Oriented Programming (OOP). The program tests all the possible combinations of workers task times. To test the different scenarios of performance allocation, the model was simulated until it produced 1800 parts (N).

For assembly line design purposes it is important to make sure that an assembly system is able to achieve a given output rate (average number of parts assembled by unit of time). However, the output variability also plays an important role on the daily planning and control of the system [7]. In this paper, the focus is given to the average inter-departure time (IDT) and the inter-departure time variability (SD) as measures of the system performance, as well as to the total required time to produce the N parts. The developed model records each time a part leaves the system, therefore the IDT is obtained by subtracting the successive times in which the parts leave the system. The SD is obtained by calculating the average of the variances obtained for each run (out of 10) of 1800 produced parts and applying the square root to this value. Since a warm-up period is not considered, the first IDT value will be larger than the following ones.

4. Results

In section 4.1 the results for all the workers with the same task time distribution are analyzed. In the following sections the focus is on the cases where task time distributions differ.

4.1. Every worker with the same task times

Fig. 2 represents the distribution of the system inter-departure times obtained with the simulation runs for all the workers with the same class of performance. In each 1 sec interval the average frequency with which a part left the system was calculated. It can be observed that the interaction of the several workers in this system configuration, introduces some positive skewness on the output time distribution due to blocking and waiting times, which increases the probability of the part spending more time in the system and therefore increasing the IDT.

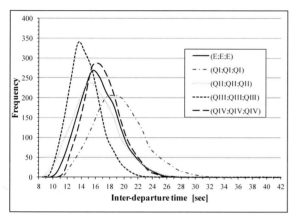

Fig. 2. Histograms – Workers with the same class of performance on every WS

The increase on the average task time results in an increase of the average system IDT (see Table 3). The same happens with the variability. If (all) the workers have a performance with higher average times and higher variability, the system IDT will also be higher and with higher variability (SD). The best possible scenario is having all workers with QIII performance and the worst is having them with QI performance. Nonetheless, if scenario (E;E;E) and (QII;QII;QII) are compared it can be concluded that the allocation of workers with QII performance (slightly faster but with task times more dispersed) increase the line performance since the system would produce the same amount of parts within less time. However, in reality workers seldom have all the same performance.

Table 3. Results – Workers with the same type of performance on every WS

Class of performances (WS1;WS2;WS3)	IDT [sec]	SD [sec]	Time Span [h:min]	Time Deviation [%]
(E;E;E)	16.55	2.88	8:17	-
(QI;QI;QI)	19.34	3.64	9:41	17%
(QII;QII;QII)	16.08	3.23	8:03	-3%
(QIII;QIII;QIII)	14.51	2.28	7:16	-12%
(QIV;QIV;QIV)	16.98	2.64	8:30	3%

4.2 The effects of allocating one worker with worst performance (QI), or best performance (QIII)

Table 4 presents the results obtained by allocating one QI or QIII worker type in the several line WS positions. Type E operators are allocated to the other two WS.

The simulation results show that, if there is one worker with the worst performance class - QI, while the others have an Expected (E) performance, the output of the assembly line depends on the workstation where he/she is allocated. The higher IDT is obtained when he/she is allocated to WS2 (E;QI;E), however (the SD is not the lowest). In this scenario, there are more chances of increasing the blocking situations on WS1 (that the worker cannot pass his/her output to WS2) and waiting in WS3 (in which the worker is waiting for inputs). On the other hand, when the results are analyzed in terms of total time to produce the 1800 units, then the time deviations, when compared with the (E;E;E) scenario, are very similar. The SD is lower for the (E;E;QI) scenario, meaning that the system output flow will be smoother when compared with the other scenarios, namely with scenario (QI;E;E).

In the case of having one worker with the best performance then it will have a more positive impact (IDT and SD reduction) when he/she is allocated in WS2, but the difference between putting him/her in WS1 or WS3 is larger than in the previous case.

Overall, if there is one worker with a worse performance among others that are representative of the overhaul population of workers (expected behavior), wherever he/she is allocated, the assembly system IDT will deviate at least 8% from the one expected when all the workers belong to E type performance (scenario E;E;E). If there's one worker among the others (E type ones) that has a better performance, the amount of time required to produce the total amount of parts will reduce a maximum of 3.7% if he/she is allocated to WS2, otherwise it's only possible to obtain a 2.3% total time reduction.

Table 4. Results – One worker QI/QIII among workers with the expected overall performance

Class of performances (WS1;WS2;WS3)	IDT [sec]	SD [sec]	Time Span [h:min]	Time Deviation [%]
(QI;E;E)	17.88	3.40	8:57	8.0%
(E;QI;E)	17.99	3.38	9:01	8.7%
(E;E;QI)	17.90	2.84	8:58	8.1%
(QIII;E;E)	16.18	2.61	8:06	-2.3%
(E;QIII;E)	15.94	2.60	7:59	-3.7%
(E;E;QIII)	16.17	2.97	8:06	-2.3%

4.3. Allocations for the lowest/highest inter-departure variability

The performance of the assembly depends not only on its output rate but also on the smoothness of its behaviour. So, to identify the workers allocation that result in smaller variability of the output several simulation runs were conducted. The results indicate that the SD is lower in the (QIII;QIII;QIV) scenario, and not with the (QIII;QIII;QIII) (see Table 5). This means that allocating a worker with a performance with more variability and slower when compared with QIII will result in lower variability ouput during the production of the 1800 parts. This happens because the worker in WS3 has a performance with a higher average time therefore the variability from the workers in WS1 and WS2 will be absorved. Note that if there was a QI worker in WS3, then the the output IDT would be higher. Also in this scenario it's required more time to produce the parts than with (QIII;QIII;QIII).

The combination with the highest output variability, is the (QI;QI;QII). In this case the worker in WS3 has a lower average time, causing the SD increasing. Again, there are gains in terms of the amount of time required to produce the 1800 parts when there's a QII performance type instead of a QI performance type on WS3(these effects are can be seen on Fig. 3).

Table 5. Results – Higher/Lower inter-departure variability

Type of performances (WS1;WS2;WS3)	IDT [sec]	SD [sec]	Time Span [h:min]	Time Deviation [%]
(QIII;QIII;QIII)	14.51	2.28	7:16	-12%
(QIII;QIII;QIV)	15.82	2.01	7:55	-4%
(QI;QI;QI)	19.34	3.64	9:41	17%
(QI;QI;QII)	18.86	4.18	9:27	14%

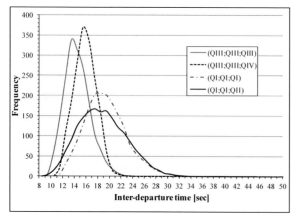

Fig. 3. Histograms – low/high SD output

5. Conclusions

Workers task times can vary significantly one from the other, both in terms of average and dispersion of task times. Considering five classes of workers performances, a three workstation assynchrounous unbuffered assembly system was simulated with different combinations of performances. Results indicate that the position in which the best/worse time performing worker is allocated has a small impact on the system output. Nevertheless, if there's one worker with a worse performance than the others, the system output can be significantly affected, specially if he/she's allocated to the middle workstation. Having one worker which outperforms the others doesn't have such a large (positive) impact. In terms of inter-departure variability, it will be reduced if the best two workers are allocated to the two initial workstations and in the third, there is a worker with a slightly larger average task times. This inter-departure variability reduction will be at the "cost" of increasing the inter-departure time, and consequently the amount of time to perform the total number of parts. Overall, the differences on workers task times can have significant impacts on the output performance of manually operated systems, and should be taken into account when managing such systems. Nonetheless, further studies should be made considering system parameters variations such as work in process buffers, work tasks transfer between workstations, among other factors. Namely, an important extension to this research will be the impact of different task times in longer assembly lines.

References

[1]   Lassila, A., Saad, S., Perera, T., Koch, T., Chrobot, J., & Camarinha-Matos, L. (2005). Modelling and Simulation of Human-Centred Assembly Systems - A Real Case Study. Emerging Solutions for Future Manufacturing Systems (Vol. 159, pp. 405-412): Springer Boston.

[2]   Butala, P., Kleine, J., Wingen, S., Gergs, H., & Leipzig, G. (2002). Assessment of assembly processes in European industry. Paper presented at the 35th CIRP-International Seminar on Manufacturing Systems, Seoul.

[3]   Doerr, Kenneth H., Arreola-Risa, Antonio. (2000). A worker-based approach for modeling variability in task completion times. IIE Transactions, 32(7), 625-636.

[4]   Smith, J. S. (2003). Survey on the use of simulation for manufacturing system design and operation. Journal of Manufacturing Systems, 22(2), 157-171.

[5]   Powell, S., Pyke, D. (1998). 1. Buffering unbalanced assembly systems. IIE Transactions, 30 (1).

[6]   Folgado, R., Henriques, E., & Peças, P. (2012). Mapping the Task Time Performance of Human-Centred Assembly Systems. Paper presented at 4th CARV, Montreal, Canada.

[7]   Kalir, A., Sarin, S. (2009). A method for reducing inter-departure time variability in serial production lines, International Journal of Production Economics, 120 (2), 340-347.

An integrated data model for quality information exchange in manufacturing systems

Y. (F.) Zhao[1], T. Kramer[2], W. Rippey[2], J. Horst[2] and F. Proctor[2]

[1] Institut de Recherche en Communication Cybernétique de Nantes (IRCCyN), Ecole Centrale de Nantes, Nantes, France

[2] National Institute of Standards and Technology (NIST), Gaithersburg, Maryland, United States of America

Abstract. Quality measurement is an integrated part of modern manufacturing systems. As the manufacturing industry has entered a digital and virtual era, information technology has become increasingly important for both machining and measurement systems. Effective information sharing and exchange among computer systems throughout a product's life cycle has been a critical issue. The quality measurement industry and standards organizations have developed several successful standard data models to enable standardized data. However, these standards have very restricted focus. From late 2010, the Quality Information Framework (QIF) project was initiated by the Dimensional Metrology Standards Consortium (DMSC) with the support of major North American quality measurement industries. This project aims to develop an integrated Extensible Modeling Language (XML) data model for the entire quality measurement chain. It consists of five application area schemas and a QIF library with nine supporting schemas. This paper introduces the scope and data model design of QIF. A pilot test project using the recently completed QMResults schema for the exchange of inspection result data is also presented.

Keywords: Quality Information Framework (QIF), manufacturing quality systems, information modeling, data model, XML schema, measurement features, characteristics, ASME Y14.5

1. Introduction

In the past two decades, the manufacturing industry has undergone drastic changes: digitization of manufacturing process chain, globalizing supply chain, fast technology development, and escalating complexity of products. For major manufactureres, such as airplane manufacturers, automobile manufacturers, etc., the technology bottleneck is shifting towards manufacturing systems integration and enterprise systems integration. The reason for this situation is that hardware and software systems generally process input and output data in their own data formats. A typical manufacturing chain often consists of (from upstream to downstream) Computer-Aided Design (CAD) software, Computer-Aided Manufacturing (CAM) software, Computer-Aided Inspection Planning (CAIP) software, and Statistical Process Control (SPC) software.

The information can not flow directly from upstream to downstream without data translation. Translation is not only time and money consuming but also results in loss of integrity of information. Furthermore, the information exchange between CAD systems from different vendors also needs data translation. There are many research efforts in standardizing data exchange within manufacturing systems. This paper will present a new effort aiming to develop a consolidated data model library for the exchange of quality data within manufacturing systems. This effort is called the Quality Information Framework. A primary benefit of this research is to reduce resources needed for systems integration in all manufacurting industries that implement dimensional metrology systems. In 2006, the automotive industry alone reported that costs due to translation of measurement data between manufacturing systems amounted to over $600 million annually [1]. Moreover, adopting QIF will facilitate commercially available components to be interoperable allowing users to buy the products that suit their individual business models.

2. Flow of information in manufacturing quality systems

Manufacturing quality systems comprise the software and hardware used for quality control in a manufacturing process. Quality control is intertwined with machining activities in a manufacturing process. As shown in Fig. 1, there are four main elements in a typical manufacturing quality system: measurement planning, inspection programming, measurement execution, and quality results analysis and reporting [2]. The solid lines in Fig. 1 represent the common information flow in typical manufacturing quality system; while the dotted lines represent alternative information exchange. Product definition information is the highest level upstream information flowing into any manufacturing quality system. It is the combined information of product design information (generated through CAD software) and multiple quality management requirements such as Product Lifecycle Management (PLM), First Article Inspection (FAI) requirements, Enterprise Resource Planning (ERP) information, etc.

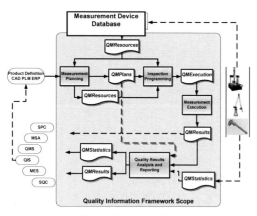

Fig. 1. The scope of QIF

The measurement planning activity receives all the information mentioned above and takes into consideration of the availability of quality measurement resources to generate measurement plans. Descriptions of all inspection resources in a facility, as well as resources assigned to specific plans are formated according to the *QMResources* data model. High-level plans, formatted according to the *QMPlans* data model, identify the measurement features, their characteristics, and the measuring sequence. The function of the inspection programming activity is to generate the machine-level measurement plan for execution based on available measurement devices. Low-level plans, formatted according to the *QMExecution* data model, provide detailed measurement operation information (i.e. probing points, scanning routes, etc.). The function of the inspection programming activity is to generate the machine-level measurement plan for execution based on available measurement devices. Most coordinate measuring machines (CMMs) require the execution commands to follow specific proprietary format for their controllers. This process is carried out by the measurement execution activity, whose responsibility is to interpret machine-level measurement plans, give equipment level commands to specific CMM control units, collect point data, fit features to data, and output feature and characteristic data formatted according to the *QMResults* data model.

Once the machine-level measurement plans are executed, the measurement data, either raw data or pre-processed data, is collected, reported, and analysed. The collection and the analysis of multiple part inspections are formatted according to the *QMStatistics* data model. The purpose of quality control is not just to judge whether a

product meets the functional requirements of the design after it is manufactured. The data gathered through quality control (both in-process and post-process) can also be used to assess the performance of the machine, to judge the quality and the efficiency of the machining process plan, and to trace any unexpected anomalies during the manufacturing process. Through analyzing measurements data, statistic process control can provide feedback to upstream processes, such as CAD design, statistical quality control (SQC), etc., to improve the production of future products.

3. QIF scope and schemas

There are multiple international standards effort trying to develop suitable data model to overcome the interoperability barrier in manufacturing quality systems. Reviews of these international standards efforts can be found in references [2-6]. Each effort focuses on a narrow slice of the overall manufacturing quality system. Most of them developed independent data models, which often overlap with other models, contributing to interoperability problems. The QIF effort was initiated by the DMSC [7] aiming to overview the existing standard data models in manufacturing quality systems to develop a consolidated data model to cover the entire manufacturing quality chain. The contents of the QIF data model was designed to encode all information in ANSI/ASME Y14.5 2009 [8] and the Dimensional Measuring Interface Standard (DMIS) 5.2 [9]. The scope of QIF covers the five interfaces shown in Fig. 1.

3.1. QIF schemas

The current release of QIF Version 0.9, has the QMResults application area schema and nine supporting schemas complete and released for beta testing. The QMPlans schema is in draft shape and the remaining three schemas have been planned to be developed in the near future. The supporting data model schemas, collectively called the QIF Library, are named as following:

- **CharacteristicTypes** - defines quality requirement information, including geometric dimensioning and tolerancing (GD&T) tolerance, attribute tolerance (i.e., Go/No-go gaging requirements), and user defined requirements.
- **FeatureTypes**- defines all 28 types of measurement features found in DMIS 5.2.
- **ConstructedFeatureTypes**-defines methods of constructing measurement features.
- **Units** - defines units for values of angle, area, force, length, mass, pressure, signed length, speed, temperature, and time.
- **MeasurementDevices**- defines the basic information regarding measurement devices such as

probe accuracy, stylus length, touch-triggering force, etc.

- **Traceability** - defines traceability information within the manufacturing quality system and with the machining process, such as part serial number, machine tool number, process id, etc.
- **Transforms** - defines coordinate systems transformation information.
- **PrimitiveTypes** - defines primitive common information that is used by other application area schemas but is not covered in the above supporting schemas, such as degree of freedom, notes, measure point, etc.
- **QIFTypes** - defines non primitive common information used by application area schemas but not covered by the above supporting schemas, such as file type, model entity, software type, etc.

One of the five application area schemas, QMResults, has been built and is believed to be technically complete for the reporting purposes it was built to serve. It is expected that testing and debugging will add only minor items related to those purposes. Testing the QMResults schema began in October 2011, when sample test files were first built. A second application area schema, QMPlans, has been drafted but, as of February 2012, is not technically complete.

3.1.1. Semantic Connecting Types Using Identifiers

In QIF each characteristic and feature is defined using four *aspects*: definition, nominal, actual, and instance. These have identifiers and are connected by references to the identifiers. Notes may be attached to any of the four aspects. Take features as an example:

- A **feature definition** includes information that is independent of the position of the feature - the diameter of a circle, for example. A single definition can be referenced by many nominal features. Only nominal features reference feature definitions.
- A **feature nominal** defines a nominal feature by referencing a feature definition and providing position information - the center of a circle and the normal to the plane of the circle, for example.
- A **feature actual** defines an actual feature that has been measured or constructed. A feature actual may optionally refer to a feature nominal and is expected to do so if there is a nominal feature. There may not be a nominal feature if the actual feature is built during a reverse engineering process.
- A **feature instance** represents an instance of a feature at any stage of the metrology process - before or after a feature has been measured. The feature instance must reference a nominal feature or an actual feature. If an actual feature is referenced, the corresponding nominal feature (if there is one) may be found through the actual feature. If a feature is measured several times, it is expected that a feature instance will be defined for each measurement and will have a different actual feature for each measurement.

In XML data files, relationships between the four aspects are expressed using strongly typed links, whose types are related to the data elements they are linking. This use of typed links, or references, will allow standard XML data file validation rules to enforce integrity of the relationships. A more technically detailed discussion of this technique can be found in the design and usage guide of QIF, which will be published soon by NIST.

3.1.2. QIF Validation to ANSI/ASME Y14.5 2009

The CharacteristicTypes schema represents the entire GD&T requirements defined in ANSI/ASME Y14.5 1994 standard. In order to make QIF compatible with the newly published 2009 version, a validation effort was carried out to update CharacteristicTypes with all the new GD&T requirements. Table 1 lists some of the new items defined in the 2009 version. Some of the new items are accompanied with new symbols. This validation process is still ongoing.

4. Case Studies for validating QIF

In September 2011, the QMResults schema Version 0.9 was released by the QIF QMResults working group. Since then, a series of validation tests have been carried out to evaluate correctness and completeness of the specification. The first step of the validation was to create QMResults data files containing simulated measurement results for sample parts. The validation requirement is that the data files contain all measurements described by part drawings, and that the XML data files be correct according to the QMResults schema. Part drawings shown in Fig. 2 are:

- A CMM calibration master ball
- Advanced Numerical Control (ANC) 101 example part
- Sheet metal scanning measurement example part.

5. Conclusions

Based on the development of QIF to date, the DMSC believes that a complete set of specifications will facilitate simple integration of commercial software solutions for manufacturing quality systems. Exchange of quality data in standard formats is judged to be a good solution to achieving interoperability of multi-vendor software components. Benefits accrued to manufacturers should include flexibility in configuring quality systems and in choosing commercial components, and effortless and accurate flow of data within factory walls as well as with contractors and suppliers. In order to facilitate and encourage software applications that manipulate QIF formated data, DMSC proposes, in the near future, to create software libraries that write data into QMResults XML files (serialization), and/or read data from QMResults XML files (deserialization). This software development will be an open source, public domain project so members of the manufacturing quality community can suggest improvements or improve the code for the benefit of all users.

Table 1. New GD&T information in ASME Y14.5 2009

New Items	Symbols/ Indicator
All over	
Movable datum targets	⌖ A1
Unequally disposed profile	Ⓤ
Independency	Ⓘ
Continuous feature	ⒸⒻ
Datum translation	▷
Maximum Material Boundary (MMB)	Ⓜ
Least Material Boundary (LMB)	Ⓛ
Regardless of Material Boundary (RMB)	N/A
Nonmandatory (MFG DATA)	N/A
Datum feature simulator	[ ] ,BASIC or BSC
Explicit degrees of freedom for datum reference frames	[u,v,w,x,y,z]
More than two tier composite position control frames	N/A
More than two tier composite profile feature control frames	N/A

References

[1]   IMTI, A Roadmap for Metrology Interoperability 2006, Integrated Manufacturing Technology Initiative (IMTI, Inc.).

[2]   Zhao, Y.F., et al., Information Modeling for Interoperable Dimensional Metrology. 2011, London: Springer.

[3]   Zhao, Y.F., et al., Dimensional Metrology Interoperability and Standardization in Manufacturing Systems. Computer Standards & Interfaces, 2011. 33: p. 541-555.

[4]   Kramer, T.R., et al., A feature-based inspection and machining system. Journal of Computer Aided Design, 2001. 33(9): p. 653-669.

[5]   Proctor, F., et al. Interoperability testing for shop floor measurement. in Performance Metrics for Intelligent Systems (PerMIS) Workshop. 2007.

[6]   Zhao, Y.F., F.M. Proctor, and W.G. Rippey, A Machining and Measurement Process Planning Activity Model for Manufacturing Systems Interoperability Analysis. 2010, National Institute of Standards and Technology; Available from: http://www.dmis.org/.

[7]   DMSC. Dimensional Metrology Standards Consortium. 2011; Available from: http://www.dmis.org/.

[8]   ANSI, ASME Y14.5-2009: Dimensioning and Tolerancing. 2009.

[9]   ANSI, Dimensional Measuring Interface Standard, DMIS 5.0 Standard, Part 1, ANSI/DMIS 105.0-2004, Part 1. 2004.

a) a CMM calibration master ball

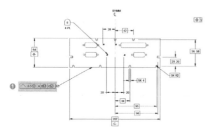

b) ANC101 example part

c) Sheet metal scanning measurement example

Fig. 2. QIF validation case studies

Bayesian network modeling of machine breakdowns

E. M. Abogrean and Muhammad Latif
School of Engineering, Manchester Metropolitan University, Manchester UK

Abstract. This paper considers a common problem that all industries contest with in practice i.e. the breaking down of machines that influence production and cost directly. In a majority of industrial applications, acquiring optimum utilisation of all available resources for existing and future predicted demand is a major function throughout all the levels of management. This paper uses a number of analytical tools and software that support one another i.e. Witness Simulation, Bayesian Network Modelling and Hugin Software. The use of expert experience and knowledge has been incorporated throughout the study as it is vital to model building and greater understanding of machine breakdown. This study uses discrete event simulation and Bayesian network modelling collectively to understand machine breakdowns to increase efficiency within a cement manufacturing plant i.e. the Crusher Machine. The Bayesian network modelling implemented by the Hugin Software is used to generate probabilities which are transferred into a discrete event simulation model using Witness Software based on the historical data, expert knowledge and opinions. The model simulates the three parameters of the machine life based on consumption of each parameter. This is translated into a probability failure rate that changes as the model is running. The model demonstrates decisions based on the probability of failure from the Bayesian model and based on life consumption of the different variables in the simulation model.

Keywords: Witness Simulation; Hugin Software; Bayesian Network Modelling, Cement.

1. Introduction

This paper considers analytical tools such as the Bayesian Network Modelling that is aided by the Hugin Software and Witness Simulation to understand machine breakdowns. This paper aims to model and reduce the effects of breakdowns that occur within a single crusher machine using the analytical tools. The development of the model will result in different probabilities of failures and to explore different approaches in calculating the most likelihood of a failure occurrence.

2. Methodology

Bayesian network modelling (Fig. 1) is a mathematical tool used to model uncertainty in a chosen area or a system, it can help identify and highlight links between variables [1]. The recognition of important variables as well as consideration of other influencing factors that seem to exist within the system is integral to the Bayesian approach. The Hugin software (Fig. 2-5) interprets historical data and expert knowledge into a probabilistic figure that shows when the likelihood of machine failure may occur, according to the parameters that exist based on the nodes and their dependencies. Hugin applies snapshot results i.e. the results indicate specific circumstances based on parameters, they are not continuously changing data but have to be changed manually to derive different results.

The role of Witness Simulation is to evaluate alternatives that either support strategic initiatives, or support better performance at operational and tactical levels [2]. Simulation provides information needed to make these types of decisions. The simulation approach supports multiple analyses by allowing rapid changes to the models logic and data, and is capable of handling large, complex systems such as a manufacturing facility. [3]

The Bayesian and Hugin approach is outlined in six steps that need to be followed in order to achieve the best results with the collated data. The six steps are as follows:

1. Establishing relevant and accurate information
2. Establishing nodes with dependencies
3. Establishing of CPT (Conditional Probability Table)
4. Normalise Probability
5. Propagate Evidence
6. Model Validation

The crusher machine is dependent on 3 parameters that exist and hence have influencing affects on the generated probability. In order to calculate the probability of the 'failure' of the crusher machine the chain rule must be applied. Given the above nodes and dependencies, the outcome or probability generated by the Hugin software is as follows shown in Fig. 6.

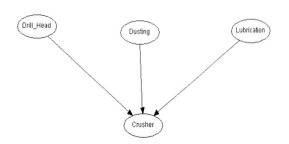

Fig. 1 Bayesian Network Modelling of a Crusher Machine and Parameters

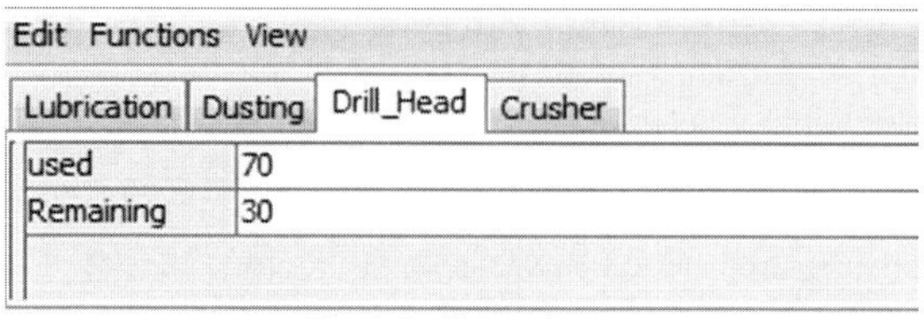

Edit Functions View

Lubrication	Dusting	Drill_Head	Crusher
used	70		
Remaining	30		

Fig. 2. Drill Head state

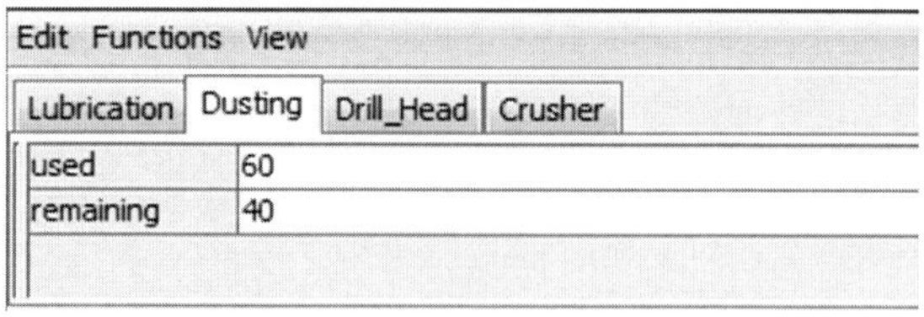

Edit Functions View

Lubrication	Dusting	Drill_Head	Crusher
used	60		
remaining	40		

Fig. 3.. Dusting state

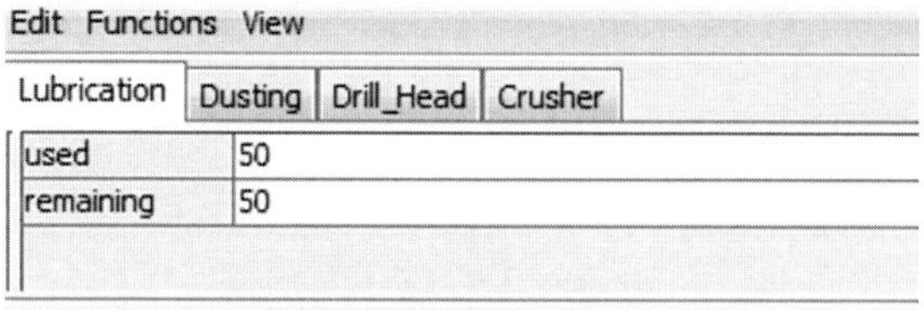

Edit Functions View

Lubrication	Dusting	Drill_Head	Crusher
used	50		
remaining	50		

Fig. 4.. Lubrication state

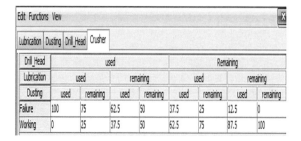

Fig. 5. Crusher Machine Parameters CPT

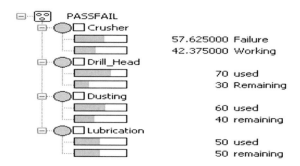

Crusher	57.625000	Failure
	42.375000	Working
Drill_Head		
	70	used
	30	Remaining
Dusting		
	60	used
	40	remaining
Lubrication		
	50	used
	50	remaining

Fig. 6. Hugin results for Drill Head, Dusting, Lubrication and Crusher.

3. Witness Simulation

The Witness simulation software uses basic elements, together with distributions and timing inputs to replicate an accurate model. The distributions are then used to influence certain factors to acquire a certain result. The basic elements to simulate the models are as follows and can be seen in Fig. 7.

Entities represent parts that flow through the model, in this case, it will be the existing parameters. *Activities* represents a station where a task is completed. *Queues* are a point where an entity is held until the entity is required or needed, or even a point where a desired waiting time can be applied. *Resources* are the labour required to perform a desired activity.

For the ease of implementing, the three stated parameters into witness, all three parameters are now based on an estimated **LIFE**, "*USED*" and "REMAINING", this represents each parameters usage that has been derived from historical data and with the use of expert knowledge. This usage rate has been attributed with "life span", this is the estimated life a parameter has and can be utilised until a change or repair is needed.

Once the overall life span has been reached, then only certain tasks need to be carried out i.e. the drill head according to historical data needs to be changed every 7 days hence the life span is 7 days/10080 minutes, the dust builds up within the machine and can cause disruption or unexpected errors if not cleared every 2 days/2880 minutes hence this is the life span of the dusting. The lubrication of the required mechanisms which exists starts to dry and cause friction if not lubricated every 3 days/4320 minutes.

The model developed aids logical understanding by the creation of the two variables per parameter, "Used" and "Remaining", as time/usage is being consumed, elapsed time will increase and the remaining time will decrease simply because the lifespan of the parameters are coming to an end. The model is shown in Fig.7 as well as the variables developed to generate the probability of failure.

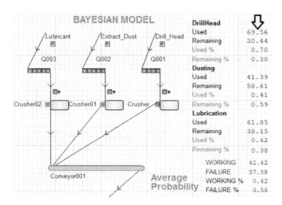

Fig. 7. Bayesian Model

The parameters do however require attention as they reach maximum life span and need adhering to; however, if two or three of the parameters almost reach the end of the life span simultaneously, this added together as a whole can easily cause breakdowns as more than one parameter has reached near the maximum usage. This is based on historical and expert opinion of the machinery.

4. Conclusion

After implementing and making use of the above analytical tools, the model developed by Witness Simulation shown helps consider many factors when simulating and approaching breakdowns. These tools aid understanding and enable better judgement to be made by the user, almost like a decision making tool as all the parameters and variables that do exist are based on legitimate reasoning that has been aided via many consultations with experts within the field. The applications of the above analytical tools result in an increase in logic being applied to the model and therefore increasing the validity of the model. These tools also help to generate and investigate different scenarios by using the different tools against each other. The main purpose of the paper is to achieve a reduction in breakdowns via the use of the above tools as described, the mean time between failure (MTBF) according to historical data is between 9000 and 9500 minutes. This means the crusher machine failure occurs approximately every 6.25 days to 6.59 days, a model was developed to understand the affects thereof. This was not a logical approach according to experts in the field, as the processing of cement and the site has many machines and cannot function effectively if machines breakdown once a week on a constant basis.

The probability of failure developed by the Bayesian approach in practice, where the probability has surpassed the 95% threshold indicated by the implementation of the rule, and this 95% threshold was decided with the consultation of experts and research. This 95% enabled a slight flexibility as different parameters had their own consumption rates as well as enabled deterioration of parameters to be considered. The results of the simulation

model can be seen in Table 1 and 2. Table 1 shows the results of the model after 30 days and Table 2 shows the same result but upto the time a breakdown occurs according to the Bayesian approach to show the difference.After running the model for 61444 minutes a breakdown occurs compared to 7 breakdowns of the MTBF model.

The Bayesian model takes into consideration all the existing parameters, here the model does not breakdown according to the mean time between failures (MTBF), and this has now been totally eliminated. The Bayesian model, models the information that has been derived from the research and experts. [5]. Now the breakdown occurs when all three parameters combined usage rate is above the 90% threshold mark, this is according to expert knowledge where the deteriorations of individual parameters and the reduction in quality starts to appear. [6, 7]. This does not result in the machine breaking down however when all three parameters simultaneously reach the threshold it should develop into a breakdown.

The average probability of failure in the model is represented by ***WORKING*** shown in Fig. 7, this is to enable the machine to breakdown when the average is above the 90% threshold and below 95%. However as discussed above, this was not so logical simply because two parameters increase in usage rate can easily increase the average, this may mean that at no particular time all three parameters surpass the 90% threshold but rather they help each other to reach the average threshold. The machine breaks down according to the average rule as implemented however, when the parameters are considered it can be seen, the first parameter that has the longest life span i.e. Drill Head has only been used 83.80%, the other two parameters have helped increase the average, Dusting has been used 91.22% and Lubrication at 95.07%.

This also had to be verified and validated with the Bayesian model to see if the average worked in correlation to the failure developed by the Hugin software. shows the results, when exactly implemented into the software; the average failure is 90.01% in the simulation model whereas the failure rate developed by the Hugin software is 86.15%. This showed a significant difference in the results that were developed.

Table 1. 30 Day MTBF Results

30 Days/ 43200 minutes	MTBF Model	Bayesian Model
Drill Heads replaced	4	4
Dusting carried out	14	14
Machine Lubricated	9	9
Total BREAKDOWNS	5	2
Total inspection time	1111.7	1553
Total Repair Time	1518	589
Total Time Lost	2629.7	2142

From the results in Table 1 it can be seen, although the number of tasks carried out remain the same, total breakdowns differ i.e. MTBF has five breakdowns whilst the Baysain has two, further the inspection times differ indicating the bayesain model takes extra time in general inspection and welfare of the machine.

Table 2 shows the correlation of the influencing factors to that of the breakdown. Here, the five breakdowns of the MTBF model have been used and the parameter usage percentages can be seen in order to examine whether the breakdown occurs for a due course according to influencing parameters or simply due to the historical data from which the MTBF has been derived. Table 2 shows clearly, when all five breakdowns occur, the influencing parameters usage percentages are far from being consumed fully. Only breakdown four shows that dusting has been used 100%, indicating dusting is required as the other parameters still have majority of the lifes reamining. The bayesian model in Table 2 indicates a very high usage rate for all three parameters indicating the occurance of a breakdown, all three paramters ar above the 90% threshold as discussed earlier on in the conclusion.

Table 2. Usage Rate of MTBF & Bayesian Parameters

MTBF	Time	Drill Head Used %	Dusting Used %	Lubrication Used %
Breakdown 1	4544	45	57	5
Breakdown 2	13578	31	67	13
Breakdown 3	22612	16	78	21
Breakdown 4	32005	5	100	37
Breakdown 5	41212	88	17	49

Bayesian	Time	Drill Head Used %	Dusting Used %	Lubrication Used %
Breakdown 1	8623	90.55	97.33	99.14
Breakdown 2	32456	97.04	90.03	99.65

The bayesian model shows an absolute change in results from that of the MTBF model based on the consumption of parameters. This results in the machine breaking down from once a week according to the MTBF model to a possible breakdown of once every 2 weeks based on Bayesian model. This indicates a definite reduction in machine breakdowns based on the condition of key influencing factors that is not considered by the MTBF appraoch.

References

[1] Nadkarni S. and Shenoy p. (2004). A Bayesian network approach to make inferences in causal maps. European Journal of operational Research .Vol.128.pp479-498.

[2] Witness Training Manual, Lanner Group Ltd, http://www.lanner.com/21/FEB/2008

[3] McGrayne, Sharon Bertsch. (2011). The Theory That Would Not Die in the history up applied Mathematics and Statistics. Hunted down Russian Submarines and Emerged Triumphant from two Centuries of Controversy. The Book Depository (Guernsey, GY, United Kingdom)

[4] Fienberg, Stephen E. (2006).When Did Bayesian Inference Become William Sealy Gosset. 1876-1937 in Heyde, and Seneta, E. (Eds), Statisticians of the Centuries.Newyork: Springer- Verlag.

[5] Jones BJ, Jenkinson I, Yang Z, Wang J. The use of Bayesian network modelling for maintenance planning in a manufacturing industry. Reliability Engineering & System Safety 2010; 95: 267-277.

[6] Theresa M. Korn; Korn, Granino Arthur, 2003. Mathematical Handbook for Scientists and Engineers: Definitions, Theorems, and Formulas for Reference and Review. New York: Dover Publications

[7] Uri D. Nodelman, June 2007, Continuous Time Bayesian Networks, PhD Thesis, Stanford University Press, Stanford University, USA.

[8] Kim GJ, 2002, Three Essays on Bayesian Choice Models, PhD Thesis, Joseph L. Rotman School of Management, University of Toronto, USA

[9] Jones BJ. (2009). A risk-based maintenance methodology of industrial systems. PhD thesis, school of engineering, Liverpool John Moors University, UK.

Exposing human behaviour to patient flow modeling

Entisar K. Aboukanda and Muhammad Latif
School of Engineering, Manchester Metropolitan University, Manchester, UK

Abstract. Patient flow models have been universally used for planning health services for both acute and chronic patients. These models invariably assume patients are homogenous and events follow traditional queuing models. These techniques are useful for examining patient flow in large population groups where Markov assumptions, or simple extensions of these, can be made. However it is realised that such assumptions are not necessarily representative in cultures and communities that do not adhere to queuing policies particularly in developing countries. This paper explores the need to consider human behaviour within a patient flow model and reports on a study that identifies some of the critical factors that a patient flow system must take into account when planning and implementing health services in developing countries. The case study explores behavioural attitudes from hospital staff and patients and its impact on the patient flow system using telephone interviews and questionnaires with staff who work in the Emergency Department (ED) at Tripoli Medical Centre (TMC). The most important result of the study is staff believe that the presence of difficult behaviour of patients is a main reason for the delay and weakening of service quality levels. Other interesting observations have been found that will be useful to health service planners in developed and developing countries.

Keywords: Patient flow; human behaviour; emergency department ED overcrowding.

1. Introduction

An Emergency Department (ED) is a medical treatment facility, specialising in the acute care of patients who attend without a prior appointment having been made [1]. The department deliver a range of treatments covering a vast arena of different injuries and illnesses; importantly, some of these may be considered life-threatening, and may therefore necessitate immediate, urgent action. Nevertheless, EDs are experiencing many different obstacles and issues owing to the fact that there has been a significant surge in patient demand; this has subsequently induced the need to ensure services and their overall quality is improved. Within hospital emergency departments, one of the most significant, urgent operational difficulties is patients overcrowding, which is recognised as threatening public health and patient safety [2]. Markedly, overcrowding is commonly considered as a circumstance wherein there are a greater number of patients than treatment beds and staff, and also where waiting times necessitate patients to endure long delays. The increase in the ED overcrowding problem has motivated researchers to delve into the issues surrounding the causes and effects, as well as how to establish a solution to this problem. Previous studies have listed the most common causes leading to overcrowding: an overall increase in patient volume, increased complexity and acuity of patients to the ED, a lack of beds for patients admitted to the hospital, avoiding inpatient hospital admissions by intensive assessment and treatment in the ED, delays in the service provided by radiology, laboratory, and ancillary services, shortage of nursing staff and/or physician staff, and a shortage of physical space within the ED [3,4,5]. Furthermore, there has been much published in the academic literature surrounding the consequences of ED overcrowding, such as increased risk of clinical deterioration, prolonged patient wait times, subsequently leading to pain and suffering, increased patient complaints, decreased staff satisfaction and decreased physician productivity, increased the pressure in terms of managing the hospital effectively, and poor service quality [6,7,8]. Various different studies that have been referred to earlier were focused on the patient flow system in order to study overcrowding within EDs. A patient flow system is a valuable tool for examining and evaluating hospital performance. In fact, the patients in ED usually require the utilisation of various different resources, namely beds, examining rooms, medical procedures, nurses, and physicians. This therefore suggests that the overall patient flow system may be described as a network [9]. In part, patient delays depend on how they physically flows through the network, and also on the ways in which information, equipment and other objects flow through it [10]. In actual fact, previous studies found that the problem in hospitals is that such movements throughout the patient flow network have been stopped or otherwise progress slowly owing to many different reasons. However, previous research into patient flow has assumed the ideal behaviours of patients. Owing to the technical challenges involved in modelling realistic human behaviours, existing work has considerable limitations in regard to its domain of applicability and, to some extent, on the validity of the results across a wide application spectrum. It is know that, under stressful conditions, human behaviour deviates substantially from the ideal; particularly when rules and regulations are not enforced, as found in less developed countries, such as in Libya [11,12,13]. The proposed study hypothesises that human behaviour is a key driver of overcrowding within Libyan hospitals; therefore, this paper has been focused on exploring human behaviour at a level to realistically predict a patient flow system within a critical hospital environment

2. Field study

To explore the effects of difficult patients' behaviour on the service time in the ED, Tripoli Medical Centre (TMC) was selected as the hospital for this study. TMC is one of the largest hospitals in Libya, located at the Eastern entrance of the city of Tripoli, with 1,438 beds in total. The emergency department of TMC is considered one of the most important emergency departments in the country, because it was founded to serve the residents of Tripoli, at an estimated one and a half million.

2.1. Methodology

This study has conducted telephone interviews and questionnaires with a number of doctors and nurses working within the ED at TMC; this was decided upon so that the patients' behaviour factors listed through a review of previous literature could be discussed. It is recognised that, by holding such interviews and discussion, the behaviour factors most important for delaying services within the department can be established. Telephone interviews have been conducted with group of eight nurses and four doctors. Through the telephone interviews, a series of questions that were prepared in advance were discussed. In fact, those interviews were a preliminary stage for building a questionnaire with 10 questions all about patients' behaviour within the ED. The questionnaire was distributed to all doctors (16 doctors) and nurses (20 nurses) who were working in the ED to find their opinions on the impact of the patients' behaviour on service times. The study explored issues related to patients' behaviour, such as the behaviour considered to cause weakness in the patient flow system; the effects of that behaviour on waiting time

3. Results

From the questionnaire sent to the ED staff in TMC: there were 11 doctor respondents, who represent 68.8% of the total questionnaires sent, and 14 nurses respondents, who

represent 70%. The staff were asked about how long they had worked in the ED and most staff who responded (doctors and nurses) had been in their jobs over five years. Figure 1 shows the distribution of staff by years of experience.

When asked about the existence of the overcrowding in the ED, 70.5% of the doctor respondents agreed, joined by approximately eighty percent (79.7%) of the nurses. The staff were also asked to indicate whether they felt that difficult patients' behaviour issues had a negative impact on the work of their ED, and almost eighty percent (78.3%) of doctors and 80 % of nurses felt that it did. In addition, they were asked about the most important behaviour that negatively affected their ED. Table 1 shows the results of their answers.

The staff were asked to indicate the negative effects caused by the difficult patients' behaviour on the work in the TMC emergency department, and it was found that doctors (approximately 75%) and nurses (approximately 80%) agreed that the most important negative impact on the work in the ED was difficult patient behaviour, which increased patient waiting time for service. Sixty-four percent of doctors and 70% of nurses believed that unacceptable patient behaviour disturbs the patient flow system. In addition, 51.4% of doctors and 55% of nurses responded that difficult behaviour contributes of staff dissatisfaction, which negatively affects the quality of service provided (see Fig. 2).

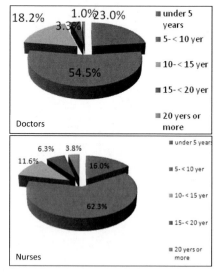

Fig. 1. Length of time staff had worked in ED

Respondents were asked for more service areas experiencing repeated unacceptable behaviour. The results show that there were six service areas experiencing delays in service because of the difficult behaviour of some patients. Table 2 shows those areas, and also shows the average real-time service, and the estimated average time that respondents believed that the difficult patient takes in addition to real service time.

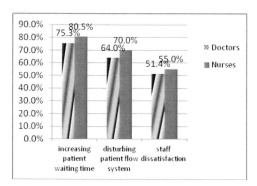

Fig. 2. The three negative effects on work at ED in TMC

Table 1. Staff opinion 0f behaviour factors affecting overcrowding

Behaviour Factors	Doctors %			Nurses %		
	M.E	Mi.E	N.E	M.E	Mi.E	N.E
Illness Believes	70.6	23	6.4	63	22.9	14.1
Over-involvement	65.3	30.1	4.6	65.8	32.1	2.1
Demanding	65.4	24.8	9.8	85.3	12.2	2.5
Arguing	62.5	27	10.5	51.8	35.8	12.4
Aggression	62	30.8	7.2	50.1	15.8	9.1
Lack of respect	58.1	27.6	14.3	75.1	34.3	15.6
Intoxication (alcoholism)	50.6	33.8	15.3	26.9	37.2	35.9
Cultural influence	30.6	40.8	28.6	59.8	30	10.2
Comm. difficulties	26.7	37.2	36.1	70.5	22.3	7.2
Interfering	22.8	46.7	30.5	35	40.6	24.4
Breaks taken	16.7	50.8	32.5	18	55.3	26.7
Where, M.E = Major effect, Mi.E = Minor effect, N.E = No Effect						

4. Discussion

Previous studies carried out on attempting to improve the flow of available ED resources—including physical resources (beds, equipment, etc.) and human resources (nursing staff, diagnostic staff, physician staff, etc.)—have sought to establish good solutions to the problem. In actual fact, there have been several proposed solutions for the overcrowding and length of the waiting time problem: fast-tracking is one of these proposed solutions. Fast-tracking helps to reduce patient waiting times through the mechanism of sending non-urgent patients to a specific station as opposed to waiting for emergency services with other urgent patients. Other solutions have been determined through studying queue theory, queue stricture, hospitals facilities available, and so on [14, 15].

Despite the importance of previous studies, there has been the neglect to consider difficult patient's behaviour and its impacts on the patient flow system. Therefore, the study hypothesis has been outlined, which assumes that patient behaviours whilst waiting in a service queue is

Table 2. Represents areas more exposed to the presence of unacceptable behaviour, the average real-time service, and the average estimated for the extra time caused by the behaviour of the patient

Service area	Repeated behaviour	T-1 (mins)	T-2 (mins)
Triage	• Interfering • Arguing • Communication difficulties • Illness Believes • Aggression	6	7.5
Reception	• Demanding • Aggression • Lack of respect • Cultural influence • Communication difficulties • Alcoholism	3.5	4
Exam.	• Communication difficulties • Arguing • Illness Believes	15.5	7
Diagnosis	• Interfering • Arguing	6	6
Discharge	• Cultural influence • Demanding • Demanding • Lack of respect • Over-involvement • Arguing	35	10.5
Queues	• Cultural influence • Demanding • Demanding • Lack of respect • Over-involvement • Breaks taken	60	30
Where, T-1= Mean time to provide the service T-2= Mean estimate extra time			

one of the reasons impeding the smooth operation of the patient flow system. To prove the hypothesis, the study relied on a survey of staff working in the ED of (TMC) to find their opinion on the issue of patients' behaviour and its impact on service.

The results show that over half of all respondents had spent more than six years working in the ED, which indicates that they have gained good experience to be able to give reliable views. Over 70% of doctors and 79% of nurses agreed that overcrowding is a significant problem in the ED at TMC. This result obviously shows the importance of conducting such a study in order to discover the reasons for this problem and work on finding a suitable solution. Doctors rated serious behaviours, shown in Table 1, which they believe to be undesirable in their emergency department, with effects on service delivery. Doctors considered illness beliefs the most important behavioural factor to cause confusion at work. This behaviour, as interpreted by the doctors surveyed, means that patients go to the ED from a belief that he/she is an emergency condition when he/she is not. According to the rules of the ED, staff cannot refuse to treat the patient who appears in the department, regardless of his/her condition; instead, staff must provide emergency services for each patient. Therefore, the illness belief behaviour leads to an increased number of patients, which causes.

5. Conclusion

This study shows that the staff who work in the ED at TMC believe that the presence of difficult behaviour of patients is a very important reason for the delay and weakening of the quality of department services delivery. It identifies the most important negative effects and service areas affected by these behaviours, as well as the staff's estimation for the extra waiting time caused by those difficult patients' behaviour. From these results, we conclude that it is necessary to find scientific and logical solutions to the problem. The most important of these solutions is the restructuring of the patient flow system to address the most common difficult behaviours within the ED.

References

[1]   Bloor. K, Barton. G, Maynard. A.The future of hospital services. London: Stationary office, 2000. ISBN 011702483x.

[2]   Trzeciak S, Rivers E. Emergency department overcrowding in the United States: an emerging threat to patient safety and public health. Emerg Medical Journal. 2003 Sep; 20(5):402-405.

[3]   Moskop JC, Sklar DP, Geiderman JM, Schears RM, Bookman KJ. Emergency department crowding, part 1-concept, causes, and moral consequences. *Annals of Emergency Medicine.* 2009;53(5):605–611

[4]   Lambe S, Washington DL, Fink A, et al. Waiting times in California's emergency departments. Ann Emerg Med 2003;41:35-44.

[5]   Andersson G, Karlberg I. Lack of integration, and seasonal American College of Emergency Physicians. Boarding of admitted and intensive care patients in the emergency department. *Annals of Emergency Medicine.* 2011;58(1):p.110

[6]   Pines JM. The left-without-being-seen rate: an imperfect measure of emergency department crowding. *Academic Emergency Medicine.* 2006;13(7):807–808.

[7]   Ahmed.M, Alkhamis.T. Simulation optimization for an emergency department healthcare unit in Kuwait. European Journal of Operational Research.2009 Nov;198(3):936-942.

[8]   Peck.J, Kim.S. Improving patient flow through axiomatic design of hospital emergency departments. CIRP Journal of Manufacturing Science and Technology.2010; 2(4):255-260.

[9]   Huw T.O. Davies, Ruth Davies. Simulating health systems: modelling problems and software solutions. European Journal of Operational Research. 1995(87): 35-44.

[10]   Jensen .K. Leadership for smooth patient flow: improved outcomes, improved service, improved bottom line. Health Administration Press, 2006. ISBN 1567932657.

[11]   Milliken. E.M . understanding human behaviour; A Guide for healthcare providers. Delmar, 1987. ISBN 0-8273-2798-6.

[12]   Duxbury. J . Difficult Patients. Oxford,2000. ISBN 0750638389.

[13]   Eric.S . The difficult patient . MedMaster, inc 1996. ISBN 0940780275.

[14]   Robert W. Derlet, MD and John R. Richards MD. Ten Solutions for Emergency Department Crowding: West J Emerg Med. 2008 January; 9(1): 24–27.

[15]   Viccellio P. Emergency department crowding: an action plan. *Acad Emerg Med.* 2001;8:185–18.

A new simulation-based approach to production planning and control

Gyula Kulcsár, Péter Bikfalvi, Ferenc Erdélyi and Tibor Tóth
Department of Information Engineering, University of Miskolc, H-3515 Miskolc-Egyetemváros, Hungary

Abstract. One of the most important means for manufacturing companies to succeed with increasing competition in fulfilling customers' orders is by planning and executing their activities in the most efficient way. In discrete manufacturing, production planning and control tasks lead to complex optimization problems. The paper presents a new comprehensive approach inspired from real-world manufacturing environments. The two-phase procedure starts in its first phase with local search based on fast execution-driven simulation and evaluation with multi-purpose assessment, while a detailed discrete event-driven simulation (DES) model is applied in the second phase to analyze selected solutions to controlling the stability, robustness and performance. The method can be used both for aggregate production planning and for shop floor production management in solving tasks. Efficiency can be further improved with detailed modelling of uncertainty and of various properties of production and business processes.

Keywords: simulation, performance indices, production planning and control, multi-objective optimization.

1. Introduction

In essence, production planning and control (PPC) systems deal with the allocation of limited resources to production activities in order to satisfy given customer demands over a well-defined time horizon. Both planning and control tasks lead in fact to optimization problems, in which the main goal is to create such plans that meet actual demands (constraints) and maximize production performance (i.e. profit). In practice, models of solving these optimization problems cover a very wide range due to varying characteristics of different production systems and their business environments. However, these models usually are very complex and difficult to cope with them.

One way of solving such planning problems is to use a hierarchical approach. It means that the planning process will run in a hierarchical way by ordering the corresponding decisions according to their relative importance. Hierarchical production planning uses a specific optimization model at each level of the hierarchy. Each model extends the constraints of the problem at the lower level of the hierarchy (i.e. [1], [6]).

In order to reduce complexity, production planning problems are usually solved at an aggregate level. It means that individual products (which are distinct but similar) are combined into aggregate product families that can be planned together. On the other hand, production resources (i.e. various machines or labour pools) can also be aggregated into presumable resource groups. The aggregation technique must assure that the aggregate plan can be disaggregated into feasible production (manufacturing) schedules whenever it is necessary.

Production planning process typically runs according to the rolling horizon principle. It means that a candidate plan is created for the actual time horizon and the decisions in the first few periods of the horizon are executed accordingly. Then, the candidate plan is revised and for good cause re-planned. The actual plan must be periodically revised due to the uncertainties that may occur in the demand list and in the production processes.

According to the rolling time horizon, PPC systems cover two main stages: the master planning and master scheduling are realized in the first stage, while fine or detailed scheduling and execution control is done in the second stage. The main goal of the first stage is to balance the capacity and demand. The PPC system includes the master function that produces production plans for material and capacity requirements. The second stage deals with sequencing of production orders that have already been released for production in the actual time frame. It is also decided exactly when and on which machines or workplaces the dependent jobs should be executed. The main goals of this second stage are to avoid tardiness of jobs and orders, to minimize flow times of parts, and to maximize utilization rates of machines/workplaces. For solving such detailed or fine scheduling tasks, advanced scheduling models and methods are needed. A detailed survey for the multi-objective scheduling problems is given in [5].

This paper proposes a new approach to support planning and control activities at the shop floor level of discrete manufacturing systems in a flexible, efficient and effective way. Typical problem-solving decisions include generating dependent orders, batching, resource allocation, work force assignment, task sequencing and timing issues. The main focus is set on simulation-based models used for detailed production scheduling in a discrete-parts manufacturing environment. The new, integrated fine scheduling approach is based on hierarchical and layered simulation.

2. Computer integrated application systems

Nowdays, production engineers and managers utilize more and more computer integrated application systems to support their decision making. Software systems applied to management of discrete production processes can be classified into four hierarchical groups according to the different supported fields and time horizons:

1. Enterprise Resources Planning (ERP)
2. Computer Aided Production Engineering (CAPE)
3. Manufacturing Execution Systems (MES)
4. Manufacturing Automation (MA).

This paper focuses only on the detailed scheduling method applied at shop floor production management level, i.e. as a function of the MES. However, it can similarly be used for aggregate production planning (ERP level). At the MES level, the main goal of the fine scheduler is to initiate detailed schedules that meet the master plan goals defined at the ERP level. The scheduler gets the actual data of dependent orders, products, resource environment and other technological constraints (tools, operations, buffers, and material handling issues and so on). The shop floor management defines the manufacturing goals and their priorities. Obviously, management from time to time may declare various goals. The scheduler has to provide a feasible sequence of jobs which meet the prescribed goal. As result of the scheduling process, a detailed production program is obtained, which declares the releasing sequence of jobs, and assigns all the necessary resources to them and proposes the starting time of operations. This program must not break any of the hard constraints but has to meet the predefined goals.

The computation time of the scheduling process is also an important issue that has to be taken into consideration especially when large number of internal orders, jobs, operations, resources, technological variants and constraints are involved in the problem. It is worth also mentioning again that the optimization problem, despite reducing its order at the MES level, still remains the most challenging one due to the varying characteristics of the manufacturing system over the considered time horizon as well as due to the very short time horizon available for decision making.

3. Simulation-based fine scheduling

Usually, all data of a specific manufacturing order (identifier, priority, product type, product quantity, due date, etc) are available, and can be downloaded from the

ERP system database. On the other hand, all the related information on product, technology and resources (bills of materials, technological process plans, available resources, processing intensities, set-up times, etc) are available in the MES database.

The proposed fine scheduling process consists of two phases (Fig. 1). In the first phase (top layer) a wide-range fine scheduling based on deterministic data models and fast execution-simulation is used to generate several, near-optimal feasible fine schedules. In the second phase (bottom layer) a narrow-range, more precise stochastic model-based event-driven simulation is used for sharp tuning. The given set of low number candidate fine schedules is evaluated according to viewpoints of stability and behaviour in uncertain environment. That is, simulation is also performing fast and stochastic issues can also be considered. At the top layer, the focus is set on creating some near-optimal feasible schedules by considering detailed constraints and capabilities of resources, while at the bottom layer the model adjusting for various uncertainties plays the primary role.

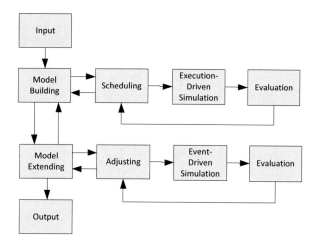

Fig. 1. Two-layered simulation-based fine scheduling.

The proposed two-phase simulation-based model supports the flexible usage of multiple production goals and requirements simultaneously. The elaborated approach helps to solve the complex, detailed scheduling problem as a whole without its decomposition. In this way, all issues like batching, assigning, sequencing and timing are answered simultaneously in much shorter time.

Moreover, simulation may easily lead to evaluation of several production performance indices already before the manufacturing execution. That may serve to corrections at MES or even at ERP level.

Based on above-mentioned two-phase modelling and simulation approach and in order to implement in practice and solve the production fine scheduling problems, special developments had to be done. To implement the first phase (top layer), a knowledge intensive searching algorithm and the corresponding software have been developed. The algorithm is based on execution-driven fast simulation, overloaded relational operators and multiple neighbouring operators. The core of the engine implemented explores iteratively the feasible solution space and creates neighbour candidate solutions by modifying the actual resource allocations, job sequences and other decision variables according to the problem space characteristics. The objective functions concerning candidate schedules are evaluated by production simulation representing the real-world environment with capacity and technological constraints. Items, parts, units and jobs are passive elements in the execution-driven simulation, and they are processed, moved, and stored by active system resources such as machines, material handling devices, humans and buffers. The numerical tracking of product units serves the time data of the manufacturing steps. The simulation process extends the pre-defined schedule to a fine schedule by calculating and assigning the time data, too. Consequently, the simulation is able to transform the original searching space to a reduced space by solving the timing sub-problem. This part of the approach encapsulates the dependency of real-world scheduling problems. Successful adaptation of the approach into practice is highly influenced by the efficiency of the simulation algorithm.

The performance analysis of the created fine schedule can be performed by calculating some objective functions based on the data of units, jobs, production orders, machines and other objects of the model. In order to express the shop floor management's goals as criteria of a multi-objective optimization problem, the software includes an evaluator module, too. This module has many such objective functions implemented, as for example:

- the number of tardy jobs;
- the sum of tardiness;
- the maximum tardiness;
- the number of set-up activities;
- the sum of set-up times;
- the average waiting rate of machines; and
- the average flow time of jobs.

It is worth mentioning here that the shop floor may be loaded with unfinished running tasks when generating schedule for the actual time horizon. So, the input data set has to include the actual state variables of the system. It means that the effects of the last confirmed schedule must be considered when creating new schedule. As described in [2, 3, 4], we have successfully applied this approach for solving extended flexible flow shop scheduling problems in practice.

However, after the first phase evaluation only very few schedule candidates have to be further analyzed, which shortens considerable the more time-consuming fine-tuning process. This is the main role of the bottom layer. Fortunately, for this purpose one may find commercial software applications available. Most of them are based on an event-driven simulation kernel, which assures flexibility and a wide spectrum of mathematical

formalism in order to perform deeper analysis. The most important aspect of this analysis is related to including of uncertainty as well as of the possibility of long-term runs. The detailed event-driven simulation model captures the relevant aspects of the control problem, which cannot be represented in the deterministic execution-driven fast simulation model. The most important issues in this respect are the uncertainty of the manufacturing system, such as (for example) uncertain processing, moving and set-up times, uncertain quality and waste product rates of operations, failures or breakdowns of machines, material handling equipment and other devices. The event-driven simulator is used as a component of the scheduling system taking the role of the real production system. The control logic of the simulator ensures that the schedule must be executed according to the predefined fine schedule. The evaluation of a given candidate schedules is measured over several runs in which uncertainty can be represented by different stochastic dispersions or even by different random numbers. In addition, the commercial simulation software helps in visualization, in statistical and/or other performance evaluation and in verifying the results of the candidate fine schedules.

4. Comparison of candidate schedules

Comparing of different schedule instances regarding to different performance measures is one of the most important tasks of the evaluation of the set of feasible schedules for the same problem. In cases where two or more candidate schedules are available, one of the main tasks is to decide which is the better solution. In complex problems it is typical that different schedules can perform better according to different performance measures.

Regarding the evaluation problem at top and bottom level of the proposed approach, a mathematical model for relative comparison of individual schedules is presented in this section.

The scheduler module creates candidate schedules by systematically modifying the values of the decision variables of the initial schedule. The objective functions concerning candidate schedules are evaluated by both simulation and evaluation modules. The relational operators (i.e. "<") have been overloaded and used to compare the generated schedules according to the multiple objective functions (some examples were enumerated in the previous section). All objective functions are supposed to be given in such a form that their minimum has to be computed according to the following formula:

$$f_k : S \to \Re^+ \cup \{0\}, \forall k \in \{1, 2, \cdots, K\} \quad (1)$$

Coefficients $w_k, k = 1, 2, \cdots, K$ as input parameters may support the user in order to adjust the actual priority of each f_k objective function independently. Each w_k is an integer value within a pre-defined close range $w_k \in \{1,2,\cdots,W\}$ and expresses the importance of f_k.

Let $s_x, s_y \in S$ be two candidate schedule solutions. A comparison function F in order to express the relative quality of solution s_y compared to s_x as a real number was generally defined as:

$$F : S^2 \rightarrow \Re, \; F(s_x, s_y) = \sum_{k=1}^{K} \left(w_k \cdot D\big(f_k(s_x), f_k(s_y)\big)\right), \quad (2)$$

where the function D compares the schedules s_x and s_y corresponding to f_k, according to the following:

$$D : \Re^2 \rightarrow \Re, \; D(a,b) = \begin{cases} 0, & if \; \max(a,b) = 0 \\ \dfrac{b-a}{\max(a,b)}, & otherwise \end{cases}. \quad (3)$$

Using definition (2) the relational operators are overloaded by the following decision:

$$\big(s_x \; ? \; s_y\big) := \big(F(s_x, s_y) \; ? \; 0\big). \quad (4)$$

Any of relational operators (i.e. the ones used in C++ programming: <, >, <=, >=, ==, !=) can be used between two candidate schedule solutions to compare them like two real numbers. For example, s_y is a better solution than s_x (i.e. $s_y < s_x$ is true) if $F(s_x, s_y)$ is less than zero. These definitions of the relational operators are suitable for applying them in meta-heuristics like taboo search, simulated annealing and genetic or evolutionary algorithms to solve multi-objective combinatorial optimization problems of production planning.

5. Conclusions

Production planning and control of discrete manufacturing systems becomes a complex and difficult task already for production systems of low and medium size (20-200 workplaces). The production entities (jobs and machines/workplaces), the human and technological resources, the logistical and technological constraints, the variations of operation intensities and of capacities, as well as the uncertainties of market and production demands and goals make the production planning and control decisions very difficult. Computer aided applications present scalable models at higher (ERP, MRP, SCM) hierarchical levels for helping functions of master planning and control. At lower (MES, MA) levels control decisions are model-dependent to such an extent that using of predefined or standard models is very limited. This is probably the main reason why the up-to-date commercially available MES applications still need important customized implementation, while the use of ISA-95 standard performs below the expected results.

Practical experience showed that production planning and control at MES level performs proactive, short term scheduling and reactive, on-line controlling, eventually re-planning tasks day by day. These activities can be carried out either by decisions of experienced humans, or by using of model-based decision support applications.

On the other hand, evaluation of production processes leads to multi-objective optimization problems. However, without continuous monitoring of several key performance indices (KPIs) the control of production is of lower quality of an open-loop control. Simulation models are suitable for evaluation. Based on practical experience, a two-phase hierarchical problem solving technique suits the best. In the first phase, at the top layer, creating jobs, allocating resources and filtering alternative feasible schedules has to be performed by a customized, very fast simulation model-based algorithm, which outputs only few candidate schedules. In the second phase, at the subordinated layer these schedules are evaluated more detailed and with including of stochastic or uncertain elements, which permit analysis of different variations on intensity rates, on allocations of resources, on priorities and goals. In this way, production management decisions support production results more effectively.

Acknowledgements: This research work was carried out as part of the TAMOP-4.2.1.B-10/2/KONV-2010-0001 project with support by the European Union, co-financed by the European Social Fund.

References

[1] Erdélyi F, Tóth T, Kulcsár Gy, Mileff P, Hornyák O, Nehéz K, Körei A, (2009) New Models and Methods for Increasing the Efficiency of Customized Mass Production. Journal of Machine Manufacturing, XLIX (E2):11-17

[2] Kulcsár Gy, (2011) A Practice-Oriented Approach for Solving Production Scheduling Problems. XXV microCAD International Scientific Conference, Miskolc, Hungary, 61-66

[3] Kulcsár Gy, Erdélyi F, (2007) A New Approach to Solve Multi-Objective Scheduling and Rescheduling Tasks. International Journal of Computational Intelligence Research, 3 (4):343-351

[4] Kulcsár Gy, Kulcsárné FM, (2009) Solving Multi-Objective Production Scheduling Problems Using a New Approach. Production Systems and Information Engineering, A Publication of the University of Miskolc, 5:81-94

[5] Lei D, (2009) Multi-objective production scheduling: a survey. The International Journal of Advanced Manufacturing Technology, 43 (9-10):926-938

[6] Venkateswaran J, Son YJ, (2005) Hybrid system dynamic—discrete event simulation-based architecture for hierarchical production planning. International Journal of Production Research, 43 (20):4397–4429.

Complexity in manufacturing supply chain applied to automotive industry: Modelling, analysis and a case study

Kanet Katchasuwanmanee and Kai Cheng
Advanced Manufacturing & Enterprise Engineering (AMEE) Department, School of Engineering and Design, Brunel University, Uxbridge, Middlesex UB8 3PH, UK

Abstract. In today's world, supply chain management becomes a key factor to achieve the goals for business. The basic principles that the organisation should take to achieve the highest potential of supply chain management, is explained in this research. The theory of "complexity" is a controversial and challenging issue for the automotive industry because this theory can be adapted to be used to solve complexity problems in the engineering field. In this project, regarding to identify and solve complexity in Toyota supply chain problem which has the inefficient process, a three dimensional complexity concept is introduced and explained to illustrate the major causes of a supply chains' complexity, which are procurement process, manufacturing system and distribution network complexity. In order to solve the complexity problems, Arena Simulation programme will be the method of this research, which is used to simulate and evaluate the existing supply chain based on the case study. After that, the proposal of improvement will be provided. Then, Katcha medel, which is an implementing model, will be given. The result comparisons indicate that Katcha system provides higher performance than the existing one in every factor including waiting time, work-in-process (WIP) and throughput time by reducing complexity of interaction of suppliers and manufacturing process and batch size. As a result, the company can achieve higher potential in order to survive in the competitive business.

Keywords: Manufacturing Supply Chain, Complexity, Simulation, Automotive Manufacturing

1. Introduction

Nowadays, business is highly competitive [1]. Every company is trying to compete better in their business by considering globalisation, customers' expectations, cost reduction, shorter allowed lead-time and shorter product life cycle [1]. In order to achieve these goals, internal and external processes must be improved [2]. Ayral [3] claims that the automotive industry is suffering from an increasing agility and flexibility in order to cope with the unpredictability and variety in customers' demand. The effects of uncertainty and unpredictability also appear at the interfaces between customers and suppliers throughout the supply chain [3]. Business management is

introduced as the age of inter network competition [4]. In the last decade, the competition was brand versus brand or store versus store, but now supply chain versus supply chain or suppliers-brand-store versus suppliers-brand-store is introduced [4].

Supply Chain Management refers to the management of multiple relationships throughout the supply chain [4]. "The supply chain is not a chain of business with one-to-one, business-to-business relationships, but a network of multiple businesses and relationships" stated by Lambert and Cooper [4]. Besides, a Supply Chain is a link of materials, information and financial flows between two or more organizations [2]. Chopra and Meindl [5] also insisted that a supply chain consists of all activities including directly or indirectly activities in order to satisfy a customer requirement such as inventory, distributions, and retailers. Thus, these issues have had a considerable impact on manufacturing structures and processes which are increasing complexity level in a supply chain.

A supply chain is not a single linear process of connections, but it is a complex web-link structure [2]. Hence, major issues of complexity in a supply chain have to be identified clearly. "Complexity Theory" is an analytical method which can be used to define the problems in complicated situations [6]. Importantly, it can be adapted to solve complex problems in the "Manufacturing Supply Chain" [6]. The complexities in the automotive manufacturing supply chain attempt to deal with the manufacturing processes and links with suppliers and customers such as procurement, manufacturing and distribution issues. Validating complexity in the supply chain seems to be difficult; however, computer simulation provides a way to validate the information-theoretic model developed and to identify the impact of these problems [3].

Most engineers are always dealing with the decision-making process which usually requires understanding of the complexity in the system [7]. Modelling is a practical technique to assist decision making in order to implementing effective manufacturing systems [8]. An existing system or an implementing system can be represented by simulated model, also investigated and compared with manipulation of the system. Therefore, in manufacturing systems, simulation programmes have been widely used to illustrate the effective of incomplete decisions, such as a new schedule [3]. In this report, the Arena Simulation programme is used to identify the supply chain and performance measurement in order to optimize solutions based on a case study which is Toyota Motor Thailand Co., Ltd.

Toyota, a well-known automotive company, is the best-selling brand in Thailand [9]. The Gateway plant is considered to be one of the most famous factories because it produces and exports about an half of overall production globally. Furthermore, Toyota Motor Thailand: Gateway plant is changing continuously in their product type because of customer demands. The complexity of the supply chain is increased because it deals not only with over 13 suppliers and 300 raw materials and 14 distributors, but also faces with mixed-model assembly lines. For the case company, Arena Simulation applied on Toyota's supply chain will be a useful tool to meet customer satisfaction.

2. Complexity in automotive manufacturing supply chain

Currently, the knowledge of the components of supply chain management usually depends on experimental study; for example, in order to control a daily activity in manufacturing supply chain, companies often apply certain simple rules to deal with common problems [10]. In the real situation, many decisions are made easily by real time controlling of the supply chain activities. Obviously, the companies whose target is to raise their reactivity capacity that assists to solve problems by reducing complexity level. Mostly, the problems are eliminated by reducing its complexity [10].

Nevertheless, from the over view, the highly complex and unsolved problems are still existing because extraordinarily complex issues cannot be solved [10]. For instance, the firms cannot predict exactly the number of inventories alone in supply chain. There are many causes of dysfunction, which are production and distribution flexibility, delivery time reduction, total quality control, new investments, have already overwhelmed the limit of its efficiency [10].

In automotive manufacturing supply chain, complexity implies the number of elements of sub-systems, level of connectivity and interaction among the elements, unpredictability, unstability, and variety in product and system states. However, the "Complexity Theory" can be applied to identify complexity of manufacturing supply chain [6]. As Gottinger [11] claimed "the more we are able to describe the complexity of an objective, the more we are able to learn about the

objective". Likewise, the understanding of system's complexity is the first phrase to solve the problems in the system [12].

Mostly, automotive industries obtain raw materials from many different suppliers and sell products to many different customers; hence, each product type has a different supply chain as Waters [13] stated "Every product has its own unique supply chain which can be both complicated and long sequence".

The complexities in the automotive industry supply chain usually attempt to deal with many sources of complexity such as network complexity, manufacturing complexity, product range complexity, customer complexity, procurement complexity, organisational complexity, information complexity and distribution complexity. Wisner [14] defined 3 foundation elements of complexity in a supply chain, which are "Purchasing", "Operations" and "Distribution" complexity as shown in Fig.1. These complexities can be examined in followings:

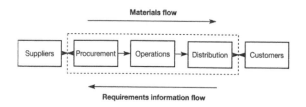

Fig. 1. Complexity in an automotive manufacturing supply chain [13]

2.1. Procurement complexity

Procurement, which is a process between manufacturer and supplier (Upstream activities), is the key link between the company and its suppliers to make sure that firms get "the right material, from the right suppliers, in the right quantity, in the right place, at the right time, with the right quality" [15]. This difficulty can be separated into three topics:

1) Number of Suppliers and Level of Its Interaction

Kauffman [16], Waldrop [17] and Dooley [18] insisted that supplier complexity bases on a factor of the number of suppliers in the supply chain and the level of supplier interaction. The complexity level of the supply base is a critical factor for the level of transaction cost, the risk of supply base and the respond of the suppliers [19].

Mostly, suppliers have their own suppliers called sub-suppliers who provide small components to the suppliers [20]. If a sub-supplier cannot delivery raw material on time, the supplier cannot operate and delivery the component on time. As a result, the company is not able to respond to customer demand [13]. This issue is a combination complexity which occurs when the system range changes with time which is changing suppliers' demand [6].

2) Differentiation of suppliers

In the supply base, differentiation of suppliers can be resulted in the level of different characteristics, which exist in the supply base, such as technical capability (e.g. Supplier does not have enough knowledge or specific tool to manufacture the raw materials), operational practices (e.g. One supplier manufactures in a push system, but another supplier manufactures in a pull system), Cross-border barriers (e.g. They are speaking in different languages, or the raw materials have to travel a long distance) [19].

3) Non-standard raw materials

A distinction can be made between standard product raw material and nonstandard raw materials. Standard raw materials are defined as materials which do not specify suppliers [20]. Typically, they are the availability of several sources of supply which enlarges the possibilities in the source selection process. For non-standard raw material, the company, who buys the raw material, is forced to obtain these parts from one supplier only [20]. Non-standard raw material is an imaginary complexity which has to be eliminated [6].

4) Raw material variety

These days, customer requires more and more customized goods, which means a high number of varied products has to be provided [21]. Therefore, complex and wide product structures are consisted of a numerous number of components [21].

Vaart [20] stated that a large variety of products will result in a lower predictable demand on the level of production types. A vast range of products means that the average demand per variant is low. Hence, the difficulty of forecasting at the individual variant level builds up as a result of forecast error [1]. Consequently, the complexity as experienced by the procurement system is related to the number of product families. However, in order to decrease real complexity, the products have to be various to achieve functional requirement which is customer demand [6].

5) Order process

Once the order is made, it needs to spend a time for waiting the actual orders. Besides, the manufacturer always takes a consideration on the order priorities to create a procurement schedule [20]. However, there is a delay for waiting the orders to be converted into the lists due to checking and allocation process. Moreover, in just-in-time method, suppliers sometimes get notice only 8 to 10 hours before "final call-off" [15]. This issue is a combination complexity which occurs when the system range changes with time which is changing the orders from the manufacturer [16].

2.2. Manufacturing complexity

This complexity is an internal procedure in firms (Internal Supply Chain). At this process, the main complexity in industrial manufacturing is needed to be considered in many aspects, which can be divided into three components as shown below:

1) Multiple operation

In the manufacturing process, lengthy processes including many different activities will not only extend lead times, but also increase unsteadiness in performance of the system [1]. The expanding steps in a process will increase the complexity in the system, and the greater possibility that there will be frequent discrepancies between planned and actual outcomes [1]. This difficulty is an imaginary complexity which should be identified and eliminated.

2) Product variety

A numerous variety in components and assemblies needs a precise responding in term of production planning and control [22] and [23]. Besides, due to the changing of production orders, the manufacturing process increases its complexity [21]. This affects to expanding cycle time and lead time, consequently [21].

This complexity has been defined in an analytical form for manufacturing systems as a measure of how product variety complicates the process; for instance, mixed-model assembly line [24] (See Fig.2). This assembly line may cause longer lead-time due to difference operation time in each model [24]. Hu [24] stated "The high number of variety undoubtedly presents enormous difficulties in the design and operation of the assembly systems and supply chain". For example, Toyota Company Thailand: Gateway Plant is producing five types of cars, and there are around 10 colours in each type. The manufacturers face with a combination complexity which occurs when the system range changes with time which is various products. However, in order to achieve customer demand, the products need to be various which can reduce real complexity in the system [6].

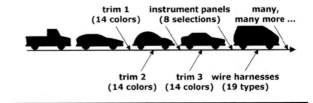

Fig. 2. Automobile mixed-model assembly line [24]

3) Lack of Employees' Knowledge

Knowledge is particularly crucial for firms with can lead and conduct the employees to achieve the works easily and accurately. Without knowledge, worker cannot assemble a simple product or even use a machine [25].

Suh [6] stated that when lack of employees' knowledge occurs, imaginary complexity will be happened.

4) Production Planning and Control

This issue deals with customers' demand undifferentiated flexibility [15]. Lauff and Werner [11] addressed "The complexity of scheduling problems in dealing with variety and uncertainty. Uncertainty comes not only from the customer, but also from the shop floor control which is production planning and control". "The disturbances and the complexity of scheduling cause deviations from a plan that is often overoptimistic" stated by Stoop and Wiers [12]. This issue is a combination complexity which occurs when the system range changes with time which is changing volume and variety of products [6].

2.3. Distribution complexity

Distribution is a process between customer and retailer (Upstream activities), is the key link between the company and its distributors to make sure that their customers receive products on time [16]. This complexity can be separated into three issues:

1) Delivery Time

Timeline and consistency of delivery is requested by customers [26]. The company has to deliver products on time in order to satisfy the customers. Moreover, some customer need higher service levels and greater product availability; for example, they want the product to be posted and delivered within 24 hours. Hence, it is exceedingly difficult to forecast the demands. This problem is a combination complexity which occurs when the system range changes with time which is changing of customer demand [6].

2) Inventory Level

Demand changes are affecting directly to usual behaviour in the supply chain such as a decreasing or increasing in demand. Many manufacturers end up with the highly inventory cost with their products in warehouses because they do not link to the actual demand from customers via retailers or dealers [26]. They should reduce its level of inventories maintaining the same level of service or increasing the level of service [26]. This issue is an imagine complexity which occurs because lack of understanding about customer demand [6].

3. Development of modeling and simulation

3.1. Data for model building

The specified data will be collected in order to transform the existing system data into the simulation model. The data for modeling building can be divided into 2 steps which are experimental & key performance factors and

input parameters. These steps will be explained in the paragraph below:

Experimental factors

All the input factors are experimental factors. These factors would be changed in the implementing system to improve the performance of the existing system and complete the objectives.

1) Lot Size of Production

The number of lot size could be decreased which means a number per arrival will be reduced.

2) Schedule of Part Arrival

After the lot size is reduced, the overall quantity should be maintained by increasing the number of lots.

3) Manufacturing Quantity

The manufacturing quantity should be calculated by customer demands in order to reduce inventory and work-in-process.

Key performance factors

After the model is created completely, the key performance factors should be defined in order to demonstrate the system performance, analyse, compare and make decisions. These factors will be examined in the following:

1) Waiting time

Waiting time shows the bottle-neck of the system which should be eliminated by increasing resources or improving efficiency of the process. This factor should be reduced in order to minimise throughput time.

2) Queuing number

According to waiting time, the number of the queue will increase when the products are waiting in front of production line or somewhere in supply chain.

3) Work-in-process (WIP)

Askin and Standridge (1993) stated "Work-in-process is the number of partially completed units in processing at any given time". When the number of WIP is high, it may cause delay of delivery.

4) Throughput time

Throughput time should be as low as possible in order to quickly responding customer demands. "Throughput time composes of the processing time, set up time, move (material handling) time and plus wait time" noted by Askin and Standridge [27].

3.2. Modeling building

In this research, ARENA Simulation (version 13.5) is used to build a simulation of the supply chain model based on Toyota Company in Thailand –a case study. There are four main steps that need to be achieved. Those steps are problem analysis, modeling and testing, experimentation and analysis and Implementation as shown in Fig.3. and Fig.4.

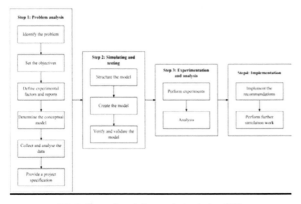

Fig.3. Step of modeling and simulation [28]

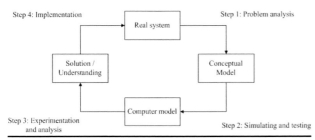

Fig.4. Modeling process [10]

The structural model is significant for the modeler before creating the model, which including the defining elements and the data logically, because it helps users understanding clearly with the system. The layout diagram of Toyota supply chain system for the simulation model is illustrated in Fig.5. The diagram indicates the raw materials are being moved from the suppliers to the manufacturing process, and delivered to distributors and the end-customer.

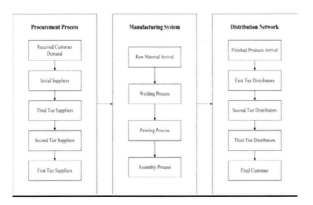

Fig. 5. Layout diagram of Toyota supply chain system

3.3. Arena simulation model

This section shows coding process for modelling the supply chain system in Toyota Company, which is translated by Arena Simulation programme (version 13.5). The model will be built logically according to the layout diagram. The completed model is divided into 3 main processes as shown in Appendix 1, Appendix 2 and Appendix 3.

4. A case study

4.1. Comparison of toyota procurement process results

Fig.6 illustrates the waiting time of components comparison in Toyota procurement process. The waiting times of all components throughout the existing system are extremely high, which are more than 45 minutes in each component. On the other hand, the waiting times of all components in Katcha system are around 20 minutes in each component.

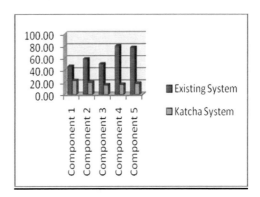

Fig. 6. Toyota procurement process waiting time of the components comparisons

According to applying the proposal improvement (See Table 1), the waiting times of Katcha procurement process are reduced dramatically in all components due to reducing batch size from 6 to 3 per batch and decreasing complexity of interaction suppliers. Consequently, the waiting times of all components in Katcha system are mostly halved comparing with existing system; for example, the waiting time of component 4 in existing is higher than the waiting time in Katcha system by 63 minutes or about 80 percentages.

Table 1. Proposal Improvement of the Katcha system

		Existing System	Katcha System
Procurement Process	Number of Suppliers	13	20
	Batch Size	6	3
Manufacturing System	Number of Welding Process	12	15
Distribution Network	Inventory Level (Parts)	25	15

Fig. 7 shows the waiting time of the processes comparisons in Toyota procurement process. Importantly, the waiting times of all processes in the actual system are higher than the waiting times of all processes in the proposed system.

After apply the proposal into the system, the waiting time in the processes are decreased dramatically. For instance, adding one more supplier to produce component 3 and 4 in the first tier supplier section, separately. Thus, the waiting times of component 3 and 4 in the first tier supplier are considerably dropped by almost 80 percents. Moreover, the total waiting time of the processes in Toyota procurement system is reduced by about 148 minutes or 75 percents.

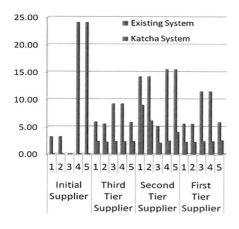

Fig. 7. Toyota procurement process waiting time in the processes comparisons

Fig. 8 demonstrates the work-in-process (WIP) comparisons in Toyota procurement process. Obviously, the numbers of work-in-process (WIP) of all components in the existing system are higher than the work-in-process (WIP) in the proposed system.

In the proposed system, the numbers of work-in-process (WIP) of the components are decreased significantly because of reducing batch size by 3 to 3 per batch and eliminating complexity by removing the bottle necks. The numbers of work-in-process (WIP) are dropped considerably compared with the actual system. For instance, there is a noticeable decrease in the number

of work-in-process (WIP) of component 4 by approximately 13 parts or 70 percents.

According to the numbers of work-in-process (WIP) of the components, the performance factor of the actual system has been improved by decreasing the total number of work-in-process (WIP) in the procurement process by 44 components or 60.66 percents.

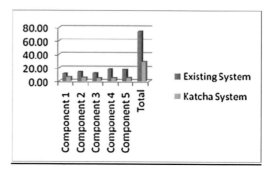

Fig. 8. Toyota procurement process work-in-process (WIP) comparisons

Fig. 9 illustrates the throughput time comparisons in Toyota procurement system. The average throughput time of the components in the existing system is higher than the Katcha system.

After applying the proposal improvement, the level of the average throughput is decreased significantly; for example, there is a noticeable drop in the throughput time of component 4 from about 90 minutes to around 26 minutes. As a result, in term of throughput time, the Katcha system provides a better performance factor by reducing the total throughput time in Toyota procurement system by approximately 220 minutes or 60 percents.

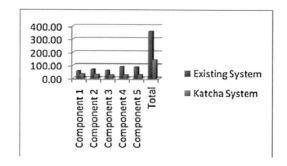

Fig. 9. Toyota procurement process throughput time comparisons

4.2. Comparison of toyota manufacturing system results

As shown in Fig.10, the waiting times of products in existing manufacturing system in Toyota Company are significantly higher than the waiting times of products in the Katcha system. For instance, the number of waiting of Camry products is plummeted from 0.185 minutes to 0.029 minutes.

As a result, the proposed system improves the performance of the actual system in term of waiting time of products by reducing the total waiting time by about 60 percents.

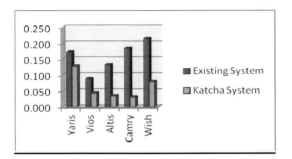

Fig. 10. Toyota manufacturing system waiting time of the products comparisons

Table 2 illustrates the waiting time of the manufacturing comparisons in Toyota procurement process. Overall, the waiting times of every process in the actual system are higher than the waiting times of all processes in the proposed system. In the existing system, the complexity that causes high waiting time of the processes is multiple operations. This complexity affects the potential of the system by producing bottle necks in the process. However, the proposal of improvement has been applied to the system, which is adding more station in the interaction processes in order to reduce complexity of the system by distribution work load.

As a result, the proposed system increases the performance of the system by reducing the total waiting time of the processes in Toyota manufacturing system by more than 70 percents (See Table 1.).

Table 2. Toyota manufacturing system waiting time in the processes comparisons

		Existing System (Minutes)	Implementing System (Minutes)	% Difference
Yaris Welding Process	1	0.0218	0	-
	2	0.0248	0.0022	-
	3	0.0967	0.0957	-
Vios Welding Process	1	0.0218	0	-
	2	0.0320	0.0074	-77.00
	3	0.0060	0.0064	7.61
Altis Welding Process	1	0.0031	0.0032	2.66
	2	0.0452	0	-
	3	0.0560	0.0023	-95.95
Camry Welding Process	1	0.0397	0	-
	2	0.0452	0	-
	3	0.0706	0.0006	-99.17
Wish Welding Process	1	0.0397	0	-
	2	0.0862	0.0011	-98.73
	3	0.0554	0.0468	-15.62
Painting Process	1	0.0263	0.0269	2.49
	2	0.0022	0.0022	2.94
	3	0.0005	0.0005	2.20
Assembly Process	1	0.0002	0.0002	-1.87
	2	0.0003	0.0003	-3.80
	3	0.0001	0.0001	-0.07
Total		0.674	0.196	-70.92

Fig.11 presents the work-in-process (WIP) comparisons in Toyota manufacturing process. The numbers of work-in-process (WIP) of all components in the proposed system are slightly lower than the work-in-process (WIP) in the existing system.

According to adding more processes in the system, the complexity, which are the bottle necks, are eliminated. As a result, the performance factor of the actual system, which is work-in-process (WIP), has been improved by reducing the total number of work-in-process (WIP) in the manufacturing system by approximately 11 percents.

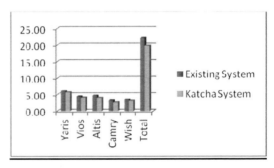

Fig. 11. Toyota manufacturing system work-in-process (WIP) comparisons

Fig.12. demonstrates the total time comparisons in Toyota manufacturing system. The average throughput time of the components in the existing system is slightly higher than the Katcha system.

The actual system has been improved in many ways such as reducing waiting time of components and in processes and decreasing the number of work-in-process (WIP) by reducing and eliminating the complexity in the system. Therefore, in term of throughput time, the proposed system improves the performance factor by reducing the total throughput time in Toyota manufacturing system by approximately 13 percents.

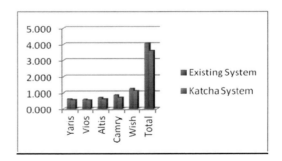

Fig. 12. Toyota manufacturing system throughput time comparisons

4.3. Comparison of toyota distribution network results

Table 3 illustrates the waiting time of the processes in Toyota distribution network. Obviously, all of the waiting times in the existing process are significantly higher than the proposed ones.

In the actual system, the complexity that causes high waiting time of the processes is the high level of inventory. This complexity affects the performance of the system by increasing waiting time. Nevertheless, the applying of the proposal improvement reduces complexity in the distribution system by decreasing the inventories. As a result, the total waiting time of the processes in Toyota distribution network is dropped noticeably by more than 55 percents (See Table 3).

Table 3. Toyota distribution network waiting time in the processes comparisons

Inventory		Existing System (Minutes)	Implementing System (Minutes)	% Difference
First Tier Distributor	1	46.89	27.75	-40.81
	2	46.39	27.26	-41.24
Second Tier Distributor	1	63.08	28.26	-55.20
	2	65.00	28.00	-56.92
	3	60.72	26.16	-56.92
	4	68.46	28.64	-58.17
Third Tier Distributor	1	69.57	29.77	-57.21
	2	72.90	32.82	-54.97
	3	75.02	29.07	-61.25
	4	68.75	30.77	-55.24
	5	64.57	28.95	-55.16
	6	68.75	26.30	-61.74
	7	72.18	33.61	-53.44
	8	76.98	30.78	-60.01
Total		919.24	408.14	-55.60

Fig. 13 illustrates the work-in-process (WIP) comparisons in Toyota distribution network. Importantly, the number of work-in-process (WIP) in all products of the actual system is higher than the work-in-process (WIP) in the Katcha system. In addition, the performance factor of the actual system has been improved by decreasing the total number of work-in-process (WIP) in the procurement process by 49 parts or about 53 percents.

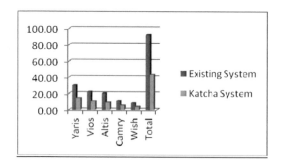

Fig. 13 Toyota distribution network work-in-process (WIP) comparisons

Fig.14 illustrates the throughput time comparisons in Toyota distribution network system. The average throughput time of the components in the actual system is considerably higher than Katcha system.

The actual system has been improved by the proposal of improvement in term of reducing waiting time of products and decreasing the number of work-in-process

(WIP) by decreasing and eliminating the complexity in the system. Thus, the average throughput is declined significantly. There is a considerable drop in the throughput time of Wish products from about 180 minutes to approximately 83 minutes.

As a result, in term of throughput time, Katcha system gives a better performance factor by reducing the total throughput time in Toyota distribution network by approximately 475 minutes or 53 percents.

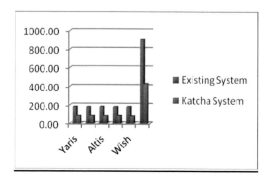

Fig. 14. Toyota distribution network throughput time comparisons

5.   Discussions

Supplier complexity is based on a factor of the number of suppliers in the supply chain system. Hence, the number of suppliers in the system should be reduced to be a minimum number in order to minimize the complexity in the supplier process. However, in this case, there are 13 suppliers throughout the procurement process, but the number of suppliers should be increased to be 20 suppliers because there are some bottlenecks, which are increasing lead time and queuing in the process.

The level of supplier interaction is one of the crucial complexities in the supply chain, which must be reduced. In the procurement process, the status of interaction between suppliers is extremely high due to the link between components. To solve this complexity, decentralization suppliers have been applied throughout the Toyota procurement system. Therefore, each supplier in each state will produce only a process for only one component.

Batch size reduction is a useful technique to leveling of supplier process. Large batch size causes an increasing of the number of queuing, work-in-process (WIP), throughput time in the procurement process; hence, batch size in each supplier must be reduced to be minimum.

In manufacturing system, lengthy processes with many different activities will increase total time and unsteadiness in performance of the system. Thus, the processes in the manufacturing system should be shortened in order to decrease throughput time of the system. Moreover, the interaction processes, which are the processes that operate for two or more different products, should be separated. In this situation, there are some interactions in the processes, which should be

reduced. Thus, a number of processes will be added into
the system in order to distribute the work load and reduce
complexity in the system.

Mixed-model assembly line may cause longer lead
time due to difference operation time in each model
because the high number of variety causes complexity in
the production system. Therefore, the variety of the
products should be reduced in order to decrease the
complexity in the system. Nonetheless, in fact, the
product variety is very important for sales, which provide
profits for the company by responding customer needs;
hence, the variety of products cannot be reduced. Then,
the solution is to improve the efficiency of the system by
eliminating the wastes instead.

Due to the fluctuating of the customer demand, the
company ends up with high inventory with their products
in warehouses. The high level of inventory is a crucial
complexity in Toyota distribution network because it
increases extremely costs for the company to hold the
inventory; therefore, the links between customers and
manufacturer should be clearly defined, and the company
should reduce the level of inventories with maintaining
the same level of service. In this case, there are some high
levels of inventory in the distribution network, which
should be reduced. Hence, the level of inventories is
decreased throughout the system.

The result comparisons indicate that Katcha system
provides higher performance than the existing one in
every factor including waiting time, work-in-process
(WIP) and throughput time. The total waiting time is
reduced about 70 percents in the procurement process, 60
percents in the manufacturing system and 55 percents in
the distribution network. The total number of work-in-
process (WIP) is dropped about 61, 11 and 53 percents in
the procurement process, the manufacturing system and
the distribution network, respectively. The total
throughput time is declined approximately 61 percents in
the procurement process, 12 percents in the
manufacturing system and 53 percents in the distribution
network. As a result, if Toyota Company applies the
proposal of improvement to its supply chain, the supply
chain system will obtain greater performance.

6. Conclusions

The main objective of supply chain management is to
respond customer need by delivery on time with low cost
and high quality. It is very important for Toyota
Company to identify and evaluate its supply chain clearly
in order to achieve the objective. Mixed-model assembly
system, which brings many benefits such as saving
investment cost by sharing resources in the same
production line and absorbing demand fluctuation, has
been used in Toyota Company: Gateway plant. However,
the high number of interaction between suppliers, system
integration and product variety including mixed-model
assembly system increase the complexity of Toyota

supply chain system. Hence, simulation techniques are
playing an important role to evaluate and simulate the
complexity in the supply chain.

In this research, Arena simulation programme is used
for modeling the existing supply chain. After the proposal
of improvement is proposed, an implementing system
called Katcha system will be simulated. The supply chain
system in Toyota Company can be divided into 3
sections, which are procurement process, manufacturing
system and distribution network. The key performance
factors of simulated results are compared between the
actual supply chain system and the proposed supply chain
system.

In conclusion, an effective supply chain can be
achieved by consideration of all complexities in the
system. In order to reduce or eliminate them, the
simulation should be used to evaluate and improve the
supply chain system. This would lead to several
improvements for responding to the customer needs and
the success of the company to survive in the competitive
business.

References

[1] Christopher, M. and Towill, D. (2001). An Integrated Model for the Design of Agile Supply Chains, International Journal of Physical Distribution & Logistics Management, 31 (4), pp. 235-246.

[2] Allesina, S., Battini, D. and Persona, A. (2007). Towards a Use of Network Analysis: Quantifying the Complexity of Supply Chain Networks, International Journal of Electronic Customer Relationship Management, 1(1) pp. 75-90.

[3] Ayral, L., Frizelle, G., Marsein, J., Van de Merwe, E., Wu, Y. and Zhou, D. (2000). A Simulation Study on Supply Chain Complexity in Manufacturing Industry, Logistics Information Management, 10(2) pp. 102-112.

[4] Lambert, M. and Cooper, C. (2000). Issue in Supply Chain Management, Industrial Marketing Management, 29(3), pp. 65-83.

[5] Chopra, S. and Meindl, P. (2007). Supply Chain Management: Strategy, Planning & Operations, 3rd edition, Pearson Prentice Hall, New York, the United States of America.

[6] Suh, Nam P., (2005). Complexity: Theory and Applications. Oxford University Press, New York, the United States of America.

[7] Blanchard, B. S. and Fabrycky, W. J. (2006). Systems Engineering and Analysis, 4th edition, Pearson Prentice Hall, New Jeysey, the United States of America.

[8] Johansson, B. and Grunberg, T. (2001). An Enhanced Methodology for Reducing Time Consumption in Discrete Event Simulation Project, Proceedings of the 13th European Simulation Symposium: Simulation in Industry, Gialbiasi and Frydman (Eds.), SCS Europe Bvba, Marseille, pp. 61-64.

[9] Nopprach S., (2006), Supplier Selection in the Thai Automotive Industry, TSB Production, Bangkok, Thailand.

[10] Bel, G., Thierry, C. and Thomas, A. (2008). Simulation for Supply Chain Management, 1st edition, ISTE Ltd and John Wiley & Sons, Inc., the United States of America.

[11] Gottinger, H. W., (1983). Coping With Complexity: Perspectives for Economics, Management and Social Sciences. D. Reidal, Dordrecht, Holland.

[12]   Choi, T. Y., Dooley, K., Rungtusanatham, M. (2001). Supply Chain Networks and Complex Adaptive Systems: Control Versus Emergence, Journal of Operations Management, 19 (1), pp. 351-366.

[13]   Waters, D. (2009). Supply Chain Management: An Introduction to Logistics, 2nd edition, Palgrave Macmillan, the United Kingdom.

[14]   Wisner, D. J., Tan, K. and Leong, K. G. (2008). Principle of Supply Chain Management: A Balanced Approach, South western, the United States of America.

[15]   Giunipero, L. C., Handfield, R. B., Monczka, R. M., Patterson, J. L. and Water, D. (2010). Purchasing and Supply Chain Management, 4th edition, Seng Lee Press, Singapore.

[16]   Kauffman, S. (1993). The Origins of Order: Self-Organization and Selection in Evolution, Oxford University Press, the United Kingdom.

[17]   Waldrop, M. (1992). Complexity: The Emerging Science at the Edge of Order and Chaos, Touchstone, New York, the United States of America.

[18]   Dooley, K. (2001). Organizational Complexity. International Encyclopaedia of Business and Management, Warner, M. (Ed.), London, the United Kingdom.

[19]   Choi, T. Y., Dooley, K., Rungtusanatham, M. (2001). Supply Chain Networks and Complex Adaptive Systems: Control Versus Emergence, Journal of Operations Management, 19 (1), pp. 351-366.

[20]   Vaart, J. T., Vries, J. and Wijngaard, J. (1996), Complexity and Uncertainty of Materials Procurement in Assembly Situations, Production Economics, 46(3), pp. 137-152.

[21]   Blecker, T., Kersten, W. and Mayer, C. (2005). Development of an Approach for Analyzing Supply Chain Complexity, Proceedings of International Mass Customization, Blecker, T and Friedrich, G. (Ed.), Gito Verlag, Berlin, Germany, pp. 47-59.

[22]   Kestal, R. (1995). Variantenvielfalt and Logistictiksysteme, Wiesbaden: Gabler, Gabler Verlag, Germany.

[23]   Westphal, J. (2000). Komplexitatsmanagement in der Produktionslogistik, Wiesbaden: Universitat Wiesbaden, Germany.

[24]   Hu, S. J., Koren, Y., Martin, S. P., Zhu, X., (2008), Modeling of Manufacturing Complexity in Mixed-Model Assembly Lines, Journal of Manufacturing Science and Engineering, 130(3), pp. 1-10.

[25]   Womack, J.p., Jones, D.T. and Rose, D. (1990). The Machine that Changed the World, Rawson Associates, New York, the United States of America.

[26]   Bard, J. F. and Nananukul, N. (2009). The Integrate Production-Inventory-Distribution-Routing Problem, International Journal of Operations & Production Management, 12(3), pp. 257-280.

[27]   Askin, R. G. and Standrige, C. R. (1993). Modelling and Analysis of Manufacturing Systems, 1st edition, ISTE Ltd and John Wiley & Sons, Inc., the United States of America.

[28]   Robinson, S. (1994). Successful Simulation: Practical Approach to Simulation Projects, McGraw-Hill, the United Kingdom.

[29]   Kauffman, S. (1993). The Origins of Order: Self-Organization and Selection in Evolution, Oxford University Press, the United Kingdom

Appendices

Appendix 1. Flow chart of the existing Toyota procurement process model

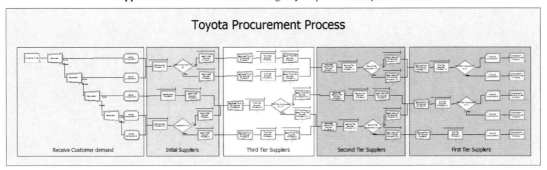

Toyota Procurement Process

Appendix 2. Flow chart of the existing Toyota manufacturing system model

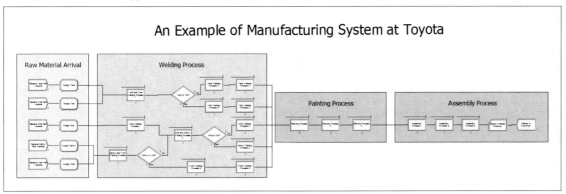

An Example of Manufacturing System at Toyota

Appendix 3. Flow chart of the existing Toyota distribution network model

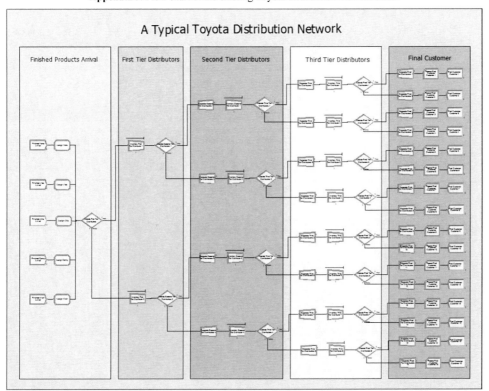

A Typical Toyota Distribution Network

Metrology

Dynamic measurement on structures using tracking-interferometers

R. Schmitt and M. Harding

Chair of Metrology and Quality Management, Laboratory for Machine Tools and Production Engineering (WZL), RWTH Aachen University, Steinbachstrabe 19, 52074 Aachen, Germany

Abstract. The measurement of structure dynamics is a common technique to characterise the dynamic behaviour of structures, analysing reasons of disturbing vibrations, design optimization and as comparison data to check and improve simulation results. Depending on the measurement task, different measurement devices are used such as laser vibrometers and accelerometers. A recent approach is to perform modal analyses using a tracking-interferometer. Firstly the fundamentals of vibration measurement and modal analysis with tracking-interferometers are discussed. Due to the adaption of the measurement device from geometry measurement to vibration measurement several issues have to be considered. Two of them are focused in this paper. The first discussed issue is the influence of the connection of the targets and nests to the vibrating structure. Moreover, for an automation of modelling of the machine it is necessary to measure the absolute distance between tracking-interferometer and the target mounted at the structure. Due to the fact that the used LaserTRACER does not have an absolute distance measurement unit, alternative methods to acquire the absolute distance information out of the signals of the available position sensitive diode are presented.

Keywords: Modal analysis, optical measurement, structural dynamics, vibration, tracking-interferometer, laser tracer

1. Modal analysis of structures

In general, modal analysis studies the dynamic properties of structures under vibrational excitation. The main applications of modal testing are the determination of material properties, the verification of theoretical models and the assessment of vibration response levels [1]. On machine tools these vibrations can affect the process and product quality negatively; therefore a detailed modal analysis can clarify the dynamic stability of the machine and improvement potentials [2]. The approach of the conventional modal analysis is now largely unified and occurs in three phases of model development, measurement and evaluation of the results [3]. Based on a model consisting of multiple nodes, sensors for vibration measurement are attached to the machine structure. Then the structure is excited with a shaker or impulse hammer, for machine tools usually the tool center point is used to

the measurement of dynamic properties [4]. The transfer behavior in the frequency range as the relocation of the mechanical system response to the spatial force excitation is evaluated for each node [5]. Finally, the positions and the results are analyzed offline in a program and any additional points measured.

Today, for modal testing usually accelerometers or laser vibrometers are used [6]. For the measurement of open surfaces and thin structures as in automobile and aerospace industries frequently vibrometers are used. Especially the 3D-scanning vibrometers need a rather broad corridor of undisturbed sight to measure simultaneously with three laser beams [7]. Because of the line of sight problems in housed structures, modal testing of machine tools is usually conducted with tri-axial accelerometers which generally provide better results [8].

For measurements on machine tools in practice almost exclusively tri-axial acceleration sensors are used that are attached with magnets or with wax at the machine structure. A major disadvantage is that due to the costs of sensors and the necessary amplifiers generally only a few acceleration sensors can be used simultaneously. The sensors have to be relocated steadily to perform a complete measurement, which leads to a high expenditure of measurement time. Moreover, the acquisition of the sensors positions to each other, which is usually done with hand measurement tools, is very time consuming too. Both, the migration of the sensors and the absolute position measurement, consume a large part of the measurement period, leading to high measurement and downtime costs. For a standard modal testing of a machining center with around two hundred measurement points using tri-axial accelerometers, the time distribution can be assumed as shown in Fig. 1.

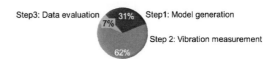

Fig. 1. Time distribution for a modal analysis of a standard machining center [8]

The first step of a modal analysis, i.e. the generation of the model consists of the following activities: Approximation of the structure by discrete points:

- Measuring the absolute position and entering the coordinates in a modal analysis software
- Connection of the points to a wire-frame structure as a geometric model

The second step is the vibration measurement through all the defined positions. This consists of the following activities:

- Fixing the sensors
- Aligning the sensors matching to the defined coordinate system
- Doing the vibration measurement

The activities "fixing and aligning the sensors" determine the time needed for this second step of a modal analysis.

The third step "the evaluation of the results" is supported by highly automated routines and functions of commercial software. It consists of:

- Building a math. model based on measured data
- Visualisation of the vibration modes

The first two steps of the model generation with its highly manual activities are the most time consuming and consist of mainly manual activities which makes them the target for automation and optimization in order to improve the process of modal testing [9].

2. Tracking interferometers

Tracking interferometers (TI) are optical measuring systems which measure the position of a reflector in the spherical coordinate system. The distance is measured relatively with a laser interferometer or absolutely according to the time of flight principle. The angles are measured with incremental angle encoders. The movement of the reflector results in a shift between the incident and reflected laser beam detected by a position-sensitive diode (PSD) and uses its signal for the automatic tracking of the laser beam and the reflector position. TI are used as portable coordinate measuring devices for measuring large volumes with a diameter of up to 50 m as well as for volumetric calibration of machine tools [10].

3. Concept of vibration measurement with traking interferometers

The followed approach describes how acceleration sensors conventionally used for modal testing can be substituted through the use of TI. Instead the accelerometers, reflectors are applied on the machine structure, whose position on the machine can be

determined quickly with the tracking-interferometer. Therefor the rotation encoders and an absolute distance measuring device (ADM) can be used. After teaching the reflector positions, the tracking interferometer is able to aim each position automatically, which is suitable for repeated measurements.

To measure the vibrations, firstly the reflector has to be locked; secondly the tracking function is shut off. Then, the machine is excited and the signals of the PSD for the lateral vibration and the interferometer for the vibrations in beam direction are read out, so that the vibration can be measured in three dimensions. Due to the symmetric geometry of the reflector a lateral movement or rotation of the reflector does not affect the beam length and consequently even the measurement done by the interferometer [11].

Fig. 2 shows a frequency response measurement done with a LT in beam direction in comparison with a measurement done with an accelerometer. The results of the interferometer (X_{int}/h) match very good with these of the accelerometer (X/h). The vibrations in beam direction are measured by the frequency stabilized He-Ne laser inteferometer of a LaserTRACER (LT).

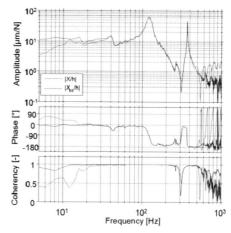

Fig. 2. Comparision of a vibration measurement done with a LT and a accelerometer

The sine-cosine signals of the interferometer allow a good resolution of the displacement below the wavelength of the laser (approx. 633 nm) depending what interpolation method is used. At low frequencies the interferometer shows even better results for the coherency than the accelerometer.

Fig. 3 shows a measurement of one point on a test structure. The beam direction are measured with the LTs interferomenter. The lateral directions measured by the PSD. The curves PSDh and PSDv show the horizontal and vertical displacement of the reflector. The quality of the lateral measurements depends strongly on the proper estimation of the diode's sensitivity and its linearity. The intesivity including the amplifiers varies for the used LTs within ranges between 2.19 mm/V and 2.60 mm/V. The signals show a good linearity for a range of at least 1 mm, but are affected by noise caused by interference, the used power supply and the A/D-converters. The reduction of the signal noise can improve the quality of the measurements with the PSD and is part of further research.

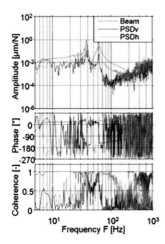

Fig. 3. Frequency response function in 3D measured with a LT

The TI records the measurement in its local coordinate system, so the measurement result should be tranformed in global coordinates. Due to the fact, that for relative displacement, it is sufficient to transform the coordinates only by rotation without translation of the origin. In combination with the assumption the basis of the TI stays parallel to the horizontal plane of the global coordinate system, only 2 degrees of freedom remain for the coordinate transformation the angle θ and Φ as shown as the matrix in Fig. 4 [12]. The rotation matrix is given in the figure. Multiplied with the measured vector in TI coordinate system (beam direction b, vertical direction v and horizontal direction h of the two dimensional PSD) the result is the vibration in global coordinates.

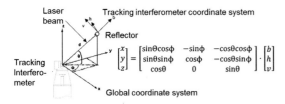

Fig. 4. Coordinate transformation from TI to global coordinates [12],[13]

4. Influences of the connection to the structure of reflectors and nests on the measurement

The usage of TI induces the use of reflectors, which have to be attached to the structure which dynamic behavior has to be characterised. The current approach is to use reflectors developed for geometric measurements. For dynamic measurements the dynamic characteristics of the nest-SMR-system has to be considered due to the fact the modal behavior of the structure should be measured and not the behavior of the reflector. To investigate this issue test measurements with a laser vibrometer has been done to compare the frequeny response of nests and spherically mounted reflectors (SMR) mounted on a structure.

Fig. 5 shows exemplarily the result of a measurement to show the difference between the applications of a nest on the structure in different directions. The direction horizontal means the nest is attached on the upper side of the structure so that the main vibration leads to shearing stress of the adhesive connection between nest and structure. The direction vertical means, the nest is attached on the side wall of the structure which leads to less shearing stress on the adhesive. This leads to better compliance of the frequency response function compared which the structure. The influencing effects of the reflectors attachment shouldn't be ignored and will be researched further in order to improve the modal testing with TI.

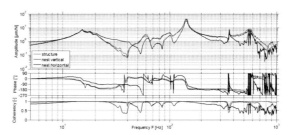

Fig. 5. Comparision of different attachment directions of nests

5. Usage of the position sensitive diodes' signals for distance measurements

For the automated geometric modeling of the structure which modal behavior should be analyzed it is useful to determine the absolute distance between LT and reflector. To compensate the absence of an absolute distance measuring unit (ADM) on the LT partially for shorter distances up to 4-5 m approaches to measure the distance with the PSD are developed. These efforts intend to avoid the implantation of an ADM unit at the LT which is a rather complex engineering task not addressed in the funded project. Two approaches are tested. Both use the precise kinematics of the tracer.

The first approach (Fig. 6) calculates the distance (l) with the geometric relation between the dislocation of the laser beam detected on the PSD (d) caused by a known incremental rotation (α) of one of the LT axis and the sensitivity of the PSD (S_{PSD}).

The maximum angle which can be used for this measurement declines with increasing distance. To maximize the used angle the LT is rotated in a manner that the signal on the PSD looks like a meander with increasing amplitude. The distance between LT and reflector is calculated by a multiplication of the measured voltage amplitude and the PSD's sensitivity.

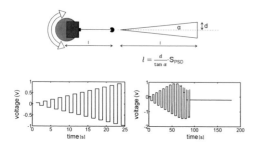

$$l = \frac{d}{\tan \alpha} \cdot S_{PSD}$$

Fig. 6. Distance measurement based on incremental rotation

The second approach uses the possibility to rotate the LT with a constant angle velocity (Fig. 7). For this measurement the laser beam is firstly centered on the reflector. Then it is moved defined over the edges of the reflector. In the third step the LT moves the laser beam over the reflector. Due to the constant angle velocity of this movement there is a linear relationship between the measured dU/dt and the distance between the LT and the reflector. [12], [13].

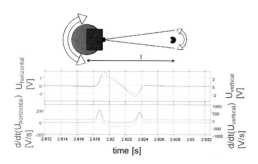

Fig. 7: Distance measurement based constant angle velocity

Both approaches show first results with a deviation below 100 mm at 4 m distance. This is not yet precise enough for the modeling purpose, but planned improvements in the firmware of the LT to reduce the angle velocity, improvements in the power supply and AD-converters with a higher resolution will lower the deviations.

6. Conclusions

This paper shows a new approach for modal testing which intends to reduce the time needed for a modal analysis. Due to the use of reflectors developed for geometric measurements it has been shown that the attachment of the reflectors has to be considered carefully in order not to affect the measurement results. The described approaches for distance measurements with the PSD meet the challenge to do absolute distance measurements without an ADM device and show the proof of principle but have to be improved. The next steps for the development in the current project intend on the improvement of the measurements with the PSD for better vibration and distance measurements. The measurement strategy to perform modal testing in industrial reality has to be developed in detail and validated. Finally the measurement uncertainty has to be evaluated in detail.

Acknowledgements: The content is based on the research and development project "DynaTrac" funded by the German Federal Ministry of Education and Research (BMBF) within the framework concept "Research for Tomorrow's Production" and managed by the Project Management Agency Karlsruhe (PTKA). The authors highly appreciate the support by Heinrich Schwenke and Mark Wissmann from etalon AG and Heiko Paluszek and Daniel Berk from sigma3D GmbH.

References

[1]   Ewins DJ (2000) Modal Testing, Theory, Practice and Application. Research Studies Press LTD, Baldock
[2]   Brecher, C.; Weck, M.: Werkzeugmaschinen – Messtechnische Untersuchung und Beurteilung, Band 5, Springer-Verlag, Berlin, 2006.
[3]   Brecher, C.; Guralnik, A.; Bäumler, S.: Measurement of structure dynamics using a tracking-interferometer in: Production Engineering - Research and Development 6 (2011), 1, ISSN 0944-6524, p. 89-96
[4]   Weck, M, Dynamisches Verhalten spanender Werkzeugmaschinen Einflussgrößen, Beurteilungsverfahren, Meßtechnik Aachen 1971
[5]   Brecher, C.; Esser, M., Witt, S.: Interaction of Manufacturing Process and Machine Tool, CIRP Annals – Manufacturing Technilogy, Jg., 58, Nr. 2, 2009, p. 588-612
[6]   Tönshoff, H.K.; Ahlborn, D.: Statistische Sicherheit bei der Modalanalyse von Werkzeugmaschinen. In: Engineering Research, Jg. 1992, H. 58 Nr.1/2, p. 1-5.
[7]   Pingle P, et al. (2009) Comparison of 3D Laser Vibrometer and Accelerometer Frequency Measurements
[8]   Schmitt, R.; Jatzkowski, P.; Schwenke, H.; Warmann, C.: Advances in the error mapping of machine tools and coordinate measurement machines (CMMs) by sequential multilateration. In: Laser metrology and machine performance IX. 9th Int. Conf. & Exh. on Laser Metro. Machine Tool, CMM & Rob. Perfor., LAMDAMAP 2009, 30th June-2nd July 2009, Brunel Uni., West Lon. 452-61.
[9]   Wendt, K.; Schwenke, H.; Bosemann, W.; Dauke, M.: Inspection of large CMMs by sequential multi-lateration using a single laser tracker. In: Laser metrology and machine performance VI, p. 121-130.
[10]   Patentschrift, DE 000019947374 A1 (01.10.1999)Schwenke, H.; Wendt, K.: Verfahren zur Ermittlung geometrischer Abweichungen von Koordinatenmessgeräten und Werkzeugmaschinen.
[11]   Jatzkowski, P.: Ressourceneffiziente Kalibrierung von 5-Achs-Werkzeugmaschinen mit Trac. Inter. 2011, p. 25
[12]   Brecher, C.; Bäumler, S.; Wissmann, M.; Guralnik, A.: Modal Testing Using Tracking-Interferometers, In: R. Allemang et al. (eds.), Topics in Modal Analysis II, Volume 6, Conference Proc. of the Soc. for Exp. Mech. S. 31, 2012
[13]   Brecher, C.; Guralnik, A.; Wissmann, M.; Berk, D.; Schwenke, H.; Paluszek, H. (submitted): Untersuchung der dynamischen Strukturverformung von Werkzeugmaschinen mit Tracking-Interferometern. Scientific conference "Sensoren und Messsysteme", 22nd - 23rd May, 2012, Nuremberg.

Model for a CAD-assisted designing process with focus on the definition of the surface texture specifications and quality

Q Qi[1], X Jiang[1], P J Scott[1] and W Lu[1, 2]

[1] EPSRC Centre for Innovative Manufacturing in Advanced Metrology, School of Computing and Engineering, University of Huddersfield, Huddersfield, HD1 3DH, UK

[2] School of Mechanical Science and Engineering, Huazhong University of Science and Technology, Wuhan, 430074, China

Abstract. In the early stage of engineering design, the optimal control of engineering surface is of essential importance to control the quality of products and to reduce the cost. A model for a CAD-assisted designing process with focus on the definition of the surface texture specifications and quality is presented for the purpose of assisting designers assign an optimal surface texture specification to fulfil the requirements of function, lower costs and shorter product lifecycles. The model is the basic philosophy to develop a surface texture intelligent information system which can be integrated with CAD systems.

Keywords: Surface texture, Geometrical Product Specifications (GPS), CAD.

1. Introduction

Surface texture plays a significant role in determining the function performance of a workpiece because of the sensitivity of surface texture to change in the process. It is of significance to assign an optimal surface texture specification for an engineering surface in the early stage of design. However, in current engineering practices, surface texture has been more commonly used only for compliance and not for manufacturing process monitoring or function correlation. It is also considered insufficient that for engineering drawing if only surface texture is provided without information of metrology as it is rather difficult to truly express the functional requirements. This can be seen in Table 1, most current commercial CAD systems still employ old profile (2D) surface texture standards or do not completely conform to the standards. This leads to large specification uncertainty (up to 300%)[1, 2] compared with the latest surface texture specification standard ISO 1302:2002 [3]. Because of the massive GPS standards [4, 5] and intricate related knowledge on design, manufacture and measurement in the field of profile and areal (3D) surface texture, it is arduous and time consuming to finish an unambiguous surface texture specification for designers, especially when there is no surface texture specification support tool in current commercial CAD systems. The latest profile [3] and areal [6] surface texture specification standards give the tools to control the surface texture by an unambiguous specification on technical drawings. The two standards make it possible for the designers to indicate the intended surface texture with the least possible effort, also making it possible for the designer of a given surface texture specification to understand, implement or verify the requirement without mistakes. Recently, a category theory based GPS surface texture knowledge platform has been developed to facilitate fast and flexible manufacturing [4, 7]. The knowledge model in these papers is the foundation to develop a design and measurement information system in profile and areal surface texture for manufacturing industry. In this context, a model for a CAD-assisted designing process with focus on the definition of the surface texture specifications and quality is presented. The model in this paper is developed to utilize the category theory based knowledge model for the development of a surface texture intelligent information system which can be integrated with CAD systems.

2. Model for a CAD-assisted designing process with focus on the definition of the surface texture specifications and quality

Based on the Geometrical Product Specifications (GPS) philosophy [8], surface texture specification should be designed based on the function requirements, lower costs and shorter product lifecycles. The essential functions of the model are to help designers find a balance point between function and total costs (include design, manufacture and measurement), and to select an optimal surface texture specification elements; also to help engineers and metrologists for monitoring manufacturing process and guiding the measurement procedure precisely.

Table 1. The status of surface texture specification design in commercial CAD systems

Commercial CAD Systems	Surface Texture Specification Design	Surface Texture Standards	
		Versions	Indications
AutoCAD	None	None	None
CATIA	Roughness symbol tool	ISO 1302: 1965	3.2
SolidWorks	Surface finish symbol menu	Simplified version of ISO 1302:2002	milling Ra 3.2
Pro/Engineer	Surface finish tool menu	ISO 1302: 1965	3.2
NX	Surface finish symbol Tool	ASME Y14.36M-1996	Mill 2.5/Rz 6.3

2.1. The integrated surface texture design conception

In contrast to traditional tolerance systems, the GPS based design process of the surface texture specification is mapped to and receives feedback from the manufacture and measurement. The specification of surface texture is assigned to transfer more manufacture and measurement

information. Fig. 1 shows the integration between design and measurement in surface texture. In the design phase, function requirements and others such as manufacturing processes and component types should be considered for a function design of surface texture. All of the specification control elements defined in ISO 1302:2002 and ISO/CD 25178-1 (currently under development) [6] can be established according to the inputs and the inference of relationships. After the inference procedure, all the inferred specification elements can be combined into a complete surface texture specification. Then the specification, callout and related specification data can be generated and saved in CAD systems. In the measurement phase, the metrologist firstly analyzes the specification, and translates it to measurement specification which will take into account the measurement conditions. Following the measurement strategy, the metrologist carries out the measurement and obtains the measurement data. In this step, the metrologist selects different options in the form removal and filtration parts. According to the data treatment selection, the software calculates the numerical result of the specified parameter in the last step. Based on the numerical result and uncertainty estimation, the metrologist should provide conformance or non-conformance with the specified specification. Finally, the measurement result and the measurement procedure can be fed back to the design stage to compare with the desired function and estimate the measurement cost to help improve the design process.

2.2. Finding the balance point between function requirements and total cost of design, manufacture and measurement

After the integration of design and measurement of

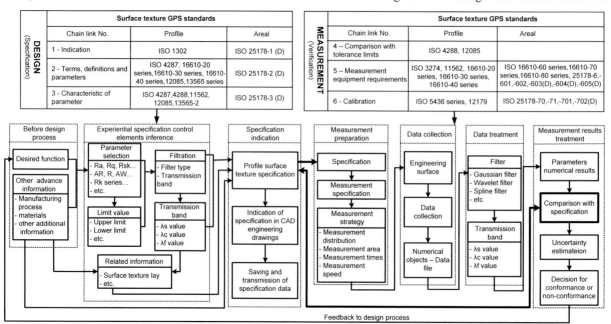

Fig. 1. The integration between surface texture design and measurement

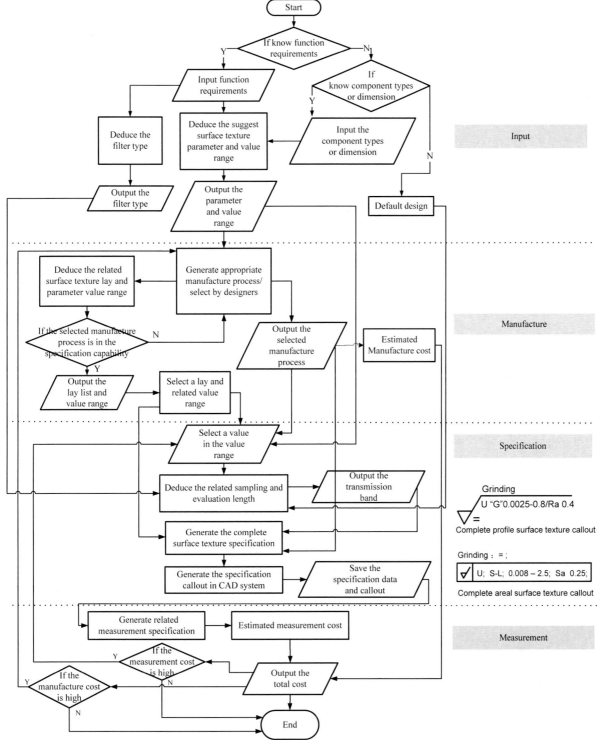

Fig. 2 The designing process for a complete and functional surface texture specification

surface texture, the corresponding measurement requirements for a specified surface texture specification can be determined. Within the design phase, the manufacture and measurement cost can be estimated according to the assigned surface texture specification. In Fig. 2, the flowchart of the cost optimal surface texture design process shows integration between input, manufacture, specification and measurement.

In the design input phase, the most crucial part is the selection of the right surface texture parameter and limit value according to the function requirements and other inputs. As in many applications surface texture is closely allied to function, for instance where two surfaces are in close moving contact with each other their surface textures will affect their sealing or wear properties. This might suggest that it is a case of "the smoother the better", but this is not always true as other factors may be involved. The financial aspect has to be considered: it costs a lot of money to produce very smooth surfaces and the expense of this exercise can considerably add to the bill without gaining a great deal of performance. However, identifying very specific parameters of the surface texture with function is fraught with problems, usually because of time and expense. One point to notice is that it is rarely the individual parameter Ra or Rq for example which is important but often the type of parameter. It is a fact that often combinations of different parameter types are needed to get the best correlation between parameter and function [9].

The manufacturing process of the specified surface can be determined by the function requirements and/or the component type of the surface. Whether the manufacturing process is assigned by the designer or deduced by the system automatically, the manufacture cost, related surface texture lay and parameter value range can be estimated accordingly.

In the specification phase, all other specification elements can be deduced according to the inputs and generated specification elements. For example, the sampling and evaluation length can be determined according to the assigned parameter and value. Then a series of complete ten/eleven specification elements for profile and areal surface texture can be generated. In the CAD system, an integrated application program will be developed to analyze the generated specification; a surface texture indication will be generated and related specification data will be saved.

In this model, the designer can gain the measurement information. The related measurement requirements for the assigned specification can be inferred and the measurement cost can be estimated. The measurement cost then will be added to the total cost which can be used to balance the design and measurement details. As the complete specification can be generated by the category model, the design cost will be decreased. If the manufacture cost is high, the designer is required to choose lower cost manufacturing process which is still fulfilling the function performance. If the measurement cost is high when the suggested measurement instrument is expensive; or the procedure requires multiple measurement (≥ 12), the designer is required to modify the specification with a larger limit value which is sufficient to convey the design intent.

3. Discussion and conclusion

A model for a CAD-assisted designing process with focus on the definition of the surface texture specifications and quality is presented. The model is the basic philosophy to utilize the category theory based knowledge model for the development of a surface texture intelligent information system which can be integrated with CAD systems. In the model, although the utilized latest profile and areal standards still have a certain specification uncertainty, the specification elements of them are considered to provide enough information for manufacturers and metrologists. When all elements are specified in one surface texture specification, the symbol may appear much longer than traditional ones. Then more drawing space is needed. A simplified version or reference symbol can be applied but should be without any significant information loss.

Acknowledgements: The authors are grateful for support from project ERC-2008-AdG 228117-Surfund and NSFC 51005089.

References

[1]   ISO/TS 17450-2:2002 Geometrical product specification (GPS) - General concepts - Part 2: Basic tenets, specifications, operators and uncertainties. International Standardization Organization, Geneva, Switzerland

[2]   Bennich P, Nielsen H, (2005) An overview of GPS, A Cost Saving Tool. http://www.ifgps.com

[3]   ISO 1302:2002 Geometrical product specification (GPS) - Indication of Surface Texture in Technical Product Documentation. International Standardization Organization, Geneva, Switzerland

[4]   Qi Q, Jiang X, Liu X, et al, (2010) An unambiguous expression method of the surface texture. Measurement 43:1398-403

[5]   Qi Q, Jiang X, Blunt L, Scott PJ, (2010) Modeling of the concepts in ISO standards for profile surface texture. Future Technologies in Computing and Engineering: Proceedings of Computing and Engineering Annual Researchers' Conference 2010: CEARC'10. University of Huddersfield, Huddersfield, 172-176

[6]   ISO/CD 25178-1 Geometrical product specifications (GPS) – Surface texture: Areal – Part 1: Indication of surface texture. International Standardization Organization, Geneva, Switzerland

[7]   Qi Q, Jiang X, Scott PJ, (2012) Knowledge modeling for specifications and verification in areal surface texture. Precision Engineering 36:322-333

[8]   ISO/TR 14638:1995 Geometrical product specification (GPS) – Masterplan. International Standardization Organization, Geneva, Switzerland

[9]   Whitehouse DJ, (2001) Function maps and the role of surfaces. International Journal of Machine Tools & Manufacture 41:1847-1961.

A generalised featured-based inspection framework for dimensional inspection of individual machined parts

Liaqat Ali, Mushtaq Khan, Khurshid Alam, Syed Hussain Imran and Mohammad Nabeel Anwar
NUST School of Mechanical and Manufacturing Engineering (SMME), Sector H-12, 44000, Islamabad, Pakistan

Abstract. Inspection and dimensional measurement of machined parts plays a vital role in manufacturing. Machined parts are inspected to much tighter tolerances in order to achieve the highest quality finished products. The dimensional inspection process of discrete components has been developed and automated with time and has come a long way, from the early use of gauge blocks, dial indicators, micrometers to today's computer controlled coordinate measuring machines (CMMs) with touch trigger probes. Despite this rapid advancement, bespoke inspection methods are used for dimensional measurement of machined parts either on a CMM or a CNC machining centre have touch trigger probes. A generalised dimensional inspection framework is required and to achieve this STEP and STEP-NC standards provide a convenient plate-form. In this paper the concept of a framework which gives a generalised feature-based inspection and measurement plan for prismatic machined parts based on STEP and STEP-NC is presented. This generalised STEP-NC compliant inspection plan could be a direct input to an intelligent controller of a CNC machining centre through an interface or can be interpreted into an inspection code for a CMM. The inspection framework uses the information regarding manufacturing features of a machined part and touch probing as given in ISO14649 (STEP-NC).

Keywords: Coordinate Measuring Machine (CMM), CNC, STEP(ISO10303), STEP-NC(ISO14649), STEP-NC compliant, work plan, Inspection

1. Introduction

The new standard STEP-NC[1] provide standards for automatic and consistent CNC component manufacture. STEP-NC formally known as ISO14649 is a departure from the current NC programming standard ISO 6983. It provides an object oriented data model for CNC machines that has detailed and structured information such as the feature to be machined, tool types used, the operations to perform, and the work plan. ISO14649 has many parts that are responsible for providing information such as general process data, manufacturing features, machining processes (milling, turning etc), set up and tooling, and inspection. A lot of work has been done in the area of CAD-CAM frameworks and interoperability in global

manufacturing [2], process control [3], mill-turn [4], software interfaces [5], composite manufacture [6] and dimensional inspection [7] regarding STEP-NC.

This paper presents an inspection framework for individual prismatic components that is STEP-NC compliant. This framework gives a generalised feature-based inspection and measurement plan for a prismatic part. This generalised STEP-NC compliant inspection plan could be a direct input to an intelligent controller of a CNC machining centre through an interface or can be interpreted into an inspection code for a CMM. The probing facility that is used for setup on a CNC machining centre can also be used for in-process inspection of simple geometric features. So the advantage is that a part to be machined can be inspected in-process to have an initial idea of dimensional deviations especially during rough milling before it goes as a finished part for final inspection at a CMM.

The main parts of this framework include a product information model and a manufacturing/inspection model based on STEP(ISO10303-21)[8] and STEP-NC standards.

2. STEP-NC compliant inspection framework for prismatic parts

Inspection planning for measuring a machined part on different measuring machines is currently different based on the measuring capabilities and the bespoke inspection routines used in each case. Three scenarios presented in the current situation for a machined part inspection are given as:

(a) Dimensional inspection of the machined part on a coordinate measuring machine using vendor-specific inspection software.
(b) Part inspection at a CNC machining centre using G and M codes for measuring part with a touch trigger probe.

(c) Part inspection at a CNC machining centre with a state of the art CNC controller using built-in probing cycles for feature measurement.

A major feature of the STEP-NC compliant framework for inspection planning of components is the inclusion of high-level and detailed information in terms of an inspection workplan, workingsteps and a mechanism to feedback inspection results. This has been achieved through the use of STEP-NC (ISO14649 part 10 &16) and AP219 (ISO10303-219)[9] as the basis for representation of product and manufacturing and inspection models for the component [7].

2.1. Objective and Function

The objective of the framework is to eliminate the need for a component's inspection planning specific to the machine tool and generate a generalised interoperable inspection plan file. The functioning of the framework is illustrated by Fig. 1. This inspection plan generated is independent of the machine tool used for inspection of the part and the information included is (a) Definition of the component, (b) Measurement of the component and (c) Method of measurement of the components.

2.2. Definition of the component

A prismatic component to be inspected is first defined in terms of its geometry and main dimensions e.g. a raw material with its main dimensions (length, width etc.). The features present in the part are then specified using the definition of the manufacturing features in the STEPNC standard (ISO14649-10). The next step is to add geometric and dimensional tolerances to the main part as well as the individual features present in it.

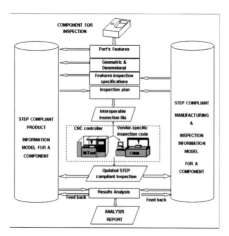

Fig. 1. STEP-NC compliant framework for inspection of discrete components [7].

2.3. Measurement of the component

The geometric elements to be inspected are specified as inspection items which include dimensions along with its dimensional tolerances, shape along with applied shape tolerances and reference dimensions from the inspection datum along with position tolerances. The entity "inspection items" according to the STEP-NC standard definition provide storage for nominal data. It is divided into three catagories namely toleranced dimension items, toleranced spanning dimension items and toleranced shape items.

The inspection items defined for the prismatic block describe the part's geometry including toleranced dimension items, toleranced shape items and toleranced pose items. Similarly inspection items in each individual feature are defined. All this information is provided by a STEP-NC compliant product information model.

The STEP-NC compliant inspection framework uses the object oriented data structure of ISO14649 for defining its entities and generating an inspection plan file. The activities of the framework are illustrated by Fig. 2. The main characteristics of the STEP-NC compliant inspection framework are as follows:

- Defining the component using the information in the STEP-NC standard
- Specifying inspection requirements for the part,
- Generating a generalised inspection plan, and conversion of the inspection file into inspection code.

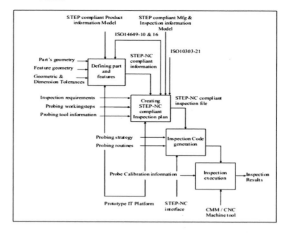

Fig. 2. Structural Definition of the Activities of the framework

3. Characteristics of the STEP-NC compliant inspection framework

3.1. Defining the component using the STEP-NC standard information

The part and its features are defined by the Product information model that includes geometrical shape, dimensions tolerances and the reference datum for inspection. The main sources of the product information are ISO14649 part10 and ISO14649 part 16, with additional information that specifies feature-based inspection requirements. Defining the part and its feature's geometric shape Specifying inspection requirements Generating a generic inspection plan. Generating a STEP-NC compliant inspection file (shown in Fig 3). Conversion of the STEP-NC compliant inspection file into inspection code (shown in Fig 4)

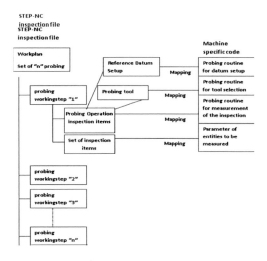

Fig 4. Manual mapping of STEP-NC inspection file onto a machine specific inspection code

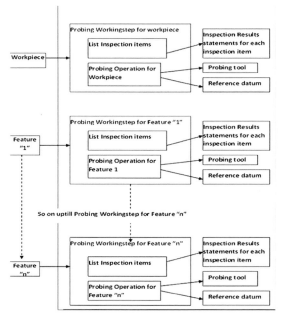

Fig 3. Inspection Workplan for workpiece having "n" number of features

4. Conclusions

In this paper the author has discussed the concept and functionality of a STEP-NC compliant inspection framework. The main objective of the framework is to prove the feasibility of STEPNC standard in generating a generic inspection plan for dimensional inspection of machined parts. The component information is provided by the product information model. Interacting with this model the manufacturing and inspection model describes inspection items, inspection datum information, probing working-steps that combine in a workplan and probing tool information. A STEP-NC compliant file for generalised inspection planning of the component is generated. This file is interpreted by mapping it into an inspection code for a CNC machine or a CMM with the measurements of inspection items the results being added to the file. However it has a few limitations these are; firstly it is only feasible for prismatic parts with simple features having regular geometry e.g. cylindrical, conical or rectangular etc. and not for parts with 3D features. Secondly it is mainly based on ISO14649-16 that defines only inspection by touch trigger probes.

References

[1] ISO14649-10, 2002, International Standards Organization, TC184/SC1/WG7, ISO14649/FDIS, "Industrial automation systems and integration — Physical device control —Data model for computerized numerical controllers — Part 10: General process data"

[2] Newman, S.; Nassehi, A.; Xu, X.; Rosso Jr., R.; Wang, L.; Yusof, Y.; Ali, L.; Liu, R.; Zheng, L.; Kumar, S.; Vichare, P. and Dhokia, V. (2008). Strategic advantages of interoperability for global manufacturing using CNC technology. Rob. & CIM, 24(6):699-708. Elsevier.

[3] Kumar, S.; Nassehi, A.; Newman, S.; Allen, R. and Tiwari, M. (2007). Process control in CNC manufacturing for discrete components: A STEP-NC compliant framework. Robotics and Computer-Integrated Manufacturing,

[4] Yusof, Y.; Newman, S.; Nassehi, A. and Case, K. (2009). Interoperable CNC system for turning operations. Proc. World Acad. Sci. Eng. and Tech. 34:941-947. WASET.

[5] Nassehi, A.; Liu, R. and Newman, S. (2007). A new software platform to support feature-based process planning for interoperable STEP-NC manufacture. International Journal of Computer Integrated Manufacturing, 20:669-683.

[6] Cantoni, S.; Carrino, L.; Nassehi, A.; Newman, S.; Tolio, T.; Valente, A. and Vitiello, C. (2009). A STEP Compliant System for Manufacturing of Composites in the Aerospace Industry. In 19th International Conference on Flexible Automation and Intelligent Manufacturing (FAIM), pp.1356-1364, Teeside, UK, 6-8 July 2009

[7] Ali, L.; Newman, S.T.; Petzing, J., 2005, "Development of a STEP Compliant inspection framework for discrete compone nts", Proceedings of IMechE, Vol 219

[8] ISO10303-21:1994, "Industrial automation systems and integration. Product data representation and exchange, implementati on methods, Clear text encoding of the exchange structure" (International Organization for Standardization, Geneva, Switzerland)

[9] ISO10303-219 TC184/SC4/WG03-N15 15,ISO/CD,2005, "Product data representation and exchange: Application protocol: Dimensional inspection information exchange" (International Organization for Standardization, Geneva, Switzerland).

Practical in-situ calibration method for the non-linear output from a low cost eddy current sensor

D. Clough[1,2], S.Fletcher[1], A.P.Longstaff[1] and P. Willoughby[2]

[1] Centre for Precision Technologies, School of Computing & Engineering, University of Huddersfield, Queensgate, Huddersfield, HD1 3DH, UK

[2] Machine Tool Technologies Ltd, 307 Ecroyd Suite, Turner Road, Lomeshaye Business Village, Nelson, Lancashire, BB 9 7DR, UK

Abstract. Increasing demand on manufacturing industry to produce tighter tolerance parts at a consistent rate means it is necessary to gain a greater understanding of machine tool capabilities, error sources and factors affecting asset availability. The machine tool spindle can be a significant contributor to both machine tool errors and failures resulting in a requirement for spindle error measurement. The use of eddy current non-contact displacement transducers is currently a popular method for measuring spindle error in a manufacturing environment. This is due to their resistance to harsh conditions where dust and coolant may be present. Unfortunately, many eddy current sensors have non-linear outputs that vary with target material and dimension. Typically, adjustments in the signal conditioning are provided to linearise the output, and calibrate the sensors for a specific target material. It is the purpose of this paper to assess current sensor calibration methods and highlight the potential for error in practical situations. To solve the problem, a method of in-situ calibration is presented, which uses the short range positional accuracy of the machine as a reference and least squares best fit of a low order polynomial. Validation is provided through the use of calibration test results and a practical example.

Keywords: Non-contact measurement, eddy current sensors, calibration methods, spindle error measurement

1. Introduction

Spindle error measurement is becoming increasingly important to machine tool users in high precision manufacturing, as they look to charcterise their spindle performance capabilities.

The requirement to measure and quantify these errors has resulted in the production of an international standard. ISO 230 part 7 'Geometric accuracy of axes of rotation' specifies test procedures and test equipment type for determining the effects of spindle errors [1].

The standard describes tests for measuring spindle radial, axial and tilt error motion which can be categorised into synchronous and asynchronous error motion. Synchronous errors are once per revolution and essentially result in the out of roundness of the spindle rotation. Asynchronous errors are non-repeating with spindle rotation and are directly responsible for part surface finish quality.

1.2. Non-contact sensing technology

The standard suggests the use of non contact measurement to analyse the spindle behaviour. The ability to monitor a target without physical contact offers several advantages over contact measurement, including the ability to achieve higher measurement resolution and increased dynamic response to moving targets. They are also virtually free of hysteresis and there is limited risk of damaging fragile targets because of contact with a measurement probe. Three typical non contact measurement sensing technologies (as suggested by ISO) are [2]:

- Capacitance
- Eddy Current
- Laser triangulation

Each of these technologies has its advantages and disadvantages for use in spindle analysis [3]. Laser triangulation sensors have a large offset so there is a reduced risk of damage during setup. However, they are susceptible to environmental influences such as humidity and material in the gap between the sensor and the target.

Capacitance sensors can achieve nanometer resolutions with high accuracy measurement however; they too are affected by use in dirty environments where dust, oil and coolant may be present.

Eddy current or inductive sensors, unlike the other sensors, are not affected by material in the gap between the sensor and the target and so are well adapted to use in hostile environments. Although the spindle in Fig.1 is not typical, it does show a typical manufacturing environment where spindle analysis might take place.

Fig. 1. Typical Manufacturing Environment

When compared to the other sensing technologies, eddy current sensors are also the cheaper option. The sensors used in this paper are from a simple low cost system. They are however not without their disadvantages and it is the purpose of this paper to address these disadvantages and offer a simple solution through the use of real case studies.

2. Eddy current technology

Eddy current sensors operate on a principle based on Lenz's law [4]. Most eddy current sensors are constructed with a sensing coil, which is a coil of wire in the head of the probe. When an alternating coil is passed through the coil it creates an alternating magnetic field. When a metallic target is present in this magnetic field the electromagnetic induction causes an eddy current in the target material in a perpendicular plane to the magnetic field of direction. This induced eddy current generates an opposing magnetic field which resists the field generated by the sensing coil. The interaction of the two magnetic fields is sensed using electronics and converted into an output voltage that is directly proportional to the distance between the sensor and the target.

Fig. 2. Principle of eddy current sensor [5]

As previously mentioned, eddy current sensors are not without their disadvantages. The main challenge is that the output changes with the use of different target

materials. This leads to a requirement for careful calibration of the sensors.

2.1 Sensor calibration

It is normal practice that the sensors are calibrated by the manfacturer before they are delivered to the customer. Each sensor is calibrated to its own individual signal conditioning unit, these cannot be interchanged between sensors without effecting the output. The length of cable used to connect the sensor to the signal conditioning unit also has an effect on the output, so once calibrated cannot be exchanged for a cable of differing length. The main challenge for high-end specifiction eddy currenet sensors with linear outputs in the region of 0.2% is that they must be calibrated to a specific target material. There is a significant difference in the output of the sensors when used with a ferrous target compared to a non-ferrous target and as such the sensors are usually calibrated to either one or the other.

This leaves room for a certain amount of measurement uncertainty between materials of a similar nature. For example the output from an aluminium alloy target would be different to that of pure aluminium. Some sensor manufactures provide a method for self calibration for fine adjustment [6]. However, if it is necessary to use the sensors on both ferrous and non-ferrous target then either a new set of sensors will be required or they will continuously need to be sent back to the manufacturer for re-calibration to a new target material.

Alternatively low cost eddy current sensors are available that are not calibrated to a specific target material. However, the output from these sensors is very non-linear and as such requires careful calibration.

From practical experience, it is not always possible to use the same target material when performing spindle measurements. In many cases different machines will have differing test bar material. In certain situations it may be necessary to perform a measurement against a component mounted in a spindle.

With this in mind a simple in-situ method for calibrating the sensor to the necessary target material is extremely beneficial. The next section of this paper offers a possible solution to this challenge.

3. Impementation and results of new calibration method

For the calibration of the eddy current sensor to a specific target material to take place, a linear profile of the output must first be established. This paper proposes setting up the sensor and required target material on a machine tool to measure the output when the machine is moved in one micron steps over the range of the sensor.

The linear scale of a machine tool is specified with linear errors on a typical modern machine in the region of $5\mu/m$, so when calibrating over 0.1 mm, any error can be considered negligible. There is a possibility of stiction error when moving the machine in small increments but experience shows that this is also negligible. This is not necessarily true for older machines and as such, calibration, where possible, should take place on newer machine tools.

To ensure the stability and accuracy of the measurement the following is taken into consideration:

- The accuracy and responsiveness of the axis over the region to be used must first be established by standard methods.
- Before any measurement is taken, it is important to move the axis of the machine over the region in which the measurement is to be taken in to ensure oil flow to the linear guide ways to counter the stiction.
- All calibration measurement is taken when approaching the target from the same direction so that no reversal error is introduced; an axis over-run [7] before the test ensures unidirectional calibration.

The measurement range of the eddy current sensor is 500μm. However, only the first 100μm was required for the case study in this paper. This was due to the most sensitive section of the sensor appearing in this part of the range. The blue trace in Fig. 3 shows the output over the first 100μm and as can be seen the output is very non-linear.

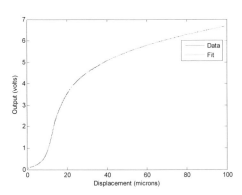

Fig. 3. Comparison of eddy current sensor output and calibration

The output data file from this test was then input into Matlab and a curve fitting tool used to fit to the data. A low order polynomial was used to ensure robustness of the fit and to enable the fit to be performed to within 0.1μm the curve was chopped into sections. The end result is a calibration file for each individual part of the curve to within 0.1μm and can be seen from the green trace in Fig. 3.

The data in Fig. 3 was captured using a 16-bit National Instruments (NI) data acquisition device and a bespoke computer application which uses standard NI APIs. The low order polynomial calibration file can then be loaded into the software whenever the sensor is being utilised with the target material for which it has been calibrated.

With the calibration file input into the windows application the repeatability of the calibrated sensor can be tested. The axis was moved in steps of 10 μm in the same direction from the same starting position for five separate runs and the deviation from linearity plotted in Fig. 4. As can be seen the sensor measured repeatably to within 0.2 μm over the 100μm range.

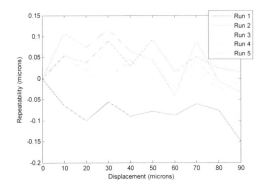

Fig. 4. Repeatability reuslts for eddy current sensor output

4. Practical validation

To ensure the calibration method worked in a practical application, a dyanmic measurment was taken using a high resoultion piezo platform flexure rig. The eddy current sensor was set up to measure against the same target and the software loaded with the calibration file captured on a standard milling machine using the proposed method. A Renishaw XL80 laser interferometer was used as a traceable reference device (see Fig. 5).

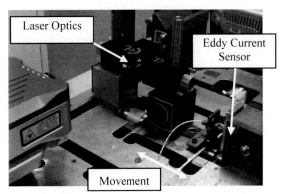

Fig. 5. High resolution piezo platform test setup

The rig was set to vibrate at 100Hz and the magnitude of the oscillation adjusted to be approximately 0.008 μm. Figure 6 shows the output from the Renishaw XL80 laser interferometer. There are some imperfections of the sine wave but this is to be expected at this level of resolution and can be attributed to the stability of the setup.

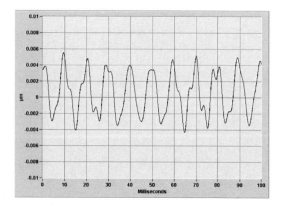

Fig. 6. Renishaw XL80 laser output on vibration rig

Figure 7 shows the output from the eddy current sensor sampling at 10 kHz. As is expected at this resolution there is some noise on the output signal but a sine wave can still clearly be seen at a magnitude of between 0.008μm and 0.01μm.

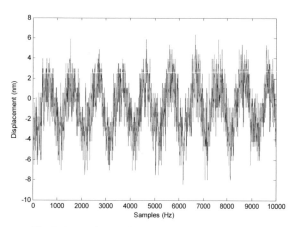

Fig. 7. Output from eddy current sensor on vibration rig

The standard deviation of the data in Fig. 7 shows the noise level to be at approximately 0.0037 μm. Since the data was sampled at 10 kHz averaging could be applied to clean up the signal, if this does not impinge on the required sampling frequency of the application.

5. Conclusions

The need of the manufacturing industry to understand machine tool capabilities and error sources has led to the production of an international standard for measuring spindle error motion. Measuring a spindle dynamically

requires the use on non-contact displacement sensors. This paper provides evidence to support the use of eddy current non-contact sensors for use in a manufacturing environment and assesses current calibration methods. The problem of needing to calibrate the sensors while working on the shop floor is highlighted. A new methodology for in-situ calibration of low cost eddy current sensors is presented which allows the sensor to be calibrated to a specific target material, while in the field, to a linear accuracy of 0.3μm and a repeatability of 0.2μm. A practical validation of the sensors is described and demonstrates the good resolution capability in the nanometre range. Due to the non-linear output of such sensors, they are very sensitive over a small range. This offers other measurement capabilities for the sensor, such as vibration monitoring.

Acknowledgements: This resource is the result of a Knowledge Transfer Partnership. The KTP project is a collaboration between Machine Tool Technologies Ltd and University of Huddersfield and was co-funded by Technology Strategy Board and Northwest Regional Development Agency

References

[1]   ISO 230-7, Test code for machine tools – Part 7: Geometric accuracy of axes of rotation.2006.

[2]   ISO 230-3, Test code for machine tools – Part 3: Determination of thermal effects.2007.

[3]   Wilson, JS. "Sensor technology handbook", Elsevier, 2005.

[4]   Nabavi, MR "A novel interface for eddy current displacement sensors", IEEE transactions on instrumentation and measurement, vol. 58, no.5, May 2009.

[5]   Lion Precision, "Difference between capacitive and eddy current sensors", TechNote LT05-0011, 2009.

[6]   Lion Precision, "Calibration of eddy current sensors", Eddy current TechNote LT02-0013, 2007.

[7]   ISO 230-2, Test code for machine tools – Part 2: Determination of accuracy and repeatability of positioning numerically controlled axes.2006.

[8]   Lai. Y, "Eddy current displacement sensor with LTCC technology," Ph.D. dissertation, Universität Freiburg, Switzerland, 2005.

[9]   Tian. G. Y, Zhao. Z. X and Baines. R.W, "The research of inhomogeneity in eddy current sensors", Sens. Actuators A, Phys., vol. 69, no. 2, p148– 151, Aug. 1998.

[10] Zhang, H. "An approach of eddy current sensor calibration in state estimation for maglev system", Electrical machines and systems, p1955-1958, 2007.

[11] Rao, BPC. "Practical eddy current testing", Oxford: Alpha Science, 2007.

Information required for dimensional measurement

J. Horst[1], C. Brown[2], K. Summerhays[3], R. Brown[4], L. Maggiano[5], and T. Kramer[6]

[1] NIST, 100 Bureau Drive, MS 8230, Gaithersburg, MD 20899 USA

[2] Honeywell FM&T, P.O. Box 419159, Kansas City, MO, 64083 USA.

[3] MetroSage, 26896 Shake Ridge Road, Volcano, CA 95689 USA.

[4] Mitutoyo America, 965 Corporate Blvd., Aurora, IL 60504 USA.

[5] Mitutoyo America, 965 Corporate Blvd., Aurora, IL 60504 USA.

[6] NIST, 100 Bureau Drive, MS 8230, Gaithersburg, MD 20899 USA.

Abstract. Much work has been done by standards organizations to model dimensional measurement information in digital formats, but the work done has revealed that standard digital formats from upstream processes are currently insufficient to enable the automatic generation of a fully-specified measurement plan for many use cases. In order to assist upstream standards committees, this paper summarizes information required for dimensional measurement, and in particular, identifies which information elements are missing in the upstream standard formats. The upstream information required to perform a dimensional measurement parses logically into four classes of information: design information, quality directives, measurement resources, and measurement rules. A high level description of the information in these four categories is presented, with emphasis on the information modeling work which needs to be performed to enable more efficient and less error prone dimensional metrology processes.

Keywords: Dimensional Metrology, Quality Measurement Systems, Geometric Dimensioning and Tolerancing, Computer-Aided Design, Information Standards

1. Introduction

Model-based design and manufacturing is on the rise, even while paper-based approaches persist in many dimensional metrology departments. If metrology departments are to benefit from a model-based approach, dimensional measurement information models, in digital formats (*i.e.*, computer-readable formats), are needed. Furthermore, *standard* information models are needed to enable more efficient and less error-prone quality measurement processes, and to enable greater agility and freedom of choice for metrology systems end users.

Standards organizations, such as the Dimensional Metrology Standards Consortium (DMSC), have been developing dimensional measurement information standards [1]. This information modeling effort has revealed that *upstream standard information models lack definitions for certain key information elements* [2].

Upstream activities feeding into the dimensional measurement activity, as shown in Fig 1, fall into a few categories: 1) design information describing the part (size and shape constraints), 2) quality management information about the part (*e.g.*, part name, part number, version, author, security requirements, measurement priorities), 3) measurement resource information (equipment and software availability and capability to perform the measurement on the part), and 4) measurement rules information (details on how to perform the measurement). When this information is complete, the quality measurement system is then able to determine if individual parts are within specified tolerance. Measurement trends can also be identified which can be used to improve the manufacturing process.

Current digital Computer-Aided Design (CAD) standards do not contain information sufficient for all types of measurements. For example, the information defined in the CAD standards does not allow full semantic association with other pieces of information, and the types of features common in these CAD formats lack some of the function-based features needed in dimensional metrology. Consequentially, a large amount of intensive engineering effort is required at the quality measurement planning workstation to define new features and make important associations.

Much of the design information required for automated dimensional measurement has already been defined within a variety of human-readable standards for geometric dimensioning and tolerancing (GD&T) (*e.g.*, ASME Y14.5 [3], ASME Y14.1 [4], ISO 1101 [5]), and some companion CAD+GD&T digital standards (*e.g.*, ISO 10303 STEP AP203 [6]). These digital standards generally focus on "design features," requiring "measurement features" to be identified downstream, usually on a paper drawing or within dimensional measurement planning software. Information added in "balloons" to drawings specifying critical information (such as key characteristics and criticality levels) for complete inspection via quality requirements is also not well defined in standard digital formats.

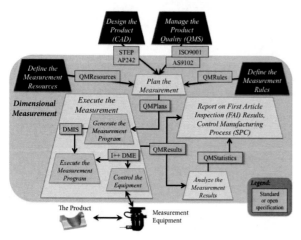

Fig. 1. Dimensional measurement activities are shown with data flow connections from four upstream activities (in black): Quality Management Systems (QMS), Computer-Aided Design (CAD), measurement resources, and measurement rules. Various standards (and open specifications) enabling data exchange between those activities are shown in gold. All standards with the prefix "QM," are part of the DMSC's Quality Information Framework (QIF) [2].

In the case of upstream Quality Management Systems (QMS) (as in Fig. 1), modeling is often in non-digital (paper-based) formats. QMS standard paper-based formats exist (*e.g.*, AS9102 [7], PPAP [8]), but all software for handling digitally formatted data employs *proprietary* exchange formats. Standard, digital QMS formats are needed to perform automated first-article inspections and other important quality manufacturing functions, such as in-process and on-machine inspection.

There are no standards, either human-readable or digital, for either measurement rules or measurement resources (see Fig. 1), but standard information models are in the planning phase within the DMSC [2].

2. Metrology-Relevant digital cad standards

There are currently two types of digital CAD standards, "lightweight" and "heavyweight." Though there is increasing overlap between the capabilities of both types of standards, the lightweight standards target visualization use cases and the heavyweight standards target CAD model development, manufacturing process development, and long-term archival use cases.

Heavyweight standards include the suite of related ISO 10303 (STEP) standards (AP242 [9], AP203 [6], and AP214 [10]). Lightweight standards include the ISO 32000-1 Portable Document Format (PDF) [11] and ISO/PAS 14306 (JT, after the original Jupiter format) [12] standards. AP242 (under development) is intended to

supersede AP203 and AP214. AP242 is planned to be a richer digital representation of existing paper-based standards for GD&T, such as ASME Y14.5 and ISO 1101. Therefore, the timing is right to define all upstream information necessary to perform dimensional metrology correctly and cost-effectively.

3. Requirements for CAD + PMI data

Beyond part geometry, the additional information required to make and measure a part has been called Product Manufacturing Information (PMI). From the standpoint of dimensional metrology, GD&T information is the most critical and complex component of PMI. We will address GD&T mostly, while also briefly discussing metrology-critical non-GD&T information.

3.1. Requirements for information models of GD&T

Correct and complete GD&T definitions and assignments enable the manufacture of parts that fit into assemblies, perform required functions, allow ease of production, assembly and measurability, and enable cost-effective manufacturing. With precisely and richly specified GD&T, design errors have the potential of getting fixed before anything is manufactured or delivered to the customer.

Correct and efficient use of GD&T is hindered when 1) a GD&T description is incorrectly defined, and so does not enable fit and function as required, 2) a correctly defined GD&T specification is incorrectly implemented by downstream processes like dimensional measurement, or 3) digital GD&T standards are incompletely specified or incompletely implemented. This paper largely addresses this third hindrance.

Successful measurement planning and execution requires access to all necessary CAD+PMI information on a part, which must include geometry, topology, features, tolerances, dimensions, datum references, and the precise associations between these elements, *e.g.*, a feature will have dimension, tolerance, and datum annotations associated with it.

Here are some key modeling requirements dimensional metrologists have articulated for upstream digital CAD+PMI standards developers [13]:

- Define "tolerance features" (not merely "geometric features") which can include multiple boundary-representation CAD faces. Tolerance features are not, in general, the same thing as design and/or manufacturing features, though there are significant overlaps.
- Carefully define associations between tolerances and tolerance features and, where appropriate, between their relevant reference features and datums & datum reference frames (DRFs). For example, in the case of dimensions like

"distance/angle between," the association of the tolerance would be to a *pair* of tolerance features.

- Model critical feature characteristics (*e.g.*, non-GD&T PMI) as commonly communicated via association to numbered "balloons" on drawings.
- Model datum definitions tailored for quality measurement and not just tailored for production.
- Represent certain features as datum features, with the added possibility that multiple features may constitute a single (so-called compound) datum.
- Allow definition of aggregates of datums in the form of DRF. For datum features that are not nominally perpendicular to the datum feature of higher precedence, reference to a basic angle is required. The order of precedence (primary, secondary, tertiary) for these datum features in the definition of the full DRF must be clearly specified.

These associations must be strong, *i.e.*, the full association must be retrievable without requiring further human intervention downstream.

How well these requirements are met by the data structures of various standard CAD+PMI formats, and particularly the in-progress ISO 10303 AP242 standard, has not been determined, but is seen as an important consideration for any efforts beyond the short-term.

3.2. Requirements for non-GD&T PMI

Some part information needed for measurement planning and execution, but not directly related to a part's physical dimensions, such as hardness, material type, color, cosmetic features, security markings, section views, reflectance, opacity, and local surface anomalies (*e.g.*, waviness, roughness, and discontinuities [14]), needs precise definition. Also, for the appropriate interpretation of measurement results, feature and part labeling conventions are needed to provide traceability of downstream interface boundaries. Finally, there will likely be a call for the definition of unconstrained "notes" or "annotations," though this is not recommended and tends to be abused, and may hinder interoperability.

3.3.   Requirements for QMS

For dimensional measurement plans to be generated automatically from a QMS, the quality management directives must contain the following information:

- First-article inspection plans, both for performing measurements and reporting them.
- The location within the measurement process of each measurement, *e.g.*, in-process, *in situ*, and/or post process [2].
- The quantity of parts in a batch.
- Statistical process requirements for each feature.
- Measurement schedules as part of the overall production scheduling process.

- Specific feature with associated tolerance, which may be specially required for FAI, for example.
- Specific levels of criticality associated with each element within a set of parts or a set of features.
- Specific gages (or classes of gages) that are needed to perform a dimensional measurement.
- Measurement criteria, consisting of the manufacturer's unique rules and regulations on measurement for measurement verification and/or product acceptance.
- List of quality standards (external or internal) or guidelines required for compliance.
- Required instructions for inspecting parts, including tolerances and dimensions to be verified and acceptable pass/fail errors.
- A plan specifying the quality requirements for a part (or parts), typically coming from an (external or internal) quality standard.
- A reaction plan for each of several possible measurement outcomes.

4.   Requirements for measurement rules

Manufacturers commonly have internal standards, called measurement rules, which specify details of *how* to perform each type of measurement on each type of characteristic feature. Here are a couple of elements anticipated in a measurement rules information model:

- A set of measurement probe sampling plans specifying point locations and densities for all types of features for a given tolerance specification and measurement system. On the part level, a sampling plan describes the frequency of measuring manufactured parts or the quantity of parts to be measured in a batch.
- A specification for the method of transformation from measured points to substitute geometric elements, *e.g.*, lines, planes, and cylinders, including directions for handling outlier points.

5.   Requirements for measurement resources

Descriptions of available measurement resources, their operational status, their calibration history, and their capabilities such as throughput, accuracy, and measurement volume, are all critical factors in the planning of any measurement. This is a very rich and layered information domain, and no attempt at a complete presentation will be attempted here. A brief sampling of major data categories will serve to illustrate the scope involved for coordinate metrology.

- Specific types of measurement devices: laser trackers; Cartesian, serial-link coordinate measuring machines (CMMs); articulated, serial-link CMMs (*e.g.*, arms); and parallel-link (*e.g.*, Stewart-platform-based) CMMs.

- Specific types of sensors for surface sampling: contact sensors (switching, analog, and scanning); non-contact sensors (laser, video, capacitance-based, *etc.*) and their geometric relationships to the basic measuring device.
- Accessories pertinent to set-up and/or operation: fixturing tools, rotary tables, probe heads, tool changers, and their geometric relationships to the basic measuring device.
- Operational environmental conditions and performance specifications for applicable combinations of all of the above. (*e.g.*, conditions related to temperature, humidity, atmospheric pressure; load-limits, working volume dimensions, speeds and accelerations; performance test and calibration results).
- Software characteristics (*e.g.*, collision avoidance in data gathering, error compensation schemes, feature fitting algorithms).

6. Conclusions

There is no single CAD+PMI standard presently available with the level of robustness desirable for data transfer from CAD to dimensional measurement. Of the various candidates for a non-proprietary format, ISO 10303 AP203, and its successor AP242, offer the most plausible route to success. This is in large part because major CAD vendors have already implemented AP203 1st Edition. It is anticipated nonetheless that AP203 2nd Edition will not meet all the requirements for automated quality measurement planning. The following are recommendations for CAD+PMI and dimensional measurement standards committees:

- Assess how well standard CAD+PMI formats (like AP242) satisfy requirements herein stated, 1) to ensure data integration and 2) to ensure work done higher upstream will reduce costs.
- Offer these requirements on robust data structures to developers on the AP242 committee.
- Distribute these proposed requirements to dimensional metrologists for refinement.
- Evaluate ISO 10303 AP242 by mapping GD&T-related aspects of this application profile to the data structures of certain GD&T information models that relate GD&T data to pertinent CAD shape geometry representations.
- Monitor the two "lightweight" CAD+PMI standards, PDF and JT, and advise their standards development committees regarding the requirements specified in this research.
- Merge results into a final requirements document.
- Ensure that final requirements be integrated by ISO 10303 developers into AP242.

- Ensure that pilot demonstrations occur.

A successful solution to this problem can be expected to bring significant benefits [15], for if these standards are widely implemented in commercial software and if certified implementations are mandated by end users, then "dimensional metrology ready" CAD standards will 1) facilitate the establishment of sound metrology data structures within proprietary CAD+PMI software, 2) permit robust metrology checking (*e.g.*, ensure that tolerancing is complete, consistent and unambiguous) at the design level, where errors can be caught and corrected at minimal expense, 3) greatly reduce development costs for communications of key data between CAD systems and quality measurement planning software, 4) reduce corresponding licensing costs to end-users, and 5) realize the digital factory, where every piece of important information is represented in a computer-readable format.

References

[1]   Squier BH, (2009) History of the DMIS standard and the Dimensional Metrology Standards Consortium. CMM Quarterly:4

[2]   Zhao Y, Brown R, Kramer T, Xu X, (2011) Information Modeling for Interoperable Dimensional Metrology. 1st ed. London: Springer

[3]   ASME (2009) Y14.5-2009: Dimensioning and Tolerancing

[4]   ASME (2003) Y14.41-2003: Digital Product Definition Data Practices

[5]   ISO (2004) ISO 1101: Geometrical Product Specifications (GPS) -- Geometrical tolerancing -- Tolerances of form, orientation, location and run-out

[6]   ISO (2011) ISO 10303-203: Configuration controlled 3D design of mechanical parts and assemblies

[7]   SAE (2004) SAE AS9102A: Aerospace First Article Inspection Requirement

[8]   AIAG (2006) PPAP-4: Production Part Approval Process

[9]   ISO (2011) ISO/NP 10303-242: Application protocol: Managed Model-based 3D Engineering

[10]   ISO (2010) ISO 10303-214: Core data for automotive mechanical design processes

[11]   ISO (2008) ISO 32000-1: Document management – Portable document format – Part 1: PDF 1.7

[12]   ISO (2011) ISO/PAS 14306: Industrial automation systems and integration -- JT file format specification for 3D visualization

[13]   IMTI (2006) A Roadmap for Metrology Interoperability. IMTI, Inc.

[14]   Henzold G, (2006) Geometrical Dimensioning and Tolerancing for Design, Manufacturing and Inspection. 2nd ed. Oxford, UK: Elsevier

[15]   Horst JA, (2009) Reduce costs and increase quality with information exchange standards for manufacturing quality. CMM Quarterly:12

Rapid Prototyping

Material switching system for multi-material bio-structure in projection micro-Stereolithography

Kwang-Ho Jo, In-Baek Park, Young-Myoung Ha and Seok-Hee Lee
School of Mechanical Engineering, Pusan National University

Abstract. Projection microstereolithography(PµSL) has some advantages such as precision of several µm, fast fabrication time, and simple fabrication process for complete 3D micro-structure. However, the application of PµSL has limitation due to the liquid material, compared with other additive manufacturing (AM) technology such as fused deposition modeling (FDM) and selective laser sintering (SLS). In this study, we propose the material switching system (MSS) to fabricate a multi-material bio-microstructure in PµSL. The system consists of three control modules; resin level control (RLC), resin dispensing control (RDC), vat level control (VLC). To evaluate the performance of the MSS, the accuracy of resin level has been measured before and after resin exchange. Then, several fabrication methods to reduce the error due to the MSS have been presented. For the application of this system, biodegradable and biocompatible materials have been synthesized and the mechanical and curing properties have been investigated. Several bio-structures of multi-material such as scaffold and transdermal drug delivery system (TDDS) have been fabricated.

Keywords: Projection microstereolithography, Material switching system, bio structure,

1. Introduction

In microstereolithography (µSL), which is based on conventional stereolithography, a commercial photocurable resin or diluted (for low viscosity) photocurable resin is used [1]. The merit of a single resin is simplification of the fabrication process, but this also makes it difficult to fabricate a microstructure with high functionality and strength. Recently, µSL technology has been applied to various research fields such as biomaterials, functional materials, and biodevices [2-4]. Studies have been carried out to develop biocompatible materials [5], nanoparticles to enhance strength [4, 6], and high-strength materials including organic matter [7]. Applications of µSL include the fabrication of biodegradable microdevices, microactuators, and microlenses [2].

Projection µSL (PµSL), a type of µSL, achieves fast fabrication with a simple system configuration [4]. The photocurable resin for PµSL is usually a mixture of several monomers and a photo initiator or a commercial resin with a diluent blended in for low viscosity. Such diluted commercial resin has weak material properties, and so some substances such as nanoparticles and organic matter are mixed in. However, it is hard to fabricate a homogeneous microstructure due to the difficulty of content control.

In this paper, we propose a material switching system (MSS) to enable exchange of the photocurable resin in the PµSL system. A microstructure with various materials can be fabricated through the use of the MSS. Parts under local stress can be strengthened through the use of an appropriate material. When a constant pattern or microstructure array is fabricated, various distinct materials can be applied to each structure. In scaffold fabrication for bioengineering, the period of biodegradation can be controlled.

2. Material Switching System(MSS)

2.1. Projection microstereolithography

PµSL is divided into free-surface and constrained-surface methods according to the generation of the liquid layer [5]. In the free surface method, each polymerized layer is made by a patterned beam focused on the resin surface; thus, building is easier than with the constrained surface method. However, it is difficult to control the layer thickness due to the waiting time for leveling the resin surface. In the constrained surface method, a transparent window is used to generate a liquid layer, so it is easy to control the layer thickness. However, a cured layer can stick to the transparent window, which will partially or totally destroy the microstructure.

Figure 1 (a) shows the free-surface-type PµSL system used in this study. The flowchart for the microstructure fabrication is shown in Fig. 1(b). A 3D CAD model is converted into the STL file format and sliced into layers with a constant thickness. A series of 1-bit cross-sectional images corresponding to each layer is then generated. Each cross-sectional image is transferred to a digital micromirror device (DMD) for the generation of the dynamic light pattern. The light emitted from the light source is shaped into a cross-sectional image by the DMD. The patterned beam is then focused onto the resin surface through several optical devices. A liquid layer is polymerized by the patterned beam, and the layer is then completed. The z-stage moves up and down to generate the next layer. Due to the viscosity of the resin, some waiting time is required for leveling the resin surface. After the new liquid layer is stabilized, the above procedure is repeated until the top layer is completed.

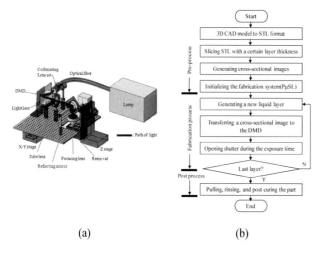

(a)   (b)

Fig. 1. PµSL system using free-surface method: (a) schematic diagram, (b) flowchart of conventional fabrication process

2.2 Material Switching System(MSS)

The MSS is used for the exchange of photocurable resin in the PµSL apparatus. Figure 2 shows the configuration of the MSS. Using this system, a photocurable resin can be exchanged during fabrication, and the microstructure array can be produced with different resins. When the vat is filled with a new resin, the gap between the resin surface and top layer of the built part should be the layer thickness exactly. For array fabrication, the level of the new resin should be the same as the first layer of the built microstructure. To meet the requirements, the system consists of three modules: resin level control (RLC), vat level control (VLC), and resin dispensing control (RDC).

The level of the resin surface in the vat is measured by the RLC module, which includes a CCD camera (Pro 9000, Logitech Korea Ltd., Seoul, Korea) that supports a video call of 30 frame/sec and two laser pointers with adjustable sizes. After a circular laser beam is illuminated onto the resin surface, the beam is reflected and formed on the focus film fixed at the z-stage. At this time, the shape of the formed beam is an ellipsoidal image due to the laws of reflection. The image on the CCD camera

fluctuates due to the influence of the resin flow. Hence, two laser pointers are used to compensate for the fluctuation.

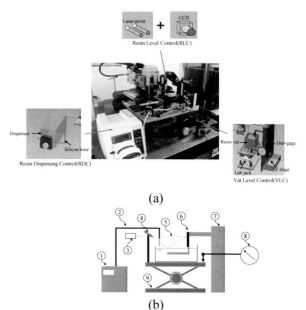

(a)

(b)

Fig. 2. Material switching system (MSS): (a) photograph; (b) schematic diagram showing 1) dispenser 2) silicon tube 3) CCD 4) laser pointer 5) beam path 6) focus film 7) z-stage 8) dial gauge and 9) electronic labjack

Figure 3 shows the principle of measurement for the resin level. The coordinate of the midpoints for two ellipsoidal images obtained by the CCD camera indicates the current level of resin. During this procedure, the system is blocked with an acrylic cover to prevent external interference.

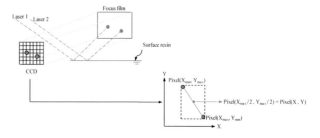

Fig. 3. Measurement of resin level

The level of the vat is measured and controlled in the VLC module. This module includes a stepping motor, electronic labjack, and dial gauge with a resolution of 10 µm. This system is used to return the vat to its lowered position for exchanging the resin. The dial gauge is used to check and control the vat position in real time. Even if the vat position is different from the previous one, the level of the new resin in the vat is controlled by the RLC and RDC modules. Hence, the fabrication for the new resin is accomplished.

The RDC module is used to provide a liquid resin during the resin exchange. It consists of a controllable peristaltic pump (WT600-1F, Longerpump Co.) with a flow rate of 0.7 ml/min, a silicon tube, and a 33 gauge syringe (ILS Co, Stützerbach, Germany). The amount of provided resin is controlled by the measured value in the RLC module.

2.3. Fabrication procedure

Figure 4 shows the procedure for the fabrication of a microstructure in MSS-based PμSL. The schematic for the resin exchange procedure is shown in Fig. 4 (b). After finishing the fabrication for the first resin, the current level of the resin measured by the RLC module is stored as pixel(x,y). The current vat position is set as zero in the dial gauge, and the labjack lowers the vat position. The first resin is removed, and the vat is cleaned and dried. The empty vat is moved onto the labjack and lifted up to the zero position in the dial gauge. The RDC module is used to fill the vat with a second resin. The level of the second resin is stored as pixel(x′,y′) and compared with pixel(x,y) in real time. When the two values are the same, the RDC module ends the filling process. Afterwards, the fabrication of the second resin is resumed.

Fig. 4 Fabrication procedure: (a) flowchart, (b) schematic diagram for changing resin

3. Photocurable resins

In this study, three photocurable resins constituted of a photoinitiator and several monomers were used. Isobornyl acrylate (IBOA) is a monofunctional monomer with the viscosity of 8 cps at room temperature and low shrinkage. Next, 1,6-hexanediol dimethacrylate (HDDA) is a monofunctional monomer with the viscosity of 7 cps at room temperature and high reactivity. Bisphenol-A-ethoxylated (4) diacrylate (BP40) is used to provide strength and low shrinkage. Tri(propylene glycol) diacrylate (TPGDA) is a trifunctional monomer with the viscosity of 121 cps at room temperature and high flexibility after curing. Trimethylolpropane triacrylate (TMPTA) is also a trifunctional monomer with the viscosity of 100 cps, fast cure response, and resistance against abrasion. As a photoinitiator, 2, 2-dimethoxy-2-phenylacetophen one (DMPA) was used. Table 1 shows the components and weight ratios of each resin.

Table 1 Resin composition

Resin	Monomers (weight ratio)	PI
IHB	IBOA : HDDA : BP40 = 8:1:1	
IHTPA	IBOA : HDDA : TPGDA = 6:2:2	5 wt.%
IHTMA	IBOA : HDDA : TMPTA = 6:2:2	

4. Fabrication examples

For the complicated microstructure shown in Fig. 5, the fabrication precision for the proposed MSS in PμSL was compared with the conventional method.

Figures 5 (a) and (b) show the respective results for the two methods. In Fig. 5 (b), the body part and crown part were fabricated with IHB and IHTMA resins, respectively. In both microstructures, the size of the crown part is almost same. However, the total height shown in Fig. 5 (b) was taller than that shown in Fig. 5 (a) by about 20 μm due to an error in the junction part.

Table 2. Fabrication conditions for Fig. 5

	Resin	Exposure_E. (mJ/cm^2)	Layer_T. (μm)	Temp. (°C)	Exposure_T (sec)
(a)	IHB				1
(b)	IHB, IHTMA	16.4	10	35	1.5

Figure 6 shows a scaffold applied to tissue engineering. The scaffold is a latticed microstructure and has to be fabricated with high connectivity between grids. Three types of resins were used to fabricate the scaffold shown in Fig. 6 (a).

Table 3 Fabrication conditions for Fig. 6

Resin	Exposure_E. (mJ/cm^2)	Layer_T. (μm)	Temp. (°C)	Exposure_T. (sec)
IHB	16.4			1
IHTPA	13.2	10	35	1.5
IHTMA				1.5

The support structure was designed to release the residual resin in the inside of the scaffold and to reduce the deformation of the lattice due to capillary force. The hole in the support is a space for the proliferation of a cell, and it also increases the connectivity ratio inside the scaffold. The grid structure on the support was fabricated with IHTPA and IHTMA. For precise fabrication of the grid shape, the still-motion method was used in PμSL. As a result, the scaffold was fabricated with an accurate shape, as shown in Fig. 6 (a). The parts of IHTPA and IHTMA were also fabricated exactly, as shown in Fig. 6 (b).

(a)

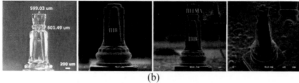

(b)

Fig. 5 Fabrication of micro-king in PμSL using
(a) conventional method, (b) MSS

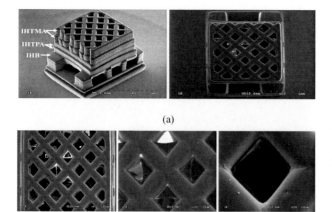

(a)

(b)

Fig. 6 Fabrication of scaffold using MSS: (a) isometric and top views, (b) magnified lattice

5. Conclusion

A material switching system (MSS) for projection microstereolithography (PμSL) was proposed that allows the use of multiple materials in microstructure fabrication. A single microstructure, microstructure array, and complex microstructure were fabricated with single and multiple materials. Although the fabrication precision of the MSS is slightly lower than that of the conventional method, the system can be applied to various research fields such as bio- and micromechanical engineering. In particular, when a microdevice is fabricated from a single material, resistance to local stresses is difficult to achieve without modifying the original design. On the other hand, using multiple materials with different properties in MSS-based PμSL enables microfabrication without having to change the design. If the proposed system is applied to scaffold fabrication in bioengineering, the biodegradation time of the scaffold can be controlled. However, the system also has disadvantages such as a longer fabrication time due to the resin exchange and lower precision due to the error in leveling the resin surface. This problems can be resolved by using the constrained-surface method in PμSL, which improves control in generation of the liquid layer.

Acknowledgement: This research was supported by Basic Science Research Program through the National Research Foundation of Korea(NRF) funded by the Ministry of Education, Science and Technology(No.2011-0010790)

References

[1]   Ha, Y. M., Park, I. B., Lee, S. H., (2010) Three-dimensional Microstructure Using Partitioned Cross-sections in Projection Microstereolithography, International Journal of Precision Engineering and Manufactureing, Vol. 11, No. 2, pp. 335-340

[2]   Park, I. B., Ha, Y. M., Kim, M. S., and Lee, S. H., (2010) Fabrication of a micro-lens array with a non-layered method in projection microstereolithography, International Journal of Precision Engineering and Manufacturing, Vol. 11, No. 3, pp. 483-490.

[3]   Lee, S. D., Choi, J. W., Park, I. B., Ha, C. S and Lee, S. H., (2007) Improvement of mechanical properties of UV-curable resin for high-aspect ratio microstructures fabricated in microstereolithography," Journal of Korean Society Precision Engineering, Vol. 24, No. 12, pp. 119-127.

[4]   Varadan, V. K., Jiang, X., Varadan, V. V., (2001) Microstereolithography and other fabrication techniques for 3D MEMS, John Wiley & Sons, Chichester.

[5]   Lu, Y., Mapili, G., Suhali, G., Chen, S., and Roy, K., (2006) A digital micro-mirror device-base system for the microfabrication of complex, spatially patterned tissue engineering scaffolds, Journal of Biomedical Material Research Part A, Vol. 77A, No. 2, pp. 396-405.

[6]   Mai, Y. M., Yu, Z. Z., (2006) Polymer nanocomposites. CRC, NewYork.

[7]   Lungu, A., Mejiritski, A., Neckers, D. C., (1998) Solid state studies on the effect of fillers on the mechanical behavior of photocured composites," Polymer, Vol. 39, No. 30, pp. 4754-4763.

Turbine blade manufacturing through Rapid Tooling (RT) process and its quality inspection

Aamir Iftikhar, Mushtaq Khan, Khurshid Alam, Asim Nisar, Syed Hussain Imran and Yasar Ayaz
School of Mechanical and Manufacturing Engineering, National University of Sciences and Technology, Islamabad, Pakistan

Abstract. Rapid Prototyping (RP) technologies have played vital role in product development and validation. Another aspect of RP is Rapid Tooling. The development and manufacturing of conventional tools (die & molds) takes considerable amount of time towards production. Rapid Prototyping technologies could be used to shorten the development time of these tools. Rapid tooling is the process of manufacturing tool for customized and small batch production of a product. Some times it is also used as "reduced time to market" strategy as it has become vital for new products to reach the market as quickly as possible.In product development, time pressure has been a major factor in determining the direction of the development and success of a new product. This research work focuses on the development of turbine blade through RT technique. Three methods of Rapid Tooling techniques are considered here i.e. with Room Temperature Vulcanization (RTV) silicon mold, Polyurethane mold and Plaster mold. Master pattern was developed using Stereo lithography and Fused Deposition Modeling (FDM) process. The quality inspection of both master patterns was done by using Coordinate Measuring Machine, in which the profile curves at different blade heights were inspected. The surface finishes of both the master patterns were analyzed using pertho meter surface tester. The SLA master pattern was selected and used for RTV, Polyurethane and Plaster mold. This selection was based on less deviation from CAD model of the turbine blade. The wax blades are then used for investment casting. Quality inspection was perfomed at prototype pattern, wax and metals stage of the product development.

Keywords: Rapid Tooling, RTV Mold, Polyeurathan Mold.

1. Introduction

The conventional method of developing tool (punch and cavity) takes more time to be ready for production [1]. Rapid Prototyping (RP) and Rapid Manufacturing (RM) technologies have shown to shorten the development time of these tools [2]. Rapid Tooling (RT) is the process of manufacturing tool for customized, small batch or trail production of a products [3].

In effect, the rapid tool produces production parts in the period between the launch of production tooling and the receipt of finished goods [5]. When applying rapid prototyping to tooling solutions, the goal is to decrease the time to produce the tool and capture the market before competitors. Presently the trend in the industries is to use short production runs in the manufacture of products [6]. As a result, product development cycle has to be shortened. This is where rapid prototyping aid the industry in delivering products quickly to the market. In recent years, advancement in RP technologies has led to considerable amount of research activities in the area of rapid tooling (RT). The basic idea of RT is to produce prototypes parts by using prototype tools so that parts truly represent the future production. Using the actual material product in real time situation, any modification or change in design and function can be detected well before mass production; it is also a good way to get customer feedback [7]. Methods of RT are reletively new [8] and thus needs validation before it can become part of the production process.

Three different types of RT molds are considered in this research work, that includes Room Temperature Vulcanizing (RTV) [4], Polyurethane (PU) and Plaster Mold (PM). These molds are developed using indirect method of tool development. Wax patterns obtained from these mold were inspected using reverse engineering techniques for their qualification. Final casting was perfomed in metal and metal part was inspected at three different stages during the product manuacturing process.

2. Procedure

2.1. RT processes

Figure 1 presents a flowchart of the part development process through indirect tool development method. Quality inspection of parts at prototype, wax and metal stage were carried out as indicated by A, B and C in Fig. 1.

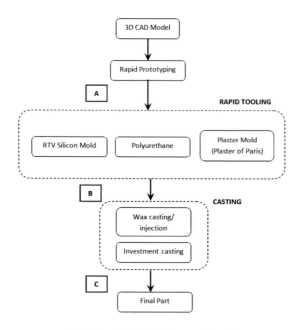

Fig 1. Rapid Tool Development Flow Chart

2.2. Selection of CAD data.

The part considerd for this research work was a turbine blade. A 3D CAD model of a turbine blade was taken, for which rapid tool was to be developed. The reason for selecting turbine blade was that it had complex curved profiles, which can be achallenge for any manufacturing process. The overall size of this bale was 30.2 x 44.2 x 88.5 mm.

2.3. Development of master pattern.

As indirect RT method was adopted, there was requirement of master pattern for tooling development [9, 10]. Two master pattern were developed through Stereolithography (SLA) and Fused Deposition Modeling (FDM). Quality inspection of the surface of both these patterns was perfomed using Coordinate Measuring Machine (CMM) (DEA, Global Image) as shown in Fig. 2. Surface roughness of these pattern was also measured using Pertho-meter. Based on the quality of surface, SLA pattern was selected for further processing.

3. Development of rapid tooling and casting

3.1. RTV silicone rubber

Room Temperature Vulcanizing (RTV) silicone mold was developed by placing the pattern in a box with gate and riser attached to the pattern. A parting line was marked with red marker and liquid silicone, ESIIL 296 with its hardener of ration 10:1, was poured over the pattern. The box was then placed in an air circulating

heating oven for curing for about 10 to 12 hours. After curing, the mold was cut and pattern was extracted and cavities were packed for wax casting. Differential pressure vacuum casting machine was used for wax casting in RTV mold. The wax pattern for RTV silicon mold is shown in Fig. 3.

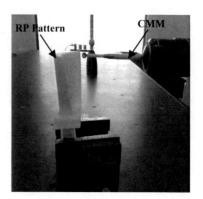

Fig 2. Blade profile measurement Through CMM

Fig. 3. Wax pattern of turbine blade through RTV silicon Mold

3.2. Polyurethane Mold (PU)

In PU mold, the key structure is made around the pattern. The purpose of this mold key structure is to make punch and cavity self alignment and locking system. This key structure also acts as parting line of the mold. The pattern with this structure is placed in a box with in-gate attached. This method of mold making is also called two half mold. Liquid PU silicone, heated at 80°C to 100°C, is poured over the patter in the box. The box is placed in vacuum chamber for degassing to remove the air being trapped in PU. The box is then placed in oven at 70°C for about 1 hour for curing. As the one half is cured, the mold box is opened and inverted to back side. After curing of single side mold the key structure is removed. The mold is again packed in the box with rear side upwords. Again liquid PU is poured, cured and baked. After both sides are cured, the pattern is remove and then mold is baked for 1 hour at 100°C. This PU mold (Fig. 4) can be used in wax injection machine. The hardness of this PU (88HRA) was almost double the RTV Silicone mold (42 HRA).

Fig. 4. Wax pattern by PU Mold

3.3. Plaster Mold (PM)

Plaster Mold (Fig. 5) is made following the same steps as in PU mold described earlier. In PM mold the same mold key structure is made and placed in a box and plaster of paris mixture is poured over it. After settling of material the mold is dried in oven at 100 ^{0}C for approximately 10 hours to remove moisture.

Fig 5. Wax pattern through Plaster Mold

3.4. Investment casting of wax patterns

The molds developed in RTV silicon, PU and PM were used to make wax patterns. After obtaining wax pattern, the blades were casted in Aluminium. Fig 6 shows different stages of the casting process.

Fig 6. Investment casting of parts

4. Results and discussion

4.1. Quality Inpection at Stage A

Results from quality inpection of five different sections of SLA and FDM pattern are presented in Fig. 7. The values shown are the difference between the patterns and the 3D CAD model. The manufactured SLA pattern shows less deviation from the original 3D CAD model as compared to FDM model both in concave and convex surface of the blade profile. The convex surface presented a higher deviation in both SLA and FDM patterns. The higher deviation in FDM parts can be associated with the expansion and contraction of the polymer during heating and cooling cycles in the process. Also, the average surface roughness of SLA pattern was found to be 0.866 μm and that of FDM was 1.322μm. Therefore, SLA pattern was selected and used for mold manufacturing.

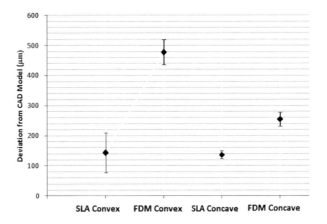

Fig 7. Quality inspection data by CMM

4.2. Quality inpection at stage B

3D digitizer was used for quality inspection of wax patterns to see the percentage of data with in the defined limit tolerance. The over all tolerance range selected was ± 0.5mm. Table 1 shows comparison of measurements at wax stage for RTV, PU and PM molds. RTV mold has overall better results (for concave and convex surface) than PU and PM, where more then 98% of data is in the defined tolerance. The convex side of the PU and PM molds are less 80% in the desired limtis. The high deviation for PU mold is because of air entrapment in the material during the processing stage which resulted in spot/bubbles on convex surface. whereas in PM mold, shrinkage of plaster due to evaporation of moisture was uncontrollable. Also, PM mold could not be clamped tightly as it was fragile and required a lot of care.

A. Iftikhar , M. Khan, A. Nisar, S. H Imran, Y. Ayaz

Table 1. Comparison of data with all RT methods at wax stage

S.No	RT Mold	Wax concave	Wax convex
1	RTV	98.3%	98.5%
2	PU	93.3%	73.4%
3	PM	89.4%	76.25%

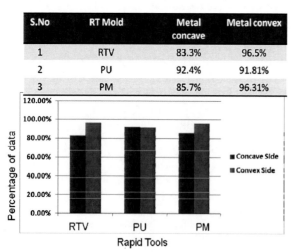

4.3. Quality Inpection at Stage C

The results for metal parts after casting were observed to be better for all three patterns where PU mold presented consistent results for both concave and convex surface of the blade (Table 2.). But taking into account the deviations at the wax stage for PU and PM molds, the total deviation of for these two molds would be unacceptable for manufacturing of shapes similar in complexity to the turbine blade. For the existing materials and method presented, RTV silicon mold presented an overall less deviation from the 3D CAD model.

Table 2. Comparison of data with all RT methods at metal stage

S.No	RT Mold	Metal concave	Metal convex
1	RTV	83.3%	96.5%
2	PU	92.4%	91.81%
3	PM	85.7%	96.31%

5. Conclusions

Three different types of Rapid Tools were developed and analyszed. The pattern manufactured through SLA was selected for better surface finish and high accuracy as compared to FDM pattern. The wax patterns were successfully produced with all three tools. Quality inspection was made by 3D non-contact digitizer. It was observed that RTV mold has better accuracy than Polyurethane mold and Plaster mold, that is more than 98% of the data was with in the tolerance. The wax patterns from Rapid tools were investment casted in Aluminum and inspected at this stage. For accurate patterns in short lead time for prototype production, RTV tool is preferred because of its high accuracy.

References

[1] Sadehg Rahmati, Phill Dickens, Rapid Tooling   analysis of Stereo lithography injection mould tooling, International Journal of Machine Tools & Manufacture 47 (2007) 740-747

[2] Rahmati Sadegh, Rezaei Mohamad Reza,   Akbari Javed, Design and Manufacturing   of a wax injection tool for Investment casting using rapid tooling, Tsinghua Science and Technology, ISSN 1007-0214 18/38 pp 108-115.(2009)

[3] Haiou Zhang, Guilan Wang, Yunhua Luo, Takeo Nakaga, Rapid hard tooling by plasma spraying for injection molding and sheet metal forming, Thin Solid Films 390 (2001) 7-12

[4] M. Vaezi, D. Safaeian, C.K. Chua, (Gas Turbine Blade Manufacturing by Use of Epoxy resin tooling and silicone rubber molding techniques). (2009)

[5] User Guide to rapid Prototyping, Todd Grim. Rapid Prototyping Association of SME. @ 2004.

[6] Prasad K.D.V.Yarlagadda, Ismet P. Iiyas, Periklis Chritodoulou, Development for sheet metal drawing using nickel electroforming and streaolithograpy processes, Journal of Materials Processing Technology 111 (2001) 286-294

[7] Rapid Manufacturing, An Industrial Revolution for the Digital Age, N. Hopkinson, R.J.M Hague, P.M. Dickens. Jhon Wiley & Sons, Ltd. 2006

[8] Wohlers Report 2010, Terry Wohlers.

[9] Rapid Prototyping, Principles and Applications, Chua C. K, Leong K.F, Lim C.S. World Scientific, 2005

[10] Rahmati Sadegh, Rezaei Mohamad Reza,   Akbari Javed, Design and Manufacturing   of a wax injection tool for Investment casting using rapid tooling, Tsinghua Science and Technology, ISSN 1007-0214 18/38 pp 108-115.(2009).

Mechanical characterization of polyamide porous specimens for the evaluation of emptying strategies in Rapid Prototyping

Andrea Cerardi, Massimiliano Caneri, Roberto Meneghello and Gianmaria Concheri
DICEA-Lin (Laboratory of Design Tools and Methods in Industrial Engineering) - University of Padova - Via Venezia, 1 - 35131 - Padova - Italy

Abstract. Rapid-prototyping is usually considered as a powerful tool in geometric and functional optimization of a product. In such an approach, focus is exclusively on the real component, and not on the rapid prototype which represents it. This work takes part on a wider study which focuses on a rational, systematic approach in obtaining an "optimized rapid prototype", with particular regard to emptying strategies without loss of structural and geometric properties. In detail, a set of tensile tests have been performed on different types of specimen, which reproduce a set of corresponding emptying strategies: each type of specimen is characterized by a different percentage of porosity (40, 60 and 80%), obtained by the combination of particular values of two parameters: reticular structure and its density. As a result, the correlation between mechanical strength and geometric structure has been evaluated, allowing the identification of a profitable emptying strategy, in terms of cost and weight.

Keywords: Rapid Prototyping, Selective Laser Syntering, Polyamide Tensile Properties, Light Weight Structures, Cellular Structures

1. Introduction

Rapid Prototyping (RP) allows to obtain end-use parts directly from three-dimensional CAD models. RP is widely used in the biomedical field, in particular for the fabrication of external prosthesis and internal implants [1], due to the ability of the process to manufacture parts with complex geometry, performing both in terms of geometrical requirements and mechanical properties.

Given the innovative characteristics of these technologies, during the design phase of both prosthesis and implants the minimization of the weight and the optimization of the material volume distribution are often overlooked, which deeply influence the behavior of the medical device and thus the patient lifestyle.

A design approach to this problem is to compose designed structures using cellular elements [2][3][4]. Moreover the resulting structures ensure the reduction of implant stiffness and the mismatch between prosthesis and human bone stiffness, which can have a positive influence on the quality of lifestyle [5]. The mechanical properties of the final structures depend on their relative density (e.g. porosity of the structure) and on the shape of the unit-cell. As stated in [6, 7], polyamide is used as a matrix in Selective Laser Sintered porous bioactive structures, which are adopted in biomedical applications.

The present work is focused on the selective laser sintering (SLS) process of polyamide. This process seems to be the most versatile among RP processes [8] and polymers are of the most promising laser sintering materials [9]. Aim of this work is to investigate mechanical behavior of specimens characterized by three different cellular shapes and different percentages of porosity. Results are presented in terms of elastic properties and tensile strength.

2. Method

The investigation consists in three steps:

1. designing the CAD model of several tensile specimens, composed by different cellular elements, and characterized by different percentage of porosity;
2. producing them by SLS process;
3. performing a mechanical characterization.

2.1. Unit-cell design

Three different shapes of lattice structures were designed and modelled in SolidWorks 2011 (Fig. 1). The topology of lattice structures has been chosen to obtain a predominant axial stress in their truss members: in such a way, when loaded, they grant superior mechanical performance compared to stochastic foams [3,10].

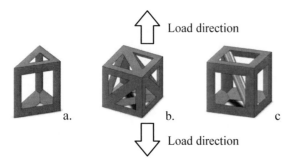

Fig. 1. Unit-cell: a. first structure topology,
b. second structure topology, c. third structure topology
and orientation of lattice structures with respect to load direction

Height and length of all cells were set to 5mm. According with [11] cell truss members had triangolar cross-sections of thickness t (Fig.1 and Fig. 4). The porosity of the structure is defined as follows:

$$\overline{\rho} = 1 - \frac{V_{ls}}{V_{env}} \qquad (1)$$

where is the volume of lattice structure and is the total volume occupied by the solid envelope.

To control the porosity of the structure only thickness (t) of cross-section has been changed; in particular, 40%, 60% and 80% porosity values were set for each structure topology. Fig. 2 shows an example of the change in the resistant area A* over the height of the different unit-cell for the specific porosity values of 40%.

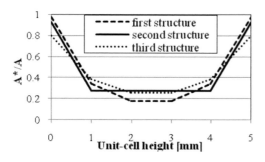

Fig.2. Difference in distribution of the material
over the height of the different unit-cell structures
for the porosity value sets to 40%.

2.2. Specimens design

Different kinds of specimens were designed to evaluate mechanical properties.

Unit-cells designed were assembled in SolidWorks to obtain the shapes of test specimens illustrated in Fig. 3, where the most important dimensions are summarized. The specimens consist in a regular 3D array of unit-cell, along X, Y and Z directions. The lattice structures were oriented like in Fig. 1, with respect to global load direction (represented by the arrows).

2.3. Specimens fabrication

Five specimens for each configuration (structure topology and porosity) were fabricated, resulting in a total of 45 specimens. Moreover 3 standard specimens (0% of porosity value) were manufactured to obtain a mechanical characterization of the original material.

All of the parts were manufactured according to the CAD templates by 3Dfast Srl on a Formiga P 100 system (EOS GmbH) in polyamide EOSINT P/PA2200 [Fig. 4 (a), (b)]. Plastic material had an average particle size of 60μm.

Dimensions of the final parts were verified using an OGP Smartscope flash CNC 300 optical system [Fig. 4 (c)]. The average error between dimensions of manufactured parts and virtual models, due to manufacturing accuracy, are smaller than 2% in the neck of truss members, while some material accumulations can be noticed at the corners.

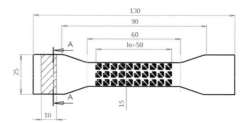

Fig. 3. Shape of specimen for tensile test

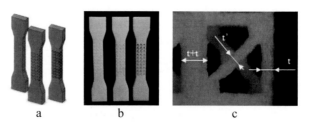

Fig. 4. Examples of a. CAD specimens,
b. fabricated parts, c. measured part

2.3. Mechanical characterization

Fig. 5. Tensile test configuration

The tensile properties are determined according to standard ISO 527 1-2:1993 using the Galdabini SUN2500 uniaxial testing machine; force and elongation were evaluated with a load cell and a Galdabini MICRON extensometer, until the failure of each specimen. The gage length is 50mm and the testing speed was set to 2mm/min. An example of test configuration is shown in Fig. 5.

3. Experimental results

Tensile stress is calculated according to equation (3320.2):

$$R = \frac{F}{A_{\min}} \qquad (2)$$

where R is the tensile stress in MPa, F is the measured force in N, and $A_{\min}$ is the minimum area in the specimen which depends on shape and porosity of the unit-cell (Fig. 2). The resulting stress-strain curve was used to estimate Young's Module of the structure and the tensile strength.

Fig 6 shows examples of curves resulting for specimens realized using the first structure at 40%, 60%, 80% porosity rates, compared to the standard specimen (0%). Mechanical properties obtained from standard specimens match the data provided by the material producer [12]. All specimens exhibit ductile failure except specimens having a porosity value of 80%. These specimens present curves with sets of steps, which highlight a progressive failure of beams in unit-cells.

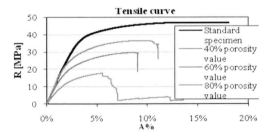

Fig. 6. Examples of test results for first structure topology at various porosity values

In Fig. 7 and 8, the estimated values of mean, maximum and minimum tensile stiffness and tensile strength for different porosity rates are plotted. It can be seen that increasing the porosity value of the manufactured structure led to a decrease in both mechanical parameters for all kind of geometries. Moreover both tensile properties exhibit a linear dependence for porosity value between 40% and 80%. It means that it's possible to formulate a linear predictive model for the porosity-stiffness and porosity-tensile strength relationships for each structure, as follows:

$$E = C1 \cdot \overline{\rho} + C2 \qquad (3)$$

$$R_m = C3 \cdot \overline{\rho} + C4 \qquad (4)$$

where C1, C2, C3 and C4 [MPa] are constants that depends on geometry (topology and porosity rate), on the orientation of the unit-cell with respect to load direction and on manufacturing parameters. For tested geometries, values of these constants are summarized in Table 1. Constants evaluated can be used in the future to approximate behavior of porous structure without mechanical testing.

Table 2, 3 and 4 summarized mean results and standard deviation obtain for the test specimens for each kind of configuration designed.

The analysis highlights that best tensile properties were observed for the first structure that presents beams parallel to the load direction. Worst properties were obtained for the third structure when porosity values are lower than 60% and for the second structure if porosity is bigger than 60%.

Table 1. Costants for porosity-stiffnessand porosity-tensile strength relationships

	First structure	Second structure	Third structure
C1	-1381	-1289	-695
C2	1926	1657	1230
C3	-38	-42	-20
C4	50	47	32

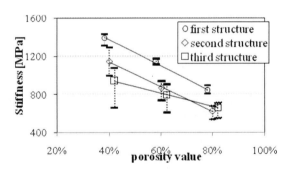

Fig. 7. Mean, maximum and minimum tensile stiffness at various porosity values

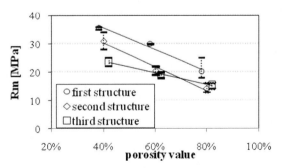

Fig. 8. Mean, maximum and minimum tensile strenght at various porosity values

Table 2. Tensile properties for the first structure

Porosity value [%]	E [MPa]	F [N]	Rm [MPa]
40	1392 ± 70	1727 ± 24	36 ± 0.6
60	1145 ± 30	804 ± 7	30 ± 0.01
80	839 ± 50	243 ± 49	20 ± 4.0

Table 3. Tensile properties for the second structure

Porosity value [%]	E [MPa]	F [N]	Rm [MPa]
40	1143 ± 207	1857 ± 250	31 ± 4.2
60	867 ± 80	860 ± 55	21 ± 1.3
80	624 ± 60	250 ± 24	14 ± 1.5

Table 4. Tensile properties for the third structure

Porosity value [%]	E [MPa]	F [N]	Rm [MPa]
40	941 ± 168	1540 ± 70	23 ± 1.1
60	795 ± 114	750 ± 39	19 ± 0.8
80	663 ± 63	262 ± 11	15 ± 0.9

4. Conclusion

Aim of this work is the mechanical characterization of several polyamide specimens, manufactured by SLS RP method, and differentiated by porosity rate and basic cellular topology. In particular, three unit-cell geometries were adopted to manufacture different tensile specimens. For each structure topology, three different rates of porosity have been adopted (40, 60 and 80%), varying a unique geometric parameter (t, thickness of beams of unit-cells), and for each couple of unit-cell and porosity rate the experimental tensile curve was plotted. For both 40 and 60% porosity rates, experimental curves highlight ductile failure, while for 80% rate successive single beam failures led to a set of steps in tensile curve at high strains. Tensile strength and stiffness showed a similar trend, depending on combination of unit-cell and porosity rate adopted: in particular, the effect of porosity rate is bigger in the first and the second geometries, and the worst mechanical properties can be observed in the third structure for a porosity value smaller than approximately 70%, and in the second structure for bigger values. Conversely, the first structure features the best mechanical performances: it can be noticed that in this structure beams are parallel to load direction. Experimental results have been summarized in a predictive model: for both tensile stress and stiffness a

linear dependency on porosity rate has been observed, and a mathematical formulation has been proposed.

Acknowledgements: The work is part of the work included in the research projects:
"CPDA105479 – Università degli studi di Padova, Progetti di Ricerca di Ateneo, Bando 2010" and
"POR CRO 2007-2013 Regione del Veneto Azione 1.1.2 - INNPRO-DENT".
The authors acknowledge the support of 3Dfast Srl in providing specimens.

References

[1] Goodridge RD, Tuck CJ, Hague RJM, (2012) Laser sintering of polyamides and other polymers. Progress in Materials Science 57:229-267
[2] Vesenjak M, Krustolovic-Opara L, Ren Z, Domazet Z, (2010) Cell shape effect evaluation of polyamide cellular structures. Polymer Testing 29:991-994
[3] Moongkhamklang P, Elzey DM, Wadley HNG, (2007) Titanium matrix composite lattice structures. Composites: Part A 39:176-187
[4] Ryan G, McGarry P, Pandit A, Apatsidis D, (2009) Analysis of the Mechanical Behavior of a Titanium Scaffold with a Repeating Unit-cell Substructure. Journal of Biomedical Materials Research Part B: Applied Biomaterials 90:894-906
[5] Harrysson OLA, Cansizoglu O, Marcellin-Little DJ, Cormier DR, West II Harvey A, (2008) Direct metal fabrication of titanium implants with tailored materials and mechanical properties using electron beam melting technology. Materials Science and Engineering C 28:366-373
[6] Zhang Y, Hao L, Savalani M M, Harris R A, Tanner K E, (2007) Characterization and dynamic mechanical analysis of selective laser sintered hydroxyapatite-filled polymeric composites. Wiley InterScience. DOI: 10.1002/jbm.a.31622
[7] Savalani M M, Hao L, Zhang Y, Tanner K E, Harris R A, (2007) Fabrication of porous bioactive structures using the selective laser sintering technique. Proceedings to IMechE Vol. 221 Part H: J. Engineering in Medicine
[8] Kruth JP, Levy G, Klocke F, Childs THC, (2007) Consolidation phenomena in laser and powder-bed based layered manufacturing. Annals of the CIRP 56/2:730-759
[9] Goodridge RD, Shofner ML, Hague RJM, McClelland M, Schlea MR, Johnson RB, (2011) Processing of a Polyamide-12/carbon nanofibre composite by laser sintering. Polymer Testing 30:94-100
[10] Cansizoglu O, Harrysson O, Cormier D, West II Harvey A, Mahale T, (2008) Properties of Ti-6Al-4V non-stochastic lattice structures fabbricated via electron beam melting. Materials Science and Engineering A 492:468-474
[11] Chua CK, Leong KF, Cheah CM, Chua SW, (2003) Development of a Tissue Engineering Scaffold Structure Library for Rapid Prototyping. Part 2: Parametric Library and Assembly Program. The International Journal of Advanced Manufacturing Technology 21:302-312
[12] EOS GmbH. [Online] [Cited: November 11, 2011.] http://eos.materialdatacenter.com/eo/de

Welding

Increasing the tolerance to fit-up gap using hybrid laser-arc welding and adaptive control of welding parameters

C. M. Allen, P. A. Hilton and J. Blackburn
TWI Ltd., Granta Park, Gt. Abington, Cambridge, CB21 6AL, United Kingdom
Copyright © TWI Ltd 2012

Abstract. Adaptively controlled hybrid laser-arc welding has been demonstrated using a 5kW 6mm.mrad Yb fibre laser. ISO 13919-1 class B (stringent) quality butt welds have been made in 4mm Al alloy, 6mm stainless steel and 8mm thickness steel plates. A laser vision sensor, used to track the joints robotically during welding, sends joint fit-up information (gap width and mismatch height) to a controller. This adjusts, in real time, the welding parameters to increase the net tolerance of hybrid welding, particularly to joint gap. Stringent quality welds can be made over a wider range of joint fit-up cases than using fixed conditions.

Keywords: laser, arc, hybrid, weld, gap, mismatch, tolerance, tracking, adaptive, control

1. Introduction

Laser welding requires precise workpiece fit-up and accurate alignment of the beam with the joint line, if high quality welds are to be made. The emergence of high brightness, fibre-delivered, lasers, with ever tighter focusable beams, makes attention to these requirements even more critical. Nevertheless, the demands that these requirements put on welding fixtures, edge preparation methods and fit-up tolerances can act to dissuade potential users of laser welding technology, particularly if large components or fabrications, assembled from a number of smaller sub-components, are to be welded.

Hybrid laser-arc welding offers a potential solution, being more tolerant to joint fit-up, owing to the wire feed addition from the arc. Nevertheless, notwithstanding the benefit of higher welding speeds with the hybrid process, this tolerance does still not match that of arc welding. In addition, material preparation and welding fixtures costs are higher, if weld defects are to be avoided.

In the current work, the capacity of a laser vision sensor to relay details of joint fit-up to the welding equipment, enabling real time adaptive control of parameters to cope with changes in joint fit-up, has been evaluated. Square edged butt joints in 4mm Al alloy or 6mm stainless steel, as well as 8mm thickness S355 steel

(in this case with a broad root-faced butt joint configuration) have been adaptively welded as test cases. Previously, adaptive control of hybrid welding has only been reported for butt joints between steel plates [1,2] using CO_2 or Nd:YAG lasers.

2. Experimental method

Robotic welding trials were carried out using a Kawasaki FS-060L robot with an IPG 5kW Yb fibre laser and ESAB synergic MIG/MAG arc welding equipment. Suitable shielding gas/wire combinations of Ar/1.2mm A18 (for steel), Ar-2%O_2/1.2mm 308LSi (for stainless) or Ar/1.2mm 5356 were used. The weld qualities achieved were evaluated with respect to ISO 13919-1 or -2 [3,4]. For the highest quality (class B) butt welds, in a plate thickness, t, these standards require:

- That the welds are free of cracks;
- The porosity content is ≤0.7% (steels) or ≤3% (Al), and of a maximum diameter ≤0.3t (max. 2mm);
- Weld root and cap undercut is ≤0.05t (max. 0.5mm (steels) or 1mm (Al);
- Excess penetration (root) and excess weld metal (cap) ≤0.15t+0.2mm (max. 5mm).

Test welds were made along 300mm long joints, with either perfect fit-up, or with varying amounts of joint gap (0-2mm) and/or joint mismatch (0-2mm). Following trials making fixed changes to welding parameters, to identify which gave the greatest fit-up tolerance, real-time adaptive changes in the robot speed or position, or the arc welding parameters, were made automatically, by feeding joint fit-up information, from a Servo-Robot Digi-I/S seam tracking sensor as shown in Fig. 1, to control programs coded in to the welding hardware.

S. Hinduja and L. Li (eds.), *Proceedings of the 37th International MATADOR Conference*,
DOI: 10.1007/978-1-4471-4480-9_9, © Springer-Verlag London 2013

Fig. 1. Laser vision sensor, ahead of hybrid welding head. Reproduced courtesy of TWI Ltd

3. Results

Figures 2, 3 and 4 show cross-sections of the Class B weld profiles made over close fitting, flush, butt joints, with the optimum conditions developed at the start of the welding trials for each of the three materials. These welds were made at 4.5, 2.5 and 1.6m/min, respectively; using wire feed rates of 9, 8 and 7m/min.

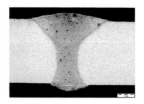

Fig. 2. Optimum hybrid weld in 4mm Al Alloy (Reproduced courtesy of TWI Ltd.)

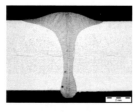

Fig. 3. Optimum hybrid weld in 6mm stainless steel (Reproduced courtesy of TWI Ltd.)

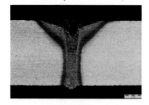

Fig. 4. Optimum hybrid weld in 8mm steel (Reproduced courtesy of TWI Ltd.)

Joints with gaps of up to 2mm in width, and/or mismatches (or hi/lo) of up to 2mm in height, were then welded with these same sets of conditions. This demonstrated that these conditions were only optimum for welding flush, close fitting joints. Figure 5, 6 and 7 (not to same scale) show selected cross sections from these experiments on the Al alloy, stainless steel and steel, respectively. Class B weld profiles were achieved up to a mismatch of 0.3mm or a gap of ~0.3mm, when welding the 4mm Al alloy, 0.5mm or 0.5mm,

respectively, when welding the 6mm stainless steel, and 0.6mm or 0.3mm, respectively, when welding the steel.

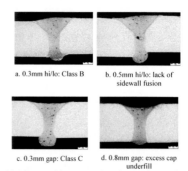

a. 0.3mm hi/lo: Class B   b. 0.5mm hi/lo: lack of sidewall fusion

c. 0.3mm gap: Class C   d. 0.8mm gap: excess cap underfill

Fig. 5. Al Alloy welds over a, b. mismatches and c, d. gaps (Reproduced courtesy of TWI Ltd.)

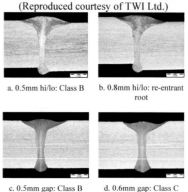

a. 0.5mm hi/lo: Class B   b. 0.8mm hi/lo: re-entrant root

c. 0.5mm gap: Class B   d. 0.6mm gap: Class C

Fig. 6. Stainless steel welds over a, b. mismatches and c, d. gaps (Reproduced courtesy of TWI Ltd.).

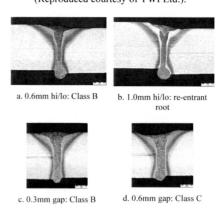

a. 0.6mm hi/lo: Class B   b. 1.0mm hi/lo: re-entrant root

c. 0.3mm gap: Class B   d. 0.6mm gap: Class C

Fig. 7. Steel welds over a, b. mismatches and c, d. gaps (Reproduced courtesy of TWI Ltd.).

As Figs. 5-7 show, the weld cap under-fill became unacceptably deep as the gap increased. Conversely, as mismatch increased, the weld root toe blend angles became re-entrant on to the higher plate and/or radiography detected lack of sidewall fusion. One or more fixed changes in the welding conditions were then tried, to improve the weld profile along joints with mismatch or gap, including:

- Reducing the welding speed;
- Increasing the wire feed rate and/or arc voltage trim;

- Changing the stand-off height of the welding head;
- Deliberately off-setting the laser beam off of the joint line (by off-setting the welding head).

Table 1. Optimum changes for 4mm Al alloy (Reproduced courtesy of TWI Ltd.)

Change	Result
Reducing head stand-off by 2mm, with increasing mismatch	Lack of fusion avoided but root still re-entrant on higher plate
Wire feed rate increased (e.g. to 13m/min), with increasing gap	Tolerance limit increases from ~0.3 to 1mm

The optimum results from these trials are summarised in Tables 1-3, for the three materials, respectively.

Table 2. Optimum changes for 6mm stainless steel (Reproduced courtesy of TWI Ltd.)

Change	Result
Welding speed reduced to 2m/min, with increasing mismatch	Tolerance limit increases from 0.5 to 0.7mm
Wire feed rate increased (e.g. to 13m/min), with increasing gap	Tolerance limit increases from 0.5 to 1mm

Table 3. Optimum changes for 8mm steel (Reproduced courtesy of TWI Ltd.)

Change	Result
Welding speed reduced to 1.2m/min, and head stand-off by 4mm, with increasing mismatch	Tolerance limit increases from 0.6 to 0.8mm
Welding speed reduced to 1.2m/min and wire feed rate increased (e.g. to 9m/min), with increasing gap	Tolerance limit increases from 0.3 to 0.5mm

These results suggested that appropriate changes in welding conditions would double, at least, the tolerance to joint gap, and give rise to more modest improvements in mismatch tolerance when butt welding the steel and stainless steel plates. The mismatch tolerance of the aluminium butt joints did not appear to be improved, but reducing the welding head stand-off did at least avoid lack of sidewall fusion defects.

To confirm the positive effects of these changes, adaptively controlled trials were carried out on butt joints where the joint preparation, in terms of gap and mismatch, increased linearly as the weld progressed. To accomplish this, appropriate control responses were programmed in to the welding equipment controllers connected to the Kawasaki robot and the arc welding equipment. These control programs then implemented appropriate changes in the welding parameters automatically, as a function of the joint gap and mismatch detected by the Servo-Robot seam tracking device.

Some of the most successful results (not to same scale) are summarised in Fig. 8 (for Class B welds, unless indicated otherwise). The overall benefits (in terms of increased tolerance) gained by using adaptive control, particularly in terms of increased gap tolerance, can be seen by comparing Fig. 8 with Fig. 5-7.

Figures 9-11 show the extents to which tolerances have been increased for making either Class B or C welds, including those over a combination of mismatch and gap, for these three different butt joint test cases. In these Figures, the red and green boxes indicate the estimated tolerances of autogenous and hybrid welding without adaptive control, respectively. Bold symbols are from validation experiments carried out in this work. The dotted lines represent the limits for Class B and Class C welds with adaptive control. Other welds were left unclassified ('X').

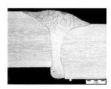 

a. Re-entrant root over 1mm mismatch but lack of fusion avoided, **by reducing head stand-off by 2mm**.

b. 0.8mm gap tolerated (Class C), **by increasing wire feed rate to 13m/min**.

c. 1.3mm mismatch tolerated, **by reducing speed to 2m/min**.

d. 1.3mm gap, **by increasing wire feed rate to 13m/min**.

e. 1.2mm mismatch tolerated, **by reducing welding speed to 1.2m/min**.

f. 0.6mm gap tolerated, **by reducing welding speed to 1.2m/min, and increasing wire feed rate to 13m/min**.

Fig. 8. Adaptively controlled butt welds in a, b: 4mm Al alloy, c, d: 6mm stainless steel and e, f: 8mm steel (Reproduced courtesy of TWI Ltd.)

These current results appear broadly in agreement with previous data using other laser types. The maximum gap tolerance reported in other work is up to 1.5-1.6mm for CO_2 laser-based hybrid welding [2,5], by increasing the wire feed rate by ~55-60%, or to 1.2mm for Nd:YAG laser-based hybrid welding [1], by reducing the welding speed by ~15%.

In terms of tolerance to mismatch, Thomy et al. [6], have reported full fusion through a butt joint between 11.2mm thickness plates with a 1.4mm mismatch using a higher power 15kW fibre laser. However, in this case the weld profile quality class achieved was not reported.

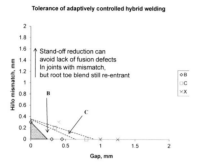

Fig. 9. Increase in fit-up tolerance possible when hybrid welding butt welds in 4mm Al alloy plate with adaptive control (Reproduced courtesy of TWI Ltd.)

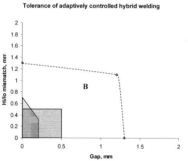

Fig. 10. Increase in fit-up tolerance possible when hybrid welding butt welds in 6mm stainless steel plate with adaptive control (Reproduced courtesy of TWI Ltd.)

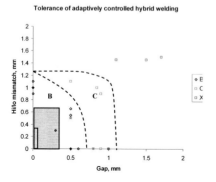

Fig. 11. Increase in fit-up tolerance possible when hybrid welding butt welds in 8mm steel plate with adaptive control (Reproduced courtesy of TWI Ltd.)

4. Conclusions

Adaptively controlled flat position hybrid laser-MIG/MAG butt welding of Al alloy, stainless steel and steel plates has been carried out, for plate thicknesses in the range 4-8mm, using a high brightness 5kW Yb-fibre laser. ISO 13919-1/2 Class B and Class C welds have been made, over joints with a number of different fit-up conditions. The main conclusions of these welding trials are; Class B weld profiles can be produced, at welding speeds of up to 4.5m/min (depending on plate thickness), over perfectly fitting joints. Welding without adaptive control can be used to tolerate joint gaps and mismatches to a limited extent (depending on plate material and thickness). Off-line trials can identify those parameters which, if changed, can increase these tolerances. Using joint fit-up data from a seam tracking sensor, subsequent adaptive control of welding parameters is then useful in combatting unacceptable levels of weld cap underfill in joints over wider gaps, sometimes in excess of 1mm (depending on plate material and thickness). Adaptive control also increase give modest tolerance increases to mismatch, when welding steel or stainless steel butt joints, avoiding re-entrant weld root toe blend angles. Adaptive control is more useful in avoiding lack of fusion defects in aluminium butt joints with mismatch, but does not improve the weld root profiles achieved in those cases.

References

[1]   Shi S G and Hilton P A, 2005: 'A comparison of the gap-bridging capability of CO2 laser and hybrid CO2 laser MAG welding on 8mm thickness C-Mn steel plate', Welding in the World, 49, pp75-87.

[2]   Shi S G, Hilton P A, Mulligan S J and Verhaeghe G, 2005: 'Hybrid Nd:YAG laser-MAG welding of thick section steel with adaptive control', Welding and Cutting, 4, 6, pp345-350.

[3]   BS EN ISO 13919-1:1997: 'Welding - Electron and laser beam welded joints - Guidance on quality levels for imperfections - Part 1. Steel'.

[4]   BS EN ISO 13919-2:2001: 'Welding - Electron and laser beam welded joints - Guidance on quality levels for imperfections - Part 1. Aluminium and its weldable alloys'.

[5]   Kim H S, Lee Y S, Park Y S, Kim J K and Shin J H, 2003: 'Study on the welding variables according to gap tolerance of butt joint in laser hybrid arc welding of carbon steel', LIM Proc pp165-169, 24-26 June, Munich, Publ. D-70331 Stuttgart, Germany.

[6]   Thomy C, Seefeld T, Vollertsen F, Vietz E, 2006: 'Application of fibre lasers to pipeline girth welding', Welding J., 85, 7, pp30-33.

Characteristics of microstructures in laser spot welds of a sintered NdFeB permanent magnet under different welding modes

Chenhui Yi[1], Baohua Chang [1], Chengcong Zhang[1], Dong Du[1], Hua Zhang[1], Yihong Li[2]
[1] Department of Mechanical Engineering, Tsinghua University, Key Laboratory for Advanced Materials Processing Technology, Ministry of Education, Haidian District, Beijing 100084, PR China
[2] Taiyuan Tongli Magnetic Materials Co., Ltd., Taiyuan 030032, PR China

Abstract. Laser spot welds were made on sintered NdFeB permanent magnets under deep penetration and heat conduction modes using 2kW continuous fiber laser. The microstructures of the welds show that the HAZ(heat affected zones) of both welds are rich of cracks caused by the melting of grain boundary phase, and the fusion zones are both composed by $Nd_2Fe_{14}B$ and α-Fe, moreover the nuggets of both welds are composed by submicron ultrafine equiaxed grains of $Nd_2Fe_{14}B$. Differently, the α-Fe columnar dendrites exist in the upper surface of the nugget of deep penetration weld but not on that of heat conduction weld. Besides, Nd-rich phase is found in the periphery of the upper surface of the nugget for the deep penetration spot welding, while it is found in the center of the upper surface of the nugget for the heat conduction spot welding. The reasons that lead to such microstructures are analyzed briefly.

Keywords: laser spot welding, NdFeB permanent magnet, microstructure

1. Introduction

The sintered NdFeB permanent magnet has a wide range of applications in microelectronic devices thanks to its high magnetic energy density[1]. Adhesive bonding and mechanical fastening are currently the most common joining methods for sintered NdFeB permanent magnets. However, there are a lot of shortcomings of the methods above, such as low production efficiency and easily aging for the adhesive bonding, and complicated structure for mechanical fastening, which is unfavorable to the miniaturization and lightweight for the device. So it is necessary to develop a new joining method for sintered NdFeB permanent magnets. Laser welding is one of the most commonly used methods in micro welding thanks to the high quality, high precision, high efficiency and low deformation associated with it[2,3].

Researches on the micro welding of sintered NdFeB permanent magnet and SPCC steel indicate that laser welding can realize an effective connection between the magnet and the steel, but the joint is prone to hot cracks and porosities and other metallurgical defects, and demonstrates low stress brittle fracture[4,5]. Therefore, it is necessary to take an in-depth research on the metallurgical behavior of the sintered NdFeB permanent magnet in laser welding to provide guideline for the manufacture of high strengh joints.

So far, the research on the metallurgical behavior of laser irradiated NdFeB permanent magnet is very little. The existing studies are limited to the modification of surface microstructure of NdFeB permanent magnet by laser scanning, trying to improve the corrosion resistance and magnetic properties[6-8]. However, the research on the microstructure of NdFeB permanent magnet after laser welding has not been reported. In this paper, deep penetration and heat conduction laser spot welding on a sintered NdFeB permanent magnet were carried out, microstructures of the welds were observed and the welds formation mechanism were analyzed, in order to provide reference for the improvement of the laser welding process of sintered NdFeB permanent magnet.

2. Experimental materials and methods

Sintered NdFeB permanent magnets N48 without coating were used in this study. The dimensions of the magnets specimens were 7.5mm×3.5mm×0.7mm, and the chemical compositions are listed in Table 1. A YLS-2000 type continuous fiber laser with maxmium power of 2kW was used in the welding experiments.

Table 1. Chemical compositions of the sintered NdFeB permanent magnet N48(wt%)

Elements	Nd	Fe	B	Pr	Dy	Co	Cu	Nb
Contents	20.63	66.75	1.00	6.88	2.99	1.50	0.15	0.10

The setup for deep penetration and heat conduction spot welding is shown in Fig. 1. The process parameters for deep penetration spot welding were: power P=120W, welding time t=0.05s and defocusing distance z=+1mm. The process parameters for heat conduction spot welding were: power P=60W, welding time t=0.3s and defocusing distance z=+1mm. The surfaces of the magnets were polished with sandpaper prior to laser welding to remove oxide layers. A side-blown argon above the magnet was used to prevent the welds from oxidation during the welding processes, and the argon flow rate was 10L/min. After the welding, the welds were mounted, ground, polished, and then etched with 4% nitric acid alcohol. SEM(scanning electron microscope) was adopted to observe the microstructures and analyze the chemical compositions.

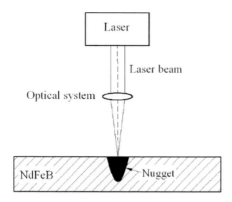

Fig. 1. Schematic diagram of laser spot welding

3. Experimental results

3.1. Cross-section morphologies

The cross-section morphologies of deep penetration spot weld and heat conduction spot weld are shown in Fig. .2. A measurment shows that the depth to width ratio of deep penetration weld is 1.2:1, which is much larger than 0.4:1 of heat conduction weld. It can be seen that the HAZ (heat affected zones) of both welds are rich of cracks caused by the melting of grain boundary phase, while the cracks in deep penetration weld are wider and much longer than those of heat conduction weld.

The melting point of the Nd-rich phase at grain boundaries is much lower than that of the main phase $Nd_2Fe_{14}B$. In the HAZ, the grain boundary phase is melted under laser heating while the main phase is not. Therefore, the HAZ are vulnerable to crack under the solidification stress of weld pool. Because of the bigger molten pool, higher superheat and greater shrinkage stress associated with deep penetration welding, the cracks in the HAZ of deep penetration weld are much more serious. Meanwhile, there are many solidification cracks in the nugget of the deep penetration weld, which extending

from the upper surface to the internal of the nugget. In contrast, the solidification cracks of heat conduction weld exit in the periphery of the upper surface of the nugget, which are shorter than those of deep penetration weld. This is also cuased by the different solidification stresses for different laser welding modes.

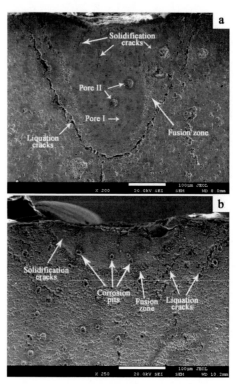

Fig. 2. The cross-section morphologies of deep penetration type spot weld(a) and heat conduction type spot weld(b)

In addition, two different types of porosities were observed in the nugget of deep penetration weld. Type I porosities are small, irregular in shape , and evenly distributed in the nugget. In contrast, type II porosities are bigger, spheroidal in shape, and with smooth inner surface. Sintered NdFeB permanent magnet is a powder metallurgy material, and there are many pores in the base metal. When the base metal melting, the gas in the pores will remain in the weld pool. Because of the fast solidification of weld pool in laser welding, the gas can not completely escape from the molten pool or combine into big bubbles and forms type I porosity. Type II porosities should be caused by the instability of the keyhole, which encapsules the metal vapor, shielding gas and air into the molten pool. Differently, the gas pores in the molten pool of heat conduction welding have enough time to grow and escape because of the longer welding time of heat conduction welding. Moreover, there is no keyhole in the molten pool of heat conduction welding. Therefore, neither types of porosities is observed in the nugget of heat conduction weld.

3.2. Fusion zone microstructures

Fusion zone is the transition part from nugget to HAZ. As can be seen in Fig. 2, the width of the fusion zone of deep penetration weld is uneven, the middle is wider than the bottom. However, the width of the fusion zone of heat conduction weld is basically the same. The higher magnification photos of the fusion zone of two different welds, as shown in Fig. 3, indicate that the fusion zone of deep penetration weld is constituted by three types of grains: the lamellar grains A epitaxially growing from the $Nd_2Fe_{14}B$ grains in HAZ, the cellular crystals B growing on the grains A, and the dendrites C, which are close to the nugget. Differently, the fusion zone of heat conduction weld is constituted by two type of grains: the lamellar grains D epitaxially growing from the $Nd_2Fe_{14}B$ grains of HAZ, and the cellular crystals E growing on grains D.

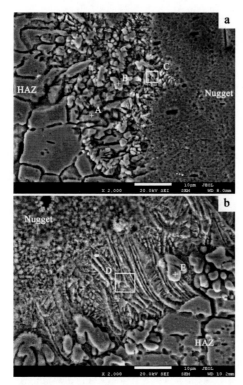

Fig. 3. The microstructures of the fusion zones of deep penetration type spot weld(a) and heat conduction type spot weld(b)

The compositions of the grains A-E described above are listed in Table 2. It can be seen that the compositions of grains A and D are very close to those of $Nd_2Fe_{14}B$ in base metal. In addition, the A and D types of grains grow from the $Nd_2Fe_{14}B$ grains in HAZ. Therefore, it can be inferred that grains A and D should be $Nd_2Fe_{14}B$ phase (a small quantity of Pr2Fe14B and $Dy_2Fe_{14}B$ are present as well). The contents of Fe are dominant in grains B, C and E, so they should be primary phase α-Fe [1].

It is thus clear that the fusion zones of deep penetration weld and heat conduction weld are both constituted by the primary α-Fe phase and the lamellar $Nd_2Fe_{14}B$ phase epitaxial growing from the $Nd_2Fe_{14}B$ grains in HAZ. It also can be seen from Fig. 3 (a) that in the fusion zone of deep penetration weld, the α-Fe grains close to the HAZ are basically cellular while they are dendrites close to the nugget. The closer to the nugget, the narrower of the dendrite arms spacing; this is the same as the phenomenon observed in reference [7].

Table 2. The contents of main elements in grains A-E in fusion zone (wt%)

	Nd	Fe	Pr	Dy
A	22.12	70.80	7.08	—
B	3.83	96.17	—	—
C	2.31	97.69	—	—
D	21.34	65.54	7.50	5.62
E	7.48	92.52	—	—

3.3. Nugget microstructures

During laser spot welding, the base metal melts, mixes and then solidifies to form a nugget. Fig. 4 (a) shows the microstructures of the upper suface of the nugget of deep penetration weld. It can be seen that the upper suface of the nugget is mainly composed of columnar dendrites, and the closer to the center of the nugget, the more narrow of the columnar dendrite arms spacing. The columnar dendrites in the periphery of the upper surface of the nugget are covered by a layer of gray tissue, while the center of the nugget surface are columnar dendrites only. The microstructures of the upper suface of the nugget of heat conduction weld are so different from those of deep penetration weld, as shown in Fig. 4 (b). It is mainly composed of massive $Nd_2Fe_{14}B$ that epitaxially growing from the HAZ around the periphery to the center of the nugget. A small amount of gray tissue can be found in the center of the upper surface of the nugget for conduction type weld.

The contents of the main chemical elements in the gray tissues and columnar dendrites shown in Fig. 4 are listed in Table 3. It can be seen that the contents of Nd in the gray tissues are higher than that in $Nd_2Fe_{14}B$, and the gray tissues are also rich of Pr and Dy. Therefore, they should be Nd-rich phase.The columnar dendrites are mainly composed of Fe, so they should be α-Fe phase. In summary, the upper suface of the nugget of deep penetration weld is mainly composed of α-Fe columnar dendrites and Nd-rich phase that distributes in the periphery of the upper surface. The upper suface of the nugget of heat conduction weld is mainly composed of $Nd_2Fe_{14}B$ phase, in addition to a small amount of Nd-rich phase that located in the center of the upper surfac

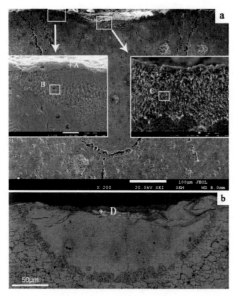

Fig. 4. The microstructures of the nugget surface of deep penetration type spot weld(a) and heat conduction type spot weld(b)

Table 3. The contents of main elements in grains A-D in upper surface of nuggets (wt%)

	Nd	Fe	Pr	Dy
A	60.49	9.17	20.06	10.29
B	3.18	96.82	—	—
C	2.34	97.66	—	—
D	63.02	4.9	20.21	11.87

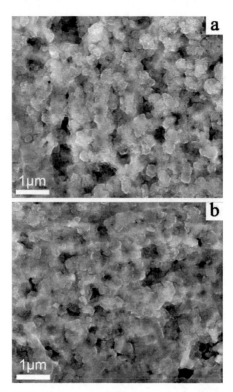

Fig. 5. The internal microstructures of the nuggets of deep penetration type spot weld(a) and heat conduction type spot weld(b)

It can be seen from Fig. 2 and Fig. 3 that the internal microstructures of the nuggets are uniform for both deep penetration and heat conduction welds, and the grain characteristics are not clear in low magnification photos. Composition analysis indicates that the compositions of the internal microstructure of the nuggets are nearly the same as those of the base metals. The high magnification photos for the nuggets of deep penetration and conduction modes welding show that the nugget microstructures of the two welds are quite similar, as shown in Fig. 5. It can be seen that the grains of the inner part of the nuggets are basically ellipsoidal and with rough surfaces, the diameter of the grains is about 200-300nm, and neighboring grains form larger size grain groups, which have voids between each other.

Some scholars [9, 10] found an interesting phenomenon in the preparation of nano-crystalline $Nd_2Fe_{14}B$ by rapid quenching method, which was that submicron grains of $Nd_2Fe_{14}B$ were formed in the NdFeB alloys solidified with a certain fast cooling rate. The cooling rate of the weld pool in laser welding is very fast, which can provide the cooling rate for the formation of submicron grains of $Nd_2Fe_{14}B$. Hence, it can be concluded from the compositions and microstructure characteristics that the inner part of the nuggets should be submicron ultrafine equiaxed grains of $Nd_2Fe_{14}B$.

4. Discussion

The microstructures of welds depend on the composition of the base metal, the highest temperature, the cooling rate and the molten pool flow modes during welding. According to the phase diagram of NdFeB alloy[1], the $Nd_2Fe_{14}B$ phase would be precipitated under equilibrium solidification conditions for the NdFeB alloy with composition listed in Table 1. However, in the conditions of two different modes of laser welding, the melts are seriously overheated and the short-range order atomic groups of $Nd_2Fe_{14}B$ are reduced significantly during the preliminary stage of solidification, which lower the crystallization temperature of $Nd_2Fe_{14}B$ phase, so the γ-Fe phase can be precipitated [1]. Moreover, the unmelted $Nd_2Fe_{14}B$ grains in HAZ provide a base for the growth of $Nd_2Fe_{14}B$, from which the lamellar-type $Nd_2Fe_{14}B$ grains are growed epitaxially and preferentially for the low crystallization energy [11].

The overheating of the melt in deep penetration welding is big enough for the precipitation of γ-Fe phase in the upper surface of deep penetration weld. However, there is no base for the epitaxial growth of $Nd_2Fe_{14}B$ phase in those area, so the γ-Fe phase will be precipitated before the $Nd_2Fe_{14}B$ phase the during the preliminary stage of solidification in the upper surface of the nugget. The overheating of the melt in heat conduction welding is low, so neither the γ-Fe phase nor the $Nd_2Fe_{14}B$ phase is precipitated.

At the beginning of the solidification, the γ-Fe phase grows in cellular crystal morphology. With the solidification of γ-Fe phase, the constitutional supercooling before the solid-liquid interface is increased gradually, and the spacing between secondary dendrite arms is decreased. With the advancing of solid-liquid interface to the center of the molten pool, the temperature, temperature gradient and cooling rate of the melt are gradually reduced, and to certain conditions the γ-Fe phase will no longer precipitate.

When the temperature is reduced to the critical point of the nucleation of $Nd_2Fe_{14}B$ phase, the $Nd_2Fe_{14}B$ phase nucleates and grows explosively.

There is a keyhole in the molten pool of deep penetration welding, in which the plasma and metal vapor erupt and lead to an outward flow of liquid metal to form a forced convection, as shown in Fig. 6 [12]. The density of Nd-rich phase is less than that of $Nd_2Fe_{14}B$ phase and γ-Fe phase [1]. Therefore it is more easily to be taken to the periphery of the upper surface of the moten pool with the flow of liquid metal. There is no keyhole in the moten pool of heat conduction welding, and the flow rate of liquid metal is small, so the low melting point Nd-rich phase is pushed to the center of the upper surface of the nugget with the advancement of the solid-liquid interface from the fusion zone to the upper surface.

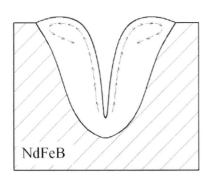

Fig. 6. The flow of liquid metal in deep penetration spot welding

5. Conclusions

In this paper, both deep penetration and heat conduction modes of laser spot welding were carried out on a sintered NdFeB permanent magnet, and the characteristics of the microstructures of the welds were analyzed, conclusions are drawn as follows: 1) The HAZ of both welds are rich of cracks caused by the melting of grain boundary phase, but the cracks of deep penetration weld are wider and much longer than those of heat conduction weld. Two different types of porosities are observed in the nugget of deep penetration weld, while neither is observed in the nugget of heat conduction weld. 2) The fusion zones of both welds are constituted by the primary α-Fe phase and the lamellar $Nd_2Fe_{14}B$ phase

epitaxially growing from the $Nd_2Fe_{14}B$ grains in HAZ. In the fusion zone of deep penetration weld, the α-Fe phase close to the HAZ are basically cellular crystals, those close to the nugget are dendrites, while the α-Fe phase in the fusion zone of heat conduction weld are all dendrites. 3) The upper suface of the nugget of deep penetration weld is mainly composed of the α-Fe columnar dendrites, and the Nd-rich phase distributed in the periphery of the upper surface. In contrast, the upper suface of the nugget of heat conduction weld is mainly composed of $Nd_2Fe_{14}B$ phase, in addition to a small amount of Nd-rich phase that located in the center of the upper surface. 4) The internal microstructures of the nuggets of deep penetration weld and heat conduction weld are basically the same, which are both composed by submicron ultrafine equiaxed grains of $Nd_2Fe_{14}B$.

References

[1]   Zhou S, Dong Q, (2004) Super Permanent Magnet. Metallurgical Industry Press, Beijing, 1-5

[2]   Schmitt F, Mehlmann B, Gedicke J, Olowinsky A, Gillner A, (2010) Laser beam micro welding with high brilliant fiber lasers. Journal of Laser Micro Nanoengineering 5: 197-203

[3]   Rohde M, Markert C, Pfleging W, (2010) Laser micro-welding of aluminum alloys: experimental studies and numerical modeling. International Journal of Advanced Manufacturing Technology 50: 207-215

[4]   Chang B, Bai S, Li X, Ding Y, Zhang H, Du D, Zhou Y, (2008) Laser spot welding of SPCC steel to NdFeB magnets. J Tsinghua Univ 48: 1728-1731

[5]   Chang B, Bai S, Du D, Zhang H, Zhou Y, (2010) Studies on the micro-laser spot welding of an NdFeB permanent magnet with a low carbon steel. Journal of Materials Processing Technology 210: 885-891

[6]   Pang J, Liu X, (2004) Microstructures in transition zone of a laser remelting/solidificaiton pool of sintered Nd-Fe-B magnets. Joural of the Chinese Rare Earth Society 22: 219-224

[7]   Pang J, Liu X, (2004) Arrangement of easy magnetization axis of cellular column Nd2Fe14B in laser melting/solidification pool on sintered Nd15Fe77B8 magnets. 14: 1183-1187

[8]   Bradley JR, Mishra RK, Pinkerton FE, Schroeder T, (1990) Microstructure and magnetic-properties of CO_2-laser surface melted Nd-Fe-B magnets. Journal of Magnetism and Magnetic Materials 86: 44-50

[9]   Shi Y, Zhou J, Wei P. The distribution of the magnetic properties and microstructures for $Nd_2Fe_{14}B$ permanent magnetic nanomaterials. Journal of Functional Materials 10: 440-443

[10]   Xiao W, (1994) Nanocrystalline Nd-Fe-B base alloys by rapid quenching. Metallic Functional 1: 1-5

[11]   He Y, Xiong K, Gao X, Zhang M, Zhou S, (1999) Crystal growth features of Nd-Fe-B ingot. Acta Metallurgica Sinica 35: 271-274

[12]   Wang H, Shi Y, (2005) Study on dynamic characteristics of molten pool during laser deep penetration welding. Aeronautical Manufacturing Technology 47: 64-6.

Study on full penetration stability of light alloys sheet laser welding

L. Chen, S. L. Gong and J. Yang
Science and Technology on Power Beam Processes Lab., BAMTRI, Beijing, China

Abstract. YAG laser welding was conducted to investigate the weld shape of BT20 titanium alloy and 5A90 Al-Li alloy sheet in this paper. The results show that laser welding mode and weld shape features depends on laser power density as well as heat input, not only rely on laser power density. And weld shape features can reveal full penetration keyhole welding stability. Both laser power density and heat input determine the keyhole stability and melt pool behavior. According to the research on effect of welding parameters on weld shape,the quality of full penetration laser welding light alloy sheet can be evaluated by the weld width ratio (Rw), that is ratio of weld back width to weld surface width. And the laser energy density and heat input threshold curve based on Rw can be used to optimize parameters for producing stable full penetration laser welding of light alloy sheet.

Keywords: laser welding, weld shape, heat input, laser power density, full penetration

1. Introduction

The light alloys are one of the most important materials for aerospace industry structures because of its light weight, superior strength-to-weight ratio, and excellent corrosion resistance. Laser welding with higher energy density offers remarkable advantages over conventional fusion welding, such as minimal component distortion and high productivity, and is specially suitable for joining aerospace structures. However, the process of laser penetration welding involves complex reactions among the laser and plasma and material, which influnce the microstructures and mechanical properties of joints.

While the laser beam acts on materials, different laser power density will produce different physical phenomenon, such as temperature rising, metal melting and vaporing, and plasma forming. It leads to different laser welding mode. It is usually thought that the penetration laser welding could perform when laser power density is larger than 10^6W/cm^2. A laser welding mode is determined by a group of parameters. But some researchs found that the unstable laser welding can occur along with stable parameters, as a result, weld width presents unstable. When laser power density close to the threshold for producing penetration laser welding, the unstable processing varies sharply between conductive laser welding and penetration laser welding, and results in untable changing in weld width and depth. A. Matsunawa thought that it results from light induced plasma screening laser [1]. Chen and Zhang also identified unstable laser welding existing when they studied stainless steel CO_2 laser welding [2]. Wang found the same results on HE130 titanium alloy YAG laser welding [3]. Based on their researches, they built up dual U curve to evaluate the parameters of stable laser welding respectively for stainless steel and HE130 titanium alloy according to the effect of focus position, laser power and welding speed on laser welding mode and weld shape. But their researches did not mention unstable full penetration laser welding, and also did not consider the effect of heat input on laser welding stability. Seto et al. [4] observed the difference of melt pool between penetration laser welding and full penetration laser welding. It should be noted that there will be unstable full penetration laser welding for sheet laser welding because of plasma or plume. In this paper, titanium alloy and aluminium lithium alloy sheets are used to investigate weld shape features for YAG laser welding, and put forward to a method of parameters optimizing.

2. Experimental investigation

The BT20 titanium alloy with thickness of 2.5mm and 5A90 aluminium lithium alloy with thickness of 3mm were used in this research. The test specimens were cut from sheet in rolling and annealing condition. The bead-on-plate weld was performed in a fixture, and the welding direction was perpendicular to plate's rolling direction. Three lasers were employed. They are AM356-YAG laser, HL2006D laser and HL3006D laser. The laser head with 200mm focal length lens were driven by robot. The laser weld nozzle and gas passage in the fixture offered shielding gas to protect the pool and back weld respectively. Argon was used as shielding gas. The gas trailing shoe was choose as additional shielding. The gas flow rates were 15l/min, 10-20l/min, and 20-25l/min respectively. The specimens' surface was chemically cleaned before laser welding. Cleaning procedures are alkali cleaning, water cleaning, pickling, water cleaning, and drying.

3. Experimental results and discussion

3.1. Effect of parameters on weld shape

Observing on the cross-sections of welds of different welding conditions, at least three different kinds of weld shapes were found for BT20 titanium alloy and 5A90 aluminium lithium alloy sheets laser welding, semi-spherical weld shape from conductive laser welding mode, nail-head weld shape and hour-glass weld shape from part and full penetration laser welding. Table 1 shows three typical weld shapes for BT20 titanium alloy and 5A90 aluminium lithium alloy laser welding under different paramenters.

According to the research, different matches in parameters, such as laser power, welding speed and defocus distance, can result in different welding shapes. However, it is found that the penetration laser welding mode and weld shape feature can be contributed to the laser power density and heat input (laser power/welding speed). When the laser energy density is higher than the threshold for keyhole formation, the penetration laser welding can not occur although heat input is large enough. When heat input is small, the penetration laser welding can also not occur even if the laser power density is overlarger than power density threshold for keyhole formation. This phenomenon reveals that the penetration laser welding mode depend on both laser energy density threshold and heat put threshold, but not on laser power density threshold alone. The result shows that the full penetration laser welding is also relative to the laser energy density and heat input.

3.2. The unstable full penetration laser welding

The previous research [5] observed that the keyhole presents periodical open and close, and that the pool back size varies along with keyhole changing. That means keyhole break intermittently through the pool during the full penetration laser welding.

When laser power density is close to threshold for penetration laser welding, there will be the unstable welding shifting between the conductive laser welding mode and penetration laser welding mode. And weld surface width changes abruptly. When laser power density and heat input is unsuitable and results in unstable full penetration laser welding, it is found that part penetration and full penetration occurs alternately on the weld back, meanwhile, the weld surface quality is perfect. Therefore, it should be noted that unstable laser welding can be caused not only from laser welding mode changing, but also from the unstable keyhole changing

during full penetration laser welding. In order to obtain stable and perfect full penetration weld shape, it is necessary that laser beam energy reaching the bottom of keyhole should be enough to produce full penetration keyhole during the sheet of laser welding.

Table 1. BT20 titanium alloy and 5A90 aluminium litjium alloy weld shape features of YAG laser welding

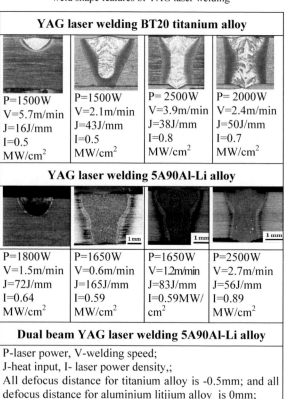

YAG laser welding BT20 titanium alloy			
P=1500W V=5.7m/min J=16J/mm I=0.5 MW/cm^2	P=1500W V=2.1m/min J=43J/mm I=0.5 MW/cm^2	P= 2500W V=3.9m/min J=38J/mm I=0.8 MW/cm^2	P= 2000W V=2.4m/min J=50J/mm I=0.7 MW/cm^2
YAG laser welding 5A90Al-Li alloy			
P=1800W V=1.5m/min J=72J/mm I=0.64 MW/cm^2	P=1650W V=0.6m/min J=165J/mm I=0.59 MW/cm^2	P=1650W V=1.2m/min J=83J/mm I=0.59MW/cm^2	P=2500W V=2.7m/min J=56J/mm I=0.89 MW/cm^2
Dual beam YAG laser welding 5A90Al-Li alloy			

P-laser power, V-welding speed;
J-heat input, I- laser power density,;
All defocus distance for titanium alloy is -0.5mm; and all defocus distance for aluminium litjium alloy is 0mm;

3.3. The definition of weld width ratio (Rw)

It is assumed that both laser energy density and heat input have effects on heat action and pool behavior of laser welding. With the increase of laser energy density and heat input, the base metal absorbs laser energy from keyhole instead of from pool surface. When laser energy density and heat input are large enough, full penetration laser welding will occur. During the full laser penetration welding, alloy evaporates strongly from keyhole and the vapour breaks forth from keyhole up and down, indicated as in Fig 1. And then the plasma/vapour plume forms over the keyhole opening, which radiates part of its energy to the pool. At the same time, weak convection appears in the part of pool in the middle of the keyhole because of the vapour expanding. As a result, it could be thought there is an additional heat source acting at the keyhole opening. The heat action of the full penetration laser welding can be imaged to be that of one lineal heat source plus two point heat sources and results in hour-glass weld in which weld width in middle is more narrow than that of surface and back weld. It means obviously that weld shape for penetration laser welding can reflect the laser welding heat action, and indicate the full penetration keyhole stability.

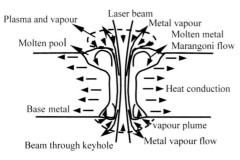

Fig 1. The sketch of heating effect for full penetration laser welding

Based on the heat action of penetration laser welding mentioned above, the weld shape can be sketched as Fig.2. The middle part width of weld (MW) depends on the action of the linear heat source, which can be illustrated as rectangle. And the upper and lower parts of the weld look like trapezoid. The weld width at top (TW) and the weld width at root (RW) reflect the action of the upper and lower point heat source. Rw is defined as the ratio of RW to TW. When Rw is larger than certain value, the unstable full penetration laser welding can be avoided. Thus Rw can show the stability of full penetration laser welding, and reveals the thresholds of laser energy density and heat input.

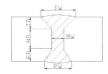

Fig. 2. the sketch for simplified weld for penetration laser welding

3.4. Influence of parameters on weld width ratio (Rw)

3.4.1 Laser power and welding speed
Fig.3 and Fig.4 show the relationship between weld width ratio (Rw) and laser power and welding speed for BT20 titanium alloy and for 5A90 Al-Li alloy respectively. When the welding speed is constant, the Rw increases rapidly at first along with laser power increasing and then change slowly. According to the reaults, Rw must be larger than 0.4 for stable full penetration laser welding for 2.5mm thick titanium alloy, and. be larger than 0.6 for 5A90 Al-Li alloy. Moreover, the larger welding speed is, the higher the threshold of laser power for full penetration laser welding is.

When laser power is constant, the Rw show a decrease tendency with the increase of weld speed. With different laser power, the maximum welding speed for full penetration welding is different. The lower the laser power, the faster the decrease of the Rw.

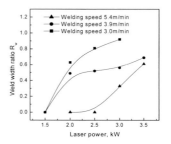

a Laser power

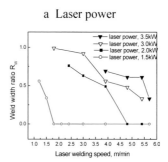

b welding speed

Fig. 3 The effects of laser power and welding speed on weld width ratio for BT20 titanium alloy

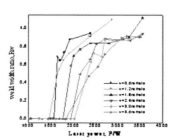

a Laser power

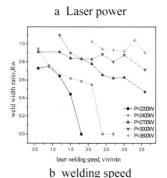

b welding speed

Fig. 4. The effects of laser power and welding speed on weld width ratio for 5A90 Al-Li alloy

When the defocus distance is constant, weld shape features can be controlled by mere the matching of laser power and welding speed. With definite the material and plate thickness, for each laser power of full penetration welding, there will be a welding speed range to match it. Fig.6 is the parameter window of laser power – welding speed (P-V) of YAG laser welding. Combining the detection of weld porosity, undercut, weld surface undercut, the better quality weld can be obtained when the Rw is between 0.4-0.9. The matching of laser power and welding speed can be divided into three zones. In the zone below curve II the weld is not penetration. The curve II, therefore, can be named critical curve for stable full penetration welding. In the zone above curve I, the weld shape is very poor quality. The curve I can be defined as critical curve for weld shape quality. The zone between those two curves is optimizing window for laser power and welding speed, in which weld with proper width and good quality.

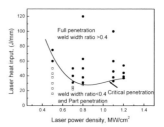

a BT20 titanium alloy

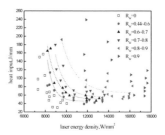

b 5A90 Al-Li alloy

Fig. 6. The relationship between laser power density and heat input for YAG laser welding

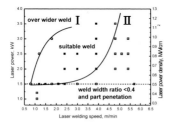

a BT20 titanium alloy

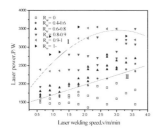

b 5A90 Al-Li alloy

Fig. 5. The optimizing window for laser power and welding speed (P-V window)

3.4.2 Heat input and laser power density

The P-V parameter window can not reveal the relation between laser power density and heat input. Statistical analysis of all experimental results illustrated the matching relationship between laser energy density and heat input, as shown in Fig.6. From the results, for titanium alloy, Rw 0.4 can be used as the critical condition for stable full penetration welding to determine the dual thresholds relationship of laser energy density and heat input of stable full penetration welding, and for Al-Li alloy, Rw should be larger than 0.6.

4. Conclusions

The features of the light alloy sheet weld shape of laser welding revealed that laser welding mode depends not only on laser power density but also on heat input. When both laser power density and heat input are close to the thresholds for throughout keyhole formation, the unstable full penetration laser welding of light alloy sheet will occur. The weld width ratio Rw in non-dimensional is a feature factor describing the sheet weld shape of laser welding and can be used to estimate the processing stability of full penetration laser welding.

References

[1]   Matsunawa A., (2001) Understanding physical mechanisms in laser welding for mathematical modeling and process monitoring. Proceedings of the first international WLT-conference on laser in manufacturing, Munich: 79-93

[2]   Zhang Xudong, Chen Wuzhu, Ren Jialie, (1997) Influence of thermal focusing on mode transition and process stability in laser welding. Journal of Tsinghua University(Science and Technology), Vol.37(8):101-104

[3]   Wang Jia chun,(2001) Laser welding process and properties and microstructure of joint for HE130 alloy.General Research Institute for Nonferrous Metals, Doctorial Dissertation.

[4]   Seto N., Katayama S.,Matsunawa A., (2002) Porosity formation mechanism and reduction method in CO2 laser welding of stainless steel. Welding Internation,Vol.16(6), 451-460

[5]   Chen Li, （2005）A Study on the Full Penetration Stability and Physical Metallurgy of the Laser Welding of Aeronautic Titanium Alloys. Huazhong University Of Science & Technology, Doctorial Dissertation.

Challenges of investigating distortion effects in laser micro welding

P. Woizeschke and F. Vollertsen
BIAS – Bremer Institut für angewandte Strahltechnik GmbH, Klagenfurter Str. 2, 28359 Bremen, Germany

Abstract. Macro effects can probably not simply be transferred to the micro range proportionally to sheet thickness because of possible influences of size and scaling effects. In order to investigate process parameter dependencies in the micro range reproducible experimental conditions have to be realized. The capability of a developed welding set-up to maintain those conditions is shown. It consists of a laser scanning head, a novel clamping device and integrated deformation measuring systems. A single-mode fiber laser beam is focused to a focus diameter of 18 μm by an F-theta lens. This results in a Rayleigh length of less than 0.2 mm. Therefore an accurate specimen positioning including clamping, in-plane prestressing and avoiding of wrinkles and buckles without damaging the specimen is mandatory. Thus, the clamping system is equipped with adjustable clamping and prestressing devices and a mechanism to flatten the specimens before welding. Due to low stiffness of thin metal foils, handling the specimen can influence its shape and properties significantly. Therefore, influences on the welding process are minimized by appropriate sample preparation, handling, and the employment of non-contact optical systems for measuring deformations. The suitability of the welding set-up for deep penetration laser micro welding is proven by investigating the reproducibility of welding results. Typical distortion effects in laser micro welding are identified and finally the effect of the needed in-plane prestress on the transverse stress behavior is investigated.

Keywords: single-mode fiber laser, micro welding, deep penetration laser welding, welding distortion, aluminum foils.

1. Introduction

The production of micro parts developed to an important industrial branch over the last decades. Thus, the importance of price-wise competitive production technologies in mass-production is essential to a growing number of customers [1]. The growing trend towards miniaturisation requires precise production and especially joining processes complying with tolerance requirements in the micrometer range [2]. An increasing number of manufacturers choose laser welding technologies in assembling due to its abundant advantages [3]. Structural dimensions smaller than 100 μm originate in tolerance challenges of a few μm. Therefore, despite of relativly low energy input with decreasing sheet thickness challenges by potential scaling effects get more severe [4].

2. Experimental

2.1 Requirements

Laser system: The high beam quality of single-mode fiber lasers enables the joining of micro parts with low heat input by deep penetration welding with foil thicknesses of 20 μm to 300 μm [5]. For deep penetration welding a high power density has to be realized. In case of aluminum the high thermal conductivity results in a threshold density of 10^7 W/cm², one magnitude order higher than for stainless steel. In addition, in micro welding of stainless steel foils was observed that this critical power density was about 2 orders larger [6] than in the macro range [7]. This requires a small focus diameter. Additionally measuring distortion demands an adequate working plane distance. Overall a high beam quality laser source and an optical system with minimized focus shift and absorption effects are needed [8].

Clamping: The clamping device has to ensure reproducible positioning of specimens on the one hand and accessibility from the upper and lower side for measuring on the other hand. Avoiding damages of the foils whilst handling is mandatory to exclude a potential influence of welding and distortion behaviors by defects like wrinkles or buckles. Especially overlap welding in the micro range demands high accuracy requirements in clamping [9, p.60f]. In foil overlap welding homogeneous low gap width is important to avoid defects like interruptions of gap bridging or cutting of the upper foil [9, p.60f]. Thus, eliminating foil sag before welding by in-plane pre-stressing is obligatory. Due to the expected short Rayleigh length accurate positioning is also required to guarantee comparable conditions by a constant focus position. For investigations on influences of geometrical parameters the adjustment of clamping distance and specimen dimensions has to be considered.

Distortion measurement: Distortion effects can be classified into categories of transverse shrinkage, longitudinal shrinkage, angular distortion, longitudinal bending, in-plane rotation and buckling [10]. To analyze these effects in micro welding non-contact measurement systems are essential due to the low stiffness of the specimens which prevent prior damaging or influencing foil properties. Furthermore, the expected size of distortion effects requires a resolution of a few μm. Thereby, the foil surface, having a characteristic rolling texture and a high reflectivity, could be a difficulty for optical measuring principles. In overlap welding measuring possibilities from the upper and the lower side have to be implemented.

2.2. Realization of a capable welding set-up

Laser system: The micro welding experiments were performed using a Trumpf TruFiber 300 single-mode fiber laser with an output power of up to 300 W with a beam propagation factor M²=1.1. Using a fiber with a diameter of 20 μm, the laser source was coupled to a Scanlab hurrySCAN25 scanner system with a collimation length of 200 mm and a focussing length of the quartz F-theta lens of 163 mm. Thus, theoretically a focus diameter of 15.4 μm and a Rayleigh length of 157 μm are expected.

    Clamping device: A frame was used for placing the specimens into the set-up to avoid damages of the foils by handling. An integrated camera system controlled the reproducible positioning of the specimens. It could be observed in preliminary tests that often only one foil was flattened by pre-stressing because of different initial sags [Fig. 1(a)]. Therefore, to create reproducible initial conditions a novel mechanism was integrated. Two rotating rolls covered with abrasive paper transferred friction forces to flatten both single foils prior final clamping and pre-stressing as illustrated in Fig.1 (b).

a) Observed failure without prior flattening

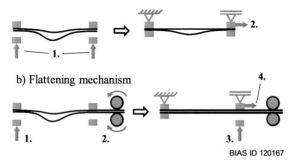

b) Flattening mechanism

BIAS ID 120167

Fig. 1. Sketch of the flattening mechanism and its benefit

Distortion measurement: A system comprising a force sensor, two laser triangulation sensors and a deflectometry system was used for the distortion category identification and distortion measurements. The force sensor measured the averaged transverse stress. One triangulation sensor above the specimen measured in-process displacement of a single spot on the surface. A measurement of the final upper surface shape was done by deflectometry after welding. Thereby, the prior

expected adverse high reflectivity of the foil surface enabled a shape measurement by the deflectometry principle in the first place. A second triangulation sensor below the surface installed on a linear stage measured the deviation of a line shape on the lower surface to compare the behaviors of lower and upper foil and the two measuring methods. Detailed reports of laser triangulation and deflectometry principles are given in [11.12].

2.3. Experimental program

As base material, commercially pure Al99.5 foil with a thickness of 50 μm was used. The clamping distance was 30 mm and the specimen width 10 mm (Fig. 2). First a set of nine overlap welding experiments with 10 N/mm² prestress at constant conditions and parameters were carried out to prove the usability of the set-up. Finally, because of distortion behavior can be affected by clamping conditions themselves [13] the influence of the initial in-plane prestress on the transverse stress behavior was investigated by performing additional experiments with 5 N/mm² and 15 N/mm². The focus spot was positioned on the surface of the workpiece using 133 W laser ouput power and a welding speed of 500 mm/s.

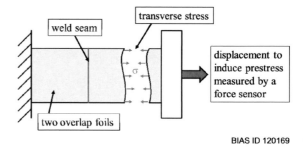

BIAS ID 120169

Fig. 2. Sketch of the specimen and the weld seam position

3. Results

3.1. Beam quality at workpiece

The beam properties were investigated using a Primes HighPower MicroSpotMonitor. For the system a real focus diameter of 17.4 μm, a Rayleigh length of 176 μm and an M² of 1.26 were measured, which are close to the expected values. The power density averaged over the spot area at 133 W output power is $5*10^7$ W/cm². Due to the Gaussian beam profile the peak intensity in the center is even higher. An averaged deviation of less than five percent between upper and lower surface seam width confirms the aimed deep penetration welding mode.

3.2. Out-of-plane distortion

First the final out-of-plane distortion was investigated by deformation measurements. A comparison of the measurements of the upper foil by deflectometry and the lower foil by triangulation showed a maximum deviation of the resulted transversal line shapes of about 10 μm. The result of a deflectometry measurement of the surface shape is illustrated in Fig. 3.

BIAS ID 120174

Fig. 3. Final surface shape measured by deflectometry of a typical weldment

In longitudinal direction a bowing of the surface is conserved while transversal to the weld two buckles and a peak on the weld seam is observed. The buckles were caused by compression stresses in that areas. Because of the insignificant difference of the weld seam width between upper and lower surface the angle distortion effect should be negligible. However, the direction of the out-of-plane displacement might be influenced initially by even a low angle distortion effect.

The time-dependent behavior of the out-of-plane distortion is plotted in Fig. 4.

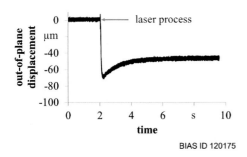

BIAS ID 120175

Fig. 4. Time-dependent out-of-plane displacement behavior

The single point time-dependent measurement 4 mm left of the weld center line on the upper surface shows a maximum displacement a few milliseconds after the laser process and a relaxation to its final value over the next couple of seconds which is caused by the increasing thermal contraction during cooling.

3.3. Transverse stress behavior and the influence of the prestress

As illustrated in Fig. 2 the foils were prestressed after flattening by an in-plane displacement of one clamping jaw. The induced prestress was measured by a force sensor. The force sensor enabled a measuring of the time-dependent behavior of the averaged transverse stress to investigate the distortion effect of transverse shrinkage. A typical measurement for a pre-stress of 10 N/mm² is shown Fig. 5.

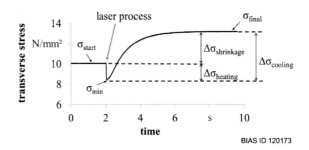

BIAS ID 120173

Fig. 5. Time-dependent transverse stress behavior of a typical weldment with a prestress of 10 N/mm²

First the transverse stress drops ($\Delta\sigma_{heating}$) from the initial prestress value σ_{start} to a minimum σ_{min} during the laser welding process which takes 20 ms and starts at t = 2 s. Afterwards the transverse stress increases to a maximum value σ_{final}. This is based on the decreasing thermal expansion during cooling. The final shrinkage $\Delta\sigma_{shrinkage}$ is given by the difference between σ_{start} and σ_{final}. The transverse shrinkage in welding is normally caused by the plastic compression of the weld seam region induced by the thermal expansion of the heated base material. Thus, the observed decrease $\Delta\sigma_{heating}$ during the heat input results from the prestress. The pre-stress value was variied to investigate its influence on the transverse stress behavior (Fig. 6).

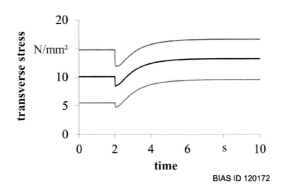

BIAS ID 120172

Fig. 6. Transverse stress behavior for three different prestress values (grey: 5 N/mm², black: 10 N/mm², red: 15 N/mm²)

The relaxation of the transverse stress $\Delta\sigma_{heating}$ from the prestress level is proportional to the initial prestress value (Fig. 6 and 7). However, the post-process increase of the transverse stress $\Delta\sigma_{cooling}$ stays constant (Fig. 6 and 7). This could be explained by the same heat input in all configurations which results in an equal thermal contraction of the specimen material during cooling. Consequentially the final shrinkage $\Delta\sigma_{shrinkage}$ decreases with increasing prestress.

The linear regression of the results (Fig. 7) shows on the one hand that theoretically no decrease of the transverse stress during heating should be expected without any prestress ($\sigma_{start} = 0$). The induced thermal expansion of the base material generates a maximum plastic compression of the weld seam region. This shortening results in a transverse shrinkage after cooling. On the other hand a maximum of $\Delta\sigma_{heating}$ should be resulted for a prestress close to the yield strength σ_Y of 29 N/mm² because of $\Delta\sigma_{heating} = \Delta\sigma_{cooling}$ and $\Delta\sigma_{shrinkage} = 0$. In this case the total thermal expansion initiates relaxing of the transverse stress during welding without any plastic compression of the weld seam.

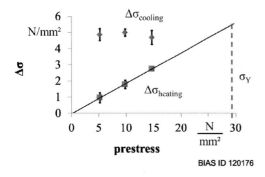

Fig. 7. Influence of prestress on the transverse stress behavior

3.4. Reproducibility

The reproducibility of the results is shown by the standard deviations of three characteristic parameters of the investigated set of experiments (Table 1).

Table 1. Standard deviations of characteristic parameters

Parameter	Standard deviation
Weld seam width	5 %
Transverse shrinkage stress	6 %
Buckling amplitude	18 %

BIAS ID 120171

The buckling amplitude has the highest standard deviation. The buckling amplitude might be the most sensitive value for influences by deviations of the material properties and process conditions.

4. Summary

The presented novel experimental set-up offers the possibility to investigate distortion effects in deep penetration laser micro welding. The non-contact measuring principle of deflectometry is suitable to identify and quantify distortion effects at micro welding of aluminum foils. A higher initial in-plane prestress results in a lower final transverse shrinkage.

Acknowledgements: The authors gratefully acknowledge financial support of this work by DFG (VO 530/35-1). This work was accomplished within the Center of Competence for Welding of Aluminum Alloys - Centr-Al.

References

[1]   Alting, L., Kimura, F., Hansen, H. N. & Bissacco, G. 2003. Micro Engineering. CIRP Annals - Manufacturing Technology 52(2), 635–657.

[2]   Hoving, W. 2004. Product Miniaturisation, a Challenge for Laser Technology. In Proc. Laser Assisted Net Shape Engineering 4 (LANE'04). Eds.: Geiger, M., Otto, A. Meisenbach Bamberg. 67-78.

[3]   Bley, H., Weyand, L. & Luft, A. 2007. An Alternative Approach for the Cost-efficient Laser Welding of Zinc-coated Sheet Metal. CIRP Annals - Manufacturing Technology 56(1), 17–20.

[4]   Vollertsen, F. 2008. Categories of size effects. Production Engineering 2(4), 377–383.

[5]   Grupp, M. 2006. Neue Schweißanwendungen mit dem Faserlaser. In Proc. 5. Laser-Anwenderforum - Laserstrahlfügen: Prozesse, Systeme, Anwendungen, Trends. Eds.: Vollertsen, Seefeld. BIAS-Verlag. Bremen. 81-87.

[6]   Miyamoto, I., Park, S. & Ooie, T. 2003. Ultrafine keyhole welding processes using single-mode fiber laser. LMP Section A. 203–212.

[7]   Miyamoto, I., Maruo, H. & Arata, Y. 1984. The role of assist gas in CO2 laser welding. In Proc. of ICALEO 1984. Department of Welding Engineering and Welding Research Institute. Osaka. Japan. Vol. 44. 68–75.

[8]   Wedel, B. & Niedrig, R. 2006. Anforderungen an die Laserbearbeitungsköpfe beim Schweißen mit hoher Strahlqualität. In Proc. 5. Laser-Anwenderforum - Laserstrahlfügen: Prozesse, Systeme, Anwendungen, Trends. Eds.: Vollertsen, Seefeld. BIAS-Verlag. Bremen. 81-87.

[9]   Brockmann, R. 2003. Beitrag zum Mikronahtschweißen von Edelstahlfolien mittels diodengepumpten Nd:YAG-Laser. PhD TU Chemnitz.

[10]   Masubuchi, K. 1980. Analysis of Welded Structures. Pergamon Press. Chapter 7.

[11]   Blais, F. 2004. Review of 20 Years of Range Sensor Development. Journal of Electronic Imaging 13(1).

[12]   Bothe, T., Li, W., Kopylow, C. von, Jüptner, W.P., Osten, W. & Takeda, M. 2004. High-resolution 3D shape measurement on specular surfaces by fringe reflection: Proc. SPIE. Optical Metrology in Production Engineering 5457(1), 411–422.

[13]   Schenk, T., Richardson, I.M., Eßer, G., Kraska, M. & Ohnimus, S. 2008. Welding Distortion of DP600 Overlap Joints and Influence of Clamping and Phase Transformation. Thermal Forming and Welding Distortion (IWOTE'08). Eds.: Vollertsen, Sakkiettibutra. BIAS-Verlag. Bremen. 83-95.

Characteristics of YAG-MIG hybrid welding 6061 Aluminum alloy with different states

Xu F.[1], Chen L.[1], Gong S. L.[1], Guo Y. F.[2] and Guo L. Y.[1]

[1] Science and Technology on Power Beam Processes Laboratory, Beijing Aeronautical Manufacturing Technology Research Institute, Beijing 100024, China

[2] The First Aircraft Institute, Xi'an 710089, China

Abstract. With the diversification of manufacture methods, joining the same materials with different states becomes indispensable in practical application. The hybrid welding has broad application prospects. It not only reduces the strict assembled gap before welding, but also improves the joint properties by filling wire during the welding process. 6061 aluminum alloys with different states were welded by YAG-MIG hybrid welding. The characteristics of welded joint, including microstructures, the tensile properties, microhardness and fracture, were investigated. The results show that the microstructures are different significantly in different states. Besides, the grain boundaries of the joint microstructures become unclear after heat treating. The strength and the elongations of welded joints could reach to that of the base metal. The tensile fracture occurs in the fusion zone and near 6061-O alloy. And the fracture presents ductile rupture. Meanwhile, the effect of the small pores in the weld on the joint's tensile properties is slight.

Keywords: Hybrid welding, 6061 aluminum alloy, Different states, Microstructure, Tensile properties

1. Introduction

6061 aluminum alloy has been widely applied in areas of aerospace, modern architecture, household appliances and trimming, due to its advantages of moderate strength, high corrosion resistance, good forming and processing properties[1-2]. In recent years, with the diversification of modern manufacture methods and the complication of product structures, it becomes indispensable to join the same materials with different states in practical application. The laser- arc hybrid welding could not only keeps the advantage of laser beam welding (such as high energy density and welding speed, low heat input and thermal damage, and low residual stress), but also makes assembled gap less severe before welding, improves the weld appearance and mechanical performance by filling wire during the welding process, etc [3-7]. So it has broad application prospects. The weldability of the 6061 aluminum alloy was studied widely [8-9]. However, there is little previous literature on the study of the welding characters for the same aluminum alloy with different states. Based on the YAG-MIG hybrid welding process of 6061 aluminum alloy with different states, the microstructure before and after the heat treating are investigated, and mechanical properties of the joints are studied systematically.

2. Experimental

The 6061-T651 sheet with 2.5mm thickness, extracted from an aluminum plate with 100mm thickness, and 6061-O aluminum alloy sheet with 2.5mm thickness are used for butt-welding. Their chemical compositions are shown in Table 1. The size of workpieces is 200mm×100mm. And the surfaces of the workpieces were cleaned by chemical method before welding.

A high power YAG laser and a MIG arc equipment with push-pull type wire feeder were assembled to realize the YAG-MIG hybrid welding. The six-axis robot is applied to the mechanical systemic part. The laser is transferred through fiber to the workpiece after being reflected and focused. The focal distance of the lens is 150mm. The laser head has 15° leaning to the normal direction of the workpiece.

Table 1. Chemical compositions of 6061 aluminum alloy and filler wire (wt %)

Material	Si	Fe	Cu	Mg	Zn	Mn	Cr	Ti	Al
6061-T651	0.66	0.42	0.27	0.93	0.05	0.08	0.08	0.04	Balance
6061-O	0.57	0.46	0.20	1.04	0.05	0.05	0.14	0.02	Balance
ER5356	0.25	0.40	0.10	4.50~5.50	0.25	0.05~0.20	0.05~0.20	0.15	Balance

The workpiece is static in the course of welding, while the laser head and MIG blowtorch are brought by the robot to realize the welding process. The samples were protected by argon.

The heat treatment, including solution heat treatment and artificial aging, are performed after the hybrid welding. The solution heat treatment is carried out of 530 ^{0}Cfor 50 minutes. Artificial aging is carried out of 165^{0}C for 8 hours. The samples are extracted from the workpieces before and after the heat treatment, respectively. The butt-joints of 6061 alloy are studies after specific treating.

3. Results and discussion

3.1. Microstructure of 6061 Aluminum alloy with different states by hybrid welding

The optimized welding parameters, including the welding speeding of 1.8m/min, laser power of 2.8kW, and MIG welding current of 50A are chose in the hybrid welding for the butt-joints. Standard metallographic practices for grinding and polishing of the joint were used. All the specimens were etched in the 2% NaOH alkaline liquor and then purged in the 20% HNO$_3$ acid liquor to clean out the remainder after etching. The joint cross-section macrograph of the 6061 alloy with different states is shown in Fig. 1.

The microstructures of different zones of joints before and after the heat treatment are also obtained, shown as Fig. 2 and Fig. 3. The grains of the base metal with O state present as typical rolling structure, while the grains of the base metal with T651 state is approximate regular shape. The strengthen phases always generate and centralize in HAZ before the heat treating. However, those phases are fine or dispersed in the HAZ after the heat treating.

Fig. 1. Cross-section of butt-joint

The HAZ and the fusion zone of the joints are very narrow. And the grains of the fusion zone are melted partially. Besides, the columnar microstructure could be observed near the fusion zone of the weld, and the equiaxed dendritic grains could be observed in the center of the weld. The columnar grains grow from the fusion zone to the centre of the weld. The mass of fine

precipitation strengthening phases disperse homogeneously, which causes unclear for the boundaries of the columnar and equiaxed dendritic grains after the heat treating. Therefore, it infers that the mechanical properties of the joint would be improved obviously after the heat treating.

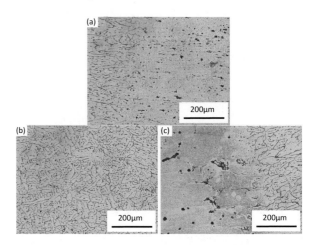

Fig. 2. Microstructure of hybrid welding joint before the heat treating (a) A zone near fusion zone; (b) Weld center of B zone; (c) C zone near fusion zone

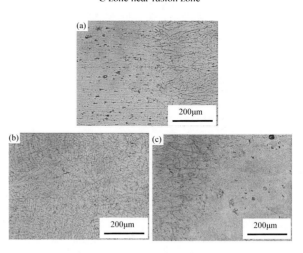

Fig. 3. Microstructure of hybrid welding joint after the heat treating (a) A zone near fusion zone; (b) Weld center of B zone; (c) C zone near fusion zone

The HAZ and the fusion zone of the joints are very narrow. And the grains of the fusion zone are melted partially. Besides, the columnar microstructure could be observed near the fusion zone of the weld, and the equiaxed dendritic grains could be observed in the center of the weld. The columnar grains grow from the fusion zone to the center of the weld. The mass of fine precipitation strengthening phases disperse homogeneously, which causes unclear for the boundaries of the columnar and equiaxed dendritic grains after the heat treating. Therefore, it infers that the mechanical properties of the joint would be improved obviously after the heat treating.

3.2. Microhardness of 6061 Aluminum alloy with different states by hybrid welding

The 6061 aluminum alloy is the typical ageing strengthening material, which causes that the joint softening phenomenon will appear in 6061-T651 alloy after welding. However, the strength of the joint in O state could be improved in the course, for which the O state is softest of all states. Fig. 4 shows the distribution of microhardness of joint by YAG-MIG hybrid welding before and after the heat treating.

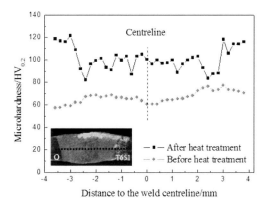

Fig. 4. Distribution of microhardness of joint before and after heat treatment

Compared with the joint before the heat treating, the microhardness of the joint would be improved significantly after the heat treating. The main reason is that the heat treatment could refine the grains of the joint a little and cause a large quantity of ageing strengthening particles to precipitate dispersedly, as shown in Fig. 3. Therefore, for the joint after the heat treating, the average microhardness value of the 6061-O alloy ($118.32HV_{0.2}$) is equal to that of the 6061-T651 alloy ($113.69HV_{0.2}$). And the average microhardness value of the weld (96.50 $HV_{0.2}$) could reach to 83.2% of the base metal.

3.3. Tensile properties of 6061 Aluminum alloy with different states by hybrid welding

The transverse direction tensile test at room-temperature was carried out. Table 2 shows the tensile properties of the joints after the heat treating. It can be seen that the ultimate strength, the yield strength of 0.2% and the elongation of the joints reach the level of the base metal. The main reason is that a large quantity of ageing strengthening particles (such as Mg_2Si, Al_2CuMg, and so on) precipitate dispersedly during the process of the heat treating. All tensile samples fracture in fusion zone and nearby the side of 6061-O alloy. It should be related to significant change of the microhardness value of joint near the side of 6061-O alloy before and after the heat treating.

The scanning electron microscope (SEM) images of fracture surface of welded joints by YAG-MIG hybrid welding are shown in Fig. 5. The tensile fracture of joints presents ductile rupture. And there have many small and shallow dimples, shown in Fig. 5 (a). Meanwhile, there have a large quantity of small pores, and their size is less than 0.2mm, shown in Fig. 5 (b) and Fig. 5 (c). The results show that the effect of the small pores in the weld on the joint's tensile properties is slight.

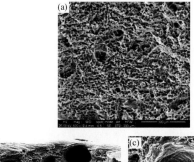

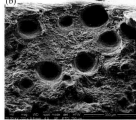 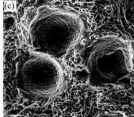

Fig. 5. Fractural SEM images of weld joint (a) Dimple; (b) and (c) oles in the weld zone

Table 2. Tensile test average results of LBW and base metal

After heat treatment	Ultimate strength R_m/MPa	Yield strength of 0.2% $R_{p0.2}$/MPa	Elongation after fracture A/%	Fracture position (four tensile samples)
Hybrid welding	311.25	268.75	6.75	All of the fractures occur in fusion zone nearby the base metal of 6061-O alloy
Base metal	≥290	≥240	≥6	–

4. Conclusions

The main characteristics of the joint (6061-T651 or 6061 O) are columnar microstructure near the fusion zone and the equiaxed dendritic grains in the center of the weld. And the boundaries of those grains become unclear after the heat treating. The average microhardness value of the 6061-O alloy is equal to that of the 6061-T651 alloy. And the average microhardness value of the weld could reach to 83.2% of the base metal. The tensile properties of the joint reach to the level to the base metal after the heat treating, and the tensile fracture presents typical ductile rupture. The effect of the small pores in the weld on the joint's tensile properties is slight.

References

[1]   Xiao YQ, Xie SS, Liu JA, Wang T, (2005) Practical handbook of aluminum fabrication technology. Beijing: Metallurgical Industry Press: 165-166

[2]   Chang YJ, Dong JH, Zhang Y, (2006) A study on the microstructures and properties of the welded 6061 aluminum alloy joint. Welding & Joining (1): 21-26

[3]   Graf T, Staufer H, (2003) Laser-hybrid welding drives VW improvements. Welding Journal 82(1): 42-48

[4]   Xu F, Chen L, Gong SL, Yang J, Zhao XM, He EG, (2011) Weld appearance and mechanical properties of aluminum-lithium alloy by YAG-MIG hybrid welding. Journal of Materials Engineering (10):28-33

[5]   Bagger C, Olsen FO, (2005) Review of laser hybrid welding. Journal of Laser Application 1(17): 2-14

[6]   Yang J, Li XY, Gong SL, Chen L, Xu F, (2010), Characteristics of aluminum-lithium alloy joint formed by YAG-MIG hybrid welding. Transactions of the China Welding Institution 31(2): 83-86

[7]   Gao ZG, Huang J, Cai Y, Li GH, (2006) Laser hybrid welding technology of light metals. Welding & Joining (9): 35-39

[8]   Che HY, Zhu L, Chen JH, Xu WF, Lv XF, (2008) Investigation of tension behavior of butt weld on plate of aluminum alloy 6061. Journal of Lanzhou University of Technology 34(2): 27-30

[9]   Tong JH, Li L, Deng D, Wang FR, (2008) Friction stir welding of 6061-T6 aluminum alloy thin sheets. Journal of university of science and technology Beijing 30(9): 1011-1017.

Pulsed laser welding of thin Hastelloy C-276: High-temperature mechanical properties and microstructure

D. J. Wu, G. Y. Ma, F. Y. Niu and D. M. Guo
Key Laboratory for Precision and Non-traditional Machining Technology of Ministry of Education, Dalian University of Technology, Dalian, Liaoning Province, 116024, PR of China,

Abstract. Based on the welding requirement during the coolant pump manufacture in the nuclear industry, the pulsed laser welding of 0.5 mm thickness Hastelloy C-276 was investigated and the well defect-free weld joint of less than 1mm width was obtained. According to the using temperature request of welding structure, the tensile test of as-welded samples at the high-temperature (200^0C, 300^0C and 400^0C) and the scanning electron microscope observation of fracture were conducted. It was found that some tensile samples were broken in the base metal and others were broken in the weld joint. The results indicated that the high-temperature yield strength, ultimate tensile strength and elongation of as-welded sample satisfied the demand compared to those of base metal, and the high-temperature in the as-welded sample was the same as that of base metal. Compared to the base metal, the plastic deformation behavior of pulsed laser weld joint was not obviously changed, and just the plastic instability in the weld joint was restrained to some extent. The high-temperature fracture morphology in the weld joint indicated the weld joint fracture type belonged to the ductile fracture. At elevated temperature, the dislocation movement and reinforced element segregation resulted in the larger size of voids at the fracture. The same voids characteristic between 200^0C to $400\ ^0$C was due to the unobvious influence of the temperature on the dislocation movement and element segregation.

Keywords: pulsed laser welding, High-temperature Mechanical properties, Microstructure

1. Introduction

Hastelloy C-276 with the excellent corrosion-resistant property is comprehensively applied in the nuclear plant equipment recently. So, aiming at the special working environment, it is important to improve the welding method to ensure the working life of Hastelloy C-276 structure. So far the argon arc welding has been widely adopted to join the sheet of C-276 in the huge structure. Since 1986, M.J. Cieslak et al. conducted the 3mm thickness Hastelloy C-276 joining experiment by the arc welding, and found five mechanisms of the equilibrium phase transformation during the welding. But the brittle phases, which were great damage to the corrosion-

resistant property of Hastelloy C-276, were also found in the weld joint [1-3]. In additional, the weldability of Hastelloy series alloy was investigated according to the Varestraint Testing, Mechanical Testing and Weld Metal Corrosion Testing by M. D. Rowe et al. [4]. Besides the traditional arc welding, the high energy beam welding was also a welding method to join Hastelloy C-276. M. Ahmad investigated the electron beam welding of 3 mm thickness Hastelloy C-276, and analyzed the microstructure of the melting zone and heat affected zone (HAZ). It was not found the detrimental phases in the melting zone [5]. D.J. Wu investigated the pulsed laser welding of 0.5 mm thickness Hastelloy C-276, and obtained the smooth weld joint without defects and HAZ. Meanwhile, he also found that the trend of brittle phase formation was weakened in the pulsed laser welding compared with that in the other welding method [6-9]. However, in the advanced passive nuclear plant, the thin Hastelloy C-276 has been applied in the manufacture of coolant pump can, and the working temperature of coolant pump can is between ambient to nearly 400 degree Celsius. So, the high-temperature mechanical property of welding structure is an important performance for evaluating the feasibility of welding method. But, so far few opening literatures were published on the high-temperature mechanical property of weld joint.

Although it was proved that the pulsed laser welding could decrease the trend of brittle phase formation to be beneficial to holding the mechanical property and corrosion-resistant property of Hastelloy C-276, the mechanical property of weld joint at high temperature was not analyzed. In the current work, on the basis of the investigation of the thin Hastelloy C-276 of pulsed laser welding, it was to evaluate the tensile property of 0.5 mm thickness Hastelloy C-276 weld joint at the 200 ^{0}C, 300 ^{0}C and 400 ^{0}C and then analyze the fracture feature by SEM.

2. Experiment procedure

In the experiment, the 0.5 mm thickness Hastelloy C-276 treated at 1170 degree Celsius for 0.3 h before quenching in water was employed. The pulsed 1064 nm Nd:YAG laser (focal length 200 mm and focal beam diameter approximately 600 um) was used to join the Hastelloy C-276 with the pure Ar as the gas-shielding. The laser pulse was the multimode gauss type at the space domain and the even type in the time domain. The laser welding parameters included the single pulse energy, duration, repetition, welding velocity and defocus. According to the previous investigation, the well smooth weld joint with free defect could be obtained under the 1.5 J, 6 ms, 30 Hz, 100 mm/min and -1 mm defocus condition [7-9]. Hence, the tensile test sample was prepared based on the welding parameters above.

The tensile tests were carried on Instron 5500R equipment with gauge length 60 mm and tensile velocity 5 mm/min at 200 °C, 300 °C and 400 °C, respectively. According to the GB/T 2651-2008, the tensile sample (shown in Fig.1) was machined by the precision wire-electrode cutting. After the tensile test, the fracture morphology was observed by JEOL JSM series scanning electron microscope (SEM).

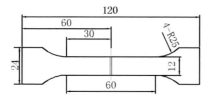

Fig. 1. Size of tensile sample (mm)

3. Analysis and discussion

3.1. Weld joint profile

Because the weld joint profile would impact the tensile result directly to some extent, so the cross-section profile of weld joint was observed to ascertain the flatness of weld joint. Fig. 2 showed the typical weld joint cross-section profile under the 1.5 J, 6 ms, 30 Hz, 100 mm/min and -1 mm defocus condition. It was found that the weld joint with no obvious HAZ was smooth. Like this, during the tensile test, the tensile stress in the weld joint would be the same as that in the base metal due to the same thickness. Hence, the tensile test result could show the real mechanical property of weld joint.

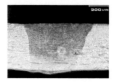

Fig. 2. Typical weld joint cross-section profile

3.2 Tensile test

Fig.3 showed the images of stretched samples. It was seen, regardless of at 200 °C, 300 °C or 400 °C, the fracture happened in the weld joint or the base metal. The necking-down with the near 45 degree rupture direction was visible at the base metal fracture. But, at the weld joint fracture, there was no obvious necking-down, and the fracture ran through the weld joint. The necking-down at the base metal fracture indicated that the base metal occurred the plastic instability, and the near 45 degree rupture direction was due to the rupture characteristic of single-phase alloy. On the other hand, at the weld joint fracture, the plastic deformation of stretched section of sample was even, so the tensile characteristic of weld joint was basically considered as the same as that of base metal.

Table 1 showed the results of tensile test. It was found, regardless of at 200 °C, 300 °C or 400 °C, the Yield Strength (YS), Ultimate Tensile Strength (UTS) and Elongation (EL) did not show obvious difference at the different fracture position under the same temperature condition. Combining with the results in Fig.3, it was concluded that the YS and UTS of pulsed laser weld joint

Table 1. Normal composition of Hastelloy C-276 (wt, %)

Temperature	Sample No.	Yield Strength (MPa)	Ultimate Tensile Strength (MPa)	Elongation (%)	Fracture Position
200°C	1#	288	737	51	Base Metal
	2#	308	765	43	Weld joint
	3#	323	740	44	Base Metal
300°C	1#	310	744	48	Weld joint
	2#	314	765	44	Weld joint
	3#	314	741	45	Base Metal
400°C	1#	292	730	55	Weld joint
	2#	268	727	45	Weld joint
	3#	296	726	50	Base Metal

were not decreased compared with those of base metal. In addition, it was found that the YS, UTS and EL at 200 °C were nearly the same as those at 300 °C. But at 400 °C, the YS, UTS were declined and the EL was increased compared to those at 200 °C and 300 °C. These characteristics of weld joint were similar with the result of A.K. Roy' investigation on the base metal [10] and the difference of value may be related to the different strain rate and the fabrication process of base metal. The results indicated that the pulsed laser welding could joining the thin Hastelloy C-276 well, and the tensile properties of weld joint between 200 °C to 400 °C were not impacted by the welding process.

Fig. 3. Images of stretched samples at (a) 200 °C, (b) 300 °C and (c) 400 °C

3.3. Stress strain analysis

For in-depth understanding the plastic deformation behavior during the tensile process, the curves on Engineering Stress VS Strain and True Stress VS True Strain (log coordinate) were shown in Fig.4 and 5, respectively.

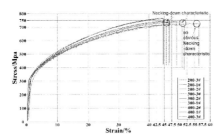

Fig. 4. Engineering Stress VS. Strain

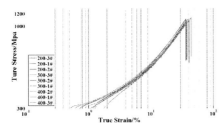

Fig. 5. True Stress VS. True Strain

In Fig.4, it was found, no matter what was the temperature, the stretched samples presented the even plastic deformation, and just in the samples of weld joint fracture there was no obvious necking-down characteristic, which agreed with the observed results in Fig. 4. But, in the sample 3# at 400 °C, the necking-down was also unobvious. This may be because the fracture was close to the weld joint so that the weld joint impacted

the curve on the necking-down expression. The even plastic deformation indicated the weld joint and the base metal both presented the similar tensile process at the given temperature. The unobvious necking- down characteristic in the samples of weld joint fracture demonstrated that the plastic instability in the weld joint was restrained compared to the base metal during the tensile process.

In Fig. 5, it was noted that all curves presented the similar non-linear relationship, and the curve slope at large strain was also nearly identical. According to the Ludwigson Relation which was given by Eq. (1), the curve slope at large strain in Fig.6 could be used as the value of n [10-11]. Based on that, it was concluded that the abilities of the ultimate even strain in the weld joint made almost no difference compared with those in the base metal.

$$\sigma = K\varepsilon^n + \Delta \qquad (1)$$

where, σ is the true stress. K, ε, n and Δ are the strain hardening coefficient, true strain, strain hardening index and Ludwigson correction at low strains, respectively. The n decides the ability of the ultimate even strain in forming of material.

The analysis results about the stress-strain curves indicated, compared to the base metal, the plastic deformation behavior of pulsed laser weld joint was not obviously changed, and just the plastic instability in the weld joint was restrained to some extent.

3.4. Fracture microstructure analysis

The Fig.6 showed the fracture morphology in the base metal. It was found, the dimple appearance was in the fracture at all temperature, and the number and size of voids at elevated temperature were increased compared to that at the room temperature, but there were not obvious difference on the voids characteristic at 200 °C, 300 °C and 400 °C. The dimple characteristic in the fracture indicated the fracture type of stretched samples was the ductile fracture. The increased number and size of voids were attributed to the dislocation movement and element segregation at elevated temperature, but under the condition of between 200 °C to 400 °C, the same voids characteristic was because the influence of the temperature on the dislocation movement and element segregation was not obvious.

In addition, the fracture morphology of weld joint was shown in Fig.7. Compared to the base metal fracture, it was seen that the weld joint fracture showed the tiny deep dimple with large density, and the size of voids was larger than that at the room temperature. At 200 °C, 300 °C and 400 °C, the voids showed the same characteristic. The tiny deep dimple with large density was due to the grain refinement and element segregation resulted from the rapid solidification during the pulsed laser welding [12], and the fracture type was still ductile fracture. The larger size of voids was also due to the dislocation movement and reinforced element segregation at elevated temperature. However, between 200⁰C to 400 °C, the same voids characteristic in weld joint fracture was due to the unobvious influence of the temperature on the dislocation movement and element segregation, which was similar with that in the base metal fracture.

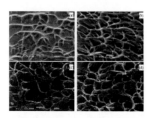

Fig. 6. Fracture at the position of base metal at (a) Room Temperature, (b) 200 °C, (c) 300 °C and (d) 400 °C

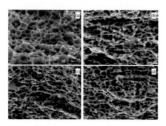

Fig. 7. Fracture at the weld joint at (a) Room Temperature, (b) 200 °C, (c) 300 °C and (d) 400 °C

4. Conclusion

The pulsed laser welding could joining the thin Hastelloy C-276 well, and the tensile properties of weld joint between 200 °C to 400 °C were not impacted by the welding process compared to the base metal. Between 200 °C to 400 °C, compared to the base metal, the plastic deformation behavior of pulsed laser weld joint was not obviously changed, and just the plastic instability in the weld joint was restrained to some extent. At the base metal fracture, the dimple characteristic indicated the fracture type of base metal was the ductile fracture. The change of voids at elevated temperature was attributed to the dislocation movement and element segregation. However, between 200 °C to 400 °C, the same voids characteristic was because the influence of the temperature on the dislocation movement and element segregation was not obvious. At the weld joint fracture, the ductile fracture was also presented. At elevated temperature, the dislocation movement and reinforced element segregation resulted in the larger size of voids. Between 200⁰C to 400 °C, the same voids characteristic at the weld joint fracture was also due to the unobvious influence of the temperature on the dislocation movement and element segregation.

Acknowledgments: This research was supported by the National KeyBasic Research Program of MOST of P.R. China (Grant No. 2009CB724307) and the National Natural Science Foundation of China (Grant No. 51175061).

References

[1]   M.J. Cieslak, G. A. Knorovsky, T.J. Headley, A. D. Romig, Jr. (1986) The Use of New PHACOMP in Understanding the solidification microstructure of Nickel base alloy weld metal. METALLURGICAL TRANSACTIONS A, 17A, 2107-2116

[2]   M.J. Cieslak, T.J. Headley, A. D. Romig, Jr. (1986) The Welding Metallurgy of HASTELLOY Alloys C-4, C-22, and C-276. METALLURGICAL TRANSACTIONS A, 17A, 2035-2047

[3]   J.S. Ogborn, D.L. Olson, M.J. Cieslak. (1995) Influence of solidification on the microstructural evolution of nickel base weld metal. Materials Science and Engineering A, 203, 134-139

[4]   M. D. Rowe, P. Crook, G. L. Hoback.(2003) Weldability of a Corrosion-Resistant Ni-Cr-Mo-Cu Alloy. Welding J., 313s-320s

[5]   M. Ahmad, J.I. Akhter, M. Akhtar, M. Iqbal, E. Ahmed, M.A. Choudhry. (2005) Microstructure and hardness studies of the electron beam welded zone of Hastelloy C276. Journal of Alloys and Compounds, 390, 88-93

[6]   G.Y. Ma, D.J. Wu, D.M. Guo.(2011) Segregation Characteristics of Pulsed Laser Butt Welding of Hastelloy C-276. METALLURGICAL TRANSACTIONS A, 42A, 3853-3857

[7]   D.J. Wu, G.Y. Ma, Y.Q. Guo, D.M. Guo. (2010) Study of weld morphology on thin Hastelloy C-276 sheet of pulsed laser welding. Physics Procedia, 5, 99-105

[8]   Y.Q. Guo, D.M. Guo, G.Y. Ma, D.J. Wu.(2011) Numerical and Experimental Investigation of Pulsed Laser Welding of Hastelloy C-276 Alloy Sheets. Advanced Materials Research, 154-155, 1468-1471

[9]   G.Y. Ma, D.J. Wu, Z.H.Wang, Y.Q. Guo, D.M. Guo. (2011) Weld Joint Forming of Thin Hastelloy C-276 Sheet of Pulsed Laser Welding. Chinese journal of lasers, 38, 0603014

[10] A.K. Roy, J. Pal, C. Mukhopadhyay.(2008) Dynamic strain ageing of an austenitic superalloy-Temperature and strain rate effects. Materials Science and Engineering A, 474, 363-370

[11] K. G. Samuel. (2006) Limitations of Hollomon and Ludwigson stress–strain relations in assessing the strain hardening parameters. J. Phys. D: Appl. Phys., 39, 203-212

[12] G.Y. Ma, D.J. Wu, D.M. Guo. (2011)Segregation characteristics of pulsed laser butt welding of Hastelloy C-276. Metallurgical and Materials Transactions A, 42A, 3853-3857.

Basic phenomena in high power fiber laser welding of thick section materials

Salminen, A.[1,2], Lappalainen, E.[1] and Purtonen, T.[1]

[1]   Lappeenranta University of Technology, Tuotantokatu 2, FI-53850 Lappeenranta

[2]   Machine Technology Centre Turku Oy, Lemminkäisenkatu 2, FI-20540 Turku

Abstract. The laser welding process is gaining ever growing attention. It has been found to be of particular interest in welding of automotive applications, where lasers have so far mostly been used for welding of thin sheet. A new generation of high power lasers has entered the market during the last couple of years. The easy availability of higher power brings the laser welding also to applications of thick section welding. Compared to old generation lasers, e.g. the CO_2 laser, new lasers have better absorption, better beam quality and higher electrical efficiency. They are not as sensitive to metal vapor on top of the keyhole as CO_2 laser in respect of process stability and resulting weld shape. This study investigates the performance, potential and problems of the deep penetration laser welding of S355 EN 10025 structural steel. Thicknesses up to 25 mm have been show potential in welding with laser power between 15–30 kW. Visual examinations of the macrographs for all and hardness tests of part of the welded specimens were made. First versions of quality windows were drawn based on the results of the experiments.

Keywords: welding, laser welding, low alloyed steel, weld quality

1. Introduction

Laser welding has been considered to be an ideal welding method for thin sheet welding especially in car manufacture. Only recently the real potential of laser welding has been realized also in welding of thicker plates. New high power lasers for welding, the fiber and the disk laser, allow use of fiber optics and high power with high beam quality simultaneously. This enables possibilities for system and production automation and easy change between processes. When welding is carried out with speed of over one meter per minute and penetration reached simultaneously is several millimeters, even up to 30 mm, the productivity improvement to conventional multipass arc welding is considerable.

Many of the really potential applications still have some restrictions in use of laser welding e.g. standard procedures and accepted practices for production and testing assuring acceptable weld quality. These procedures and practices are based on long experience of the behavior of arc welding and the resulting weld. Since output power over 10 kW has become available for fiber

and disk lasers, the mechanical engineering companies using thick steel sections have shown increasing interest in this new technology. Even though CO_2 laser reached 40 kW laser power in mid 1990s, the new lasers seem to allow, for the first time, a really scalable laser. Modular construction in fiber and disk laser gives potential to reach high power with numbering up the basic components of low power to scale the power.

The new lasers also reach considerably higher wall plug efficiency to lower the need of electricity and lower the cost of use. Furthermore, overall efficiency of them in process is further increased by the fact that they are not as sensitive to metal vapor on top of the keyhole as CO_2 lasers. One of the first evaluations was carried out against Nd:YAG laser by Verhaege et al. from TWI. Already that time the improvement in weld productivity was found out to follow the improvement of beam quality [1]. Detailed information about the actual performance of these lasers is, however, still lacking [1, 2]. Lot of effort is currently put in development of laser-arc-hybrid welding (LAHW) for pipeline construction. The new generation lasers have shown good performance in this application by various authors [3, 4, 5, 6, 7]. The discussions with laser manufacturers also reveal that the sales of lasers for welding applications are increasing.

In the keyhole laser welding the high laser intensity is vaporizing and melting the material. Typically the vaporization causes metal vapor, often called the plume, above the formed keyhole. It has been reported in case of welding with laser close to 1000 nm wavelength e.g. YAG, disk or fiber laser, that there is a visible tall plume and its upper part has low refraction index which should exert an effect of refraction, defocusing and Rayleigh scattering to reduce penetration. Spattering and humping phenomena are observed to occur due to the effect of laser-induced plume [8].

The weldability of high power lasers has been studied to some extent, but most of the reports deal with hybrid welding. A comparison of effect of focal point has been carried out by Katayama et al. [8] with austenitic stainless steel. Figure 1 illustrates this case. Figure shows clearly the fact that smaller focal point enables typically higher welding speed for welding of given penetration. This effect is fading out only when the welding speed is lowered down to 1 m/min and lower. This is probably the intensity range in which the plasma starts to play an important role in absorption.

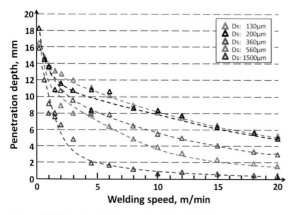

Fig. 1. Welding of stainless steel 1.4301 with different focal point diameters, bead on plate joint configuration, laser power 10 kW, focal point position 0 mm, and shielding gas Ar 30 l/min. [8]

When going into further details for certain material and thickness, different performance can be shown e.g. for different laser power levels. A rather large collection of data published about laser welding with high power lasers using fiber optics is published by author [9]. There are only few published weldability lobes of this kind available yet. Figure 2 shows an example of weldability lobe with 15 kW fiber laser for S355 steel of 8 mm in thickness. From this figure it can be seen that with certain laser power welding speed should be high enough to achieve full penetration. For example with 15 kW laser power, full penetration is attained with a welding speed range between 1.8 and 5.5 m/min.

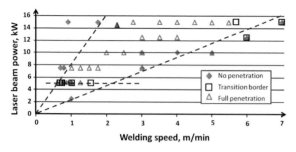

Fig. 2. The welding performance window for S355 steel of 8 mm in thickness in respect of welding speed reachable with certain laser power. Laser 15 kW fiber laser, fiber core diameter 200 μm, collimation length 150 mm, focal length 300 mm [10]

One of the major advantages of laser welding is the fact that compared to traditional arc welding methods, laser welding gives a narrower heat-affected zone, due to the lower energy input applied per unit length, reducing work piece thermal distortions. This is all due to very efficient

joining process. One reason for laser welding not being very successfully applied in industry is the way of productivity thinking. Typically productivity of welding is calculated by efficiency to melt filler material. Autogenous laser welding has almost no productivity with this consideration and LAHW can reach poor productivity. When the real figure for this calculation should be the joined joint cross section per time unit, the utilization of laser welding partly requires total change of thinking. In thick section welding most efficient processes are so called narrow gap (NG) welding techniques with TIG, MAG or SAW, the latter being considered to be the most productive. NG SAW can produce for example full penetration weld in 50 mm thick low alloyed steel in speed of 0.49 h/m, which means average welding speed of 20.4 mm/min. 60 mm thickness can be welded in 0,6 h/m i.e. speed of 16.7 mm/min. This calculation does not take into account the welding of root pass and the machining after root pass welding prior to actual welding [9]. Another, actually quite a logical, way to compare the actual productivity is to measure the speed of joining i.e. what is the speed on which a certain thickness of material can be joined to another part of equal thickness. This comparison, the joining speed (V_J), was presented by professor Kutsuna [12] and it is calculated according to equation (1), where t is the material thickness in mm and V_W the welding speed in mm/min.

$$V_J = t*V_W \qquad (1)$$

With this equation the joining speed of CO_2 laser was calculated to be with 12 kW laser power 2 times higher and fiber laser (10 kW) to be 5.6 times faster than SAW. The examples of NG SAW shown above give joining speed value of 0.1 mm²/min.

There are few different error types identified for high power welding, which have not been defined for CO_2 laser welding. The root side ball shaped formation is the one; bead side aggressive material removal is the second one. The root side ball formation is in connection of laser power, with inadequate power the keyhole reaches to the bottom, but is disturbed producing uneven quality. Bead side aggressive material removal, the groove formation of top side, is in connection with power density.

Previous studies have shown that the welding process produces good quality in case the focal is point below the top surface of the work piece. If the focal point position is above the surface, it causes strong material removal behavior to the weld. From the Schlieren images it can be seen that a lot of material is removed from the weld as spatters aiming to backwards in an opening angle of 45 degrees. Typically this happens in upper side whereas the root side is formed to have extensive amount of material in continuous and regular pearl-like shapes [13]. When the actual weld flaws are added into speed, penetration, power diagram, the actual weldability windows are created. Figure 3 shows a good example of this kind of window.

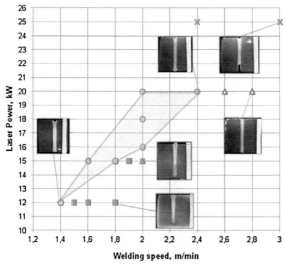

OAcceptable Weld ✕Cut through △Imperfections ■Incomplete penetration

Fig. 3. The weldability window for 20 mm thick S355 steel. Laser 30 kW fiber laser, fiber core diameter 200 µm, collimation length 140 mm, and focal length 300 mm. [14]

This paper investigates thick section laser welding of S355, EN 10025-2 structural steel at high laser power. The aim of the study is to identify the processing parameters suitable for the autogenous laser welding and made comparison between joint manufacture and productivity of different parameters.

2. Experimental procedure

2.1. Motivation

The aim of this study was to find out the best combination of laser welding parameters and equipment setup leading to high-quality full-penetration welds with a minimum amount of weld defects in butt-joints of thick steel plates. Also finding the limits for welding these material thicknesses was one aim for the experiments.

Welding with high power laser was performed on S355 EN 10025-2 structural steel with following variables: laser power, welding speed, focal point position and material thickness and joint configuration. In one setup also the diameter of working fiber was varied.

2.2. Material and equipments

The steel used in experiments, S355 EN 10025, is common steel for structural fabrication: from offshore structures and mining equipment to wind mill components. Minimum yield strength is 355 MPa and tensile strength 470–630 MPa. Nominal chemical composition of the steel is given in Table 1. Test plates of S355 10, 12, 15, 20 and 25 mm in thickness were manufactured into test pieces of 200 mm x 75 mm in size.

The joint preparation was carried out with four different means: plasma cutting, oxygen laser cutting, machining and abrasive water jet cutting. Mainly oxygen laser cutting, machining and abrasive water jet cutting were used. Some of the joints were prepared with grit blasting to clean the groove surface. Also protective paint was removed by grit blasting from most of the test pieces. In some cases the work pieces were before welding tack welded together such that the set air gap in the joint was not altering during welding

Table 1. Nominal chemical composition of steel S355 EN10025

C max	Si	Mn	S max	P max	Cr max	Mo max
0.1	0.1/ 0.55	1.00/ 1.65	0	0	0.25	0.1
Nb max	**V max**	**Ti max**	**Ni max**	**Cu max**	**Al total**	**N max**
0	0.1	0.03	0.5	0.3	0.015/ 0.055	0

Totally four different laser set ups were tested (Table2), two of them being fiber laser systems and two disk laser based system. Typically the laser welding head was mounted on an industrial robot. The robot moved the laser welding head over the clamped work pieces at the programmed welding speed. The samples were tightly fixed on a flat position on the fixture such that welding optics was perpendicular to direction of welding and work piece surface. In test system 2 the angle between laser beam and work piece was 83 degrees, to protect optical equipment from possible back reflection of the laser beam. Argon with a flow rate of 20 l/min, delivered to the weld via side copper tube was used as a shielding gas. Power levels on work piece between 10–30 kW were used.

Table 2. The systems used in experiments.

Test system	1	2	3	4
Laser type	disk	fiber	disk	fiber
Max. power, kW	12	15	16	30
Fiber ⌀, µm	0.2	0.2	0.2,0.4,0.6	0.2
Collimation length, mm	200	150	150	140
Focal length, mm	300	300	280	300

2.3. Experimental design

The laser power was tested at several different power levels, partly overlapping between different laser setups and laser types. The goal was to achieve acceptable weld for each thickness. In view of the numerous parameters affecting the laser welding process, a procedure of trial and error was mainly used to test a range of parameters suitable for acceptable weld. Based on literature and previous experiments preliminary experiments were carried out for each material thickness. Then the optimum focal point position and power-speed combinations were tested. Also part of the previous experiments was repeated with different laser set up. For example best of the fiber laser results were tested also with disk laser. In next Fig. 4 used welding speed – laser power combinations for different material thicknesses are presented. As from this figure can be seen, the area of 15 kW laser power was mostly tested.

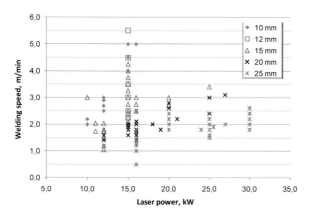

Fig. 4. The laser power, welding speed combinations used for welding trials of different material thicknesses.

Focal point position was changed according to testing results from top of the workpiece downwards below the surface. Typical focal point position was – 7.5 mm which was noticed in preliminary tests to be the best one especially for thicknesses of 20 mm. Focal point position was not widely varied during experiences to get more information from influence of other parameters.

2.4. Weld quality assessment

All the welds were evaluated visually after welding. Weld bead and root sides were photographed. A set of samples were selected for metallographic evaluation. The samples were sectioned, grinded, polished and etched prior to photography. The acceptance criteria in visual inspection and evaluation of the cross-sections of macrographs for all selected samples were according to EN ISO 13919-1 standard (15).

The correlation between consumed energy and weld performance was also investigated. Laser line energy (E_L) was calculated based on the equation (2), where P_L is laser output power (kW) and V_W is welding speed (m/min):

$$E_L = P_L / V_W \qquad (2)$$

3. Results and discussion

3.1. Inspection of the welds

Visual inspection showed that it is possible to reach acceptable quality for the welds. The parameter windows are occasionally quite narrow. The preliminary evaluation was carried out based on this. The typical flaws seen were those of predicted by the literature. Occasionally beam was engraving the top of weld bead leaving a deep groove on top of the weld. In many cases weld with good results from visual testing was not acceptable quality from inside the weld. It was noticed that it is important to make cross section macrographs during the parameter testing before defining parameters for the next experiment. For example in focal point position testing, certain focal point positions lead to good visual quality in bead and root side of the weld but there was a crack inside the weld, so that those parameters were not potential for further studies.

Typical weld flaws are: lack of penetration, incompletely filled groove and cracking. Typically the cracking, in case of butt joints with machined groove sides, was the weld joint midsection cracking in cases with partial penetration. The crack is typically horizontal and close by the weld root.

3.2. Weld penetration

Weld penetration depth is an important feature to measure from the welds in order to justify the performance on laser welding. Welding speed achieved with different laser powers is typically dependent on the material thickness. In next Fig. 5 it is presented how these experiments are placed in this welding speed – penetration depth – window. In Fig. 5 are presented both disk and fiber laser tests in all used focal point positions and focal spot sizes. Both laser types were used in 12 kW and 15 kW power levels.

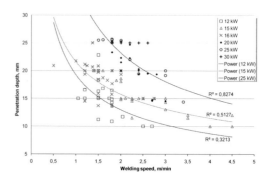

Fig. 5. The effect of welding speed on penetration depth with various laser powers.

The welding speed seems to vary a lot within different parameter combinations. The correlation coefficients (R^2 values) of the power curves fit are poor due to low number of experiments and large variation of experiments. This is caused by different other variations not shown in this figure e.g. the joint manufacture varies and the focusing conditions are different between same laser power tests. However we can see that e.g. the difference between 15 and 16 kW is not considerable if conditions are equal.

3.3. Effect of joint manufacture

The typical set up for laser welding is to use minimum or practically no air gap in a butt joint. To have very high quality joint edge manufacturing is in order to ensure the proper joining and to avoid the lack of material in joint. Usually machining is being used to get good enough quality, but also abrasive water jet cutting or laser oxygen cutting are used. Influence of joint edge manufacturing to penetration depth is presented in next Fig. 6.

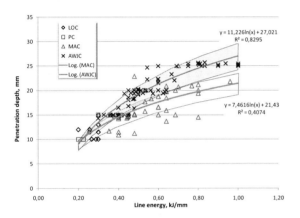

Fig. 6. The penetration depth as function of line energy introduced achieved with different joint edge manufacture. LOC is laser oxygen cutting, PC plasma cutting, MAC machining and AWJC abrasive water jet cutting. Painted areas represent ±10 % area from logarithm trendlines.

From Fig. 6 it can be seen that with same amount of line energy with abrasive water jet cut penetration depth can be higher than with edges made by machining. This seems to be true in all tested thicknesses. In this point it has to be remembered that abrasive water jet cutting was made in different, ordinary, workshops and the machined edges were all done with same tools, machines and methods. Although from quality of welds it was also noticeable that in lower thicknesses laser welding requires better quality cutting to achieve results with high enough quality.

In next Fig. 7 it is presented how the joint edge manufacturing influences the cross section shape of the weld. From the figure it can be seen that with abrasive water jet cutting and machining with grit blasting the penetration depth is almost the same. When altering the machined surface with grit blasting to the machined surface without grit blasting, penetration depth decreases but the weld width increases. This explains partly the difference in penetration depths presented in Figure 6. From these figures it can be noticed also that laser

welding does not require the highest possible quality for the groove surface to achieve the deepest penetration. At the same time it can be seen also that the deeper the penetration the more there is cracking in the bottom of the partial penetration.

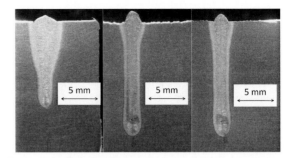

Fig. 7. Cross sections of welds with a) MAC b) MAC + grit blasting and c) AWJC joint edges. Test system 3, fiber diameter 0.4 mm, welding speed 1,8 m/min, laser power 12 kW, focal point position -6 mm, material thickness 25 mm.

3.4. Productivity

As mentioned earlier the typical welding productivity evaluation does not work with favor to the autogenous laser welding. Therefore the performance was evaluated with terms of joining speed developed to give information of the speed achieved to join the two parts together. Results gained with various welding set ups show really high joining speed, see Fig. 8. The joining speed is increasing with the laser power and seems to show some optimum spot size for achieving the maximum speed. As the figure also shows there is quite a large variation in the values. This is likely to be due to the considerable variations in test arrangements and joint preparations.

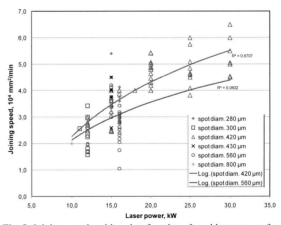

Fig. 8. Joining speeds achieved as function of used laser power for various optical set ups, spot diameters used.

The laser beam focal point diameter has an influence also to the weld cross section. With larger spot the weld width increases but the penetration depth decreases. Respectively the quality of the weld seems to become better. Top surface of the weld is wider and smoother and the amount of cracks in the cross section is smaller.

4. Conclusion

This study of laser welding of structural steels with high-power fiber lasers was able to achieve certain maximum welding speeds for different material thickness and laser power levels for S355 steel with extremely narrow welds of a good visual appearance. It can be concluded that; Too high quality joint edge can decrease the penetration depth in high power laser welding. With abrasive water jet cut joint edge it is possible to reach even 20 % deeper penetration with the same amount of line energy than with machined joint edge. When comparing grit blasted and not grit blasted machined joint edge, with grit blasting the penetration is deeper and without grit blasting weld is wider. Very high productivity is possible in laser welding with high power fiber and disk laser because of the good beam quality and small focal spot size. Especially improving the quality of the weld and finding reasons for different kind of flaws in the weld need more research.

Acknowledgements: The authors are grateful to Finnish Academy for funding of Pamowe project. Authors are also grateful for laser teams in University of Lulea, SLV Rostock, Winnova in Laitila and IPG Photonics in Burbach for arranging an opportunity to carry out the experimental. The authors also want thank Mr. Antti Heikkinen from Lappeenranta University of Technology for performing the metallography.

References

[1]   Veerhaeghe G, Hilton P. Battle of the sources – using a high power Yb-fibre laser for welding steel and aluminium. In: Junek L, editor. Proceedings of the IIW Conference on Benefits of New Methods and Trends in Welding to Economy, Productivity and Quality, 10–15 July, Prague, Czech Republic; 2005. pp. 188–200.

[2]   Quintino, L.; Costa, A.; Miranda, R.; Yapp, D.; Kumar, V.; Kong, C. J. Welding with high power fiber lasers - A preliminary study. Materials & Design. 2007, 28, pp. 1231-1237.

[3]   Rethmeier, M.; Gook, S.; Gumenyuk, A. Perspectives of application of Laser-GMA-hybrid girth welding for pipeline construction. In: Prof. G. Turichin, editor. Proceedings of the VI International Conference "Beam Technologies & Laser Applications". September 23-25, Saint-Petersburg, Russia, 2009; 278-288.

[4]   Miranda, R.; Costa, A.; Quintino, L.; Yapp, D.; Iordachescu, C. Characterization of fiber laser welds in X100 pipeline steel. Materials & Design. 2009, 3: 2701–2707.

[5]   Lappalainen, E., Purtonen, T., and Salminen, A., Effect of arc parameters in high brightness laser hybrid arc welding of structural steel. In: 30th International Congress on Applications of Lasers & Electro-Optics: ICALEO 2011, 24-27 October, Orlando, Florida, USA, Laser Institute of America, 10 pp

[6]   Howse, D. S., Scudamore, R. J., Booth, G. S. Yb Fibre Laser/MAG Hybrid Processing for Welding of Pipelines. IIW IV-880-05. 2005.

[7]   Yapp, D., Kong, C-J. Hybrid Laser-Arc Pipeline Welding. IIW XII-1887-06. 2006. Cranfield. Welding engineering research centre Cranfield university.

[8]   Katayama, S., Kawahitoa, Y. and Mizutan, M., Elucidation of laser welding phenomena and factors affecting weld penetration and welding defects, Physics Procedia 5 (2010) 9–17

[9]   Salminen, A., Piili, H., and Purtonen, T., The characteristics of high power fibre laser welding. Proc. IMechE Vol. 224 Part C: J. Mechanical Engineering Science, Professional Engineering Publishing, Volume 224, Number 5 / 2010, ISSN 0954-4062, pages 1019-1029.

[10] Kaplan, A. and Wiklund, G. Advanced welding analysis methods applied to heavy section welding with a 15 kW fiber laser. Proceedings of the International Conference on Welding of the IIW. July 12-17, 2009, 62, pp. 295-300.

[11] Lukkari, J., Efficient SMA welding for manufacturing of the parts of energy industries. Welding technology days, Jyväskylä 22.4.1010, The Welding Society of Finland, 36 pages, In Finnish

[12] Kutsuna, M., Advanced Laser Based Processes for Improving Manufacturing Systems, Proc. Int. Conf. Shipyard Applications for Industrial Lasers SAIL2008. 18th and 19th August 2008 Williamsburg, Virginia. PennState University Applied Research Laboratories. 72 pp.

[13] Salminen, A., Lehtinen, J. and Harkko, P. The effect of welding parameters on keyhole and melt pool behavior during laser welding with high power fiber laser. In: 27th International Congress on Applications of Lasers & Electro-Optics ICALEO 2008, 20-23 October, Temecula, California, USA; 354-363.

[14] Sokolov, M., Salminen, A., Kuznetsov, M. and Tsibulskiy, I., Laser welding and weld hardness analysis of thick section S355 structural steel, Materials and Design, 2011, vol. 32, nro. 10, p. 5127-5131,

[15] EN ISO 13919-1:1996, Welding – Electron and laser-beam welded joints- Guidance of quality levels for imperfections – Part 1: Steel.

Laser Technology – Additive Manufacturing

Characterisation of test results of the performance of additive manufacturing parts obtained by selective laser sintering processes incorporating glass and carbon fibres

F. Roure i Fernández[1], M.M. Pastor i Artigues[1], M. Ferrer Ballester[1], J. Minguella i Canela[2], R. Uceda Molera[2], A. Arjona Mora[2]

[1] Dept. de Resistència de Mat. i Estructures a l'Eng. i Dept. d'Eng. Mecànica, Universitat Politècnica de Catalunya

[2] Fundació Privada Centre CIM

Abstract. Selective Laser Sintering, and in general Additive Manufacturing Processes are becoming mature technologies; in the sense that after many years of market introduction there are some well-known manufacturers that provide fully operative systems that are even utilized for direct parts manufacturing. However, the parts final users are becoming more and more demanding, so they can integrate the AM parts in complex assemblies; thus increasing the materials requirements and the final part's properties. Building into previous works on AM materials, the purpose of the present paper is to explore the properties yielded by the introduction of Glass short fibre randomly to polyamide and to assess the performance of the probe parts taking into consideration the AM manufacturing direction in the building platform and the Energy Density applyied to the operations. Folowing to a literature review, the paper first describes the statistical approach and the samples preparation. Then, the samples are characterized by tensile tests and the properties of the material are described for the different manufacturing strategies adopted. The outcomes of the work are highly applicable to the AM parts manufacture process, as the construction of parts in the building platform is usually dictated by the objective of optimizing the use of the total volume; thus implying that not all parts are manufactured in the same main construction direction.

Keywords: Selective Laser Sintering, Additive Manufacturing, Materials, Properties, Performance

1. Introduction

The opportunity to work in virtual environments, from design to testing and planning of production, gives engineers the power to do a wide series of tests and modify quickly several design proposals. However, it is necessary to fully understand how the factors involved in the production AM processes affect the parts obtained in order to take the most out of the prototyping process.

The objective of the present work is to undertake a series of experimental tests with composite materials in Selective Laser Sintering process. For this reason, test specimens of a PA220 base material and different weight fractions of glass-fiber are analysed, in order to assess the performance in terms of mechanical properties of manufactured prototypes and so to determinate the optimal configuration of parameters for SLS equipment for the studied composite material combination.

The results provide valuable information to ETSEIB's scholars for further studies and FEA simulation models and also have practical application for the improvement of the capabilities portfolio of Fundació Centre CIM.

1.1. Literature review

Selective Laser Sintering is one of the AM processes most widely applied in the industry. With this method, a powder base material is preheated to a temperature near its fusion point and a high power laser acts on a layer of material following the profile of the corresponding 2D section. Then, a new layer of material is added and the procedure restarts until the model is complete.

The commercially available SLS equipments have operating parameters established on their configuration. In most cases these presetting provide good results. Nevertheless, there is research work like Raghunath and Pandey [1], H.J. Yang et al. [2] that study the effects of each variable in order to optimize an ad-hoc configuration for specific applications. The main construction parameters in SLS process are the energy density applied, the build orientation of layers, the curves of temperature in pre-heating, sintering and cooling stages, the temperature gradients inside the build chamber, layer thickness; others variables are material dependant such as the type of material, consistency of powder, size and geometry of the powder particles. Most authors agree in evaluating the level of sintering as function of the energy density (ED) applied in the process. Nelson's [3] definition calculates the ED on the work surface per unit of area and it depends on the laser's parameters:

S. Hinduja and L. Li (eds.), *Proceedings of the 37th International MATADOR Conference*,
DOI: 10.1007/978-1-4471-4480-9_10, © Springer-Verlag London 2013

$$ED = \frac{LP}{SS \cdot LS} \qquad\qquad (1)$$

were, LP is laser power in Watts, SS is hatch spacing in mm, LS is laser speed in mm/s and therefore ED is energy density in J/mm2.

Caulfield et al. [4] experimental research analyzes the relation between mechanical properties and construction's parameters in terms of energy density. Their results show an increase of relative density, Young's modulus, yield and fracture strength when increasing the energy density. The dimensional accuracy shows better results at low energy levels, due to shrinkage from temperature gradients. Geometry compensation factors are usually applied to the original CAD model when it is transformed into CAM instruction files. Some authors, like Raghunath et al. [1] and Yang et al.[2], had focus research in study the shrinkage compensation in SLS. Their observations show that deviations in the sintering XY plane are less sensitive than in Z-direction (average differences of 0.6%, 0.84% and 3.24% in X, Y and Z-direction respectively), because of layer's thickness and the energy transfer between sintering levels.

Energy consumption is a very important issue in every manufacturing process. Authors like Franco et al. [5] researched to determinate the range of energy for optimal level of sintering and found a better degree of sintering in a polyamide material for an energy density range between 0,02 and 0,08J/mm2, where the use of energy is maximize to merge the powder particles of material; above this range the productivity of laser decreases because an excessive exposure of laser could burn them and cause a degradation of the powder particles, specially for plastic materials. A similar trend curve can be observed in Caulfield et al. [4] work.

Nowadays, applications in engineering require an exploration for new and improved materials. The characteristics of AM techniques allow expanding the portfolio of materials typically used to composite mixes. In SLS the secondary material is usually added within the powder particles (like fiber, bead and powder) or support structures (like mesh or rod). Authors like Chung et al. [6] and Majewski et al. [7] have studied the effects of reinforce plastic materials with different proportions of glass fiber, in order to analyze the final mechanical properties. In both cases, they observed an increment of rigidity in the test specimens with the presence of glass fiber.

2. Experimental set-up

The geometry of the test specimens has been chosen according to the specifications of the ASTM D638, standard method for tensile properties of plastics, as shown in Fig. 1.

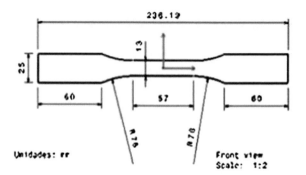

Fig. 1. Geometry of test specimens.

All specimen samples were manufactured on a Vanguard HS©(3D Systems) SLS machine, property of Fundació Centre CIM, using Polyamide powder as base-reference material and glass fiber as allied material in 25% and 50% weight fractions of the composite. The standard parameters used by Centre CIM were applied: laser speed 7500mm/s, scan spacing 0,33m and laser power as a proxy value for ED in a range from 38W to 50W.

In order to determinate the influence of the construction direction, 4 main sample orientations were considered: XY-0º, YZ-0º, YZ-45º and YZ-90º. Fig. 2 shows the planes of construction for the test specimens (black lines indicate construction layers).

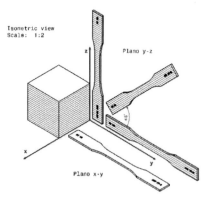

Fig. 2. Samples construction orientations.

Mechanical properties were obtained from tensile tests (under ASTM D638 regulations), using an Instron 3366 Tensile Test machine. Six samples were tested for each series of experiments to verify the consistency of data. Strain and force were measured using the equipment's software and exported into DAT files.

The optimal SLS processing parameters are defined by the configuration that produces a near-full dense part, dimensionally accurate and best mechanical properties according to the application desired. The study focused three of the highest influence in accordance with the literature reviewed: material composition, energy density and layer orientation. In case of this SLS machine, the setting for hatch spacing and laser speed are fixed per work session, laser power has a more flexible condition; in order to optimize time and resources this study uses the laser power as the controlled variable to test variations of energy density. Table 1 shows the parameters and the levels evaluated for each case.

Table 1. Parameter levels considered for experimentation

Parameter	Levels of each parameter			
	Level 1	Level 2	Level 3	Level 4
Material composition	PA100%	PA75%+ CF25%	PA50%+ GF50%	—
Laser power	38W	44W	50W	—
Layer orientation	XY-0°	YZ-0°	YZ-90°	YZ-45°

2.1 Experimentation and assessment

The samples produced were assessed in terms of performance according to the following criteria.

2.1.1 Parts dimensions
Dimensional accuracy for part width has a similar correlation, as a function of ED. The deviations are minimal; measures do not exceed 2,3% versus the 13mm reference. Part width has a smaller sensitivity comparing to part thickness in this case, due to the magnitude of each value, a 40:1 in part width and 10:1 in part thickness comparing to operating parameters like scan spacing used (0,33mm).

The level of dimensional accuracy of each sample fulfills the tolerances; therefore there is no significantly effect. The differences in the dimensions can be compensated through improving the shrinkage scale factors. In both cases ED is the variable of highest contribution on the overall dimensions.

2.1.2 Yield stregth
The main factors to determine the maximum yield strength in a SLS process are the type of material and the level of union between particles, which depends on the ED applied. The results data suggest material as the most determinant factor, but in a more distributed scenario for this case. Type of material shows a 49% contribution to the maximum yield strength, ED a 37% and layer orientation a 14%. Fig. 3 shows a comparison of these impacts, with an average calculation of the results in terms of the yield strength of test specimens due to each variable.

In case of parts build in the XY-0° and YZ-0° planes, the direction of the applied force in the tensile test has the same direction as the construction layers; thus achieving the best performance. In the XY plane, the stratum is the thickness of the part and in the YZ plane, it forms the front face of the part. The extreme case is the YZ-90°

orientation, with construction layers perpendicular to the direction of the applied force, where the yield strength depends mainly on the level of sintering between layers.

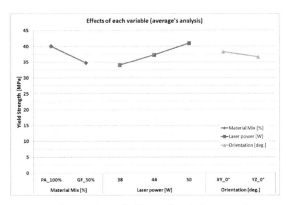

Fig. 3. Comparison of effects of each factor analysed

Also, the influence of the specific material utilised it is clearly demonstrated when undertaking stress tests to the samples. Being the PA100% composition the most elastic material with a compromise in maximum stress, Fig. 4 shows how GF50% restricts the elongation with a very fragile behaviour and CF25% achieves the highest stress limit with an intermediate elongation level.

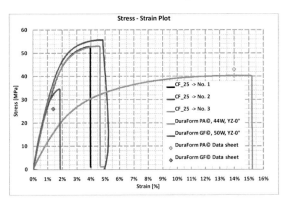

Fig. 4. Stress-Strain plot for different samples and reference values

2.1.3 Elongation at break
The relation between the orientation of layers and the direction of the applied force is important to predict the maximum stress level and the type of fracture. Models in AM processes are not isotropic; depending on the level of sintering achieved the union between working layers may have less strength than between powder particles, because the laser acts on an already merged section of powder in the previous layer while sintering a new one on top. An excess of ED could burn the material and cause its degradation.

In case of test specimens with the orientation in the XY-0° and YZ-0° planes, the stress is applied along each layer and it takes on the material mainly rather than the joints between layers. Therefore the type of fracture has irregular shape because even when the properties are assumed to be the same in the plane of analysis, the microstructure of the sintering between powder particles and layers could have a slight difference due to small defects and voids between particles (the shape of the grain has an irregular shape). In case of samples YZ-45° and YZ-90°, the force applied has a different direction to the built orientation, the joints between layers would yield first and the rupture shape will follow the layer's pattern.

3. Results overview

Results showed that at higher ED levels the mechanical properties are improved due to a higher grade of sintering. The positive effect of higher ED levels depends on the combination of parameters and the properties of a specific material, thermoplastics have a physical limit of the energy than can be applied, after that turning point the material suffers degradation, it burns out and the mechanical properties of the final work pieces decrement; the higher consumption of energy in the process is wasted. There are studies analyzing a wider range of ED in the literature reviewed; the interval analyzed in the present work was established using the actual operation in Centre CIM as reference, with the characterization chart of the materials (base and composite) and the refinement of parameters as practical application objectives along with the research study.

The addition of fiber-glass into the base polyamide material provides more rigidity to the final work-piece; the elastic modulus increments considerably, the maximum tensile stress and the elongation at break decrease, the last one does it substantially. Depending on the application and the required functions of the design, the proportion of allied material in the prototype can be modified in order to fulfill the requirements.

In general, the XY orientation plane provides better mechanical properties because the laser transfer energy directly to a larger area, as result a higher level of sintering is achieved. From a tensile resistance perspective, the XY-0° and YZ-0° planes share a similar performance and mechanical properties due to the fact that the applied force and the construction layers have the same orientation.

4. Conclusions and future work

The results show that ED has a major role on the final properties of the work-piece. Also, the bigger the area of sintering, the better mechanical properties of the final model are obtained. The grade of sintering between the particles is better than between layers. Hence, it is highly recommendable to maximize the area in the XY working plane and set the building orientation depending on the

direction of the stress-concentration vectors. This is particularly critical for prototype models with practical testing or direct application purposes. The optimal selection of the studied construction factors -material composition, ED parameters, part build orientation-, can help to develop a more ad-hoc setting for specific applications. Future work will consist in further analysing the composite GF-PA utilising other mixture percentagesm as well as the introduction of different compounds in mix such as Carbon Fiber. This further work will enable a fully characterisation of the spectrum and so to analyse the benefits in mechanical properties compared to the cost of the supply materials. Also, it will be possible to explore the requirements and properties of polyamide composites and to model with finite-element models of non-isotropic materials in order to predict the behaviour of virtual parts in different materials.

References

[1]   Raghunath, N.; Pandey, Pulak M.; 2006. "Improving accuracy though shrinkage modeling by using Taguchi method in selective laser sintering". International Journal of Machine Tools & Manufacture, 47 (2007), pp. 985-995.

[2]   Yang, H.-J.; Hwang, P.-J.; Lee, S.-H.; 2002. "A study on shrinkage compensation of the SLS process by using the Taguchi method". International Journal of Machine Tools & Manufacture, 42 (2002), pp. 1203-1212.

[3]   Nelson, J.C.; "Selective laser sintering: a definition of the process and an empirical sintering method". PhD dissertation, University of Texas, Austin, 1993

[4]   Caulfield, B.; McHugh, P.E.; Lohfeld, S.; 2006. "Dependence of mechanical properties of polyamide components on build parameters in the SLS process". Journal of Materials Processing Technology, 182 (2007), pp. 477-488.

[5]   Franco, Alessandro; Lanzetta, Michele; Romoli, Luca; 2010. "Experimental analysis of selective laser sintering of polyamide powders: an energy perspective". Journal of Cleaner Production, 18 (2010), pp. 1722-1730.

[6]   Chung, Haseung; Das, Suman; 2006. "Processing and properties of glass bead particulate-filled functionally graded Mylon-11 composites produced by selective laser sintering". Materials Science and Engineering A, 437 (2006) pp. 236-234.

[7]   Majewski, Candice; Hopkinson, Neil; 2010. "Effect of section thickness and build orientation on tensile properties and material characteristics of laser sintered nylon-12 parts". Rapid Prototyping Journal, 17/3 (2010), pp. 176-180.

Laser deposition of Ti-6Al-4V wire with varying WC composition for functionally graded components

P. K. Farayibi[1], A.T. Clare[1], J.A. Folkes[1], D.G. McCartney[2], T.E. Abioye

[1] Division of Manufacturing
[2] Division of Materials, Mechanics and Structures
Faculty of Engineering, University of Nottingham, University Park, Nottingham, NG7 2RD, UK

Abstract. Laser deposition provides a suitable means for metallic coating and surface alloying for many applications. This process allows new functionally graded components to be built and surface properties to be locally modified. Tailoring to meet location specific demands such as modified tribological, corrosion and thermal properties may be achieved. This paper investigates cladding of Ti-6Al-4V wire and WC powder concurrently fed into the laser generated melt pool on a Ti-6Al-4V substrate. The addition of WC particles promotes its hardness and improves wear resisitance. Details of the laser deposition conditions of the titanium alloy coating with varying WC composition are presented. Results obtained showed the micrograph of the functionally graded area coating of Ti-6Al-4V/WC matrix in which WC particles are dissolved or have irregular reaction layer in the matrix. SEM, EDX, XRD and microhardness characterisation is presented for the coating. Fibre laser deposition is demonstrated as a means to produce titanium matrix ceramic composite.

Keywords: Laser cladding, Ti-6Al-4V, WC, Hard facing, Microstructure, Hardness.

1. Introduction

Laser cladding is a flexible manufacturing technology which allows deposition of materials which may improve wear, corrosion, and thermal properties [1]. In this process, a laser generated melt pool is steadily fed with the desired materials to form a clad. Repeated clad multipass gives an area coating. This process is characterised by rapid cooling and solidification [2] which make it an excellent means for fabrication of discontinuously reinforced metal matrix composites (MMC).

Metal matrix composites are desirable for various applications which may include wear surfaces, thermal management, and structural stiffness [3]. Various manufacturing processes such as stir casting, liquid metal infiltration, and powder metallurgy have been employed in the past for the production of MMCs [4], however, owing to the rapid cooling and solidification which is peculiar to laser deposition process, functionally graded components can be produced easily. Thus, compositionally graded engineering components can be manufactured for functionality in thermal, wear, and structural applications [5]. This process can be considered flexible and cost effective, as deposits with tailored properties can be customised to meet location specific demands, allowing new components and repairs to be made.

Metal-ceramic systems have been developed for aerospace, automotive, and marine applications and the most widely used of these MMCs is Al based [3]. Ti alloys are used in aerospace and the chemical processing industries, but have limited tribological applications due to high friction coefficients and low hardness [6]. However, Ti alloy MMC have not gained much popularity due to cost of Ti alloy production. It has only been used in areas where performance is of priority compare to cost. as discontinuously reinforced Ti alloys have found use in the automotive industry for engine valves [7].

Succcessful attempts have been made to deposit Ti and its alloy with SiC, TiC and WC, and emphasis is made on powder deposition and surface laser melt injection[8, 9]. However, the laser deposition of Ti-6Al-4V wire with varying WC flow rate is yet to be performed. This will allow compositionally graded structures or coatings to be formed with excellent utilisation of the costly Ti-6Al-4V alloy. Thus, the stiffness, hardness and wear properties of the alloy maybe improved with the injection of WC particles. Tungsten carbide (WC) is considered a suitable reinforcing ceramic, as it has been discovered to have smaller reaction layer thickness and thermal expansion coefficient closer to Ti-6Al-4V when compared to silicon carbide, which reduces the MMC crack tendency [10].

In this study, cladding of concurrently fed Ti-6Al-4V wire and WC powder for production of tracks and area cladding was investigated using a range of powder feed rates. This aims at determining geometric characteristics of single tracks with varying WC powder flow rate, and microstructure formation.

2. Experimental

2.1. Materials

The 1.2 mm diameter Ti-6Al-4V wire used was supplied by VBC Group (Loughborough, UK). WC powder with size range of 40-160 μm was supplied by Technogenia, France. The powder particles have a near spherical morphology, as shown in Fig. 1. The mean particle size, determined by laser diffractometry (Mastersizer S, Malvern Instruments, UK) was 124 μm. The substrate material was Ti-6Al-4V in form of rectangular coupons (180 mm x 100 mm x 5 mm) which were grit-blasted with alumina grit prior to deposition to improve laser absorptivity and degreased with acetone to remove surface contaminants.

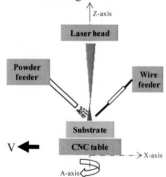

Fig. 1 Scanning electron microscopic image of powder morphology.

2.2. Cladding system setup and process parameters

Laser deposition was undertaken using a 2-kW Ytterbium-doped, CW fibre laser (IPG Photonics) operating at 1070nm wavelength coupled with a beam delivery system (125 mm collimating lens and a 200 mm focussing lens), and a Precitec YC 50 cladding head.

Fig. 2 Schematic configuration of the process setup.

A front feeding nozzle was used for wire delivery, coupled with a REDMAN wire feeder mechanism (Redman Controls and Electronic Ltd, England), and a rear feeding nozzle for powder delivery coupled with a Model 1264 powder feeder (Praxair Surface Technologies), and a computer numerically controlled (CNC) table (4-axis) were used. The laser remained stationary and the CNC table traversed in the direction of the negative x-axis. Figure 2 shows the schematic configuration of the process setup. The deposition process was conducted in a flexible chamber. The chamber was flushed with argon, Ar, for 10 minutes prior to the start of deposition and continuously flushed during the experiment. Table 1 summarises the parameters employed in this study.

Table 1 Process condition for the laser deposition

Parameter	Value
Laser Power, W	1800
Beam spot area, mm^2	7.5
Traverse speed, mm.min^{-1}	172
Wire feed rate, mm.min^{-1}	800
Powder feed rate, g.min^{-1}	10 – 40
Carrier gas flow rate, l.min^{-1}	10
Shielding gas flow rate, l.min^{-1}	30

2.3. Single track experiments

Experiments were performed to examine the effect of powder feed rate (PFR). All other process parameters were held constant as listed in Table 1. The PFR was varied between 10 and 40 g.min^{-1}.

2.4. Deposition of clad layers

In deposition of clad layer, the same parameters as in Table 1 were used with range of PFR. The clad layer was deposited discontinuously in the pattern shown in Fig 3. The pitch for depositing successive tracks was 0.55w, where w is the track width.

2.5 Deposition characterisation

Cross section dimensions were measured with a Talysurf CLI 1000 surface profiler (Taylor Hobson Precision Ltd, UK) in laser mode. This was used to determine height and width of tracks. Measurements were taken at three (3) different positions along each track. Circular samples of 36 mm were cut from clad plates by electro-discharge machine (EDM) and surfaces were finished by milling and ground with SiC papers and polished with 6 μm and 1μm diamond pastes.

Vickers microhardness tests were performed on these polished surfaces using 200 gf for 15 sec with a LECO M-

400 Hardness Tester. Ten (10) indents were made across the diameter, and mean and standard deviation values were calculated for the different PFR values employed.

X-ray diffraction (XRD) was also performed on these top surfaces to identify phases formed. This was done using a Siemens D500-1 XRD machine which was operated at 40kV and 25mA to generate a Cu K (α) radiation. A step size of 0.05° and 4 s counting time in the range 20-90° was employed.

Scanning electron microscope (SEM) was used to examine microstructure of track cross section and top surfaces of clad layers. Samples were ground and polished prior to examination using back scattered electron (BSE) to study images. Energy dispersive X-Ray analysis (EDXA) was also used to aid phase identification.

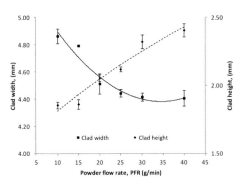

Fig. 4 Dependencies of clad width and height on powder flow rate.

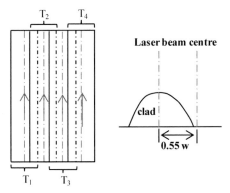

Fig. 3 Deposition strategy for clad layer

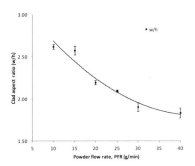

Fig. 5 Variation of clad aspect ratio with increasing powder flow rate.

3. Results

3.1. Clad characteristics

The geometrical characteristics of the single tracks were obtained and plotted against powder flow rate, PFR. Fig. 4 shows the dependence of clad width and height on powder feed rate in the deposition process. It was observed that the clad width is inversely dependent on the PFR, while clad height positively varies with the feed rate. This observed trends can be attributed to having a constant energy per unit length with increasing material delivery into the melt pool.

As material delivery increases, there is more attenuation of the laser energy delivered onto the substrate which leads to reduction of the width size. Increase in clad height is favoured by more material delivered available and becoming trapped in the melt pool [11]. Thus, the aspect ratio of the single track was observed to have an inverse relationship with the powder feed rate, as shown in Fig. 5. This can be used to predict the tendency of having inter-run porosity when the parameters used for a single track is used for area coating.

3.2. Microstructural characterisation

Fig. 6 shows the XRD spectrum of the clad layer deposited using 20g/min WC PFR. The highest peak was found at $2\theta = 40.3°$ which corresponds to W (110) (BCC) phase, other W peaks are found at $2\theta = 58.3°$, 73° and 87°. TiC (FCC) phase was observed in the spectrum with major peaks at 2θ positions 36.2°, 42°, and 61°. β-Ti (BCC) phase with highest peak at 39.7° was also identified. In summary, the phases present are W, WC, TiC and β-Ti. As the WC PFR increased from 20 to 40 g.min[-1], there was no change to the phases that were present.

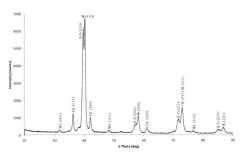

Fig. 6 XRD spectrum of coating deposited with 20g/min PFR

However, as determined from the relative intensity peak height ratios of β-Ti : W, as WC PFR increases from 20 to

40 g.min^{-1}, the ratio decreases from 0.92 to 0.45. There was no significant change in the peak height ratio value of TiC : W, as the difference is ±0.03.

Fig 7 shows the cross sectional image of a single track from the SEM. Most of the WC particles assimilated into the melt pool have been transported to the clad periphery by Marangoni convection leaving fewer particles in the central region of the pool. These fewer particles experience intense heat compare to the ones at the periphery. This makes central particles to be either partially or fully dissolved leaving them with irregular reaction layer and continuously dissolved edges, while majority of the particles conveyed to the periphery are encapsulated in a regular reaction layer.

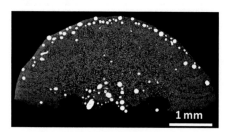

Fig. 7 BSE image of a track cross section showing bright WC particles at the track periphery

Fig 8 shows the clad periphery section with un-dissolved WC particles maintaining their spherical morphology as seen in Fig 1. These particles are surrounded by a dark phase layer.

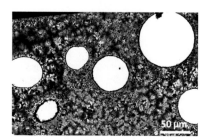

Fig. 8 BSE image from the periphery region in Fig 8 showing magnified region of WC and matrix

Fig. 9 shows points where EDX data were conducted in the central region with partially dissolved WC particles and the clad periphery with more un-dissolved WC particles. As W and C go into the Ti alloy solution, an exothermic reaction occurs to producing TiC, which are seen as black precipitates marked point A and are dispersed in the matrix. This also corresponds to the reaction layer formed around the WC particles present at the clad periphery. As TiC is formed, the Ti solution is β-stabilised, and enriched to a saturation level by increasing W, V, and C contents. This makes W to precipitate out of the solution. Thus, the diffusion of W and C into the Ti solution resulted into the formation of TiC and W as XRD pattern confirms. The bright white precipitates (point B) are W solid solution (W ss) with fair amount of other

elements (Ti, Al, V, C) dissolving in it, as these points have the highest value of W at% in the matrix. Table 2 presents the details of the elemental composition in atomic percentages (at. %) for the elements present. The point labelled C is identified as β-Ti solid solution (Ti ss). The diffusion of W and C promotes the precipitation of W into bigger two-phase nodules formed in the central regions, as seen in red box in Fig. 9 (a). These nodule features were observed in all the samples. Also similar phases were observed in the microstructure of the track edges as seen in Fig 9 (b).

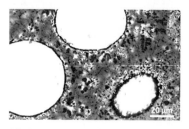

Fig. 9 Magnified view of the track section; a) central region with EDX points, b) edges with similar phases

Table 2 Elemental composition (at.%) of phases in Fig. 13

Pt	Ti	Al	V	W	BSE contrast	Probable phase
A	81.5	8.0	3.0	7.5	Black	TiC
B	36.7	-	2.8	60.5	Bright	W ss
C	82.3	9.3	3.2	5.2	Grey	Ti ss

The SEM examination of the polished flat-top surface, as seen in Fig. 10, shows how the WC particles were distributed across the coating surface. The surface has bands of overlap region which possess higher concentration of WC particles. This resulted from the WC particles that were transported to the track edges and when tracks were overlapped, the regions are enriched with particles shelled by a reaction layer.

Fig. 11 shows a magnified view of the WC particles present in the central region with continuous dissolution of the particle edges until solidification sets in. The WC particles in the overlap regions were mostly surrounded by the reaction layer.

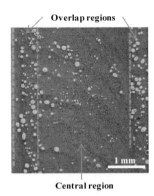

Fig. 10 BSE image of the planar view of area coating deposited using 20g/min PFR

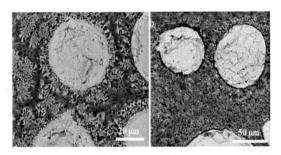

Fig. 11 Micrographs of WC particles at a) central region with continuous dissolution; and b) Overlap region with reaction layer

Thus, as the PFR increases, the amount of partially dissolved WC particles seen in the central region of track increases, and also the amount of the particles present at the clad periphery and are shelled in a reaction layer increases. This resulting surface microstructure morphology of the coatings is synonymous to the schematics in Fig. 12.

Fig. 12 Schematic illustration of single tracks overlapping to form an area coating

3.3. Micro-hardness tests

Table 3 shows the microhardness test results obtained across the coating surface. The results showed that the hardness of the surfaces increases with increasing PFR used for deposition. The hardness of Ti-6Al-4V about 350 H_V, but it was enhanced to mean values between 600-1030 $H_{V0.2}$. Besides the partially dissolved or un-dissolved WC present in the coatings, the presence of the two-phase nodules of W and TiC precipitates enhance the hardness of the central region which makes the hardness value of the region to be less far from the overlap region. Some indents give as high as 2117 $H_{V0.2}$ which must have been on a WC particle or a region near it. Such values were not used in the evaluation of the mean hardness values.

Table 3 Microhardness variation across the coating surfaces (SD – Standard Deviation; SE – Standard Error)

$H_{v0.2}$ (kgf.mm^{-2})						
WC, PFR	10 g/min	15 g/min	20 g/min	25 g/min	30 g/min	40 g/min
Mean	601.4	745.3	758.9	806.9	894.4	1027.1
SD	96.4	144.0	95.2	79.4	158.9	323.0
SE	30.5	48.0	30.1	25.1	50.2	107.7

4. Conclusions

The cladding of Ti-6Al-4V wire with varying WC PFR resulted in a graded composition with central regions having more dissolved particles, and more WC particles are concentrated in the overlap regions in an area coating. The clad width decreases while height increases with increasing PFR. With the aid of XRD, SEM and EDAX, phases identified are W, WC, TiC and β-TiC. Hardness across the coating surface shows a direct and positive variation with increasing WC PFR. Thus, hardness properties are improved with increasing WC PFR.

References

[1] Ion, J.C., (2005) Laser Processing of Engineering Materials. Principle, Procedure and Industrial App. Elsevier Butterworth Heinemann, Linacre House, Jordan Hill, Oxford, UK.

[2] Hofmeister, W., Griffith, M., Ensz, M., Smugeresky, J., (2001) Solidifcation in Direct Metal Deposition by LENS Processing. J. of Minerals, Metals and Materials Society. 53(9): p. 30-34.

[3] Miracle, D.B., (2001)Aeronautical applications of metal matrix composites. ed. D.B. Miracle and S.L. Donaldson. Vol. 21., ASM International, Materials Park, OH, USA.

[4] Miracle, D.B., (2005)Metal matrix composites – From science to technological significance. Composites Science and Technology,65(15–16): p. 2526-2540.

[5] Toyserkani, E., Khajepour, A., Corbin, S., (2005) Laser Cladding. CRC Press, Florida, USA.

[6] Huang, C., Zhang, Y., Vilar, R. (2011) Microstructure characterisation of laser clad TiVCrASi high entropy alloy coating on Ti-6Al-4V substrate. Advanced Materials Research. 154-155: p. 621-625.

[7] Hunt, W.H., Miracle, D.B., (2001) Automotive applications of metal matrix composites. ed. D. S.L. Vol. 21., ASM International, Materials Park, OH.

[8] Pleshakov, E., Senyavs'kyi, Y. Filip, R. (2002) Laser Surface Modification of Ti-6Al-4V alloy with Silicon Carbide. Materials Science,. 38(5): p. 37-42.

[9] Folkes, J.A., Shibata, K. (1994) Laser cladding of Ti-6Al-4V with various carbide powders. Journal of Laser Applications,. 6(2): p. 88-94.

[10] Chen, Y., Liu, D., Li, F., Li, L., (2008) WCp/Ti-6Al-4V graded metal matrix composites layer produced by laser melt injection. Surf. and Coati. Technology. 202(19): p. 4780-4787.

[11] Mok, S.H., Bi, G., Folkes, J., Pashby, I., (2008) Deposition of Ti-6Al-4V using a high power diode laser and wire, Part I: Investigation on the process characteristics. Surface and Coatings Technology. 202(16): p. 3933-3939.

Laser cladding of rail steel with Co-Cr

A. T. Clare, O. Oyelola, T. E. Abioye and P. K. Farayibi
Division of Manufacturing, Faculty of Engineering, The University of Nottingham, University Park, Nottingham NG7 2RD, UK.

Abstract. Degradation and subsequent failure of rail tracks is commonly caused by rolling contact fatigue amongst other mechanisms of wear. Rail crossings are known to exhibit more of these failures due to increased localised traffic and environmental conditions. A high proportion of the costs associated with the repair of rail tracks is due to the RCF phenomenon. In order to mitigate these costs, a laser cladding process of worn regions has been proposed for the repair of worn tracks in-situ to limit the need for them to be replaced and for the protection of newly rolled rails. The ability of a laser beam to give localised heating with good deposition rates and good metallurgical bonding is being exploited in the use of the laser beam to deposit powders in this study. Stellite 6 is chosen to demonstrate repair and also surface coating/ protection of rail steels. Samples with good metallurgical bonding were produced with laser cladding. Characterisation of the resulting deposits and tribological properties of the materials from a material compatibility view is presented. A laboratory test which gives preliminary information about the performance of the material selected to the widely used rail material is also presented. Laser cladding is demonstrated to be a viable solution to repair worn track.

Keywords: Laser Cladding, Stellite 6, Rail steel, Microstructure, Wear

1. Introduction

Railway track degradation and the resulting repair is a significant cost burden upon rail network providers. The cost to the environment is also significant as continual grinding, replacement, and melting of worn material are energy intensive processes. Increased tonnages, the development of faster trains and the limited time frames available for maintenance of rail track has led to an increase in the amount of material being sent to land fill. Rolling contact fatigue (RCF) amongst other forms of degradation in rails has made it important that new materials capable of withstanding the attendant problems be developed for use as rail materials and coatings. Degradation in rail tracks due to RCF can be observed in various ways, which include; Spalling, head checks and gauge corner cracks amongst others.

RCF defects are caused by a combination of high normal and tangential stresses between the wheel and rail. This leads to shearing through plastic deformation (ratchetting) of the surface layer of the rail material and reduction in the ductility of the material [1]. Defects can lead to rail fracture and ultimately derailment if not addressed. Advances in rail technology and metallurgy have led to the use of premium materials which are more wear resistant and less RCF prone. Rail steel hardness has been closely related to the wear; although as manufactured hardness does not give a total indication of wear performance [2]; this has led to the development of premium rail materials that have high hardness and also good workhardening ability. Over time, in the quest for the development of a more wear resistant material, it was found that R260 gave better wear performance compared to the R200 grade. Since then, it has been widely used in most rail networks [3]. Also, more extensive work has shown that R350HT head hardened rail offers more improved wear and RCF performance when compared to both grades mentioned earlier [4,5]. A number of materials have been developed to achieve less maintenance problems and costs. However, the use of these materials does not completely rid the issues that arise with the problems highlighted.

The capital cost of any rail network is significant. Premium work hardening and corrosion resistant materials are available whose properties outstrip those of the 260 grade. However, due to the higher cost of these, manufacturing entire rail sections in these materials is not feasible

Particularly sections where the tracks are most prone to damage include crossings and areas of high curvature. Also, in networks where high speed tilting trains are used, the nature of the design dictates the inner wheel is at a greater angle to the track. This invariably leads to loss of more material due to the high banking forces. In the repair and protection of wear prone materials, surface modification techniques have been used successfully in other engineering applications. Areas of potential application of the surface modification technique to be used would include; areas with a high track curvature, areas close to train stations (where rail squeal is an issue), crossings and areas which have been known to have high rate of rail degradation per unit time.

Lasers have been widely used in surface modification of specialist and wear prone surfaces. Also, surface repair has been achieved using laser beams. Researchers have attempted laser glazing of railroad tracks with an aim towards reducing the friction at the top layer. They were able to achieve a 40% reduction in the friction coefficient and also were able to reduce crack propagation through the bulk material [6]. Laser cladding offers a route for the repair of worn track and protective coating of new track. In laser cladding, a laser beam is scanned over the surface to be cladded whilst powder is simultaneously injected into the melt pool that is created. This causes the powder material to melt and fuse to the substrate material creating an intricate bond. Laser cladding has been used widely in the addition of varied powder materials to regions where material has been lost due to wear or corrosion.

Advances in powder metallurgy and the development of high power lasers; which are capable of achieving higher deposition rates has led to an increase in the application of laser cladding in surface modification. It is important for any technology to be used in repair of rail tracks to be robust enough to deliver high rates of material per unit time as the time frame for in situ repair is limited. It is also not feasible to provide surface modification for entire length of a rail network.

In this study, a premium work hardening alloy is selected as a potential candidate for pre-service rail cladding and for potential repair and preventative maintenance. This is aimed at providing protection for the top surface of the bulk rail when in service. This method would allow for the retention of the bulk material and also the use of a low cost alloy with a protective coating at sites prone to wear. Clad/substrate interface and microstructures are examined in detail and hardness profiles are presented. A preliminary laboratory wear test is also presented. Data presented is important when assessing the compatibility of hardfacing materials for use in rail applications.

2. Experimental

2.1. Materials

The work hardening ability of a material is known to be important in material selection for rail components. Advances in rail technology have led to the wide use of the pearlitic grade steels. These, when compared to bainitic grades in the rail industry have been found to possess better workhardening ability. This has ensured its use in most rail networks worldwide, although grades vary [7]. In these materials a ferritic phase and an alternating cementite phase are observed. The composition thus ensures the material has properties intermediate between the soft, ductile ferrite and the hard, brittle cementite. During loading, the hardness of the material increases as the interlamellar spacing between the constituents of the microstructure reduces. The plastic ferrite and the hard cementite thus combine to give this effect. For the present study, a pearlitic grade of steel, widely used in most rail networks is used (R260), its composition is shown in Table 1.

Table 1. Material composition in wt %

Ele.	*Stellite 6*	*R260*
C	-	0.74
Si	1.5	0.25
Cr	30.4	0.04
Mn	0.7	1.3
Fe	2.1	balance
Co	balance	-
Ni	2.4	-
W	5.8	-
Al	-	0.003
S	-	0.008

In selecting a hardfacing material with suitable work hardening ability, a number of properties were considered; the ability of the material to be deposited without cracks, the wear resistance and toughness of the material, the environmental and financial cost implication of using the material. These chercteristics were assessed in a prior study. Stellite 6 was selected as it met most of these considerations. The material is a cobalt based hard facing alloy which combines a complex mixture of carbides in a CoCr based matrix. Stellites have been widely used in industry to resist wear in hostile environments and also as structural materials, particularly in areas of high temperature [8,9]. These properties thus make the material relevant in the development of a coating to be used in rail applications. The material composition is shown in Table1

Particle size range for the Stellite 6 powder used was found to be between 75μm - 275μm with a mean size of 144μm and the particles were spherical.

2.2. Cladding set up

The laser used in cladding is a 2 kW IPG Ytterbium-doped, continuous wave, Fibre laser operating at a wavelength of 1070 nm with a Gaussian beam profile at focus. The optics arrangement delivers a beam spot size of 3.1mm at a defocused distance of 20mm. This spot size is used for cladding so as to achieve higher deposition rates and also to replicate an industrial application methodology. The laser is equipped with a CNC table which has four axis of movement. The delivery of powder to the melt pool created by the laser beam was through a lateral feed nozzle and the powder was delivered using a Miller Thermal (Model 1264) powder feeder. The powder feeder was calibrated with the powder to deliver known quantities of material per unit time. Argon gas is used as both the carrier and shielding gas; to create an inert environment around the melt pool. Plates cut from rail steel sections were used as substrates and were grit blast to improve the laser absorptivity of the surface, and thereafter degreased with acetone to remove contaminants before processing.

2.3. Cladding trials

In the deposition of Stellite 6 on rail substrates, Single (clad passes) layer tracks as well as multilayer tracks were deposited. Parameters used in the deposition of the material are shown in Table 2.

Table 2. Conditions for Stellite 6 laser deposition

Condition	Parameter range
Laser power	1.6kW
Traverse speed	400-800mm/min
Powder flow rate	15-30g/min

During deposition, as expected, it was found that the parameters required to clad suitable single clads were not necessarily suitable for the cladding of multi-layer clads. Samples were obtained from single, overlap and multilayer deposits by sectioning them into mountable sizes using Electrical Discharge Machining (EDM). The resulting samples were mounted in a conductive resin and polished to a 1μm surface finish. Clads were characterized with a scanning electron microscope (SEM) with Electron dispersive X-ray spectroscopy (EDS) functionality. X-Ray Diffraction (XRD) analyses were performed to identify the phases present in the clad microstructure. Multi-pass clad layers were used in XRD analysis due to their wider exposed area to the X-ray beam.

Micro Hardness tests were carried using a LECO hardness tester. Hardness values were taken from the top of the clad at equidistant positions of 0.25 mm and extend into the substrate material. This was done after the microstructure was evaluated and the phases present were compared to determine the effects of changes in processing parameters on the type of structures present. A load of 0.3N (300gf) was used with indent duration of 15 seconds. The hardness of the substrate rail is confirmed to be 310HB (Hardness of R260). Wear test were also conducted using the ball on disc tribometer. The use of this laboratory test in replicating and assessing rail wear has been described by other authors [10]. Sliding tests are not capable of completely replicating wheel /rail contact as the rolling effect is not present [11]. This test was

however useful in determining the work hardening behaviour of the materials involved and in the creation of a benchmark to determine the suitability of the material for use in the application before performing field tests. Cobalt based Tungsten Carbide ball with a diameter of 9.5mm and hardness 1510 HV is used for the experiments. The normal load on the ball was kept constant at 50N with a rotating speed of 240 rpm. The total distance covered for each trial is 1000 m which corresponds to 14,400 cycles.

3. Results and discussions

In the deposition of Stellite 6 on rail substrates, it was found that cracking was predominant in samples with a high energy density .This is thought to be due to the non-uniform dissipation of the heat energy during cooling leading to a high coefficient of thermal expansion mismatch between the clad and the substrate. However, clads showed intricate bonding to the substrate with no obvious signs of distortion. Samples when etched with a solution containing 60 ml HCl, 15ml H_2O, 15ml HNO_3, 15ml CH_3COOH[i] revealed a dendritic structure with interdendritic eutectics. In the structure observed, due to the high concentration of Co, a Co rich phase forms the dendrite with the interdendritics and eutectics formed afterwards from Cr and C. this observation is similar to the results discussed in [ii]. In all, it is found the clad thus consists of the Co rich dendrite, eutectics of Cr, Co, W carbides and the Co based matrix. The dendrite structures at the edges of the clad orient in the expected direction of heat flow away from the root of the clad towards the clad edge. Dilution from the substrate surface into the clad is effectively controlled as the deposited clad has a composition similar to the parent powder material. Figure 1 shows the clad structure and the microstructure being discussed.

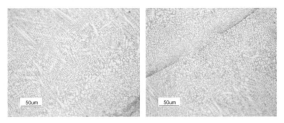

Fig.1. Columnar and dendritic growth within clad structure

From the figure, it is observed there is a mix of cellular and columnar dendrite growth within the structure. The boundaries of the clad and the area between subsequent clads during multicladding show the cellular growth due to the rapid cooling at these regions and the inability of the cells to grow to larger sizes during solidification.

EDS analysis however reveals both constituents have a similar composition. Table 3 shows the composition of both clad and unprocessed powder.

Table 3. Material composition of clad after processing

	Si	Cr	Mn	Fe	Co	Ni	W
Powder	1.5	30.4	0.7	2.1	57	2.1	5.8
Clad	1.1	27	0.5	2.4	58	2.2	8.5

3.1. XRD analysis

In the analysis of the clads using X-ray diffraction, patterns generated from both the unprocessed powder and the clad are taken into consideration. From Figure 2, the Co peak is seen to have the most intensity in the clad layer at $2\theta = 44°$. There is an increase in the presence of Cobalt Carbide (CoC_x) when comparing between both plots. New phases after processing include; a Chromium carbide (Cr_7C_3) phase, a mixed carbide phase of Co_2W_4C formed within the cobalt matrix. The presence of these carbide phases is expected to give the deposit a high hardness and thus high wear resistance.

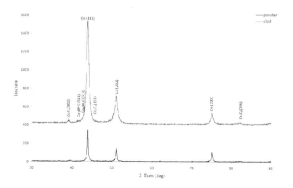

Fig. 2. Stellite 6 powder and Clad - XRD spectra

3.2. Tribological tests

In selecting a sample for the Microhardness test, a single bead sample which showed an obvious HAZ and whose structure was considered desirable for multi-cladding was used (1.6kW power, 400mm/min traverse speed, 15g/min powder feed rate). Microhardness tests gave the average hardness of the clad layer as 565 HB (602 HV). Within the dilution zone, the hardness was 620 HB and in the heat affected zone, the hardness was 906 HB. The hardness value when compared to the hardness of the rail steel material (310 HB) is much greater. The drastic increase in hardness of the substrate at the HAZ region is due to the formation of Martensite in this region. The rapid cooling of the substrate during cladding leads to a diffusionless process which favours Martensitic transformation. In Figure 3, the regions marked P correspond to the region of pearlitic structure and the region marked M is the martensitic structure. The presence of complex carbides and tungsten within the material account for the high hardness values observed.

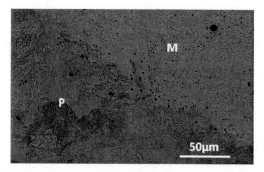

Fig.3. Boundary between HAZ and unaffected substrate showing Martenstic transformation

In the sliding tests, it is observed that for up to 14400 revolutions of the ball, the bulk rail material wears at a rate slightly higher than the clad surface as the worn volume removed was more in the rail material. Comparing results from both profiles after the tests, it is seen that the substrate has a marginally greater area worn away during the experiments. To demonstrate the work hardening of the materials, hardness results within the wear scar and unaffected substrate are compared. Results obtained are shown in Table 4.

Table 4. Parameters within wear grove

	Width/mm	Depth/μm	Hardness before wear test(HB)	Hardness after wear test(HB)
Clad	1.63	0.049	790	988
Substrate	2.11	0.045	422	700

There is a marked increase in the hardness of the materials after the test which shows the work hardening ability of both materials. The substrate had a percentage increase of 39% and the cladded surface had a 30% increase. Figures 4 and 5 show the wear scars when observed with a SEM, it is seen that the mechanism of wear is dissimilar in both cases. The Stellite surface wears at a uniform rate with no sign of accumulation of debris within the wear scar.

Fig. 4. (a) Wear scar of Stellite 6, (b) exploded view of boxed region in (a)

The rail substrate however wears in a non uniform manner with accumulation of material within the scar. Accumulation of debris leads to micro cutting on the surface being worn. Excessive plastic deformation is noticed within the worn region. Since plastic deformation leads to increased rate of RCF, it is expected the Stellite 6 clad would perform better in service against RCF.

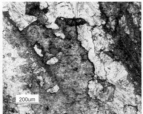

Fig. 5. (a) Wear scar of Rail material, (b) exploded view of boxed region in (a)

4. Conclusion

Laser cladding of Stellite 6 powder on rail substrates has been carried out. Microstructure results produced show that the deposits maintain their compositional integrity as there is minimal dilution from the substrate. Preliminary hardness results show that clad has elevated hardness values when compared to the substrate which indicates the clad is potentially suitable for the protection of the substrate material. Wear results indicates both materials have comparable wear resistant properties. Wear mechanism in the substrate material is however less desirable as it leads to failure of the material ultimately. Further tests including a twin disc test and field tests are still to be carried out to determine the efficacy of using the material as a protective coating for rail materials.

References

[1]   D. F Cannon, K.O Edel, SL Grassie, K Sawley; Rail defects: An overview (2003) Fatigue & Fracture of Eng. Mat.& Struct. 26, Issue 10, pages 865–886

[2]   F C. Robles Hernández, N. G. Demas, K. Gonzales, A. A. Polycarpou(2007) Mechanical properties and wear performance of premium rail steels. Wear, Vol. P. 263 766-772

[3]   European Standard, final draft   prEN 13674-1(November 2002) Railway applications -Track-Rail, Part 1: Vignole railway rails 46kg/m and above

[4]   Gregor Girsch, A Jorg Wolfgang Schoech (2010) Technology Track, Railway Gazette International August 2010, p-45-48

[5]   R. Heyder, G Girsch (2005) Testing of HSH® rails in high speed tracks to minimise rail damage. wear 258, p; 1041-1021

[6]   R.J DiMelfi, D. Sanders, B Hunter, J.A Eastman, K.J Sawley, K.H Leong , J.M Kramer (1998) Mitigation of subsurface crack propagation in railroad rails by laser surface modification. Surface and coatings Vol 106, p; 30-43

[7]   F.J. Franklin, J.E. Garnham, D.I. Fletcher, C.L. Davis , A. Kapoor (2008) Modelling rail steel microstructure and its effect on crack initiation. Wear 265: Pg 1332–1341

[8]   J. L. De mol Van Otterloo and J. Th. M. De Hossont -1997. Microstructural features and mechanical properties of a cobalt-based laser coating: Acta mater. Vol. 45. No. 3. no. 1225-1236

[9]   S. Niderhauser, B. Karlsson (2003) Mechanical properties of laser cladded steel. Mat. Sci. and tech ;Vol 19, p 1611- 1616

[10] F C. Robles Hernández, N. G. Demas, K. Gonzales, A. A. Polycarpou (2011) Correlation between laboratory ball-on-disk and full-scale rail performance tests. Wear 270, 479-491

[11] Ki Myung Lee, Andreas A. Polycarpou (2005) Wear of conventional pearlitic and improved bainitic rail steels ; Wear 259, p;391–399

[12] D'Oliveira Ana Sofia, C.M., Se´rgio, P. C., Vilar R.C. (2002) Microstructural features of consecutive layers of Stellite 6 deposited by laser cladding. Surf. and Coatings Tech. 153, 203–209

[13] Zhonga M, Liu W., .Yao K, Goussain, J. C Mayer C.., Becker A. (2002) Microstructural evolution in high power laser cladding of Stellite 6+WC layers Surf. and Coa.Tech. 157, 128–137

Concurrent Inconel 625 wire and WC powder laser cladding on AISI 304 stainless steel

T.E Abioye[1], J. Folkes[1], D.G. McCartney[2] and A.T Clare[1]
[1]Manufacturing Division, [2]Division of Materials, Mechanics and Structures,
Faculty of Engineering, University of Nottingham, NG7 2RD, United Kingdom

Abstract. WC-Ni based superalloy metal matrix composite coatins deposited by laser cladding have the potential to significantly increase the lifetime of metallic components particularly in intense corrosive–wear environments. Abrasive wear performance of the coatings largely depends upon their un-dissolved WC contents after solidification. This work investigates the concurrent deposition of WC powder and Inconel 625 wire single tracks for composite coating using a range of laser processing parameters with the aim of defining the process characteristics and studying the WC contents of the clad. A fibre laser with a lateral feeding system for the concurrent wire and powder single track deposition on AISI 304 stainless steel plate was used in this study. Optical microscopy and scanning electron microscopy were used to evaluate clad microstructure. The un-dissolved WC content was determined by image analysis. The results include the generation of a process map that predicts the characteristics of the deposition process at varying cladding conditions. The volume fraction of un-dissolved WC particles is found to decrease with increasing laser power, transverse speed and wire feed rate.

Keywords: Laser cladding; Metal matrix composites; Abrasive wear performance; Composite Coatings; Inconel 625.

1. Introduction

Laser deposition of WC-Ni based superalloy metal matrix composite coatings is becoming a prominent technique for improving the wear and corrosion performance of engineering components [1]. The established methods of producing these coatings involve lateral [2, 3] and coaxial [4, 5] feeding of pre-blended powders, combined coaxial powder and lateral wire feeding systems [6], and concurrent lateral wire and lateral powder feeding systems [7]. Previous studies established that concurrent wire and powder feeding systems improve the material deposition efficiency, produce composite clad of better surface finish and allow for an independent control of the feeding rate of the additive materials [8].

The abrasive performance of the WC-Ni based composite coatings reportedly shows positive dependency on the volume fraction and even distribution of un-dissolved WC in the coatings [9, 10]. Efforts directed towards increasing the WC volume fraction of the composite coatings include composite cladding of pre-blended powders of different WC shapes [10] and compositions [11, 12]. The result shows that spherical shaped WC produced clads of lower WC content because they dissolve more rapidly than the ball milled (crushed) type while higher volume fraction was found for clad made with powders of higher WC concentration. Laser induction rapid cladding of WC-Ni alloy coating performed at varying laser specific heat energy by Zhou et al. [13] shows that minimal WC dissolution occurred in clads deposited with lower specific heat energy. Also, WC reinforced plasma transfer arc welding (PTAW) composite cladding with three NiCr alloys of varying Cr content has been investigated [14]. It was found that WC dissolution increases with increasing Cr content of the matrix alloys. The result was attributed to the strong Cr-C affinity which enhances the formation of (W, Cr) C phases all over the matrix. Usually, the volume fraction of un-dissolved WC particles in solidified clad deposits is traced to the degree of WC dissolution that occurs during the cladding process. To date, a study of the variation of the WC content of WC-Ni based superalloy coatings with the main cladding parameters using concurrent wire and powder lateral feeding system has not yet been reported.

This work investigates the concurrent deposition of WC powder and Inconel 625 wire single tracks using a range of laser processing parameters. The aim is to define the deposition process characteristics and study the variation of the WC volume fraction with the cladding parameters. Inconel 625 wire matrix was selected because it exhibits a combination of high temperature strength and high resistance to corrosion and oxidation in marine environments [15]. The high ductility of the alloy, due to its face centred cubic (FCC) crystal structure, improves its resistance to cracking that may occur due to contraction after welding [16]. WC is known to be very hard (1780 kgf/mm^2) and possesses a high melting point (2720 °C). Compared to the other carbides such as SiC,

WC also combines favourable properties such as high density (15630 kg/m^3), low thermal expansion coefficient and a good wettability with molten Ni alloys[17].

2. Methodology

2.1. Materials and laser process

Fig 1 shows a schematic diagram of the concurrent wire and powder laser deposition system used in this study. Deposition was performed using a 2-kW Ytterbium doped fibre laser (IPG Photonics) operating at 1070 nm wavelength. The beam was focused to a small round spot of about 3.1mm at 20mm away from focus giving a 212 mm working distance with a Gaussian energy distribution. Inconel 625 wire of 1.2mm diameter (d) supplied by VBC group, Loughborough, UK was "front fed" (ahead of the laser) at an angle of 42°±1 to the horizontal so as to aim the wire tip at the centre of the meltpool. A WF200DC wire feeder (Redman Controls and Electronic Ltd) was used. Spherical WC powder (40-160 μm; mean size = 124 μm) commercially named spherotene was back-fed into the meltpool through a Praxair (Model 1264) powder feeder. The powder nozzle was oriented at 67° to the horizontal so as to improve WC deposition rate.

A scanning electron microscope image of the morphology of the WC powder is shown in Fig 2 (a). Table 1 gives the chemical compositions of the wire and substrate used in this work, as quoted by the manufacturers. Plates of dimension 100mm X 180mm X 6mm were machined out of austenitic stainless steel AISI 304 and used as substrate material. These were grit blasted and cleaned with acetone before the deposition runs so as to improve substrate surface laser absorptivity and remove contaminants respectively.

Single track ceramic-metal composite depositions were performed at varying laser cladding parameters on several AISI 304 stainless steel plates inside a transparent enclosure (bag) which was evacuated and back-filled with high purity argon gas supplied at 25 l/min. The process parameters were selected from the process map previously developed for laser deposition of Inconel 625 wire only. This was done with the aim of understanding the effect of simultaneous injection of WC powder on the laser deposition characteristics of Inconel 625 wire. Table 2 gives the details of the process parameters used for the deposition process.

In order to understand the variation of WC content with the cladding parameters, powder feed rate was kept constant while each of other parameters was varied one at a time. A total of 48 tracks were deposited. The observed process characteristics were recorded for each of the deposited track.

2.2. Microstructural characterisation

Track sample was transversely sectioned. As shown in Fig 2 (b), a total of 14 micrographs, each at 200X magnification, taken with a scanning electron microscope at different parts of the sample were analysed using image analysis. The volume fraction of un-dissolved WC contained in each micrograph was determined. The average of the values obtained from the 14 micrographs was taken as the un-dissolved WC volume fraction of the track sample. The process was repeated for all the deposited tracks.

$$E_L = 60\frac{P}{V} \tag{1}$$

$$A_C = \frac{W_{FR} \times \prod d^2}{4V} \tag{2}$$

Energy/unit length (E_L) in J/mm is defined by Equation 1. The deposited composite clad area (A_{CC}) in mm^2 was obtained from image analysis of the scanning electron microscope (SEM) images of the transversely cross-sectioned tracks. Equation 2 defines the cross-sectioned area (A_C) in mm^2 of Inconel 625 wire laser deposited clad.

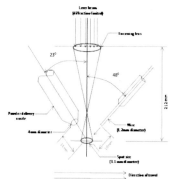

Fig 1. A schematic diagram of the laser deposition system

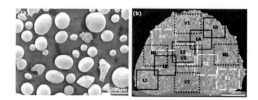

Fig 2: Scanning Electron Micrographs of (a) WC powder (b) a cross-sectioned WC-Inconel 625 wire composite track

Table 1. Chemical compositions of Inconel 625 wire and 304 stainless steel in wt.%

Element	Ni	Cr	Mn	Si	Al	Ti	Fe	C	Mo	Nb	P	S
Inconel 625	Bal	22.46	-	-	0.26	0.26	0.14	0.02	8.84	3.46		-
304 Stainless steel	7.86	18.58	1.78	0.42	-	-	Bal	0.08	-	-	0.10	0.03

Table 2. Cladding parameters

Name	Symbol	Value	Unit
Laser power	P	1000, 1200, 1400, 1600 and 1800	W
Wire feed rate	W_{FR}	400, 600, 800 and 1000	mm/min
		3.8, 5.7, 7.6 and 9.6	g/min
Transverse speed	V	100, 200, 300 and 400	mm/min
Powder feed rate	P_{FR}	25	g/min

3. Results and discussion

Fig 3 (a) is a process map developed for laser deposition of Inconel 625 wire. The process map shown in Fig 3 (b) is valid for the fibre laser deposition of single track WC powder-Inconel 625wire composite coating for the range of parameters used in this experiment (see Table 2).

Except for the constant powder feed rate of 25 g/min, the process map parameters correspond to those represented by the data points in the regions 1, 2 and 3 of the process map for Inconel 625 wire only. The two maps define three characteristic regions namely wire dripping (region 1), smooth wire transfer (region 3) and wire stubbing (5). A study of the two process maps shows that the top left corner (region 1) is characterised with wire dripping effect while the bottom right corner is characterised with wire stubbing effect (see Fig 3). Therefore, laser cladding at any process condition at these corners will result in track of poor surface quality. Wire dripping is found to be caused by feeding too little material volume for a given heat energy/unit length of track pass (J/mm) however, the map show that the effect can be avoided by either reducing the laser heat energy (J/mm) or increasing the material volume/unit length of track pass. On the other hand, wire stubbing effect caused by feeding excessive material volume for a given heat energy/unit length of track pass can be eliminated by either increasing the energy/unit length or decreasing the material volume/unit length of the track pass.

Additionally, the two process maps predict that smooth wire transfer may not be practicable when cladding below 200 J/mm. When compared with the process map developed for Inconel 625 wire only, a rightwards shift is noticeable in the map developed for WC powder-Inconel 625 wire laser deposition. The shift in the process map is considered to be caused by the reduction in laser heat energy available to melt the feed wire due to absorption by the powder. With combined wire and powder lateral feeding system, the injected powder acted as a thermal barrier and absorbed a portion of the laser heat energy before it reached the wire tip. Therefore, for smooth wire transfer to occur, wire feed rate (W_{FR}) must be lowered since wire tip will now require more time to interact with the laser heat energy before it can melt. Accordingly, wire dripping and wire stubbing effects which are due to too low W_{FR} and excessive W_{FR} respectively for a given laser heat energy/unit length will now occur at lower W_{FR} thus shifting the process map to the right.

Fig. 4 shows the variation of volume fraction of undissolved WC in the deposited tracks with the laser cladding parameters. It is found that the undissolved WC volume fraction in the deposited tracks decreases with increasing laser power, transverse speed and wire feed rate. The increase in the volume fraction of undissolved WC in a deposited track with decreasing laser power can be traced to the low dissolution of WC particles caused by decreasing laser heat energy/unit length. For example, 1080 J/mm energy available at 1.8 kW power and 100 mm/min transverse speed produced a very hot meltpool. Consequently, the WC particles suffered intense heating causing the smaller particles to melt completely while the larger particles melted to smaller sizes, as shown in Fig 5a and 5b. At 1.2 kW, but with the same transverse speed, the laser heat energy/unit length of track pass has significantly reduced to 720 J/mm producing lower meltpool temperature hence, there is little melting of WC particles. Eventually, the WC particles appear larger and are of high number density in the clad, as shown in Fig. 5c and 5d. In these micrographs the undissolved WC appear bright contrast and near circular in appearance. The Inconel matrix is dark and the light phase with a dendritic morphology is believed to be M_6C (Fe_3W_3C) which has precipitated during solidification of the clad from a Ni-rich molten phase which contained significant quantities of dissolved W and C.

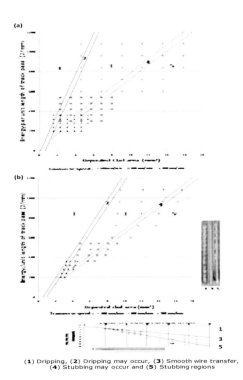

(**1**) Dripping, (**2**) Dripping may occur, (**3**) Smooth wire transfer, (**4**) Stubbing may occur and (**5**) Stubbing regions

Fig 3. Plot of energy per unit of length of track pass versus clad area for tracks of (a) Inconel 625 wire (b) WC powder-Inconel 625 wire coatings.

Previously, an increase in WC volume fraction with increasing transverse speed reported by Anandan et al. [18] on laser cladding of pre-blended WC-Ni alloy powder is related to low WC dissolution as the speed increases. However, the current result suggests that the reverse is the case with concurrent wire and powder laser cladding. A possible explanation is that efficiency of injecting WC powder into the melt pool plays a role. Laser heat energy/unit length of track pass decreases with increasing transverse speed. Therefore, smaller meltpool size is produced at higher transverse speed. It would be expected that all feed wire entered the meltpool but not all powder is trapped in the laser meltpool. Consequently, powder catchment efficiency is reduced as transverse speed increases. Also, WC dissolution is found to decrease with increasing transverse speed however, the overall decrease in undissolved WC content with increasing speed shows that the effect of powder catchment is more significant than WC dissolution.

The decrease in un-dissolved WC volume fraction with increasing W_{FR} is due to lower assimilation of the WC particles at higher W_{FR}. Increasing the W_{FR} means more wire volume is delivered to the meltpool per unit time. Since the laser power and transverse speed is constant, the thermal mass per unit length of the meltpool is also constant. So, at higher W_{FR}, the energy content of the meltpool is rapidly consumed thus reducing the fluidity and activeness of the meltpool in a short time. Hence, clad solidification is quickened. As a result, only small volume of powder particles could make it into the meltpool before the meltpool solidifies while others are bounced/reflected to the surrounding cladding area.

4. Conclusion

A process map predicting the characteristics for the concurrent laser deposition of WC powder-Inconel wire 625 single tracks has been developed. Compared with the process map developed for Inconel 625 wire only, the rightward shift in the map is a direct product of WC powder injection into the deposition process. The effects of process parameters on the volume fraction of the undissolved WC particles in the deposited composite tracks were studied. It can be concluded that; WC volume fraction decreases with increasing laser power, transverse speed and wire feed rate. A good control of WC powder dissolution and catchment efficiency is highly significant to obtaining WC powder–Inconel 625 wire composite coating of desirable WC content. WC particle dissolution is principally accountable for the inverse effect of laser power whereas powder catchment efficiency is mainly responsible for the negative effects of transverse speed and wire feed rate. For the range of parameters used in this experiment, a maximum of 30 vol. % of undissolved WC was found to be possible for the concurrent WC powder–Inconel 625 wire composite laser cladding.

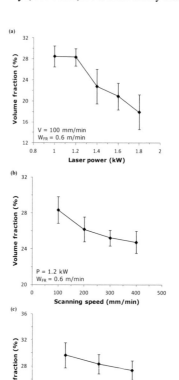

Fig 4. Effects of cladding parameters on the WC volume fraction.

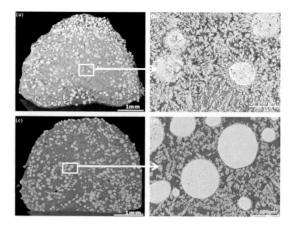

Fig 5. SEM images of track cross sections for laser power of (a) and (b) 1.8 kW, (c) and (d) 1.2 kW.

References

[1]   St-Georges L, (2007) Development and characterization of composite Ni–Cr +WC laser cladding. Wear, 263(1-6): 562-566.

[2]   Zhou S, Dai X, Zeng X, (2009) Effects of processing parameters on structure of Ni-based WC composite coatings

during laser induction hybrid rapid cladding. Applied Surface Science, 255(20): 8494-8500.

[3] Zhong M, Liu W, Zhang Y, Zhu X, (2006) Formation of WC/Ni hard alloy coating by laser cladding of W/C/Ni pure element powder blend. International Journal of Refractory Metals and Hard Materials, 24(6): 453-460.

[4] Balla V.K, Bose S, Bandyopadhyay A, (2010) Microstructure and wear properties of laser deposited WC-12%Co composites. Materials Science and Engineering: A, 527(24-25):6677-6682.

[5] Farayibi P.K, Folkes J, Clare A, Oyelola O, (2011) Cladding of pre-blended Ti–6Al–4V and WC powder for wear resistant applications. Surface and Coatings Technology, 206(2-3): 372-377.

[6] Syed W.U.H, Pinkerton A.J, Li L, (2006) Combining wire and coaxial powder feeding in laser direct metal deposition for rapid prototyping. Applied Surface Science, 252(13):4803-4808.

[7] Syed W.U.H, Pinkerton A.J, Liu Z, Li L, (2007) Coincident wire and powder deposition by laser to form compositionally graded material. Surface and Coatings Technology, 201(16-17): 7083-7091.

[8] Syed W.U.H, Pinkerton A.J, Li L, (2005) Simultaneous wire- and powder-feed direct metal deposition: An investigation of the process characteristics and comparison with single feed methods. Journal of Laser Applications, 18(1):65-72.

[9] Wu P, Zhou C.Z, Tang X.N, (2003) Microstructural characterization and wear behavior of laser cladded nickel-based and tungsten carbide composite coatings. Surface and Coatings Technology, 166(1):84-88.

[10] Huang S.W, Samandi M, Brandt M, (2004) Abrasive wear performance and microstructure of laser clad WC/Ni layers. Wear, 256(11-12):1095-1105.

[11] Van Acker K, Vanhoyweghen D, Persoons R, Vangrunderbeek J, (2005) Influence of tungsten carbide particle size and distribution on the wear resistance of laser clad WC/Ni coatings. Wear, 258(1-4):194-202.

[12] Zhou R, Jiang Y, Lu D, (2003) The effect of volume fraction of WC particles on erosion resistance of WC reinforced iron matrix surface composites. Wear, 255(1–6):134-138.

[13] Zhou S, Huang, Y, Zeng X, Hu Q, (2008) Microstructure characteristics of Ni-based WC composite coatings by laser induction hybrid rapid cladding. Materials Science and Engineering: A, 480(1-2):564-572.

[14] Liyanage T, Fisher G, Gerlich A.P, (2012) Microstructures and abrasive wear performance of PTAW deposited Ni–WC overlays using different Ni-alloy chemistries. Wear, volumes 274–275:345-354.

[15] Cooper K.P, Slebodnick P, Thomas E.D, (1996) Seawater corrosion behavior of laser surface modified Inconel 625 alloy. Materials Science and Engineering A, 206(1):138-149.

[16] Reed R.C, (2006) The superalloys Fundamentals and Apllications, Cambridge, UK: Cambridge University Press.

[17] Tobar M.J, Álvarez C, Amado J.M, Rodríguez G, Yáñez A, (2006) Morphology and characterization of laser clad composite NiCrBSi–WC coatings on stainless steel. Surface and Coatings Technology, 200(22–23):6313-6317.

[18] Anandan S, Pityana S, J. Dutta Majumdar J, (2011) Structure–property-correlation in laser surface alloyed AISI 304 stainless steel with WC + Ni + NiCr. Materials Science and Engineering: A.

Evaluation of effect of heat input in laser assisted additive manufacturing of stainless steel

M. Eskatul Islam[1], Antti Lehti[1], Lauri Taimisto[1], Heidi Piili[1], Olli Nyrhilä[2], Antti Salminen[1,3]

[1]   Laser Processing Research Group, Lappeenranta University of Technology, P.O. Box 20
FIN-53851 Lappeenranta, Finland.
[2]   EOS Finland, Lemminkäisenkatu 36, FI-20520 Turku, Finland.
[3]   Machine Technology Center Turku Ltd., Lemminkäisenkatu 28, FI-20520 Turku, Finland.

Abstract. Laser based additive manufacturing (AM) has risen a lot interest in industry as whole manufacturing method has become more reliable with improved accuracy and quality of final products. Laser sintering is additive manufacturing process where laser beam is used to melt powder material layer by layer and this way even very complex 3D geometries can be manufactured from metallic, ceramic, composite or polymer powder. The aim of this study was to characterize laser based AM of stainless steel powder with the assistance of pyrometer and active illumination imaging system by finding the correlation between the parameters. This enables understanding of basic phenomena occurring during process but also development of process control system. Material used in this study was EOS Stainless Steel PH1. Sintering of single layer of stainless steel on the powder bed was performed with a self-made trial set-up with IPG 200 W SM CW laser, a scanner (focal length of 1000 mm) and sintering chamber. Pyrometer used in this study was by Temperature-Control-System (TCS) manufactured by Thyssen-Laser-Technik and active illumination system by Cavitar. It was found out that balling phenomena is causing many defects in laser sintering process and also in quality of final product. It was concluded that with constant heat input the simultaneous increase in laser power or decrease in scanning speed within specified range, resulted sintered powder bed surface with minimum balling effect. The variation and stability for this phenomenon is characterized and segmented according to the heat input. It was also noticed that heat input has a strong effect to process stability. All these findings refer that process control via heat input is a key issue in achieving high quality of end product in laser sintering.

Keywords: Laser based additive manufacturing, additive manufacturing, AM, pyrometer, active illumination imaging, direct metal laser sintering, DMLS, selective lasersintering, SLS.

1. Introduction

Laser based AM is a manufacturing process which uses laser beam to fabricate work pieces layer by layer by following 3D CAD based design. When this technology is used, need for designing a suitable manufacturing process can be avoided, due to complex geometry can be produced "in one print" [1,2,3]. Laser based AM (to

which Direct metal laser sintering (DMLS) belongs to) can utilize metal powder to manufacture metal pieces[3]. A laser beam scans over a layer of powder to melt metal powder point by point. This process is repeated by spreading, leveling, and melting of additional layers of powder to produce a complete structure of the manufactured part. The metal powder during laser based AM process heated to a temperature slightly less than the melting point [4]. Nearly any 3D geometry can be fabricated using AM and the melting mechanisms can be defined by stages [5] as in Fig. 1. There is no melting if laser energy is not high enough to cause any physical change of the powder material. In the partial melting stage, a suitable amount of laser power with low scanning speed indicates melting in the liquid phase. Melting of particles occurs with balling effect in the next stage in high laser power level and high scanning speed. Complete melting in final stage was high enough to produce continuous tracks of molten liquid. This forms a dense surface after solidification [5, 6].

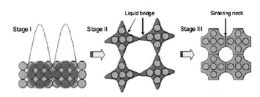

Fig. 1. Formation mechanism by DMLS in stainless steel. [5]

Scanning line distance between parallel scan has an impact in the heat distribution and lower distance than necessary can create surface deterioration by increasing excessive heat generation [7]. Balling effect is a defect associated with DMLS which is a complex physical metallurgical process. According to the perturbation theory [8] laser irradiation of the melt decreases due to a larger cylindrical melt formed creates the capillary instability. This phenomena leads to the formation of coherently bonded track free of any balling phenomenon after sintering. There are two kinds of balling phenomena observed during DMLS operation [5, 9]. In the first kind, using a low laser power give an increase to the balling characterized by highly coarsened balls possessing an interrupted dendrite structure in the surface layer of balls [6]. In the second kind, balling phenomena featured by a large amount of micrometer-scaled (10 μm) balls on laser sintered surface are formed at a high scan speed. The formation mechanism was developed to laser-induced melt splashes caused by a high capillary instability of the melt. Increasing laser scan speed on the powder surface impact and after affect as cracks can be found [6].

2. Experimental analysis

In this study stainless steel EOS PH1 powder was used which has pre-alloyed stainless steel in fine powder with a lowest possible layer thickness of 20 μm. (EOS datasheet) A trial set-up of IPG 200 W SM CW fiber laser was used having a maximum laser power level of 200 W and a wavelength of 1070 nm. A straight oriented scanner was performing sintering with a focal length of 1000 mm and a spot diameter of 100 μm. Hatch spacing maintained with 0.1 mm. Square shaped part dimension calibrated 5 x 5 mm space. Laser power was set to values between 60-200 W and scan speed to values between 0.005-1 m/s. A nitrogen atmosphere was maintained in the work chamber (see Fig. 2) by feeding nitrogen in and out and covering the top of chamber with float glass such that laser beam transmits. A steel backed rectangular mould (8.5cm x 8.5cm) was build to pour the powder material. To focus and scan perfectly a small cavity was build on that surface. Thyssen Laser-Technik made pyrometer for temperature control and monitoring was used into two ranges, 1200-1400 nm and 1400-1700 nm. Active illumination imaging system by Cavitar (with camera system) was used to capture images during laser process and operated in a specific wavelength range. Cavilux HF Illuminator consists of a 500 W diode laser with wavelength of 809 nm. Camera was used in a speed of 25 frames/s (fps). Oven test was done to define the flaw in measured temperature in case of using float glass. The error found to be 30-35°C and thus negligible. Heat input Q [J/m] was used on these experiments to demonstrate the amount of laser energy used in process per unit meter of processed sinter layer (laser power P [W or J/s], scan speed v [m/s]) Equation defines heat input [10, 11] as in equation 1,

$$Q = \frac{P}{v} \qquad (1)$$

3. Result and discussions

Melting point of low carbon stainless steel, equivalent with material properties as EOS Stainless Steel PH1, is

around 1400°C (1232-1532°C). The melting point of the martensitic and ferritic metal stainless steel is around 1405-1440°C and low carbon stainless steel is 1232-1530°C which is a closure point of the sintering temperature.(ASM materials) Temperature achieved in this test during 2000 J/m processing was 1390°C. From 1 J/mm to 3.35 J/mm the maximum temperature is in a range of 1315-1390°C. It can be concluded that, high enough heat input is necessary to reach melting temperature of stainless steel during this AM process. Phase selection was based on heat input calculated using equation 1 and arranged in Table 1. During the first experiment, it was noticed that when heat input values equivalent or over 0.5 J/mm is used, strong balling effect of molten stainless steel powder was observed. At 0.8 J/mm balling phenomena increased gradually and during 1000 J/m balling effect found as the maximum among all the procedures. All most in all cases, molten material tends to form ball as with the increase of laser energy in this specific range of 250-2000 J/m. With a heat input of 2500 J/m the surface showed formation of homogenous and non porous evenly distributed surface. At the beginning, temperature rise to around 1343°C from a temperature of 770°C.

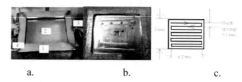

Fig. 2. ((a) sintering chamber, 1. Chamber, 2. Float glass, 3. Gas outlet, 4. Gas inlet. (b) mould for layer operation of PH1 powder (c.) hatch pattern.)

It created a predominant melted surface at the middle of the experiment. The surface has fewer pores in temperature range of 1050-1270°C. With heat input 5000 J/m; the surface in the initial and middle was in a partially heat affected and in the end, smooth surface. Initially the temperature rose to 1270°C and declined to a range of 1000-1100°C during the operation of 10-20 s. From 20-32 s the temperature increased up to 1200°C. At high heat input surface shows partial reflective nature instead of high absorption of the laser beam as it should for stainless steel. The layer formed is the most uniform of all results, still got some unevenly distributed porosity. It can be concluded that, at this heat input, the heat penetration of heat to this semi molten surface of the stainless steel powder start decreasing but can produce a denser surface with less balling phenomena. Heat input of 10000 J/m on stainless steel made the surface to collapse at the end of formation. At the beginning the temperature raised to 1033°C. After reaching the peak temperature at 1043°C, the temperature starts to decline from its original position. The temperature dropped to a temperature region of 580-870°C from a region of 1043-800°C within 15s. During this time some balling was created but it disappears soon after the heat input starts to decline. The experiments with this heat input on stainless steel suggest lower heating impact by lower absorption. It can be stated that high heat input greater than 5000 J/m and below 10000 J/m does not have a good impact to sintered layer rather it deteriorates the surface. At 2500 J/m heat input, the heat input is more evenly distributed causing the formation of a uniformly melted even surface. The interesting finding in 5000 J/m from microscopic and active illumination image, suggests a more reflective nature than the absorptive nature throughout the experiment. Unique heat input in the surface and uniform outlook from the experiment is found at the end of the experiment. As a matter of high heat input from the laser energy at this 5000 J/m point start to make significant changes in the heat affected zones due to the heat conduction. At 10000 J/m heat input, the surface starts to deteriorate with the maximum porosity found and instead of bonding the material, the semi molten pool reflects.

Table 1. Experimental result and analysis.

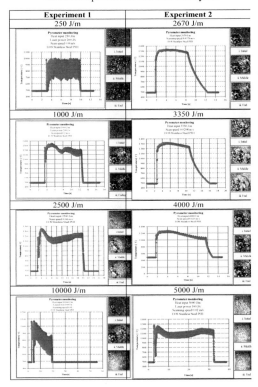

Two postulates to define the heat input mechanism from 2670-4450 J/m for second test, can be stated as following: The process temperature reaches close by the melting point in the beginning. Soon after sintering starts, process temperature starts declining and surface becomes smoother in 3000-3350 J/m. The surface at higher heat input might act like a mirror like a mirror to the laser beam and cannot withstand the heat input above 4450 J/m. Microscopic analysis was used to analyze the impact of heat input on the sintered layer, as shown in Fig. 3. When heat input of 250 J/m was used, there are some uniformly molten areas and surface consists of un-melted

sintered parts. With the increase of heat input the metallic bond start to form at 300 J/m. The surface shows the rising of balling and porous surface from 300-800 J/m to be bigger than before and formed a balling affected hemispheric shaped metallic neck. At 400 J/m, the outer surface of the sintered layer starts to form more regular shapes but in the inner section, this portion continues hemispheric neck building, i.e. balling. This phenomenon followed in result case of 500 J/m and became denser. At 800 J/m the maximum porosity can be seen.

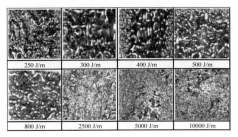

Fig 3. Microscopic images analysis.

At 2500 J/m, the surface starts to melt more homogenously excluding balling syndrome. Still some porosity remains but it starts declining. Thus when heat input of 2500-5000 J/m is used, micrograph of sintered piece looks more uniform. In case of 5000 J/m heat input value, micrograph is seen with finer and uniform formation at the sintered parts piece. The structure starts to deteriorate from 10000 J/m with high porosity and lack of metallic bonding. The main controlling characteristics for this experiment was heat input derived temperature of the sintered layer. As seen from Fig. 4, the total maximum temperature against heat input can be divided in three regions and corresponding trends as I, II and III covers the heat input of 0-2200 J/m, 2200-5000 J/m and 5000-10000 J/m with balling, best sintering and reflective zone. Extend the trend lines of regions corresponds to 1418 and 1470°C. It can be concluded that, temperature corresponding to the trend intersecting point from 1418°C demonstrate a decreasing order which can give the good quality surface in a range of 2500-5000 J/m. On the contrary, for 1470°C, the absorption can be lower than in range of 5000-10000 J/m heat input regions. The evaluation between comparison of microscopic and active illumination based image based on the finding table II, the impact of the parameters working for sintering the single layer with EOS PH1 stainless steel can be characterized. The first experiment (a) covers a wide range of heat input up to 10000 J/m. This result suggests the impact of the heat input on the layer surface and the closure range of finest formation of the sintered layer in between 2500 J/m to 5000 J/m. Table 2 (a., b.) suggests an inner microstructure phenomena for the best formation of layer possible with heat input of 2500 J/m to 10000 J/m whereas the conditions at 10000 J/m is quite unstable as it deteriorates the material inner formation due to high input of 2500 J/m to 5000 J/m and temperature gradient development in low scan speed between 0.04m/s to 0.02

m/s gives the smooth layer with low porous environment on the layer. On the other hand from active illumination system point of view heat input in 2500 J/m and 5000 J/m gives the best visual structure where all others show balling phenomena impact which cannot be fully visualized with microscopic analyze.

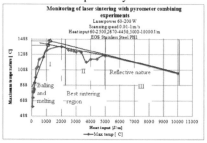

Fig. 4. Maximum temperature via heat input characteristics curve.

Table 2. Comparative analysis on microscopic and active illumination image. (a. 250-2500 J/m, b.5000-1000 J/m, c. 2670-4450 J/m)

Processing parameters			Microscopic image	Active illumination image
Laser power [Watt]	Scan speed [m/s]	Heat input [J/m]		
200	0.8	250		
200	0.76	299		
200	0.5	400		
200	0.4	500		
200	0.1	2000		
100	0.04	2500		

a.

Processing parameters			Microscopic image	Active illumination image
Laser power [Watt]	Scan speed [m/s]	Heat input [J/m]		
100	0.0374	2670		
100	0.035	2850		
100	0.033	3000		
100	0.0298	3350		
100	0.027	3700		
100	0.025	4000		
100	0.0225	4450		

c.

Processing parameters			Microscopic image	Active illumination image
Laser power [Watt]	Scan speed [m/s]	Heat input [J/m]		
100	0.02	5000		
100	0.01	10000		

Therefore it can be suggested that the best layer formation, possible for this EOS PH1 stainless steel powder, lies on the heat input region from 2500 J/m to 5000 J/m with 100 W Laser power and a scan speed between 0.04 and 0.02 m/s. Table 2 (c.) shows the comparative analysis based on processing parameters of a heat input ranged from 2670-4450 J/m with microscopic and active illumination image of the sintered layer from 0.022-0.037m/s. The impact of heat input on the layer [see Table 2 (c.)] Can be found best in the 3000 and 3350 J/m with scanning speeds of 29.8 and 33 mm/s. It should be noted that in all cases except the last one laser power was 100 W; this is why most of the surface suffers balling

syndrome in the beginning of layer formation and in the rest of the surface area become smoother.

4. Conclusions

Objective of this study was to examine and co-relate the best possible sintering surface determination and characteristics categorizing with heat input impact. The main challenge was to interpret the comparison these two processes monitoring and controlling in a numeric interpretations and evaluation from a wide range of experimental data. It can be concluded from the experimental analysis procedures that, high balling phenomenon and porosity is the main restriction from creating smooth surface. Throughout this analysis, the monitoring and comparison suggests that, instead of using individual parameters (like laser power or scan speed) heat input is the controlling and characterizing parameter.

Acknowledgements: This project was done as part of InnoBusiness-ArvoBusiness project consortium (TEKES). Authors would also like kindly to thank EOS Finland for all knowledge for this study and to the staff of LUT Laser for their support.

References

[1]   Steen W.M, Majumder J., (2010) Laser Material Processing,4th ed. London, Springer-Verlag. pp.349,353,358.

[2]   I. Gibson, D.W. Rosen, and B. Stucker, (20109 Additive Manufacturing Technologies. Springer, LLC, pp1,2.

[3]   M. Shellabear, O. Nyrhilä, J. (2004) DMLS-development history and state of the art, presented at LANE 2004 conference, September 21-24, Erlangen, Germany.

[4]   Narendra B. Dahotre, Sandip P. Harimkar, (2008) Laser Fabrication and Machining of Materials; Springer Science and Business Media, LLC; ISBN 978-0-387-72343-3, 9.3.2.1.

[5]   Dongdong Gu, Yifu Shen; (2008) Processing conditions and microstructural features of porous 316L stainless steel components by DMLS, App. Surface Science 255, 1880–87.

[6]   Dongdong Gu, Yifu Shen; (2009) Balling phenomena in direct laser sintering of stainless steel powder: Metallurgical mechanisms and control methods, Mat. & Des.30, 2903-10.

[7]   Y. Ning, Y. S. Wong, J. Y. H. Fuh, and H. T. Loh, (2006) An Approach to Minimize Build Errors in Direct Metal Laser Sintering, IEEE Tran. on Autom. Sci.& Engg. Vol. 3, No. 1..

[8]   Niu H.J. & Chang I.T.H. (1999) Instability of scan tracks of selective laser sintering of high speed steel powder. Scripta Mater; vol.41, pp.1229–1234.

[9]   Simchi A. & Asgharzadeh H. (2004) Densification and microstructural evaluation during laser sintering of M2 high speed steel powder. Mat.Sci.. Tech. vol.20, pp.1462-1468.

[10] A. Lehti, L. Taimisto, H. Piili, A. Salminen, O. Nyrhilä; (2011) Evaluation of different monitoring methods of laser assisted additive manufacturing of stainless steel, ECerS XII, Stockholm, Sweden, pp.2.

[11] M. Eskatul Islam , (2011)   Laser assisted additive manufacturing of stainless steel, Lappeenranta University of Technology, Lappeenranta, pp. 137-138.

Comparison of theoretical and practical studies of heat input in laser assisted additive manufacturing of stainless steel

M. Eskatul Islam[1], Lauri Taimisto[1], Heidi Piili[1], Olli Nyrhilä[2], Antti Salminen[1,3]

[1]   Laser Processing Research Group, Lappeenranta University of Technology, P.O. Box 20, FIN-53851 Lappeenranta, Finland.

[2]   EOS Finland, Lemminkäisenkatu 36, FI-20520 Turku, Finland.

[3]   Machine Technology Center Turku Ltd., Lemminkäisenkatu 28, FI-20520 Turku, Finland.

Abstract. Laser based AM (additive manufacturing) is cutting edge technology in industry, since industrial AM machines has become more reliable through executing fast, flexible and cost effective production. Commercial machines using laser beam can create solid very complex 3D geometries for metallic, ceramic or composite components in tooling and direct manufacturing of parts. The aim of this study was to characterize measurable outputs of laser based AM by defining the correlation between theoretical and practical study of effect of heat input to success of process. The process was monitored with active illumination imaging system and pyrometer. Material used in this study was EOS Stainless Steel PH1. The used fiber laser was IPG 200 W SM CW. Comparative theoretical study was carried out with own setup consisting of laser, scanner, sintering chamber and powder unit. The commercial EOS sintering machine was used for practical studies. The earlier assumption that heat input had a strong effect to the sintering process stability was verified in test of this study. It was noticed that as in theoretical case, also in practice most of instabilities of sintering process and defects in quality of end product are dependent on heat input and its control.

Keywords: Laser assisted additive manufacturing, AM, DMLS, fiber laser, stainless steel, heat input, linear energy density.

1. Introduction

AM is typically based on using a model, initially generated with 3D CAD, in production system to produce 3D products. Therefore, AM technology can significantly simplify the production of complex 3D objects from CAD data to final product. [1, 2] The system can generate a three-dimensional part by selectively melting thermoplastic, ceramic or metallic powders with a localised heating by laser, usually now a day with a fibre laser. [3] There are various names for the typical process of laser based metal AM. The basic principle is however about the same. Direct metal process is based on a method where layer of metal powder is deployed and a laser beam is scanned over a layer of powder to sinter or melt it from programmed locations according to 3D model. After this the laser treaded layer is lowered, new

layer deployed and melted/sintered with laser beam. The powder deployment can be carried out with a roller (see Fig. 1) or with a blade. This process is repeated by spreading, levelling, and sintering of additional layers of powder to produce a complete structure of the manufactured part is often called selective laser sintering [4]. Fig. 1 shows the basic systems for laser based sintering mechanism.

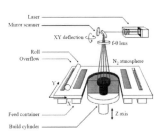

Fig. 1. Schematic of a laser based AM system principle. [5]

Porosity is yet a problem and can be reduced by post sintering applications. [6] Direct Metal Laser Sintering (DMLS) got flexibility in materials and shapes in case of producing complex shaped porous metallic components. Limitation of the powder sintering method to be consider is the pore characteristics (e.g., size, shape, and distribution of pores), which are usually difficult to control. [7] The oxidation has also another problem i.e. oxidation or excessive heating damage can cause burning of the powder material. This is usually avoided by filling the chamber with nitrogen. [5] To speed up the process pre-heating can be a good choice if the total heating can be controlled. The un-sintered/melted particles can be re-used and the raw material can be anything capable to be sintered or melted. [3] The wider areas are treated by melting zones beside each other with so called hatch. The parameters of hatch are type and line distance. Hatch distance has an impact in the heat distribution and too short a hatch line distance (hatch distance) can lead into a short scanning interval between each scan can create surface deterioration by increasing excessive heat generation. [8] The parameters to vary respect of heat input are the hatch distance, speed of laser beam, laser power and overall heating strategy. Even heat distribution requires control of those parameters. Another flaw in melted/sintered layers is so called balling i.e. ball formation. Material is not forming constant layer but large balls. This effect is unfavourable since it ruins quality. According to the perturbation theory [9] laser irradiation of the melt decreases due to a larger cylindrical melt formed creates capillary instability. This phenomena leads to the formation of coherently bonded track free of any balling phenomenon after sintering. Besides an increase in laser energy intensity leads to a larger degree of under cooling of the melt, favouring the rapid growth of dendrites in a refined morphology during solidification. [7, 10]

2. Experimental procedure

2.1. Materials and equipments

The material used was EOS PH1 powder is pre-alloyed stainless steel in powder form with excellent mechanical properties. The modified research machine of laser sintering used in LUT Laser fabricated by EOS, is comparable to commercial equipment and consists of a laser, scanner, material building platform and recoating unit, computer control, feeder, and material overflow cartridge. The used Yb-fiber laser was 200 W IPG SM CW, which produces 200 W of optical power at a wavelength of 1070 nm. Stainless steel structure manufacturing is performed in nitrogen atmosphere. Used EOSint M type machine possess DMLS technology by fusing metal powder into solid parts by melting with a focusable laser source. Laser works through scanner on the material recoating unit surface to build the layer of the part. [11, 12, 13] The protective glass of the machine is replaced with a float glass to operate the camera and active illumination system properly. The pyrometer sensor was installed inside the sintering chamber to avoid collision with re-coater blade. Diode laser and camera for active illumination imaging, was set to cover the sintering area in the powder bed, from outside the float glass of the sintering chamber. Thyssen Laser-Technik made temperature control and monitoring system was used with a measurement range of 1200-1400 and 1400-1700 nm. Active illumination imaging system manufactured by Cavitar was designed to capture images from the bright object by pouring illumination to the focus beam centred area and capable to be operated in a specific wavelength range. Cavilux HF Illuminator was used with a 500 W diode laser with a wavelength of 809 nm to illuminate the powder material.

2.2. Calculation of thermal features

Heat input was used to demonstrate the amount of laser energy processed per unit meter of processed sinter layer. Heat input Q [J/m] can be expressed by laser power P [W or J/s] and scan speed v [m/s or mm/s] and seen [13, 14] from equation 1,

$$Q = \frac{P}{v} \tag{1}$$

Average processing temperature T_s of the sintered zone is the calculated average temperature, for an entire settling region instead of total process temperatures. It interprets only the area where sintering operation tends to stability between the maximum T_{max} and minimum T_{min} temperature [°C] from a number of occurring points N_t. This can be interpreted [14] as equation 2.

$$Ts = \frac{\sum Tmax + \sum Tmin}{\sum Nt} \tag{2}$$

Instead of using multilayer, in this experiments volumetric energy density is used combining the influences of laser power P [W], scan speed v [mm/s], scan line spacing h [mm] and layer thickness d [μ] for a single layer and is defined [15] as volumetric linear energy density (volumetric LED) ϵ [J/m³] for single layer as in 3.

$$\epsilon = \frac{P}{vhd} \tag{3}$$

Hatch pattern was designed in CAD software and converted to STL format to activate sintering machine command. Hatch pattern was used with horizontal and inclined surface. Contour based hatch structure was monitored along with three different hatch spacing. The data series used during this experiment categorized for monitoring in constant maximum laser power of machine (200 W) and with best result found during theoretical studies [11] (100 W) region, variable region (100-200 W), variable scan speed from the maximum (1000 mm/s) to minimum (50 mm/s) machine range, optimum machine constant scan speed (700 mm/s), hatch spacing (0.10 mm, 0.15mm and 0.20 mm), hatch geometry (inclined or parallel with or without contour) rectangular shape. Single layer thickness used for the sintering layer used was constant to 150 μm.

3. Result and discussion

At constant laser power process at 200 W (see Table 3.1.a.) with decreasing scanning speed from 1000 mm/s to 50 mm/s gives an increase in the heat input value from 200 to 4000 J/m according to equation 1. The layer formation for this decreasing scanning speed and increasing heat input gives raise the increasing order of denser layer formation accordingly. At the beginning from 200 to 667 J/m, the surface formed shows a porous pattern with balling phenomenon. This balling phenomenon was heavily prevailing from 667 to 1000 J/m and started decreasing after this point. From 1000 to 2222 J/m, the porosity starts decreasing. First pattern (see Table 1.c.), with 0.15 mm hatch spacing, the heat input shows a porous surface with balling affect. The heat affected surface found with heat input from 143 to 286 J/m was in a decreasing pattern of porosity.

But in the second pattern, using lower hatch distance of 0.10 mm, shows a significant change with a denser surface than the previous one. Balling phenomena was more severe than before but porosity decrease. At a hatch spacing of 0.20 mm, more porosity and balling phenomena prevailed. Three different hatch geometries (see Table 1.) were evaluated. Laser power with a constant 200 W and scan speed varied between 900 to 500 mm/s. The better heat affected surface found in inclined without contour phase. Both porosity and balling prevailed in all cases, highest presence found in parallel with contour phase. This gives a significant understanding of the usage of contour on HAZ and surface pattern. The characteristics curve (see Fig. 2 a.) for constant power zone shows a linear pattern of increase in linear energy density via average processing temperature of the sintered zone (1 a.) using equation 2 and 3.

Table 1. Active illumination image analysis.

Speed [mm/s]	Heat input [J/m]	Active illumination image	Speed [mm/s]	Heat input [J/m]	Active illumination image
A			B		
Laser power 200 W Hatch distance 0.15 mm, without contour			Laser power 100 W Hatch distance 0.15 mm, without contour		
1000	200		1000	100	
800	250		800	125	
600	333		600	167	
400	500		400	250	
200	1000		200	500	
100	2000		100	1000	
80	2500		80	1250	
60	3333		60	1667	

Laser power [W]	Heat input [J/m]	Hatch distance [mm]	Active illumination image	Speed [mm/s]	Heat input [J/m]	Hatch pattern	Active illumination image
C				D			
Scanning speed 700 mm/s Parallel without contour				Laser power 200 W Hatch distance 0.15 mm			
200	286	0.15		900	222		
150	214			700	286		
100	143			500	400	inclined without contour	
200	286	0.10		900	222		
150	214			700	286		
100	143			500	400	inclined with contour	
200	286	0.20		900	222		
150	214			700	286		
100	143			500	400	parallel with contour	

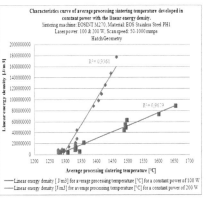

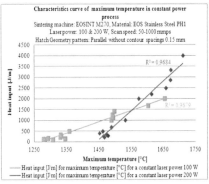

Fig. 2. Summary of constant laser power sintering process.

The significant understanding is at higher laser power the linear energy density increases more rapidly which affects the surface formation already shown in active illumination images. For the heat input versus maximum temperature (1 b.) using equation 1, both the initial temperature and rate of temperature increase is higher with 200 W than 100 W. This supports the HAZ shown. Maximum temperature rose to from 1307°C to 1654 and from 1553°C to 1710°C in case of 100 and 200 W respectively. The optimum machine scan speed was used to check the machine efficiency (3 a.) with a varying laser power and three different hatch distances, showing linear increment in linear energy density. The significant result is the change of hatch distance gives a linear decrease of linear energy density for average sintering temperature using 2 and 3.

from the volumetric linear energy density, is more appropriate in case of commercial machine environment.

Acknowledgements: This project was done as part of InnoBusiness-ArvoBusiness project consortium (TEKES). Authors would like humbly to thank EOS Finland, for all knowledge for this study and to the staff of LUT Laser for their support.

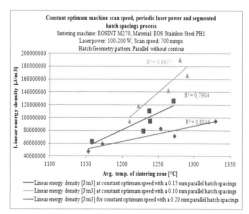

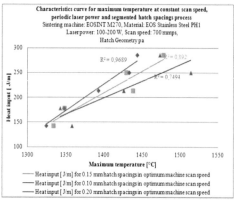

Fig. 3. Summary of hatch spacing sintering pattern.

Similar pattern is shown also in case heat input via maximum temperature (3 b.) using equation 2.1. In both cases the rate of increasing linear energy density or heat input and average processing temperature or maximum temperature shows linearity in their behaviour pattern.

4. Conclusions

It can be concluded that, the best method for evaluation can be achieved by using maximum temperature and average processing temperature of the sintered zone via heat input and linear energy density for single layer. The optimization of the HAZ for creating a better surface depends on the processing parameters and hatch geometry. This conclusion supports the theoretical idea of controlling heat input is the best way to manipulate the DMLS surface for EOS Stainless Steel PH1. The new understanding which differ from theoretical perspective from these experiments, gives the comprehensive conclusion of using linear energy density for single layer

References

[1]   Gibson I., Rosen D.W. & Stucker B. (2010) Additive Manufacturing Technologies, Rapid prototyping to Direct Digital Manufacturing. Springer Science and Business Media, LLC: Newyork, p. 1-2.

[2]   Nyrhilä O. & Shellabear M. (2004) DMLS-development history and state of the art. LANE 2004 conference, Erlangen, p. 1-12.

[3]   Steen W. M. & Majumder J. (2010) Laser Material Processing, 4th ed. London: Springer-Verlag, p. 358.

[4]   Dahotre N. B. & Harimkar S. P. (2008) Laser Fabrication and Machining of Materials. Newyork: Springer Science and Business Media, LLC, p.357.

[5]   Yasa E., Deckers J. & Kruth J.P. (2011) The investigation of the influence of laser re-melting on density, surface quality and microstructure of selective laser melting parts, Rapid Prototyping Journal, vol. 17 iss. 5, p. 2.

[6]   O'Neill W., Sutcliffe C.J., Morgan R., Landsborough A. & Hon K.K.B. (2007) Investigation on Multi-Layer Direct Metal Laser Sintering of 316L Stainless Steel Powder Beds. [online document] Available at http://www.sciencedirect.com/, 1-2.

[7]   Dongdong G. & Yifu S. (2008) Processing conditions and micro-structural features of porous 316L stainless steel components by DMLS, Applied Surface Science 255, p. 1880–1888.

[8]   Y. Ning, Y. S. Wong, J. Y. H. Fuh, and H. T. Loh, (2006) An Approach to Minimize Build Errors in Direct Metal Laser Sintering, IEEE Transactions on automation science and engineering, Vol. 3, No. 1.

[9]   Niu, H.J. & Chang, I.T.H. (1999). Instability of scan tracks of selective laser sintering of high speed steel powder. Scripta Materialia, Vol. 41, No. 11, pp. 1229–1234.

[10] Simchi A. & Asgharzadeh H. 2004. Densification and microstructural evaluation during laser sintering of M2 high speed steel powder. Material Science Technology; vol.20, pp.1462-1468.

[11] Delgado J., Ciurana J., Reguant C. and Cavallini B. (2010) Studying the repeatability in DMLS technology using a complete geometry test part. Taylor and Francis group, London, p. 349.

[12] EOS (Electro Optical Systems), [online document] Available at http://www.rmsiberia.com/Producto/eosint_m270_en.pdf .

[13] A.Lehti, L.Taimisto, H.Piili, A.Salminen, O.Nyrhilä;(2011) Evaluation of different monitoring methods of laser assisted additive manufacturing of stainless steel, ECerS XII, Stockholm, pp.2.

[14] M.Eskatul Islam, (2011) Laser assisted additive manufacturing of stainless steel, Lappeenranta University of Technology, Lappeenranta pp. 52,137,138.

[15] Dongdong G. & Yifu S. (2009) Balling phenomena in direct laser sintering of stainless steel powder. Metallurgical mechanisms and control methods Materials and Design, Elsevier, vol. 30, pp.2909.

Balling phenomena in Selective Laser Melting (SLM) of pure Gold (Au)

Nadeem Ahmed Sheikh[1*], Mushtaq Khan[2], Khurshid Alam[2], Hussain Imran Syed[2], Ashfaq Khan[3], Liaqat Ali[2]

[1] Department of Mechanical Engineering, Muhammad Ali Jinnah University, Islamabad, Pakistan.
[2] School of Mechanical and Manufacturing Engineering (SMME), National University of Sciences and Technology (NUST), Islamabad, Pakistan.
[3] Department of Mechanical Engineering, University of Engineering and Technology (UET) Peshawar, Pakistan.

Abstract. Selective Laser Melting (SLM) is the process of melting metal powder using laser to produce intricate and difficult to manufacture metallic components for wide variety of applications. SLM is a highly nonlinear process and the understanding of melt pool flow physics and associated heat transfer is important. This research work focuses on the numerical simulation of heat flux driven flow in the melt pool of 24 carat gold (Au) powder using 50W fibre laser. Balling phenomenon was observed at low as well as high scan speeds in the experimental work. The size of the droplets formed during balling was observed to increase at high laser power with lower scan speeds, however no clear trend was observed for the changes in the spatial gap between the droplets. It was observed that these patterns indicate zones of melt pool splitting and can be linked to the negative surface tension gradient of pure metals. An initial simulation was performed using equations suitable for incompressible flow coupled with heat transport in porous media showing thermal gradients within the meltpool.

Keywords: Numerical methods, Selective Laser Melting (SLM), Balling, Melt pool, Surface Tension.

1. Introduction

Laser based Additive Manufacturing (AM) processes use laser energy to process metals, polymers and ceramics. Some of the most commonly used laser based AM processes for research and industrial product development include Laser Engineered Net Shaping (LENSTM), Direct Light Fabrication (DLF), Direct Metal Laser Sintering (DMLS), Selective Laser Melting (SLM) and Selective Laser Sintering (SLS) [1-5]. The advantages of these AM processes are geometrical freedom, reduced time and steps to design and most importantly mass customization.

The SLM process, which is a laser melting process, involves complex Marangoni [6] as well as buoyancy driven convection within the small sized melt pool. Despite small size of melt pool, conductive heat transfer is by far less influential compared to convective mode due to fluidic motion [7]. Due to flow currents in the melt

pool, the meltpool may split in to droplets. This instability is related to heat transfer mechanism of viscous molten metal on a solid substrate [8] and becomes complex for non-wetting underlying surfaces [9]. Thermo-capillary or Marangoni flows in the melt pool induces a surface tension gradient and convective fluid flow within the melt pool [6].

Empirical as well as computational methods have been employed to analyze the physics of SLM process. For meltpool flow simulations often Computational Fluid Dynamics (CFD) along with conjugate heat transfer models are used [10]. However uncertainties related to the values of surface tension co-efficient, heat flux absorption and viscosity of material in the fluid state [11] may lead to erroneous results. A number of efforts have been reported in this regard, such as enhanced thermal conductivity approach to account for melt pool convection using energy equation only [10, 12, 13].

Here, a coupled field method is proposed for numerical simulation of settling flow in the melt pool. Initial simulation results are correlated with the experimental observations.

2. Experimentation

The experimental work was performed on SLM 100 system from MTT on the parameters and ranges defined in the Table 1. The detailed experimental work has been presented elsewhere [14]. The material selected for processing was 24 carat gold (Au) powder. The layer thickness and substrate temperature of each Au powder layer was fixed at 100μm and 100°C respectively.

Table 1. Laser parameters and their range [17]

S.no	Parameters	Values
1	Laser Power (W)	10, 15, 20, 25, 30, 35, 40, 45 and 50 W
2	Scan Speed (mm/s)	10, 15, 25, 35, 45, 55, 65, 75, 100, 130, 150, 160, 200, 250, 300, 350, 400, 450 & 500
3	Layer thickness (μm)	100
4	Substrate temperature (°C)	100

3. Mathematical Model

The flow is governed by the incompressible Navier-Stokes equations along with coupled energy balance that includes convection and conduction modes. The Boussinesq approximation is used to include the effect of temperature on the velocity field [15]. The governing equation for the viscous incompressible flow can be described using,

$$-\eta \nabla^2 U + \rho U \cdot \nabla U + \nabla p = F \tag{1}$$

$$\nabla \cdot U \tag{2}$$

The first and second terms relate the rates of momentum gain/lost through viscous transfer and convection respectively and the third relates the pressure (p) force along with other body forces F. η is the dynamic viscosity and ρ is the density of the fluid having U velocity. While the heat equation is given by,

$$\nabla \cdot (-\kappa \nabla T + \rho C_p T U) = Q \tag{3}$$

The left hand side includes heat flux vector due to conductive and convective modes. κ is the thermal conductivity and Cp is the heat capacity of the fluid.

While Q is the term accounting for the heat source/sink. The above set of equations is coupled through the coupling of shear stress with the temperature gradient. For marangoni effect the flow variations in temperature is related to the velocity gradient through equation 4.

$$\eta \frac{\partial u}{\partial y} = \gamma \frac{\partial T}{\partial x} \tag{4}$$

The variations in the temperature field with in the melt pool produce buoyancy force of the form shown in Eq.5 that may displace the fluid when coupled with Eq. 4.

$$F = \alpha g \rho \nabla T \tag{5}$$

The set of equations is solved using Comsol multiphysics [15].

3.1. Laser Beam Model

The heat imparted on the granular/powder bed is through Laser at specific scan speeds. The total heat flux per unit, 'E', width can be calculated using the laser power and the spot size.

$$E(J/m) = \frac{Laser\ power\ (J/s)}{Laser\ speed\ (m/s)} \tag{6}$$

This allows the steady heat load calculations for the laser processing. However the distribution of heat flux within the laser spot is spatially non-uniform [6]. This can be modelled using a Gaussian distribution using Eq. 7.

$$E = E_o e^{-(x-x_o)^2 / a} \tag{7}$$

Where E_o is the peak amplitude of beam, a is the area of zone under the laser spot and x_0 is the centre line of laser spot.

3.2. Simulation Geometry and Model Parameters

The geometry selected for the simulations of melt pool is a 2D rectangle of 0.2 x 0.1 m for the melt pool of the 100 μm depth, which indicates the area under laser spot. The domain is modelled by assuming symmetric along the centre line of the laser focus spot on the powder bed and the material in liquid phase. The parameters used for simulation are listed in Table 2.

Table 2. Simulation parameters and their values

Para	Value	Description
μ	17300[kg/m^3]	Molten gold density
γ	-8e-5	Temperature derivative of the surface tension
η	5.13 [mPa*s]	Dynamic viscosity
κ	318[W/(m*K)]	Thermal conductivity
Cp	25.41[J/(mol*K)]	Heat capacity
g	9.8[m/s^2]	Acceleration due to gravity

4. Heat transfer in meltpool

Splitting of meltpool and balling formation is a direct consequence of heat transfer mechanism in the liquid melt pool. The three possible heat transfer methods for molten material are (a) heat loss to the atmosphere, (b) heat loss through the powder layer and (c) heat loss to the substrate underneath the melt pool (Fig. 1.).

The maximum heat transfer occurs through the substrate, then through the powder layer and the least through the top surface to the atmosphere, thus creating a natural temperature gradient in the melt pool and influencing flow of material.

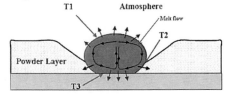

Fig. 1. Heat transfer and temperatures during the laser melting process

5. Results and discussion

5.1. Experimnetal results

In the experimental work, a trend in the size of the droplets (balling) formed at various scan speeds is observed (Fig. 2.). With the increase of scan speed the droplet size decreases. It is apparent that at low scan speeds and high laser power, the energy input into the powder material is high. This provides ample time for the molten state of material to split and form droplets before the re-solidification. This increase in time results in the formation of larger droplets which are far apart from each other as also observed elsewhere [5].

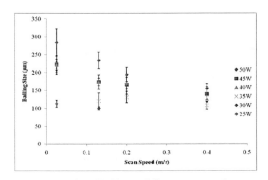

Fig. 2. Size of balling at different scan speeds.

On the other hand at higher scan speeds a smaller melted and heat affected zone is expected. Shrink in the width of the melt pool with increase in its length can lead to discontinuity of melt bead. Under these conditions balling occurs due to the destabilization of the melt pool and results in smaller balling size as observed in this case. Presence of droplets away from the path of the laser scan is also observed at lower scan speeds and low energy densities (Fig. 3.). This could occur due to the presence of a negative surface tension gradient, which induces an outward flow of molten metal from the centre of the melt

pool [5]. As gold had almost no adhesion to substrate, the liquid metal moved away from the path of the laser and created large droplets on both sides.

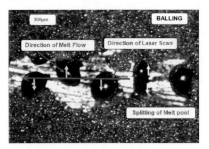

Fig. 3. Microscope image of single line scan of gold: melt pool splitting at 25W laser power and 25 mm/s scan speed

5.2. Simulation results

In order to study the influence of temperature gradient on the flow physics, two types of boundary conditions are used for simulation.

For the first case shown in Fig. 4, the side and lower boundaries are assumed to be insulated. So the heat transfer occurs through the top surface. This resembles the case where the melt pool is wider as compared to the depth of the pool. Here a single large flow of the molten material is observed where the shearing line is at the centre of the meltpool. The molten material flows away from the centre of the meltpool on both sides of the meltpool, similar to that shown in Fig. 3. Also, as this occur for low laser power, the heat transfer to the underlying substrate and the powder in close vicinity of the meltpool will be almost negligible, which supports the insulated boundary condition used for simulations. The surface plot (Fig. 4.) shows the temperature gradient with reference to the peak reference temperature of 1000K. With low temperature difference, the high density molten metal is driven downward with the rolling occurring away from the centre of the meltpool.

Fig. 5 presents the second case of boundary conditions where heat conduction occurs from the bottom and left side of the meltpool. This is a more realistic representation of the heat transfer as also indicated in Fig. 1. Clearly the higher temperature difference is observed in the domain and the resultant flow field is significantly different from the Fig. 4. Fig. 5 shows that the two flow currents in the melt pool. In the high temperature zone, strong rolling can be observed where flow is upward from bottom at the centre of pool, while another roll away from the centre of melts pool shows counter rotation as also observed in the experimental work [14]. The shearing effect at the boundary of the two types of rolling splits the meltpool. The exact size of melt pool is difficult to measure, however the results shown in Fig. 4 and Fig. 5 show qualitatively similar trends as observed in the experimental work with different heat flows from the boundaries of melt pool.

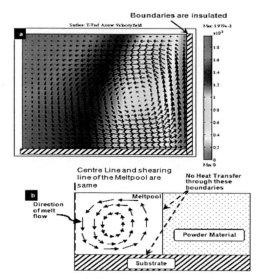

Fig. 4. (a) The difference in temperature with velocity field, (b) schematics of the melt flow within the meltpool.

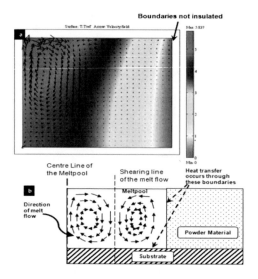

Fig. 5. (a) The colour map with velocity field (b) schematics of the melt flow within the meltpool.

6. Conclusions

Experimental results show formation of larger balling in the melt pool at low scan speed and high laser power which is associated with increase in phase change time from molten to solid. Smaller billing was observed at high scan speed i.e. at low energy input which is because of the instability of size of meltpool. In simulation, a single large flow of material is observed in the case when insulated boundary conditions are considered that may lead to large size balling at high energy intensity. However, two different flows are observed in the simulation case when heat transfer is considered through the substrate and the powder bed leading to small droplets

at low energy intensities. The insulated boundary condition in simulation was found valid for the case when meltpool splits from the centre and droplets move away from the line of laser scan on both sides, while the non-insulated boundary condition were found valid for the case when meltpool splits at a shearing line other than the centre of the meltpool, thus creating balling on the line of laser scan.

References

[1]   Liao HT, Shie JR, (2007) Optimization on selective laser sintering of Metallic Powder via Design of Experiments Method. Rapid Prototyping Journal 13 (3):156-162

[2]   Keicher D M., Miller WD, Smugeresky J.E, Romero JA, (1998) Laser Engineered Net Shaping (LENS), Beyond Rapid Prototyping to Direct Fabrication, Proceedings of the TMS Annual Meeting, San Antonio. Texas. USA. 369–377

[3]   Lewis GK, Nemec R, Milewski J, Thoma DJ, Cremers D, Barbe M, (1994) Directed Light Fabrication", Proceedings of the International Congress on Applications of Lasers and Electro-Optics, Orlando. Florida. USA. 17–26

[4]   Childs THC, Hauser C, Badrossamay M, (2005) Selective Laser Sintering (Melting) of Stainless and Tool Steel Powders: Experiments and Modelling, Proceeding of IMechE, Part B: Journal of Eng. Manufacture. 219(4): 339–357

[5]   Rombouts M, Kruth JP, Froyen L, Mercelis P, (2006), "Fundamentals of Selective Laser Melting of alloyed steel powders", CIRP Annals - Manufacturing Technology, Vol. 55 No. 1, pp. 187–192

[6]   Khan M, (2010) Selective Laser Melting (SLM) of Gold (Au), PhD Thesis. Loughborough University. UK

[7]   Mills KC, Keene BJ, Brooks RF, Shirali A, (1998) Marangoni Effects in Welding. The Royal Society :911-925.

[8]   Morgan R, Sutcliffe CJ, O'Neill W, (2004) Density Analysis of Direct Metal Laser Re-Melted 316L Stainless Steel Cubic Primitives, Journal of Materials Science, 39(4): 1195-1205

[9]   O'Neill W, Sutcliffe CJ, Morgan R, Hon KKB, (1998) Investigation of Short Pulse Nd:YAG Laser Interaction with Stainless Steel Powder Beds, Proceedinf of 9th Solid Freeform Fabrication Sym. Austin. Texas. USA: 147-160

[10] Toyserkani E, Khajepour A, Corbin S, (2004) 3-D Finite Element Modeling of Laser Cladding by Powder Injection: Effects of Pulse Shaping on the Process, Optics and Laser in Engineering. 41: 849-867.

[11] Eustathopoulos N, Dervet B, Ricci E, (1998) Temperature Coefficient of Surface Tension for Pure Liquid Metals, Journal of Crystal Growth.191: 268-274.

[12] Song B, Dong S, Liao H, Coddet C, (2011) Process parameter selection for selective laser melting of Ti6Al4V based on temperature distribution simulation and experimental sintering. The International Journal of Advanced Manufacturing Technology [Published Online]

[13] Rombouts M, Kruth JP, Froyen L, (2009) Impact of Physical Phenomena During Selective Laser Melting of Iron Powders, Proceedings of The Minerals, Metals & Materials Society (TMS): Fabrication, Materials, Processing and Properties, 1: 397-404

[14] Khan M, Dickens P, (2012) Selective Laser Melting (SLM) of Gold (Au), Rapid Prototyping Journal. 18(1): 81-94.

[15] Comsol (2008), Comsol User Manual, [www. Comsol.com].

Formability of micro material pre-forms generated by laser melting

H. Brüning and F. Vollertsen

BIAS - Bremer Institut für angewandte Strahltechnik GmbH, Klagenfurter Str. 2, 28359 Bremen

Abstract. A laser-based micro upsetting process is presented which takes advantage of scaling effects in order to optimise the conventional multi-stage upsetting process. Due to the fact that the surface tension exceeds the gravitation force with increasing miniaturisation, in the micro-range an accumulation of molten material forms a nearly perfect sphere. This material accumulation can be finally formed in a secondary process step. In former investigations it has been shown, that the laser-based free form heading process improves the upset ratio up to 250 instead of 2.3 commonly achieved by conventional multi-stage upsetting processes. In this article the second stage of the upsetting process is examined by investigating the deformation behavior of laser assisted free formed austenitic chromium-nickel steel (1.4301) with a rod diameter of 0.5mm and head diameters ranged from 1.1mm to 2.1mm using a pneumatic forming machine. The experiments show that the total maximum achievable upset ratio is much higher compared to a conventional multi-stage upsetting process. The punch force and the height of the preforms after forming operation were measured, so that flow curves are calculated.

Keywords: laser based free form heading, upsetting, formability, size effect, scaling effect, micro preform, new processes, microforming, laser melting.

1. Introduction

Accompanied by cost reduction, newly invented products are expected to either have an increased functionality by given size or a similar amount of functions by smaller size compared to the preceding product. These boundary conditions are the main forces which lead in most cases to function compaction in many fields of technology, e.g. automotive electronics and telecommunication. In the area of fabrication of metallic components in high quantities, cold forming has been established not only due to its high cost efficiency but also due to its economic utilisation of material. Functional compaction inevitably leads to miniaturization of components. Concerning forming of metallic parts, knowledge and experience gained in so-called "macro-range" is not, or only with limitations, directly transferrable into "micro-range" due to size effects [1]. Components belong to micro-range if at least two dimensions do not exceed the length of 1mm [2].

In many mass production chains free form heading is an important process because material needed for each step of the shaping process is provided. Depending on the basic material and the sample diameter d_0, a certain length l_0 of the sample, called the upsetting length, can be accumulated in one step by conventional upsetting process. The process is limited in two dimensions: the maximum natural strain φ as a limit for formability of basic material and the upset ratio s as a limit against buckling [3].

The upset ratio decreases with decreasing sample diameter, and thus with increasing miniaturisation, the conventional upsetting process becomes more and more inefficient as stated by Vollertsen [4]. Especially in case of work-hardening material, the conventional upsetting process has to be interrupted by several heat treatment cycles which increase process time, process complexity and overall costs. In order to still be able to manufacture small sized metallic components, a laser based free form heading process has been demonstrated [5] which takes advantage of the fact that in micro-range forces due to surface tension become dominant compared to gravitation [6]. Besides conventional upsetting, it is also possible to form the preforms by rotary swaging as stated in [7].

Previous investigations on microstructure of preforms generated by laser melting showed, that under conditions as mentioned above a dendritic microstructure is observed [9]. It is known that this type of microstructure does generally not allow a high formability [10] especially in combination with cold work hardening steels such as 1.4301. In this paper, the formability of micro material preforms generated by laser melting is investigated.

2. Method

2.1. Generation of material preforms by laser melting

In the laser based free form heading process the laser beam energy is focused on the bottom surface of a rod inducing the end of the rod to melt. The molten mass forms a nearly perfect sphere due to surface tension.

The investigations were focused on rods of austenitic chromium-nickel steel 1.4301 (X5CrNi18-10) with a diameter of 0.5mm. The longitudinal axes of the rods are orientated parallel to gravitational force so that the laser irradiation of bottom surface takes place in a right angle. The focus layer of laser beam is coincident with initial position of the bottom surface of the rod. In order to prevent oxidation of specimen, Argon as shielding gas is used. Due to the fact that molten rod material forms a sphere and no feeding is applied, the irradiated surface moves out of focus layer thus leading to a defocussing effect [5]. A schematic view of both, the conventional upsetting process and the experimental setup, is shown in Fig. 1 and the corresponding laser parameters are shown in Table 1. In Fig. 1 F_P demonstrates the punch force while the associated arrow shows its direction.

Table 1. Specifications of the used laser system

Laser type	Fibre laser, cw
Wavelength	1085nm
Focal distance	100mm
Beam radius	0.02mm
Divergence angle	40mrad
Max. power	300W

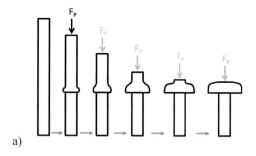

a)

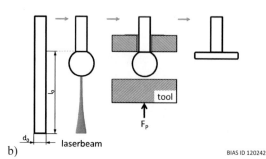

b)   BIAS ID 120242

Fig. 1. (a) Conventional and (b) laser based free form heading

2.2. Upsetting of material preforms

The generation of material accumulation is followed by a compression test. This requires two coplanar tool surfaces moving towards each other so that the initial height of the preform is reduced. In case of upsetting preforms as stated above, a radial fixation of the shaft of the preform is necessary to avoid axial movement.

3. Experiment

Three sets of preforms with d_0=0.5mm were generated using irradiation power P_L=64W and pulse durations t_P of 500ms, 1500ms and 4000ms resulting in different average head diameters d_H. Every set consists of 20 preforms. The head diameters were measured using a micrometer calliper. The results are shown in Table 2. It can be seen that the standard deviation of the head diameter of the smaller preforms lies within or at least very close to measuring inaccuracy of the measuring instrument so that it seems reasonable to state that generating material preforms by laser melting is a process resulting in products with well defined volume. Forming work pieces in closed forging dies necessitates semi-finished parts with a well determined volume in order to be able to form products with very little probability of defects such as incomplete die filling [8]. Preforms generated by laser melting achieve these standards making them a good opportunity as wrought material for a following forming processes.

Table 2. Laser parameters and resulting head diameters with standard deviation

L_P [W]	t_P [ms]	d_H [mm]	Std.Dev. [mm]	Std.Dev. [%]
64	500	1.136	0.004	0.39
64	1500	1.598	0.005	0.34
64	4000	2.135	0.010	0.49

The second process step, which is a free form upsetting process, is carried out using a pneumatic press "Dynamess" with maximum punch force F_P of 50kN thus enabling a precise adjustment of maximum punch force as well as force increase. The required tools are made of heat-treated steel 42CrMo4 (1.7225) so that a hardness of 55HRC is reached. Both, the lower and the upper die have flat, coplanar and polished surfaces. The lower die is divided into two parts parallel to the longitudinal axis of the preform allowing on one hand the deformed preform to be removed easily. On the other hand a clamping effect on the shaft of the preform is achieved by the condition, that the diameter of the resulting hole is smaller than the shaft diameter. A radius of 0.2mm forms the fillet from hole to die surface. A 3d-cross sectional view of lower and upper die in correspondace with a material preform (head diameter 1.1mm) is shown in Fig. 2.

The maximum formability of material preforms generated by laser melting is investigated by fixing the preform in the lower die and progressively increasing the punch force thus resulting in decreasing final head heights h_{Hf} and increasing final flattened surface area A_{Hf}. The punch forces were defined for preforms with $d_H = 2.135$mm and have been scaled down to smaller preforms as equation 1.

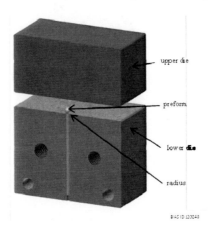

Fig. 2. Cross sectional view on upper and lower die with material preform

$$F_{P2} = F_{P1} \cdot (d_{H2}^2/d_{H1}^2). \qquad (1)$$

After the punch force has been applied and the preform was unloaded, the flattened surface area A_{Hf} as well as the head height h_{Hf} (see Fig. 2.) were measured so that the average natural strain φ_H of the head can be determined.

Fig. 3. Left: Preform, $d_0 = 0.5$mm. Right: Preform after forming process

4. Results

Figure 4 shows the yield stress k_f in dependence of the average natural strain φ_H. The yield stress is calculated as follows [11]:

$$k_f = \frac{F_p}{A_{Hf}\left[1 + \frac{1}{3}\mu \dfrac{d_{Hf}}{h_{Hf}}\right]} \qquad (2)$$

with, μ as the friction coefficient between preform and die. A friction coefficient of $0.05 < \mu < 0.15$ [3] is

generally assumed for cold heading with lubricated interacting surfaces depending on surface preparation. Due to the fact that no lubrication but polished surface have been used in these experiments, a friction coefficient of $\mu = 0.15$ is used for further calculations. The increase of punch force was $F_p = 100$N/s for all experiments carried out.

The average natural strain φ_H is calculated as follows:

$$\varphi_H = ln\ (h_{Hf}/d_H). \qquad (3)$$

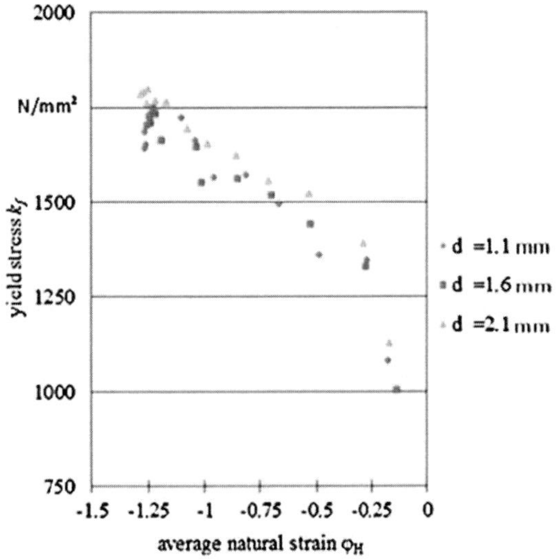

Fig. 4. Yield stress k_f in dependence of average natural strain φ_H for different head diameters d_H

Overall, it can be stated that the yield stress k_f increases linear with increasing absolute value of average natural strain φ_H. This behavior is generally expected but beyond that it is to be noticed that there is a very good accordance in yield stress between the three different sizes of preforms being formed. Compression tests in micro range with cylindrical specimens of 1.4301 have been carried out by Mebner [12]. A comparison of the values of yield stress in dependence of natural strain shows, that the deviations in yield stress are approximately ten percent. A reason for this might be, besides the geometry of specimens, the difference in grain size.

In general, the maximum formability of preforms is reached as soon as cracks appear. If cracks appear at all, it is most likely that they come into existence at the fillet between circumference and flattened surface of the preform as this is a strong deformed area. Figure 5 shows a cross-section polish of a formed preform with $\varphi_H = -1.7$ as well as a section of the strong formed area. It can be seen that neither the maximum applied force in the experiments resulting in $\varphi_H = -1.3$ (see Fig. 4) nor an increasing of maximum punch force enabling $\varphi_H = -1.7$ leads to cracks in the preform. A further exaltation of the punch force was not committed in order to not exceed the mechanical properties of the die.

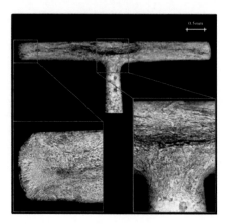

Fig. 5. Cross-sectional view of deformed preform $\varphi_H = -1.7$

5. Conclusion

The presented two-staged upsetting process is based on generating preforms by laser melting which are afterwords formed in a secondary process step. It is shown, that preforms can be reproducably generated with a standard deviation in diameter of less than 0.5%. There will be a focus on this value to be further reduced in future investigations. Even though a dendritic microstructure is observed after free form heading process, average natural strains of $\varphi_H = -1.7$ can be achieved. Due to the fact that larger punch forces would lead to plastic deformation of the used forming tool it is assumed that mechanically more resistant forming tools might enable natural strain $|\varphi_H| \geq 2.0$.

Acknowledgement: The authors gratefully acknowledge the financial support by Deutsche Forschungsgesellschaft (DFG, German Research Foundation) for Subprojekt A3 "Stoffanhäufen" within the SFB 747 (Collaborative Research Centre) "Mikrokaltumformen – Prozesse, Charakterisierung, Optimierung".

References

[1]   Geiger, M., Kleiner, M., Eckstein, R., Tiesler, N., Engel, U. (2001) Microforming. Annals of the CIRP 50:445-462.
[2]   Masuzawa, T. (2000) State of the Art of Micromachining. Keynote Paper. Annals of the CIRP 49:473-488.
[3]   Lange, K., Umformtechnik, Band 2: Massivumformung (1988) Springer-Verlag, Berlin.
[4]   Vollertsen, F. (2008) Categories of size effects. Prod. Eng.-Res. Dev. 2, p. 377-388.
[5]   Vollertsen, F., Walther, R. (2008) Energy balance in laser-based free form heading. Annals of the CIRP 57:291-294.
[6]   Vollertsen, F., Größeneffekte – eine systematische Einordnung. In: Größeneinflüsse bei Fertigungsprozessen p. 1-9 (2009). BIAS Verlag. Bremen.
[7]   Kuhfub, B., Piwek, V., Moumi, E., Vergleich charakteristischer Einflussgrößen beim Mikro- und Makrorundkneten. In: 4. Kolloquium Mikroproduktion p. 219-228 (2009). BIAS Verlag. Bremen.
[8]   Ilschner, B., Singer, R.F., Werkstoffwissenschaften und Fertigungstechnik (2001) Springer-Verlag, Berlin.
[9]   Stephen, A., Vollertsen, F., (2010) Upset ratios in laser-based free form heading. Physics Procedia 5:227-232.
[10] Weißbach, W., Werkstoffkunde. Strukturen, Eigenschaften, Prüfung (2010) Vieweg & Teubner, Wiesbaden.
[11] Siebel, E., Die Formgebung im bildsamen Zustand (1932). Verlag Stahleisen, Düsseldorf.
[12] Mebner, A., Kaltmassivumformung metallischer Kleinstteile: Werkstoffverhalten, Wirkflächenreibung, Prozessauslegung (1998). Meisenbach Verlag. Bamberg.

Modeling of coaxial single and overlap-pass cladding with laser radiation

N. Pirch[2], S. Keutgen[2], A. Gasser[2], K. Wissenbach[2], I. Kelbassa[1,2]

[1]   Chair for Laser Technology, RWTH Aachen University, Germany

[2]   Fraunhofer Institute for Laser Technology ILT, Steinbachstr. 15, Aachen, Germany

Abstract. The repair with laser cladding of single crystalline Ni-base superalloy gas turbine blades requires specific solidification conditions in order to realize an epitaxic dendritic growth front. For each superalloy there exists a process window concerning solidification conditions within the formation of new grains ahead of the epitaxic dendritic growth front can be prevented. In order to determine this process window a three-dimensional finite element model for laser cladding has been developed. This model emphasizes on a precise calculation of free surface shape of melt pool because the microstructure as processed depends essentially on the solidification conditions around the solidus/liquidus temperature. So due to the small distance between the melt pool surface and the growth front all errors concerning the approximation of melt pool surface and the implementation of boundary conditions will be mapped directly onto the solidification conditions. This model does not use any geometrical approximation of the melt pool surface but determines the melt pool surface shape by the balance equation for capillary forces.

Keywords: Laser cladding, modeling, FEM, superalloy, single crystal

1. Introduction

Laser cladding has been successfully applied or developed for fabrication of aircraft structural parts, gas turbine engine component repair, remanufacturing of parts with manufacturing errors, repair of tools and dies, and medical implants [1-3]. It has been shown during the last decade that single crystal repair by laser cladding [4, 5]. The process requires close control of processing parameters, i.e. the solidification conditions must stay inside a small window which depends on the alloy. For process development purposes this processing window concerning solidification conditions has to be mapped to a process window concerning processing parameters. This relation between microstructure and processing parameters can be determined by modeling both the temperature field of the process and the solidification behavior of the alloy. A review and state of the art respectively of modeling of microstructure development for single crystal superalloy under solidification conditions during laser cladding (temperature gradient

$\sim 10^5$-10^6 K/m, interface velocity 10^{-3} -10^{-1} m/s and cooling rate 10^2-10^5 K/s) is given by Mokadem [6].

This paper addresses a precise computation of temperature field by a further development of a three-dimensional quasi-stationary finite element model of laser cladding by coaxial powder injection [7]. The further development applies to the computation of the melt pool surface shape. In the previous model this was done by mass balance whereby the resulting melt pool surface shape is completely determined by the powder density distribution. The process has now been monitored with high speed videography. The results concerning the melt pool surface shape in a longitudinal side view disagree with the computed ones. This means that the formation of melt pool surface shape is governed by the capillary forces and this physically process can't be described by a mass balance. Within the new model the pressure balance equation is used for the computation whereby the track cross section from the metallographic analysis is used as input for the computation. The projection of the triple line onto the surface is part of the solution and determined by an iterative process.

The subroutines for the calculation of powder density distribution, particle temperature and laser power attenuation due to the laser beam particle interaction and local inclination angle of surface element against beam propagation direction is taken from the previous model.

Because the track cross section is used as input for the computation the three dimensional finite element quasi-stationary model for laser cladding is not completely self-consistent. But the model represents a significant progression concerning the accuracy of the temperature field computation and makes thereby possible a mapping between the processing parameters and the solidification conditions.

2. Mathematical analysis

Figure 1 shows the laser cladding process. A moving laser beam strikes the surface with an top hat intensity distribution. The absorbed and thermalized optical energy melts the surface region of substrate and additive material and consecutively the clad forms by solidification as shown in the figure.

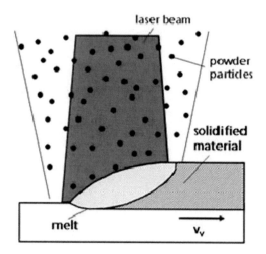

Fig. 1. Schematic representation of the laser cladding process

The quasi-stationary temperature field T is obtained from the heat equation

$$\frac{\partial(\rho c_p T)}{\partial t} - v_v \frac{\partial(\rho c_p T)}{\partial x} = div(\lambda \cdot grad T).$$

ρ : density, λ : thermal conductivity,

c_p : heat capacity, v_v : scan speed

The heat equation is complemented by the boundary condition.

$$-\lambda \frac{\partial T}{\partial n}_{\substack{\text{heat flux in the}\\\text{substrate}}} = \underbrace{<\vec{n}, \vec{q}_{Laser, trans}>}_{\substack{\text{thermalized optical}\\\text{energy}}}$$

$$-\underbrace{n_x \cdot v_v \cdot \rho_P c_p (T - T_P)}_{\substack{\text{energy for temperature}\\\text{equalisation powder par-}\\\text{ticle melt surface}}} - \underbrace{\sigma_o \varepsilon T^4}_{\text{thermal radiation}}$$

$q_{Laser, trans}$: laser intensity distribution transmitted through the powder gas beam , ρ_P : particle density distribution, T_P : powder particle temperature, σ_o : Stefan Boltzmann constant, ε : emissivity
$n^{tr} = (n_x, n_y, n_z)$: normal vector at melt pool surface which is derived from the balance of heat flux at the surface. The calculation of bead shape h(x,y) is based on the pressure balance equation whereby the integration is performed within the domain restricted by the projection of the triple line onto the x,y-plane (Fig. 2). The geometry

of the bead cross section and res-pectively the projection of the triple line onto the y,z-plane is used as input for the calculation of melt pool shape (Fig. 2). If the triple line along solidification front lies outside the laser intensity the isotherm for the melt temperature ends at right angle and the melt pool surface of the melt pool h ends horizontal in the solidified region (Fig. 2). This boundary condition is used to determine the right hand side of the pressure balance equation. The two subroutines for the self-consistent calculation of bead shape and temperature field are strongly interrelated. The calculation of bead shape is based on the geometry of the triple line which is deduced from the temperature field. For the calculation of the temperature field the track geometry is assumed. For this reason an iterative approach is used for the self-consistent calculation of bead shape and temperature field.

$$\gamma(T) \cdot \mathbf{div} \frac{\nabla h}{(1+<\nabla h, \nabla h>)^{1/2}} = p_{ext} - p_{int}$$

The iterative solution starts with an approximative geometry in the x,y-plane of the triple line, which correlates with the dimension of laser intensity distribution, calculates the melt pool surface and subsequently the quasi-stationary temperature field. From the temperature field the next approximation of the projection of the triple line onto the x,y-plane is deduced.

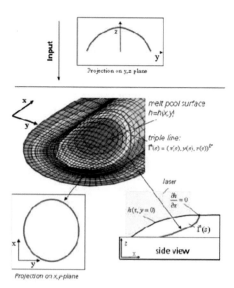

Fig. 2. Schematic for mathematical model for melt pool surface calculation during laser cladding process area

The iteration stops when the difference between two successive approximations of the triple line falls below a given limit.

3. Results

The 3D finite element model of laser metal deposition (LMD) by coaxial powder injection is extended for modeling of overlapping tracks. Figure 3 shows the result for the temperature field for single and two overlapping tracks. The maximum process temperature differs slightly and

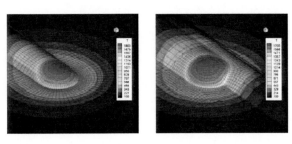

Fig. 3. 3D temperature field for single and overlapping track (Top Hat, D= 1.5mm, A*P$_W$=107W, v$_v$=300mm/min, m$_P$=1.357 g/min.

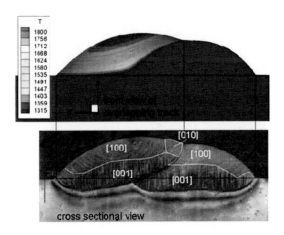

Fig. 4. Comparison of experimental and computed melt pool surface in cross sectional view of overlapping tracks (Top Hat, D= 1.5mm, A*P$_W$=107W, v$_v$=300mm/min, m$_P$=1.357 g/min.

the geometry of the melt pool is inclined against the z-axis (Fig.4). The comparison of experimental and computed melt pool geometry shows an excellent agreement (Fig. 8).

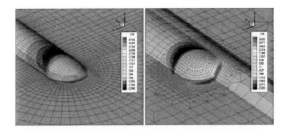

Fig. 5. 3D cooling rate for single and overlapping track (Top Hat, D= 1.5mm, A*P$_W$=107W, v$_v$=300mm/min, m$_P$=1.357 g/min.

The maximum heating rate is reduced up to 25 % for the overlapping track compared to the single track (Fig. 9). Concerning the cooling rate only marginal differences are encountered (Fig. 5, 6).

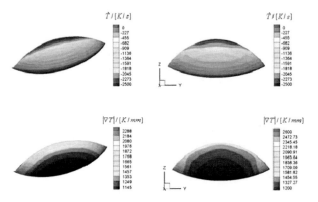

Fig. 6: 3D cooling rate and absolute value of temperature gradient for single (right) and overlapping (left) track on the solid liquid interface projected onto the y,z-plane (Top Hat, D= 1.5mm, A*P$_W$=107W, v$_v$=300mm/min, m$_P$=1.357 g/min.

During the growth competition between the different dendrite growth directions the growth direction with the smallest angle to the temperature gradient will overgrow the other one. Due to the local variation of the orientation of the temperature gradient on the solid liquid interface different dendrite growth directions are initialized during solidification. Fig. 7 shows that for overlapping tracks the [010] growth direction will be found in the border region of the second to the first track (Fig. 7). This effect is verified by the experimental investigations (Fig. 4). If the columnar to equiaxed transition criterion is applied to the local solidification conditions the results for the single and overlapping tracks (fig. 8) differ only slightly concerning the minimum values and the distribution. Thus the region at risk for the columnar to equiaxed transition that is the region with low values for the criterion is only slightly higher in the single track.

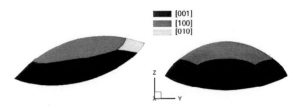

Fig. 7: 3D local dendrite growth direction for single (right) and overlapping (left) track on the solid liquid interface projected onto the y,z-plane (Top Hat, D= 1.5mm, A*P$_W$=107W, v$_v$=300mm/min, m$_P$=1.357 g/min.

Fig. 8: 3D columnar to equiaxed transition criterin according to Gäumann /1/ for single (right) and overlapping (left) track on the solid liquid interface projected onto the y,z-plane (Top Hat, D= 1.5mm, $A*P_W$=358W, v_v=300mm/min, m_P=1.357 g/min.

4. Conclusion

A three-dimensional quasi-stationary finite element model of laser metal deposition (LMD) by coaxial powder injection as given in the first report has been extended through a new program module for the calculation of melt pool surface. Now the experimental and computed shape of the melt pool surface agree in view of cross section and longitudinal section. The results of the investigations show that the shaping of melt pool surface is governed by the capillarity forces and can not be modeled in general by the local mass balance. Furthermore the model is advanced for modeling of overlapping tracks. Concerning cooling rate and the criterion for columnar to equiaxed transition according to Gäumann /1/ the results for single track and overlapping tracks differ only slightly. But varying dendrite growth directions are encountered for overlapping tracks compared to the single track. It has been shown during the last decade that single crystal repair by laser cladding [2,3] requires an exact control of the processing parameters, i.e. the solidification conditions must be kept in a small window depending on the alloy. For process development the model can be used to derive appropriate process parameters for the required solidification conditions. The present model will be extended in the future to a three-dimensional transient finite element model which will cover more complex geometries. In addition through comparison between experimental and computational results it will be investigated whether the model has to be completed with a module for the calculation of the Marangoni convection.

References

[1]   J.L. Miller, K-H. Grote, "Solid Freeform Manufacturing technologies as an important step in development process", Computers in Industry, 28(1), 11-16 (1995).

[2]   G.K. Lewis, E. Schlienger, "Practical considerations and capabilities for laser assisted direct metal deposition", Materials and Design, 21(4), 417-423 (2000).

[3]   G. Backes, E.W. Kreutz, A. Gasser, E. Hoffmann, S. Keutgen, K. Wissenbach, R. Poprawe, "Laser-shape reconditioning and manufacturing of tools and machine parts", Proc. ICALEO'98, Orlando, FL USA, Vol. E, 48-56 (1998).

[4]   M. Gäumann, P.-H. Journeau, J.-D. Wagnière et W. Kurz, "Epitaxial Laser Metal Forming on a single Crystal Superalloy", Laser Assisted Net shape Engineering 2, Proceedings of the LANE'97, eds.: M. Geiger, F. Vollertsen, Meisenbach Bamberg 1997, pp. 651-657.

[5]   S. Mokadem, C. Bezençon, J.-M. Drezet, A. Jacot, J.-D. Wagnière, W. Kurz, "Microstructure control during single crystal laser welding and deposition of Ni-base Superalloy", In "Solidification Processes and Microstructures", eds. M Rappaz, C Beckermann, R Trivedi, TMS 2004, p.67-75

[6]   S. Mokadem, C. Bezençon, A. Hauert, A. Jacot, W. Kurz, „Laser Repair of Superalloy Single Crystals with Varying Substrate Orientations", Metallurgical and Materials Transactions A, Volume 38, Number 7, July 2007 , pp. 1500-1510.

[7]   N. Pirch, S. Mokadem, S. Keutgen, K. Wissenbach, E.W. Kreutz, „3D-Model for Laser Cladding by Powder Injection", Proceedings of the LANE 2004, September 21 - 24, 2004,Erlangen, Germany M. Geiger, A. Otto eds., Laser Assisted Net Shape Engineering 4, Volume 2, pp.851-858.

High speed LAM

I. Kelbassa[1,2], A. Gasser[1,2], W. Meiners[2], G. Backes[1], B. Müller[1]

[1]   Chair for Laser Technology, RWTH Aachen University, Steinbachstrasse 15, 52074 Aachen, Germany
[2]   Fraunhofer Institute for Laser Technology, Steinbachstrasse 15, 52074 Aachen, Germany

Abstract. Laser Additive Manufacturing (LAM) of parts and components fascinates due to process specific advantages such as nearly unrestricted geometrical freedom, material freedom and achievable properties of the parts built-up. Hence, adaptive parts and components can be manufactured, repaired and modified by LAM. This article presents the advances in speeding up the LAM techniques Selective Laser Melting (SLM) and Laser Metal Deposition (LMD) towards higher speeds and deposition rates. In the first part build up increase for (SLM) is presented. The main approach for higher build up rates is the use of higher laser power up to 1kW. A machine concept and process strategies for the efficient transfer of laser power into build up rate for different materials (aluminum ans steel) will be presented. The second part focuses on investigations on LMD at high processing speeds of up to 600 m/min. Process basics and achieved results in terms of layers are presented. This new process enables to produce dense layers with thicknesses ranging from 10 μm up to 0.3 mm in one pass with very low dilution at high speeds, resulting in completely new advantages regarding economic efficiency of LMD processes.

Keywords: Selective Laser Melting, SLM, Laser Metal Deposition, LMD, build-up rate, High Speed SLM and LMD, LAM.

1. SLM (Selective Laser Melting)

The additive manufacturing technology Selective Laser Melting (SLM) makes it possible to manufacture metal components layer by layer according to a 3D-CAD volume model. Thereby, SLM enables the production of nearly unlimited complex geometries without the need of part-specific tooling or preproduction costs [1].

Figure 1 illustrates the principle of the SLM process and the steps the process can divided into. First, the 3D-CAD volume model is broken down into layers and transferred to the SLM machine. Subsequently, the powder material (grain fraction 10-45 μm) is deposited as a defined thin layer on a substrate. The geometric information of the individual layers is transmitted by the laser beam to the powder bed wherein the regions to contain solid material are scanned under an inert atmosphere, leaving a solid layer of the piece to be produced. After lowering the substrate by one layer thickness, the process steps are repeated until the part is finished. Since standard metallic powders are used, which melt completely, the part has a density of approximately 100%, thus assuring mechanical properties that match or even beat those of conventionally manufactured parts (cutting, casting).

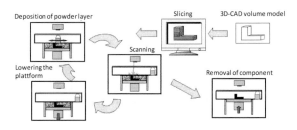

Fig. 1. Principle of the SLM process.

1.1. Current applications

Currently, SLM is used to manufacture functional prototypes and to build up final parts directly. In this case the field of commercial applications is limited to single parts or parts in small batches.

The tool- and mold-making industry is a typical example of a branch producing final parts in small batches of approximately 1 to 8. Because of the almost infinite geometrical freedom, SLM is applied to manufacture tooling inserts containing conformal cooling channels (see Fig. 2, left). Thanks to SLM, an improved tool cooling can be attained, resulting in reduced cycle times and improved part quality. As a result the rapid manufacturing method SLM offers massive cost savings in combination with better functionalities despite the higher manufacturing costs for small batch production [2].

Medical technology is another area applying the infinite geometrical freedom and variability of SLM. According to the current state of the art individual implants in a batch size of one are manufactured with SLM. Typical examples of application are hip implants or surgical instruments out of titanium alloys as well as dental restorations out of cobalt chromium (see Fig. 2

middle). Compared with conventional manufacturing methods like for example casting, SLM can significantly decrease the processing time and the production costs. Furthermore, the given geometric freedom can be used to manufacture implants with new functionalities such as hollow structures, graded porosity, adapted rigidity or surface structure [3].

Fig. 2. Current applications of SLM (left: tooling insert with conformal cooling channels, middle: dental restorations (CoCr) and hip implant (titanium), right: pneumatic valve (Festo AG & Co. KG)

An application for the manufacturing of functional prototypes is represented by the pneumatic valve out of the aluminum alloy AlSi10Mg produced in cooperation with Festo AG & Co. KG (see Fig. 2, right). In comparison to conventional manufacturing methods, such as casting, the pneumatic valve can be processed in small series efficiently and economically with the use of SLM machines according to the current state of the art (200 W). In order to enable SLM to enter series production with higher lot sizes, an increased productivity for the manufacturing process is essential while maintaining the requirements for constant component quality of the manufactured parts.

1.2. Improving the productivity of the SLM process

In order to qualify SLM for series production, the Fraunhofer Institute for Laser Technology ILT has been conducting scientific research and development within the Cluster of Excellence "Integrative Production Technology for High-Wage Countries". This project aims to reduce the production time and maintain the requirements for constant component quality. The production time of the SLM process can be divided into primary and auxiliary processing times. The primary processing time is the time required by the laser beam to melt the powder layer. The secondary processing time is the sum of times required for the process chamber to be prepared for the primary period. These include, for example, time needed to equip the SLM machine and the powder deposition. The main influencing variables to decrease the primary processing time and thus, manufacture parts economically, are hatch distance Δy_s, layer thickness D_s and scanning velocity v_{scan}. The layer thickness and scanning velocity are limited amongst other factors by the available laser power. The hatch distance is limited by the diameter of the beam and typically equals approximately 0.7 times the beam diameter.[1] A benchmark to measure the productivity of the SLM process is given by the process-related build-up rate,

which is determined by the product of hatch distance, layer thickness and scanning velocity according to the following equation:

$$\dot{V} = D_s \cdot \Delta y_s \cdot v_{scan} \qquad (1)$$

To enable SLM to enter series production, manufacturing higher lot sizes in an economical way, the process-related build-up rate has to be increased significantly by increasing the laser power. According to the current state of the art, SLM machines are equipped with laser power up to 200 W (max. 400 W) and a focus diameter of approximately 100 µm; process-related build up rates of 1-4 mm³/s can be achieved. In order to increase the process-related build up rate a TrumaForm LF250 SLM machine was completely redesigned and rebuilt. Therefore, the SLM machine is equipped with a 1 kW laser source and a redesigned optical system that allows changing the beam diameter between 200 and 1000µm during the process. As a first approach the increased laser power can be used to increase the scanning velocity resulting in an enhanced process-related build-up rate.

When aluminum alloys (AlSi10Mg) are processed, a constant beam diameter (200 µm) can be employed, thereby allowing the laser power to be increased up to 1 kW. This can be accomplished because of the increased heat conductivity of aluminum alloys, in comparison to steels or nickel-based alloys, which does not result in spattering at the point of processing, thus maintaining a constant focus diameter.

In contrast, processing steel materials by increasing the laser power while maintaining a constant beam diameter (200µm) has the effect of increasing the intensity at the point of processing. This, in turn, leads to a higher evaporation rate resulting in a higher incidence of spattering, impedes process stability and component quality. To avoid this problem, the redesigned SLM machine allows an increase of the beam diameter up to 1000 µm.

The following investigations offer an overview of actual research topics in High Power Selective Laser Melting (HP-SLM). In these cases, an attempt is made to increase the process-related build-up rate which indicates how productive the SLM process is.

1.3. Increasing the productiviy for Aluminium alloys

The condition that limits the manufacture of SLM parts with an increased process-related build-up-rate and, thus, series-identical functional characteristics, is the density: It has to be approximately 100% (≥99,5%). To test density, cubic test components (10x10x10 mm³) are built and examined by light microscopy in order to examine an appropriate process window for the fabrication of real-life components. Figures 3 illustrates the results for the measured density according to the processed scanning velocity and hatch distance employing a constant beam diameter of 200 µm and a laser power between 300 and 1000 W. If a laser power of 300 W is used, components can be produced with a density of approximately 100% at a scanning velocity up to 500 mm/s. Increasing the laser power up to 500 W links with a scanning velocity of 1200 mm/s which represents an increase of the process-related build up rate of 200% with the respect to the present state of the art (see Fig. 3, left). Further increases with a laser power up to 1000 W enable scanning velocities up to 2200 mm/s resulting in a process-related-build up rate of 16 mm³/s.[4]

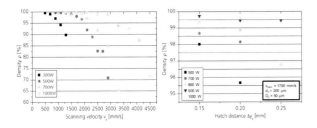

Fig. 3. Current Density according to scanning velocity (left) and hatch distance (right) for SLM parts out of AlSi10Mg

Besides the variation of scanning velocity the impact of increased hatch distance on the achieved density and process-related build-up rate is investigated. Therefore, the hatch distance is varied between 150, 200 and 250 µm. The results in Fig. 3 (right) illustrate that the densities of approximately 99.5% are still reached with a hatch distance of 200 µm. After the hatch distance, in combination with the scanning velocity, is adapted for 900 W laser power, the process-related build-up rate could be increased up to 21 mm³/s [4].

The tests conducted show that when increased laser power and adapted process parameters (scanning velocity, hatch distance) are used, the process-related build-up rate can be increased by a factor of five with respect to the current state of the art employing SLM machines with 200 W laser power.

1.4. Increasing the productiviy for steel alloys

In addition to processing aluminum alloys, SLM is used to manufacture components out of steels such as 1.2343, 1.2709 or 1.4404 in a wide range of applications. As described before increasing the laser power and maintaining a constant beam diameter (200µm) for steel materials leads to process instabilities. The intensity at the point of processing is increased whereby the evaporation rate rises and a higher incidence of spattering occurs [5]. To avoid these process instabilities the redesigned and rebuilt SLM machine is equipped with a variable focus diameter in order to change the focus diameter between 200 and 1000µm during the process.

However the accuracy and detail resolution of additive manufactured parts are hampered by larger melt pools which grow with larger beam diameters (1000 µm).

For this reason the skin-core strategy has to be taken into consideration. According to this strategy, the manufactured part is divided into an inner core and a skin which forms the outer core of the part. Different process parameters and focus diameters can be designated to each area. The core area does not have strict limitations or requirements concerning the accuracy and detail resolution. Therefore, the core area can be processed with an increased beam diameter (1000 µm) and an increased laser power, thus resulting in an increased process-related build-up rate. In contrast the skin area is manufactured with the small beam diameter (200 µm) in order to assure the accuracy and surface quality of the part. The skin core principle is illustrated in Fig. 4 in detail.

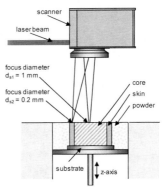

Fig. 4. Skin-core principle

To test the application of the skin-core principle, an injection moulding tool with conformal internal cooling channels, which cannot be processed conventionally (for example by cutting) is manufactured. To accomplish this, the component tool is subdivided into an inner and outer shell (see Fig. 5 middle). The outer shell (skin) is manufactured with a layer thickness of 50 µm employing the small beam diameter (200 µm) to assure the required accuracy. In contrast the inner core is processed with a beam diameter of 1000 µm and a layer thickness of 200 µm, resulting in layer thickness ratio between skin and core of 1:4. To ensure metallurgical bonding between the skin and core areas, an overlap depending on the thickness ratio is maintained. The results show that the skin can be built with a scanning velocity of 400 mm/s resulting in a process-related build up rate of 3mm³/s. The core is processed with an increased hatch distance and layer thickness, whereby a process-related build-up rate of 16 mm³/s is achieved. For this reason the process-related build-up rate for the whole part can be calculated to 12 mm³/s. This is a significant increased process-related build-up rate in comparison to parts processed on SLM machines at the current state of the art. To guarantee series-identical properties, the specimen is cut horizontally as well as vertically after processing and examined with light microscopy as to the density (see Fig. 5, right). The skin and the core of the injection moulding insert show a density of approximately 100%.

In addition the overlap between skin and core is processed without pores or binding errors.

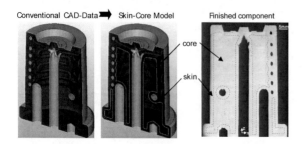

Fig. 5. Current Density Injection moulding tool with internal conformal cooling channels (left: CAD model, middle: skin-core model, right: cross-section of finished SLM component)

In summary, the investigations conclude that real-life components can be manufactured with an increased process-related build-up rate and, therefore, increased productivity by using the rebuild machine setup. A laser power of 1000 W and a focus diameter change between 200 and 1000 μm can be realized. Based on these results a multi-beam SLM system has been designed and realized in cooperation with SLM Solutions (SLM 280 HL). The availability of commercial High Power SLM systems is an important step for integrating this technology into industrial use and series production.

2. LMD (Laser Metal Deposition)

Laser Metal Deposition is a well known and established technology for the repair, wear and corrosion protection of high added value components [6]. Figure 6 illustrates the LMD process. A powder additive material is fed into the interaction zone, melted and metallurgically bonded to the substrate. By subsequent overlapping of tracks and layer by layer 3D geometries can be built up.

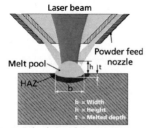

Fig. 6. Principle of the LMD process

For example, it has been proven, that complete compressor blades for Blisks (Blade Integrated Disks) can be built up with deposition rates of up to 1,2 kg/h [7].

2.1. High speed LMD

Aim of the current development is to speed-up the LMD process far beyond the current state-of-the-art. The approach is shown scematically in Fig. 7.

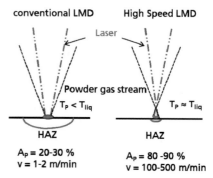

Fig. 7. Approach for High-speed LMD

In conventional LMD processes the laser radiation and the powder gas jet are focused onto the substrate material. The amount of absorbed energy in the powder A_p is in the range of 20 to 30 %. Typical processing speeds v range between 0,5 m/min and 2 m/min. The temperature of the particles T_p is below the meltig point T_{liq} of the particles. The overall amount of absorbed energy is in the range of 60 to 70 % (Figure 7, left side). 30 to 40 % of the irradiated laser energy is lost due to reflection.

In the case of High-speed LMD the laser radiation as well as the powder focus is placed slightly above the substrate material. The powder is focused to a diameter below 1 mm, compare figure 1.8, left side. Most of the laser energy is deposited in the powder (up to 90 %), leading to particle temperatures in the range of the melting point (Figure 7, right side). The remaining transmitted energy is used to produce a metallurgical bond of the molten particles to the base material by generating a welding bath.

Figure 8, right side, shows the experimental set-up for measuring the transmitted power through the powder focus.

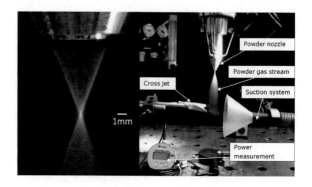

Fig. 8. Powder focus and experimental set-up for tranmission measurements in the powder gas stream

The transmitted power depends on the used grain size distribution and the powder mass flow. For a grain size distribution between 20 and 32 µm and a powder mass flow of 14 g/min the transmission is reduced to 10 %. Figure 9 shows the measured transmission values for Inconel 625 powder.

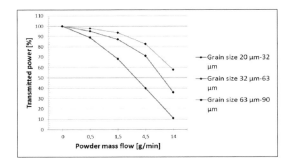

Fig. 9. Transmitted power as a function of powder mass flow for different grain size distributions of the powder (Inconel 625)

High-speed LMD trials with processing speeds of up to 600 m/min at laser powers of up to 1800 W were performed. The produced layers were cut, inspected metallographically and analysed regarding hardness, dilution and porosity. Figure 10. shows a cross section of a multiple layer built up consisting out of 30 single layers at a processing speed of 200 m/min (laser power 1400 W). The additive material used is Inconel 625. Each layer has a thickness of approximatively 11 µm.

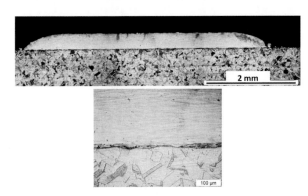

Fig. 10. Cross section of a layer built out of 30 single layers out of Inconel 625 at a processing speed of 200 m/min

Figure 11 shows the hardness of Inconel 625 layers achieved for different processing speed. For comparison, the hardness of a layer produced at conventional processing speed (1 m/min) is also shown. With High-speed LMD a higher hardness is achieved due to significantly higher cooling rates.

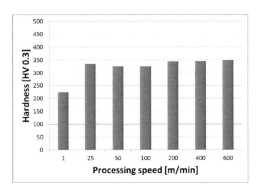

Fig. 11. Hardness of Inconel 625 layers produced at different processing speeds

In summary, in this first approch it has been demonstrated that LMD can be speeded up to levels of 600 m/min producing sound layers. The potential of such layers for wear and corrosion applications will be investigated in the future.

References

[1] Meiners, W.; Direktes Selektives Lasersintern ein-komponentiger metallischer Werkstoffe. PHD Thesis RWTH Aachen, 1999.

[2] Michaeli, W. and Schönfeld, M.; Komplexe Formteile kühlen", Kunststoffe, Vol. 8, 2006, pp. 37 – 41.

[3] Höges, S.; Wirtschaftliche Herstellung individueller Implantate, Maschinenmarkt, Vol. 26, 2010, pp. 20 – 23.

[4] Buchbinder, D.; Schleifenbaum, H.; Heidrich, S.; Meiners, W.; Bültmann, J.: High Power Selective Laser Melting (HP SLM) of aluminium parts. Phys Procedia 2011, Vol. 12, 1-8

[5] Schleifenbaum, H.; Diatlov, A.; Hinke, C.; Bültmannn, J.; Voswinckel, H.: Direct photonic production: towards high speed additive manufacturing of individual goods. Production Engineering – Research and Development 2011, Vol. 5, No. 4, pp. 359 – 371.

[6] Gasser, A.; et al.: Laser Additive Manufacturing, Laser Technik Journal 4 2010, S. 58-63

[7] Johannes Witzel et al.: Additive Manufacturing of a Blade Integrated Disk By Laser Metal Deposition, ICALEO 2011, Orlando, USA.

Supersonic Laser Deposition of Corrosion and Wear Resistant Materials

A Cockburn, R. Lupoi, M Sparkes, W O'Neill

Institute for Manufacturing, University of Cambridge, 17 Charles Babbage Road, Cambridge, CB3 0FS

Abstract. Supersonic Laser Deposition (SLD) is a new coating and fabrication process which combines the supersonic powder stream found in Cold Spray with laser heating of the deposition zone. SLD combines some advantages of CS: solid-state deposition and high build rate (≤ 8 kghr^{-1}) with the ability to deposit materials which are either difficult or impossible to deposit using cold spray alone. A system has been developed in house which can impact metallic powder particles onto a substrate which is locally heated using a 4kW IPG fibre laser. A pyrometer and control system is used to record and maintain impact site temperature. This paper describes the deposition of titanium, 316L stainless steel and Stellite 6 powders, the conditions which were found to be optimal, the build rates wqhich were achieved, and the structure and hardness of the coatings produced.

Keywords: laser, coating, titanium, Stellite 6®

1. Introduction

The protection of components via the application of corrosion and / or wear resistant layers of additional material is a well established route to enhancing the service life of structural components. Techniques which are employed to achieve this include, weld overlaying, laser cladding and various thermal spray techniques such as plasma spray and HVOF. The processes listed above all involve either total or substantial melting of the coating material and in some cases, the substrate. Disadvantages associated with liquid state deposition include increased risk of oxidation, residual stress due to solidification shrinkage, restrictions on coating-substrate combinations due to the potential formation of undesirable intermetallic phases and limited control of the microstructure of the final coating.

One coating process which has the potential to avoid the problems of oxidation, dilution and distortion is cold spray. The main advantage of cold spray is that there is no bulk particle melting, hence any materials sprayed retain their initial composition and or phases. Cold spray is most commonly used to deposit metals such and aluminium and copper but has been used to deposit a variety of metals (1-3) such as , Zn, , Ni, Ca, Ag, Co, Fe, Nb, NiCr alloys, MCrAlY alloys and Ti alloys. In the cold spray process, powder is entrained in a high pressure, high velocity gas jet so that it impacts the substrate leading to deformation and bonding, building up the coating. Since deposition is reliant on plastic deformation, it is difficult to deposit brittle or hard materials such ceramics unless they are co-deposited with a ductile matrix material. The requirement for extensive plastic deformation has meant that successful application of the process to harder materials has required the use of extensive gas heating and/ or the use of helium as an accelerating gas in order to produce satisfactory coatings.

Supersonic Laser Deposition (SLD) is a new coating and fabrication process in which a supersonic powder stream similar to that found in found in Cold Spray impinges on a substrate which is simultaneously heated with a laser. SLD retains the advantages of Cold Spray: solid-state deposition, high build rate (≤ 8 kg hr^{-1}) and the ability to deposit metals onto a range of substrates including those which are incompatible with conventional welding, while eliminating the operating costs arising from the use of extensive gas heating and large volumes of helium gas, and allowing the range of materials deposited to expand to higher strength materials which are of considerable engineering interest.

This paper presents the status of current work on the SLD process detailing the capabilities of the current deposition system and the conditions at the deposition site. Particular attention will be paid to titanium, 316L and Stellite 6, where the effect of process parameters on material deposition and coating structure and properties are described.

2. Experimental methods

2.1. SLD System Description

A high pressure (10-30 bar range) nitrogen gas supply is split and delivered to a converging-diverging (De-Laval) nozzle, Fig. 1, both directly and via a Praxair 1264HP high pressure powder feeder where metal powder particles are en-trained. The two streams recombine and pass through the nozzle where they are accelerated to supersonic speeds. The high velocity, powder stream exits the nozzle and is directed towards the substrate. The impacts site is simultaneously illuminated by an YLS fibre laser (IPG), with a maximum power of 4 kW. The deposition zone temperature is monitored by a high speed IR-Pyrometer (Kleiber) and used to control the laser power via a PID loop.

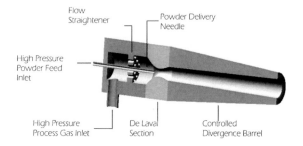

Fig. 1 Schematic diagram showing De Laval nozzle used in SLD

The supersonic nozzle, laser head and pyrometer (Fig. 2) are held stationary while the substrate can be moved using a CNC X-Y stage with an additional rotational axis allowing flat samples, cylinders and discs to be coated.

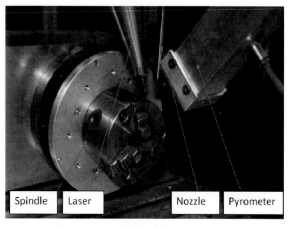

Fig. 2 View of CNC spindle and main system components

2.2. Computational Fluid Dynamics

The Computaional Fluid Dynamics (CFD) package Fluent™ was used to examine the effect of gas temperature on particle velocity. Two dimensional, axisymmetric methodology was employed to construct the nozzle flow domains using Gambit 2.2.30. The velocity and trajectory of particles in the two-phase model was computed using a drag force balance written in a Lagrangian reference frame. Particles were injected singly with particle parameters chosen to reflect the particles being sprayed.

2.3. Materials

SLD uses powder similar to that which is suitable for cold spray. Generally powders whose diameters lie between 10 and 50 μm are most suitable. A Malvern Mastersizer was used to determine the powder size distribution.

Table 1. Powder sizes used in this study

Material	Volume weighted mean (μm)
Titanium	29.7
316L	15.5
Stellite 6®	41.1

Coatings have been deposited onto both flat plates and cylinders made from mild steel.

The effect of different deposition parameters was explored through the deposition of a series of single tracks. Coatings consisting of overlapping tracks were deposited after the most successful combination of deposition parameters had been identified.

2.4 .Sample Characterization

Coating density and structure were investigated through the optical examination of polished cross sections in combination with A4i image analysis software. Chemical etching was used to reveal microstructure, allowing the degree of particle deformation and the scale of the coating structure to be observed.

A Scanning electron microscope was used to examine both coating fracture surfaces and polished cross sections where a high level of magnification was required. The density of the titanium coatings under optimum processing conditions was measured through an Archimedes balance tester which determines a materials density by weighing the coating in two different fluids of known density, such as air and water. Vickers hardness (Hv) measurements were made using a Mitutoyo HV-112 hardness measurement machine.

A DFD Instruments Precision GM01 Adhesion Testing machine was used to measure the adhesion strength of the coatings. 8.16 mm diameter dollys were glued to specimens which had been cut from their surrounding coating using a core drill, in order to impose the load on the interface over the area of the dolly only. The glue used was a single part heat cured epoxy, E1100s, supplied by DFD Instruments.

3. Results & Discussion

3.1 CFD

The effect of gas heating on gas and paticle velocity was modeled using CFD. Final gas and particle velocities are plotted in Fig. 3. The results indicate that a titanium particle of mean diameter 30 μm sprayed using 300 °C nitrogen rather than unheated nitrogen impact the substrate at in excess of 600 ms^{-1}.

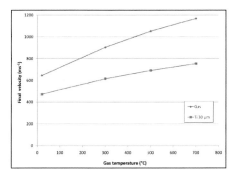

Fig. 3 Gas and particle velocities predicted by CFD modelling

3.2. Deposition characteristics

Titanium coatings have been primarily deposited using cold nitrogen gas at a pressure of 30 bar. Computational fluid dynamics predicts that this should result in an impact velocity of approximately 470 ms^{-1}. Trials on flat substrates looked at the effect of deposition site temperature on build rate while traverse rate was kept constant at 8.3 mms^{-1}. As Fig. 4 shows, build rate increases significantly with deposition site temperature.

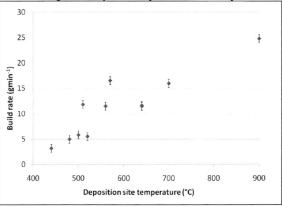

Fig. 4. The effect of temperature on build rate(4)

The use of a rotary spindle and a 4 kW laser has allowed deposition site temperature to be maintained at higher traverse rates. This allows coatings on cylindrical substrates, Fig. 5, to be deposited with

effective traverse rates of up to 500 mms^{-1} and a build rates of 46 gmin^{-1}.

Fig. 5 Thick titanium coating on 51 mm diameter mild steel tube. Shown after post deposition machining.

A similar optimisation process carried out for 316L stainless steel and Stellite 6 allowed the optimal deposition conditions given in Table 2 to be determined. In both cases the temperatures required for deposition are above 1000 °C and optimal traverse rates used were less than those which proved to be possible with titanium.

In the case of Stellite 6, no deposition occurred with room temperature gas. In order to generate satisfactory tracks, gas heating was required. The effect of gas heating on titanium deposition has also been investigated. A series of coatings were deposited using gas heating to elevate gas inlet temperature to up to 305 °C. According to CFD predictions this should have increased the impact speed of a 30 μm particle to 616 ms^{-1}. Gas heating was found to increase maximum build rate to over 8 kghr^{-1}.

Table 2 Optimal deposition parameters for the deposition of titanium, 316L steel and Stellite 6

Material	Deposition temperature (°C)	Gas temperature (°C)	Traverse rate (mms^{-1})	Build rate (gmin^{-1})
Titanium	550	20	500	46
316 L	1050	20	50	30
Stellite 6	1000	> 450	40	35

3.2. Coating Characterisation

Titanium. Little porosity is evident in a polished cross section of a typical coating but when etched with Kroll's reagent, Fig 6, relatively little particle deformation is observed and many of the particles retain an essentially circular cross section.

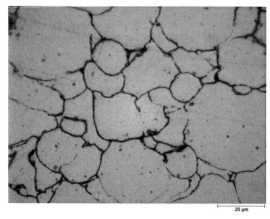

Fig. 6 Optical micrograph showing an etched titanium coating

Archimedes based density measurements show that as deposited coatings are between 87 - 89% dense. Gas heating had no significant effect on coating density but machined samples were found to have densities of up to 95% at the cost of a significant reduction in coating thickness.

316 L stainless steel. Fig. 7 shows a typical 316L stainless steel coating which is approximately 1 mm thick and has been deposited at a deposition site temperature of 1050 °C at a traverse rate of 10 mms^{-1}. Although there is very little porosity evident, the individual overlapping tracks are visible. This may be due to surface oxidation forming on each track before the next pass takes place.

Fig. 7. Micrograph showing Cross section of 316L

The etched micrograph shown in Fig. 8 shows particles which are significantly more deformed than those observed in titanium cross sections. Significant deformation of grains within the particles is also evident. This indicates that deposition has taken place in the solid state with no melting or recrystallization having taken place.

Fig. 8 316L coating etched with marbles reagent

Stellite 6 coatings, Fig. 9, showed little porosity and no signs of oxidation between overlapping tracks or at the coating substrate interface. As Fig. 10 (a) shows, individual particles in the coating have a very fine microstructure, far smaller in scale than Stellite 6 coatings deposited from the molten state, Fig. 10 (b)..

Fig. 9. micrograph showing Stellite 6 coating

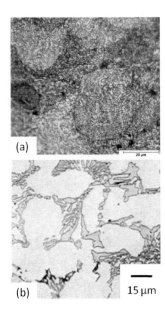

Fig. 10 (a) Stellite 6 deposited using SLD and (b) a Stellite 6 weld overlay coating (5)

3.3 Mechanical properties

Adhesion strength for titanium was found to be 77 MPa. This is an improvement over values reported for titanium cold sprayed onto steel substrates by Marraco et al (6)

Pull off data could not be obtained for the 316 L and Stellite 6 coatings using this technique since the maximum load it can apply is limited by the strength of the glue attaching the dolly to the coating. The glue failure load for each case is listed in Table 3.

Table 3. Adhesion and hardness data for the coatings

Material	Adhesion strength (MPa)	Average hardness ($Hv_{0.3}$)
Titanium	77 (± 11)	272
316L	> 70	250
Stellite 6	> 62	500

Hardness values recorded for titanium are in excess of the values found for CP2 grade titanium sheet which was measured as 180.3 Hv on the same machine. This may be due to the work hardened surface left by the machining operation. The hardness values recorded for both 316 L and Stellite 6 are slightly above literature values which are 240 and 380-490 Hv respectively (7, 8). This could be due to the fine grain structure of the coatings or the degree of deformation which the particles experience during deposition.

4. Conclusions

It has been shown that SLD provides a viable route for the deposition of titanium, 316L stainless steel and Stellite 6 coatings onto steel substrates. The SLD system can deposit titanium at up to 46 $gmin^{-1}$ without gas heating and achieve bond strengths in excess of those reported for cold spray deposits onto steel. Particle deformation on impact is limited contributes to an as deposited density of 85 %. Both 316L and Stellite 6 deposits exhibit significant particle deformation while retaining the fine grained microstructure inherent with solid state depositing from gas atomized powder. Adhesion strengths for both 316L and Stellite 6 are in excess of 70 MPa and the hardness values recorded for Stellite 6 coatings are above the literature values for the material suggesting that SLD deposited Stellite 6 may prove to be suitable in applications where both wear and corrosion resistance are required.

References

[1]   J. Pattison, S. Celotto, R. Morgan, M. Bray, W. O'Neill. Cold gas dynamic manufacturing: "A non-thermal approach to freeform fabrication", International Journal of Machine Tools & Manufacture 47 (2007) 627–634.

[2]   E. Calla, D.G. McCartney, and P.H. Shipway. "Effect of Deposition Conditions on the Properties and Annealing Behavior of Cold-Sprayed Copper", Journal of Thermal Spray Technology Volume 15(2) June 2006, 255-262.

[3]   Heli Koivuluoto, Mari Honkanen, Petri Vuoristo. "Cold-sprayed copper and tantalum coatings Detailed FESEM and TEM analysis", Surface & Coatings Technology 204 (2010) 2353–2361.

[4]   Heli Koivuluoto, Mari Honkanen, Petri Vuoristo. "Cold-sprayed copper and tantalum coatings Detailed FESEM and TEM analysis", Surface & Coatings Technology 204 (2010) 2353–2361.

[5]   Matthew Bray, Andrew Cockburn, William O'Neill "The Laser-assisted Cold Spray process and deposit characterization", Surface and Coating Technology, Volume 203, Issue 19, 25 June 2009, Pages 2851-2857

[6]   George F. Vander Voort. ASM Handbook volume 9: Metallography And Microstructures . Vol 9, ASM Handbook, ASM International, 2004, p. 762–774

[7]   T. Marrocco, D.G. McCartney, P.H. Shipway, and A.J. Sturgeon "Production of Titanium Deposits by Cold-Gas Dynamic Spray: Numerical Modeling and Experimental Characterization", Journal of Thermal Spray Technology, Volume 15(2) June 2006.

[8]   S.D. Washko and G. Aggen. ASM Handbook Volume 1: Properties and Selection: Irons, Steels, and High-Performance Alloys. Vol1, ASM Handbook, ASM International, 990, p 841–907

[9]   P. Crook, ASM Handbook Volume 1: Nonferrous Alloys and Special-Purpose Materials, Vol 2, ASM Handbook, ASM International, 1990, p 446–454.

Laser Technology – Modelling

Modelling of direct metal laser sintering of EOS DM20 bronze using neural networks and genetic algorithms

A. Singh[1], D.E. Cooper[2], N.J. Blundell[3], G.J. Gibbons[4], D.K. Pratihar[5]
[1] Dept of Mechanical Engineering, IIT Kharagpur, West Bengal, India.
[2] WMG, University of Warwick, Coventry, CV4 7AL, UK.
[3] WMG, University of Warwick, Coventry, CV4 7AL, UK.
[4] WMG, University of Warwick, Coventry, CV4 7AL, UK.
[5] Dept of Mechanical Engineering, IIT Kharagpur, West Bengal, India

Abstract: An attempt was made to predict the density and micro-hardness of a component produced by Laser Sintering of EOS DM20 Bronze material for a given set of process parameters. Neural networks were used for process-based-modelling, and results compared with a Taguchi analysis. Samples were produced using a powder-bed type ALM (Additive Layer Manufacturing)-system, with laser power, scan speed and hatch distance as the input parameters, with values equally spaced according to a factorial design of experiments. Optical Microscopy was used to measure cross-sectional porosity of samples; Micro-indentation to measure the corresponding Vickers' hardness. Two different designs of neural networks were used - Counter Propagation (CPNN) and Feed-Forward Back-Propagation (BPNN) and their prediction capabilities were compared. For BPNN network, a Genetic Algorithm (GA) was later applied to enhance the prediction accuracy by altering its topology. Using neural network toolbox in MATLAB, BPNN was trained using 12 training algorithms. The most effective MATLAB training algorithm and the effect of GA-based optimization on the prediction capability of neural networks were both identified.

Keywords: Direct Metal Laser Sintering, Genetic Algorithms, Neural Networks.

1. Introduction

Direct Metal Laser Sintering (DMLS) is an Additive Manufacturing technique, capable of constructing metallic components by depositing and selectively melting successive layers of metal powder [15]. Figure 1 shows the principle of the process, with raw material fed into a processing area by a re-coating mechanism.

In recent years, neural networks have become very useful tool in the modelling of input–output relationships of some complicated systems [1]. They have excellent ability to learn and generalize (interpolate) the complicated relationships between input and output variables. There are different training schemes for these neural networks [2]. Counter Propagation Neural Network (CPNN) and Back Propagation Neural Network

(BPNN) are two designs of neural network, with the approximation efficiency of each varying with the type of data used [11, 16]. Radial basis function network [5] was also used to check and compare the accuracy of modelling, but it did not yield appreciable results. Margaris et al. [3] discussed the implementation of CPNNs. Network optimization concerns the technique used to achieve the optimum number of hidden neurons in a CPNN [4]. BPNNs have been used for a variety of modelling tasks for complex systems [6].

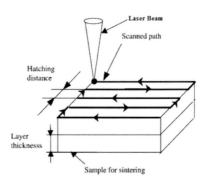

Fig.1. Schematic of DMLS Parameters [8]

The properties of components produced by DMLS depend heavily on fine control of the input parameters, so identifying the precise effect of each parameter is crucial. Wang et al. [7] explored the part shrinkage of samples manufactured by Selective Laser Sintering (SLS), by varying seven process parameters. An experimental design approach was used towards SLS of low carbon steel by Chatterjee et al. [8], where the parameters used were layer thickness and hatching distance, to consider the effects of density, hardness and porosity of sintered components. Ning et al. [9] and Wang et al. [7] used models to intelligently select the parameters for modelling the DMLS process. One of the notable studies related to the application of soft computing towards laser sintering included the estimation of build time [10]. Comparisons between the applicability of BPNN and CPNN towards manufacturing process (TIG welding) have been demonstrated by Juang et al. [11]. Apart from SLS, multiple designs of neural networks have been used in the past to model different aspects of various other manufacturing processes. Lu et al. [12] worked on modelling the Laser Engineered Net Shaping (LENS) process, where BPNN-based models were applied to control the deposition height of the prototype. For laser welding, Lim and Gweon [13] investigated the application of neural networks in estimating joint strength for pulsed laser spot welding. Balasubramanian et al. [14] discussed about the performance of BPNN for the modelling of stainless steel butt joints.

In this study, a unique comparison of CPNN and BPNN had been carried out in the context of modelling the sample hardness and cross-sectional porosity. To have a better estimation of predictive capability of the two designs, they were trained and tested with three unique data sets. Using MATLAB, several BPNN training algorithms were tested. The effect of a binary-coded GA (Genetic Algorithm) was also studied towards enhancing the predictive capability of a BPNN.

2. Methodology

The following steps were followed to carry out the experiments:-

- Sample production by ALM;
- Metallographic sectioning & polishing;
- Visual examination and hardness tests to obtain the desired output values to be fed into the Neural Network;

The values of the input parameters were Laser Power (kW): 0.75, 1.00, 1.25, 1.50, 1.75; Laser Scan Speed (m/min): 5.0, 6.5, 8.0, 9.5, 11.0; Hatch Distance (mm): 0.25, 0.50, 0.75, and 1.00.

2.1. Techniques used, and developed approaches

After the values for the various experimental runs were obtained, the entire data was assembled in input-output pairs. A total of 99 such pairs were obtained, which were divided into three sets of 89 training values and 10 testing values. These values were then used in the neural network-based modelling task. An artificial neural network (ANN) is a mathematical or computational model that is inspired by the structure and/or functional aspects of biological neural networks. The CPNN is a hybrid network, consisting of an outstar network and competitive filter network. The hidden layer is a Kohonen

network, which categorizes the pattern that was input. The output layer is an outstar array, which reproduces the correct output pattern for the category. The second kind of neural network used in the study was a BPNN, the topology of which is shown in Fig. 2. The numbers of nodes in the input and output layers are Ni and No, respectively. The use of a larger number of hidden nodes can potentially improve the accuracy and convergence of the back-propagation (BP) algorithm at the cost of computational processing time [2].

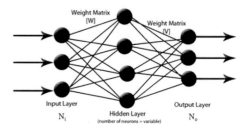

Fig. 2. BPNN Architecture

In the tests carried out, Ni = 3 and No = 2, while the number of neurons in the hidden layer was varied and tested with the help of a Genetic Algorithm (GA). Juang et al. [11], Goh et al. [6] and Margaris et al. [3] discussed the network structures and training schemes in detail. Pratihar [2] discussed the training schemes of BPNN in detail.

A binary-coded GA was used to optimize the topology of the network. Of the two types of errors described above, the GA tends to minimize the training error by choosing the best combination of network parameters, such as number of neurons of the hidden layer 'nh', coefficient of transfer function of the hidden layer 'ah', coefficient of transfer function of the output layer 'ao' (Pratihar [2]). The following steps were used in GA implementation:

- Creation of random population (size of 100). Each chromosome in the population represents a certain combination of 'nh', 'ah' and 'ao';
- Fitness evaluation of each chromosome using the mean square error (MSE) of the BPNN [2], after 10000 iterations, keeping the topology represented by that chromosome into account;
- Tournament-based selection [2] was used to select the pool of better chromosomes;
- Single point crossover and Mutation with respective probabilities of 0.9 and 0.09, forming a new pool of 100 chromosomes and indicating the completion of a generation;
- The process was repeated for 100 generations, and the fittest chromosome was finally chosen;

The modelling was conducted in C++, where codes were written for GA optimized BPNN and CPNN. Using neural network toolbox in MATLAB, analysis of 12 training algorithms was carried out for feed-forward network. C++ coding was performed on a GCC compiler (version: Dec 20 1999 15:39:08). Minitab v16 was used to perform the Taguchi L9 analysis.

3. Results and discussion

Inputs of the neural network were normalized in the scale of 0 to 1. In neural network toolbox of MATLAB, feed-forward networks were developed using 12 different training algorithms, namely traingd (Gradient descent), traingdm (Gradient descent with momentum), traingdx (Gradient descent momentum with an adaptive learning rate), trainrp (Resilient BP algorithm), traincgf (Conjugate gradient BP with Fletcher-Reeves updates), traincgp (Conjugate gradient BP with Polak-Ribiere updates), traincgb (Conjugate gradient BP with Powell-Beale restarts), trainscg (Scaled conjugate gradient method), trainbfg (BFGS quasi-Newton method), trainoss (One step secant method), trainlm (Levenberg-Marquardt optimization) and trainbr (Levenberg-Marquardt optimization with Bayesian regularization). Three combinations of 89 training and 10 testing cases have hereby been referred to as Set-1, Set-2 & Set-3. The number of neurons in the hidden layer was varied from 2 to 17, keeping the number of unknowns (5 × number of neurons in the hidden layer) lower than the number of equations (89 training cases). Tests were conducted for 'ah' and 'ao' by individually varying the ranges, and the optimum range was found to be (0.2 to 15.95, in steps of 0.25) and (0.2 to 3.35, in steps of 0.05) for ah and ao, respectively. Table 1 shows the optimum values of the parameters: nh, ah and ao.

The Mean Generalization Error (MGE) represents the mean absolute difference between the normalized values of computed and actual porosity and hardness. The training algorithm shown is the same as the traingdm algorithm discussed later in this section in the MATLAB results. For the CPNN, the number of neurons in the competition layer was varied from 2 to 89. The network was allowed to train as long as the MSE was converging towards 0. The loop was terminated the moment the MSE started diverging. Upon analysing the final MSE before divergence occurred, the best network topology was chosen.

Upon increasing the number of hidden neurons, up to a certain number the pattern was uniform. After that, there were indications of improper (over/under)-training. For all the three cases, the best values for MGE were obtained between 25 to 30 neurons. The BPNN took 10000 iterations (for NN weight modification) converge to the specified MGE, while the CPNN took only 4 iterations. Table 2 shows the best CPNN configuration (lowest MSE-based analysis).

Table 1. BPNN Results

Set	Optimum Parameters			MGE
	n_h	a_h	a_o	
Set-1	15	8.75	2.10	0.1076
Set-2	15	8.50	2.25	0.1300
Set-3	16	8.75	2.10	0.1676

Table 2. CPNN Results

Set	No. hidden neurons	MGE
Set-1	30	0.1110
Set-2	28	0.1393
Set-3	29	0.1320

Table 3. MATLAB Results

Algorithm	Set-1	Set-2	Set-3
traingd	0.1334 (17)	0.1304 (14)	0.1798 (8)
traingdm	0.1442 (16)	0.1336 (12)	0.1392 (5)
traingdx	0.1006 (10)	0.1387 (11)	0.1024 (16)
trainrp	0.0843 (10)	0.1255 (13)	0.1920 (10)
traincgf	0.0896 (10)	0.1066 (13)	0.1605 (9)
traincgp	0.1016 (17)	0.1404 (14)	0.1636 (13)
traincgb	0.1473 (16)	0.1170 (12)	0.1845 (14)
trainscg	0.1044 (11)	0.1329 (13)	0.1539 (11)
trainbfg	0.1056 (16)	0.1357 (14)	0.1577 (15)
trainoss	0.1249 (16)	0.1238 (14)	0.0964 (14)
trainlm	0.1298 (13)	0.1457 (7)	0.1659 (12)
trainbr	0.0990 (9)	0.1399 (14)	0.1594 (13)

Feed-forward back-propagation neural network was trained using 12 different training algorithms; the results have been compared in this section. As training parameters, the number of iterations/epochs was set to a maximum of 10000. The performance goal, based on the MSE, was set to 0 (zero) and rest all other training parameters were at their default values. For all the networks, the one hidden layer with a tansig activation function and one output layer with purelin activation function were used. The value of 'nh' was varied from 2 to 17. The best approximation was identified as the network topology (i.e., the number of neuron in the hidden layer) with the least MSE at the completion of the training process. Table 3 displays the values of MGE for various training algorithms. The number of neurons of the hidden layer for the same network has been indicated in the bracket. The MATLAB training algorithm named traingdx turned out to be the most accurate one with an average MGE of 0.1139 on the normalized scale.

Using the Taguchi L9 analysis, laser scan speed was found to have the maximum effect on both the outputs, while laser power and hatch distance were found to have the minimum effect on porosity and mean hardness, respectively. Figure 3 overleaf, compares the predicted values of the two outputs with their respective actual values for the test cases (using trainrp for Set-1)

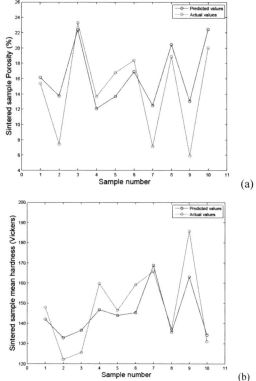

(a)

(b)

Fig. 3. Actual vs. Predicted values of (a) porosity and (b) mean hardness for the test cases

4. Conclusion

The MATLAB training algorithm "traingdx" was found to be the most accurate, with an average Mean Generalisation Error (MGE) of 0.1139 on a normalized (0 to 1) scale. Counter Propagation and Genetic Algorithm (GA) optimized-Back Propagation Neural Networks (BPNN) had average MGE values of 0.1274 and 0.1350, respectively. The training time for CPNN was much shorter than for BPNN. Laser Scan Speed was seen to have the greatest influence on both the outputs, using the Taguchi method. The use of GA-based optimization successfully reduced the MGE for BPNN trained by the gradient descent with momentum algorithm from 0.1350 to 0.1274.

References

[1] Freeman JA, Skapura DM, (1991) Neural Networks: Algorithms, Application and Programming Techniques, Addison–Wesley,

[2] Pratihar DK, (2008) Soft Computing, Narosa Publishing House, New Delhi.

[3] Margaris A, Souravlas S, Kotsialos E, Roumeliotis M, (2007) Design and Implementation of Parallel Counterpropagation Networks Using MP, Informatica, 18:79-102

[4] Dong Y, Sun C, Tai X, (2007) An Adaptive Counter Propagation Network, Eighth ACIS International Conference on Software Engineering, Artificial Intelligence, Networking, and Parallel/Distributed Computing (IEEE Transactions), Qingdao, 695 – 700

[5] Park JS, Sandberg IW, (1991) Universal Approximation using Radial-Basis-Function Networks, Neural Computation 3 (MIT Press), 246-257

[6] Goh ATC, (1995) Back-propagation neural networks for modelling complex systems, Artificial Intelligence in Engineering 9, 143-151

[7] Wang RJ, Wang L, Zhao L, Liu Z, (2007) Influence of process parameters on part shrinkage in SLS, International Journal of Advanced Manufacturing Technology 33:498 – 504

[8] Chatterjee AN, Kumar S, Saha P, Mishra K, Roy Choudhury A, (2003) An Experimental design approach to selective laser sintering of low carbon steel, JMPT 136:151 – 157

[9] Ning Y, Fuh JYH, Wong YS, Loh HT, (2004) An Intelligent parameter selection system for the direct metal laser sintering process, International Journal of Production Research 42:183 – 199

[10] Mungunia J, Ciurana J, Riba C, (2009) Neural-network-based model for build-time estimation in selective laser sintering, Proceedings of the Institution of Mechanical Engineers, Part B: Journal of Engineering Manufacture, 995 – 1002

[11] Juang SC, Tarng YS, Lii HR, (1998) A comparison between the back-propagation and counter-propagation networks in the modelling of the TIG welding process, JMPT 75:54 – 62

[12] Lu ZL, Li DC, Lu BH, Zhang AF, Zhu GX, Pi G, (2010) The prediction of the building precision in the Laser Engineered Net Shaping process using advanced network, Optics and Lasers in Engineering 48:519-525

[13] Lim DC, Gweon DG, (1999) In-Process Joint Strength Estimation in Pulsed Laser Spot Welding Using Artificial Neural Networks, Journal of Manufacturing Processes 1:31 – 42

[14] Balasubramanian KR, Buvanashekaran G, Sankaranarayanasamy K, (2010) Modelling of laser beam welding of stainless steel sheet butt joint using neural networks, CIRP Journal of Manufacturing Science and Technology 3:80 – 84

[15] Das S, Wohlert W, Beaman JJ, Bourell DL, (1998) Producing Metal Parts with Selective Laser Sintering/Hot Isostatic Pressing, JOM 50 (12):17 – 20

[16] Dharia A, Adeli H, (2003) Neural Network model for rapid forecasting of freeway link travel time, Engineering Applications of Artificial Intelligence 16:607 – 613.

A thermo-mechanical model for laser processing of metallic alloys

A. Suárez, J. M. Amado, M. J. Tobar, A. Yáñez
Universidade da Coruña, D. de Ingeniería Industrial II, Ferrol, E-15403, Spain.

Abstract. Among the wide scope of laser materials processing techniques stand the use of high power laser as an alternative tool in standard manufacturing process as cutting, welding or bending and forming. The nature of the laser beam energy profile provides with a controlled and precise heat input in the material minimizing mechanical and microstructure distortions. The development of analytical or numerical models is useful not only to get an insightful understanding of the underlying aspects but also to optimize process parameters and reduce experimental trials. Then numerical methods are often preferred over the analytical ones as they allow accommodating for variations in the process such as geometry or temperature dependent material properties. A 3D customized transient FEM model was developed to understand the main features of the metallurgical and thermo-mechanical phenomena and make accurate predictions. The metallurgical model is based on the carbon diffusion during the austenitization process through 1D diffusion models. Martensitic transformation and backtempering are included. Experimental tests for the validation of the model are also discussed.

Keywords: thermomechanical, metallurgical, phase transformations, laser cladding

1. Introduction

The laser cladding technique allows the deposition of low dilution tracks of molten material on a metal substrate, a metallurgical bond between fed material and the substrate is achieved during solidification.

One of the main concerns is the development of high residual stresses. They affect the working life of the part due to the enhancement of the susceptibility to both fatigue and stress-corrosion. Additionally workpiece will show distortion and cracks..

Experimental methods to measure residual stresses are usually complex. In contrast, computer simulations can produce an insightful understanding of the underlying aspects of the process.

Both 2D and 3D CFD models of the laser cladding have been studied [1-3], but they are not easily coupled with mechanical analysis; so heat conduction models are preferred. The effect of deposition patterns on deflection of the workpiece [4], the cladding track geometry was also evaluated inside simple plasticity models [5-7].

Phase transformations were studied mainly for laser hardening. Several authors [8,9] based their work on the simple kinetic model [10], others use data from TTT/CCT diagrams and the Johnson-Mehl-Avrami (JMA) equation [11]. Also exist 2D kinetic models coupled to 3D FEM models [12], but 1D kinetic models are faster [13]. Backtempering due to adjacent laser passings was also studied [14] inside a tempering model based on the JMA equation.

Metallurgy received less attention in laser cladding. In [15] 3D models using the software SYSWELD are developed including metallurgical transformations and coupled with the mechanical analysis.

In this paper a metallurgical model based on 1D carbon diffusion for steels was developed. It includes the tempering due to the deposition of the contiguous cladding tracks in a JMA approach. This model was coupled with a previously developed thermomechanical model [16] already published.

2. Model description

In the model [16] thermal and mechanical fields are sequentially coupled. The simulation starts with a transient thermal analysis. These results are applied as thermal load for the subsequent quasistatic mechanical analysis. Birth & Death techniques are used for the simulation of the material deposition.

Metallurgy is fully coupled with both thermal and mechanical analysis. Each thermal load step is solved starting by a pure thermal step and the subsequent metallurgical step, allowing the change of material properties during the analysis. The phase volume fractions are stored, so during each mechanical step the properties of any element can be calculated. Each element has assigned a vector with all required information. It is implemented as a main routine that controls the transition between stability regions of different phases, according to the element temperature. Specific routines containing the kinetic models are called to compute the phase evolution.

Initially the steel sample is in a ferritic-pearlitic microstructure. The carbon content of pearlite (0.8%) is much higher than in ferrite, so above austenitization temperature, austenite grains nucleate in pearlite. Ferrite will transform at higher temperatures. During the transformation the lattice changes from BCC to FCC, increasing density around 4%.

Austenitization is regarded as a two-step process [8]. The first stage is the fast intragranular carbon diffusion in pearlite. The second step is the intergranular diffusion, homogenizing the carbon distribution [2]. Each of these two steps is simulated in a different 1D diffusion model.

The first model uses the pearlite interlamellar spacing to calculate cementite and ferrite lamellae widths (Fig. 1) from the carbon concentration of these phases. The intragranular diffusion distance is divided in intervals and carbon concentration is calculated in each of them. Complete austenitization is achieved when carbon concentration reaches 90% of its value in pearlite.

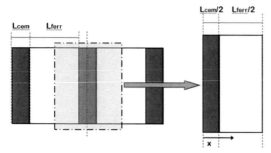

Fig. 1. Schematic view of the geometrical model for the pearlite dissolution.

The second model uses the pearlite grain size and the carbon content to calculate the diffusion distance between grains. This distance is divided in intervals, considered austenitized when carbon content is above austenite critical value. The fraction of austenitized intervals gives the austenite percentage. In between temperatures A_{C1} and A_{C3} the maximum percentage is adjusted using the C-Fe phase diagram.

Both models are based on the carbon diffusion from a homogeneous phase of width h and initially constant carbon content C_0, from which carbon diffuses to a region of lower carbon content (Fig. 1).

The 1D diffusion equation is applied [8]:

$$\frac{dC(x,t)}{dt} = D(T)\frac{\partial C(x,t)}{\partial x} \qquad (1)$$

With the previous initial condition, considering symmetry at the origin and one reflection at the end of the cell ($x=LT$), the following equation results:

$$C(x,t) = \frac{C_0}{2}\left[errf\left(\frac{(h-x)}{2\sqrt{D(T)t}}\right) + errf\left(\frac{(h+x)}{2\sqrt{D(T)t}}\right) \right.$$
$$\left. + errf\left(\frac{(h-(2LT-x))}{2\sqrt{D(T)t}}\right) + errf\left(\frac{(h+(2LT-x))}{2\sqrt{D(T)t}}\right) \right] \qquad (2)$$

The temperature dependent diffusion coefficient is calculated as follows:

$$D(T) = D_0 \exp\left(-\frac{Q}{RT}\right) \qquad (3)$$

Where D_0 is the diffusion coefficient, Q is the activation energy of carbon diffusion, R is Boltzmann's constant and T is the absolute temperature. The carbon diffusion from austenite to ferrite at an isothermal temperature and different times is represented in Fig. 1.

The main cooling mechanism is the heat conduction to the workpiece in a "self-quenching" process, producing martensite. The crystalline structure changes from FCC to BCT by means of a displacive transformation without diffusion depending only on the degree of sub cooling below the martensite start temperature (M_s). It is modeled using the Koistinen-Marburger equation:

$$f_M = 1 - \exp\left(-\alpha\left(M_s - T\right)\right) \qquad (4)$$

Where α is a constant dependent on the transformation rate and the steel composition. Usually, for steels with carbon content smaller than 1.1 % $\alpha = 0.011$ [24].

The previously formed martensite undergoes a heating process during the next laser passing causing its tempering: decomposing of martensite into a stable microstructure of ferrite with dispersed carbides.

Tempering starts at 100°C and continues up to A_{C1} temperature. Between 100°C and 200°C a finely dispersion of ε carbide and transition metal carbides are formed [18]. At temperatures between 200°C and 300°C takes place the transformation of retained austenite into ferrite and cementite. At higher temperatures only cementite is formed and the carbides coarsen, softening the microstructure [18].

The process is simulated with the using the JMA equation. Parameters were taken from Lakhkar et al. [14]

$$f_{TM} = f_M\left(1 - \exp(1 - kt)^n\right) \qquad (5)$$

where,

$$k = k_0 \exp\left(-\frac{Q}{RT}\right) \qquad (6)$$

Being f_M the maximum fraction of martensite, Q the activation energy and n and k_0 experimentally derived constants. The equation is valid for isothermal conditions, so the laser process is subdivided into a series of isothermal steps (Fig. 2), and the effect of each step is added according to the Scheil's additive rule [17]. Hardness is commonly calculated by means of the rule of mixtures [17]: a linear combination of the hardness of each phase, weighted by its volume fraction.

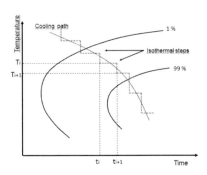

Fig. 2. Discretization of the cooling process into a series of isothermal steps.

3. Experimental

Laser surface melting tests were performed to validate the metallurgical model using carbon steel plates of AISI 1045 and a 2 kW CO_2 laser. The tests consisted of 3 laser tracks of 80 mm length on plates of 100x50x10 mm with velocity of 5 mm/s, 1200W of laser power and a 4 mm laser spot (TEM01*). Different distances between tracks were tested to study martensite tempering.

Laser cladding tests with single and multiple cladding tracks were performed using a 2.2 kW Nd:YAG laser with a coaxial laser cladding nozzle. 80 mm length tracks were deposited on plates as before with parameters: 1800 W of power, a speed of 5 mm/s and a powder flow rate of 0.48 g/s. The samples were cut, mechanically grounded, polished and etched with Nital 3 to reveal the microstructure. Microhardness was also measured.

4. Results

The FEM mesh contains hexahedral elements in the zone below the laser scans and coarse tetrahedral elements in the rest of the plate (126590 elements and 89468 nodes).

Results for 2 and 4 mm spacing are shown in Fig. 3. Absorptivity of the plate was tuned using these results. The calculated melted zones, heat affected zones (HAZ) and tempered zones are similar to the experimental ones. Hardness profiles are in good agreement with maximum errors of about 16% and typical errors below 6%. The maximum differences between the width and depth of the calculated HAZ are about 300 μm. The effect of tempering is especially important for 2 mm spacing tests, but also noticeable for the 4 mm ones. The melted zones obtained are connected when the spacing is 2 mm. The simulated melted zones are in good agreement except for

the last track of this sample: 40% underestimated in depth, due to the neglecting of Marangoni convection.

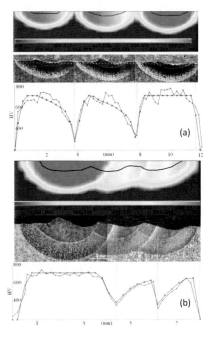

Fig. 3. Results of the laser surface melting tests, spacing between laser passings: (a) 4 mm (first track at left); (b) 2 mm (first track at right).

Four cladding tracks of AISI H13 tool steel on plates of the same steel were simulated with a mesh composed of 57329 elements and 48176 nodes. Each cladding track deposited cools very fast by heat conduction to the steel plate, producing martensite. During the deposition of the contiguous cladding track reheating produces the tempering of the generated martensite so at the end of the simulation only the last track is fully martensite (Fig. 4). The final hardness map is shown in Fig. 5 (a).

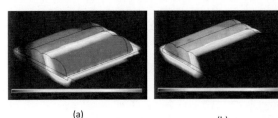

(a)   (b)

Fig. 4. Phase volume fractions calculated with the laser cladding model (after final cooling): (a) martensite; (b) tempered martensite.

Martensitic transformation has an important impact on the stresses. During the heating the contraction caused by the transformation counteracts the thermal expansion, and the opposite happens during cooling down. Thus the strains are lower than without phase transformation. However this is not true for stresses: the higher yield strength of martensite could lead to higher stresses.

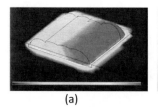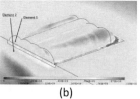

Fig. 5. (a) Final hardness profile and (b) Von Mises equivalent stress calculated with the laser cladding model.

Von Mises equivalent stress is represented in Fig. 5(b). The last clad track has a high percentage of martensite and shows low compressive stresses. The highest stresses are located in the interfaces between the tracks and the plate. The Von Mises stress history of 2 elements is shown in Fig. 6. The elements are situated in: 1) clad track; 2) plate (untransformed zone). The temperature history of 1 is also represented. Element 2 shows a typical profile without phase transformations, during the multitrack deposition process: valleys when the material is hot and peaks when it is cold. Element 1 suffers complete martensitic transformation, generating a stress reversal from tensile to compression during the cooling, producing two peaks instead of one and final stresses lower than before.

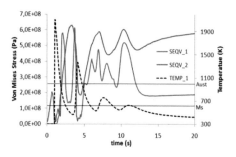

Fig. 6. Von Mises Von Mises stress and temperature histories calculated for two elements from the laser cladding model.

5. Conclusions

A 3D FEM model was developed for the thermomechanical and metallurgical simulation of laser hardening and laser cladding techniques. The metallurgical model simulates some of the phase transformations in steels: austenitization, martensite transformation and tempering. 1D carbon diffusion models were used for the simulation of the two steps of the austenitization process and tempering. Metallurgy was coupled with the thermomechanical simulations to compute the strains and stresses. Martensite has a great influence on the final stress profile, as it was expected. Experimental tests of laser surface melting show good agreement with the calculated melted zones, HAZ and hardness values

References

[1]   A.F.A. Hoadley and M. Rappaz. A thermal model of laser cladding by powder injection. Metall. Trans. B, 23:631-642, 1992.

[2]   M. Picasso, C.F. Marsden, J.D. Wagniere, A. Frenk, and M. Rappaz. A simple but realistic model for laser cladding. Metallurgical and materials transactions B, 25:281-291, 1994.

[3]   J. Choi, L. Han, and Y. Hua. Modeling and experiments of laser cladding with droplet injection. J. Heat Transfer, 127:978-987, 2005.

[4]   A.H. Nickel, D.M. Barnett, and F.B. Prinz. Thermal stresses and deposition patterns in layered manufacturing. Materials Science and Engineering, 317:59-64, 2001

[5]   R. Jendrzejewski, G. Sliwinski, M. Krawczuk, and W. Ostachowicz. Temperature and stress during laser cladding of double-layer coatings. Surface and Coatings Technology, 201:3328-3334, 2006.

[6]   A.M. Deus. A thermal and mechanical model of laser cladding. PhD thesis, Urbana, Illinois, 2004

[7]   M.J. Tobar, A.Suárez, J.C. Álvarez, J. M. Amado, and A. Yáñez. A 3D transient FEM analysis of residual stress generation during laser cladding. Proceedings of the LANE 2007.

[8]   M. Davis, P. Kapadia, J. Dowden, W.M. Steen and C.H.G. Courtney. Heat Hardening of Metal Surfaces with a Scanning Laser Beam. Journal of Physics D: Applied Physics 19 (1986).

[9]   R. Patwa and Y.C. Shin. Predictive modeling of laser hardening of AISI 5150H steels. International Journal of Machine Tools and Manufacture 47 (2007).

[10]   M.F. Ashby and K.E. Easterling. Transformation hardening of steel surfaces by laser beams-I Hypo-eutectoid steels. Acta Metallurgica, 32:1935-1948, 1984.

[11]   M.J. Tobar, C. Álvarez, J.M. Amado, A. Ramil, E. Saavedra and A. Yáñez. Laser transformation hardening of a tool steel: simulation based parameter optimization and experimental results. Surface and Coatings Technology, 200:6362-6367, 2006.

[12]   S. Skavarenina and Y.C. Shin. Predictive modeling and experimental results for laser hardening of AISI 1536 steel with complex geometric features by a high power diode laser. Surface and Coatings Technology, 201:2256-2269, 2006.

[13]   G.N. Haidemenopoulos. Coupled thermodynamic/kinetic analysis of diffusional transformations during laser hardening and laser welding. Journal of Alloys and Compounds, 320:302-307, 2001.

[14]   R.S. Lakhar, Y.C. Shin, and M.J.M. Krane. Predictive modeling of multi-track laser hardening of AISI 4140 steel. Materials Science and Engineering A, 253:501-5028, 2008.

[15]   F. Brückner, D. Lepski, and E. Beyer. Calculation of stresses in two and three dimensional structures generated by induction asisted laser cladding. Munich: Proceedings of the Fifth International WLT-Conference Lasers in Manufacturing, LiM 2009.

[16]   A. Suárez, J.M. Amado, M.J. Tobar, A. Yáñez, E. Fraga, and M.J Peel. Study of residual stresses generated inside laser cladded plates using FEM and diffraction of synchrotron radiation. Surface and Coatings Technology 204 (2010).

[17]   H. Mehrer. Diffusion in Solids Fundamentals, Methods, Materials, Diffusion-Controlled Processes. Springer, 2007.

[18]   G. Roberts, G. Krauss and R. Kennedy. Tool Steels. ASM International, 2009.

Transient numerical simulation of CO_2 laser fusion cutting of metal sheets: Simulation model and process dynamics

Stefanie Kohl[1,2], Karl-Heinz Leitz[1,2] and Michael Schmidt[1,2]
[1] University of Erlangen-Nürnberg, Chair of Photonic Technologies, Paul-Gordan-Str. 3, D-91052 Erlangen, Germany
[2] Erlangen Graduate School in Advanced Optical Technologies, Paul-Gordan-Str. 6, D-91052 Erlangen, Germany

Abstract. Simulations are a versatile tool of increasing importance for profound analysis and comprehensive understanding of laser-based material processing. Based on our simulation model of laser-matter interaction, we developed a numerical three-dimensional model for the transient simulation of fusion cutting of sheet metal. This model is based on the finite-volume method (FVM) and covers various laser-metal interaction mechanisms, such as multiple surface reflection and absorption according to Fresnel, but also more complex interactions like absorption in the metal plasma. In this study, we focused on a Gaussian shaped CO_2 laser beam applied for the cutting and steel sheets. The implemented physics include fluid and vapor dynamics according to Navier-Stokes, heat conduction, the surface tension of the liquid-vapor interface and the modeling of the enthalpies of fusion and vaporization. As accurate input parameters are crucial for any simulation, special focus was put on the implementation of material properties at temperatures around the melting point. The additional modeling of inert process gas ensures realistic blow-out of the molten material. Furthermore, adaptive meshing enables high calculation accuracy in the areas of interest while minimizing computation time drastically.

Keywords: Laser Fusion Cutting, Laser Cutting Simulation; Finite Volume Method

1. Introduction

Laser material processing attains more and more importance in production technology. Since laser cutting represents a contact and force-free as well as a precise and highly flexible cutting tool, it is the most widespread representative among all these technologies. Though laser cutting does not have to face such severe process errors like pores and humping which complicate laser welding, its simulation is still of major importance both from a scientific as well as an industrial point of view. Since simulations enable the variation of basically all the process and material parameters as well as they offer virtually unlimited accessibility of the work piece and its properties during the process, they are an indispensable tool for both examination and planning of the cutting process.

Many of the existing models available for laser cutting only focus on certain aspects of the process or disregard certain interactions. The model presented in this paper, however, aims at covering the different mechanisms and interactions of laser cutting comprehensively, facilitating high-accuracy modeling of the process based on its underlying physical equations.

2. Simulation model

All modeling was done using the open source computational fluid dynamics (CFD) software package OpenFOAM® (Open Field Operation and Manipulation), which has proofed to be a suitable environment for the simulation of a multitude of laser material interaction processes for many years [1-2]. It represents a toolbox of modules for the simulation of complex numerical problems based on partial differential equations, such as fluid flows, heat transfer, solid dynamics as well as electromagnetics. Its open and modular structure enables the flexible implementation of the different interaction mechanisms between laser and material as well as subsequent processes like heat conduction, melt and vapor dynamics and phase transitions.

All liquid-vapor surfaces within the model are regarded as free surfaces and as such they were reconstructed based on their volume fraction making use of a volume-of-fluid (VOF) approach [3]. The size of the time steps is adapted dynamically and derived from the Courant number [4], ensuring sufficiently small steps during the whole simulation.

The cubic mesh of the simulated volume allows for automated adaption within the volumes of interest using a predefined refinement function, which in case of laser material processing focuses on areas of high energy input of the laser as well as phase changes and high temperatures. This refinement and unrefinement facilitates highly accurate calculations while keeping the computation time at a very low level, compared to calculations using a constant grid.

2.1. Laser-Metal Interaction

In our simulation model, the absorbed and reflected fraction of the laser beam at incidence onto the metal was implemented according to the Fresnel equations. Therefore, a correct estimation of the complex refractive index was necessary.

However, to our knowledge, no experimental data exists for the refractive index of iron or steels in the infrared regime for temperatures around the melting point. So we calculated the complex refractive index both from the Drude model [5], and from the Dausinger model [6] taking interband absorption into account using the electron density [7] and electric conductivity [8] as input parameters. From there, the ratio of angle-dependant absorption and reflection ratio were determined using an approximation of the Fresnel formula [9]. For large wavelengths and high temperatures, the results from both methods agree well.

The energy input of the absorbed beam into the metal was implemented assuming exponential decay inside the metal according to Lambert-Beer, and the reflected beam was modeled using a computation time efficient combination of a ray tracing algorithm and a diffusion based method [10].

Additionally, the absorption in the metal plasma vapor was taken into account since its effect increases drastically with wavelength. Therefore, it was modeled introducing a temperature and density dependent absorption coefficient [11].

2.2. Implemented Physics

The main difficulty in calculating the conductive and convective heat transfer by solving the heat equation in laser material processing is posed by the need of taking the enthalpies during fusion and evaporation, as well as recondensation and resolidification into account.

In our model this problem is approached by using an iterative method based on [12][13], correcting the enthalpies within every calculation step of the heat equation multiple times. Thus, within the first step of each calculation, both enthalpies are neglected. From this first preliminary temperature solution, the enthalpies can be estimated more correctly. These improved values are then used in the subsequent step.

The fluid dynamic within the simulations is modeled incompressibly and is based on mass conservation, the Navier-Stokes equation and the phase transport equation [14]. It is solved using the PISO (Pressure Implicit with Splitting of Operators) algorithm [15].

The description of the evaporation process and the vapor dynamic are predicated on the calculation of the mass of the evaporated material as well as on its associated vapor pressure.

In fusion cutting, the use of the inert process gas is crucial. Due to the difficulties of its implementation, the gas jet is modeled making use of a simplified approach. That means that the gas flow is provided as a constant pressure field defined in the simulated volume above the metal surface rather than modeling the nozzle.

2.3. Simulation Parameters

Collecting accurate material parameters for steel at temperatures around and above the melting point, however, posed a problem that, unfortunately, remains yet unsolved. Therefore, the properties of iron were used, since for this material a consistent and complete set of material parameters around melting temperature could be found in [8,9]. The values are given in Table 1, the thickness of the simulated sheet can be varied arbitrarily, for this paper a value of 1 mm was chosen.

Table 1. Properties of iron at its melting temperature

Physical property	Symbol	Value
Molar mass	M	55.845 g/mol
Density	ϱ	7015 kg/m^3
Specific heat	c_P	841.17 J/kgK
Thermal conductivity	λ_T	34.5 W/Km
Viscosity	ν	0.784 m^2/s
Temperature of fusion	T_M	1809 K
Evaporation temperature	T_V	3133 K
Enthalpy of fusion	H_M	272.4 kJ/kg
Enthalpy of evaporation	H_V	6095 kJ/kg
Surface tension	σ	1.872 N/m
Complex refractive index at λ=10.6 μm	n+ik	14,7 + i 15,5

As inert process gas nitrogen was employed. Its thermophysical properties can be found in [16-18].

The parameters of the Gaussian CO_2 laser beam which was defined in the simulation model for this work are given in Table 2.

Table 2. Parameters of the simulated laser

Physical property	Symbol	Value
Wavelength	λ	10.6 µm
Laser power	P_L	4 kW
Beam profile		Gaussian
Waist radius	w_0	100 µm
Feed rate	v_0	0.4 m/s

3. Simulation results and discussion

In Fig. 1, a cross section through the metal sheet along the cutting direction is given for different time steps of the cutting process. The laser beam was orthogonal to the sheet irradiating the metal sheet from the top.

In the beginning of the cutting process, the laser hits the metal sheet and melts and evaporates the material [Fig. 1 (a)]. Then, droplet formation and melt spilling can be observed. Before full penetration of the sheet, the melt is expelled to the top [Fig. 1(b)]. Afterwards, due to the pressure of the process gas, the melt is expelled downwards [Fig. 1 (c-f)].

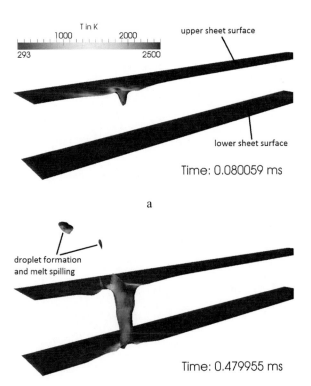

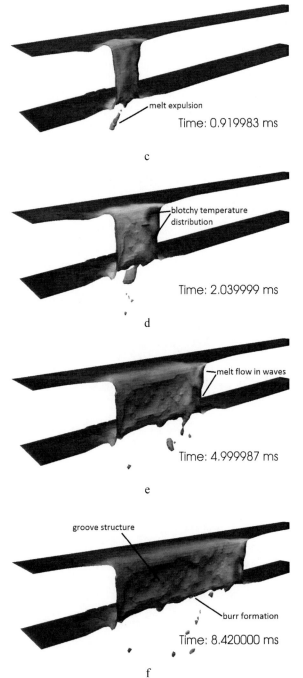

Fig. 1. Simulated laser cutting process using the given model and parameters. The illustrated surface displays the metal-air boundary and the color indicates its temperature. The different panels represent the different time steps (a-f). The color legend is identical for all panels and therefore is given only in panel a

The simulation clearly shows that the melt does not flow away from the cutting front uniformly, but rather in a wavelike process, which can be derived from the curvy cutting front and from its blotchy temperature distribution [Fig. 1 (d-f)].

After cool down of the cutted material, a realistic, rough groove structure along the cutting edge can be observed [Fig. 1 (f)]. Moreover, burr at the lower side of the cutting edge is formed [Fig. 1 (f)].

Figure 2 shows a close-up of the cutting front. For reasons of clarity, the default smoothing interpolation of the temperature field was skipped. Again, a wave within the melt flow can be observed. This wave exhibits a pronounced shadowing effect on the area underneath it, which, hence, cannot be irradiated by the laser and stays considerably cooler. Therefore, the resulting temperature distribution along the cutting front is blotchy.

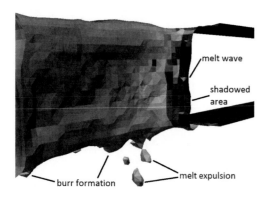

Fig. 2. Close-up of the metal-air boundary of the same cutting simulation. Here, the smoothing of the temperature field was omitted. The melt wave in the middle of the cutting front shawdows the area beneath it, which stays significantly cooler

4. Conclusion and Perspectives

In this work, the numerical simulation of the CO_2 based laser cutting process of metal sheets within the framework of OpenFOAM® could be demonstrated. The implemented laser-metal interaction and the physical description of the dynamics proofed to be an effective way for the description of this process. The melt flow and expulsion as well as the cutting edge structures showed realistic results.

Acknowledgements: The authors gratefully acknowledge the funding of the project "Thermomechanical and Fluiddynamic Interactions in Laser Deep Penetration Welding" as well as the funding of the Erlangen Graduate School in Advanced Optical Technologies (SAOT) by the German National Science Foundation (DFG) in the frame work of the excellence initiative.

References

[1]   Geiger M, Leitz KH, Koch H, Otto A, (2009) A 3D Transient Model of Keyhole and Melt Pool Dynamics in Laser Beam Welding Applied to the Joining of Zinc Coated Sheets. Prod. Eng. Res. Devel. 3:127-136

[2]   Otto A, Koch H, Leitz KH, Schmidt M, (2011) Numerical Simulations – A Versatile Approach for Better Understanding Dynamics in Laser Material Processing. Phys. Proc. 12:11-20

[3]   Leitz KH, Koch H, Otto A, Schmidt M, (2012) Numerical Simulation of Process Dynamics During Laser Beam Drilling With Short Pulses. Appl. Phys. A. 106:885-891

[4]   Hirt CW, Nichols BD, (1981) Volume of Fluid (VOF) Method for the Dynamics of Free Boundaries. J. of Comp. Phy. 39:201-225

[5]   Courant R, Friedrichs K, Lewy H, (1928) Über die partiellen Differenzengleichungen der mathematischen Physik. Math. Ann. 100:32-74

[6]   Wooten F, (1972) Optical Properties of Solids. London: Academic Press

[7]   Dausinger F, Shen J, (1993) Energy Coupling Efficiency in Laser Surface Treatment. ISIJ Int. 33:925-933

[8]   Haynes WM, Lide DR, (2011) CRC Handbook of Chemistry and Physics.[e-book] 91st ed. Internet Version 2011. Available through: < http://www.hbcpnetbase.com> [Accessed 20 June 2011]

[9]   Brandes EA, Brook GB, (1992) Smithells Metals Reference Book. 7th ed. Oxford: Butterworth Heinemann

[10]   Prokhorov AM, Konov VI, Ursu I, Mihailescu IN, (1990) Laser Heating of Metals. Bristol: IOP Publishing Ltd

[11]   Poprawe R, (2005) Lasertechnik für die Fertigung. Berlin: Springer

[12]   Voller VR, Cross M, Markatos NC, (1987) An Enthalpy Method for Convection/Diffusion Phase Change. Int. J. for Num. Meth. In Eng. 24:271-284

[13]   Brent AD, Voller VR, Reid KJ, (1988) Enthalpy-Porosity Technique for Modeling Convection-Diffusion Phase Change: Application to the Melting of a Pure Metal. Num. Heat Trans. 13:297-318

[14]   Kunkelmann C, Stephan P, (2009) CFD Simulation of Boiling Flows Using the Volume-of-Fluid Method Within OpenFOAM. Num. Heat Trans. 56:631-646

[15]   Issa RI, (1985) Solution of the Implicitly Discretised Fluid Flow Equations by Operator-Splitting. J. of Comp. Phys. 62:40-65

[16]   Kadoya K, Matsunaga N, Nagashima A, (1985) Viscosity and Thermal Conductivity of Dry Air in the Gaseous Phase. J. Phys. Chem. Ref. Data. 14:947-970

[17]   Hanley HJM, Ely JF, (1973) The Viscosity and Thermal Conductivity of Dilute Nitrogen and Oxygen. J. Phys. Chem. Ref. Data. 2:735-755

[18]   Span R, Lemmon EW, Jacobsen RT, Wagner W, (1998) A Reference Quality Equation of State for Nitrogen. Int. J. of Thermodyn. 19:1121-1132.

Numerical methods for laser path calculation for surface treatment of dies and moulds

Y. Liu, R. Ur-Rehman, D. Heinen and L. Glasmacher
Fraunhofer IPT, Steinbachstrasse 17, 52074 Aachen Germany

Abstract. Utilization of laser based processes such as alloying/dispersing are in frequent use to improve the properties of contact surfaces and consequently to increase the lifetime of moulds and dies. The process involves transmission of energy from laser beam and/or alloying materials in precise amounts and accurately at the desired positions on the part surface. The complexity in the implementation of the processes has been investigated at Fraunhofer IPT and a Computer Aided Manufacturing (CAM) solution has been established to optimize the application of powder based laser processes for surface treatment of moulds and dies. In early part of the paper the background of laser processes for surface treatment technology is discussed, based upon physics knowledge and experiment results. Accordingly, the larger part of the research focused on the analysis and optimization of the CAM strategies and specialization of CNC data for laser machine. The optimization of CAM strategies is achieved by application of numerical methods such as statistic calculation of triangulation and surface parameterization and approximation. A consistent workflow that integrates process database, laser machine simulation and specialized post-processing is employed to achieve stable CAM solution. The results from implementation of CAM solution are also demonstrated.

Keywords: laser surface treatment, CAM solution, CNC, moulds and dies, triangulation mesh, laser path generation

1. Introduction

The manufacture industry is facing constantly the challenge to reduce production costs and increase production volumes at persistent product quality. Being the most common-used tooling in most of the sectors of manufacture industry, moulds and dies are the determinative part influencing the production cost and the part quality. Industrial reports show that increasing forging dies lifetime by 100-200% can reduce the overall costs by 30% [1].

The deterioration of moulds and dies is caused by thermal and mechanical stresses and chemical exposures during their usage. All of these factors result in a high abrasive and adhesive wear of the parts, cracks on the contact surfaces and dimension deviation, which are the main criteria for useful lifetime of moulds and dies [2] and [3].

The earlier works in [4-7] investigated into the effect of laser process to metallic parts. According to Ayers and Tucker [4] and Tassin [5], laser imparts hard particles into metals and alloys to adjust the dissolution of these particles in the melted substrate, leading to desired metallurgical structures and properties with continuity in the chemical and mechanical properties. Therefore laser alloying/dispersing is an effective and economic approach to extend the useful lifetime of dies and moulds by enhancing the mechanical and chemical characteristics of contact surface material. For instance, a combined laser alloying and nitriding treatment process can produce a hard surface layer of 0.5- 1.0 mm case depth and a hardness of up to 1000 HV that ensures wear resistance and fatigue resistance [8]. In addition, the compound treated layer enhances good tribological and corrosion resistance properties of the die surface [6].

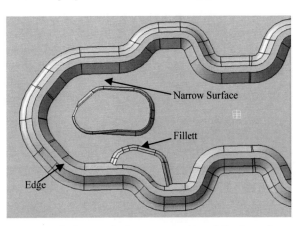

Fig. 1. Typical geometric features on moulds and dies for surface treatment

Although laser processes have advantages that surfaces can be treated with dimensional precision and efficient control of thermal energy, however there are challenges in the application of these processes on the complex geometries in industrial practice. The critical areas of moulds and dies are typically narrow and small geometric features, e.g.: which are shown in Fig. 1 of a die for forming processes. These areas normally consist of freeform surface, and surrounded by wall structure that leads to collision risk to the laser head. The investigations at Fraunhofer IPT have considered these problems during laser surface treatment and utilized 5-axis CNC laser machine to accomplish the application on all critical areas and to ensure the precision of the laser process [12].

To improve the efficiency of the surface treatment process chain, an automatic planning of laser path through computer aided manufacturing (CAM) has been strongly realized and CAM solution has been established at Fraunhofer IPT to achieve the process requirements in an integrative method. The CNC machines are programmed with NC data that is a combination of laser path and technological parameters for laser surface treatment. The NC programming is automatically accomplished by an integrated CAM solution to control the risk of process failure and to reduce the overall cost of prototypes for destructive testing as well as regular production.

2. Knowledge of laser alloying/dispersing process

The understanding of the physics of laser processes is essential for the development of an efficient surface treatment process to enhance the properties of part material. The laser beam heats up the local areas near the die's surface and melts them. Meanwhile powdered materials (e.g WC-Co-Cr, TiC or VC) are conveyed into the melting pool through an inert gas stream and completely diffuse into solution. The Fig. 2 below presents a schematic representation of the processes [12]. In typical applications the process utilizes a Nd:YAG or diode laser source that can concentrate 1.0~2.0 kws in a spot diameter/size of 0.8~2.0 mm.

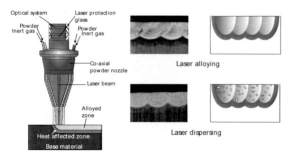

Fig. 2. Schematic arrangement of laser surface treatment process

Fraunhofer IPT carried out experimental investigation in the scope of European Commission's funded research project **CURARE** to achieve optimal surface treatment process and to identify the criteria for the stability of process. During these experiments the physics of

interactions between laser and the material has been investigated [12].

The experimental setup consisted of machine + laser has been built up to investigate into the effects of laser intensity, alloying materials and their amounts, gas flow upon the surface and the formation of heat effected zones. Fig. 3. shows the influence of the process on the surface and below the surface. The performance of the following milling and grinding processes has also been verified on the test parts. The laser process is carried out by a computer numerical controlled (CNC) machine - Alzmetall GS-1005 LOB, integrated with a Precitec laser head.

Fig. 3. Cross section of alloyed cast part, its alloyed surface and milled/grinded surface

3. CAM solution for planning for laser process

The control of surface treatment process is complex since it involves the control of simultaneous movements of laser in five degrees of freedom (5-axis CNC). The objective of CAM solution for laser process is to automatically generate precise laser path together with consideration of laser technological requirements, NC machine specialty and characteristics of geometry features on moulds and dies.

The CAM module developed at Fraunhofer IPT for laser process for surface treatment uses given geometry features on part as input for laser path calculation. The CAM solution automatically generates interpolated move positions on part surface within allowable tolerance and outputs the NC data for laser process for surface treatment. Based upon the types of data structure of input geometry, CAM strategies can be categorized into two groups: parametric surface-based and triangulated mesh-based. The specialized requirements for laser surface treatments have been investigated and handled by representing the part geometry by a triangulated mesh and integrating laser processes into advanced numerical techniques for calculation of laser path for modern CNC laser machine.

3.1. Surface-based laser path calculation

Surface-based calculation is aiming at CAD data consisting of NURBS surfaces/curves. This type of data is normally input as IGES or STEP file, and then a 3D model is built up as an intermediate of interpolation [11].

The first step is to build-up a data structure with 2-dimensional (u-v) parameterized surface model from CAD data. The u-v net combined with correspondent to the 3D coordinates positions are generated from the NURBS geometries. The procedure of laser path generation can be described as following steps:

1. Parameterize surfaces and boundary curves from 3D space into u-v space
2. Based on types of CAM strategy and laser setup, build up virtual intersection of part surface and laser
3. Calculate intersection points (moves) between the intersection geometries and parameterized surface on laser path, as in Fig. 4
4. Refine the laser path by interpolating positions between intersections within specified tolerance
5. Re-group the points and store into data structure for laser path according to CAM strategy configuration
6. Calculate orientation of laser at each position in the laser path, based upon the user's configuration of laser process and to avoid collision of laser head with die/mould and accordingly re-adjust move position when necessary

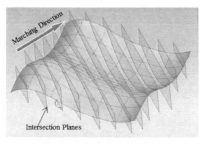

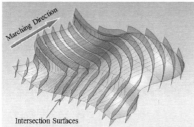

Fig. 4. (a). calculation of iso-planer laser path (b). calculation of constant offset laser path

The Fig 4 shows two typically used strategies. (a) illustrates the interpolation geometries used in the strategy "Iso-planer Parallel moving", and 4 (b) demonstrates the strategy "Constant offset morphing from curves", which means laser path offsets from the initial curve.

3.2. Triangulated mesh-laser path calculation

In some application cases, only a triangulated-mesh is given as the input geometry for laser path calculation, for example data acquired from laser metrology process. The difficulty in such applications is that we cannot directly obtain continuous and analytic information within the triangulated mesh, but only discrete data of part surface such as vertices and edges.

We propose a set of mesh-based laser path calculation methods to solve this problem. The procedure can be divided into two steps:

1. Generation of initial geometries
2. Propagation of laser path based on initial geometries

Firstly we build up an extended "graph" data structure which stores the 3D dimension information of the mesh, such as vertices, edges and faces, and also the topologic information such as query of neighbour facets. Then statistic and topologic methods are used to obtain these features from the graph, also combined with empirical knowledge of part geometry, such as estimation of the tensor of the local curvature, triangle normal voting method to detect edges and intersection between analytic geometry elements and triangulated mesh.

Afterwards a rapid algorithm generates the laser path as propagating curves from the initial geometries. A "map" structure that represents the geodesic distances from vertices to initial geometry is established on the mesh edges or vertices, by an optimized algorithm based on Fast Marching Method (FMM) [9, 10]. This method spreads the laser path on the mesh in a similar pattern as wave-front propagation, with time complexity by $O(n \cdot log(n))$, in which n is the number of vertices. The picture below demonstrates an example of mesh-based laser path generated by our method.

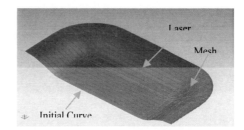

Fig. 5. Laser path generated on mesh

4. Simulation and prototype test result

A post-processor is implemented to translate the calculated laser path to NC program specialized for the laser machine. We preceded an analysis of the time-sequence and the interactions among system components including laser controller, inert gas pump, powder conveyor and machine kinematics.

A machine simulation module developed based on ModuleWorks simulation libraries [13] has been integrated into the CAM system and applied to check collision of machine kinematic components and work piece, before the generated 5-axis NC program runs on the laser machine. The following picture illustrates an example of a demonstrator for alloying testing.

Fig. 6. Simulation and real process

A testing part is made of steel 1.2379. The powder used for alloying is WC-Co-Cr. The alloyed result is shown in the picture below:

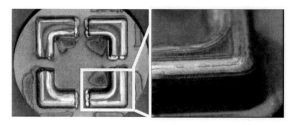

Fig. 7. Test result

A metallographic inspection has been made and showed that with proper process parameter setting, the alloyed depth on processed area can reach up to 1.5 mm and the hardness reaches to 800 HV0.3.

5. Conclusion

In this paper we introduced two sets of numerical methods for laser path calculation, respectively for parametric surface model and triangulated-mesh geometry. These methods have been integrated into a CAM system, together with a specialized post-processor to generate NC data for laser CNC machine. The CAM systems integrate a machine simulation to check the NC data and prevent any unforeseen collision of machine components. Tested parts show that the surface quality on the treated areas has been significantly improved. The CAM systems have also been utilized in several industrial applications and have demonstrated to be effective and efficient to improve the useful life of the dies and moulds.

Acknowledgement: The authors would like to express their gratitude to European Commission for supporting CURARE project (grant number 222317) of seventh Framework Program. The cooperation and support of CURARE consortium partners is highly appreciated.

Reference

[1]   Hejarsmide NN, (01/2002) Forging Company Hejarmide, Sweden. www.hejarsmide.se

[2]   Robert F, (2008) Standzeitverlängerung von Werkzeugen durch Laserbehandlung und Nitrieren. Giesserei 95 H. 5:110–111

[3]   Harksen S, Glaeser T, Bleck W, Klocke F, (06/2009) Beurteilung von Temperaturwechsel und Mechanischem Verschleiß auf duplexbehandeltem Warmarbeitsstahl. Proceedings of the 8th International Tooling Conference, 8ITC, Mainz

[4]   Ayers JD, Tucker TR, (1980) Particulate-TiC-Hardened steel surfaces by laser melt injection. Thin Solid Film 73-1:201-207

[5]   Tassin C, Laroudie F, Pons M, Lelait L, (03/1996) Improvement of the wear resistance of 316L stainless steel by laser surface alloying. Surface and Coatings Technology 80/1–2:207-210

[6]   Gläeser T, (2006) Standzeitverlängerung von Werkzeugen und Formen durch Laserlegieren/-dispergieren und Nitrieren, VDWF im Dialog. Magazin des Verbands Deutscher Werkzeug- und Formenbauer e. V., Ausgabe 03

[7]   Gläeser T, (2006) Standzeiterhöhung von Schmiede- und Druckgießwerkzeugen durch Laserlegieren/-dispergieren und Nitrieren. Laser Magazin, Ausgabe 03

[8]   Ruset C, Grigore E, Gläser T, Bausch S, (2006) Combined treatments – a way to improve surface performances. The fifth International Edition of Romanian Conference on Advanced Materials, Bucharest-Magurele, Romania, Sept. 11-14, 2006

[9]   Sethian JA, (02/1996) A Fast Marching Level Set Method for Monotonically Advancing Fronts, PNAS (Proceedings of the National Academy of Sciences of the United States of America), Vol. 93-4:1591-1595

[10] Kimmel R, Sethian JA, (07/1998) Computing geodesic paths on manifolds. PNAS (Proceedings of the National Academy of Sciences of the United States of America) vol. 95:8431-8435

[11] Ur-Rehman R, (07/2010) Numerical Techniques for CAM Strategies for Machining of Mould and Die, Proceedings of the 36th International Matador Conference 6: 259-262

[12] Klocke F, Heinen D, Ruset C, Liu Y, Arntz K, (2010) Kombinierte Oberflächenbehandlung von Werkzeugen, wt Werkstattstechnik online Jahrgang 100

[13] ModuleWorks CNC simulation and toolpath verification CAD/CAM components, http://www.moduleworks.com/ cad-cam-components/simulation-verification.asp, ModuleWorks GmbH

Laser Technology – Surface Engineering

Microstructures and performances of the WC-10Co4Cr coatings deposited by laser hybrid plasma spraying technology

S. Q. Li[1,2], S. L. Gong[2], Y. P. Duan[1], S. H Liu[1]

[1] Department of Materials Processing Engineering, School of Material Science &Engineering, Dalian University of Technology, Linggong Road 2, Ganjingzi District, Dalian 116085, Liaoning Province,PR China

[2] National Key Laboratory of Science and Technology on Power Beam Processes, Beijing Aeronautical Manufacturing Technology Research Institute, Dongjunzhuang 1,Chaoyang District,Beijing 100024, PR China

Abstract: In this paper, the WC-10Co4Cr coatings were deposited by the laser hybrid plasma spraying (LHPS) technology on the 38CrMoAl substrate. Microstructures, tribological characteristics and corrosion-resistance performance of the coatings were studied using an optical microscope, an X-ray diffraction, a scanning electron microscope, an SRV friction and wear tester and the Neutral Salt Spraying (NSS) test. Test results indicated that LHPS WC-10Co4Cr coating had good wear-resistance and corrosion-resistance performance comparing with the base material. The LHPS coating overcame many defects of the conventional spraying coatings such as poor bonding strength, many pores and many cracks. LHPS is able to improve the bonding strength as the interfaces are melted and joined by the laser simultaneously with plasma spraying. The coating achieves metallurgy bonding and its microstructure becomes more compact and therefore its performances are greatly improved.

Keywords: laser hybrid plasma spraying; WC-10Co4Cr coating; microstructure; performances analysis

1. Introduction

WC coatings deposited by plasma spraying technology are widely used to protect the components from abrasive wear less than 450^0C. However, APS ceramic coatings contain many porosities and cracks for its mechanical bonding method which reduced the coating's performances such as abrasive resistance, toughness, corrosion resistance and service life. To improve the coating's performance, laser irradiation has been tried after plasma spraying in the previous studies and it was shown that the laser treatment can improve the APS spraying coating's bonding strength and microstructure, but the formation of macro-cracks in the post-spraying configuration is very difficult to be suppressed and which can lead to the spallation of the coating from the substrate. In recent studies, laser hybrid plasma spraying (LHPS) may be an interesting method for densifying the ceramic coatings' microstructure and improving their performances. This method consists of APS with the simultaneous laser beam irradiation which enables the

spraying and remelting in a single process. LHPS seems to be effective in densifying the microstructure and improving the coating's properties for example the thermal shock resistance of the ZrO_2-Y_2O_3 coating [1,2,3], the wear resistance of the alumina-titania coating [4], the corrosion resistance of NiTi coating [5,6,7]. However, no reports can be found about the WC coating. In this study, WC-CoCr coating were deposited by LHPS, not only the microstructure and wear-resistance performance but also the bonding method and corrosion-resistance performance are investigated. Moreover, the anti-wear performance and anti-corrosion performance of the LHPS WC-CoCr coatings are compared with the 38CrMoAl substrate at the same time.

2. Materials and experimental procedures

2.1. Samples

In this paper, 38CrMoAl is used as the base material. The samples, 5mm thick, 35mm wide and 35mm long, are used to fabricate the coating for the anti-corrosion performance test. A surface area of 35 mm by 35mm is blasted by Al_2O_3. The samples for wear-resistance performance test are Φ25mm and 7.5mm long. To form the coating, the WC-10Co4Cr powder is used and the powder's particle diameter range is from 15 to 45um.

2.2. Fabrication procedure

The coating was fabricated by the laser hybrid plasma spraying system. Both the plasma gun and the laser nozzle were installed inside a chamber equipped with an exhaust system. An Nd YAG laser (HL3006D, TRUMPF，GERMANY) working in continuous wave mode at 1.06 μm was used and the laser beam was delivered by optical fibre to the laser nozzle. An air

S. Hinduja and L. Li (eds.), *Proceedings of the 37th International MATADOR Conference*,
DOI: 10.1007/978-1-4471-4480-9_12, © Springer-Verlag London 2013

plasma spraying equipment (APS2000, BAMTRI, CHINA) was utilized in this experiment. A schematic diagram of the laser hybrid plasma spraying process is depicted in Fig. 1.The plasma spraying was scanned perpendicular to the substrate, and the laser beam was set at 45^0 with respect to the substrate. Laser irradiation was superimposed on the sprayed area. The laser beam scanned up and down at 2mm spacing and the coating thickness was aimed to be 350μm. The Deposition parameters of LHPS coatings are given in Table 1.

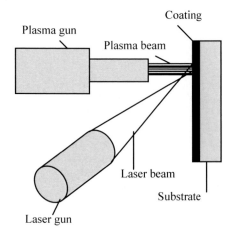

Fig. 1. The schematic diagram of the laser hybrid plasma spraying technology

Table 1. Deposition parameters of LHPS coatings

Voltage (V)	Current (A)	Powder feeding (gmin^{-1})	Feed interval (mm)	Scan speed (mms^{-1})
68	480	21	2	280
Spray distance (mm)	Primary gas (Ar), slm	Secondary gas (H$_2$) slm	Laser Power (W)	Beam size (mm)
95	60	2	2300	Φ3

2.3. The coatings' microstructures and performances test

Microstructural characterization and the porosity of WC-CoCr coatings was studied using an optical microscope JENAPHOTO2000 (made in Germany) and a scanning electron microscope (SEM) of type CamScan3400. The crystal structure of the coatings and the original powder are tested by the X-ray diffraction with 40KV operating voltage and 20mA current and CuKα radiation. The coatings' micro hardness was tested by the DMH-2

micro-hardness tester according to the standard of GB/T4340.1-1999.The bonding strength was tested by the tensile test machine according to the standard of HB5476-91. The wear tests were conducted using a standard SRV high-temperature friction and wear tester with reciprocating motion against 9.525mm sintered Al$_2$O$_3$ ball, the friction and wear results were compared with the 38CrMoAl substrate. The wear conditions were given as 10N load, 5HZ frequency, and temperature from room temperature to 800^{0}C, 1mm stroke and 0.5h test duration. And the worn surfaces of the38CrMoAl and LPHS coatings after 800^{0}C friction and wear test were analyzed using a SEM. The NSS test is processing according to the standard of GB/T 10125-1997, the salt spray liquid is 3.5% NaCl solution, PH=6.5~7.2, test temperature is 35^{0}C. All the sides of the tested samples are enwrapped by the neoprene except the definite test area.

3. Results and discussions

Figure 2 gives the result of the X-ray diffraction patterns of the LHPS WC-CoCr coating. The phase composition is mainly of the WC, W$_2$C, Co$_{25}$Cr$_{25}$W$_8$C$_2$, Co$_6$W$_6$C, and Co$_2$C. It indicated that part of WC phase decomposed under the high temperature during the spraying processing and the new phases of W$_2$C, Co$_2$C were formed in the coating. 2WC—W$_2$C+C, Co+C—Co$_2$C.The compounds of Co$_{25}$Cr$_{25}$W$_8$C$_2$ and Co$_6$W$_6$C have not only the high hardness but also the good toughness which are beneficial to the coating's anti-wear and anti-corrosion performances.

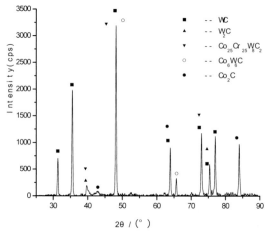

Fig.2. The XRD patterns of the LHPS coatings

Figure 3 shows a typical cross section of a LHPS WC-CoCr coating which were tested by the optical microscope. The coating's microstructure is compact and few cracks and porosities can be seen in the coating and there is no inter-space between the coating and the substrate. The LHPS coating achieved metallurgy bonding and this can be verified by a melting line between the coating and the substrate.

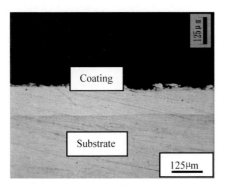

Fig.3. Microstructure of the LHPS WC-CoCr coating

Table 2 shows the bonding strength, micro-hardness and porosity of the WC-CoCr coatings deposited by LHPS. For the addition of the laser power, the LHPS coating achieved metallurgy bonding and its bonding strength is thus improved to be 78MPa. Moreover, the micro-hardness of LHPS WC-CoCr is rather high and its porosity is much lower than 2%. These above characteristics would contribute to the coating's wear-resistance performance and corrosion-resistance performance.

Table 2. The bonding strength and micro-hardness and cavity of the coatings

Coating	Bonding strength (MPa)	Micro-hardness HV0.2	Porosity
WC-CoCr	78	1077	<2%

Figure 4 shows the friction coefficient of the substrate and WC-CoCr coating at 800^0C during the 1800s test time. For comparison of friction and wear results to the LPHS WC-CoCr coating the 38CrMoAl substrate was tested under the same test environment. The average friction coefficient was 0.80 ± 0.05 for the 38CrMoAl substrate and 0.87 ± 0.09 for the LHPS WC-CoCr coating at 800^0C.

For the 38CrMoAl, its friction coefficient has a little fluctuation between 0.90 and 1.0 in the first 500 seconds, from 500s till to the end of 1800s, its friction coefficient always hold the line of 0.78 and it is very stable. Figure 5 (a) shows that after the SRV high-temperature abrasion test, there was some visible trace in the wear area and the material shed seriously and many wear products or pits take part in the next abrasion processing and make the abrasion more serious [8]. This can be explained that the substrate has good toughness but low hardness; hence the material was easy to be cut under the friction force. It can be summed up that the 38CrMoAl substrate's wear mechanism is mainly of plastic grinding.

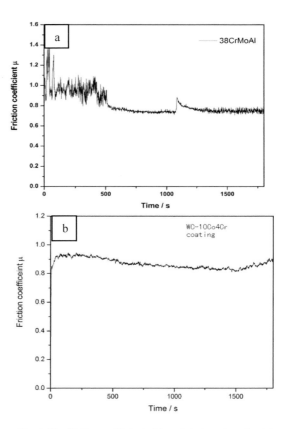

Fig. 4. The friction coefficient of the substrate and coating (a) 38CrMoAl (b) LHPS WC-CoCr coating

As is shown in Fig. 4 (b), the LHPS WC-CoCr coating's friction coefficient has a little fluctuation between 0.8 and 0.9 during the 1800s, as a whole the friction coefficient hold the line of 0.85 and its average friction coefficient was 0.86 ± 0.07 which is higher than the substrate of 0.80 ± 0.05. That is because of the bigger surface roughness of LHPS WC-CoCr coating than the 38CrMoAl. Figure 5 (b) gives the SEM configurations of LHPS coating after friction and abrasion test at 800^0C. It can be seen that there is a few of wear products and some slight wear traces on the coatings wear surface. The wear rate of the 38CrMoAl is $5.96\times10^{-5}(mm^3/N.m)$ but the LHPS WC-CoCr coating is only $6.85\times10^{-6}(mm^3/N.m)$. It indicated that the LHPS WC-CoCr coating has more excellent anti-high temperature abrasion performance than the 38CrMoAl substrate. Its main failure mechanism is a combination of mild scratching and plastic deformation. These results can be explained by the next three reasons. Firstly, the WC-CoCr powders melt more sufficiently and have good fluidity and uniformity for the addition of the laser power and hence the coating achieved metallurgy bonding. Secondly, the coating's microstructure becomes compact and there are no big pores or cracks in it (Fig. 3), the hard ceramic phases of WC, W_2C, $Co_{25}Cr_{25}W_8C_2$ are strongly fastened by the CoCr metal phase. Thirdly, the LHPS WC-CoCr coating has higher micro-hardness and good toughness, its hard ceramic phases and the metal phases form a firm framework to resist the wear force.

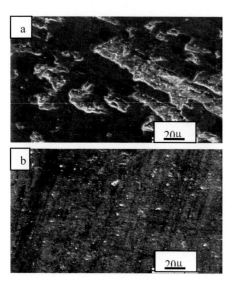

Fig. 5. The worn morphology of the substrate and WC-CoCr coating (a) 38CrMoAl substrate (b) LHPS WC-10Co4Cr coating

Figure 6 shows the corrosion configuration of the substrate and the LHPS coating. It can be seen that after 96 hours of the NSS test, the LHPS WC-CoCr coating demonstrated more excellent anti-corrosion performance than the substrate. After 1.5 hours of the NSS test, some corrosion points became visible on the surface of 38CrMoAl, and more than 5/6 surface area was corroded after 96 hours' test. However the first corrosion point on the surface coating of LHPS WC-CoCr appeared when the corrosion time increased to be 12 hours and the corrosion points germinated and increased slowly, till the 96 hours, there were only about 10 corrosion points on the surface coating. The excellent anti-corrosion performance of the LHPS WC-CoCr coating was mainly because of its high bonding strength and compact microstructure for the addition of the laser power in the processing.

Fig. 6. The corrosion configuration of the substrate and the coating after 96 hours (a) 38CrMoAl substrate (b)LHPS WC-CoCr coating

4. Conclusions

The LHPS WC-CoCr coating partly achieved metallurgy bonding for the addition of the laser in the spraying process, and the coating's bonding strength was improved and its microstructure was densified, which contributed to the coatings' anti-wear and anti-corrosion performances. Compared with the 38CrMoAl substrate, the anti-wear performance of the LHPS WC-CoCr coating is more excellent for its hard mico-hardness and appropriate ductility. The wear mechanism of the LHPS WC-CoCr coating is a combination of scratching and mild ploughing but the 38CrMoAl substrate is mainly of plastic grinding and spallation. The corrosion-resistance performance of the LHPS coating is more excellent than the 38CrMoAl substrate because of its metallurgy bonding method and the compact microstructure which can be effective to prohibit the corrosion solution from penetrating into the coating. Therefore the substrate can be protected excellently by the LHPS WC-CoCr coating from corrosion.

References

[1]   Sang Ok Chwa, Akira Ohmori. Thermal diffusivity and erosion resistance of ZrO2-8wt. %Y2O3 coatings prepared by a plasma laser hybrid spraying technique. Thin Solid Films, 2002(415):160-166.

[2]   Sang Ok Chwa, Akira Ohmori. Microstructures of ZrO2-8wt. %Y2O3 coatings prepared by a plasma laser hybrid spraying technique, J. of Surface & Coatings Technology, 2002(153):304-312.

[3]   J.H.Ouyang, S.Sasaki. Microstructure and tribological characteristics of ZrO2-Y2O3 ceramic coatings deposited by laser-assisted plasma hybrid spraying [J]. Tribology International, 2002 (35):255-264.

[4]   L. Dubourg , R.S. Lima, C. Moreaub. Properties of alumina–titania coatings prepared by laser-assisted air plasma spraying [J]. Surface & Coatings Technology, 2007 (201): 6278–6284.

[5]   Hitoshi Hiraga, Takashi Inoue, Hirofumi Shimura, Akira Matsynawa.Caviation erosion mechanism of NiTi coatings made by laser plasma hybrid spraying[J].WEAR, 1999 (231):272-278.

[6]   Hitoshi Hiraga, Takashi Inoue, Akira Matsynawa, Hirofumi Shimura. Effect of laser irradiation condition on bonding strength in laser plasma hybrid spraying [J]. Surface & Coatings Technology, 2001 (138):284-290.

[7]   Hitoshi Hiraga, Takashi Inoue, Shigeharu Kamado, Yo Kojimab, Akira Matsunawac, Hirofumi Shimura. Fabrication of NiTi intermetallic compound coating made by laser hybrid spraying of mechanically alloyed powders[J].Surface & Coatings Technology, 2001(139):93-100.

[8]   He DingY, Fu BinY, Jiang JianM. Fiction and Wear Papers [J].2007, 27(2):116-120.

A spectroscopic and theoretical photo-thermal approach for analysing the laser ablated structures

Mihai Stafe

Department of Physics, University 'Politehnica' of Bucharest, Splaiul Independentei 313, 060042 Bucharest, Romania.

Abstract. We investigated experimentally and theoretically the micro-ablation process of metallic (Al) targets by using nanosecond laser-pulses at 532 nm wavelength in atmospheric air. We analyzed experimentally the dependence of the crater depth, ablation rate, and the heat affected zone on the laser fluence and pulse number. The fluence was varied between 3 and 3000 J/cm^2 by changing the laser-energy, whereas the pulse number was varied with an increment that ensures high accuracy of the optical microscopy measurements on the ablated structures and also an approximate constant ablation rate during multi-pulse irradiation. The microscopy data indicate that the crater depth (and the ablation rate) increases approximately linearly with the 1/3 power of the fluence. For a given fluence, the crater depth increases strongly, approximately linearly during the first 30 pulses and much slowly during the next pulses. The variation of the crater depth and ablation rate with pulse number was further addressed by spectrometric analysis of the ablation-plasmas produced at high fluences on the metallic target. We found direct connection between the emission lines intensities and the crater depth, whereas the plasma electron-temperature varies similarly to the ablation rate when increasing the pulse number. The ablation threshold fluence and the ablation rate in the low fluence regime (i.e. near the ablation threshold) were also addressed theoretically within the frame of a photo-thermal model which accounts for the material heating, melting and evaporation upon irradiation with the nanosecond laser pulses. The theoretical and experimental results are in good agrrement indicating the validity of the model for low laser fluences.

Keywords: laser ablation, ablation plasma, photo-thermal modelling

1. Introduction

The efficiency of material removal upon irradiation with short and intense laser pulses is described by the ablation rate which gives the maximum thickness of the layer removed during a laser pulse. Understanding and controlling the ablation rate is essential for determining the production efficiency, dimensions, and the quality of the produced structures in laser processing [1-5]. Femto-second lasers are the most suitable lasers for obtaining high quality micrometer and sub-micrometer structures in metals due to the reduced thermal effects induced by these lasers in the irradiated area [4-6]. Nevertheless, due to the high costs for acquisition and maintenance of the femtosecond lasers, there is a continuous effort to develop new techniques for laser processing based on nanosecond lasers [5-7]. These studies propose to determine the optimum laser and ambient conditions that allow one to obtain high quality structures at high processing rates on different materials. Previous experiments on nano-second laser ablation of metals indicated that high quality structures could be obtained at small laser fluences, i.e. slightly above the ablation threshold fluence which is demonstrated to be within the range of 1 to 10 J/cm^2 [1, 2, 5-7]. In this low fluence regime, the ablation rate is very small (tens to hundreds of nm per pulse) leading to small processing speeds of large areas. The ablation rate was demonstrated to increase logarithmically with fluence in a certain range above the ablation threshold but to the detriment of the quality [1, 2].

At high fluences (i.e. above plasma ignition threshold fluence) the critical thermo-dynamical conditions come into play along with the phenomena associated to the ablation plasma, and the laser-plasma interaction could influence very strongly the dimensions and the quality of the structures [10-14]. Spectroscopic analysis of the laser-induced plasmas brings essential information on the ablation process and its associated phenomena such as material heating, melting, evaporation, and ionization [1, 2, 14-17]. Several studies have addressed the spatial-temporal characteristics of the ablation plasmas [20-24]. Since understanding the relationship between the characteristics of the ablation plasmas and the ablated areas could be very useful in laser processing [19-22].

Here, we investigate the dependence of the ablation rate of aluminium on the fluence and pulse number of a nanosecond laser in ambient atmosphere. We demonstrate the influence of the laser fluence and pulse number on the ablation rate, dimensions and quality of the laser produced structures. We also analyse the possibility to connect the real-time spectroscopic data from the ablation plasma to the ablation rate and dimensions of the craters ablated on metallic targets in the high fluence regime.

The experimental study is completed with a theoretical analysis of the main results regarding the ablation process at low fluences within the frame of a photo-thermal model which accounts for the fast conversion of the laser energy absorbed by the free electrons of the material into heat via electron-atoms collisions, which leads further to melting, evaporation and recession of the target surface due to evaporation [1-3,25-27].

2. Experiment

The experimental setup consists of a frequency doubled 'Quantel-Brilliant' Q-switched Nd-YAG laser system, a large dynamic range 'Newport' variable attenuator, 10x beam expander, focusing lens and aluminium plate targets placed in the focal plane of the lens perpendicularly to the laser beam. The experiments were carried out in ambient atmosphere. The laser system works in the TEM00 mode and generates fundamental pulses at 1064 nm wavelength which are frequency doubled (532 nm wavelength) with a second harmonic generator module. The second harmonic pulses are characterized by a duration of 4.5 ns, repetition rate of 10 Hz, and maximum energy of 10 mJ.

The diameter of the laser beam on the target is ~20 μ m, whereas, the fluence was varied between 3 and 3000 J/cm^2 by using the variable attenuator. The pulse number was set as follows. In order to study the influence of the laser fluence on the ablation rate we set the pulse number for drilling a crater to 10 in order to maintain approximate constant ablation rate during multiple pulses incidence and, on the other hand, to obtain deep enough craters to allow for small relative errors for the microscopy measurements. In order to study the influence of pulse number on the ablation rate we set the fluence at certain values and drilled craters with different pulse numbers (i.e. 4, 8, 12, etc). The ablated craters were analyzed with a metallographic microscope with axial resolution of 1 micron so that we estimated a maximum relative error of the measurements of ~10%. The ablation rate was calculated subsequently by dividing the crater depth by the pulse number.

The ablation plasma plume was imaged with an imaging lens on the tip of an optical fiber (10 μ m diameter) connected to a high resolution Ocean Optics spectrometer (~0.6 nm resolution). We analyzed the intensities of the 309.2 and 396.1 nm lines of neutral Al atoms [28] as a function of pulse number while maintaining constant fluence.

3. Results and discussion

The microscopy measurements on the depth of the craters drilled with 10 pulses at different laser fluences are presented in Fig. 1(a). The graphs indicate that for small fluences (e.g. 10 J/cm^2), the ablation rates is ~1 micron/pulse. In this fluence regime the craters are practically

free of rims. At high laser fluences (e.g. 1000 J/cm^2) the measurements indicate removal rates as high as 8 microns/pulse. In this fluence regime the craters are surrounded by rims and large re-deposited material. Figure 1 (a) also indicates that the ablation rate and, consequently, the depth of the craters, increase linearly with the 1/3 power of fluence. By extrapolating toward a zero removal rate the 1/3 power fitting curves we can estimate a minimum threshold fluence of ~1.2 J/cm^2 required for material ablation.

These experimental results can be interpreted in terms of photo-thermal material removal under the laser radiation. Considering the photo-thermal activation of the material removal, the excitation laser energy propagates mainly within the target material due to the poor thermal contact between the target surface and the surrounding air. Since the thermal penetration depth is at least one order of magnitude smaller than the laser spot diameter on the target surface, the temperature distribution in the axial z direction can be determined from the one-dimensional heat equation. In a coordinate system that is fixed to the irradiated surface, the heat equation describing the propagation of the heat flow and of the melting front into the target can be written as [18, 25]:

$$\rho\left[c_p + \Delta H_m \delta(T - T_m)\right]\left(\frac{\partial T}{\partial t} - v_a \frac{\partial T}{\partial z}\right) - k\frac{\partial^2 T}{\partial z^2} = Q(z,t) \quad (1)$$

where ρ, c_p, $\Delta H_m(T_m)$ and k denote the mass density, the specific heat, the melting enthalpy at normal melting point, and the thermal conductivity of the material, respectively, taken as constant, temperature independent parameters. v_a is the ablation velocity, the instantaneous position of the ablated surface being labeled with $z=0$. δ function was approximated as a Gaussian function of temperature around the melting point with standard deviation of 10-100 K depending on the temperature gradient [18].

$$Q(z,t) = (1 - R)\alpha I(t)\exp(-\alpha z) \quad (2)$$

The source term [1, 18], accounts for the reflectivity R of the target surface and for the laser energy absorbed into the sample per unit volume and time that is converted into heat (via $\alpha I(z,t)$), α being the optical absorption coefficient of the target. $I(t)$ describes the Gaussian temporal profile of a laser pulse. Evaporation at the melted surface determines the recession of the irradiated surface. The velocity of recession due to evaporation (i.e. the ablation velocity) is given by the Hertz-Knudsen equation [1, 18],

$$v_a = 0.32\left(\frac{M}{k_B T_s}\right)^{1/2}\frac{p_0}{\rho}\exp\left[\frac{M \Delta H_v(T_b)}{k_B}\left(\frac{1}{T_b} - \frac{1}{T_s}\right)\right] \quad (3)$$

where $\Delta H_v(T_b)$ is the vaporization enthalpy at the normal boiling point T_b, p_0=1atm, k_B is the Boltzmann constant, and M is the atomic mass. The correction coefficients 0.32 in (3) accounts for the over-saturation and condensation of the vapour [1]. The ablation rate is obtained by time integration of the ablation velocity over the time interval corresponding to the succeeding period of the laser pulses. Hence, the ablation rate may be determined by solving the heat equation and computing the time evolution of the surface temperature T_s. The boundary and initial conditions of the heat equation are as follows. The initial condition at the beginning of a laser pulse is $T(t=0) =T_{amb}$, T_{amb} being the ambient temperature. The constant temperature boundary condition at rear surface $(z=h)$ is determined by the small thermal penetration depth relative to the target plate thickness h, whereas the boundary condition at the irradiated surface $(z=0)$ gives the energy balance,

$$T\big|_{z=h} = T_{amb}, \ k\,\partial T/\partial z\big|_{z=0} = \rho v_a \Delta H_v(T_b) \qquad (4)$$

where the right term of the second equation accounts for the surface vaporization of the sample. We neglected here the loss of energy by radiation.

The heat equation is integrated by using the finite differences method in MATLAB, the values of the thermal and optical parameters of the metallic target being given in [1]. The numerical results regarding the dependence of the ablation rate on the fluence up to 20 J/cm^2 are presented in Fig. 1(b) along with the experimental results. The reasonable agreement between the theoretical and experimental results demonstrates that the most important phenomena are accounted for in the model proposed here for low laser fluences (i.e. up to 20 J/cm^2) which does not induce strong ionization of the ablation plume. Moreover, the model succeeds in estimating the ablation threshold fluence, the theoretical value of 1 J/cm^2 being close to the experimentally estimated threshold fluence of 1.2 J/cm^2.

The high fluence regime is addressed experimentally by spectroscopic analysis of the emission of the ablation plasma in connection to the crater depth variation when increasing the pulse number. We monitored the intensity of the 309.2 and 396.1 nm lines as a function of pulse number and observed a direct connection between the lines intensity and the crater depth. Figure 2 (a) presents the variation of the intensity of 396.1 nm line with pulse number for a fluence of 500 J/cm^2. The line intensity decays strongly during the first 30 pulses and becomes ~3 times smaller after the 30th pulse as compared to line intensity corresponding to the 4th pulse. This is consistent with the variation of the crater depth [inset plot in Fig. 2 (a)]: the crater depth increases strongly with pulse number, and becomes ~3 times larger after 30 pulses as compared to the crater depth after 4 pulses. Further increase of pulse number leads to a much slower decay of the line intensity, which is consistent with the much

slower increase of the crater depth with pulse number. This makes the spectroscopic analysis suitable for monitoring in 'real-time' the multi-pulse ablation process.

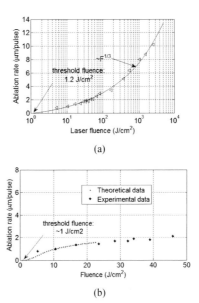

Fig. 1. (a) Ablation rate vs laser fluence. (b) Theoretical (red dots) and experimental (black *) ablation rate vs laser fluence

From the spectroscopic data we also found information on the plasma temperature which we demonstrate to be connected to the ablation rate as follows. Considering that conditions for local thermal equilibrium are fulfilled and the distribution of ablation species has a Boltzmann form, the ratio of the intensities of the 396.1 and 309.2 nm lines emitted by the neutral Al species enables us to calculate the electron temperature from the equation [14,17]:

$$\frac{I_1}{I_2} = \frac{f_1 g_1 \lambda_2^{\,3}}{f_2 g_2 \lambda_1^{\,3}} \exp\left(-\frac{E_1 - E_2}{k_B T}\right) \qquad (7)$$

where E_1 and E_2 are the energies of the upper states of the two transitions accounted here, and I, λ, g and f denote the intensity, wavelength, statistical weight and oscillator strength of the two transitions, respectively. The spectroscopic constants corresponding to the two lines can be found in [28]. The dependence of the electron temperature on the pulse number is presented in Fig. 2 (b) for a 500 J/cm^2 fluence. The temperature is practically constant when increasing the pulse number to approximately 30; being ~6000 K while further increase of the pulse number leads to approximately 10% decrease of the temperature, to ~5400 K. This is totally analogous to the behaviour of the ablation rate which is, ~3.5 microns/pulse for the first ~30 pulses, whereas it decays to ~0.5 microns/pulse for the next pulses.

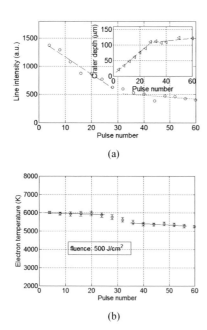

(a)

(b)

Fig. 2. (a) 396.1 nm line intensity (and crater depth in the inset plot) vs pulse number. (b) Electron plasma temperature vs pulse number.

4. Conclusion

We investigated the dependence of the dimensions of the laser ablated craters in aluminium on the fluence and pulse number of a nanosecond laser at 532 nm wavelength in atmospheric air. The laser fluence at the target surface was varied between 3 and 3000 J/cm^2 by changing the energy with a variable attenuator. The microscopy measurements indicate that the craters depth and the ablation rate increase approximately linearly with the 1/3 power of the laser fluence. The extrapolation curve gave an ablation threshold fluence of ~1.2 J/cm^2. The ablation threshold fluence and the ablation rate in the low fluence regime were also analysed theoretically within the frame of a photo-thermal model which accounts for the material heating, melting and evaporation upon irradiation with the nanosecond laser pulses. The theoretical and experimental results on the dependence of the ablation rate on laser fluence are in good agreement demonstrating the validity of the model. The ablated craters at high fluences were further addressed by spectrometric analysis of the ablation-plasmas produced during irradiation of the metallic targets. This analysis provides real-time information on the pulsed laser ablation of the targets since we found a direct connection between the lines intensities and the depth of the ablated craters, the plasma electron temperature varying similarly to the ablation rate when increasing the pulse number. Thus, we demonstrate that up to a certain number of pulses (~30th pulses) the intensity of the neutral Al spectral lines decreases strongly with the number of pulses while further increase

of the number of pulses leads to a much slower decrease of the lines intensities. By comparing these spectroscopic data to the optical microscopy measurements of the ablated craters we connected the behaviour of the lines intensity to the crater depth which increases strongly during the first ~30 pulses and much slowly afterwards. Further analyses of the optical microscopy and the spectroscopic data indicated that there is a direct connection between the ablation rate and the electron plasma temperature which remain constant until the pulse number reaches ~30th while further increase of the pulse number leads to smaller electron temperatures.

Acknowledgment: This paper is supported by the Sectoral Operational Programme Human Resources Development, financed from the European Social Fund and by the Romanian Government under the contract number POSDRU/89/1.5/S/64109.

References

[1]   D. Bauerle, Laser processing and chemistry, Springer- Verlag, Berlin- Heidelberg- New York, 2000
[2]   M.von Allmen, A. Blatter, Laser-Beam Interactions with Materials, Springer-Verlag, Berlin, 1995
[3]   A.E. Wynne, B.C. Stuart, Appl. Phys. A 76, 373–378 (2003)
[4]   A. Semerok, Appl. Surf. Sci. 138–139, 311–314 (1999)
[5]   B.N. Chichkov et al, Appl. Phys. A 63(2), 109–115 (1996)
[6]   P. Simon, J. Ihlemann, Appl. Phys. A 63, 505 -508 (1996)
[7]   M.R.H. Knowles et al, The International Journal of Advanced Manufacturing Technology 33, 95-102 (2007)
[8]   M. Henry, J. Wendland1, P. M. Harrison, D. Hand, Proceedings icaleo 2007, Paper M505
[9]   A. Stephen, T. Lilienkamp, S. Metev, and G. Sepold, RIKEN Review No. 43: Focused on 2nd International Symposium on Laser Precision Microfabrication (LPM 2001)
[10]   N.M. Bulgakova, A.V. Bulgakov, Appl. Phys. A 73, 199–208 (2001)
[11]   B. Garrison et al, Phys. Rev. E 68, art. no. 041501 (2003)
[12]   C. Porneala, D. A. Willis, Appl. Phys. Lett. 89, 211121 (2006)
[13]   L. Quanming, Phys. Rev. E 67, 016410 (2003)
[14]   S. Amoruso et al, J. Phys. B 32, 131-172 (1999)
[15]   J. Bovatsek et al, Thin Solid Films 518, 2897-2904 (2010)
[16]   H. Griem (1964) Plasma Spectroscopy, New York: McGraw-Hill
[17]   S. Dadras et al, J. Phys. D: Appl. Phys. 41, 225202 (2008)
[18]   N.M. Bulgakova et al, Appl. Phys. A 79, 1323–1326 (2004)
[19]   J.A. Aguilera, C. Aragon, Appl. Surf. Sci. 197–198, 273 (2002)
[20]   K.J. Saji et al, (2006) J. Appl. Phys. 100, 043302
[21]   S. Amoruso et al, Appl Phys A 92, 907–911 (2008)
[22]   Y. Yamagata et al, J. Appl. Phys. 88, 6861–7 (2000)
[23]   J.M. Vadillo, J.J. Laserna, Spectrochim. Acta B 59, 147 (2004)
[24]   L. St-Onge, M. Sabsabi, Spectrochim. Acta B 55, 299 (2000)
[25]   N.M. Bulgakova et al, Applied Surface Science 257, 10876–10882 (2011)
[26]   D. Marla et al, J. Appl. Phys. **109**, 021101 (2011)
[27]   M. Stafe et al, Appl. Surf. Sci. **253**, 6353–6358 (2007)
[28]   http://physics.nist.gov/PhysRefData/ASD/lines_form.html.

Process and properties research on 17-4PH stainless steel by laser synchronous hybrid hardening

Jianhua Yao[1,2], Junhao Sai[1,2], Qunli Zhang[1,2] and Fanzhi Kong[1,2]

[1] Key Laboratory of E&M (Zhejiang University of Technology), Ministry of Education & Zhejiang Province, Hangzhou, 310014, P. R. China

[2] Research Center of Laser Processing Technology and Engineering, Zhejiang University of Technology, Hangzhou, 310014, P. R. China

Abstract. Laser synchronous hybrid hardening – laser alloying and solid solution hybrid treatment – produces a compound strengthened layer on 17-4PH stainless steel with alloyed zone on the top layer and solid solution zone on the subsurface. A best laser hybrid hardening method for production has been found by comparing the performances of hardened layer with different sequences of laser treatments. The depth of the whole hardened layer is more than 1.2mm with average microhardness at 445 $HV_{0.2}$. The highest microhardness of top of the hardened layer can reach to 683.2 $HV_{0.2}$, which is 2 times of that of substrate. To evaluate the surface performances of the compound strengthened layer, friction and wear test and cavitation test had been done on the surface of the compound strengthened layer. The results show that, the wear mass loss and cavitation mass loss of strengthened layer is respectively only 17% and 46% of that of substrate.

Keywords: 17-4PH stainless steel; Laser synchronous hybrid hardening; Wear resistance; Cavitation property

1. Introduction

Coal-fired power is the main way of generating in China, despite the proportion of the total domestic power generation of clean power generation increasing year by year. It still occupies more than 70% of China's total installed capacity. Supercritical and ultra-supercritical thermal power units with thermal efficiency up to 45% have the advantages such as low coal consumption, environment friendly, high technical contents and incomparable economy. They are widely applied in China and great economic efficiency is received [1]. Turbine blades work in extremely poor conditions and are impacted by the steam and droplets which contain salts and oxygen. Solid particle erosion and fretting wear are the main failure types of the high pressure Curtis stage blades [2~3].

Silver brazing stellite pieces, integrated solid solution treatment and special surface treatments are the commonly used ways to strengthen turbine blades made of 17-4PH stainless steel[4~5]. Neither integrated solid

solution treatment nor silver brazing Stellite pieces is an ideal way to be spread and applied in industry because the deformation, expensive cost and easily exfoliating problems.

Laser green manufacturing technology with high efficiency, small deformation, no environmental pollution and low dilution rate can regenerate the failure parts due to corrosion or erosion or size out of tolerance[6]. This technique has been tried in some parts of repairing turbine blades[7]. Aiming at the characteristics of 17-4PH stainless steel, Laser synchronous hybrid hardening is proposed to produce a hardened layer with a certain depth and required performances on the surface of the material. And varieties of researches have been done on the structures and performances of the strengthened layer.

2. Experimental procedure

The steam turbine blade steel 17-4PH was used as the substrate, whose chemical composition is given in Table 1. The size of specimen is 100mm×50mm×5mm and the microhardness is 330~350 $HV_{0.2}$, delivery state. A self-made special alloying powder mainly containing Co, Cr, Ni, W, Al is used as the alloying material.

Table 1 Chemical composition of 17-4PH (wt.%)

C	Cr	Ni	Cu	Nb	Mn	Fe
0.04	16.5	4.5	3.3	0.3	0.7	Bal.

Experiments were carried out with a 7kW transverse CO_2 laser. Rectangular integral focus lens were used as the focus system. The experiment was designed to use two laser beams scanning the surface of specimen at a same speed, to imitate the synchronous hybrid strengthening treatment. Then the whole sample was under an aging treatment at a preset temperature. Scheme of experiment devices is shown in Fig.1 and the parameters are given in Table 2.

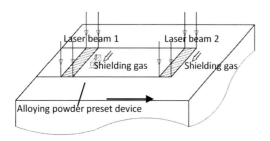

Fig.1 Scheme of experiment devices

Table 2 Test parameters of laser hybrid strengthening

No.	Velocity (mm/min)	Treatment, power density(W/mm²)	Aging
1	100	SS¹, 10.42 / A, 100	
2	200	SS, 10.42 / A, 125	450°C, 3h
3	300	SS, 16.67 / A, 150	
4	300	SS, 108.33 / A, 150	
5	300	A, 150 / SS, 108.33	

Note: 1. SS – laser solid solution treatment, A – laser alloying treatment

3. Results and discussion

3.1. Microstructure of laser hybrid strengthened layer

The laser hardened layer from surface to inner can be divided into laser alloyed zone, transition zone, heat affected zone and laser solid solution zone, shown in Fig.2. As the material is strengthened in a nonequilibrium state, the microstructure of hardened layer is quite different with that of equilibrium state. The microstructure of laser alloyed zone is mainly oriented growth dendrites with coarser types in bottom. There are coarse grains in the transition zone that are similar to retained austenite. The microstructure of heat affected zone is acicular martensite transiting to cryptocrystalline martensite, which is shown in Fig.3 (a). Low carbon lath martensites distribute in laser solid solution zone with finer structure and arrangement disorder. Besides, fine dendrimerlike precipitated phases appear in interdendritic region of the dendrites of alloyed zone. That is shown in Fig.3(b). The result of EDS shows that there are W, Co and other hard phase forming elements in it, given in Fig.3 (c). It plays a positive role for improving the hardness of alloyed layer.

There is no obvious difference between processes 4 and 5. The microstructure of upper of the alloyed layer by process 5 turns into free crystals and equiaxial crystals. It

can be referred that the later laser solid solution treatment stimulates the growth and development of the surface crystals and makes the component uniform.

Fig.2 Morphology of laser hybrid hardened layer

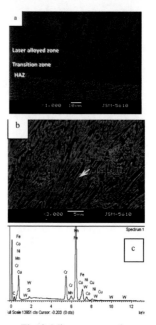

Fig. 3. Microstrure results

3.2. Microhardness analysis of laser hybrid hardened layer

Hardness may characterize the comprehensive properties like surface strength and elastic and plastic of material. The microhardness of solid solution zone after aging has greatly increased with the separation of precipitation sclerosis phases. The hardness profiles in different processes are given in Fig.4 and Fig.5.

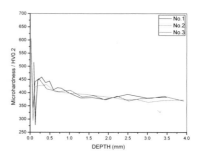

Fig.4. Microhardness curves of hardened layers by processes 1, 2

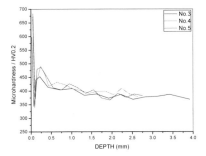

Fig.5. Microhardness curves of hardened layers by processes 3, 4, 5

From the hardness curves it can be seen, the microhardness of laser hybrid hardened layer drops from surface to substrate in gradient. It appears the lowest microhardness, left of which is laser alloyed zone and the right side solid solution zone. The depth of the hardened layer is more than 1.2mm. Thereinto, the depth of alloyed zone is about 80μm and the highest microhardness is 683.2HV$_{0.2}$. According to Fig.4, the hardness of surface layer has increased and the weakness of transition zone has been improved when laser beam running faster. As is shown in Fig.5, the following solid solution treatment didn't make the microstructure of alloyed zone change. The hardness performances of the hardened layers by these two different treatments are almost the same, which improves the feasibility of the processes.

3.3. Wear resistance of laser hybrid hardened layer

The wear test provides the basis for assessing the surface performance of hardened layer. Si$_3$N$_4$ ceramics ball is used as the grinding material for this test. The test condition is was: test load 200g, rotate velocity 800r/min, gyration radius 2.5mm, room temperature, atmospheric environment, and test for 120min. The results are listed in Fig.6 and Table 3.

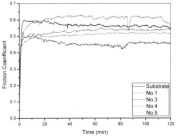

Fig.6. Friction and wear curves

The abrasion loss of hardened layers is less than that of substrate. Thereinto, friction coefficient of hardened layer made by processes 3,4 and 5 is smaller than that of substrate. It can be refer that the surface of material remelted under the treatment of laser beam, and new microstructures appeared in this area. Especially at a faster speed, the wear resistance of hardened layer had been greatly improved - abrasion loss and friction

coefficient declined obviously. Compared with the friction coefficient of hardened layer made by process 4, the friction coefficient of that by process 5 is lower. It is because the following solid solution treatment makes the surface structure denser and the surface smoother.

Table 3 Friction coefficient and abrasion loss

	Average friction coefficient	Weight before test (g)	Weight after test (g)	Weight-loss (mg)
Substrate	0.565	3.1748	3.1736	1.2
No.1	0.601	2.3279	2.3274	0.5
No.3	0.456	3.6073	3.6071	0.2
No.4	0.528	3.6519	3.6517	0.2
No.5	0.505	3.5573	3.5571	0.2

3.4. Cavitation performance of laser hybrid hardened layer

Cavitation test was operated by an ultrasonic cell disruptor so as to imitate the work condition of turbine blades. Cavitation tests were separately done on the substrate and laser hybrid strengthened specimen by process 3. The results are given in Fig.7 and Fig.8.

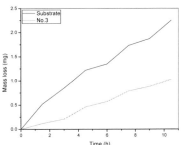

Fig.7 Cavitation mass loss-time curve

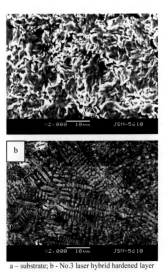

a – substrate; b - No.3 laser hybrid hardened layer

Fig.8 Cavitation morphologies

The results show that, the mass loss of laser hybrid hardened specimen is only 46% of that of substrate. Fig.8 shows the surface morphologies after cavitation tests. A large number of uneven lump cavitation pits form on the surface of substrate and spalling is much more serious, shown in Fig.8(a). There is no large area of spalling on the surface of laser hybrid hardened specimen after cavitation test. Some tiny and shallow pits appear in interdendritic region, that Fig.8(b) shows. The fine structures in this area relieve the stress concentration at the joint of grain and boundary and also improve the cavitation resistance of the surface. The alloying elements have also enhanced the cavitation resistance performance of hardened surface.

4. Conclusions

Laser synchronous hybrid hardening has produced a composite strengthened layer on 17-4PH stainless steel. The average thickness of hardened layer (microhardness $\geq 400\ HV_{0.2}$) is more than 1.2mm and the hardness of that drops in gradient along the cross section. The surface layer has a compact structure with high hardness. The average hardness of strengthened layer reaches $445HV_{0.2}$. The highest microhardness of surface layer is $683.2HV_{0.2}$, which is nearly 2 times of that of the substrate. The surface wear resistance and cavitation resistance of laser hardened layer has been greatly improved. It has the lower friction coefficient than the substrate.The faster speed of laser processing plays an active role of the improvement of surface performance and the weakness of transition zone. The following laser solid solution treatment hasn't aroused the transformation of alloyed structure. Additionally, the advantages of the alloyed layer such as high hardness and wear-resistance have been preserved.

Acknowledgements: The authors would like to appreciate financial support from National Science Foundation of China (50971117) and Zhejiang Provincial Commonweal Technology Application Research International Cooperation Project (2011C24006).

References

[1]   TANG Fei, DONG Bin, ZHAO Min. USC Unit Development and Application in China[J]. Electric Power Construction, 2010, 31(1): 80-82.

[2]   JIANG Pu-ning. Technology of Stress Corrosion Cracking Resistance in Nuclear Steam Turbines. Thermal Turbine, 2010, 39(2): 89-92.

[3]   LUO Da-yong. Analysis on Failure blade of Steam Turbine. Equipment Manufacturing Technology, 2010, 10: 161-162.

[4]   Kochmańsk I P, Nowacki J. Activated gas nitriding of 17-4PH stainless steel[J]. Surface and Coatings Technology, 2006, 200: 6558-6562.

[5]   A. Gholipour, M. Shamanian, F. Ashrafizadeh. Microstructure and wear behavior of stellite 6 cladding on 17-4 PH stainless steel[J]. Journal of Alloys and Compounds, 2011, 509(14): 4905-4909.

[6]   Ricciardi G, et al. Surface treatment of automobile parts by RTM[J]. Laser Beam Surface Treating and Coating, 1988, 957: 66-74.

[7]   Y.P. Kathuria. Some aspects of laser surface cladding in the turbine industry[J]. Surface and Coatings Technology, 2000, 132: 262-269.

Fiber laser surface texturing of TiN coatings and their resistance to scratch test

A.G. Demir[1], N. Lecis[1], B. Previtali[1] and D. Ugues[2]

[1] Politecnico di Milano, Dipartimento di Meccanica, Via La Masa 1, 20156 Milan, Italy
[2] Politecnico di Torino, DISMIC, C.so Duca degli Abruzzi 24, 10129 Torino, Italy

Abstract. Laser surface texturing (LST) is an established technology that involves generation of micro features, usually in the form of dimples (or craters), on a relatively large surface, most commonly to increase the tribological performance. Recently LST has been applied on thin hard coatings to further improve the wear resistance. In the realized patterns on thin coatings, one important aspect is that the texturing should not damage or alter the coating layer. Under certain conditions, the LST pattern can cause local detachments of the coating or induce cracks. These aspects should also be considered when the LST pattern to be applied has to be chosen. Fortunately they can be evaluated in relatively harsher conditions such as non lubricated contact. The present work reports the study of LST of TiN coatings. The tribological performance of the patterns is evaluated by scratch tests under non lubricated conditions in order to measure the coating adhesion and toughness. LST is performed on TiN coating of about 3 μm thickness making use of a pulsed active fiber laser working in *ns* pulse regime. Laser process parameters are varied to obtain several patterns with different dimple diameter, depth and pitch, which are essential to control the tribological behaviour. Scratch tests are applied to the obtained patterns to evaluate the friction coefficient, critical load for crack generation and adhesion. Finally the most suitable LST pattern is identified and proposed to be used also in application under lubricated conditions.

Keywords: Laser surface texturing, TiN, fiber laser, scratch test

1. Introduction

Laser surface texturing (LST), among other processes, stands out as a more flexible and precise option in order to change and optimize surface properties of a material. LST has found applications in many fields ranging from biomedical to optical; whereas the vast majority of the research covers the study of the tribological improvements it provides. It has been shown that micro dimples generated on the surface of the material can decrease the friction coefficient, potentially decreasing the energy use and increasing the useful life of a component [1]. The dimples allow debris entrapment, act as lubricant reservoirs thus decrease the possibility of third body wear, and generate static force to contribute to load support. Recent applications showed potential of LST to further improve the tribological performance of hard surface coatings [2]. Such studies have demonstrated LST could be applied to the substrate prior to coating, or inversely after the coating.

LST is applied with pulsed laser sources operating in different laser wavelengths and pulse regimes. Both determine the interaction between the beam and material. Potentially shorter wavelengths and shorter pulse durations are preferable in view of high quality machining. However, in the case of LST, productivity and system robustness play a key role, as the process is required to be applied to relatively large surfaces compared to the micro dimensions of the dimples. On the other hand, LST of hard coatings, require a better control of the process as these coatings are characterized by a thickness of only a few μm in most cases [3]. Excessive laser ablation can result in reaching the substrate, coating loss, and a local mixture of coating and substrate. Overall, LST can damage the coating rather than increasing its performance, which can be primarily observed as detachment under working conditions.

The present work presents the study of LST of hard TiN coatings with a pulsed fiber laser system operating in *ns* pulse regime, which is a suitable solution to high productivity industrial applications. Different process parameters are investigated in order to relate them to geometrical attributes of dimples. Obtained patterns are subjected to scratch testing to identify the best condition in terms of coating adhesion and cohesion in non lubricated conditions. This harsh evaluation allows the most successful conditions to be rapidly determined. Moreover useful insights to the subsequent design of the tribological properties of the laser textured surface in real application can be inferred.

2. Experimental

TiN cathodic arc PVD coatings were deposited on 42CrMo4 substrate. The thickness was 2.65 ± 0.13 μm. LST was applied with a 50 W pulsed active fiber laser operating in *ns* pulse regime (IPG Phonics). Laser source was coupled to a scanning head with a 100 mm focal lens (Sunny TSH8310D), which produced a calculated waist diameter of 39 μm at focal plane. General characteristics of the laser are reported in Table 1.

Table 1. General characteristics of the laser system

Maximum average power	50 W
Wavelength	1064 nm
Pulse repetition rate (PRR)	20-80 kHz
Pump current %	10-100%
Pulse energy range	0.5-1.2 mJ
Minimum pulse duration	100 ns
M^2	1.7
Collimated beam diameter	5.9 mm
Focal length	100 mm
Beam waist diameter	39μm

Actually LST operation can be carried out by on the fly machining with the use of a single laser pulse per dimple. This requires an accurate synchronization of pulse repetition rate with scanning velocity to impose the desired distance between dimples (pitch), while the ablation depth is limited to the available pulse energy. On the other hand, LST can be performed in a similar way to percussion drilling, with a train of pulses being shot to a certain position to generate a single dimple, and then move on to the next position to do the same. This method is relatively less productive, yet enables better control on the process, especially in terms of dimple depth, which is highly required for LST of thin coatings. Therefore, this point-to-point strategy was employed for the texturing of TiN coatings. The laser system used in this work is controlled by a modulation signal with a chosen duration, rather than imposing a determined number of pulses. Therefore the effective control parameter becomes the modulation duration (t_{mod}), which is the duration of the pumping diodes being injected with current. Preliminary experiments showed that in order to carry out a gentle, superficial ablation, the required modulation duration is supposed to be less than the laser emission rise time. This in fact, results in obtaining an emission profile ramping up and down without reaching a regime (see Fig.1)

Designed experiments were used to study the effect of different LST conditions on the dimple geometry and the tribological performance of the textured surface. Pumping current and pulse repetition rates were fixed, whereas modulation time (t_{mod} *[μs]*) was varied to obtain different energy conditions. Focal position (*f [mm]*) was also varied to obtain different spot diameters. These two parameters allow obtaining different dimple diameter and depth conditions. Distance between dimples, namely pitch (*p [μm]*), was the third varied parameter to obtain

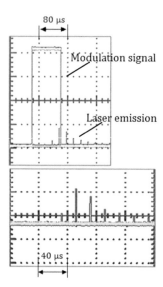

Fig. 1. Emission characteristics of the laser with 80 μs modulation duration.

different texturing densities. With these parameters a 2^3 factorial plan with central point was executed, resulting in 9 different LST patterns. The fixed and varied parameters are listed in Table 22.

Table 2. Fixed and varied parameters in the 2^3 factorial plan

Fixed parameters			
Pulse repetition rate (PRR)	50 kHz		
Pump current %	100%		

Varied parameters	**High**	**Mid**	**Low**
Modulation duration, t_{mod} (μs)	65	72.5	80
Focal position, f (mm)	0	0.75	1.5
Pitch, p (μm)	100	150	200

The results were analyzed by SEM to check the basic geometrical attribute, back scattering mode was used for revealing if the substrate was reached during ablation by identifying change of material density. Dimple radii (r *[μm]*) and the area of substrate coating mix (A_{mix} *[μm²]*) were measured from these images (see an example in Fig. 2). Pitch, as an ineffective parameter in laser machining, was excluded from geometrical analysis. In other words, different pitch conditions with the same laser parameter combination correspond to same dimples with different distance. Eventually, five replicates were taken for each combination.

Scratch tests were performed using a scratch tester (Microcombi Platform - CSM Instruments) with an increasing load from 0.3 to 30 N. Scratches of 3 mm length were made using a diamond Rockwell indenter with a spherical tip radius of 200 μm sliding at a constant speed of 1.26 mm/min. The tester was also equipped with acoustic sensors to detect the nucleation of the first cracks corresponding to the coating critical load. Both the critical loads for crack nucleation around the dimples

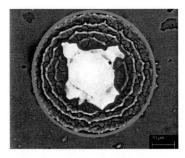

Fig. 1. Representation of the measures taken with back scattered SEM images. The red circle represents the circle fit to measure the dimple radius; and the highlighted area in blue is where substrate and coating mix occurs.

(LC_I) and for the detachment of TiN coating (LC_{II}) were measured for all conditions tested. During all tests the friction coefficient was also measured.

3. Results and discussion

3.1. Geometrical aspects

The SEM images of the laser drilled dimples revealed that within the given range of parameters it is possible to control the process to achieve texturing without damaging the coating (see Fig. 3). As a matter of fact, working in defocused conditions, which is intuitively contradictory for laser surface processing, is found to be favourable in this case. Having the focal plane 1.5 mm above, the process is also controllable to vary the dimple radius, without the coating damage, by increasing the t_{mod}, thus the energy content. On the other hand, working in focal plane results in deep drilling conditions due to the high energy intensity. Moreover, the diameter change is less evident in different energetic conditions (t_{mod}).

Measured values of A_{mix} evidently show that there is a drastic increase of the mixed alloy amount, as the energy content increases working in the focal plane on the surface (see Fig. 4). For $f=0$ mm, mixed alloy content doubles at the more energetic condition, from 814.4 μm^2 ± 110.7 to 1670.7 μm^2 ± 134.6. Moving the focal plane away from the surface, the area decreases and finally reaches zero, meaning at $f=1.5$ mm no contamination to the TiN coating occurs. On the other hand, a more peculiar behaviour in terms of the measured radii is observed (see Fig. 4). A strong difference is present in different focal position conditions. At fixed f conditions, the increase in dimple radius with the increase of laser energy, is more steep for highly defocused condition $f=1.5$ mm, compared to focal plane on surface. This can be simply associated to the fact that a larger beam diameter is projected to the surface in this condition. With lower energetic condition, a good portion of the projected beam area lays under the ablation threshold, thus a small dimple is obtained, whereas increasing the energy content increases the area of the beam that is

above the ablation threshold. The two parameters that determine the energy content (t_{mod}) and the projected beam diameter (f) are strongly interacting, as an inverse behaviour is present for different energetic conditions, when focal position is varied.

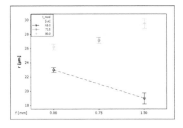

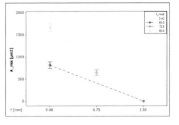

Fig. 4. Interval plots of A_{mix} [μm^2] and r [μm] (error bars represent 95% confidence interval for the mean)

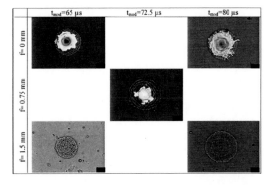

Fig. 3. Back scattering mode SEM images of the dimples obtained with different process conditions

3.2. Tribological performance

Under non lubricated conditions, no significant change was observed in terms of friction coefficient for the different texture conditions, as it was measured to be 0.22±0.02 with an applied load of 30N. The same type of observation was valid for the critical load (LC_{II}) to nucleate adhesion cracks on TiN, which did not vary significantly for the different texturing conditions. The measured LC_{II} was 22.3 N ±1.7. However, patterns exhibited different behaviour in terms of crack nucleation around the dimples (LC_I). Severe detachments were observed in dimples with a large amount of mixed alloys, which is probably due to the embrittlement of the material; and due to the form of the spatter in these dimples, which increases stress concentration. Worst performance was observed for dimples made with highest energetic conditions, thus t_{mod} at the higher, and f at the lower value. In this condition cracks occurred at $LC_I=12.8$ N $\pm$ 3.9 (see Fig. 5).

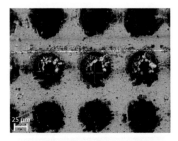

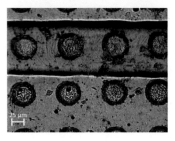

Fig. 5. Optical images taken during the scratch test, belonging to the worst and best LST conditions (applied load is 10N at the frame instance).

Best performance was observed for the exact opposite end of the same parameters, thus for the least energetic conditions (t_{mod} at the lower, and f at the higher value), where the critical load LC_I was never reached and no cracks were observed up to LC_{II} [see Figure 5 (b)]. The dimples obtained with t_{mod} at the higher, and f at the higher value exhibited an intermediate behaviour, as detachments were observed in 50% of the executed runs. It is important to note that no significance was found for the parameter that determines distance between the dimples, pitch (p). This points to the fact that, even if there is partial detachment of the dimples occurring, no connection of cracks will be present. Therefore, it can be presumed that under lubricated conditions, it is more likely to find partial dimple detachment, which can contribute to third body wear in the form of debris, whereas coating failure due to crack propagation is not expected. Moreover, in order to increase the proportion of the textured surface compared to the all over surface area it can be recommended to use smaller, shallow dimples with smaller pitch.

4. Conclusions

LST of TiN coatings was studied in terms of process parameters, which were then associated to the dimple geometry and tribological performance to maintain coating integrity under scratch test. It was found out that

shallow dimples without substrate contamination can be potentially applied in smaller distances to allow better tribological performance in real application conditions, which involve the use of lubricant.

References

[1]   Etsion I, (2005) State of the Art in Laser Surface Texturing. Journal of Tribology 127:248-253
[2]   Voevodin AA, Zabinski JS, (2006) Laser surface texturing for adaptive solid lubrication. Wear 261:1285-1292
[3]   Vandoni L, Demir AG, Previtali B, Lecis N, Ugues D, (2012) Wear behaviour of fiber laser textured TiN coatings in heavy loaded sliding regime in press.

Parallel direct patterning of ITO thin films using ultra short pulses

P. Fitzsimons, Z. Kuang, W. Perrie, S. Edwardson, G. Dearden and K. G. Watkins
School of Engineering, University of Liverpool, Brownlow Hill, Liverpool, L69 3BX

Abstract. Indium-tin oxide (ITO) is a commonly used material in the manufacture of several photonic devices such as flat panel displays, solar cells and electronic ink. The conventional method of patterning ITO involves etching; these methods require the use of harmful chemicals and suffer from low reproducibility due to surface contamination, under and over etching; as a direct result manufacturing costs are high. Direct laser patterning of ITO films offers a viable alternative to these methods. An ultra short pulse laser system with pulse duration of 10 picoseconds (ps) and variable repetition rate (5 to 50 kHz) was applied to ITO removal from a glass substrate. Removal of the film was achieved through determination of the ablation threshold (Φth) of both the ITO and glass. By varying the traverse speed, fluence and offset a processing window was identified. In addition, a spatial light modulator (SLM) was introduced into the optical path and used for diffractive multiple beam processing; this increased throughput by increasing the "effective" repetition rate.

Keywords: Laser direct write, Laser patterning, picosecond, Indium Tin Oxide, transparent conductive oxide, parallel processing, multiple beam processing, SLM.

1. Introduction

First manufactured in the 20th [1] century transparent conductive oxides (TCOs) which are electrically conducting and transparent have undergone continuous development resulting in the prodution of several different types including: Al-doped ZnO, SnO_2 and F-doped In_2O_3. TCO products have found widespread use in optoelectronic devices. Since the 1960s Indium Tin Oxide (ITO) has been the most widely utilised TCO in manufacturing as it provides good conductivity, transmission and stability in most environments. These favourable properties have seen it used in the manufacture of flat panel televisions, computer monitors, touch screens, radio frequency identification (RFID) and solar cells.

The traditional approach to TCO patterning consists of dry or wet etching; this process suffers from low throughput, reproducibility and requires the use of hazardous chemicals [2]. Concordantly manufacturing costs are high, whilst the chemicals used in manufacture raise health and environmental concerns. More recently,

long pulse ($\geq$ns) laser systems have been employed as a tool for direct writing (DW) patterns on ITO [3]; this process is maskless, produces high precision structures and requires no further tooling or chemical usage. The direct write process is only restricted by the design capabilities of the software employed and the laser system. However, thermal loading and shockwaves generated with these systems can reduce machining quality and damage substrates [4]. In contrast ultra short pulse lasers induce very little thermal effects and due to their small spot size (microns in diameter) and high intensities allow for precision manufacture of simple and complex shapes. Such desirable properties have led to interest in the application of both femtosecond and picosecond pulses in ITO removal [5]. The drawback of these systems, from a manufacturing perspective, had previously been attributed to high investment cost, low power and repetition rate. These high investment costs required to purchase and maintain can still be present with femtosecond (fs) systems; although high repetition rate systems are available [6]. Ps lasers with multi-Watt powers and MHz repetition rates are commercially available.

The high intensities generated with ps lasers enable removal of most materials including ITO; normally only a fraction of the available power is required to achieve this. When processing with a single beam, consequently the majority of available output power is wasted. One method of utilising the available energy is through the generation of multiple beams using a SLM with a computer generated hologram (CGH). This group has previously presented work on the application of SLM in micro-processing [7-11].

As electronic devices become cheaper and more widely available there is a market requirement upon businesses to be able to produce these devices as quickly and as efficiently as possible.

In this paper, parallel processing of ITO with multiple beams generated using a phase only SLM was demonstrated. Initially the processing window was determined using a single beam output before moving to a multi-beam set-up. Using this optical arrangement the throughput of manufactured ITO devices can be increased, improving efficiency whilst maintaining the flexibility, precision removal and reliability of the laser DW process.

2. Experimental

The High-Q picosecond laser (model: IC-355-800 ps) used here is a class IV laser capable of emitting three output wavelengths of 1064, 532 and 355nm. The system can operate at four different frequencies of 5, 10, 20 and 50 kHz; the maximum pulse energy at 1064nm was 170μJ. The output beam has an M^2 value of ≈1.1 with pulse duration of 10ps (FWHM). After passing through a 3x beam expander the beam was directed to a scanning head (Nutfield XLR8), controlled by Scanner Application.

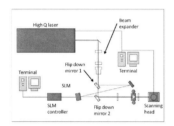

Fig. 1. Experimental arrangement. The blue path shows the single beam processing line. The red path shows the new optical arrangement once the SLM was integrated into the system.

Software (SCAPS), where it was focussed using a 100mm *f*-theta lens to give a spot size of ≈23μm. Samples of ITO coated and uncoated glass (Knight Optical WIW2000-I02) were mounted on to a custom designed mount attached to a 5-axis Aerotech precision motion control system; this system was controlled via the NView MMI software with a repeatability of 0.5μm. The pulse energy was varied between 2.5 – 60μJ for ITO and 20 – 90μJ for glass at a frequency of 10 KHz; this corresponds to a peak fluence range of 1 - 25 Jcm^{-2} and 12 – 57 Jcm^{-2} respectively.

Fig. 2. Circuit used during single and multi beam processing. The light blue areas were processed to leave behind the circuit shown in dark grey.

Initially, single pulses were used to determine the ablation threshold (Φ_{th}) of both ITO and glass through fast scanning of the galvanometers; subsequently the scan speed and hatch spacing were varied to identify an

operating window. Using these parameters a small functional circuit was fabricated. The conductivity of the single beam circuit was tested to ensure no charge movement across boundaries was possible. The circuit design used throughout processing is shown in Fig. 2 for the purposes of this work a simple design was used but more complex patterns could be implemented.

In the second stage of testing a Hamamatsu phase only spatial light modulator (SLM) was used in conjunction with a binary grating to divide the beam and used to fabricate small functional circuits (Fig. 1). The machined areas where examined using a Nikon microscope with CCD camera and a WYKO NT1100 optical surface profiler.

3. Results and discussion

The absorption coefficients of ITO and glass at 1064nm were determined to be 4.9×10^4 cm^{-1} and 9 cm^{-1} respectively; this was in good agreement with literature [12]. This difference in absorption gives rise to a distinct difference in ablation thresholds; this separation is highlighted in Fig. 3. As the beam was near Gaussian, the diameter of the single pulse ablated holes can be used to determine the spot size and ablation threshold as a function of fluence as described by Mannion et al [13] (equation 1). By plotting the square of the diameter against the pulse energy the effective focussed spot size was calculated to be ≈23μm for both materials.

$$D^2 = 2\omega_0 \ln\left(\frac{\theta_o}{\theta_{th}}\right) \qquad (1)$$

Figure 3 shows a plot of the measured diameters squared (D^2) versus fluence (ϕ) for both ITO and glass; by extrapolating to zero the threshold values of ITO (Φ_{th}^{ITO}) and glass (Φ_{th}^{GL}) were determined as 0.70 Jcm^{-2} and 10 Jcm^{-2} respectively. By measuring the depth of selected ablated craters the ITO film thickness was recorded as ≈0.1μm.

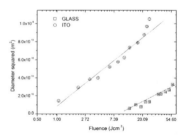

Fig. 3. Separation between ablation thresholds of ITO and glass. The threshold for ITO represents the lower boundary for the operating window. The upper boundary will be affected by incubation and cannot be attributed to the glass threshold.

The separation between ablation thresholds represents the processing window, within which ITO removal without substrate damage is possible. However, fabricating a complete circuit required pulses to be overlapped.

Previously reported results from Rosenfeld et al. [14] indicated a lowering of ablation threshold in dielectric materials irradiated with multiple pulses. In these studies the dielectrics tested were uncoated, therefore to minimise incubation effects on the substrate ITO processing was performed with as few scans as possible.

Figure 4 shows optical microscopy images of the post-processed surface at x80 magnification. Image 4(a) shows the effect of scanning at a relatively high speed; ITO film removal was incomplete and remained adhered to the surface. The maximum traverse speed identified as producing tracks equivalent to the film thickness with a single scan, similar to those in images 4(b) and 4(c), was 37.5 mm/s. In 4(c) the effect of low speeds was observed to induce surface modification at the centre of the track, damage was attributed to thermal accumulation; despite steps to minimise the effect. Finally, 4(b) represents the best processing quality obtainable; the track width was consistent and no damage attributed to thermal build-up can be observed.

Fig. 2. Images of ablated ITO tracks made with a fluence of 4.24 Jcm^{-2}. (a) 112mm/s (b) 32mm/s and (c) 22.5mm/s. Using the highest traverse speed produced inconsistent tracks whilst in (c) the lowest traverse speed led to surface modification attributed to thermal loading. 4(b) represents the best balance of fluence and traverse speed to produce consistent tracks with no surface modification

Table 1 shows the fluence and traverse speed combinations that resulted in damage to the substrate. At speeds above those indicated film removal was either complete or partial. Results for partial removal where included as part of the processing window because in the next step the hatched areas produced required a low offset [5]. The low offset resulted in the remaining ITO being removed on the next pass.

The pulse offset required to fully remove the ITO was determined by generating a series of 2mm^2 hatches with decreasing offset. A fluence of 3.39 Jcm^{-2} was used to produce the hatched squares as there were several different traverse speeds available for processing and also because the low offset required reduced the likelihood of substrate damage. These areas were subsequently examined under optical microscope and white light profiling to check for remaining ITO or surface modification. Figure 4 shows that with low offset (low separation) it was possible to fully remove the film from the surface. This is shown in Figure 4 where the surface roughness is comparable to glass.

Table 1: The minimum traverse speed that can used at each flunce is shown below, at speeds below these damage to the substrate, as shown in 3(C), occurred. For Fluences where no damage was visible at tested speeds the < was used to indicate that damage could occur below this value.

Fluence (Jcm^{-2})	Traverse speed (mm/s)
5.94	37.5
5.52	37.5
5.09	32
4.67	32
4.24	32
3.82	32
3.39	25
2.97	< 22.5
2.55	< 22.5
2.12	< 22.5

3.1. Parallel patterning of circuits

In this section parallel patterning of an ITO layer with two beams was undertaken; the optical setup was modified to include an SLM displaying a computer generated hologram (CGH) to divide the beam. In this case a simple binary grating was used, although other methods are available [4 - 9].

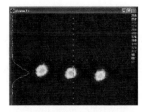

Fig 3. Image showing the multiple beams produced with a binary grating (displayed on the SLM) [15]. The zero order is the central spot, whilst the spots either side are first order.

The optimum grating period and grey scale setting were previously determined to be 2 and 80 respectively [15], figure 4 above shows the CGH displayed on the SLM (4a) to produce three spots of similar intensity (4b). The zero order was blocked after the SLM to prevent circuit overlapping; parallel processing was carried out using two beams.

Using these SLM parameters in conjunction with the removal parameters determined previously the simultaneous fabrication of two circuits was completed. The circuits are shown in figure 6(a) and (b). During post-processing, samples were tested for conductivity across circuit boundaries and to compare the conductance of each fabricated circuit. These measurements were taken at random intervals over a period of 48 hours;

during this time no conductance across either boundary was observed. The conductance of the fabricated circuits was found to be $\approx 4 \times 10^5 \mathrm{Sm}^{-1}$ for both. A surface roughness of 16nm and 15.6nm was recorded for hatched areas of both circuits; comparable to unprocessed glass.

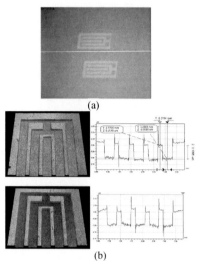

(a)

(b)

Fig. 4. Circuits produced using parallel processing. Image (a) shows the two parallel processed circuits scribed into the ITO layer. The white light profile for both these circuits are shown in (b); the uniformity of both circuits can be observed. These were produced simultaneously using a fluence of 3.39 Jcm^{-2}, traverse speed of 32 mm/s and a 10% offset.

4. Conclusions

Using the parameters identified in the study several small functional circuits were fabricated through micro processing of a thin ITO film with picosecond pulses. By utilising the difference in ablation thresholds of ITO and glass, removal of the thin layer was possible with no observable damage to the substrate; confirmed by optical microscope and white light profiling. The lower threshold for processing was taken as the ablation threshold of ITO (0.70 Jcm^{-2}). The upper limit for processing was dependent on the traverse speed used. Both single and multi-beam circuits were tested for conductivity with no cross boundary current detected. The circuit conductivity recorded for single and parallel processed circuits highlight the repeatability of this process. Through the introduction of an SLM and the application of a CGH, the time taken to manufacture the circuits was reduced. The application of more complex holograms could allow for the fabrication of greater numbers simultaneously. This enables the fabrication process to be more energy efficient, quicker and cheaper, especially when compared to current methods. Future work will focus on using the SLM and CGH combination to introduce beam shaping during processing; this will allow for more precise control whilst simultaneously reducing the risk of damage to the substrate by redistributing the intensity.

References

[1] Inzelt G (2008) Conducting polymers a new era in electrochemistry. Monographs in electrochemistry DOI: 10.1007/978-3-540-75930-0

[2] Huang CJ, Su YK, Wu SL (2003) The effect of solvent on the etching of ITO electrode. Materials Chemistry and Physics 84, 1: 146 – 150

[3] Lee YJ, Bae JW, Han HR, Kim JS, Yeom GY (2001) Dry etching characteristics of ITO thin films deposited on plastic substrates. Thin Solid Films 383, 1 -2: 281 – 283

[4] Chen M, Chen Y, Hsaiao W, Gu Z (2007) Laser direct write patterning technique of indium tin oxide film. Thin Solid Films 515, 24: 8515 – 8518

[5] Raciukaitis G, Brikas M, Gedvilas M, Rakickas T (2007) Patterning of indium-tin oxide on glass with picosecond lasers. Applied Surface Science 253: 6570 – 6574

[6] Choi HW, Farson DF, Bovatsek J, Arai A, Ashkenasi D (2007) Direct-write patterning of indium-tin-oxide film by high pulse repetition frequency femtosecond laser ablation. Applied Optics 46, 23: 5792 – 5799

[7] Kuang Z, Perrie W, Leach J, Sharp M, Edwardson SP, Padgett M, Dearden G, Watkins KG (2008) High throughput diffractive multi-beam femtosecond laser processing using a spatial light modulator. Applied Surface Science 255: 2284 – 2289

[8] Kuang Z, Liu D, Perrie W, Edwardson SP, Fearon E, Dearden G, Watkins KG (2009) Fast parallel diffractive multi-beam femtosecond laser surface micro structuring. Applied Physics A 255: Issue 13 – 14: 6582 – 6588

[9] Liu D, Kuang Z, Shang S, Perrie W, Karanakis D, Kearsley A, Knowles M, Edwardson SP, Dearden G, Watkins KG (2009) Ultrafast parallel laser processing of materials for high throughput manufacturing. Proc. Of LAMP

[10] Kuang Z, Perrie W, Liu D, Edwardson SP, Cheng J, Dearden G, Watkins KG (2009) Diffractive multi-beam surface micro-processing using 10ps laser pulses. Applied Surface Science 255: 9040 – 9044

[11] Liu D, Kuang Z, Perrie W, Scully PJ, Baum A, Edwardson SP, Fearon E, Dearden G, Watkins KG (2010) High-speed uniform parallel 3D refractive index micro-structuring of poly(methyl methacrylate) for volume phase gratings. Applied Physics B 101: 817 – 823

[12] Yavas O, Takai M (1999) Effect of substrate absorption on the efficiency of laser patterning of indium tin oxide thin films. Journal of Applied Physics 85, 8

[13] Mannion PT, Magee J, Coyne E, O'Connor GM, Glynn (2004) The effect of damage accumulation behaviour on ablation thresholds and damage morphology in ultrafast laser micro-machining of metals in air. Applied Surface Science 233: 275 – 287

[14] Rosenfeld A, Lorenz M, Stoian R, Ashkenasi D (1999) Ultrashort laser pulse damage threshold of transparent materials and the role of incubation. Applied Physics A 69, 7: S373 – S376.

Laser resonant excitation of precursor molecules in chemical vapor deposition of diamond thin films and crystals

Y. F. Lu, Z. Q. Xie, L. S. Fan and Y. S. Zhou

1 Department of Electrical Engineering, University of Nebraska-Lincoln, Lincoln, NE 68588-0511, USA

Abstract. Fast growth of diamond thin films and crystals in open air was achieved by combustion synthesis with resonant laser energy coupling. A pre-mixed $C_2H_4/C_2H_2/O_2$ gas was used as precursors for chemical vapor deposition of diamond. Through the resonant excitation of the CH_2 wagging mode of the ethylene (C_2H_4) molecules using a CO_2 laser tuned at 10.532 μm, high-quality diamond thin films and crystals were grown on various substrates with a high growth rate. The effects of the resonant laser energy coupling were investigated using optical emission spectroscopy and mass spectroscopy. Excitations of precursor molecules by different laser powers were studied. The density of the incident laser power was adjusted to modify diamond crystal orientation, optimize diamond quality, and achieve high-efficiency laser energy coupling. The wavelength-tunable CO_2 laser steers the chemical reactions and promotes proportion of intermediate oxide species, which results in preferential growth of (100)-oriented diamond. The wavelength-tunable CO_2 laser was also used to resonantly excite the vibration modes of ammonia molecules to synthesize N-doped diamond. The laser wavelength was tuned to match frequencies of the NH wagging mode of the ammonia molecules. Vibrational excitation of the ammonia molecules promotes nitrogen concentration in the deposited diamond films. This study opens up a new avenue for controlled chemical vapor deposition of crystals through resonant vibrational excitations to steer surface chemistry.

Keywords: diamond, resonant laser excitation, laser-assisted combustion flame, chemical vapour deposition.

1. Introduction

Diamond has long been the ultimate fantasy and avid pursuit due to its extraordinary properties such as high hardness, high thermal conductivity, low friction coefficient, low electrical conductivity, optical transparency and chemical inertness and due to its wide range of applications. Since the first report of commercially successful synthetic diamond in 1955 using the high-pressure high-temperature technique [1], remarkable advancements have been made in the synthesis of diamond crystals. Currently, diamond synthesis at low pressure by chemical vapor deposition (CVD) including thermal-assisted (hot filament) CVD, microwave plasma-assisted CVD, plasma torch CVD, and combustion CVD has been widely investigated [2-10]. Combustion synthesis of diamond using an oxyacetylene combustion flame was firstly demonstrated by Hirose and Kondo in 1988 [11]. It has been argued that combustion synthesis of diamond is the most flexible of the CVD alternatives because of its scalable nature, minimum utility requirement, and significantly reduced capitol costs relative to plasma-aided process [12].

All diamond CVD methods, including plasma CVD, hot-filament CVD, and combustion-flame CVD, hinge on chemical reactions among precursor gases near thermal equilibrium [2-10]. Consequently, it is not possible to achieve selectivity among various competing chemical processes for controlled diamond growth. Many attempts have been made to achieve controlled growth of diamond, including utilization of bias-enhanced nucleation (BEN) [13,14] and introduction of nitrogen into the reaction precursors [15,16]. Both methods are capable of producing (100)-textured diamond films. However, to the best of our knowledge, there has been no research work on laser resonant vibrational excitation of molecules for controlled diamond growth.

The application of diamond in electric devices is severely limited due to the low conductivity of undoped diamond. Nitrogen addition during diamond CVD has been widely investigated in order to obtain n-type diamond [17-22]. However, a high level N-containing additive usually results in a loss of faceting and a reduction of diamond phase purity. Several radical species, including CN, HCN, NH and NO, are proposed to be responsible for the observed change [23-26]. However, few efforts are made on limitation of the deterioration induced by nitrogen-containing additive.

In this study combustion synthesis is assisted with bond-selective resonant laser energy coupling. A laser energy assisted combustion synthesis process was developed for growing high-quality diamond crystals with high growth rates in open air. Ethylene (C_2H_4) is introduced into the oxyacetylene combustion flame in order to couple the laser energy into the combustion reaction through the resonant excitation of the ethylene molecules. Diamond crystals with an average length of 5 mm and a diameter of 1 mm were obtained on silicon substrates. The effects of the vibrational energy-coupling approach through laser-induced resonant excitation of precursor molecules are investigated.

Laser resonant excitations of precursor molecules with different powers were investigated in diamond growth. The wavelength of the CO_2 laser was tuned to 10.532 μm in order to match a vibrational mode of ethylene, one of the precursor molecules. By adjusting the incident laser power density from 0 to 2.7×10^4 W/cm^2, the effect of laser energy coupling on diamond surface morphology, diamond quality, as well as energy coupling efficiency, were studied. At certain incident laser power densities, uniform (100)-textured diamond particles were obtained in the center areas of the deposited diamond films.

CO_2 laser-assisted, ammonia-added oxyacetylene combustion flames were used for synthesizing N-doped diamond films. With wavelength-matched resonant laser excitation of ammonia molecules, a drastic morphological transition from amorphous or nanocrystalline ball-like structures to columnar microcrystalline diamond structures in the deposited films was observed. N-doped diamond films with a nitrogen content of 1.5×10^{20} N atoms/cm^3 were achieved by tuning the laser wavelength to 9.219 μm to match the N-H wagging mode. The effects of the laser-induced resonant excitation of the precursor molecules on the diamond growth were investigated using optical emission spectroscopy and mass spectrometry.

2. Experimental section

2.1. Growth of diamond Crystals

Figure 1 shows the experimental setup for the CO_2 laser-assisted fast growth of diamond crystals in open air. A commercial oxygen–acetylene torch with a 1.5 mm orifice tip was used to generate the flame. The fuel was a mixture of ethylene (C_2H_4) and acetylene (C_2H_2) with a gas flow ratio of 1:1, which was then mixed with oxygen (O_2) with a volume ratio of 1.03 (($C_2H_4+C_2H_2$)/O_2). A wavelength-tunable CO_2 laser (PRC, spectrum range from 9.2 to 10.9 μm) was used in the synthesis process to achieve resonant excitation of the C_2H_4 precursor molecules. Since there is no obvious vibrational mode within the laser spectrum range for the acetylene molecules, ethylene was added into the precursor mixture to couple the laser energy into the diamond growth process [27]. The CH_2 wagging mode (v_7, 949.3 cm^{-1}) of the ethylene molecules corresponds to a wavelength of 10.534 μm, which has a close match with the CO_2 laser line at 10.532 μm. Laser energy was then coupled into the reactions through resonant excitation of the C_2H_4 molecules.

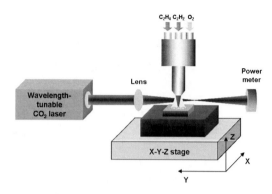

Fig. 1. Illustration of the experimental setup for the CO_2 laser-assisted fast growth of diamond crystals in open air.

Silicon (100) wafers with a dimension of $10\times10\times0.6$ mm^3 were used as substrates. The silicon substrates were pre-seeded with 100,000-grit diamond powders with an average size of 0.25 μm. The substrate was placed on a water cooling box mounted on a motorized X-Y-Z stage. The CO_2 laser, tuned to 10.532 μm with a power of 800 W, was directed in parallel with the substrate surface but perpendicularly to the flame axis. The laser beam was focused using a ZnSe convex lens (focal length = 254 mm) to 2 mm in diameter, which was around the same as the average diameter of the inner flame. The original length of the flame was around 6 mm, which shrunk to about 3.6 mm when the laser beam was introduced into the flame right underneath the nozzle. The distance between the inner flame tip and the diamond growth site was maintained at about 0.5 mm, by the program-controlled X-Y-Z stage. The temperature of the growth site was monitored by a pyrometer (Omega Engineering, Inc., OS3752). The temperature was maintained at 760 - 780 ℃ by adjusting the flow rate of the water-cooling box. To understand the effects of laser-induced energy coupling, diamond samples were also prepared without laser irradiation, but under the same deposition condition including the same gas flow ratio and deposition temperature.

2.2. Growth of diamond films

The experimental setup for the growth of diamond film is similar with that of the growth of diamond crystals, as shown in Fig. 1, and described in Section 2.1. Differences are the substrates and incident laser powers. In the growth of diamond films, tungsten carbide (WC) plates (BS-6S, 6 wt. % Co, Basic Carbide Corp.) with a dimension of $12.7\times12.7\times1.6$ mm^3 were used as substrates. The surface roughness of the WC substrates was 400 nm. The incident laser power of the CO_2 laser was varied. The laser power was tuned from 0 to 800 W to study the laser power dependence in diamond growth.

2.3. Growth of Nitrogen-doped diamond films

The experiment system for the CO_2 laser-assisted combustion for N-doped diamond deposition is similar to that used in Section 2.1, as illustrated in Fig. 2. A commercial oxyacetylene torch with a 1.5-mm orifice tip was used to generate oxyacetylene flames. Acetylene (C_2H_2, 99.6%), oxygen (O_2, 99.996%), and ammonia (NH_3, 99.99%) gases were mixed in a torch through three gas-flow meters (B7920V, Spec-Air Gases & Technologies). The NH_3 gas acted as a nitrogen source. The gas flow rates of C_2H_2, O_2, and NH_3 were 1200, 1180, and 150 standard cubic centimeter per minute (sccm), respectively. Wavelengths of 9.219, 10.35, 10.719 μm (matching the N-H wagging mode of the NH_3 molecule), and 10.591 μm (wavelength from common commercial CO_2 lasers) were used in this study. A tungsten carbide (WC) substrate (BS-6S, Basic Carbide Corp.) with dimensions of $25.4 \times 25.4 \times 1.6$ mm^3 and a cobalt composition of 6% was placed on a hollow brass block with water cooling.

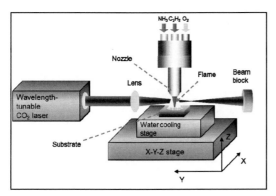

Fig. 2. Illustration of the experimental setup for the CO_2 laser-assisted combustion flame method for deposition of nitrogen-doped diamond in open air.

2.4. Characterization of the diamond crystals and films

Surface morphologies and dimensions of the grown diamond crystals and films were characterized by a scanning electron microscope (SEM; XL-30, Philips Electronics). Diamond qualities were evaluated both by Raman spectroscopy (inVia H 18415, Renishaw) and X-ray diffraction (XRD; AXS D8, Bruker). The exciting source of the Raman system is an argon-ion laser with a wavelength of 488 nm and a power of 100 mW, which was operated in the multichannel mode with the beam focused to a spot diameter of approximately 5 μm. Prior to and after the Raman characterization of the diamond samples, the Raman system was calibrated using a single crystal silicon (100) wafer. The nitrogen doping level were measured using a CAMECA IMS-4f double focusing magnetic sector instruments equipped with oxygen and/or Cs primary beam sources. The distribution

of chemical species in the flames was studied using a mass spectrometer (JEOL, AccuTOF, Direct Analysis Real Time).

2.5. Optical emission spectra of the combustion flames

Optical emission spectroscopy (OES) was used to study the effects of the resonant laser energy coupling. OES spectra of the flames during the film depositions were collected in a direction perpendicular to the flame axis. The setup for OES study is hown in Fig. 3. The optical emission of the flame was introduced into a spectrometer (Andor Shamrock SR-303i-A) coupled with an ICCD (Andor iStar DH-712) through two lenses and an optical fiber which were all made of UV-grade quartz. All the spectra were taken at a point 0.3 mm above the tip of the inner flame to compare the deposition conditions. The dimension ratio of the image and the flame was 1:1. The optical fiber, with a diameter of 50 μm, was precisely positioned in the projected image of the inner flame. A background spectrum taken before the collections of the emission spectra was subtracted in all results.

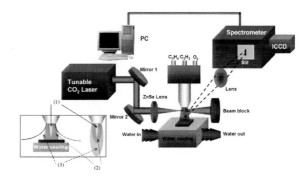

Fig. 3. Optical emission spectroscopy of the flame.

2.6. Mass spectrometry of the combustion flames

Positive ions in the $C_2H_4/C_2H_2/O_2$ flame were detected using a time-of-flight mass spectrometer (MS, AccuTOFTM, JEOL USA, Inc.), as shown in Fig. 4. A stainless steel orifice with an inner diameter of 400 μm on the mass spectrometer collects ions in open air. The combustion torch was fixed on a motorized X-Y-Z stage, and the relative position of the flame to the orifice was precisely adjusted so that the tip of the flame inner cone was right in front of the orifice under all excitation conditions. The MS data were analyzed using the TSSPro software (Shrader Analytical and Consulting Laboratories, Inc. Version 3.0) and the MS Tools software (JEOL USA, Inc.).

Fig. 4. Mass spectrometry of the flame. (a) Picture of the mass spectrometer and the flame setup. (b) A close view of the relative position of the MS orifice and the flame nozzle. (c) Picture of flame analysis using the mass spectrometer.

3. Results and discussion

3.1. Characterization of the diamond crystals and films

3.1.1. Characterization of the diamond crystals

Diamond crystals were grown using a continuous one-step approach. In the initial 1 hour of the diamond deposition, a diamond film was deposited on the silicon substrate with a preferential growth within the central area of the film. This was similar to our previous work of diamond film deposition on tungsten carbide [27]. As the film thickness increased, diamond grains under the inner flame tip exhibited a much higher growth rate and yielded diamond crystals. The diamond crystal shown grown for 5-hour in Figs. 5 (a) (side view) and 5 (e) (top view) illustrates how the crystal started to grow on the diamond film. The diamond crystal reached 0.6 mm long and 0.6 mm in diameter. When the growth time increased to 15 hour, the diamond crystal grew to a length of around 2 mm and a diameter of 0.8 mm, as shown in Figs. 5 (b) (side view) and 5 (f) (top view). Since the primary growth area was always maintained at the top of the diamond crystal, it is easy to understand that the vertical growth rate was much higher than the horizontal growth rate, leading to the formation of a pillar-like shape. A diamond pillar of 5 mm long and 1 mm in diameter grown for 36 hours is shown in Figs. 5 (c) (side view) and 5 (g) (top view). The average growth rate for the 36-hour grown crystal was 5 mm / 36 hr which equals to 139 μm/hr. This is more than twice that of the conventional combustion CVD mentioned in the above paragraphs.

It has been reported that for the longer-time deposition, the crystal surface temperature plays a critical role in the uniform growth of diamond [28]. Such a phenomenon was also observed in this work. In the growth longer than 36 hours, the temperature of the diamond surface became higher than 850 ℃ when the diamond grew higher than 5 mm. The higher temperature led to the increase of secondary crystallization, which degraded the uniformity of the diamond crystals. It is believed that the surface temperature could be maintained

constant by decreasing the total gas flow and introducing the multi-step deposition method [28].

Compared with diamond particles obtained in the 15-hour growth without laser irradiation as shown in Figs. 5 (d) (side view) and 5 (h) (top view), laser-assisted combustion synthesis yielded diamond crystals with better uniformity and larger size. The diamond particles grown without laser irradiation were more like a pile up of polycrystalline diamond grains and graphitized carbons.

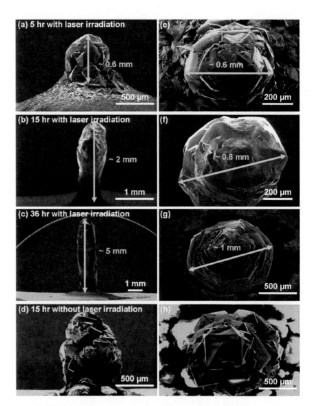

Fig. 5. SEM images of diamond crystals grown with laser excitation at 10.532 μm for (a) 5, (b) 15, (c) 36 hours, respectively; and (d) diamond grown without laser irradiation for 15 hours. Figures (e), (f), (g), and (h) are the top-viewed images of the samples in (a), (b), (c), and (d), respectively.

Figure 6 shows the Raman spectra of the four diamond samples shown in Figs. 5 (a) - (d). Intense and sharp Raman peaks at 1332 cm⁻¹ were observed in all of the four samples. Diamond crystals obtained from the laser-assisted combustion synthesis only show the 1332 cm⁻¹ Raman peak (Figs. 3.2(a)-(c)) with narrow FWHM values of 4.5 - 4.7 cm⁻¹. No other signals were observed in the Raman spectra of these three samples. Both strong Raman peak and narrow FWHM value indicate the high quality of the diamond crystals grown with resonant laser energy coupling. The diamond sample obtained without laser energy coupling shows a Raman peak at 1332 cm⁻¹ (Fig. 6 (d)) but with a much wider FWHM of 9.2 cm⁻¹. A broad band from 1100 to 1600 cm⁻¹ was observed in Fig.

6 (d), indicating the existence of nondiamond carbon species, such as sp^2 amorphous carbon.

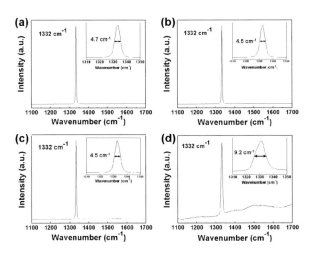

Fig. 6. Raman spectra of the diamond crystals grown with laser excitation at 10.532 μm for (a) 5, (b) 15, (c) 36 hours, respectively; and (d) diamond grown without laser for 15 hours. Insets are the expanded Raman peaks of each spectrum, showing the full width at half maximum.

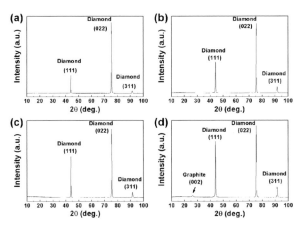

Fig. 7. X-ray diffraction spectra of the diamond crystals grown with laser excitation at 10.532 μm for (a) 5, (b) 15, (c) 36 hours, respectively; and (d) diamond grown without laser for 15 hours.

Figure 7 shows the XRD spectra of the four samples. Discrete peaks from the (111), (022), and (311) diamond facets were observed from all of the four samples. The signals are in good agreement with the data from the ASTW 6-675 table for natural diamond, indicating the formation of diamonds. However, a 2θ signal at 26° is observed from sample (d), as shown in Fig. 7 (d). This signal is ascribed to the (002) plane of graphite, clearly indicating the existence of the graphite phase.

Both Raman and XRD spectra demonstrate the high quality of the diamond crystals grown with the 10.532 μm CO_2 laser resonant excitation. In the mean while, the 15-hr grown sample without laser irradiation exhibits

poor diamond crystallization with the extensive existence of non-diamond carbon structures. Both of them indicate the advantage of the laser-induced resonant excitation.

3.1.2. Characterization of the diamond films
Morphologies and grain sizes of the deposited diamond films are shown in Fig. 8. Diamond films deposited without laser excitations and with low power densities less than 1.7×10^3 W/cm² have randomly-oriented grains with small sizes. However, (100)-oriented facets began to dominate in the center areas of the deposited diamond films when the laser power density increased to 3.3×10^3 W/cm². The preferential growth of (100) facets became more dominant when the laser power increased. The size and uniformity of the (100) facets reached its maximum at an incident power density of 1.0×10^4 W/cm². The orientations of the diamond facets became more randomly distributed with further increased laser power density, as indicated in Figs. 8 (i) and (j). The SEM images indicate that the laser resonant excitation of C_2H_4 molecules plays an important role in modifying the morphology of diamond films. Within a certain range of incident laser power densities, uniformly distributed diamond (100) facets can be obtained. The SEM images also show that the grain sizes of the diamond crystals increased with increasing laser power.

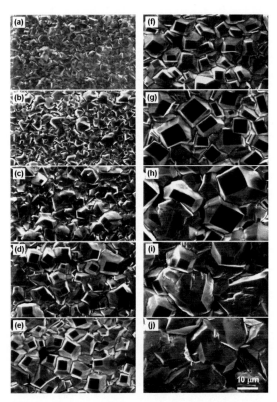

Fig. 8. SEM images of diamond films deposited (a) without laser excitation and with laser excitations at (b) 8.3×10^2, (c) 1.7×10^3, (d) 3.3×10^3, (e) 5.0×10^3, (f) 6.7×10^3, (g) 1.0×10^4, (h) 1.3×10^4, (i) 2.0×10^4, and (j) 2.7×10^4 W/cm².

Figure 9 shows the Raman spectra for the diamond films deposited without laser excitation and with laser excitations at different incident laser power densities. Sharp diamond peaks around 1332 cm^{-1} and broad amorphous carbon bands around 1500 cm^{-1} were observed in all samples. The sample deposited without laser excitation exhibits a strong band of amorphous carbon and a relatively weak diamond peak. When the laser excitation was used, the band of amorphous carbon was suppressed and the diamond peak became stronger. This trend continued until the laser power density was increased to 5.0×10^3 W/cm^2, as shown in Fig. 9 (a). Figure 9 (b) indicates that Raman spectra of the diamond samples deposited using the laser excitations with laser power densities higher than 5.0×10^3 W/cm^2 showed similar relative height of diamond peaks and amorphous carbon bands.

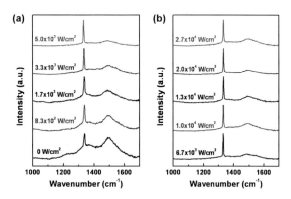

Fig. 9. Raman spectra of diamond films deposited without laser excitation and with laser excitations at different laser power densities.

Figure 10 shows the corresponding Raman shifts (left, solid squares) and full width at half maximum (FWHM; right, solid triangles) values of the diamond peaks shown in Fig. 9. Diamond peak shifts are correlated to residual stresses in the diamond films, whereas FWHM values of the peaks represent diamond qualities. It is observed that the diamond peak shifted from ~1338 to ~1332 cm^{-1} and then slightly increased to ~1334 cm^{-1}. At laser power densities of 5.0×10^3 and 6.7×10^3 W/cm^2, the diamond peak position was closest to 1332 cm^{-1}, which is the peak position of natural diamond. The FWHM curve of the diamond peaks has similar shape to that of the Raman shift curve, as shown in Fig. 10 (right, solid triangles). The shift and broadening of Raman peaks for the samples deposited without laser excitation and with low-power-density (8.3×10^2~3.3×10^3 W/cm^2) laser excitations were believed resulting from low diamond qualities due to defects, impurities and non-diamond carbon contents. The diamond quality increased as the incident laser power increased. On the other hand, when the laser power density increased beyond 6.7×10^3 W/cm^2, another factor, residual stress, became dominant in determining the position of the diamond peak. The major contribution of

the residual stress comes from the thermal stress developed during the cooling-down process from the growth temperature to room temperature, which is caused by the thermal expansion mismatch between the diamond films and the substrates [29-34]. For the diamond films deposited with high laser power density larger than 1.0×10^4 W/cm^2, diamond growth rates were much higher, which introduced higher lateral stress and resulted in the shift and broadening of the diamond peaks. Based on the Raman spectroscopy, it is believed that diamond crystals with best quality can be obtained when the incident power density is 5.0×10^3~6.7×10^3 W/cm^2.

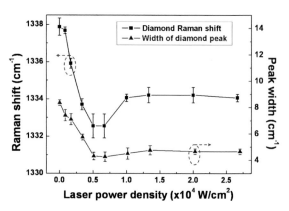

Fig. 10. Raman shifts (left, solid squares) and full width at half maximum values (right, solid triangles) of diamond peaks as functions of laser power densities.

3.1.3. Characterization of the N-doped diamond films

Figure 11 shows the scanning electron microscope (SEM) images of the deposited films. Uniform and clear diamond films have been steadily obtained with oxyacetylene flames with an oxygen to acetylene ratio of about 1:1 [see Fig. 11 (a)] [12,35-37]. The films deposited using ammonia-added flames without laser excitations have amorphous or nanocrystalline cauliflower-like structures as shown in Fig. 11 (b). The morphological change observed upon ammonia addition agrees with the results obtained by other techniques [36,37]. Substantial N-containing additive in the diamond deposition process considerably reduces diamond quality and leads to preferential sp^2 carbon deposition. With a laser irradiation at 10.591 μm, similar cauliflower-like structures were observed in the deposited films. In the films deposited at 10.719 μm, spherules consisting of poorly faceted diamond grains were observed. The ball-like structures became indistinguishable in the films deposited with laser excitation at a wavelength of 10.35 μm. Clear diamond grains with an average size of 2 μm were observed in the films. With a laser excitation at the strongest absorption wavelength, 9.219 μm, (111) faceted diamond grains with an average diameter of 5 μm were clearly observed [see Fig. 11 (f)]. From the secondary ion mass spectrometry (SIMS) result, a high N doping concentration, 1.5×10^{20} atom/cm^3, was determined for the sample. Such a high level doping is believed to be difficult to achieve when using nitrogen or ammonia as dopant sources [19].

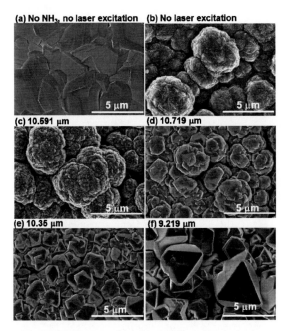

Fig. 11. SEM images of the diamond films deposited using ammonia-free flames (a), and ammonia-added flames without laser (b) and with 500 W CO_2 laser excitations at 10.591, 10.719, 10.35, and 9.219 μm (c-f).

The bonding structures in the diamond films were characterized by Raman spectroscopy as shown in Fig. 12. To analyze the diamond quality, a quality factor $Q_i = I_d / (I_d+I_g)$ was derived from the Raman spectra, Id and Ig being the intensity of the diamond peak and the graphitic (G) band, respectively [38]. The Raman spectrum of the film deposited using ammonia-added flames without laser excitation possesses a typical nanocrystalline diamond feature with a Q_i value of 0.17, as shown in spectrum (a). The tiny Raman peak centered at 1331 cm^{-1} is a typical diamond Raman signal. Besides the diamond peak, four extra peaks at 1170, 1371, 1493 and 1550 cm^{-1} are resolved. The band located at 1371 cm^{-1} (D-band) is attributed to the breathing modes of sp^2 atoms in rings, reflecting disordered carbon in the film. The broadband centered at 1550 cm^{-1} (G band) is attributed to the bond stretching of all pairs of sp^2 atoms in both rings and chains and indicates graphite-like carbon content mixed with amorphous carbon. The peaks at 1170 and 1492 cm^{-1} are a signature of trans-polyacetylene which is related to the hydrogen content in the nanodiamond lattices [39]. The film deposited under the laser excitation at 10.591 μm has similar Raman features of nanocrystalline diamonds. At 10.35 and 10.719 μm, the diamond peaks in spectra (c) and (d) obviously increase. For the film deposited at 9.219 μm, the diamond peak is much sharper and stronger associated with weaker D and G bands, which indicates an improvement in the diamond quality

of the film. The quality factor, 0.45, is twice as high as that of the film deposited without laser excitations. The diamond quality (Q_i) was plotted as a function of the ammonia absorption coefficient (α) of the corresponding laser wavelength used in the process in Fig. 13. The diamond quality linearly increases with respect to the absorption coefficient of the ammonia gas. High-quality N-doped diamond films grown with resonant laser excitation at 9.219 μm show the advantage of vibrational resonant excitations on reaction control.

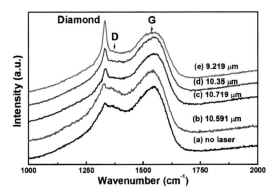

Fig. 12. Raman spectra of the diamond films deposited using ammonia-added flames with no laser excitation (a), and with laser excitations at 10.591, 10.719, 10.35, and 9.219 μm (b-e).

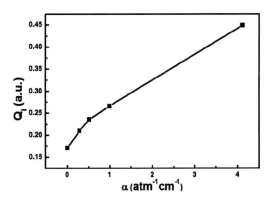

Fig. 13. Raman spectra of the diamond films deposited using ammonia-added flames with no laser excitation (a), and with laser excitations at 10.591, 10.719, 10.35, and 9.219 μm (b-e).

3.2. Mechanism discussion

3.2.1. Optical emission spectroscopy

Optical images of the $C_2H_4/C_2H_2/O_2$ flame with and without laser irradiation showed that the inner flame was shortened from 6 to 3.6 mm in length and was broadened by 80% in width [27]. The shorter inner cone is the result of the enhanced reactions in the flame induced by the laser resonant excitation. The chemical reaction becomes stronger and faster and hence expanded horizontally. A power meter was used to monitor the absorption of the laser powers by comparing the measured powers before and after the flame absorption. Laser power absorption spectra of the precursors were then measured within a spectral range from 9.2 to 10.9 µm. By irradiating the laser beam through the $C_2H_4/C_2H_2/O_2$ flame used for the laser-assisted growth of diamond crystals, an obvious absorption peak was observed at 10.532 µm, as shown in Fig. 14. However, when the C_2H_2/O_2 flame (volume ratio of 1.05) was used, no obvious absorption was observed within the same spectral range. The absorption peak at 10.532 µm by $C_2H_4/C_2H_2/O_2$ flame was ascribed to the resonant excitation of the CH_2 wagging mode (v_7, 949.3 cm^{-1}) of the C_2H_4 molecules, as illustrated in the inset of Fig. 14. At the ground state, the C_2H_4 molecule vibrates like a butterfly. By absorbing laser energy at 10.532 µm, the CH_2 wagging mode vibration is resonantly excited.

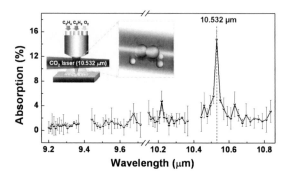

Fig. 14. Absorption of CO_2 laser power by the $C_2H_4/C_2H_2/O_2$ flame with respect to different laser wavelengths. Inset is a schematic of the CH_2-wagging mode excitation of an ethylene molecule by the 10.532 µm CO_2 laser.

It was demonstrated that by enhancing the reactions in the flame, the laser resonant excitation of the C-H bond vibration in C_2H_4 molecules results in the increased concentrations of several intermediate species [27]. According to the OES obtained from the flame with and without the laser irradiation shown in Fig. 15, obvious increases in concentrations of OH, CH and C_2 species were observed in the laser-irradiated flame. It is suggested that OH radicals play a critical role in combustion synthesis of diamond by etching the surface-bond hydrogen and stabilizing the sp^3 hybridized surface carbon bonds [40,41]. The CH radical is also believed to be helpful in diamond growth [42]. Both increments explain why both growth rate and crystal quality were improved in the laser-assisted combustion synthesis of diamond. The C_2 radical, however, has a two-sided effect on the diamond growth. On the unhydrided surface, the insertion of C_2 into a C=C bond produces a carbene, leading to the formation of the secondary nucleation. On the monohydride surface, however, the addition of a C_2 to a hydrogen-terminated diamond (110) surface is energetically very favorable, making the growth of the existing crystal proceeds readily [43,44].

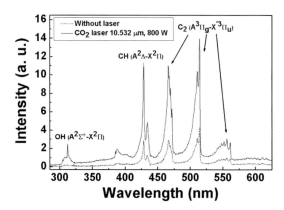

Fig. 15. Optical emission spectra of the flame before (dashed curve) and after (solid curve) the CO_2 laser irradiation at a wavelength of 10.532 µm and a power of 800 W.

Optical images of the $C_2H_4/C_2H_2/O_2$ flames without and with laser irradiations at different laser powers shown in Fig. 16 indicate that the inner flame is shortened by laser irradiation and the length keeps decreasing as the laser power density increases. The flame shrank from 6 to 3.5 mm in length and was broadened by 70% in width when laser power density was increased to 2.7×10^4 W/cm^2. The shortened inner cone is the result of the accelerated reactions in the flame induced by the laser resonant excitation. Variation of flame brightness also indicates the promotion of chemical reactions in the flames. With an increased laser power, the brightness of the flame was also increased, indicating promoted chemical reactions and higher concentrations of reacting species.

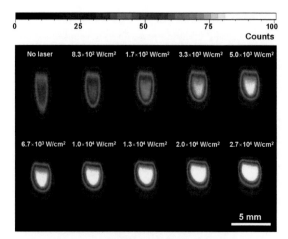

Fig. 16. Optical images of flames without laser excitation and with laser excitations at different laser power densities.

It was demonstrated that by accelerating the reactions in the flame, laser resonant excitation of the C-H bond vibration in C_2H_4 molecules results in the increased concentrations of several relevant intermediate species [27,45-46]. According to the OES spectra obtained from the flames without and with laser irradiations at different laser powers shown in Fig. 17, obvious increases in emission intensities hence concentrations of OH, CH and C_2 species were observed in the laser-irradiated flames. The emission intensities of these species kept increasing as laser power increased. It is suggested that OH radicals play a critical role in combustion synthesis of diamond by etching the surface-bond hydrogen and stabilizing the sp^3 hybridized surface carbon bonds [40,41]. The CH radicals are also believed to be helpful in diamond growth [42]. Both increments of OH and CH radicals explain the increases in growth rate and crystal quality in the laser-assisted combustion synthesis of diamond. The C_2 radical has a more important effect on the diamond growth [30,31]. On an unhydrided surface, the insertion of C_2 into a C=C dimer bond produces a carbene, leading to secondary nucleation and fast diamond growth. On the monohydride surface, the addition of a C_2 into C-H bonds of a hydrogen-terminated diamond (100) and especially (110) surface is energetically favorable, making the growth directly proceed from the existing crystal surface at a slow rate [39,40]. Under our experimental condition where hydrogen is relatively abundant, at medium laser power densities smaller than 1.0×10^4 W/cm^2, the surfaces of diamond crystallites should be almost completely covered by hydrogen atoms and the surfaces are monohydrided, thus the growth can readily proceed from the parent crystallites and the (100) orientation will dominate, considering the fact that those with higher Miller indices essentially resemble (100) surface [43]. However, when the laser power density increases to above 1.0×10^4 W/cm^2, the C_2 concentration increases to a critically high value, and the small portion of unhydrided surfaces could also increase to a critical value. Under such conditions, diamond growth through secondary nucleation becomes dominant, thus jeopardizing the crystal orientations [43,44]. At the same time, the crystal grain boundaries (consisting of π-bonded carbon atoms) also increase in proportion and contribute to the slight broadening of the Raman peak at 1332 cm^{-1} for laser power above this point.

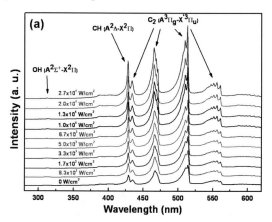

Fig. 17. Optical emission spectra of flames without laser excitation and with laser excitations at different laser power densities.

It was demonstrated that the laser resonant excitation of NH_3 molecules resulted in an enhanced reaction rate and an increase in the concentrations of several intermediate species. OES of the flames was conducted to examine the species in the flames as shown in Fig. 18. Compared with the spectra of ammonia-free flames, the ammonia-added flames showed a considerable declination of C_2 and CH radical intensities while an intensive CN peak centered at 380 cm^{-1} raised up. With laser excitations at 10.591 and 9.219 μm, the intensity of the whole spectra increased by 41.0% and 107.6%, respectively. This reveals a highly enhanced reaction with resonant laser excitation.

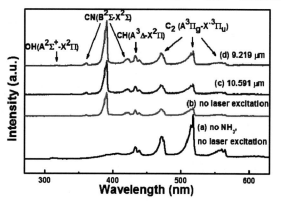

Fig. 18. Optical emission spectra of ammonia-free oxyacetylene flames (a), and ammonia-added flames with no laser excitation (b) and with 500 W CO_2 laser excitations at 10.591 and 9.219 μm (c,d).

Mass spectrometry (MS) of the flames was carried out to simultaneously detect both stable species and more reactive radicals. To provide insight into the chemistry of this process, the absolute mole fractions of several critical species were calculated from the equation $c = I_s / I_w$, with I_s and I_w being the intensities of an individual radical and a whole spectrum, respectively. Figure 19 shows the calculated mole fractions of the species in the flames for different experimental conditions. C_xH_y ($x = 3, 5, y = 3$) species were dominant in the ammonia-free flames. The extant CN and CNO radicals were from the diffused N_2 in the air or the N-impurity from the acetylene cylinder. With the addition of 150 sccm of ammonia, the concentrations of all N-containing species increased drastically while C_xH_y species decreased slightly. This is in good accordance with the OES results. Compared with the flames without laser excitations, the mole fractions of OH and H_3O radicals slightly increased with a 500 W laser excitation at 9.219 μm. An H_3O radical is believed to be formed by attaching an H atom to a H_2O molecule. Both H atoms and OH radicals play a critical role in the synthesis of diamond by etching the surface-bond hydrogen and stabilizing the sp^3 hybridized surface carbon bonds [40,41]. The mole fractions of most N-containing species increased, corresponding to a 2.29% increase of the overall N-containing species (CN, CNO, NO, and H_2CN). The N-H single bond energy of the NH_3 molecules is about 391 kJ/mol. Due to the relative low power density of the continuous wave CO_2 laser (10^4 W/cm^2) as well as the intermolecular vibrational redistribution, it is impossible to directly break the NH bond through laser vibrational excitation. However, during the course of the reaction in the combustion flame, laser vibrational excitation can play an important role by actively intervening in the reaction, which can guide the reactants by controlling the phase of their motions. The selective excitation of NH_3 molecule steers the gas phase chemistry and result in increase in N-containing species. The increase in the N-containing species explains the high doping level in the diamond films deposited at 9.219 μm. A pronounced difference, however, was the decreasing CN concentration. Hong et al. [47] proposed that CN radicals have a two-sided effect on diamond growth. Abstraction of hydrogen from the surface to form HCN or H_2CN promotes diamond growth if a small amount of N-containing species is added. With substantial N-containing additives, the high bond dissociation energy (D_0 = 180 kcal/mol) and the large dimerization energy (ΔH = -134 kcal/mol) of the CN radicals favors the formation of cyanogen (C_2N_2). The condensed paracyanonen induces the cauliflower structures. This explains the structure distortion with the ammonia addition. However, with resonant laser excitations, the chemistry in the flames has a slight modification. The increasing etching species (H and OH) suppresses the formation of paracyanonen. This contributes to the reproduction of diamond growth sites and promotes diamond deposition. Diamond deposition becomes dominant in the competition between diamond and graphite growth. The decreasing CN species accompanied by the increasing CNO and H_2CN species, as well as the increasing etchant concentration, supports this hypothesis. The gas phase chemistry redistribution induced by vibrational excitation of ammonia molecules demonstrates the advantage of laser excitations in materials synthesis.

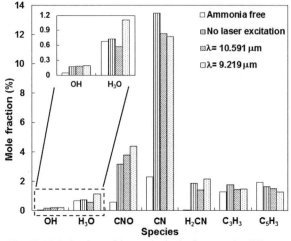

Fig. 19. Mole fractions of the species in the flames under different laser excitations.

4. Conclusions

A laser-assisted combustion synthesis was developed for fast growth of diamond crystals in open air by coupling laser energy into the combustion process through the resonant excitation of C_2H_4 molecules. High-quality diamond crystals up to 5 mm in length and 1 mm in diameter were obtained. A high diamond growth rate up to 139 μm/hr was achieved on a silicon substrate. Optical emission spectra of the flame with and without laser irradiation indicate that the laser-induced vibrational excitation enhances the reaction and increase the active radicals in the flame, which result in the fast growth of high-quality diamond crystals. This study also suggests a bond-selective energy-coupling approach through laser-induced resonant excitation of precursor molecules for materials preparation.

Resonant excitation of precursor molecules using different laser powers in laser-assisted growth of diamond crystals was studied to modify diamond surface morphology, obtain high diamond crystal quality, and high energy coupling efficiency. At a laser power density range of $5.0 \times 10^3 \sim 1.0 \times 10^4$ W/cm^2, (100)-oriented diamond crystals were grown in the center area of the deposited diamond films. According to the Raman spectroscopy of the diamond films deposited with laser excitations at different incident laser powers, best diamond qualities could be obtained when the incident power density was in a range of $5.0 \times 10^3 \sim 6.7 \times 10^3$ W/cm^2. The increment rates of diamond film thicknesses indicate that the highest efficiency of laser energy coupling could be achieved in a range of $5.0 \times 10^3 \sim 6.7 \times 10^3$ W/cm^2. Considering all the above factors, it is believed that the incident power density of 6.7×10^3 W/cm^2 is the optimal value under the growth condition used in this study. This study suggests a laser-assisted approach for modifications of surface orientations in crystal growth through laser excitations of precursor molecules.

Vibrational resonant excitations of the NH_3 precursor molecules were achieved using a wavelength-tunable CO_2 laser in combustion CVD of N-doped diamond films. The chemical reaction in the diamond forming combustion flame was steered by laser resonant excitation of NH_3 molecules. Compared with laser wavelengths of 10.35, 10.719, and 10.591 μm, the laser excitation at 9.219 μm, which corresponds to the strongest absorption peak, is more effective in exciting the NH_3 molecules and leads to high-quality N-doped diamond. The OES and MS results both revealed an enhanced reaction with the resonant laser excitation at 9.219 μm. The increase in the N-containing species and etchant concentrations in the flames with resonant laser excitations explains the high doping level and the improvement of the diamond crystallinity. CN radicals play a critical role in balancing the diamond and graphite deposition. It has proven to be a promising approach to use vibrational resonant excitations of precursor molecules to achieve high-quality N-doped diamond films.

References

[1]. Bundy FP, Hall HT, Strong HM, Wentorf RH Jr, (1955) Man-made diamonds. Nature 176: 51-55

[2]. Haubner R, Lux B, (2003) Diamond growth by hot-filament chemical vapor deposition: state of the art. Diam. Relat. Mater. 2:1277-1294

[3]. McCauley TS, Vohra YK, (1995) Homoepitaxial diamond film deposition on a brilliant cut diamond anvil. Appl. Phys. Lett. 66:1486-1488

[4]. Mokuno Y, Chayahara A, Soda Y, HorinoHa HY, Fujimori N, (2005) Synthesizing single-crystal diamond by repetition of high rate homoepitaxial growth by microwave plasma CVD. Diam. Relat. Mater. 14:1743-1746

[5]. Asmussen J, Grotjohn TA, Schuelke T, Becker MF, Yaran MK, King DJ, Wicklein S, Reinhard DK, (2008) Multiple substrate microwave plasma-assisted chemical vapor deposition single crystal diamond synthesis. Appl. Phys. Lett. 93:031502

[6]. Zou YS, Yang Y, Chong YM, Ye Q, He B, Yao ZQ, Zhang WJ, Lee ST, Cai Y, Chu HS, (2008) Chemical vapor deposition of diamond films on patterned GaN substrates via a thin silicon nitride protective layer. Cryst. Growth Des. 8:1770-1773

[7]. Terranova ML, Manno D, Rossi M, Serra A, Filippo E, Orlanducci S, Tamburri E, (2009) Self-assembly of n-diamond nanocrystals into supercrystals. Cryst. Growth Des. 9:1245-1249

[8]. Eguchi K, Yata S, Yoshida T, (1994) Uniform and large-area deposition of diamond by cyclic thermal plasma chemical vapor deposition. Appl. Phys. Lett. 64:58-60

[9]. Donnet JB, Oulanti H, Le Huu T, Schmitt M, (2006) Synthesis of large single crystal diamond using combustion-flame method. Carbon 44:374-380

[10]. Grigoryev EV, Savenko VN, Sheglov DV, Matveev AV, Cherepanov VA, Zolkin AS, (1998) Synthesis of diamond crystals from oxygen-acetylene flames on a metal substrate at low temperature. Carbon 36:581-585

[11]. Hirose H, Komaki K, (1988) Eur. Pat. Appl. EP324538

[12]. Ravi, KV, (1995) Combustion synthesis: is it the most flexible of the diamond synthesis processes? Diam. Relat. Mater. 4:243-249

[13]. Stoner BR, Sahaida SR, Bade JP, Southworth P, Ellis PJJ, (1993) Highly oriented, textured diamond films on silicon via bias-enhanced nucleation and textured growth. J. Mater. Res. 8:1334-1340

[14]. Fox BA, Stoner BR, Malta DM, Ellis PJ, Glass RC, Sivazlian FR, (1994) Epitaxial nucleation, growth and characterization of highly oriented, (100)-textured diamond films on silicon. Diam. Relat. Mater. 3:382-387

[15]. Locher R, Wild C, Herres N, Behr D, Koidl P, (1994) Nitrogen stabilized ⟨ 100 ⟩ texture in chemical vapor deposited diamond films. Appl. Phys. Lett. 65:34-36

[16]. Ayres VM, Bieler TR, Kanatzidis MG, Spano J, Hagopian S, Balhareth H, Wright BF, Farhan M, Abdul Majeed J, Spach D, Wright BL, Asmussen J, (2000) The effect of nitrogen on competitive growth mechanisms of diamond thin films. Diam. Relat. Mater. 9:236-240

[17]. Birrell J, Carlisle JA, Auciello O, Gruen DM and Gibson JM, (2002) Morphology and electronic structure in nitrogen-doped ultra nanocrystalline diamond. Appl. Phys. Lett. 81:2235-2237

[18]. Baranauskas V, Li BB, Peterlevitz A, Tosin MC and Durrant SF, (1999) Nitrogen-doped diamond films. J. Appl. Phys. 85:7455-7458

[19]. Okano K, Koizumi S, Silva SRP and Amaratunga GAJ, (1996) Low-threshold cold cathodes made of nitrogen-doped chemical-vapour-deposited diamond. Nature 381:140-141

[20]. Hu Q, Hirai M, Joshi RK and Kumar A, (2009) Structural and electrical characteristics of nitrogen-doped nanocrystalline diamond films prepared by CVD. J. Phys. D: Appl. Phys. 42:025301

[21]. Sowers AT, Ward BL, English SL and Nemanich RJ, (1999) Field emission properties of nitrogen-doped diamond films. J. Appl. Phys. 86:3973-3982

[22]. Mort J, Machonkin MA and Okumura K, (1991) Compensation effects in nitrogen-doped diamond thin films. Appl. Phys. Lett. 59:3148-3150

[23]. Corvin RB, Harrison JG, Catledge SA and Vohra YK, (2002) Gas phase thermodynamic models of nitrogen-induced nanocrystallinity in chemical vapor deposited diamond. Appl. Phys. Lett. 80:2550-2552

[24]. Wolden CA, Draper CE, Sitar Z and Prater JT, (1998) The influence of nitrogen addition on the morphology, growth rate, and Raman spectra of combustion grown diamond. Diamond Relat. Mater. 7:1178-1183

[25]. Schermer JJ and de Theije FK, (1999) Nitrogen addition during flame deposition of diamond: a study of nitrogen-enhanced growth, texturing and luminescence. Diamond Relat. Mater. 8:2127-2139

[26]. Atakan B, Beuger M and Hoinghaus KK, (1999) Nitrogen compounds and their influence on diamond deposition in flames. Phys. Chem. Chem. Phys. 1:705-708

[27]. Ling H, Xie ZQ, Gao Y, Gebre T, Shen XK, Lu YF, (2009) Enhanced chemical vapor deposition of diamond by wavelength-matched vibrational excitations of ethylene molecules using tunable CO_2 laser irradiation. J. Appl. Phys. 105:064901

[28]. Wang XH, Zhu W, Vonwindheim J, Glass JT, (1993) Combustion growth of large diamond crystals. J Cryst. Growth 129:45-55

[29]. Xu ZQ, Lev L, Lukitsch M, Kumar AJ, (2007) Analysis of residual stresses in diamond coatings deposited on cemented tungsten carbide substrates. J. Mater. Res. 22:1012-1017

[30]. Nakamura Y, Sakagami S, Amamoto Y, Watanabe Y, (1997) Measurement of internal stresses in CVD diamond films. Thin Solid Films 308-309:249-253

[31]. Fan QH, Fernandes A, Pereira E, Gracio J, (1999) Evaluation of biaxial stress in diamond films. Diamond and Relat. Mater. 8:645-650

[32]. Fan QH, Gracio J, Pereira E, (2000) Residual stresses in chemical vapour deposited diamond films. Diamond and Relat. Mater. 9:1739-1743

[33]. Kuo CT, Lin CR, Line HM, (1996) Origins of the residual stress in CVD diamond films. Thin Solid Films 290-291:254-259

[34]. Kim JG Yu J, (1998) Behavior of residual stress on CVD diamond films. Mater. Sci. and Eng. B 57:24-27

[35]. Hirose Y and Amanuma S, (1990) The synthesis of high quality diamond in combustion flames. J. Appl. Phys. 68:6401-6405

[36]. Stolk RL, van Herpen MMJW, Schermer JJ and ter Meulena JJ, (2003) Influence of nitrogen addition on oxyacetylene flame chemical vapor deposition of diamond as studied by solid state techniques and gas phase diagnostics. J. Appl. Phys. 93:4909-4921

[37]. Okkerse M, de Croon MHJM, Kleijn CR, Marin GB and van den Akker HEA, (2002) Influence of nitrogen on diamond growth in oxyacetylene combustion chemical vapor deposition. J. Appl. Phys. 92:4095-4102

[38]. Bergmann U, Lummer K, Atakan B and Kohse-Hoinghaus K, (1998) Flame deposition of diamond films: an experimental study of the effects of stoichiometry, temperature, time and the influence of acetone. Ber. Bunsen-Ges. Phys. Chem. 102:906-914

[39]. Farrari AC and Roberson J, (2004) Raman spectroscopy of amorphous, nanostructured, diamond-like carbon, and nanodiamond. Phil. Trans. R. Soc. Lond. 362:2477-2512

[40]. Komaki K, Yanagisawa M, Yamamoto I, Hirose Y, (1993) Synthesis of Diamond in Combustion Flame under Low Pressures. Jpn. J. Appl. Phys. 32:1814-1817

[41]. Miller JA, Melius CF, (1992) Kinetic and thermodynamic issues in the formation of aromatic compounds in flames of aliphatic fuels. Combust. Flame 91:21-39

[42]. Yalamanchi RS, Harshavardhan KS, (1990) Diamond growth in combustion flames. J. Appl. Phys. 68:5941-5943

[43]. Gruen DM, Redfern PC, Horner DA, Zapol P, Curtiss LA,(1999) Theoretical studies on nanocrystalline diamond: nucleation by dicarbon and electronic structure of planar defects. J. Phys. Chem. B 103:5459-5467

[44]. Redfern PC, Horner DA, Curtiss LA, Gruen DM, (1996) Theoretical studies of growth of diamond (110) from dicarbon. J. Phys. Chem. 100:11654-11663

[45]. Ling H, Sun J, Han YX, Gebre T, Xie ZQ, Zhao M, Lu YF (2009) Laser-induced resonant excitation of ethylene molecules in C2H4/C2H2/O2 reactions to enhance diamond deposition. J. Appl. Phys. 105:014901

[46]. Xie ZQ, Zhou YS, He XN, Gao Y, Park JB, Ling H, Jiang L, Lu YF, (2010) Fast growth of diamond crystals in open air by combustion synthesis with resonant laser energy coupling. Cryst. Growth Des. 10:1762-1766

[47]. Hong TM, Chen SH, Chiou YS and Chen CF, (1995) Optical emission spectroscopy studies of the effects of nitrogen addition on diamond synthesis in a CH4–CO2 gas mixture. Appl. Phys. Lett. 67:2149-2151.

Characterization of the power relationship in laser bonding of linear low density polyethylene film to polypropylene using a commercial adhesive

C. F. Dowding and J. Lawrence
School of Engineering, University of Lincoln, Brayford Pool, Lincoln, LN6 7TS, UK

Abstract. This work investigates the effectiveness of using a 10.6 μm CO_2 marking laser to activate a thermally sensitive adhesive in the bonding of a thin linear low density polyethylene film to a polypropylene substrate. The potential technical hurdle of bubbling has been witnessed and described, this prevented contact between the two and a viable bond. The Gaussian profile of the laser used caused the centre of the bond to be over irradiated whilst the edge was under-irradiated; this, limited the real area on which an effective bond was achieved. A bond strength of approximately 2.2 N/mm^2 was achieved using the laser activated adhesive bonding technique described. There was little bond strength dependency upon laser irradiance once above a threshold level. An average peel force of 1.09 N was measured and the work demonstrated this technique is highly consistency.

Keywords: Laser Marker; CO_2; Thermal Bonding; Peel.

1. Introduction

Localized bonding between polymers is already used extensively across a wide range of industrial applications, from primary manufacture to packaging and logistics [1]. This currently requires a mechanical contact process involving a permenantly heated steel plate machined to suit the specific task for which it is intended; thus, it is limited by inherent weaknesses such as limited speed, inflexible geometry, great energy consumption [2], lengthy service downtime [3] and a reliance upon thermal conduction and diffusion [4]. Typically, a mechanical bonding technique uses a threshold temperature of approximately 115°C to achieve a stated bond strength of 0.53 N/mm [5]

There are many benefits that can be brought to bear by a non-contact heating process like that offered by a laser. Use of galvanometric mirrors allows the user great scan geometry flexibility, high operation speed, continuous motion and vastly increased energy efficiency during operation [6]. Presently, such an approach has not been adopted and is little understood.

Using a 10.6 μm CO_2 marking laser on a prototype materials handling system which is designed to reliably apply the active materials together in the correct fashion, this work initiates the characterization of the laser motivated bonding mechanism. The subject of primary interest to the potential user is the irradiance required for adhesion and the relationship between irradiance and bond strength. This paper will, as a result, also define benchmark bond strength with respect to area that is achievable using laser activation of the adhesive.

2. Experimental arrangement and proceedure

2.1. The arrangement of material handling equipment

The critcal requirement of any polymer thermal bonding system is reliable and consistent contact between the film and base polymer to ensure a functional bond. This action is inherently provided by action of the contact methodlogy of traditional sealing systems; these force the substrate onto the film, allowing thermal conduction and, in some cases, an opportunity to trim unnecessary material from the edge of the item being sealed [3].

This requirement remains for any non-contact bonding methodology to allow functional activation of the adhesive. A system has therefore been developed for this series of experiments to provide reliable contact between the film, adhesive and substrate following the schematic arrangement of Figure 1. This arrangement rigidly supports the marker laser (Fenix Flyer; Synrad, Inc.) which has been modified to use a non-standard Ø48 mm single element wide area scanning lens, held in place with specially made collar to hold the lens in a suitable position for the galvanometer arrangement of the laser. A working distance of 450 mm is required which is provided by a variable lift stage in the support frame. The 75 μm thick linear low density polyethylene (LLDPE) film is supplied from a 600 mm clutched reel having been pre-coated with a thin layer of adhesive by the supplier (Dow Chemical Company, Inc.) and drawn taught by clutched reels; film tension can be varied by adjustment of reel clutches. The thin nature of the adhesive layer differentiates this technique from 'clear welding' [7]. The film passes under two crowned rollers to ensure prevention of wrinkles developing as the film is drawn through the system. The substrate material, 640 μm thick polypropylene (PPE) sheet, is pushed up into the film to a plane that coincides with the laser's focal length.

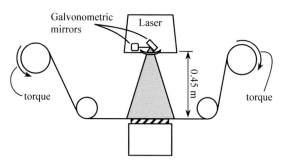

Fig. 1. Schematic diagram of material handling arrangment.

2.2. Laser control

Laser control was provided by the supplied software (Winmark Pro Version 6.2.0; Synrad, Inc), using the 200mm working-distance lens setting in conjunction with the replacment lens to achieve a wide, 335mm^2, machining area without spherical distortion. A scan speed of 115 mm/s (after lens scaling correction) was used for every bond. 11 bonds per sample were processed using identical laser scan speed (which was highly repeatable) and laser power which was found to vary by less than 2%. Three of these samples were produced per power which were: 10; 10.5; 11; 11.5; 12; 12.5; 13; 13.5; 14; and 14.5 W.

2.3. Beam analysis

The laser beam power was verfied using a laser power meter (UP25N250FH12; Gentec Electro-Optics Inc.) attached to a calibrated meter (Maestro; Gentec Electro-Optics, Inc.). The beam generated had a gaussian profile. The mean diameter of the circular laser beam was 480 μm at the focal plane.

2.4. Peel force measurment

Grouping bond tracks into sets of 11 averaged the variance in bond force across the entire sample. This mean result was then further confirmed by comparison to the other two samples for each power level. The use of three samples also averaged any random error inherent within the tensile tester (TA.XT plus; Stable Micro Systems, Ltd).

Each sample of 11 bonds was clamped into the jaws of the tensile tester in the arrangement shown in Fig. 2. The film was peeled directly against the bond direction. A key feature of this arrangement is the swivel joint that allowed the film jaws to correct against imperfect manual

alignment of the film; thus, providing an even distribution of peel force across the 11 separate bonds and ensuring the correct force vector with respect the direction of the bonds. Every sample was peeled using this arrangement at a rate of 1 mm/s.

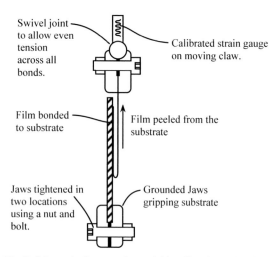

Fig. 2. Schematic diagram of material handling for peel testing.

3. Results and discussion

3.1. Result consistency

The samples produced exhibited some bubbling which appeared to be related to laser power and substrate-film contact pressure. This is merely a subjective observation that will be investigated in great detail in later work; however, this bubbling did cause some consistency issues for this analysis. Bubbles in the path of a bond track prevented contact between the film and substrate; thus, preventing a successful bond occurring at the bubble location.

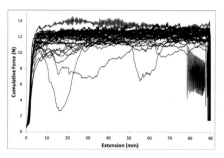

Fig. 3. Plot of peel force with respect to peel extension for all 30 samples, the features of interest are those data points that fall more than 1.02 N below the general peel force and coincide with the location of bubbles.

Figure 3 plots the bond strength of all 30 samples that were peeled. Large deviations from the normal bond strength range (cumulative force over 11 bonds) coincide with the presence of one or more bond tracks being interrupted by a bubble. The great density of bond traces between 11 and 12.5 N demonstrates that this is a highly consistent technique. This work is concerned with the bond strength with respect to laser power when the mechanical contact between film and substrate is satisfactory; thus, erroneous bubble or tensile tester related results are omitted.

Due to the considerable volume of data collected, software has been applied to filter such data from the results sampled. Data is considered in power level groupings and one of the three samples tested for 10W is given as an example in Fig. 4. In Region 1, a bubble was present (of varying size) on all three samples. In Region 2 and Region 3 the tensile tester hit a physical limit, hence the juddering force plots. The data in all three regions is useless to this work and so is filtered out; only data with a cumulative peel force greater than 10.5 N and is within ±0.51 N (the standard deviation of the acceptable data) of the mean peel cumulative force of all three 10 W samples.

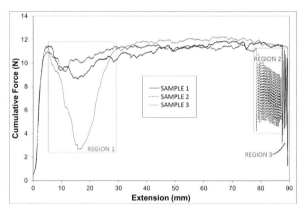

Fig. 4. Plot of peel force with respect to peel extension for the three samples laser bonded using 10 W. Regions 1, 2 and 3 all show the result of systematic error that are ommited from the analysis.

3.2. Bond physical characteristics

The plan view of a 10 W laser bond track given in Fig. 5 indicates melting. A central horizontal track is surrounded by a darker border on either side. Within this approximately 400 μm wide central track a clear pattern of material melting is evident, also exhibiting a central channel. The adhesive manufacturer (Dow Chemical Company, Inc.) has disseminated little technical detail of the adhesive; but one detail made public is the required activation temperature: 115°C [5]. Outside this central region, the laser intensity (the beam has a Gaussian profile) is insufficient to activate the adhesive.

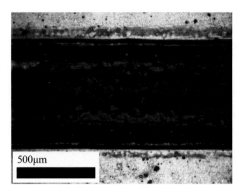

Fig. 5. A 10× Micrograph showing a plan view of a 10 W, 115 mm/s bond track.

Figure 6 shows a cross section of the same bond. This clearly shows the thermally effected area emanating radially from the top of the substrate. The dark borders evident in Fig. 5 can be explained by the ridged topography of the film at the edges of the bond that are clearly visible in Figure 6. LLPDE melts at 130°C [8], thus it is not evident in the main tract of the bond in Figure 6. This demonstrates that a Gaussian profile beam is not ideal for such work as it does not provide an even thermal profile.

Fig. 6. A 10x Micrograph showing a cross sectional view of a 10 W, 115 mm/s bond track.

3.3. Bond strength with respect to laser power

The plot shown in Fig. 7 was generated using the logic described in Section 3.1 and compares the peel force required for an average (found mathematically) single track bond with the dimensions indicated in Figs. 5 and 6. Included in Fig. 7 are two trend lines; one is fitted, using least squares, to the shape of the data and the other is plotted at a constant bond force of 1.09 N. The data plotted in Fig. 7 does not demonstrate a strong force-power relationship; however, it can be seen from inspection of the error bars (which are defined by the standard deviation of the acceptable data) that the data is consistent. A minor boost to bond strength is indicated in the 11.5 to 13 W region, although this is not conclusive evidence.

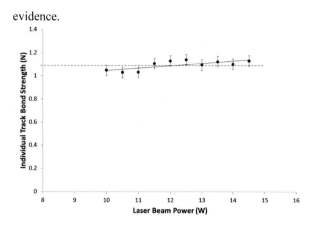

Fig. 7. Plot of the bond strength achieved in a single mean bond track with respect to laser beam energy.

The important information one can gleen from this data are:

    i) The bond strength is not greatly modified by laser power; that is, once activated, the adhesive provides a relatively constant bond; and

    ii) The technique's consistency is resilient to large power variations.

This information is a of great interest; as it suggests a threshold laser irradiance is required to activate the adhesive, generating a bond. It is important to note that without the adhesive very little, if any, bond was achieved using similar laser irradiance - unlike typical laser polymer welding [9]. Further work will be conducted to determine the threshold irradiance level.

3.4. Laser activated adhesive bond stregnth

Figure 3 demonstrates with a reasonable level of certainty *via* the consistency of all the data collated and the lack of evidence for a stong relationship between laser power and bond strength (see Fig. 7); thus, an average laser activated adhesive stength of approximately 2.2 N/mm has been determined using the data deemed acceptable by the rules defined in Section 3.1. This value is far greater than that claimed by Dow Chemical Company for traditional mechanical processing methods [5].

4. Conclusions

This work investigated the use of a 10.6 μm CO_2 marking laser to activate a thermally sensitive adhesive to bond a thin linear low density polyethylene (LLDPE) film to a polypropylene (PPE) substrate. The potential technical hurdle of bubbling has been witnessed and described: the adhesive appears to evaporate during irradiation, becoming trapped between the LLDPE film and the PPE, preventing contact between the two and preventing a

viable bond. It was proposed this was conditional upon contact pressure and laser irradiance; these variables will be objectively investigated in future work. The Gaussian profile inherent in the lightly modified commercial laser marker used for this work is not ideal for adhesive activation. The centre of the bond is over irradiated whilst the edge is under-irradiated; thus, limiting the real area on which an effective bond is achieved and reducing the user's ability to predict the bond area achieved by a process. A bond strength of approximately 2.2 N/mm^2 was achieved using the laser activated adhesive bonding technique described. This value greatly exceeds that reported to be achieved using mechanical techniques in the adhesive manufacturer's promotional material. This work has also shown that there is little bond strength dependency upon laser power (and therefore irradiance) once a threshold power is achieved. A peel force of between 1 and 1.2 N were achieved per average bond track; this small range of data spread demonstrates the high consistency of this technique. An average peel force of 1.09 N was measured from a sample population of 330 separate bond tracks. Future work will build on these results to find the threshold irradiance required to activate the adhesive and the effects of scan speed upon bond strength.

References

[1]   K. Hishinuma (2009) Heat Sealing Technology and Engineering for Packaging: Principles and Applications; DEStech Publications.

[2]   Promotional Material: FS910 Thermoforming Machine (2012) Mecapack, S.A.

[3]   M.J. Kirwan, D McDowell, R. Cole (2003) Food Packaging Technology; Blackwell

[4]   M.J. Troughton (2008) Handbook of Plastics Joining: A Practical Guide, 2nd Edition; William Andrew.

[5]   Promotional Material: Reliable and versatile performance for easy-open packaging (2010) The Dow Chemical Company, Inc.

[6]   J.C. Ion (2005) Laser Processing Of Engineering Materials: Principles, Procedure And Industrial Application; Butterworth-Heinemann, UK.

[7]   N.M. Woosman (2005) Proc IMechE Part D, 219 (9) pp. 1069-1074.

[8]   C. Mueller, G. Capaccio, A. Hiltner, E. Baer (1998) Journal of Applied Polymer Science 70 (10) pp. 2021-2030.

[9]   P.A. Hilton, I.A. Jones, Y. Kennish, in: I. Miyamoto, K.F. Kobayashi, K. Sugioka, R. Poprawe, H. Helvajian (Eds.) First International Symposium on High-Power Laser Macroprocessing, SPIE, Osaka, Japan, 2003, pp. 44-52.

Printed by Books on Demand, Germany